Lecture Notes in Artificial Intelligence 6929

Subseries of Lecture Notes in Computer Science

LNAI Series Editors

Randy Goebel
University of Alberta, Edmonton, Canada
Yuzuru Tanaka
Hokkaido University, Sapporo, Japan
Wolfgang Wahlster
DFKI and Saarland University, Saarbrücken, Germany

LNAI Founding Series Editor

Joerg Siekmann
DFKI and Saarland University, Saarbrücken, Germany

W0107277

Salem Benferhat John Grant (Eds.)

Scalable Uncertainty Management

5th International Conference, SUM 2011
Dayton, OH, USA, October 10-13, 2011
Proceedings

Springer

Series Editors

Randy Goebel, University of Alberta, Edmonton, Canada
Jörg Siekmann, University of Saarland, Saarbrücken, Germany
Wolfgang Wahlster, DFKI and University of Saarland, Saarbrücken, Germany

Volume Editors

Salem Benferhat
Université Artois, CRIL, CNRS, UMR 8188
Rue Jean Souvraz, 62307 Lens, France
E-mail: benferhat@cril.fr

John Grant
Towson University, Department of Mathematics
Towson, MD 21252, USA
E-mail: jgrant@towson.edu

ISSN 0302-9743 e-ISSN 1611-3349
ISBN 978-3-642-23962-5 ISBN 978-3-642-23963-2 (eBook)
DOI 10.1007/978-3-642-23963-2
Springer Heidelberg Dordrecht London New York

Library of Congress Control Number: 2011935865

CR Subject Classification (1998): I.2, H.4, H.3, H.5, C.2, H.2

LNCS Sublibrary: SL 7 – Artificial Intelligence

© Springer-Verlag Berlin Heidelberg 2011
This work is subject to copyright. All rights are reserved, whether the whole or part of the material is concerned, specifically the rights of translation, reprinting, re-use of illustrations, recitation, broadcasting, reproduction on microfilms or in any other way, and storage in data banks. Duplication of this publication or parts thereof is permitted only under the provisions of the German Copyright Law of September 9, 1965, in its current version, and permission for use must always be obtained from Springer. Violations are liable to prosecution under the German Copyright Law.
The use of general descriptive names, registered names, trademarks, etc. in this publication does not imply, even in the absence of a specific statement, that such names are exempt from the relevant protective laws and regulations and therefore free for general use.

Typesetting: Camera-ready by author, data conversion by Scientific Publishing Services, Chennai, India

Printed on acid-free paper

Springer is part of Springer Science+Business Media (www.springer.com)

Preface

In many applications nowadays, information systems are becoming increasingly complex, open, and dynamic. They involve massive amounts of data, generally issued from different sources. Moreover, information is often inconsistent, incomplete, heterogeneous, and pervaded with uncertainty. The annual International Conference on Scalable Uncertainty Management (SUM) has grown out of this wide-ranging interest in the management of uncertainty and inconsistency in databases, the Web, the Semantic Web, and artificial intelligence. The SUM conference series aims at bringing together researchers from these areas by highlighting new methods and technologies devoted to the problems raised by the need for a meaningful and computationally tractable management of uncertainty when huge amounts of data have to be processed. The First International Conference on Scalable Uncertainty Management (SUM 2007) was held in Washington DC, USA, in October 2007. Since then, the SUM conferences have taken place successively in Naples (Italy) in 2008, again in Washington DC (USA) in 2009, and in Toulouse (France) in 2010.

This volume contains the papers presented at the 5th International Conference on Scalable Uncertainty Management (SUM 2011) which was held in Dayton, Ohio (USA), during October 10–12, 2011. In this edition, 58 papers were submitted, of which 6 papers were withdrawn by their authors (only one during the reviewing process). Among the 52 remaining papers, 32 papers were accepted as regular papers, and 3 papers as short papers. Each paper was reviewed by at least three Program Committee members.

In addition, the conference greatly benefited from invited lectures by two world-leading researchers in artificial intelligence and the Semantic Web: Joseph Y. Halpern (on "Causality, Responsibility, and Blame: A Structural-Model Approach") and Umberto Straccia (on "Fuzzy Logic, Annotation Domains and Semantic Web Languages").

This conference revisited the idea, introduced at SUM 2010, of having discussants. Each discussant is in charge of a subset of accepted papers focusing on the same topic, and is asked to prepare a short survey on this topic for introducing the discussions. This volume contains six discussant contributions which provide an overview of a selection of topics where the research on the management of uncertainty is specially active.

We wish to thank all the authors of submitted papers, the invited speakers, the discussants, and all the conference participants for fruitful discussions. We would like to thank all the members of the Program Committee, as well as the additional reviewers, who devoted time for the reviewing process.

We would like to extend a very special thanks to the General Chair Thomas Sudkamp, from Wright State University, for his excellent local organization which made the conference a success.

Lastly, thanks are also due to the creators and maintainers of the conference management system EasyChair (http://www.easychair.org).

July 2011

Salem Benferhat
John Grant

Organization

Executive Committee

Program Chairs	Salem Benferhat and John Grant
General Chair	Thomas Sudkamp

Program Committee

Leila Amgoud (France)
Nahla Benamor (Tunisia)
Leopoldo Bertossi (Canada)
Isabelle Bloch (France)
Reynold Cheng (Hong Kong)
Carlos Iván Chesñevar (Argentina)
Laurence Cholvy (France)
Jan Chomicki (USA)
Alfredo Cuzzocrea (Italy)
Anish Das Sarma (USA)
Thierry Denoeux (France)
Jürgen Dix (Germany)
Didier Dubois (France)
Zied Elouedi (Tunisia)
Scott Ferson (USA)
Michael Fink (Austria)
Wolfgang Gatterbauer (USA)
Lluís Godo (Spain)
Gianluigi Greco (Italy)
Jon C. Helton (USA)
Anthony Hunter (UK)
Eyke Hüllermeier (Germany)
Gabriele Kern-Isberner (Germany)
Vladik Kreinovich (USA)
Weiru Liu (UK)
Peter Lucas (The Netherlands)
Thomas Lukasiewicz (UK)
Zongmin Ma (P.R. China)
Tommie Meyer (South Africa)
Cristian Molinaro (USA)
Zoran Ognjanovic (Serbia)
Francesco Parisi (Italy)
Bijan Parsia (UK)
Simon Parsons (USA)
Gabriella Pasi (Italy)
Olivier Pivert (France)
Henri Prade (France)
Andrea Pugliese (Italy)
Guilin Qi (P.R. China)
Christopher Re (USA)
Emad Saad (Kuwait)
Prithviraj Sen (India)
Xinxin Sheng (USA)
Prakash P. Shenoy (USA)
Guillermo R. Simari (Argentina)
Umberto Straccia (Italy)
V.S. Subrahmanian (USA)
Karim Tabia (France)
Sunil Vadera (UK)
Leon Van Der Torre (Luxembourg)
Jef Wijsen (Belgium)
Dan Wu (Canada)
Ronald R. Yager (USA)
Vladimir I. Zadorozhny (USA)

Additional Reviewers

Richard Booth
Martin Caminada
Weiwei Cheng
Cassio De Campos
Xiang Li
Emilia Oikarinen
Pere Pardo
Vadim Savenkov
Francesca Spezzano
Serena Villata
Shawn Yang
Yinuo Zhang
Zhizheng Zhang

Table of Contents

Probabilistic Inference

Dynamic of Beliefs

Information Retrieval and Databases

Ontologies

Possibility Theory and Classification

Logic Programming

Applications

Causality, Responsibility, and Blame: A Structural-Model Approach

Joseph Y. Halpern

Computer Science Department
Cornell University
Ithaca, NY 14853, USA
halpern@cs.cornell.edu

This talk will provide an overview of work that I have done with Hana Chockler and Judea Pearl [1,4,5] on defining notions such as *causality*, *explanation*, *responsibility*, and *blame*. I first review the Halpern-Pearl definition of causality—what it means that A is a cause of B—and show how it handles well some standard problems of causality. This definition is based on what are called *structural equations*, which are ways of describing the effects of interventions. The definition (like most in the literature) views causality as an all-or-nothing concept. Either A is a cause of B or it is not. I show how the account can be extended to take into account the degree of responsibility of A for B. For example, if someone wins an election 11–0, each person is less responsible for his victory than if he had won 6–5. Finally, I discuss more recent work [2,3] on combining a theory of normality (or defaults) with the structural equations. A slightly revised definition of causality that uses normality deals well with problems that have been pointed out in the original Halpern-Pearl definition, and helps explain different intuitions that people have regarding causality.

References

1. Chockler, H., Halpern, J.Y.: Responsibility and blame: A structural-model approach. Journal of A.I. Research 20, 93–115 (2004)
2. Halpern, J.Y.: Defaults and normality in causal structures. In: Principles of Knowledge Representation and Reasoning: Proc. Eleventh International Conference (KR 2008), pp. 198–208 (2008)
3. Halpern, J.Y., Hitchcock, C.: Graded causation and defaults (2011) (unpublished manuscript)
4. Halpern, J.Y., Pearl, J.: Causes and explanations: A structural-model approach. Part I: Causes. British Journal for Philosophy of Science 56(4), 843–887 (2005)
5. Halpern, J.Y., Pearl, J.: Causes and explanations: A structural-model approach. Part II: Explanations. British Journal for Philosophy of Science 56(4), 889–911 (2005)

S. Benferhat and J. Grant (Eds.): SUM 2011, LNAI 6929, p. 1, 2011.
© Springer-Verlag Berlin Heidelberg 2011

Fuzzy Logic, Annotation Domains and Semantic Web Languages

Umberto Straccia

ISTI - CNR, Pisa, Italy
straccia@isti.cnr.it
http://www.umberto-straccia.name

Abstract. This talk presents a detailed, self-contained and comprehensive account of the state of the art in representing and reasoning with fuzzy knowledge in Semantic Web Languages such a RDF/RDFS, OWL 2 and RIF and discuss some implementation related issues. We further show to which extend we may generalise them to so-called annotation domains, that cover also e.g. temporal, provenance and trust extensions.

Keywords: Fuzzy Logic, Semantic Web Languages, RDFS, OWL, RIF.

1 Introduction

Managing uncertainty and fuzzyness is starting to play an important role in Semantic Web research, and has been recognised by a large number of research efforts in this direction (see, e.g., [68] for a concise overview).

We recall that there has been a long-lasting misunderstanding in the literature of artificial intelligence and uncertainty modelling, regarding the role of probability/possibility theory and vague/fuzzy theory. A clarifying paper is [28]. We recall here the salient concepts for the inexpert reader. Under *uncertainty theory* fall all those approaches in which statements rather than being either true or false, are true or false to some *probability* or *possibility* (for example, "it will rain tomorrow"). That is, a statement is true or false in any world/interpretation, but we are "uncertain" about which world to consider as the right one, and thus we speak about e.g. a probability distribution or a possibility distribution over the worlds. For example, we cannot exactly establish whether it will rain tomorrow or not, due to our *incomplete* knowledge about our world, but we can estimate to which degree this is probable, possible, and necessary. On the other hand, under *fuzzy theory* fall all those approaches in which statements (for example, "the tomato is ripe") are true to some *degree*, which is taken from a truth space (usually $[0,1]$). That is, an interpretation maps a statement to a truth degree, since we are unable to establish whether a statement is entirely true or false due to the involvement of vague concepts, such as "ripe", which do not have an *precise* definition (we cannot always say whether a tomato is ripe or not). Note that all fuzzy statements are truth-functional, that is, the degree of truth of every statement can be calculated from the degrees of truth of its constituents, while uncertain statements cannot always be a function of the uncertainties of their

S. Benferhat and J. Grant (Eds.): SUM 2011, LNAI 6929, pp. 2–21, 2011.
© Springer-Verlag Berlin Heidelberg 2011

constituents [27]. More concretely, in probability theory, only negation is truth-functional, while in possibility theory, only disjunction (resp. conjunction) is truth-functional in possibilities (resp. necessities) of events. Furthermore, mathematical fuzzy logics are based on truly many-valued logical operators, while uncertainty logics are defined on top of standard binary logical operators.

We present here some salient aspects in representing and reasoning with fuzzy knowledge in Semantic Web Languages (SWLs) such as *triple languages* RDF & RDFS [19] (see, e.g. [69,70]), *conceptual languages* or *frame-based languages* of the OWL 2 family [50] (see, e.g. [45,58,62]) and *rule languages*, such as RIF [53] (see, e.g. [65,66,68]).

In the following, we overview briefly SWLs and relate them to their logical counterpart. Then, we briefly sketch the basic notions of Mathematical Fuzzy Logic, which we require in the subsequent sections in which we illustrate the fuzzy variants of SWLs.

2 Semantic Web Languages: Overview

The Semantic Web is a 'web of data' whose goal is to enable machines to understand the semantics, or meaning, of information on the World Wide Web. In rough terms, it should extend the network of hyperlinked human-readable web pages by inserting machine-readable *metadata*[1] about pages and how they are related to each other, enabling automated agents to access the Web more intelligently and perform tasks on behalf of users.

Semantic Web Languages (SWL) are the languages used to provide a formal description of concepts, terms, and relationships within a given knowledge domain to be used to write the metadata. There are essentially three family of languages: namely, *triple languages* RDF & RDFS [19] (*Resource Description Framework*), *conceptual languages* of the OWL 2 family (*Ontology Web Language*) [50] and *rule languages* of the RIF family (*Rule Interchange Format*) [53]. While their syntactic specification is based on XML [74], their semantics is based on logical formalisms, which will be the focus here (see Fig. 1): briefly,

- RDFS is a logic having intensional semantics and the logical counterpart is ρdf [47];
- OWL 2 is a family of languages that relate to *Description Logics* (DLs) [6];
- RIF relates to the *Logic Programming* (LP) paradigm [43];
- both OWL 2 and RIF have an extensional semantics.

RDF & RDFS. The basic ingredients of *RDF* are *triples* of the form (s, p, o), such as $(umberto, likes, tomato)$, stating that *subject* s has *property* p with *value* o. In *RDF Schema* (RDFS), which is an extension of RDF, additionally some special keywords may be used as properties to further improve the expressivity of the language. For instance we may also express that the class of 'tomatoes are a subclass of the class of vegetables', $(tomato, \mathsf{sc}, vegetables)$, while Zurich is an instance of the class of cities, $(zurich, \mathsf{type}, city)$.

[1] Obtained manually, semi-automatically, or automatically.

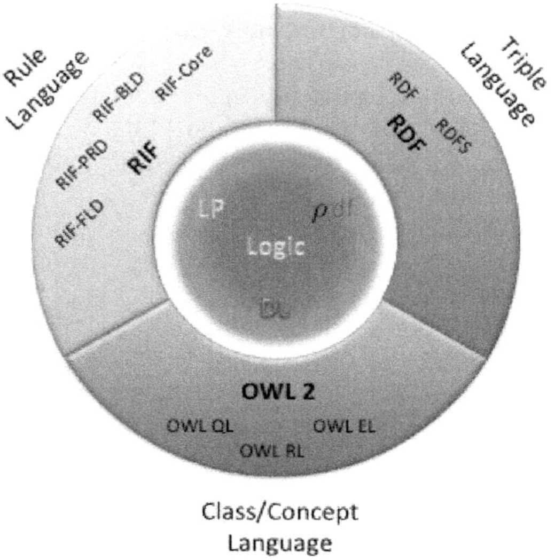

Fig. 1. Semantic Web Languages from a Logical Perspective

Form a computational point of view, one computes the so-called *closure* (denoted $cl(\mathcal{K})$) of a set of triples $\mathcal{K}$. That is, one infers all possible triples using inference rules [46,47,52], such as

$$\frac{(A, \mathsf{sc}, B), (X, \mathsf{type}, A)}{(X, \mathsf{type}, B)}$$

"if A subclass of B and X instance of A then infer that X is instance of B",

and then store all inferred triples into a relational database to be used then for querying. We recall also that there also several ways to store the closure $cl(\mathcal{K})$ in a database (see [1,37]). Essentially, either we may store all the triples in table with three columns *subject, predicate, object*, or we use a table for each predicate, where each table has two columns *subject, object*. The latter approach seems to be better for query answering purposes. Note that making all implicit knowledge explicit is viable due to the low complexity of the closure computation, which is $\mathcal{O}(|\mathcal{K}|^2)$ in the worst case.

OWL Family. The Web Ontologoy Language *OWL* [49] and its successor *OWL 2* [23,50] are "object oriented" languages for defining and instantiating Web ontologies. Ontology (see, e.g. [31]) is a term borrowed from philosophy that refers to the science of describing the kinds of entities in the world and how they are related. An OWL ontology may include descriptions of classes, properties and their instances, such as

```
class Person partial Human
      restriction (hasName someValuesFrom String)
      restriction (hasBirthPlace someValuesFrom Geoplace)
```

"The class Person is a subclass of class Human and has two attributes: hasName having a string as value, and hasBirthPlace whose value is an instance of the class Geoplace".

Given such an ontology, the OWL formal semantics specifies how to derive its logical consequences. For example, if an individual Peter is an instance of the class Student, and Student is a subclass of Person, then one can derive that Peter is also an instance of Person in a similar way as it happens for RDFS. However, OWL is much more expressive than RDFS, as the decision problems for OWL are in higher complexity classes [51] than for RDFS. In Fig. 2 we report the various OWL languages, their computational complexity and as subscript the DL their relate to [6,26].

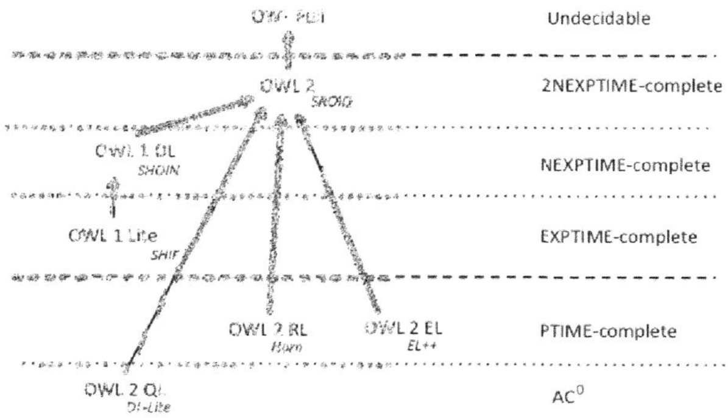

Fig. 2. OWL family and complexity

OWL 2 [23,50] is an update of OWL 1 adding several new features, including an increased expressive power. OWL 2 also defines several *OWL 2 profiles*, i.e. OWL 2 language subsets that may better meet certain computational complexity requirements or may be easier to implement. The choice of which profile to use in practice will depend on the structure of the ontologies and the reasoning tasks at hand. The OWL 2 profiles are:

OWL 2 EL is particularly useful in applications employing ontologies that contain very large numbers of properties and/or classes (basic reasoning problems can be performed in time that is polynomial with respect to the size of the ontology [5]). The EL acronym reflects the profile's basis in the $\mathcal{EL}$ family of description logics [5].

OWL 2 QL is aimed at applications that use very large volumes of instance data, and where query answering is the most important reasoning task. In OWL 2 QL, conjunctive query answering can be implemented using conventional relational database systems. Using a suitable reasoning technique, sound and complete conjunctive query answering can be performed in LOGSPACE with respect to the size of the data (assertions) [4,21]. The QL acronym reflects the fact that query answering in this profile can be implemented by rewriting queries into a standard relational Query Language such as SQL [72].

OWL 2 RL is aimed at applications that require scalable reasoning without sacrificing too much expressive power. OWL 2 RL reasoning systems can be implemented using rule-based reasoning engines as a mapping to *Logic Programming* [43], specifically *Datalog* [72], exists. The RL acronym reflects the fact that reasoning in this profile can be implemented using a standard rule language [30]. The computational complexity is the same as for Datalog [25] (polynomial in the size of the data, EXPTIME w.r.t. the size of the knowledge base).

RIF Family. The *Rule Interchange Format* (RIF) aims at becoming a standard for exchanging rules, such as

```
Forall ?Buyer ?Item ?Seller
  buy(?Buyer ?Item ?Seller) :- sell(?Seller ?Item ?Buyer)
```

"Someone buys an item from a seller if the seller sells that item to the buyer"

among rule systems, in particular among Web rule engines. RIF is in fact a family of languages, called *dialects*, among which the most significant are:

RIF-BLD The *Basic Logic Dialect* is the main logic-based dialect. Technically, this dialect corresponds to Horn logic with various syntactic and semantic extensions. The main syntactic extensions include the frame syntax and predicates with named arguments. The main semantic extensions include datatypes and externally defined predicates.

RIF-PRD The *Production Rule Dialect* aims at capturing the main aspects of various production rule systems. Production rules, as they are currently practiced in main-stream systems like Jess[2] or JRules[3], are defined using ad hoc computational mechanisms, which are not based on a logic. For this reason, RIF-PRD is not part of the suite of logical RIF dialects and stands apart from them. However, significant effort has been extended to ensure as much sharing with the other dialects as possible. This sharing was the main reason for the development of the RIF Core dialect;

RIF-Core The *Core Dialect* is a subset of both RIF-BLD and RIF-PRD, thus enabling limited rule exchange between logic rule dialects and production rules. RIF-Core corresponds to Horn logic without function symbols (i.e., Datalog) with a number of extensions to support features such as objects and frames as in F-logic [38].

RIF-FLD The *Framework for Logic Dialects* is not a dialect in its own right, but rather a general logical extensibility framework. It was introduced in order to

[2] `http://www.jessrules.com/`

[3] `http://www.ilog.com/products/jrules/`

drastically lower the amount of effort needed to define and verify new logic dialects that extend the capabilities of RIF-BLD.

3 Mathematical Fuzzy Logic Basics

Given that SWLs are grounded on Mathematical Logic, it is quite natural to look at *Mathematical Fuzzy Logic* [36] to get inspiration for a fuzzy logic extensions of SWLs. So, we recap here briefly that in Mathematical Fuzzy Logic, the convention prescribing that a statement is either true or false is changed and is a matter of degree measured on an ordered scale that is no longer $\{0,1\}$, but $[0,1]$. This degree is called *degree of truth* (or *score*) of the logical statement ϕ in the interpretation $\mathcal{I}$. In this section, *fuzzy statements* have the form $\phi\colon r$, where $r \in [0,1]$ (see, e.g. [35,36]) and ϕ is a statement, which encode that the degree of truth of ϕ is *greater or equal* r. A *fuzzy interpretation* $\mathcal{I}$ maps each basic statement p_i into $[0,1]$ and is then extended inductively to all statements:

$$
\begin{array}{rcl}
\mathcal{I}(\phi \wedge \psi) = \mathcal{I}(\phi) \otimes \mathcal{I}(\psi) & , & \mathcal{I}(\phi \vee \psi) = \mathcal{I}(\phi) \oplus \mathcal{I}(\psi) \\
\mathcal{I}(\phi \rightarrow \psi) = \mathcal{I}(\phi) \Rightarrow \mathcal{I}(\psi) & , & \mathcal{I}(\neg\phi) = \ominus \mathcal{I}(\phi) \\
\mathcal{I}(\exists x.\phi(x)) = \sup_{a \in \Delta^{\mathcal{I}}} \mathcal{I}(\phi(a)) & , & \mathcal{I}(\forall x.\phi(x)) = \inf_{a \in \Delta^{\mathcal{I}}} \mathcal{I}(\phi(a)) \ ,
\end{array}
$$

where $\Delta^{\mathcal{I}}$ is the domain of $\mathcal{I}$, and $\otimes$, $\oplus$, $\Rightarrow$, and $\ominus$ are so-called *t-norms*, *t-conorms*, *implication functions*, and *negation functions*, respectively, which extend the Boolean conjunction, disjunction, implication, and negation, respectively, to the fuzzy case [40]. Usually, the implication function $\Rightarrow$ is defined as *r-implication*, that is, $a \Rightarrow b = \sup\{c \mid a \otimes c \leq b\}$. The notions of satisfiability and logical consequence are defined in the standard way. A fuzzy interpretation $\mathcal{I}$ *satisfies* a fuzzy statement $\phi\colon r$ or $\mathcal{I}$ is a *model* of $\phi\colon r$, denoted $\mathcal{I} \models \phi\colon r$ iff $\mathcal{I}(\phi) \geq r$.

One usually distinguishes three different logics, namely *Łukasiewicz, Gödel*, and *Product logics* [36], whose combination functions are reported in Table 1. Zadeh logic, namely $a \otimes b = \min(a,b)$, $a \oplus b = \max(a,b)$, $\ominus a = 1 - a$ and $a \Rightarrow b = \max(1-a, b)$, is entailed by Łukasiewicz logic, as $\min(a,b) = a \otimes (a \Rightarrow b)$ and $\max(a,b) = 1 - \min(1-a, 1-b)$. Table 2 and 3 report axioms these functions have to satisfy. Table 4 recalls some salient properties of the various fuzzy logics. Worth noting is that a fuzzy logic satisfying all the listed properties has

Table 1. Combination functions of various fuzzy logics

	Łukasiewicz logic	Gödel logic	Product logic
$a \otimes b$	$\max(a+b-1, 0)$	$\min(a,b)$	$a \cdot b$
$a \oplus b$	$\min(a+b, 1)$	$\max(a,b)$	$a + b - a \cdot b$
$a \Rightarrow b$	$\min(1-a+b, 1)$	$\begin{cases} 1 & \text{if } a \leq b \\ b & \text{otherwise} \end{cases}$	$\min(1, b/a)$
$\ominus a$	$1 - a$	$\begin{cases} 1 & \text{if } a = 0 \\ 0 & \text{otherwise} \end{cases}$	$\begin{cases} 1 & \text{if } a = 0 \\ 0 & \text{otherwise} \end{cases}$

Table 2. Properties for t-norms and s-norms

Axiom Name	T-norm	S-norm
Tautology / Contradiction	$a \otimes 0 = 0$	$a \oplus 1 = 1$
Identity	$a \otimes 1 = a$	$a \oplus 0 = a$
Commutativity	$a \otimes b = b \otimes a$	$a \oplus b = b \oplus a$
Associativity	$(a \otimes b) \otimes c = a \otimes (b \otimes c)$	$(a \oplus b) \oplus c = a \oplus (b \oplus c)$
Monotonicity	if $b \leq c$, then $a \otimes b \leq a \otimes c$	if $b \leq c$, then $a \oplus b \leq a \oplus c$

Table 3. Properties for implication and negation functions

Axiom Name	Implication Function	Negation Function
Tautology / Contradiction	$0 \Rightarrow b = 1,\ \ a \Rightarrow 1 = 1,\ \ 1 \Rightarrow 0 = 0$	$\ominus 0 = 1,\ \ \ominus 1 = 0$
Antitonicity	if $a \leq b$, then $a \Rightarrow c \geq b \Rightarrow c$	if $a \leq b$, then $\ominus a \geq \ominus b$
Monotonicity	if $b \leq c$, then $a \Rightarrow b \leq a \Rightarrow c$	

necessarily to collapse to the Boolean, two-valued, case. As a note, [29] claimed that fuzzy logic collapses to boolean logic, but didn't recognise that to prove it, all the properties of Table 4 have been used. Additionally, we have the following inferences: let $a \geq n$ and $a \Rightarrow b \geq m$. Then, under Kleene-Dienes implication, we infer that "if $n > 1 - m$ then $b \geq m$". More importantly, to what concerns our paper, is that under an r-implication relative to a t-norm $\otimes$, we have that

$$\text{from } a \geq n \text{ and } a \Rightarrow b \geq m, \text{ we infer } b \geq n \otimes m\ . \tag{1}$$

To see this, as $a \geq n$ and $a \Rightarrow b = \sup\{c \mid a \otimes c \leq b\} = \bar{c} \geq m$ it follows that $b \geq a \otimes \bar{c} \geq n \otimes m$. In a similar way, under an r-implication relative to a t-norm $\otimes$, we have that

$$\text{from } a \Rightarrow b \geq n \text{ and } b \Rightarrow c \geq m, \text{ we infer that } a \Rightarrow c \geq n \otimes m\ . \tag{2}$$

We say $\phi\colon n$ is a *tight logical consequence* of a set of fuzzy statements $\mathcal{K}$ iff n is the infimum of $\mathcal{I}(\phi)$ subject to all models $\mathcal{I}$ of $\mathcal{K}$. Notice that the latter is

Table 4. Some additional properties of combination functions of various fuzzy logics

Property	Łukasiewicz Logic	Gödel Logic	Product Logic	Zadeh Logic
$x \otimes \ominus x = 0$	+	+	+	−
$x \oplus \ominus x = 1$	+	−	−	−
$x \otimes x = x$	−	+	−	+
$x \oplus x = x$	−	+	−	+
$\ominus \ominus x = x$	+	−	−	+
$x \Rightarrow y = \ominus x \oplus y$	+	−	−	+
$\ominus (x \Rightarrow y) = x \otimes \ominus y$	+	−	−	+
$\ominus (x \otimes y) = \ominus x \oplus \ominus y$	+	+	+	+
$\ominus (x \oplus y) = \ominus x \otimes \ominus y$	+	+	+	+

equivalent to $n = \sup\{r \mid \mathcal{K} \models \phi: r\}$. n is called the *best entailment degree* of ϕ w.r.t. $\mathcal{K}$ (denoted $bed(\mathcal{K}, \phi)$), i.e.

$$bed(\mathcal{K}, \phi) = \sup\{r \mid \mathcal{K} \models \phi: r\}\ .$$

On the other hand, the *best satisfiability degree* of ϕ w.r.t. $\mathcal{K}$ (denoted $bsd(\mathcal{K}, \phi)$) is

$$bsd(\mathcal{K}, \phi) = \sup_{\mathcal{I}}\{\mathcal{I}(\phi) \mid \mathcal{I} \models \mathcal{K}\}\ .$$

We refer the reader to [34,35,36] for reasoning algorithms for fuzzy propositional and First-Order Logics. For illustrative purpose, we recap here a simple method to determine $bed(\mathcal{K}, \phi)$ and $bsd(\mathcal{K}, \phi)$ via Mixed Integer Linear Programming (MILP) for the case of propositional Łukasiewicz logic. To this end, it can be shown that

$$\begin{aligned} bed(\mathcal{K}, \phi) &= \min x. \text{ such that } \mathcal{K} \cup \{\neg\phi: 1 - x\} \text{ satisfiable} \\ bsd(\mathcal{K}, \phi) &= \max x. \text{ such that } \mathcal{K} \cup \{\phi: x\} \text{ satisfiable}\ . \end{aligned}$$

Now, for a formula ϕ consider a variable x_ϕ (with intended meaning: the degree of truth of ϕ is greater or equal to x_ϕ). Now we apply the following transformation σ that generates a set of MILP in-equations:

$$\begin{aligned} bed(\mathcal{K}, \phi) = \min x. \text{ such that } & x \in [0,1], x_{\neg\phi} \geq 1 - x, \sigma(\neg\phi), \\ & \text{for all } \phi' \geq n \in \mathcal{K}, x_{\phi'} \geq n, \sigma(\phi'), \end{aligned}$$

$$\sigma(\phi) = \begin{cases} x_p \in [0,1] & \text{if } \phi = p \\ x_\phi = 1 - x_{\phi'}, x_\phi \in [0,1] & \text{if } \phi = \neg\phi' \\ x_{\phi_1} \otimes x_{\phi_2} \geq x_\phi, & \\ \sigma(\phi_1), \sigma(\phi_2), x_\phi \in [0,1] & \text{if } \phi = \phi_1 \wedge \phi_2 \\ x_{\phi_1} \oplus x_{\phi_2} \geq x_\phi & \text{if } \phi = \phi_1 \vee \phi_2 \\ \sigma(\neg\phi_1 \vee \phi_2) & \text{if } \phi = \phi_1 \Rightarrow \phi_2\ . \end{cases}$$

In the definition above, $z \leq x_1 \oplus x_2$ and $z \leq x_1 \otimes x_2$, with $0 \leq x_i, z \leq 1$, can be encoded as the sets of constraints:

$$\begin{aligned} & z \leq x_1 \oplus x_2 \mapsto \{z \leq x_1 + x_2\}, \\ & z \leq x_1 \otimes x_2 \mapsto \{y \leq 1 - z, x_1 + x_2 - 1 \geq z - y, y \in \{0,1\}\}\ . \end{aligned}$$

As the set of constraints is linearly bounded by $\mathcal{K}$ and as MILP satisfiability is NP-complete, we get the well-known result that determining the best entailment/satisfiability degree is NP-complete for propositional Łukasiewicz logic.

We conclude with the notion of *fuzzy set* [76]. A *fuzzy set* R over a countable crisp set X is a function $R: X \to [0,1]$. The *degree of subsumption* between two fuzzy sets A and B, denoted $A \sqsubseteq B$, is defined as $\inf_{x \in X} A(x) \Rightarrow B(x)$, where $\Rightarrow$ is an implication function. Note that if $A(x) \leq B(x)$, for all $x \in [0,1]$, then $A \sqsubseteq B$ evaluates to 1. Of course, $A \sqsubseteq B$ may evaluate to a value $v \in (0,1)$ as well.

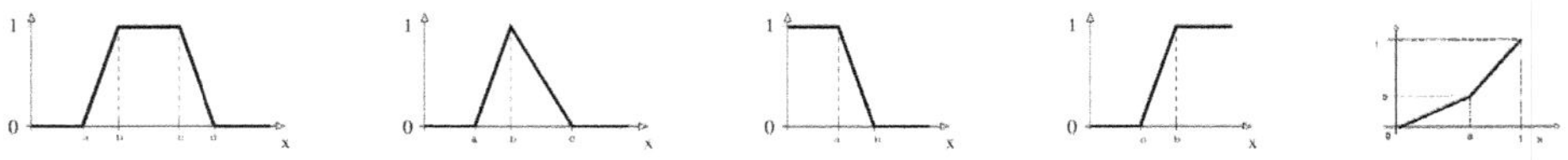

(a) (b) (c) (d) (e)

Fig. 3. (a) Trapezoidal function $trz(a, b, c, d)$, (b) triangular function $tri(a, b, c)$, (c) left shoulder function $ls(a, b)$, (d) right shoulder function $rs(a, b)$ and (e) linear modifier $lm(a, b)$

A (binary) *fuzzy relation* R over two countable crisp sets X and Y is a function $R\colon X \times Y \to [0, 1]$. The *inverse* of R is the function $R^{-1}\colon Y \times X \to [0, 1]$ with membership function $R^{-1}(y, x) = R(x, y)$, for every $x \in X$ and $y \in Y$. The *composition* of two fuzzy relations $R_1\colon X \times Y \to [0, 1]$ and $R_2\colon Y \times Z \to [0, 1]$ is defined as $(R_1 \circ R_2)(x, z) = \sup_{y \in Y} R_1(x, y) \otimes R_2(y, z)$. A fuzzy relation R is *transitive* iff $R(x, z) \geqslant (R \circ R)(x, z)$.

Eventually, the trapezoidal (Fig. 3 (a)), the triangular (Fig. 3 (b)), the L-function (left-shoulder function, Fig. 3 (c)), and the R-function (right-shoulder function, Fig. 3 (d)) are frequently used to specify membership degrees. For instance, the left-shoulder function is defined as

$$ls(x; a, b) = \begin{cases} 1 & \text{if } x \leq a \\ 0 & \text{if } x \geq b \\ (b - x)/(b - a) & \text{if } x \in [a, b] \end{cases} \tag{3}$$

4 Fuzzy Logic and Semantic Web Languages

We have seen in the previous section how to "fuzzyfy" a classical language such as propositional logic and FOL, namely fuzzy staements are of the form $\phi\colon n$, where ϕ is a statement and $n \in [0, 1]$.

The natural extension to SWLs consists then in replacing ϕ with appropriate expressions belonging to the logical counterparts of SWLs, namely ρdf, DLs and LPs, as we will illustrate next.

4.1 Fuzzy RDFS

In *Fuzzy RDFS* (see [69] and references therein), triples are annotated with a degree of truth in $[0, 1]$. For instance, "Rome is a big city to degree 0.8" can be represented with $(\mathit{Rome}, \mathsf{type}, \mathit{BigCity})\colon 0.8$. More formally, *fuzzy triples* are expressions of the form $\tau\colon n$, where τ is a RDFS triple (the truth value n may be omitted and, in that case, the value $n = 1$ is assumed).

The interesting point is that from a computational point of view the inference rules parallel those for "crisp" RDFS: indeed, the rules are of the form

$$\frac{\tau_1\colon n_1, \ \ldots, \ \tau_k\colon n_k, \{\tau_1, \ldots, \tau_k\} \vdash_{\mathsf{RDFS}} \tau}{\tau\colon \bigotimes_i n_i} \tag{4}$$

Essentially, this rule says that if a classical RDFS triple τ can be inferred by applying a classical RDFS inference rule to triples $\tau_1, \ldots, \tau_k$ (denoted $\{\tau_1, \ldots, \tau_k\} \vdash_{\mathsf{RDFS}} \tau$), then the truth degree of τ will be $\bigotimes_i n_i$.

As a consequence, the rule system is quite easy to implement for current inference systems. Specifically, as for the crisp case, one may compute the closure $cl(\mathcal{K})$ of a set of fuzzy triples $\mathcal{K}$, store them in a relational database and thereafter query the database.

Concerning the query language, *SPARQL* [55] is the current standard, but a new version (*SPARQL 1.1*) is close to be finalised [56]. From a logical point of view, a SPARQL query may be seen as a *Conjunctive Query* (CQ), or an union of them, a well-known notion in database theory [2]. Specifically, an *RDF query* is of the rule-like form

$$q(\boldsymbol{x}) \leftarrow \exists \boldsymbol{y}.\varphi(\boldsymbol{x}, \boldsymbol{y}) \ , \tag{5}$$

where $q(\boldsymbol{x})$ is the *head* and $\exists \boldsymbol{y}.\varphi(\boldsymbol{x}, \boldsymbol{y})$ is the *body* of the query, which is a conjunction (we use the symbol "," to denote conjunction in the rule body) of triples τ_i $(1 \leq i \leq n)$. $\boldsymbol{x}$ is a vector of variables occurring in the body, called the *distinguished variables*, $\boldsymbol{y}$ are so-called *non-distinguished variables* and are distinct from the variables in $\boldsymbol{x}$, each variable occurring in τ_i is either a distinguished or a non-distinguished variable. If clear from the context, the existential quantification $\exists \boldsymbol{y}$ may be omitted. In a query, built-in triples of the form (s, p, o) are allowed, where p is a built-in predicate taken from a reserved vocabulary and having a *fixed interpretation.* Built-in predicates are generalised to any n-ary predicate p. For convenience, "functional predicates"[4] are written as *assignments* of the form $x{:=}f(\boldsymbol{z})$ and it is assumed that the function $f(\boldsymbol{z})$ is safe (also non functional built-in predicate $p(\boldsymbol{z})$ should be safe as well). A query example is:

$$q(x, y) \leftarrow (y, created, x), (y, \mathsf{type}, Italian), (x, exhibitedAt, Uffizi) \tag{6}$$

having intended meaning to retrieve all the artefacts x created by Italian artists y, being exhibited at Uffizi Gallery.

Roughly, the *answer set* of a query q w.r.t. a set of tuples $\mathcal{K}$ (denoted $ans(\mathcal{K}, q)$) is the set of tuples $\boldsymbol{t}$ such that there exists $\boldsymbol{t}'$ such that the instantiation $\varphi(\boldsymbol{t}, \boldsymbol{t}')$ of the query body is true in the closure of $\mathcal{K}$, i.e., all triples in $\varphi(\boldsymbol{t}, \boldsymbol{t}')$ are in $cl(\mathcal{K})$.

Once we switch to the fuzzy setting, queries are similar as for the crisp case, except that fuzzy triples are used in the query body in place of crisp triples. A special attention is required to the fact that now all answers are graded and, thus, an order is induced on the answer set. Specifically, a *fuzzy query* is of the form

$$q(\boldsymbol{x})\colon s \leftarrow \exists \boldsymbol{y}.\tau_1\colon s_1, \ldots, \tau_n\colon s_n, s{:=}f(\boldsymbol{s}, \boldsymbol{x}, \boldsymbol{y}) \ , \tag{7}$$

where now additionally s_i is the score of triple τ_i and the final score s of triple $\boldsymbol{x}$ is computed according to a user function f applied to variables occurring in the query body. For instance, the query

$$q(x)\colon s \leftarrow (x, \mathsf{type}, SportsCar)\colon s_1, (x, hasPrice, y), s = s_1 \cdot cheap(y) \tag{8}$$

[4] A predicate $p(\boldsymbol{x}, y)$ is functional if for any $\boldsymbol{t}$ there is *unique* t' for which $p(\boldsymbol{t}, t')$ is true.

where e.g. $cheap(y) = ls(20000, 30000)(y)$, has intended meaning to retrieve all cheap sports car. Then, any answer is scored according to the product of being cheap and a sports car.

It is not difficult to see that indeed fuzzy CQs can easily be mapped into SQL as well. For further details see [69].

Annotation Domains and RDFS. We have seen that fuzzy RDFS extends triples with an *annotation* $n \in [0, 1]$. Interestingly, we may further generalise fuzzy RDFS, by allowing a triple being annotated with a value λ taken from a so-called *annotation domain* [3,20,48,70][5], which allow to deal with several domains (such as, fuzzy, temporal, provenace) and their combination, in a uniform way. Formally, let us consider a non-empty set L. Elements in L are our annotation values. For example, in a fuzzy setting, $L = [0, 1]$, while in a typical temporal setting, L may be time points or time intervals. In the annotation framework, an interpretation will map statements to elements of the annotation domain. Now, an *annotation domain* for RDFS is an idempotent, commutative semi-ring

$$D = \langle L, \oplus, \otimes, \bot, \top \rangle ,$$

where $\oplus$ is $\top$-annihilating [20]. That is, for $\lambda, \lambda_i \in L$

1. $\oplus$ is idempotent, commutative, associative;
2. $\otimes$ is commutative and associative;
3. $\bot \oplus \lambda = \lambda$, $\top \otimes \lambda = \lambda$, $\bot \otimes \lambda = \bot$, and $\top \oplus \lambda = \top$;
4. $\otimes$ is distributive over $\oplus$, i.e.$\lambda_1 \otimes (\lambda_2 \oplus \lambda_3) = (\lambda_1 \otimes \lambda_2) \oplus (\lambda_1 \otimes \lambda_3)$;

It is well-known that there is a natural partial order on any idempotent semi-ring: an annotation domain $D = \langle L, \oplus, \otimes, \bot, \top \rangle$ induces a partial order $\preceq$ over L defined as:

$$\lambda_1 \preceq \lambda_2 \text{ if and only if } \lambda_1 \oplus \lambda_2 = \lambda_2 .$$

The order $\preceq$ is used to express redundant/entailed/subsumed information. For instance, for temporal intervals, an annotated triple $(s, p, o)\colon [2000, 2006]$ entails $(s, p, o)\colon [2003, 2004]$, as $[2003, 2004] \subseteq [2000, 2006]$ (here, $\subseteq$ plays the role of $\preceq$).

Remark 1. $\oplus$ is used to combine information about the same statement. For instance, in temporal logic, from $\tau\colon [2000, 2006]$ and $\tau\colon [2003, 2008]$, we infer $\tau\colon [2000, 2008]$, as $[2000, 2008] = [2000, 2006] \cup [2003, 2008]$; here, $\cup$ plays the role of $\oplus$. In the fuzzy context, from $\tau\colon 0.7$ and $\tau\colon 0.6$, we infer $\tau\colon 0.7$, as $0.7 = \max(0.7, 0.6)$ (here, max plays the role of $\oplus$).

Remark 2. $\otimes$ is used to model the "conjunction" of information. In fact, a $\otimes$ is a generalisation of boolean conjunction to the many-valued case. In fact, $\otimes$ satisfies also that

[5] The readers familiar with the annotated logic programming framework [39], will notice the similarity of the approaches.

1. $\otimes$ is bounded: i.e.$\lambda_1 \otimes \lambda_2 \preceq \lambda_1$.
2. $\otimes$ is $\preceq$-monotone, i.e. for $\lambda_1 \preceq \lambda_2$, $\lambda \otimes \lambda_1 \preceq \lambda \otimes \lambda_2$

For instance, on interval-valued temporal logic, from $(a, \mathsf{sc}, b)\colon [2000, 2006]$ and $(b, \mathsf{sc}, c)\colon [2003, 2008]$, we will infer $(a, \mathsf{sc}, c)\colon [2003, 2006]$, as $[2003, 2006] = [2000, 2006] \cap [2003, 2008]$; here, $\cap$ plays the role of $\otimes$.[6] In the fuzzy context, one may chose any t-norm [36,40], e.g.product, and, thus, from $(a, \mathsf{sc}, b)\colon 0.7$ and $(b, \mathsf{sc}, c)\colon 0.6$, we will infer $(a, \mathsf{sc}, c)\colon 0.42$, as $0.42 = 0.7 \cdot 0.6$) (here, $\cdot$ plays the role of $\otimes$).

Remark 3. Observe that the distributivity condition is used to guarantee that e.g. we obtain the same annotation $\lambda \otimes (\lambda_2 \oplus \lambda_3) = (\lambda_1 \otimes \lambda_2) \oplus (\lambda_1 \otimes \lambda_3)$ of the triple (a, sc, c) that can be inferred from triples $(a, \mathsf{sc}, b)\colon \lambda_1$, $(b, \mathsf{sc}, c)\colon \lambda_2$ and $(b, \mathsf{sc}, c)\colon \lambda_3$.

The use of annotation domains appears to be quite appealing as

1. it applies to several domains, such as the fuzzy domain, the temporal domain, provenance, trust and any combination of them [3];
2. from an inference point of view, the rules are conceptually the same as for the fuzzy case: indeed, just replace in Rule 4, the values n_i with λ_i, i.e.

$$\frac{\tau_1\colon \lambda_1, \ \ldots, \ \tau_k\colon \lambda_k, \{\tau_1, \ldots, \tau_k\} \vdash_{\mathsf{RDFS}} \tau}{\tau\colon \bigotimes_i \lambda_i} \tag{9}$$

3. annotated conjunctive queries are as fuzzy queries, except that now variables s and s_i range over L in place of $[0, 1]$;
4. a query answering procedure is similar as for the fuzzy case: compute the closure, store it on a relation database and transform an annotated CQ into a SQL query.

From a computational complexity point of view, it is the same as for crisp RDFS plus the cost of $\otimes$, $\oplus$ and the scoring function f in the body of a query. A prototype implementation is available from `http://anql.deri.org/`.

4.2 Fuzzy OWL

Description Logics. (DLs) [6] are the logical counterpart of the family of OWL languages. So, to illustrate the basic concepts of fuzzy OWL, it suffices to show the fuzzy DL case (see [45], for a survey). Briefly, one starts from a classical DL, and attaches to the basic statements a degree $n \in [0, 1]$, similarly as we did for fuzzy RDFS. As a matter of example, consider the DL $\mathcal{ALC}$ ($\mathcal{A}$ttributive $\mathcal{L}$anguage with $\mathcal{C}$omplement), a major DL representative used to introduce new extensions to DLs: the table below shows its syntax, semantics and provides examples.

---

[6] As we will see, $\oplus$ and $\otimes$ may be more involved.

Syntax		*Semantics*	*Example*
$C, D \rightarrow$	$\top$	$\mid \top(x)$	
	$\bot$	$\mid \bot(x)$	
	A	$\mid A(x)$	$Human$
	$C \sqcap D$	$\mid C(x) \wedge D(x)$	$Human \sqcap Male$
	$C \sqcup D$	$\mid C(x) \vee D(x)$	$Nice \sqcup Rich$
	$\neg C$	$\mid \neg C(x)$	$\neg Meat$
	$\exists R.C$	$\mid \exists y.R(x,y) \wedge C(y)$	$\exists has\_child.Blond$
	$\forall R.C$	$\forall y.R(x,y) \Rightarrow C(y)$	$\forall has\_child.Human$
$C \sqsubseteq D$		$\forall x.C(x) \Rightarrow D(x)$	$Happy\_Father \sqsubseteq Man \sqcap \exists has\_child.Female$
$a{:}C$		$C(a)$	$John{:}Happy\_Father$
$(a,b){:}R$		$R(a,b)$	$(John, Mary){:}Loves$

The upper pane describes how *concepts/classes* can be formed, while the lower pane shows the form of *statements/formulae* a knowledge base may be build of. Statements of the form $C \sqsubseteq D$, called, *General Inclusion Axioms* (GCIs), dictated that the class C is a subclass of the class D, $a{:}C$ dictates that individual a is an instance of class C, while $(a,b){:}R$ states that $\langle a,b \rangle$ is an instance of the binary relation R. The definition $A = C$, is used in place of having both $A \sqsubseteq C$ and $C \sqsubseteq A$, stating that class A is defined to be equivalent to C.

Fuzzy DLs [58,64,45] are then obtained by interpreting the statements as fuzzy FOL formulae and attaching a weight n to DL statements, yielding *fuzzy DL statements*, such as

$$C \sqsubseteq D \colon n \ , \ a{:}C \colon n \ \text{ and } \ (a,b){:}R \colon n \ .$$

A notable difference to fuzzy RDFS is that one may use additionally some special constructs to enhance the expressivity of fuzzy DLs [12,15,16,60], these include

- fuzzy modifiers applied to concepts, such as

$$NiceVeryExpensiveItem = Nice \sqcap very(ExpensiveItem)$$

  defining the class of nice and very expensive items, where $Nice$ and $ExpensiveItem$ are classes/concepts and $very$ is a linear modifier, such as $ln(x, 0.7, 0.3)$;
- the possibility of defining *fuzzy concrete concepts* [60], i.e. concepts having a specific fuzzy membership function, e.g., allowing a definition for $ExpensiveItem$

$$ExpensiveItem = Item \sqcap \exists hasPrice.HighPrice$$
$$HighPrice = rs(100, 200)$$

- various forms of *concept aggregations* [15] using so-called *Aggregation Operators* (AOs). These are mathematical functions that are used to combine information [71]. The arithmetic mean, the weighted sum, the median and, more generally Ordered Weighted Averaging (OWA) [75] are the most well-known AOs. For instance,

$$Hotel \sqcap (0.3 \cdot Cheap + 0.5 \cdot CloseToVenue + 0.2 \cdot Comfortable) \sqsubseteq GoodHotel \quad (10)$$

may be used to define a sufficient condition for a good hotel as a weighted sum of being cheap, close to the venue and comfortable ($Cheap, CloseToVenue$ and $Comfortable$ are classes here).

From a decision procedure point of view, one may proceed similarly as for the best entailment degree problem for fuzzy propositional logic. That is, the decision procedure consists of a set of inference rules that generate a set of in-equations (that depend on the t-norm and fuzzy concept constructors) that have to be solved by an operational research solver (see, e.g. [14,60]). An informal rule example is as follows:

"If individual a is instance of the class intersection $C_1 \sqcap C_2$ to degree greater or equal to $x_{a:C_1 \sqcap C_2}$[7], then a is instance of C_i $(i = 1, 2)$ to degree greater or equal to $x_{a:C_i}$, where additionally the following in-equation holds:

$$x_{a:C_1 \sqcap C_2} \leq x_{a:C_1} \otimes x_{a:C_2} \text{ ."}$$

Note that for Zadeh Logic and Łukasiewicz Logic a MILP solver is enough to determine whether the set of in-equations has a solution or not.

However, recently there have been some unexpected surprises [7,8,9,22]. [9] shows that $\mathcal{ALC}$ with GCIs *(i)* does not have the finite model property under Łukasiewicz Logic or Product Logic, contrary to the classical case; *(ii)* illustrates that some algorithms are neither complete not correct; and *(iii)* shows some interesting conditions under which decidability is still guaranteed. [7,8] show that knowledge base satisfiability is an undecidable problem for Product Logic. The same holds for Łukasiewicz Logic as well [22]. In case the truth-space is finite and defined a priori, decidability is guaranteed (see, e.g. [13,11,59]).

Some fuzzy DLs solvers are: *fuzzyDL* [12], *Fire* [57], *GURDL* [32], *DeLorean* [10], *GERDS* [33], and *YADLR* [41]. There is also a proposal to use OWL 2 itself to represent fuzzy ontologies [16]. More precisely, [16] identifies the syntactic differences that a fuzzy ontology language has to cope with, and shows how to encode them using OWL 2 annotation properties. The use of annotation properties makes possible *(i)* to use current OWL 2 editors for fuzzy ontology representation, *(ii)* that OWL 2 reasoners discard the fuzzy part of a fuzzy ontology, producing almost the same results as if it would not exist; and *(ii)* an implementation is provided as a Protégé plug-in.

Eventually, as for RDFS, the notion of conjunctive query straightforwardly extends to DLs and to fuzzy DLs as well: in the classical DL case, a query is of the form (compare to Eq. (5))

$$q(\boldsymbol{x}) \leftarrow \exists \boldsymbol{y}.\varphi(\boldsymbol{x}, \boldsymbol{y}) \ , \tag{11}$$

where now $\varphi(\boldsymbol{x}, \boldsymbol{y})$ is a conjunction of unary and binary predicates. For instance, the DL analogue of the RDFS query (6) is

$$q(x, y) \leftarrow Created(y, x), Italian(y), ExhibitedAt(x, uffizi) \ . \tag{12}$$

[7] As for the fuzzy propositional case, for a fuzzy DL formula ϕ we consider a variable x_ϕ with intended meaning: the degree of truth of ϕ is greater or equal to x_ϕ.

Similarly, a *fuzzy DL query* is of the form (compare to Eq. (7))

$$q(\boldsymbol{x})\colon s \leftarrow \exists \boldsymbol{y}.A_1\colon s_1, \ldots, A_n\colon s_n, s{:=}f(\boldsymbol{s}, \boldsymbol{x}, \boldsymbol{y}) \ , \tag{13}$$

where now A_i is either an unary or binary predicate. For instance, the fuzzy DL analogue of the RDFS query (8) is

$$q(x)\colon s \leftarrow \mathit{SportsCar}(x)\colon s_1, \mathit{HasPrice}(x,y), s{:=}s_1 \cdot \mathit{cheap}(y) \ . \tag{14}$$

Annotation Domains and OWL. The generalisation of fuzzy OWL to the case in which an annotation $n \in [0,1]$ is replaced with an annotation value λ taken from an annotation domain proceeds as for RDFS, except that now the annotation domain has the form of a complete lattice [63].

From a computational complexity point of view, similar results hold as for the [0,1] case [17,18,63]. While [63] provides a decidability result in case the lattice is finite, [17] further improves the decidability result by characterising the computational complexity of KB satisfiability problem for $\mathcal{ALC}$ with GCIs over finite lattices being EXPTIME-complete, as for the crisp variant, while [18] shows that the KB satisfiability problem for $\mathcal{ALC}$ with GCIs over non finite lattices is undecidable.

4.3 Fuzzy RIF

The foundation of the core part of RIF is *Datalog* [72], i.e. a Logic Programming Language (LP) [43]. In LP, the management of imperfect information has attracted the attention of many researchers and numerous frameworks have been proposed. Addressing all of them is almost impossible, due to both the large number of works published in this field (early works date back to early 80-ties [54]) and the different approaches proposed.

Basically [43], a Datalog program $\mathcal{P}$ is made out by a set of rules and a set of facts. *Facts* are ground *atoms* of the form $P(\boldsymbol{c})$. On the other hand rules are similar as conjunctive DL queries and are of the form

$$A(\boldsymbol{x}) \leftarrow \exists \boldsymbol{y}.\varphi(\boldsymbol{x}, \boldsymbol{y}) \ ,$$

where now $\varphi(\boldsymbol{x}, \boldsymbol{y})$ is a conjunction of n-ary predicates. In Datalog it is further assumed that no fact predicate may occur in a rule head (facts are the so-called extensional database, while rules are the intentional database). A *query* is a rule and the *answer set* of a query q w.r.t. a set $\mathcal{K}$ of facts and rules (denoted $ans(\mathcal{K}, q)$) is the set of tuples $\boldsymbol{t}$ such that there exists $\boldsymbol{t}'$ such that the instantiation $\varphi(\boldsymbol{t}, \boldsymbol{t}')$ of the query body is true in *minimal model* of $\mathcal{K}$, which is guaranteed to exists.

As pointed out, there are several proposals for fuzzy Datalog (see [68] for an extensive list). However, a sufficiently general form is obtained in case facts are graded with $n \in [0,1]$, i.e. facts are of the form $P(\boldsymbol{c})\colon n$ and rules generalise fuzzy DL queries (compare to Eq. (13)): i.e., a *fuzzy rule* is of the form

$$A(\boldsymbol{x})\colon s \leftarrow \exists \boldsymbol{y}.A_1\colon s_1, \ldots, A_n\colon s_n, s{:=}f(\boldsymbol{s}, \boldsymbol{x}, \boldsymbol{y}) \ , \tag{15}$$

where now A_i is an n-ary predicate. For instance, the fuzzy GCI in Eq. (10), can be expressed easily as the fuzzy rule

$$GoodHotel(x)\colon s \leftarrow Hotel(x), Cheap(x)\colon s_1, CloseToVenue(x)\colon s_2,$$
$$Comfortable(x)\colon s_3, s := 0.3 \cdot s_1 + 0.5 \cdot s_2 + 0.2 \cdot s_3 \quad (16)$$

A *fuzzy query* is a fuzzy rule and, informally, the *fuzzy answer set* is the ordered set of weighted tuples $\langle \boldsymbol{t}, s\rangle$ such that all the fuzzy atoms in the rule body are true in the minimal model and s is the result of the scoring function f applied to its arguments. The existence of a minimal is guaranteed if the scoring functions in the query and in the rule bodies are *monotone* [68].

We conclude by saying that most works deal with logic programs without negation and some may provide some technique to answer queries in a top-down manner, as e.g. [24,39,42,73,61]. Deciding whether a wighted tuple $\langle \boldsymbol{t}, s\rangle$ is the answer set is undecidable in general, though is decidable if the truth space is finite and fixed a priory, as then the minimal model is finite.

Another rising problem is the problem to compute the top-k ranked answers to a query, without computing the score of all answers. This allows to answer queries such as "find the top-k closest hotels to the conference location". Solutions to this problem can be found in [44,66,67].

Annotation Domains and RIF. The generalisation of fuzzy RIF to the case in which an annotation $n \in [0,1]$ is replaced with an annotation value λ taken from an annotation domain is straightforward and proceeds as for RDFS. From a computational complexity point of view, similarly to the fuzzy case, deciding whether a wighted tuple $\langle \boldsymbol{t}, \lambda\rangle$ is the answer set is undecidable in general, though is decidable if the annotation domain is finite.

5 Conclusions

We have provided a "crash course" through the realm of Semantic Web Languages, their fuzzy variants and their generalisation to annotation domains, by illustrating the basics of these languages, some issues, and related them to the logical formalisms on which they are based.

References

1. Abadi, D.J., Marcus, A., Madden, S., Hollenbach, K.: Sw-store: a vertically partitioned dbms for semantic web data management. VLDB J. 18(2), 385–406 (2009)
2. Abiteboul, S., Hull, R., Vianu, V.: Foundations of Databases. Addison Wesley Publ. Co., Reading (1995)
3. Zimmermann, A.P.A., Lopes, N., Straccia, U.: A general framework for representing, reasoning and querying with annotated semantic web data. Technical report, Computing Research Repository (2011), Available as CoRR technical report, at `http://arxiv.org/abs/1103.1255`

4. Artale, A., Calvanese, D., Kontchakov, R., Zakharyaschev, M.: The DL-Lite family and relations. Journal of Artificial Intelligence Research 36, 1–69 (2009)
5. Baader, F., Brandt, S., Lutz, C.: Pushing the $\mathcal{EL}$ envelope. In: Proceedings of the Nineteenth International Joint Conference on Artificial Intelligence (IJCAI 2005), pp. 364–369. Morgan-Kaufmann Publishers, Edinburgh (2005)
6. Baader, F., Calvanese, D., McGuinness, D., Nardi, D., Patel-Schneider, P.F. (eds.): Description Logic Handbook: Theory, Implementation, and Applications. Cambridge University Press, Cambridge (2003)
7. Baader, F., Peñaloza, R.: Are fuzzy description logics with general concept inclusion axioms decidable? In: Proceedings of 2011 IEEE International Conference on Fuzzy Systems (Fuzz-IEEE 2011). IEEE Press, Los Alamitos (to appear, 2011)
8. Baader, F., Peñaloza, R.: Gcis make reasoning in fuzzy dl with the product t-norm undecidable. In: Proceedings of the 24th International Workshop on Description Logics (DL 2011), CEUR Electronic Workshop Proceedings (to appear, 2011)
9. Bobillo, F., Bou, F., Straccia, U.: On the failure of the finite model property in some fuzzy description logics. Fuzzy Sets and Systems 172(1), 1–12 (2011)
10. Bobillo, F., Delgado, M., Gómez-Romero, J.: Delorean: A reasoner for fuzzy OWL 1.1. In: Proceedings of the 4th International Workshop on Uncertainty Reasoning for the Semantic Web (URSW 2008), CEUR Workshop Proceedings, vol. 423 (October 2008)
11. Bobillo, F., Delgado, M., Gómez-Romero, J., Straccia, U.: Fuzzy description logics under gödel semantics. International Journal of Approximate Reasoning 50(3), 494–514 (2009)
12. Bobillo, F., Straccia, U.: fuzzyDL: An expressive fuzzy description logic reasoner. In: 2008 International Conference on Fuzzy Systems (FUZZ 2008), pp. 923–930. IEEE Computer Society, Los Alamitos (2008)
13. Bobillo, F., Straccia, U.: Towards a crisp representation of fuzzy description logics under Łukasiewicz semantics. In: An, A., Matwin, S., Raś, Z.W., Ślęzak, D. (eds.) Foundations of Intelligent Systems. LNCS (LNAI), vol. 4994, pp. 309–318. Springer, Heidelberg (2008)
14. Bobillo, F., Straccia, U.: Fuzzy description logics with general t-norms and datatypes. Fuzzy Sets and Systems 160(23), 3382–3402 (2009)
15. Bobillo, F., Straccia, U.: Aggregations operators and fuzzy owl 2. In: 2011 International Conference on Fuzzy Systems (FUZZ 2011). IEEE Computer Society, Los Alamitos (2011)
16. Bobillo, F., Straccia, U.: Fuzzy ontology representation using owl 2. International Journal of Approximate Reasoning (2011)
17. Borgwardt, S., Peñaloza, R.: Description logics over lattices with multi-valued ontologies. In: Proceedings of the Twenty-Second International Joint Conference on Artificial Intelligence, IJCAI 2011 (to appear, 2011)
18. Borgwardt, S., Peñaloza, R.: Fuzzy ontologies over lattices with t-norms. In: Proceedings of the 24th International Workshop on Description Logics (DL 2011), CEUR Electronic Workshop Proceedings (to appear, 2011)
19. Brickley, D., Guha, R.V.: RDF Vocabulary Description Language 1.0: RDF Schema. W3C Recommendation, W3C (2004), `http://www.w3.org/TR/rdf-schema/`
20. Buneman, P., Kostylev, E.: Annotation algebras for rdfs. In: The Second International Workshop on the role of Semantic Web in Provenance Management (SWPM 2010), CEUR Workshop Proceedings (2010)

21. Calvanese, D., Giacomo, G., Lembo, D., Lenzerini, M., Rosati, R.: Tractable reasoning and efficient query answering in description logics: The dl-lite family. Journal of Automated Reasoning 39(3), 385–429 (2007)
22. Cerami, M., Straccia, U.: Undecidability of KB satisfiability for ł-$\mathcal{ALC}$ with GCIs (July 2011) (Unpublished Manuscript)
23. Cuenca-Grau, B., Horrocks, I., Motik, B., Parsia, B., Patel-Schneider, P.F., Sattler, U.: OWL 2: The next step for OWL. Journal of Web Semantics 6(4), 309–322 (2008)
24. Damásio, C.V., Medina, J., Ojeda Aciego, M.: A tabulation proof procedure for residuated logic programming. In: Proceedings of the 6th European Conference on Artificial Intelligence, ECAI 2004 (2004)
25. Dantsin, E., Eiter, T., Gottlob, G., Voronkov, A.: Complexity and expressive power of logic programming. ACM Computing Surveys 33(3), 374–425 (2001)
26. Description Logics Web Site, `http://dl.kr.org`
27. Dubois, D., Prade, H.: Can we enforce full compositionality in uncertainty calculi? In: Proc. of the 12th Nat. Conf. on Artificial Intelligence (AAAI 1994), Seattle, Washington, pp. 149–154 (1994)
28. Dubois, D., Prade, H.: Possibility theory, probability theory and multiple-valued logics: A clarification. Annals of Mathematics and Artificial Intelligence 32(1-4), 35–66 (2001)
29. Elkan, C.: The paradoxical success of fuzzy logic. In: Proc. of the 11th Nat. Conf. on Artificial Intelligence (AAAI 1993), pp. 698–703 (1993)
30. Grosof, B.N., Horrocks, I., Volz, R., Decker, S.: Description logic programs: combining logic programs with description logic. In: Proceedings of the Twelfth International Conference on World Wide Web, pp. 48–57. ACM Press, New York (2003)
31. Guarino, N., Poli, R.: Formal ontology in conceptual analysis and knowledge representation. International Journal of Human and Computer Studies 43(5/6), 625–640 (1995)
32. Haarslev, V., Pai, H.-I., Shiri, N.: Optimizing tableau reasoning in alc extended with uncertainty. In: Proceedings of the 2007 International Workshop on Description Logics, DL 2007 (2007)
33. Habiballa, H.: Resolution strategies for fuzzy description logic. In: Proceedings of the 5th Conference of the European Society for Fuzzy Logic and Technology (EUSFLAT 2007), vol. 2, pp. 27–36 (2007)
34. Hähnle, R.: Many-valued logics and mixed integer programming. Annals of Mathematics and Artificial Intelligence 3,4(12), 231–264 (1994)
35. Hähnle, R.: Advanced many-valued logics. In: Gabbay, D.M., Guenthner, F. (eds.) Handbook of Philosophical Logic, 2nd edn., vol. 2. Kluwer, Dordrecht (2001)
36. Hájek, P.: Metamathematics of Fuzzy Logic. Kluwer, Dordrecht (1998)
37. Ianni, G., Krennwallner, T., Martello, A., Polleres, A.: A rule system for querying persistent rdfs data. In: Aroyo, L., Traverso, P., Ciravegna, F., Cimiano, P., Heath, T., Hyvönen, E., Mizoguchi, R., Oren, E., Sabou, M., Simperl, E. (eds.) ESWC 2009. LNCS, vol. 5554, pp. 857–862. Springer, Heidelberg (2009)
38. Kifer, M., Lausen, G., Wu, J.: Logical foundations of Object-Oriented and frame-based languages. Journal of the ACM 42(4), 741–843 (1995)
39. Kifer, M., Subrahmanian, V.S.: Theory of generalized annotated logic programming and its applications. Journal of Logic Programming 12, 335–367 (1992)
40. Klement, E.P., Mesiar, R., Pap, E.: Triangular Norms. Trends in Logic - Studia Logica Library. Kluwer Academic Publishers, Dordrecht (2000)

41. Konstantopoulos, S., Apostolikas, G.: Fuzzy-dl reasoning over unknown fuzzy degrees. In: Proceedings of the 2007 OTM Confederated International Conference on On the Move to Meaningful Internet Systems, OTM 2007, vol. Part II, pp. 1312–1318. Springer, Heidelberg (2007)
42. Lakshmanan, L.V.S., Shiri, N.: A parametric approach to deductive databases with uncertainty. IEEE Transactions on Knowledge and Data Engineering 13(4), 554–570 (2001)
43. Lloyd, J.W.: Foundations of Logic Programming. Springer, Heidelberg (1987)
44. Lukasiewicz, T., Straccia, U.: Top-k retrieval in description logic programs under vagueness for the semantic web. In: Prade, H., Subrahmanian, V.S. (eds.) SUM 2007. LNCS (LNAI), vol. 4772, pp. 16–30. Springer, Heidelberg (2007)
45. Lukasiewicz, T., Straccia, U.: Managing uncertainty and vagueness in description logics for the semantic web. Journal of Web Semantics 6, 291–308 (2008)
46. Marin, D.: A formalization of rdf. Technical Report TR/DCC-2006-8, Deptartment of Computer Science, Universidad de Chile (2004), `http://www.dcc.uchile.cl/cgutierr/ftp/draltan.pdf`
47. Muñoz, S., Pérez, J., Gutierrez, C.: Minimal deductive systems for rdf. In: Franconi, E., Kifer, M., May, W. (eds.) ESWC 2007. LNCS, vol. 4519, pp. 53–67. Springer, Heidelberg (2007)
48. Straccia, U., Lopes, N., Polleres, A., Zimmermann, A.: Anql: Sparqling up annotated rdf. In: Patel-Schneider, P.F., Pan, Y., Hitzler, P., Mika, P., Zhang, L., Pan, J.Z., Horrocks, I., Glimm, B. (eds.) ISWC 2010, Part I. LNCS, vol. 6496, pp. 518–533. Springer, Heidelberg (2010)
49. OWL Web Ontology Language overview, W3C (2004), `http://www.w3.org/TR/owl-features/`
50. OWL 2 Web Ontology Language Document Overview, W3C (2009), `http://www.w3.org/TR/2009/REC-owl2-overview-20091027/`
51. Papadimitriou, C.H.: Computational Complexity. Addison Wesley Publ. Co., Reading (1994)
52. RDF Semantics, W3C (2004), `http://www.w3.org/TR/rdf-mt/`
53. Rule Interchange Format (RIF), W3C (2011), `http://www.w3.org/2001/sw/wiki/RIF`
54. Shapiro, E.Y.: Logic programs with uncertainties: A tool for implementing rule-based systems. In: Proceedings of the 8th International Joint Conference on Artificial Intelligence (IJCAI 1983), pp. 529–532 (1983)
55. SPARQL, `http://www.w3.org/TR/rdf-sparql-query/`
56. SPARQL, `http://www.w3.org/TR/sparql11-query/`
57. Stoilos, G., Simou, N., Stamou, G., Kollias, S.: Uncertainty and the semantic web. IEEE Intelligent Systems 21(5), 84–87 (2006)
58. Straccia, U.: Reasoning within fuzzy description logics. Journal of Artificial Intelligence Research 14, 137–166 (2001)
59. Straccia, U.: Transforming fuzzy description logics into classical description logics. In: Alferes, J.J., Leite, J. (eds.) JELIA 2004. LNCS (LNAI), vol. 3229, pp. 385–399. Springer, Heidelberg (2004)
60. Straccia, U.: Description logics with fuzzy concrete domains. In: Bachus, F., Jaakkola, T. (eds.) 21st Conference on Uncertainty in Artificial Intelligence (UAI 2005), pp. 559–567. AUAI Press, Edinburgh (2005)
61. Straccia, U.: Uncertainty management in logic programming: Simple and effective top-down query answering. In: Khosla, R., Howlett, R.J., Jain, L.C. (eds.) KES 2005. LNCS (LNAI), vol. 3682, pp. 753–760. Springer, Heidelberg (2005)

62. Straccia, U.: Answering vague queries in fuzzy DL-Lite. In: Proceedings of the 11th International Conference on Information Processing and Managment of Uncertainty in Knowledge-Based Systems (IPMU 2006), pp. 2238–2245. E.D.K, Paris (2006)
63. Straccia, U.: Description logics over lattices. International Journal of Uncertainty, Fuzziness and Knowledge-Based Systems 14(1), 1–16 (2006)
64. Straccia, U.: A fuzzy description logic for the semantic web. In: Sanchez, E. (ed.) Fuzzy Logic and the Semantic Web, Capturing Intelligence. ch.4, pp. 73–90. Elsevier, Amsterdam (2006)
65. Straccia, U.: Fuzzy description logic programs. In: Proceedings of the 11th International Conference on Information Processing and Managment of Uncertainty in Knowledge-Based Systems (IPMU 2006), pp. 1818–1825. E.D.K, Paris (2006)
66. Straccia, U.: Towards top-k query answering in deductive databases. In: Proceedings of the 2006 IEEE International Conference on Systems, Man and Cybernetics (SMC 2006), pp. 4873–4879. IEEE, Los Alamitos (2006)
67. Straccia, U.: Towards vague query answering in logic programming for logic-based information retrieval. In: Melin, P., Castillo, O., Aguilar, L.T., Kacprzyk, J., Pedrycz, W. (eds.) IFSA 2007. LNCS (LNAI), vol. 4529, pp. 125–134. Springer, Heidelberg (2007)
68. Straccia, U.: Managing uncertainty and vagueness in description logics, logic programs and description logic programs. In: Baroglio, C., Bonatti, P.A., Małuszyński, J., Marchiori, M., Polleres, A., Schaffert, S. (eds.) Reasoning Web. LNCS, vol. 5224, pp. 54–103. Springer, Heidelberg (2008)
69. Straccia, U.: A minimal deductive system for general fuzzy RDF. In: Polleres, A., Swift, T. (eds.) RR 2009. LNCS, vol. 5837, pp. 166–181. Springer, Heidelberg (2009)
70. Straccia, U., Lopes, N., Lukacsy, G., Polleres, A.: A general framework for representing and reasoning with annotated semantic web data. In: Proceedings of the Twenty-Fourth AAAI Conference on Artificial Intelligence (AAAI 2010), pp. 1437–1442. AAAI Press, Menlo Park (2010)
71. Torra, V., Narukawa, Y.: Information Fusion and Aggregation Operators. In: Cognitive Technologies. Springer, Heidelberg (2007)
72. Ullman, J.D.: Principles of Database and Knowledge Base Systems, vol. 1,2. Computer Science Press, Potomac (1989)
73. Vojtás, P.: Fuzzy logic programming. Fuzzy Sets and Systems 124, 361–370 (2001)
74. XML, W3C, `http://www.w3.org/XML/`
75. Yager, R.R.: On ordered weighted averaging aggregation operators in multicriteria decisionmaking. IEEE Trans. Syst. Man Cybern. 18, 183–190 (1988)
76. Zadeh, L.A.: Fuzzy sets. Information and Control 8(3), 338–353 (1965)

Logic Programming and Uncertainty

Chitta Baral

Faculty of Computer Science and Engineering
Arizona State University
Tempe, AZ 85287-8809
chitta@asu.edu

Abstract. In recent years Logic programming based languages and features–such as rules and non-monotonic constructs–have become important in various knowledge representation paradigms. While the early logic programming languages, such as Horn logic programs and Prolog did not focus on expressing and reasoning with uncertainty, in recent years logic programming languages have been developed that can express both logical and quantitative uncertainty. In this paper we give an overview of such languages and the kind of uncertainty they can express and reason with. Among those, we slightly elaborate on the language P-log that not only accommodates probabilistic reasoning, but also respects causality and distinguishes observational and action updates.

1 Introduction

Uncertainty is commonly defined in dictionaries [1] as the state or condition of being uncertain. The adjective, uncertain, whose origin goes back to the 14th century, is ascribed the meanings, "not accurately known", "not sure" and "not precisely determined". These meanings indirectly refer to a reasoner who does not accurately know, or is not sure, or cannot determine something precisely. In the recent literature uncertainty is classified in various ways. In one taxonomy [38], it is classified to finer notions such as subjective uncertainty, objective uncertainty, epistemic uncertainty, and ontological uncertainty. In another taxonomy, uncertainty is classified based on the approach used to measure it. For example, probabilistic uncertainty, is measured using probabilities, and in that case, various possible worlds have probabilities associated with them.

Although the initial logic programming formulations did not focus on uncertainty, the current logic programming languages accommodate various kinds of uncertainty. In this overview paper we briefly discuss some of the kinds of uncertainty that can be expressed using the logic programming languages and their implications.

The early logic programming formulations are the language Prolog and Horn logic programs [13,23]. A Horn logic program, also referred to as a definite program is a collection of rules of the form: $a_0 \leftarrow a_1, \ldots, a_n.$ with $n \geq 0$ and where $a_0, \ldots, a_n$ are atoms in the sense of first order logic. The semantics of such programs can be defined using the notion of a least model or through the least fixpoint of a meaning accumulating operator [13,23].

For example, the least model of the program:

$a \leftarrow b, c.$
$d \leftarrow e.$

S. Benferhat and J. Grant (Eds.): SUM 2011, LNAI 6929, pp. 22–37, 2011.
© Springer-Verlag Berlin Heidelberg 2011

$b \leftarrow .$
$c \leftarrow .$

is $\{b, c, a\}$ and based on the semantics defined using the least model one can conclude that the program entails $b, c, a, \neg d$ and $\neg e$. The entailment of $\neg d$ and $\neg e$ is based on the closed world assumption [34] associated with the semantics of a Horn logic program. Thus there is no uncertainty associated with Horn logic programs.

Although Prolog grew out of Horn logic programs, and did not really aim to accommodate uncertainty, some Prolog programs can go into infinite loops with respect to certain queries and one may associate a kind of "uncertainty" value to that. Following are some examples of such programs.

P_1: $a \leftarrow a.$
$b \leftarrow .$

P_2: $a \leftarrow not\ a, c.$
$b \leftarrow .$

P_3: $a \leftarrow not\ b.$
$b \leftarrow not\ a.$
$p \leftarrow a.$
$p \leftarrow b.$

With respect to the Prolog programs P_1 and P_2 a Prolog query asking about a may take the interpreter to an infinite loop, and with respect to the program P_3 a Prolog query asking about a, a Prolog query asking about b and a Prolog query asking about p could each take the interpreter to an infinite loop.

In the early days of logic programming, such programs were considered "bad" and writing such programs was "bad programming." However, down the road, there was a movement to develop logic programming languages with clean declarative semantics, and Prolog with its non-declarative constructs was thought more as a programming language with some logical features and was not considered a declarative logic programming language. With the changed focus on clean declarative semantics, P_1, P_2 and P_3 were no longer bad programs and attempts were made to develop declarative semantics that could graciously characterize these programs as well as other syntactically correct programs. This resulted in several competing semantics and on some programs the different semantics would give different meanings. For example, for the program P_3, the stable model semantics [16] would have two different stable models $\{a, p\}$ and $\{b, p\}$ while the well-founded semantics [39] will assign the value unknown to a, b and p.

The important point to note is that unlike Horn logic programs, both stable model semantics and well-founded semantics allow characterization of some form of "uncertainty". With respect to P_3 the stable model semantics effectively encodes two possible worlds, one where a and p are true (and b is false) and another where b and p are true (and a is false). On the other hand the well-founded semantics does not delve into possible worlds; it just pronounces a, b and p to be unknown.

On a somewhat parallel track Minker and his co-authors [24] promoted the use of disjunctions in the head of logic programming rules, thus allowing explicit expression of uncertainty. An example of such a program is as follows.

P_4: $a \; or \; b \leftarrow .$
$p \leftarrow a.$
$p \leftarrow b.$

The program P_4 was characterized using its minimal models and had two minimal models $\{a, p\}$ and $\{b, p\}$. As in the case of stable models one could consider these two minimal models as two possible worlds. In both cases one can add probabilistic uncertainty by assigning probabilities to the possible models.

In the rest of the paper we give a brief overview of various logic programming languages that can express uncertainty and reason with it. We divide our overview to two parts; one where we focus on logical uncertainty without getting into numbers and another where we delve into numbers. After that we conclude and mention some future directions.

2 Logical Uncertainty in Logic Programming

Logical uncertainty can be expressed in logic programming in various ways. In the previous section we mentioned how uncertainty can be expressed using the stable model semantics as well as using disjunctions in the head of programs. We now give the formal definition of stable models for programs that may have disjunctions in the head of rules. A logic program is then a collection of rules of the form:

$$a_0 \; or \ldots or \; a_k \leftarrow a_{k+1}, \ldots, a_m, \; not \; a_{m+1}, \ldots, \; not \; a_n.$$

with $k \geq 0$, $m \geq k$, $n \geq m$, and where $a_0, \ldots, a_n$ are atoms in the sense of first order logic. The semantics of such programs is defined in terms of stable models. Given such a program P, and a set of atoms S, the Gelfond-Lifschitz transformation of P with respect to S gives us a program P^S which does not have any *not* in it. This transformation is obtained in two steps as follows:

(i) All rules in P which contains $not \; p$ in its body for some p in S are removed.

(ii) For each of the remaining rules the $not \; q$ in the bodies of the rules are removed.

A stable model of the program P is defined as any set of atoms S such that S is a minimal model of the program P^S. An atom a is said to be true with respect to a stable model S if $a \in S$ and a negative literal $\neg a$ is said to be true with respect to a stable model S if $a \notin S$. The following examples illustrates the above definition. Consider the program

P_5: $a \leftarrow not \; b.$
$b \leftarrow not \; a.$
$p \; or \; q \leftarrow a.$
$p \leftarrow b.$

This program has three stable models $\{a, p\}$, $\{a, q\}$ and $\{b, p\}$. This is evident from noting that $P_5^{\{a,p\}}$ is the program:

$a \leftarrow .$
$p \; or \; q \leftarrow a.$
$p \leftarrow b.$

and $\{a, p\}$ is a minimal model of $P_5^{\{a,p\}}$. Similarly, it can be shown that $\{a, q\}$ and $\{b, p\}$ are also stable models of P_5.

As we mentioned earlier, the logical uncertainty expressible using logic programs is due to both the disjunctions in the head as well as due to the possibility that even programs without disjunctions may have multiple stable models. However, in the absence of function symbols, there is a difference between the expressiveness of logic programs that allow disjunction in their head and the ones that do not. Without disjunctions the logic programs capture the class coNP, while with disjunctions they capture the class Π_2P [8].

In the absence of disjunctions, rules of the kind

P_6: $a \leftarrow not\ n\_a.$
$n\_a \leftarrow not\ a.$

allow the enumeration of the various possibilities and rules of the form

P_7: $p \leftarrow not\ p, q.$

allow elimination of stable models where certain conditions (such as q) may be true. The elimination rules can be further simplified by allowing rules with empty head. In that case the above rule can be simply written as: P_8: $\leftarrow q.$ When rules with empty heads, such as in P_8, are allowed, one can replace the constructs in P_6 by exclusive disjunctions [20] of the form: P_9: $a \oplus n\_a \leftarrow .$ to do the enumeration, and can achieve the expressiveness to capture the class coNP with such exclusive disjunctions, rules with empty heads as in P_8 and stratified negation. The paper [20] advocates this approach with the argument that many find the use of unrestricted negation to be unintuitive and complex. On the other hand use of negation is crucial in many knowledge representation tasks and while using them having not to worry whether the negation used is stratified or not makes the task simpler for humans.

2.1 Answer Sets and Use of Classical Negation

Allowing classical negation in logic programs gives rise to a different kind of uncertainty. For example the program

P_{10}: $a \leftarrow b.$
$\neg b \leftarrow .$

has a unique answer set $\{\neg b\}$ and with respect to that answer set the truth value of a is unknown. We now give the formal definition of answer sets for programs that allows classical negation. A logic program is then a collection of rules of the form:

$$l_0\ or \ldots or\ l_k \leftarrow l_{k+1}, \ldots, l_m,\ not\ l_{m+1}, \ldots,\ not\ l_n.$$

with $k \geq 0$, $m \geq k$, $n \geq m$, and where $l_0, \ldots, l_n$ are literals in the sense of first order logic. The semantics of such programs is defined in terms of answer sets [17]. Given such a program P, an answer set of the program P is defined as a consistent set of literals S such that S satisfies all rules in P^S and no proper subset of S satisfies all rules of P^S, where P^S is as defined earlier. A literal l is defined to be true with respect to an answer set S if $l \in S$. With the use of classical negation one need not invent new atoms for the enumeration, as was done in P_6, and simply write:

P_{11}: $a \leftarrow not\ \neg a.$
$\neg a \leftarrow not\ a.$

Moreover, since answer sets are required to be consistent, one need not write explicit rules of the kind: $\leftarrow a, \neg a.$ which were sometimes explicitly needed to be written when not using classical negation.

2.2 Other Logic Programming Languages and Systems for Expressing Logical Uncertainty

Other logic programming languages that can express logical uncertainty include abductive logic programs [21] and various recent logic programming languages. Currently there are various logic programming systems that one can be use to express logical uncertainty. The most widely used are Smodels [29], DLV [11] and the Potassco suite [15].

3 Multi-valued and Quantitative Uncertainty in Logic Programming

Beyond logical uncertainty that we discussed in the previous section where one could reason about truth, falsity and lack of knowledge using logic programming, one can classify uncertainty in logic programming in several dimensions.

- The truth values may have an associated degree of truth and falsity or we may have multi-valued truth values.
- The degree of truth or the values (in the multi-valued case) could be discrete or continuous.
- They can be associated with the whole rule or with each atom (or literal) in the rule.
- The formalism follows or does not follow axioms of probability.
- The formalism is motivated by concerns to learn rules and programs.

Examples of logic programming with more than three discrete truth values include use of bi-lattice in logic programming in [14], use of annotated logics in [6] and various fuzzy logic programming languages.

Among the various quantitative logic programming languages the recollection [37] considers Shapiro's quantitative logic programming [36] as the first "serious" paper on the topic. Shapiro assigned a mapping to each rule; the mapping being from numbers in (0,1] associated with each of the atoms in the body of the rule to a number in (0,1] to be associated with the atom in the head of the rule. He gave a model-theoretic semantics and developed a meta-interpreter. A few years later van Emden [12] considered the special case where numbers were only associated with a rule and gave a fixpoint and a sound and conditionally-complete proof theory.

A large body of work on quantitative logic programming has been done by Subrahmanian with his students and colleagues. His earliest work used the truth values [0,1] $\cup \{*\}$, where $*$ denoted inconsistency and as per the recollection [37] it was "the first work that explicitly allowed a form of negation to appear in the head." This was followed by his work on paraconsistent logic programming [6] where truth values could be from any lattice. He and his colleagues further generalized paraconsistency to generalized annotations and generalized annotated programs where complex terms could be used as annotations.

3.1 Logic Programming with Probabilities

The quantitative logic programming languages mentioned earlier, even when having numbers, did not treat them as probabilities. In this section we discuss various logic programming languages that accommodate probabilities[1].

Probabilistic Logic Programming

The first probabilistic logic programming language was proposed by Ng and Subrahmanian [27]. Rules in this language were of the form:

$$a_0 : [\alpha_0, \beta_0] \leftarrow a_1 : [\alpha_1, \beta_1], \ldots, a_n : [\alpha_n, \beta_n].$$

with $n \geq 0$ and where $a_0, \ldots, a_n$ are atoms in the sense of first-order logic, and $[\alpha_i, \beta_i] \subseteq [0, 1]$. Intuitively, the meaning of the above rule is that if the probability of a_j being true is in the interval $[\alpha_j, \beta_j]$, for $1 \leq j \leq n$, then the probability of a_0 being true is in the interval $[\alpha_0, \beta_0]$. Ng and Subrahmanian gave a model theoretic and a fixpoint characterization of such programs and also gave a sound and complete query answering method. The semantics made the "ignorance" assumption that nothing was known about any dependencies between the events denoted by the atoms. Recently a revised semantics for this language has been given in [9].

Ng and Subrahmanian later extend the language to allow a_i's to be conjunction and disjunction of atoms and the a_is in the body were allowed to have not preceding them. In presence of not the semantics was given in a manner similar to the definition of stable models.

Dekhtyar and Subrahmanian [10] further generalized this line of work to allow explicit specification of the assumptions regarding dependencies between the events denoted by the atoms that appear in a disjunction or conjunction. Such assumptions are referred to as *probabilistic strategies* and examples of probabilistic strategies include: (i) independence, (ii) ignorance, (iii) mutual exclusion and (iv) implication. While some of the probabilistic logic programming languages assume one of these strategies and hard-code the semantics based on that, the hybrid probabilistic programs of [10] allowed one to mention the probabilistic strategies used in each conjunction or disjunction. For example, $\wedge_{ind}$ and $\vee_{ind}$ would denote the conjunction and disjunction associated with the "independence" assumption and would have the property that $Prob(e_1 \wedge_{ind} \ldots \wedge_{ind} e_n) = Prob(e_1) \times \ldots \times Prob(e_n)$. Following are examples, from [10], of rules of hybrid probabilistic programs:

$$price\_drop(C) : [.4, .9] \leftarrow (ceo\_sells\_stock(C) \vee_{igd} ceo\_retires(C)) : [.6, 1].$$
$$price\_drop(C) : [.5, 1] \leftarrow (strike(C) \vee_{ind} accident(C)) : [.3, 1].$$

The intuitive meaning of the first rule is that if the probability of the CEO of a company selling his/her stock or retiring is greater than 0.6 then the probability of the price dropping is between 0.4 and 0.9, and it is assumed that the relationship between the CEO retiring and selling stock is not known. The intuitive meaning of the second rule is that

[1] Some of these were discussed in our earlier paper [4], but the focus there was comparison with P-log.

if the probability of a strike happening or an accident happening –which are considered to be independent–is greater than 0.3 then the probability of the price dropping is greater than 0.5.

Lukaciewicz [25] proposed the alternative of using conditional probabilities in probabilistic logic programs. In his framework clauses were of the form: $(H \mid B)[\alpha_1, \beta_1]$

where H and B are conjunctive formulas and $0 \leq \alpha_1 \leq \beta_1 \leq 1$, and a probabilistic logic program consisted of several such clauses. The intuitive meaning of the above clause is that the conditional probability of H given B is between α_1 and β_1. Given a program consisting of a set of such clauses the semantics is defined based on models where each model is a probability distribution that satisfies each of the clauses in the program.

Bayesian Logic Programming

Bayesian logic programs [22] are motivated by Bayes nets and build up on an earlier formalism of probabilistic knowledge bases [28] and add some first-order syntactic features to Bayes nets so as to make them relational. A Bayesian logic program has two parts, a logical part that looks like a logic program and a set of conditional probability tables. A rule or a clause of a Bayseian logic program is of the form: $H \mid A_1, \ldots, A_n$

where $H, A_1, \ldots, A_n$ are atoms which can take a value from a given domain associated with the atom. An example of such a clause is:

$highest\_degree(X) \mid instructorr(X).$

Its corresponding domain could be, for example, $D_{instructor} = \{yes, no\}$, and $D_{highest\_degree} = \{phd, masters, bachelors\}$. Each such clause has an associated conditional probability table (CPT). For example, the above clause may have the following table:

instructor(X)	highest_degree(X) phd	highest_degree(X) masters	highest_degree(X) bachelors
yes	0.7	0.25	0.05
no	0.05	0.3	0.65

Acyclic Bayesian logic programs are characterized by considering their grounded versions. If the ground version has multiple rules with the same ground atom in the head then combination rules are specified to combine these rules to a single rule with a single associated conditional probability table.

Stochastic Logic Programs

Stochastic logic programs [26] are motivated from the perspective of machine learning and are generalization of stochastic grammars. Consider developing a grammar for a natural language such that the grammar is not too specific and yet is able to address ambiguity. This is common as we all know grammar rules which work in most cases but not necessarily in all cases and yet with our experience we are able to use those rules. In statistical parsing one uses a stochastic grammar where production rules have associated weight parameters that contribute to a probability distribution. Using those weight parameters one can define a probability function $Prob(w|s,p)$, where s is a sentence, w is a parse and p is the weight parameter associated with the production rules of the grammar. Given a grammar and its associated p, a new sentence s' is parsed to w' such that $Prob(w'|s',p)$ is the maximum among all possible parses of s'. The weight parameter p is learned from a given training set of example sentences and their parses. In the learning process, given examples of sets of sentences and parses $\{(s_1,w_1),\ldots,(s_n,w_n)\}$ one has to come up with the p that maximizes the probability that the s_i's in the training set are parsed to the w_i's.

Motivated by stochastic grammars and with the goal to allow inductive logic programs to have associated probability distributions, [26] generalized stochastic grammars to stochastic logic programs. In stochastic logic programs [26] a number in [0,1], referred to as a "probability label," is associated with each rule of a Horn logic program with the added conditions that the rules be range restricted and for each predicate symbol q, the probability labels for all clauses with q in the head sum to 1. Thus, a Stochastic logic program [26] P is a collection of clauses of the form

$$p\ :\ a_0 \leftarrow a_1,\ldots,a_n.$$

where p (referred to as the the probability label) belongs to $[0,1]$, and $a_0,a_1,\ldots a_n$ are atoms. The probability of an atom g with respect to a stochastic logic program P is obtained by summing the probability of the various SLD-refutation of $\leftarrow g$ with respect to P, where the probability of a refutation is computed by multiplying the probability of various choices; and doing appropriate normalization. For example, if the first atom of a subgoal $\leftarrow g'$ unifies with the head of the stochastic clause $p_1\ :\ C_1$, also with the head of the stochastic clause $p_2\ :\ C_2$ and so on up to the head of the stochastic clause $p_m\ :\ C_m$, and the stochastic clause $p_i\ :\ C_i$ is chosen for the refutation, then the probability of this choice is $\frac{p_i}{p_1+\cdots+p_m}$.

Modularizing Probability and Logic Aspects: Independent Choice Logic

Earlier in Section 2 we discussed how one can express logical uncertainty using logic programming. One way to reason with probabilities in logic programming is to assign probabilities to the "possible worlds" defined by the approaches in Section 2. Such an approach is taken by Poole's Independent Choice Logic of [31,32], a refinement of his earlier work on probabilistic Horn abduction [33].

There are three components of an Independent Choice Logic of interest here: a choice space $\mathcal{C}$, a rule base $\mathcal{F}$ and a probability distribution on $\mathcal{C}$ such that $\Sigma_{X\in\mathcal{C}}Prob(X)=1$. A Choice space $\mathcal{C}$ is a set of sets of ground atoms such that if $X_1\in\mathcal{C}$, $X_2\in\mathcal{C}$ and $X_1\neq X_2$ then $X_1\cap X_2=\emptyset$. An element of $\mathcal{C}$ is referred to as an "alternative" and an element of an "alternative" is referred to as an "atomic choice". A rule base $\mathcal{F}$ is a logic program such that no atomic choice unifies with the head of any of its rule and it has a unique stable model. The unique stable model condition can be enforced by restrictions such as requiring the program to be an acyclic program without disjunctions. $\mathcal{C}$ and $\mathcal{F}$ together define the set of possible worlds and the probability distribution on $\mathcal{C}$ can then be used to assign probabilities to the possible worlds. These probabilities can then be used in the standard way to define probabilities of formulas and conditional probabilities.

Logic Programs with Distribution Semantics: PRISM

The formalism of Sato [35], which he refers to as PRISM as a short form for "PRogramming In Statistical Modeling", is very similar to Independent Choice Logic. A PRISM formalism has a possibly infinite collection of ground atoms, F, the set Ω_F of all interpretations of F, and a completely additive probability measure P_F which quantifies the likelihood of the interpretations. P_F is defined on some fixed σ algebra of subsets of Ω_F.

In Sato's framework interpretations of F can be used in conjunction with a Horn logic program R, which contains no rules whose heads unify with atoms from F. Sato's logic program is a triple, $\Pi = \langle F, P_F, R\rangle$. The semantics of Π is given by a collection Ω_Π of possible worlds and the probability measure P_Π. A set M of ground atoms in the language of Π belongs to Ω_Π iff M is a minimal Herbrand model of a logic program $I_F \cup R$ for some interpretation I_F of F. The completely additive probability measure of P_Π is defined as an extension of P_F.

The emphasis of the original work by Sato and other PRISM related research is on the use of the formalism for design and investigation of efficient algorithms for statistical learning. The goal is to use the pair $DB = \langle F, R\rangle$ together with observations of atoms from the language of DB to learn a suitable probability measure P_F.

Logic Programming with Annotated Disjunctions

In the LPAD formalism of Vennekens et al. [40] rules have choices in their head with associate probabilities. Thus an LPAD program consists of rules of the form:

$$(h_1 : \alpha_1) \vee \ldots \vee (h_n : \alpha_n) \leftarrow b_1, \ldots, b_m$$

where h_i's are atoms, b_is are atoms or atoms preceded by *not*, and $\alpha_i \in [0, 1]$, such that $\sum_{i=1}^{n} \alpha_i = 1$. An LPAD rule instance is of the form: $h_i \leftarrow b_1, \ldots, b_m$.

The associated probability of the above rule instance is then said to be α_i. An instance of an LPAD program P is a logic program P' obtained as follows: for each rule in P exactly one of its instance is included in P', and nothing else is in P'. The associated probability of an instance P', denoted by $\pi(P')$, of an LPAD program is the product of the associated probability of each of its rules.

An LPAD program is said to be sound if each of its instance has a 2-valued well-founded model. Given an LPAD program P, and a collection of atoms I, the probability assigned to I by P is given as follows:

$$\pi_P(I) = \sum_{P' \text{ is an instance of } P \text{ and } I \text{ is the well-founded model of } P'} \pi(P')$$

The probability of a formula ϕ assigned by an LPAD program P is then defined as:

$$\pi_P(\phi) = \sum_{\phi \text{ is satisfied by } I} \pi_P(I)$$

4 Logic Programming with Probabilities, Causality and Generalized Updates: P-log

An important design aspect of developing knowledge representation languages and representing knowledge in them is to adequately address how knowledge is going to be updated. If this is not thought through in the design and representation phase then updating a knowledge base may require major surgery. For this reason updating a knowledge base in propositional logic or first-order logic is hard. This is also one of the motivations behind the development of non-monotonic logics which have constructs that allow elaboration tolerance.

The probabilistic logic programming language P-log was developed with updates and elaboration tolerance in mind. In particular, it allows one to easily change the domain of the event variables. In most languages the possible values of a random variable get restricted with new observations. P-log with its probabilistic non-monotonicity allows the other way round too. Another important aspect of updating in P-log is that it differentiates between updating due to new observations and updating due to actions; this is especially important when expressing causal knowledge.

Elaborating on the later point, an important aspect of probabilistic uncertainty that is often glossed over is the proper representation of joint probability distributions. Since the random variables in a joint probability distribution are often not independent of each other, and since representing the joint probability distribution explicitly is exponential in the number of variables, techniques such as Bayseian networks are used. However, as pointed out by Pearl [30], such representations are not amenable to distinguish between observing the value of a variable and execution of actions that change the value of the variable. As a result prior to Pearl (and even now) most probability formalisms are not able to express action queries such as the probability that X has value a given that Y's value is made to be b. Note that this is different from the query about the probability that X has value a given that Y's value is observed to be b. To be able to address this accurately a causal model of probability is needed. P-log takes that view and is able to express both the above kind of queries and distinguishes between them.

With the above motivations we give a brief presentation on P-log[2] starting with its syntax and semantics and following up with several illustrative examples.

A P-log program consists of a declaration of the domain, a logic program without disjunctions, a set of random selection rules, a set of probability atoms, and a collection of observations and action atoms.

The declaration of the domain consists of sort declarations of the form $c = \{x_1, \ldots, x_n\}$. or consists of a logic program T with a unique answer set A. In the latter case $x \in c$ iff $c(x) \in A$. The domain and range of attributes[3] are given by statements of the form: $a : c_1 \times \cdots \times c_n \rightarrow c_0$.

[2] Our presentation is partly based on our earlier paper [4].

[3] Attributes are relational variables. In probabilistic representations, a variable such as Color can take the value from {red, green, blue, ... }. Now if we want talks about colors of cars, then color is a function from a set of cars to {red, green, blue, ... }. In that case we call "color" an attribute.

A random selection rule is of the form

$$[\, r\,]\; random(a(\bar{t}) : \{X : p(X)\}) \leftarrow B. \tag{1}$$

where r is a term used to name the rule and B is a collection of extended literals of the form l or $not\ l$, where l is a literal. Statement (1) says that *if B holds, the value of $a(\bar{t})$ is selected at random from the set $\{X : p(X)\} \cap range(a)$ by experiment r, unless this value is fixed by a deliberate action.*

A probability atom is of the form: $pr_r(a(\bar{t}) = y \mid_c B) = v.$ where $v \in [0, 1]$, B is a collections of extended literals, *pr* is a special symbol, r is the name of a random selection rule for $a(\bar{t})$, and $pr_r(a(\bar{t}) = y \mid_c B) = v$ says that *if the value of $a(\bar{t})$ is fixed by experiment r, and B holds, then the probability that r causes $a(\bar{t}) = y$ is v.* (Note that here we use 'cause' in the sense that B is an immediate or proximate cause of $a(\bar{t}) = y$, as opposed to an indirect cause.)

Observations and action atoms are of the form: $obs(l).$ $do(a(\bar{t}) = y)).$

where l is a literal. Observations are used to record the outcomes of random events, i.e., random attributes, and attributes dependent on them.

We now illustrate the above syntax using an example from [4] about certain dices being rolled. In that example, there are two dices owned by Mike and John respectively. The domain declarations are given as follows:

$dice = \{d_1, d_2\}.$
$score = \{1, 2, 3, 4, 5, 6\}.$
$person = \{mike, john\}.$
$roll : dice \rightarrow score.$
$owner : dice \rightarrow person.$
$even : dice \rightarrow Boolean.$

The logic programming part includes the following:

$owner(d_1) = mike.$
$owner(d_2) = john.$
$even(D) \leftarrow roll(D) = Y, Y\ mod\ 2 = 0.$
$\neg even(D) \leftarrow not\ even(D).$

The fact that values of attribute $roll : dice \rightarrow score$ are random is expressed by the statement

$[\, r(D)\,]\; random(roll(D))$

The dice domain may include probability atoms that convey that the die owned by John is fair, while the die owned by Mike is biased to roll 6 at a probability of .25.

Let us refer to the P-log program consisting of the above parts as T_1.

$pr(roll(D) = Y \mid_c owner(D) = john) = 1/6.$
$pr(roll(D) = 6 \mid_c owner(D) = mike) = 1/4.$
$pr(roll(D) = Y \mid_c Y \neq 6, owner(D) = mike) = 3/20.$

In this domain the observation $\{obs(roll(d_1) = 4)\}$ records the outcome of rolling dice d_1. On the other hand the statement $\{do(roll(d_1) = 4)\}$ indicates that d_1 was simply put on the table in the described position. One can have observations such as

$obs(even(d_1))$ which means that it was observed that the dice d_1 had an even value. Here, even though $even(d_1)$ is not a random attribute, it is dependent on the random attribute $roll(d_1)$.

The semantics of a P-log program is given in two steps. First the various parts of a P-log specification is translated to logic programs and then the answer sets of the translated program is computed and are treated as possible worlds and probabilities are computed for them. The translation of a P-log specification Π to a logic program $\tau(\Pi)$ is as follows:

1. Translating the declarations: For every sort declaration $c = \{x_1, \ldots, x_n\}$ of Π, $\tau(\Pi)$ contains $c(x_1), \ldots, c(x_n)$. For all sorts that are defined using a logic program T in Π, $\tau(\Pi)$ contains T.
2. Translating the Logic programming part:
   (a) For each rule r in the logic programming part of Π, $\tau(\Pi)$ contains the rule obtained by replacing each occurrence of an atom $a(\bar{t}) = y$ in r by $a(\bar{t}, y)$.
   (b) For each attribute term $a(\bar{t})$, $\tau(\Pi)$ contains the rule:
   $$\neg a(\bar{t}, Y_1) \leftarrow a(\bar{t}, Y_2), Y_1 \neq Y_2. \tag{2}$$
   which guarantees that in each answer set $a(\bar{t})$ has at most one value.
3. Translating the random selections:
   (a) For an attribute a, we have the rule: $intervene(a(\bar{t})) \leftarrow do(a(\bar{t}, Y))$. where, intuitively, $intervene(a(\bar{t}))$ means that the value of $a(\bar{t})$ is fixed by a deliberate action. Semantically, $a(\bar{t})$ will not be considered random in possible worlds which satisfy $intervene(a(\bar{t}))$.
   (b) Each random selection rule of the form
   $$[\, r \,]\ random(a(\bar{t}) : \{Z : p(Z)\}) \leftarrow B.$$
   with $range(a) = \{y_1, \ldots, y_k\}$ is translated to the following rule:
   $$a(\bar{t}, y_1) \text{ or } \ldots \text{ or } a(\bar{t}, y_k) \leftarrow B,\ \ \text{not}\ \ intervene(a(\bar{t})) \tag{3}$$
   If the dynamic range of a in the selection rule is not equal to its static range, i.e. expression $\{Z : p(Z)\}$ is not omitted, then we also add the rule
   $$\leftarrow a(\bar{t}, y),\ \ \text{not}\ \ p(y), B,\ \ \text{not}\ \ intervene(a(\bar{t})). \tag{4}$$
   Rule (3) selects the value of $a(\bar{t})$ from its range while rule (4) ensures that the selected value satisfies p.
4. $\tau(\Pi)$ contains actions and observations of Π.
5. For each Σ-literal l, $\tau(\Pi)$ contains the rule: $\leftarrow obs(l),\ \ \text{not}\ \ l$.
6. For each atom $a(\bar{t}) = y$, $\tau(\Pi)$ contains the rule: $a(\bar{t}, y) \leftarrow do(a(\bar{t}, y))$.
   The last but one rule guarantees that no possible world of the program fails to satisfy observation l. The last rule makes sure the atoms that are made true by the action are indeed true.

The answer sets of the above translation are considered the possible worlds of the original P-log program. To illustrate how the above translation works, $\tau(T_1)$ of T_1 will consist of the following:

$dice(d_1).\ dice(d_2).\ score(1).\ score(2).$
$score(3).\ score(4).\ score(5).\ score(6).$
$person(mike).\ person(john).$
$owner(d_1, mike).\ owner(d_2, john).$
$even(D) \leftarrow roll(D, Y), Y\ mod\ 2 = 0.$
$\neg even(D) \leftarrow \ \text{not}\ \ even(D).$
$\neg roll(D, Y_1) \leftarrow roll(D, Y_2), Y_1 \neq Y_2.$
$\neg owner(D, P_1) \leftarrow owner(D, P_2), P_1 \neq P_2.$
$\neg even(D, B_1) \leftarrow even(D, B_2), B_1 \neq B_2.$
$intervene(roll(D)) \leftarrow do(roll(D, Y)).$
$roll(D, 1)$ or ... or $roll(D, 6) \leftarrow B,$ not $intervene(roll(D)).$
$\leftarrow obs(roll(D, Y)),$ not $roll(D, Y).$
$\leftarrow obs(\neg roll(D, Y)),$ not $\neg roll(D, Y).$
$roll(D, Y)) \leftarrow do(roll(D, Y)).$

The variables D, P, B's, and Y's range over $dice$, $person$, $boolean$, and $score$ respectively.

Before we explain how the probabilities are assigned to the possible worlds, we mention a few conditions that the P-log programs are required to satisfy. They are:

(i) There can not be two random selection rules about the same attribute whose bodies are simultaneously satisfied by a possible world.

(ii) There can not be two probability atoms about the same attribute whose conditions can be simultaneously satisfied by a possible world.

(iii) A random selection rule can not conflict with a probability atom in such a way that probabilities are assigned outside the range given in the random selection rule.

The probabilities corresponding to each of the possible worlds are now computed in the following way:

(a) Computing an initial probability assignment P for each atom in a possible world: For a possible world W if the P-log program contains $pr_r(a(\bar{t}) = y \mid_c B) = v$ where r is the generating rule of $a(\bar{t}) = y$, W satsifies B, and W does not contain $intervene(a(\bar{t}))$, then $P(W, a(\bar{t}) = y) = v$.

(b) For any $a(\bar{t})$, the probability assignments obtained in step (a) are summed up and for the other possible values of $a(\bar{t})$ the remaining probability (i.e., 1 - the sum) is uniformly divided.

(c) The *unnormalized probability*, $\hat{\mu}_T(W)$, of a possible world W *induced by* a given P-log program T is $\hat{\mu}_T(W) = \prod_{a(\bar{t}, y) \in W} P(W, a(\bar{t}) = y)$ where the product is taken over atoms for which $P(W, a(\bar{t}) = y)$ is defined. The above measure is then normalized to $\mu_T(W)$ so that the sum of it for all possible worlds W is 1.

Using the above measure, the probability of a formula F with respect to a program T is defined as $Prob_T(F) = \Sigma_{W \models F} \mu_T(W)$.

We now show how P-log can be used to express updates not expressible in other probabilistic logic programming languages. Lets continue with the dice rolling example. Suppose we have a domain where the dices are normally rigged to roll 1 but once in a while there may be an abnormal dice that rolls randomly. This can be expressed in P-log by:

$roll(D) = 1 \leftarrow \; not\ abnormal(D)$
$random(roll(D)) \leftarrow abnormal(D)$

Updating such a P-log program with $obs(abnormal(d_1))$ will expand the value that $roll(d_1)$ can take.

Now let us consider an example that illustrates the difference between observational updates and action updates. Lets augment the dice domain with a new attribute $fire\_works$ which becomes true when dice d_1 rolls to 6. This can be expressed by the rule:

$fire\_works \leftarrow roll(d_1) = 6.$

Now suppose we observe fire works. This observation can be added to the P-log program as $obs(fire\_works)$, and when this observation is added to the P-log program it will eliminate the earlier possible worlds where $fire\_works$ was not true and as a result the probability that dice d_1 was rolled 6 will increase to 1. Now supposed instead of observing the fire works someone goes and starts the fire work. In that case the update to the P-log program would be $do(fire\_works = true)$. This addition will only add $fire\_works$ to all the previous possible worlds and as a result the probability that dice d_1 was rolled 6 will remain unchanged.

As suggested by the above examples, updating a P-log program basically involves adding to it. Formally, the paper [4] defines a notion of coherence of P-log programs and uses it to define updating a P-log program T by U as addition of U to T with the requirement that $T \cup U$ be coherent. The paper also shows that the traditional conditional probability $Prob(A|B)$ defined as $\frac{Prob(A \wedge B)}{Prob(B)}$ is equal to the $Prob_{T \cup obs(B)}(A)$ where $obs(B) = \{obs(l) : l \in B\}$.

Since the original work on P-log [3,4] which we covered in this section there have been several new results. This includes work on using P-log to model causality and counterfactual reasoning [5], implementation of P-log [19], an extension of P-log that allows infinite domains [18] and modular programming in P-log [7].

5 Conclusion and Future Directions

In this paper we have given a personal overview of representing and reasoning about uncertainty in logic programming. We started with a review of representing logical uncertainty in logic programming and then discussed some of the multi-valued and quantitative logic programming languages. We briefly discussed some of the probabilistic logic programming languages. Finally we discussed logic programming languages that have distinct logical and probabilistic components and concluded with the language of P-log that has distinct logical and probabilistic components, that allows a rich variety of updates and makes a distinction between observational updates and action updates. Our overview borrowed many examples, definitions and explanations from the book [2] and the articles [37] and [3,4]. We refer the reader to those articles and the original papers for additional details.

Although a lot has been done, there still is a big gap between knowledge representation (KR) languages that are used by humans to encode knowledge, KR languages that are learned and KR languages used in translating natural language to a formal language. We hope these gaps will be narrowed in the future, and to that end we need to develop ways to learn theories in the various logic programming languages that can express and reason with uncertainty. For example, it remains a challenge to explore how techniques from learning Bayes nets and statistical relational learning can be adapted to learn P-log theories. P-log also needs more efficient interpreters and additional refinements in terms of explicitly expressing probabilistic strategies.

References

1. Online etymology dictionary (July 2011), `http://dictionary.reference.com/browse/uncertain`
2. Baral, C.: Knowledge representation, reasoning and declarative problem solving. Cambridge University Press, Cambridge (2003)
3. Baral, C., Gelfond, M., Rushton, N.: Probabilistic reasoning with answer sets. In: Proceedings of LPNMR7, pp. 21–33 (2004)
4. Baral, C., Gelfond, M., Rushton, N.: Probabilistic reasoning with answer sets. TPLP 9(1), 57–144 (2009)
5. Baral, C., Hunsaker, M.: Using the probabilistic logic programming language p-log for causal and counterfactual reasoning and non-naive conditioning. In: IJCAI, pp. 243–249 (2007)
6. Blair, H., Subrahmanian, V.: Paraconsistent logic programming. Theoretical Computer Science 68, 135–154 (1989)
7. Damasion, C., Moura, J.: Modularity of P-log programs. In: Delgrande, J.P., Faber, W. (eds.) LPNMR 2011. LNCS, vol. 6645, pp. 13–25. Springer, Heidelberg (2011)
8. Dantsin, E., Eiter, T., Gottlob, G., Voronkov, A.: Complexity and expressive power of logic programming. In: Proc. of 12th Annual IEEE Conference on Computational Complexity, pp. 82–101 (1997)
9. Dekhtyar, A., Dekhtyar, M.: Possible worlds semantics for probabilistic logic programs. In: Demoen, B., Lifschitz, V. (eds.) ICLP 2004. LNCS, vol. 3132, pp. 137–148. Springer, Heidelberg (2004)
10. Dekhtyar, A., Subrahmanian, V.S.: Hybrid probabilistic programs. Journal of Logic Programming 43(3), 187–250 (2000)
11. Eiter, T., Faber, W., Gottlob, G., Koch, C., Mateis, C., Leone, N., Pfeifer, G., Scarcello, F.: The dlv system. In: Minker, J. (ed.) Pre-prints of Workshop on Logic-Based AI (2000)
12. van Emden, M.: Quantitative deduction and its fixpoint theory. The Journal of Logic Programming 3(2), 37–53 (1986)
13. van Emden, M., Kowalski, R.: The semantics of predicate logic as a programming language. Journal of the ACM 23(4), 733–742 (1976)
14. Fitting, M., Ben-Jacob, M.: Stratified and Three-Valued Logic programming Semantics. In: Kowalski, R., Bowen, K. (eds.) Proc. 5th International Conference and Symposium on Logic Programming, Seattle, Washington, August 15-19, pp. 1054–1069 (1988)
15. Gebser, M., Kaufmann, B., Kaminski, R., Ostrowski, M., Schaub, T., Schneider, M.: Potassco: The Potsdam Answer Set Solving Collection. AI Communications - Answer Set Programming archive 24(2) (2011)
16. Gelfond, M., Lifschitz, V.: The stable model semantics for logic programming. In: Kowalski, R., Bowen, K. (eds.) Logic Programming: Proc. of the Fifth Int'l Conf. and Symp., pp. 1070–1080. MIT Press, Cambridge (1988)

17. Gelfond, M., Lifschitz, V.: Logic programs with classical negation. In: Warren, D., Szeredi, P. (eds.) Logic Programming: Proc. of the Seventh Int'l Conf., pp. 579–597 (1990)
18. Gelfond, M., Rushton, N.: Causal and probabilistic reasoning in p-log (to appear in an edited book)
19. Gelfond, M., Rushton, N., Zhu, W.: Combining logical and probabilistic reasoning. In: AAAI Spring 2006 Symposium, pp. 50–55 (2006)
20. Greco, S., Molinaro, C., Trubitsyna, I., Zumpano, E.: NP-Datalog: A logic language for expressing search and optimization problems. TPLP 10(2), 125–166 (2010)
21. Kakas, A., Kowalski, R., Toni, F.: Abductive logic programming. Journal of Logic and Computation 2(6), 719–771 (1993)
22. Kersting, K., Raedt, L.D.: Bayesian logic programs. In: Cussens, J., Frisch, A. (eds.) Proceedings of the Work-in-Progress Track at the 10th International Conference on Inductive Logic Programming, pp. 138–155 (2000)
23. Lloyd, J.: Foundations of logic programming. Springer, Heidelberg (1984)
24. Lobo, J., Minker, J., Rajasekar, A.: Foundations of disjunctive logic programming. MIT Press, Cambridge (1992)
25. Lukasiewicz, T.: Probabilistic logic programming. In: ECAI, pp. 388–392 (1998)
26. Muggleton, S.: Stochastic logic programs. In: De Raedt, L. (ed.) Proceedings of the 5th International Workshop on Inductive Logic Programming, Department of Computer Science, Katholieke Universiteit Leuven, p. 29 (1995)
27. Ng, R.T., Subrahmanian, V.S.: Probabilistic logic programming. Information and Computation 101(2), 150–201 (1992)
28. Ngo, L., Haddawy, P.: Answering queries from context-sensitive probabilistic knowledge bases. Theoretical Computer Science 171(1–2), 147–177 (1997)
29. Niemelä, I., Simons, P.: Smodels – an implementation of the stable model and well-founded semantics for normal logic programs. In: Dix, J., Furbach, U., Nerode, A. (eds.) Proc. 4th International Conference on Logic Programming and Non-Monotonic Reasoning, pp. 420–429. Springer, Heidelberg (1997)
30. Pearl, J.: Causality. Cambridge University Press, Cambridge (2000)
31. Poole, D.: The independent choice logic for modelling multiple agents under uncertainty. Artificial Intelligence 94(1-2), 7–56 (1997)
32. Poole, D.: Abducing through negation as failure: Stable models within the independent choice logic. Journal of Logic Programming 44, 5–35 (2000)
33. Poole, D.: Probabilistic horn abduction and bayesian networks. Artificial Intelligence 64(1), 81–129 (1993)
34. Reiter, R.: On closed world data bases. In: Gallaire, H., Minker, J. (eds.) Logic and Data Bases, pp. 119–140. Plenum Press, New York (1978)
35. Sato, T.: A statistical learning method for logic programs with distribution semantics. In: Proceedings of the 12th International Conference on Logic Programming (ICLP 1995), pp. 715–729 (1995)
36. Shapiro, E.: Logic programs with uncertainties: A tool for implementing expert systems. In: Proc. IJCAI (1983)
37. Subrahmanian, V.S.: Uncertainty in logic programming: some recollections. ALP Newsletter (May 2007)
38. Tannert, C., Elvers, H., Jandrig, B.: The ethics of uncertainty. in the light of possible dangers, research becomes a moral duty. EMBO Report 8(10), 892–896 (2007)
39. Van Gelder, A., Ross, K., Schlipf, J.: The well-founded semantics for general logic programs. Journal of ACM 38(3), 620–650 (1991)
40. Vennekens, J., Verbaeten, S., Bruynooghe, M.: Logic programs with annotated disjunctions. In: ICLP, pp. 431–445 (2004)

Evaluating Probabilistic Inference Techniques: A Question of "When," not "Which"

Cory J. Butz

Department of Computer Science
University of Regina, Canada
butz@cs.uregina.ca

Abstract. Historically, it has been claimed that one inference algorithm or technique, say A, is better than another, say B, based on the running times on a test set of Bayesian networks. Recent studies have instead focusing on identifying situations where A is better than B, and vice versa. We review two cases where competing inference algorithms (techniques) have been successfully applied together in unison to exploit the best of both worlds. Next, we look at recent advances in identifying structure and semantics. Finally, we present possible directions of future work in exploiting structure and semantics for faster probabilistic inference.

1 Introduction

Bayesian networks [20] are an established framework for uncertainty management in artificial intelligence. A Bayesian network consists of a directed acyclic graph and a corresponding set of conditional probability tables. The *probabilistic conditional independencies* [23] encoded in the directed acyclic graph indicate that the product of the conditional probability tables is a joint probability distribution. Approaches to exact inference in Bayesian networks, the complexity of which has been shown by Cooper [8] to be NP-hard, can be broadly classified into two categories. One approach performs inference directly in a Bayesian network [1]. Two common algorithms are variable elimination (VE) [25] and arc-reversal (AR) [19,21]. A second approach to Bayesian network inference is join tree propagation [4], which systematically builds and passes messages in a join tree constructed from the Bayesian network and then computes posterior probabilities for each variable. Zhang's [24] experimental results indicate that VE is more efficient than the classical join tree methods when updating twenty or less non-evidence variables, given a set of twenty or fewer evidence variables. Madsen and Jensen [15] suggested a join tree algorithm, called *Lazy propagation* (LP), and empirically demonstrated a significant improvement in efficiency over previous join tree methods. More recently, Madsen [16,17] examined hybrid approaches to Bayesian network inference. Inference is still conducted in a join tree, but direct methods are utilized for message construction. Of the three hybrid approaches tested, LP with AR (LPAR) tended to be no worse than the other two approaches and was sometimes faster. All of the above studies, however, exclusively applied the same inference algorithm on a test set of Bayesian networks.

S. Benferhat and J. Grant (Eds.): SUM 2011, LNAI 6929, pp. 38–51, 2011.
© Springer-Verlag Berlin Heidelberg 2011

In order to determine *when* one inference algorithm is better than another, one needs to be able to characterize the semantics of the probability information on hand [6,7]. In [7], we gave the first join tree approach that labels the probability information passed between nodes in terms of conditional probability tables rather than *potentials* [9,12]. This information allowed us [3] to use AR to determine the messages to be propagated during inference in a join tree, but to call VE to build them. Thus, [3] was the first paper to use one inference algorithm to identify messages and another to build messages. Madsen [18] took the next step by using AR to build messages and using VE to compute posterior probabilities. We extended [18] by selectively choosing VE or AR to build messages [2]. Similarly, four cost measures s_1, s_2, s_3, s_4 were recently studied in [18] for sorting the operations in LPAR. It was suggested in [18] to use s_1 with LPAR, since there is an effectiveness ranking, say s_1, s_2, s_3, s_4, when applied in isolation. In [5], we also suggested to use s_1 with LPAR, but to use s_2 to break s_1 ties, s_3 to break s_2 ties, and s_4 to break s_3 ties. The important point is that the above works [2,3,5,18] attempt to choose "when" to apply a particular inference method rather than simply choosing "which" algorithm should be considered the best overall.

This paper is organized as follows. Section 2 contains background knowledge. In Section 3, we show the benefits of applying multiple algorithms in probabilistic inference. Recent advances in identifying semantics during probabilistic inference are discussed in Section 4. The conclusion and future work are presented in Section 5.

2 Background Knowledge

We review Bayesian networks and common approaches to probabilistic inference.

2.1 Bayesian Networks

The following draws from [22]. Let $U = \{v_1, v_2, \ldots, v_n\}$ denote a finite set of discrete random variables. Each variable v_i is associated with a finite domain, denoted $dom(v_i)$, representing the values v_i can take on. For a subset $X \subseteq U$, we write $dom(X)$ for the Cartesian product of the domains of the individual variables in X. Each element $x \in dom(X)$ is called a *configuration* of X. A *potential* [10] on $dom(X)$ is a function ψ on $dom(X)$ such that $\psi(x) \geq 0$, for each configuration $x \in dom(X)$, and at least one $\psi(x)$ is positive. For brevity, we refer to a potential as a probability distribution on X rather than $dom(X)$, and we call X, not $dom(X)$, its domain. A joint probability distribution on U, denoted $p(U)$, is a potential on U that sums to one. Given $X \subset U$, a *conditional probability table* for a variable $v \notin X$ is a distribution, denoted $p(v|X)$, satisfying the following condition: $\sum_{c \,\in\, dom(v)} p(\ v = c \mid X = x\ ) = 1.0$, for each configuration $x \in dom(X)$.

A *Bayesian network* [20] on U is a pair (D, C). D is a *directed acyclic graph* on U. C is a set of conditional probability tables defined as: for each variable

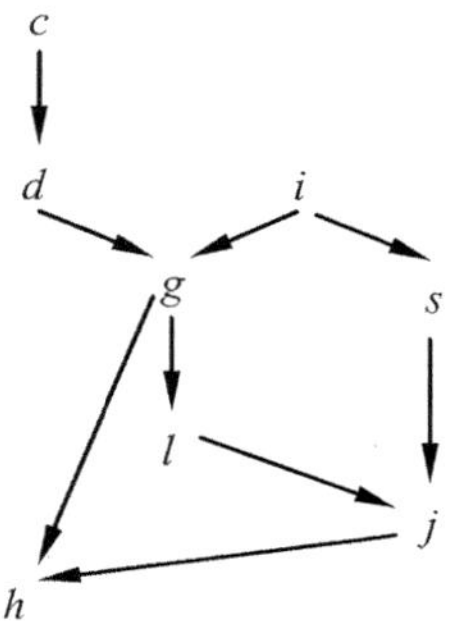

Fig. 1. The directed acyclic graph of the ESBN

Table 1. Conditional probability tables for the ESBN in Figure 1

c	$p(c)$
0	0.20

i	$p(i)$
0	0.75

g	j	h	$p(h\|g,j)$
0	0	0	0.25
0	1	0	0.65
1	0	0	0.50
1	1	0	0.85

c	d	$p(d\|c)$
0	0	0.40
1	0	0.70

g	l	$p(l\|g)$
0	0	0.30
1	0	0.60

i	s	$p(s\|i)$
0	0	0.40
1	0	0.80

d	i	g	$p(g\|d,i)$
0	0	0	0.90
0	1	0	0.20
1	0	0	0.50
1	1	0	0.40

s	l	j	$p(j\|s,l)$
0	0	0	0.10
0	1	0	0.60
1	0	0	0.45
1	1	0	0.50

$v_i \in D$, there is a conditional probability table for v_i given its parents P_i in D. Based on the *probabilistic conditional independencies* [23] encoded in D, the product of the conditional probability tables in C is a joint distribution $p(U)$.

For example, the directed acyclic graph in Figure 1 is called the *extended student Bayesian network* (ESBN) [13]. We give conditional probability tables in Table 1, where only binary variables are used in examples, and probabilities not shown can be obtained by definition. By the above,

$$p(U) = p(c) \cdot p(d|c) \cdot p(i) \cdot p(g|d,i) \cdots p(h|g,j). \tag{1}$$

We will use the terms Bayesian network and directed acyclic graph interchangeably if no confusion arises. A *topological ordering* [13] is an ordering $\prec$ of the variables in a Bayesian network B so that for every arc (v_i, v_j) in B, v_i precedes v_j in $\prec$. For example, $c \prec d \prec i \prec g \prec s \prec l \prec j \prec h$ is a topological ordering of the directed acyclic graph in Figure 1, but $d \prec c \prec i \prec g \prec h \prec l \prec j \prec s$ is not.

2.2 Variable Elimination

In inference, $p(X|E = e)$ is the most common query type, which are useful for many reasoning patterns, including explanation, prediction, inter-causal reasoning, and many more [13].

Here, X and E are disjoint subsets of U, and E is observed taking value e. We describe a basic algorithm for computing $p(X|E = e)$, called *variable elimination* (VE), first put forth by [25]. Inference involves the elimination of variables. Algorithm 1, called *sum-out* (SO), eliminates a single variable v from a set Φ of potentials, and returns the resulting set of potentials. The algorithm *collect-relevant* simply returns those potentials in Φ involving variable v.

Algorithm 1. SO(v,Φ)
Ψ = collect-relevant(v,Φ)
ψ = the product of all potentials in Ψ
$\tau = \sum_v \psi$
return $(\Phi - \Psi) \cup \{\tau\}$

SO uses Lemma 1, which means that potentials not involving the variable being eliminated can be ignored.

Lemma 1. [22] *If ψ_1 is a potential on W and ψ_2 is a potential on Z, then the marginalization of $\psi_1 \cdot \psi_2$ onto W is the same as ψ_1 multiplied with the marginalization of ψ_2 onto $W \cap Z$, where $W, Z \subseteq U$.*

The *evidence potential* for $E = e$, denoted $1(E = e)$, assigns probability 1 to the single value e of E and probability 0 to all other values of E. Hence, for a variable v observed taking value λ and $v \in \{v_i\} \cup P(v_i)$, the product $p(v_i|P(v_i)) \cdot 1(v = \lambda)$ keeps only those configurations agreeing with $v = \lambda$.

Algorithm 2, taken from [13], computes $p(X|E = e)$ from a discrete Bayesian network B. VE calls SO to eliminate variables one by one. More specifically, in Algorithm 2, Φ is the set C of conditional probability tables for B, X is a list of query variables, E is a list of observed variables, e is the corresponding list of observed values, and σ is an elimination ordering for variables $U - XE$, where XE denotes $X \cup E$.

Algorithm 2. VE(Φ, X, E, e, σ)
Multiply evidence potentials with appropriate conditional probability tables
While σ is not empty
  Remove the first variable v from σ
  Φ = sum-out(v, Φ)
$p(X, E = e)$ = the product of all potentials $\psi \in \Phi$
return $p(X, E = e)/\sum_X p(X, E = e)$

As in [13], suppose the observed evidence for the ESBN is $i = 1$ and $h = 0$ and the query is $p(j|h = 0, i = 1)$. The weighted-min-fill algorithm [13] can yield $\sigma = c, d, l, s, g$. VE first incorporates the evidence:

$$\psi(i=1) = p(i) \cdot 1(i=1),$$
$$\psi(d,g,i=1) = p(g|d,i) \cdot 1(i=1),$$
$$\psi(i=1,s) = p(s|i) \cdot 1(i=1),$$
$$\psi(g,h=0,j) = p(h|g,j) \cdot 1(h=0).$$

To eliminate c, the SO algorithm computes

$$\psi(d) = \sum_c p(c) \cdot p(d|c).$$

SO computes the following to eliminate d

$$\psi(g,i=1) = \sum_d \psi(d) \cdot \psi(d,g,i=1).$$

To eliminate l,

$$\psi(g,j,s) = \sum_l p(l|g) \cdot p(j|l,s).$$

SO computes the following when eliminating s,

$$\psi(g,i=1,j) = \sum_s \psi(i=1,s) \cdot \psi(g,j,s).$$

For g, SO can compute:

$$\begin{aligned} &\sum_g \psi(g,i=1,j) \cdot \psi(g,i=1) \cdot \psi(g,h=0,j) \\ &= \sum_g \psi(g,i=1,j) \cdot \psi(g,h=0,i=1,j) \\ &= \psi(h=0,i=1,j). \end{aligned}$$

Next, VE multiplies all remaining potentials as

$$p(h=0,i=1,j) = \psi(i=1) \cdot \psi(h=0,i=1,j).$$

Finally, VE answers the query by

$$p(j|h=0,i=1) = \frac{p(h=0,i=1,j)}{\sum_j p(h=0,i=1,j)}.$$

2.3 Arc Reversal

Arc reversal (AR) [19,21] eliminates a variable v_i by reversing the arcs (v_i, v_j) for each child v_j of v_i, where $j = 1, 2, \ldots, k$. With respect to multiplication, addition, and division, AR reverses one arc (v_i, v_j) as a three-step process:

$$p(v_i, v_j|P_i P_j) = p(v_i|P_i) \cdot p(v_j|P_j), \tag{2}$$

$$p(v_j|P_iP_j) = \sum_{v_i} p(v_i, v_j|P_iP_j), \tag{3}$$

$$p(v_i|P_iP_jv_j) = \frac{p(v_i, v_j|P_iP_j)}{p(v_j|P_iP_j)}. \tag{4}$$

Suppose the variable v_i to be removed has k children. The distributions defined in (2) - (4) are built for the first $k-1$ children. For the last child v_k, however, only the distributions in (2) - (3) are built. When considering v_k, there is no need to build the final distribution for v_i in (4), since v_i will be removed as a barren variable. Therefore, AR removes a variable v_i with k children by building $3k-1$ distributions. However, AR only outputs the k distributions built in (3).

For example, consider eliminating variable a in Fig. 2 (i). There are two arcs (a, c) and (a, d) to be reversed. Suppose arc (a, d) is reversed first:

$$p(a, d|b) = p(a) \cdot p(d|a, b),$$
$$p(d|b) = \sum_{a} p(a, d|b),$$
$$p(a|b, d) = \frac{p(a, d|b)}{p(d|b)}.$$

The resulting directed acyclic graph is shown in Fig. 2 (ii). The reversal of the other arc (a, c) gives Fig. 2 (iii) by computing:

$$p(a, c|b, d) = p(a|b, d) \cdot p(c|a),$$
$$p(c|b, d) = \sum_{a} p(a, c|b, d).$$

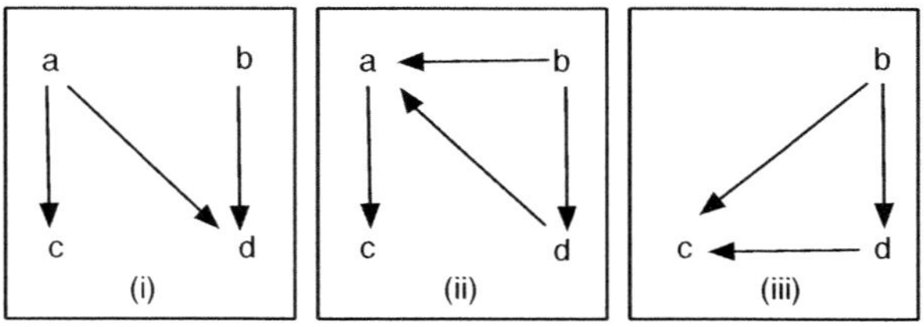

Fig. 2. Eliminating a in (i) by reversing arc (a, d) (ii) followed by arc (a, c) (iii)

2.4 Join Tree Propagation

Shafer [22] emphasizes that *join tree propagation* is central to the theory and practice of probabilistic expert systems. A join tree [20,22] is a tree with sets of variables as nodes, with the property that any variable in two nodes is also in any node on the path between the two. The *separator* S between any two neighbour nodes N_i and N_j is $S = N_i \cap N_j$. The task of transforming a directed acyclic graph into a join tree has been extensively studied in probabilistic reasoning literature.

LP [15] maintains structure in the form of a multiplicative factorization of potentials at each join tree node and each join tree separator, as illustrated in Figure 3. Maintaining a decomposition of potentials offers LP the opportunity to exploit barren variables and independencies induced by evidence. Doing so improves the efficiency of join tree propagation remarkably as the experimental results in [15] clearly emphasize. We refer the reader to [4] for a recent study on probabilistic inference using join tree propagation.

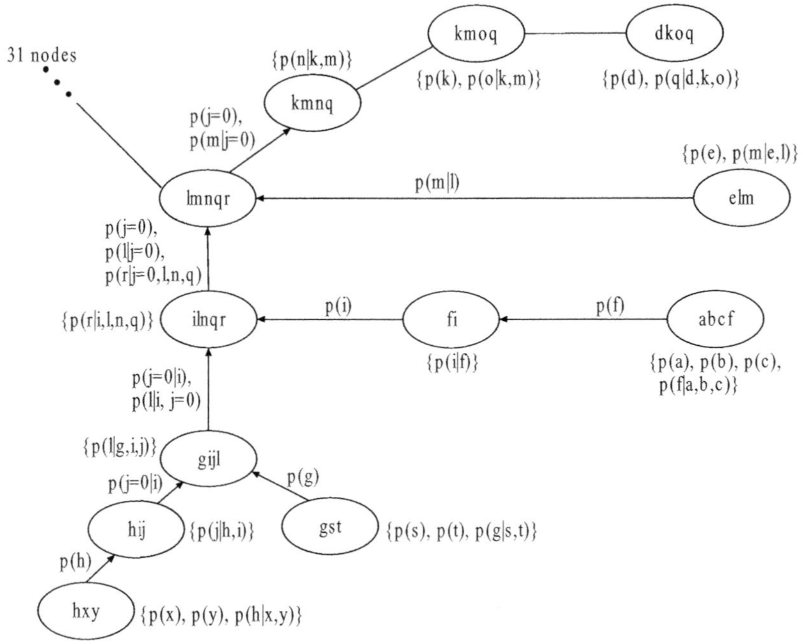

Fig. 3. A join tree partially depicted. Nodes include $\{a, b, c, f\}$ and $\{e, l, m\}$. Node $\{e, l, m\}$ was assigned two conditional probability tables $\{p(e), p(m|e, l)\}$ from the Bayesian network. Some edges have been directed to illustrate the direction of propagation. Node $\{i, l, n, q, r\}$ will pass three messages $\{p(j = 0), p(l|j = 0), p(r|j = 0, l, n, q)\}$ to its neighbour $\{l, m, n, q, r\}$.

3 "When" versus "Which"

Our purpose here is to report on two studies [2,5] that have shown faster inference by exploiting structure and semantics.

3.1 Using Multiple Algorithms for Message Construction

We first review [2], which selectively applies AR or VE to build messages in join tree propagation, compared with LPAR, which exclusively applies AR to build messages.

Table 2. Description of four benchmark Bayesian networks and constructed join trees

Bayesian network	# variables	# evidence variables	# CPT rows	# Join tree nodes	# Join tree messages
Alarm (BN0)	100	26	974	68	798
Water (BN28)	24	8	18029184	12	27
ISCAS 85 (BN43)	880	10	5096	326	1742
CPCS (BN78)	54	10	1658	30	193

The evaluation in [2] was carried out on four benchmark Bayesian networks taken from the 2006 UAI probabilistic inference competition. The elimination ordering is determined using the min-fill criteria [13], while the ordering of the children of the variable being eliminated when using AR is determined by a fixed topological ordering of the variables in the Bayesian network. Table 2 describes the Bayesian network name and number from the 2006 UAI competition, the number of variables in each Bayesian network, the number of evidence variables in each Bayesian network, the number of rows in the conditional probability tables of the Bayesian network, the number of nodes in each join tree, and the number of messages passed in the join tree when no evidence is involved.

The experiments not involving evidence were conducted as follows. Load the Bayesian network ignoring the given evidence variables and build a join tree. The inward and outward phases of join tree propagation are performed to compute a factorization of $p(N)$ for every join tree node N. For experiments involving collected evidence, the evidence $E = e$ is stated in description of the Bayesian network and was determined by the competition organizers. In this case, the directed acyclic graph of the Bayesian network is pruned based on the given evidence. Next, a join tree is constructed from the pruned directed acyclic graph. Finally, inward and outward phases of join tree propagation are performed to compute a factorization of $p(N - E, E = e)$ for every join tree node N.

Table 3 reports on Bayesian inference not involving evidence processing. Running times for LPAR and for DataBayes are listed in milliseconds and are the average of three runs. The last column shows the speed-up percentage of DataBayes over LPAR. The average percentage gain is 46%.

Next, we measure the runtime of inference involving evidence. Table 2 indicates the number of evidence variables as specified in the 2006 UAI probabilistic

Table 3. The performance of LPAR and DataBayes not involving evidence processing in four benchmark Bayesian networks

Bayesian network	LPAR inward	LPAR outward	DataBayes inward	DataBayes outward	Net inward	Net outward	Net total
Alarm	1866	24602	1696	19610	9%	20%	20%
Water	355	1653	15	111	96%	93%	94%
ISCAS 85	5845	24420	5380	13212	8%	46%	39%
CPCS	1129	1976	1039	1133	8%	43%	30%

Table 4. The performance of LPAR and DataBayes involving evidence processing in four benchmark Bayesian networks

Bayesian network	LPAR inward	LPAR outward	DataBayes inward	DataBayes outward	Net inward	Net outward	Net total
Alarm	2387	12694	1310	9392	45%	26%	29%
Water	2131	7857	1766	2598	17%	67%	56%
ISCAS 85	3700	13494	3419	13103	8%	3%	4%
CPCS	1372	3034	1213	1918	12%	37%	29%

inference competition. The times reported in Table 4 are given in milliseconds and are the average of three runs. Note that, once again, DataBayes is always faster than LPAR. The average percentage gain is 30%.

The results in Tables 3 and 4 empirically demonstrate that by selectively applying VE and AR for message construction, join tree propagation can be performed faster.

3.2 Using Multiple Heuristics for Determining AR Orderings

Madsen [18] has demonstrated that the order in which arcs are reversed in LPAR can affect the amount of computation needed. Experimental results suggest that *cpt-weight* (*cptw*), denoted s_1, is the best of four measures s_1, s_2, s_3, s_4 [18]. When using cost measure s_1 to reverse arcs in LPAR, [18] will reverse the first arc tied for the lowest score in the event of a tie. Instead, we suggest breaking ties with other cost measures s_2, s_3, s_4.

Recall the elimination of variable a in Figure 4, where a, b, d are binary and c's domain has four values. Here the s_1 scores of arcs (a, c) and (a, d) corresponding to the children c and d of a are the same. Since $s_1(a, c)$ is equal to $s_1(a, d)$, s_1 does not distinguish between arcs (a, c) and (a, d). Instead of randomly reversing an arc, say (a, d), we can let the other heuristics s_2, s_3, s_4 decide. In this case, s_3 suggests to reverse (a, c) and then (a, d), as depicted in Figure 4.

The experiments in [5] use the LPAR method in [18], namely, AR is applied to build all messages and VE is applied to compute posterior marginals. Here, we

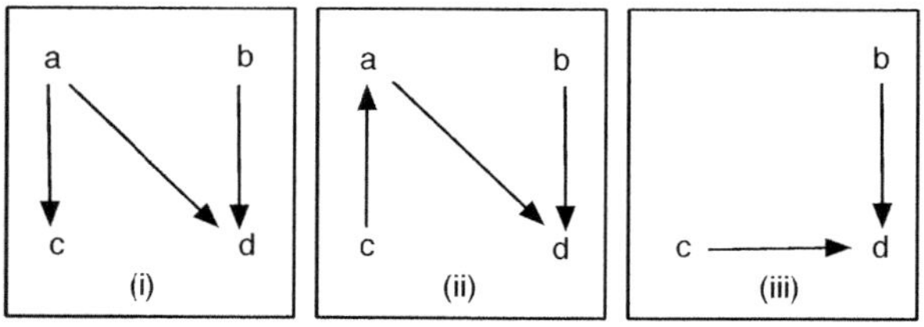

Fig. 4. Eliminating a in (i) by reversing arc (a, c) (ii) followed by arc (a, d) (iii)

Table 5. Description of test Bayesian networks and corresponding join tree nodes $\mathcal{N}$

Bayesian network	$\|U\|$	$\|\mathcal{N}\|$	max $\|dom(\mathcal{N})\|$	total size
Barley	48	36	7,257,600	17,140,796
KK	50	38	5,806,080	14,011,466
ship-ship	50	35	4,032,000	24,258,572

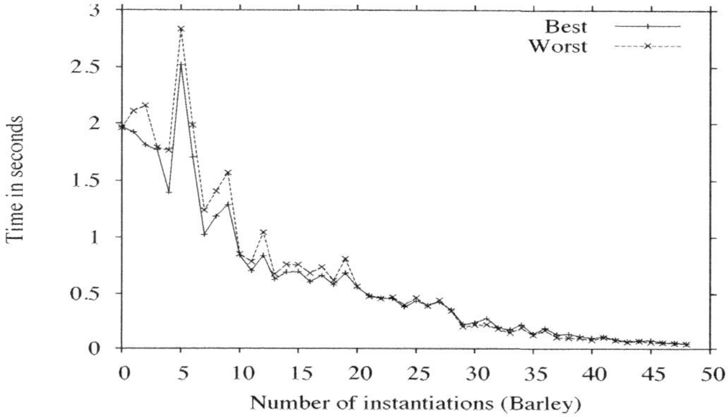

Fig. 5. Time savings of breaking ties by reversing the best and worst arcs as determined by the next cost measure on Barley

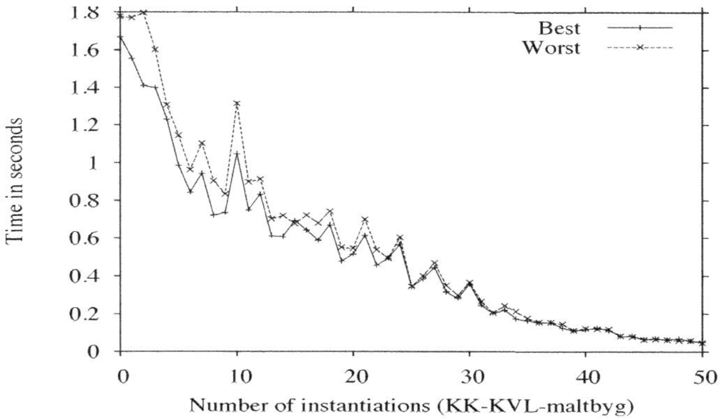

Fig. 6. Time savings of breaking ties by reversing the best and worst arcs as determined by the next cost measure on KK

only review results on three real-world networks, called Barley [14], KK [14][1], and ship-ship [11], which are described in Table 5. For each size of evidence set, ten sets of evidence are generated, with the same evidence used in different runs. To reflect the potential time savings of breaking ties, the *best* and *worst*

[1] KK is a preliminary version of Barley.

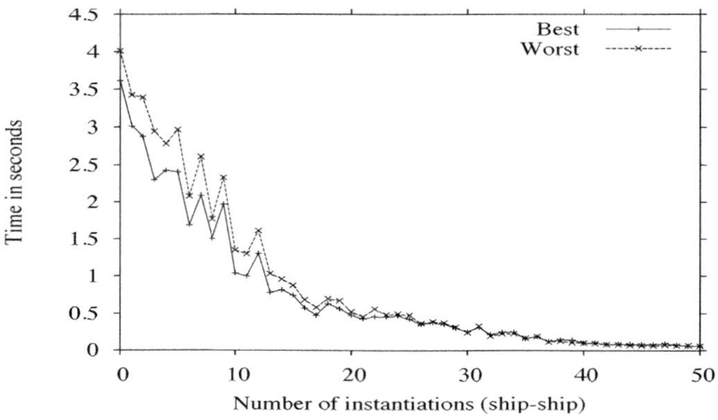

Fig. 7. Time savings of breaking ties by reversing the best and worst arcs as determined by the next cost measure on ship-ship

arcs are reversed based on the next cost measure. Figs. 5 - 7 show running times in seconds on our three Bayesian networks.

The results in Figs. 5 - 7 empirically demonstrate that probabilistic inference can be performed faster by applying multiple heuristics.

4 Knowing "When:" Semantics of Intermediate Factors

In [7], we gave a method for deciding semantics in join tree propagation. We have more recently shown in [6] how to determine semantics of the intermediate factors constructed during probabilistic inference.

Let $\psi(\cdot)$ be any potential constructed by VE on a Bayesian network B. If the semantics of B ensure the $\psi(\cdot) = p(\cdot)$, then $\psi(\cdot)$ is denoted as $p(\cdot)$; otherwise, it is denoted as $\phi(\cdot)$. Thereby, the semantics of every potential ψ constructed by VE can be denoted with a $p-$label or a $\phi-$label.

Recall evidence $i = 1$ and $h = 0$ in the Bayesian network in Figure 1, which Koller and Friedman [13] call non-trivial. In Section 2.2, all intermediate distributions were denoted as potentials in the computation of $p(j \mid i = 1, h = 0)$. However, even with evidence, structure and semantics can still be identified.

Example 1. Computing $p(j \mid i = 1, h = 0)$ in the Bayesian network of Figure 1 involves, in part,

$$\sum_{c,d,g,s,l} p(c) \cdot p(d|c) \cdot p(i) \cdot p(s|i) \cdot p(g|d,i) \cdot p(l|g) \cdot p(j|l,s) \cdot p(h|g,j) \cdot 1(i = 1) \cdot 1(h = 0).$$

Eliminating variables c and d requires

$$\sum_{d} p(g|d,i) \cdot 1(i = 1) \cdot \sum_{c} p(c) \cdot p(d|c) \qquad (5)$$

$$= \sum_d p(g|d, i = 1) \cdot \sum_c p(c, d) \tag{6}$$

$$= \sum_d p(g|d, i = 1) \cdot p(d) \tag{7}$$

$$= \sum_d p(d, g|i = 1)$$

$$= p(g|i = 1).$$

Variable g can be eliminated as:

$$\sum_g p(g|i = 1) \cdot p(l|g) \cdot p(h|g, j) \cdot 1(h = 0)$$

$$\sum_g p(g|i = 1) \cdot p(l|g) \cdot p(h = 0|g, j)$$

$$= \sum_g p(g, l|i = 1) \cdot p(h = 0|g, j) \tag{8}$$

$$= \sum_g \phi(g, l, h = 0|i = 1, j) \tag{9}$$

$$= \phi(l, h = 0|i = 1, j). \tag{10}$$

The remainder of the example is omitted.

In contrast to Section 2.2, Example 1 shows that all intermediate distributions have structure and semantics, regardless of: the involvement of evidence potentials (5); the side or sides of the bar on which evidence appears (8), (9); marginalization operations (6); and p-labels (7) or ϕ-labels (10).

5 Conclusion and Future Work

Rather than exclusively applying what is considered to be the "best" inference algorithm or technique, we have reviewed two cases where competing methods were successfully applied together. The rationale is to take full advantage of what each technique has to offer. However, in order to know "when" to apply each method, the importance of recognizing and exploiting structure and semantics in probabilistic inference is underscored.

We have reviewed current work on identifying the semantics of intermediate factors in discrete Bayesian network inference [6]. One direction of future work in probabilistic inference is to exploit semantics of intermediate factors for processing subsequent queries in VE. That is, some calculations performed when answering a given query in VE may be reused when processing a subsequent query.

Our work on semantics in [7] allowed us to suggest the idea of prioritized messages. Current join tree algorithms treat all propagated messages as being of equal importance. On the contrary, it is often the case in real-world Bayesian networks that only some of the messages propagated from one join tree node to another are relevant to subsequent message construction at the receiving node. In Figure 3, for building messages $p(j = 0)$ and $p(m|j = 0)$ from node $\{l, m, n, q, r\}$, messages $p(j = 0)$ and $p(l|j = 0)$ from $\{i, l, n, q, r\}$ are respectively *relevant*, whereas the message $p(r|j = 0, l, n, q)$ is *irrelevant* in both cases. In [4], we proposed the first join tree propagation algorithm that identifies and constructs the *relevant* messages first. In other words, join tree propagation is conducted at a "message-to-message" level rather than a "node-to-node" level.

Lastly, while [5] focuses on finding better child orderings in AR, it does not address the problem of finding good elimination orderings for AR. That is, the elimination orderings used in [5] for AR are determined by a standard method for determining elimination orderings for VE.

References

1. Butz, C.J., Chen, J., Konkel, K., Lingras, P.: A formal comparison of variable elimination and arc reversal in Bayesian network inference. Intell. Dec. Analysis 3(3), 173–180 (2009)
2. Butz, C.J., Konkel, K., Lingras, P.: Join tree propagation utilizing both arc reversal and variable elimination. Intl. J. Approx. Rea. (2010) (in press)
3. Butz, C.J., Hua, S.: An improved Lazy-AR approach to Bayesian network inference. In: Proc. of Canadian Conference on Artificial Intelligence, pp. 183–194 (2006)
4. Butz, C.J., Hua, S., Konkel, K., Yao, H.: Join tree propagation with prioritized messages. Networks 55(4), 350–359 (2010)
5. Butz, C.J., Madsen, A.L., Williams, K.: Using four cost measures to determine arc reversal orderings. In: Proc. 11th European Conference on Symbolic and Quantitative Approaches to Reasoning with Uncertainty, pp. 110–121 (2011)
6. Butz, C.J., Yan, W.: The semantics of intermediate CPTs in variable elimination. In: Proc. 5th European Workshop on Probabilistic Graphical Models, pp. 41–49 (2010)
7. Butz, C.J., Yao, H., Hua, S.: A join tree probability propagation architecture for semantic modelling. J. Int. Info. Sys. 33(2), 145–178 (2009)
8. Cooper, G.F.: The computational complexity of probabilistic inference using Bayesian belief networks. Art. Intel. 42(2-3), 393–405 (1990)
9. Castillo, E., Gutiérrez, J., Hadi, A.: Expert Systems and Probabilistic Network Models. Springer, New York (1997)
10. Hájek, P., Havránek, T., Jiroušek, R.: Uncertain Information Processing in Expert Systems. CRC Press, Ann Arbor (1992)
11. Hansen, P.F., Pedersen, P.T.: Risk analysis of conventional and solo watch keeping. Research Report, Department of Naval Architecture and Offshore Engineering, Technical University of Denmark (1998)
12. Kjaerulff, U.B., Madsen, A.L.: Bayesian Networks and Influence Diagrams: a Guide to Construction and Analysis. Springer, New York (2008)
13. Koller, D., Friedman, N.: Probabilistic Graphical Models: Principles and Techniques. The MIT Press, Cambridge (2009)
14. Kristensen, K., Rasmussen, I.A.: The use of a Bayesian network in the design of a decision support system for growing malting barley without use of pesticides. Computers and Electronics in Agriculture 33, 192–217 (2002)

15. Madsen, A.L., Jensen, F.V.: Lazy propagation: A junction tree inference algorithm based on lazy evaluation. Artif. Intell. 113(1-2), 203–245 (1999)
16. Madsen, A.L.: An empirical evaluation of possible variations of lazy propagation. In: Proc. 20th Conference on Uncertainty in Artificial Intelligence, pp. 366–373 (2004)
17. Madsen, A.L.: Variations over the message computation algorithm of lazy propagation. IEEE Trans. Sys. Man Cyb. B 36, 636–648 (2006)
18. Madsen, A.L.: Improvements to message computation in lazy propagation. Intl. J. Approx. Rea. 51(5), 499–514 (2010)
19. Olmsted, S.: On representing and solving decision problems. Ph.D. thesis, Department of Engineering Economic Systems. Stanford University, Stanford, CA (1983)
20. Pearl, J.: Probabilistic Reasoning in Intelligent Systems: Networks of Plausible Inference. Morgan Kaufmann, San Francisco (1988)
21. Shachter, R.: Evaluating influence diagrams. Oper. Res. 34(6), 871–882 (1986)
22. Shafer, G.: Probabilistic Expert Systems. Society for Industrial and Applied Mathematics (1996)
23. Wong, S.K.M., Butz, C.J., Wu, D.: On the implication problem for probabilistic conditional independency. IEEE Trans. Syst. Man Cybern., A 30(6), 785–805 (2000)
24. Zhang, N.L.: Computational properties of two exact algorithms for Bayesian networks. Appl. Intell. 9(2), 173–184 (1998)
25. Zhang, N.L., Poole, D.: A simple approach to Bayesian network computations. In: Proc. 10th Canadian Conference on Artificial Intelligence, pp. 171–178 (1994)

Probabilistic Logic Networks in a Nutshell

Matthew Iklé

Adams State College, Alamosa CO

Abstract. We begin with a brief overview of Probabilistic Logic Networks, distinguish PLN from other approaches to reasoning under uncertainty, and describe some of the main conceptual foundations and goals of PLN. We summarize how knowledge is represented within PLN and describe the four basic truth-value types. We describe a few basic first-order inference rules and formulas, outline PLN's approach to handling higher-order inference via reduction to first-order rules, and follow this by a brief summary of PLN's handling of quantifiers.

Since PLN was and continues to be developed as one of several major components of a broader and more general artificial intelligence project, we next describe the OpenCog project and PLN's roles within the project.

Keywords: probabilistic logic, probabilistic networks, artificial general intelligence, PLN, OpenCog.

1 Introduction: What is PLN?

First introduced as a probabilistic reasoning system within the Webmind Artificial Intelligence project by Ben Goertzel and the late Jeff Pressing, the Probabilistic Logic Networks (PLN) system has evolved and grown considerably. PLN now serves as the probabilistic reasoning system within the open source OpenCog AI engine[7], which has replaced Webmind. The primary focus of PLN is to serve as a systematic, comprehensive, and pragmatic system to manage uncertainty: to handle and reason about imprecise, uncertain, incomplete, and inconsistent data, and reasoning involving uncertain conclusions.

Perhaps one of PLN's most striking characteristics is its dual nature. Designed as part of a broader artificial intelligence system, PLN is very practical, encompassing heuristic approaches as necessary. At the same time, considerable effort has been made to ground as much of PLN as possible upon solid theoretical and mathematical foundations. A result of this duality is the following list of desired characteristics:

- PLN should enable uncertainty-savvy versions of all known varieties of logical reasoning: including, for instance, higher-order reasoning involving quantifiers, higher-order functions, and so forth;
- PLN should reduce to crisp 'theorem prover" style behavior in the limiting case where uncertainty tends to zero;
- PLN should encompass inductive and abductive as well as deductive reasoning;

S. Benferhat and J. Grant (Eds.): SUM 2011, LNAI 6929, pp. 52–60, 2011.
© Springer-Verlag Berlin Heidelberg 2011

- PLN should agree with probability theory in those reasoning case where probability in its current state of development provides solutions within reasonable calculational effort based on assumptions that are plausible in the context of real-world embodied software systems;
- PLN should gracefully incorporate heuristics not explicitly based on probability theory, in cases where probability theory, at its current state of development, does not provide adequate pragmatic solutions;
- PLN should provide "scalable" reasoning, in the sense of being able to carry out inferences involving at least billions of premises and carry out more intensive and accurate reasoning when the number of premises is fewer;
- PLN should easily accept input from, and send input to, natural language processing software systems.

2 Relationship of PLN to Other Uncertain Inference Engines

It is clear that uncertain inference is hardly a new idea. What is new within PLN is its focus on bridging the theoretical and practical, and how it incorporates and integrates ideas from a variety of sources. PLN borrows heavily upon other approaches to uncertain inference and in many ways represents an amalgam of a large number of these ideas, including such standard approaches as Bayesian probability theory, and fuzzy logic, as well as from more unusual ideas including Pei Wang's Non-Axiomatic Reasoning System (NARS)[12], algorithmic information theory, and Walley's theory of imprecise probabilities[11]. One of the key differences between PLN and other approaches to probabilistic logic lies in PLN's foundation upon "term logic." As we shall see later, this foundational choice allows one to reduce PLN's higher-order inference rules to more basic first-order rules.

Overall, PLN owes the most to Pei Wang's NARS system and Walley's theory of imprecise probabilities. Pei Wang pioneered the use of uncertain term logic in his NARS system, and in large measure provided the motivation for the development of PLN. Indeed PLN began as part of a collaboration with Wang as an attempt to create a probabilistic analogue to NARS, though there remain many conceptual and mathematical differences between the two, and PLN has long ago diverged from these roots.

Peter Walley's theory of imprecise probabilities provided motivation for the development of our "indefinite probabilities" approach. Essentially a hybridization of Walley's imprecise probabilities with Bayesian credible intervals, "indefinite probabilities" provide a general and mathematically sound method for calculating the "weight-of-evidence" underlying the conclusions of uncertain inferences. Moreover, both Walley's imprecise beta-binomial model and standard Bayesian inference can be mathematically viewed as limiting cases of the indefinite probability model.

Of the wide array of uncertain inference methods, Bayes' nets represent perhaps the most similar approach to PLN, although the graph structures themselves are quite dissimilar. While both methods succeed at embodying probability theory in a set of date structures and algorithms, PLN was designed with different purposes in mind. As a pragmatic approach with an eye towards interaction with an integrative artificial intelligence system, PLN was designed to interface with other cognitive processes and with other kinds of inference, including intensional inference, fuzzy inference, and higher-order inference using quantifiers, variables, and combinators.

While PLN utilizes fuzzy set membership as the semantics for Member relationship truth-values, it maintains a clear distinction between uncertainty and partial membership. For many of the purposes commonly associated with fuzzy membership, PLN uses intensional probabilities, giving the advantage of keeping more things within a probabilistic framework.

3 Knowledge Representation within PLN

Declarative knowledge representation within PLN is handled by a weighted labeled hypergraph called the Atomspace, which consists of multiple types of nodes and links, generally weighted with probabilistic truth values and attention values PLN is divided into first-order and higher-order sub-theories (FOPLN and HOPLN). These terms are used in a nonstandard way drawn from NARS. We develop FOPLN first, and then derive HOPLN therefrom. FOPLN is a term logic, involving terms and relationships (links) between terms. It is an uncertain logic, in the sense that both terms and relationships are associated with truth value objects, which may come in multiple varieties ranging from single numbers to complex structures like indefinite probabilities[3]. Terms may be either elementary observations, or abstract tokens drawn from a token-set T.

3.1 Core FOPLN Relationships

"Core FOPLN" involves relationships drawn from the set: negation; Inheritance and probabilistic conjunction and disjunction; Member and fuzzy conjunction and disjunction. Elementary observations can have only Member links, while token terms can have any kinds of links. PLN makes clear distinctions, via link type semantics, between probabilistic relationships and fuzzy set relationships. Member semantics are usually fuzzy relationships (though they can also be crisp), whereas Inheritance relationships are probabilistic, and there are rules governing the interoperation of the two types.

3.2 Auxiliary FOPLN Relationships

Beyond the core FOPLN relationships, FOPLN involves additional relationship types of two varieties. There are simple ones like Similarity, defined by

$$Similarity\ A\ B$$

We say a relationship R is simple if the truth value of $R\ A\ B$ can be calculated in terms of the truth values of core FOPLN relationships between A and B.

There are also complex ones like IntensionalInheritance, which measures the extensional inheritance between the set of properties or patterns associated with one term and the corresponding set associated with another.

3.3 PLN Truth Values

Truth-values come in four basic types. In order of increasingly information about the full probability distribution they are

- strength truth-values, which consist of single numbers; e.g., $< s >$ or $< .8 >$. Usually strength values denote probabilities but this is not always the case.
- SimpleTruthValues, consisting of pairs of numbers. These pairs come in two forms: $< s, w >$, where s is a strength and w is a "weight of evidence" and $< s, N >$, where N is a "count." "Weight of evidence is a qualitative measure of belief, while "count is a quantitative measure of accumulated evidence.
- IndefiniteTruthValues, which quantify truth-values in terms of an interval $[L,\ U]$, a credibility level b, and an integer k (called the lookahead). IndefiniteTruthValues quantify the idea that after k more observations there is a probability b that the conclusion of the inference will appear to lie in $[L,\ U]$. See [3] for more details.
- DistributionalTruthValues, which are discretized approximations to entire probability distributions.

This gradation of truth-value types serves several purposes. Strength and Simple truth values can be used when speed is of the essence or when one simply has little information. When accuracy is most important and when we have additional information concerning an Atom's full probability distribution, then Indefinite and Distributional truth values may be more pertinent.

3.4 PLN Rules and Formulas

A distinction is made in PLN between rules and formulas. PLN logical inferences take the form of "syllogistic rules," which give patterns for combining statements with matching terms. Examples of PLN rules include, but are not limited to, are

- the deduction $((A \rightarrow B) \wedge (B \rightarrow C) \Rightarrow (A \rightarrow C))$,
- induction $((A \rightarrow B) \wedge (A \rightarrow C) \Rightarrow (B \rightarrow C))$,
- abduction $((A \rightarrow C) \wedge (B \rightarrow C) \Rightarrow (A \rightarrow C))$,
- inversion rules $((A \rightarrow B) \Rightarrow (B \rightarrow A))$.

Related to each rule is a formula which calculates the truth value resulting from application of the rule. As an example, suppose s_A, s_B, s_C, s_{AB}, and s_{BC} represent the truth values for the terms A, B, C, as well the truth values of the relationships $A \rightarrow B$ and $B \rightarrow C$, respectively. Then, under suitable conditions imposed upon these input truth values, the formula for the deduction rule is given by:

$$s_{AC} = s_{AB}s_{BC} + \frac{(1 - s_{AB})(s_C - s_B s_{BC})}{1 - s_B},$$

where s_{AC} represents the truth value of the relationship $A \rightarrow C$. This formula is directly derived from probability theory given the assumption that $A \rightarrow B$ and $B \rightarrow C$ are independent. Using a combination of probability theory and heuristics, PLN also effectively handles cases in which independence is not a valid assumption.

4 Higher-Order PLN

Higher-order PLN (HOPLN) is defined as the subset of PLN that applies to predicates (considered as functions mapping arguments into truth values). It includes mechanisms for dealing with variable-bearing expressions and higher-order functions.

A predicate, in PLN, is a special kind of term that embodies a function mapping terms or relationships into truth-values. HOPLN contains several relationships that act upon predicates including Evaluation, Implication, and several types of quantifiers. The relationships can involve constant terms, variables, or a mixture.

PLN supports a variety of quantifiers, including traditional crisp and fuzzy quantifiers, plus the AverageQuantifier defined so that the truth value of

$$AverageQuantifier\ X\ F(X)$$

is a weighted average of $F(X)$ over all relevant inputs X [3]. AverageQuantifier is used implicitly in PLN to handle logical relationships between predicates, so that e.g. the conclusion of the above deduction is implicitly interpreted as

AverageQuantifier X
    Implication
        Evaluation is_Fluffy X
        Evaluation is_cat X

4.1 Reducing HOPLN to FOPLN

In [3] it is shown that in principle, over any finite observation set, HOPLN reduces to FOPLN. The key ideas of this reduction are the elimination of variables via use of higher-order functions, and the use of the set-theoretic definition of function embodied in the *SatisfyingSet* operator to map function-argument relationships into set-member relationships.

As an example, consider the Implication link. In HOPLN, where X is a variable

Implication
    R_1 A X
    R_2 B X

may be reduced to

Inheritance
    SatisfyingSet(R_1 A X)
    SatisfyingSet(R_2 B X)

where e.g. *SatisfyingSet*$(R_1\ A\ X)$ is the fuzzy set of all X satisfying the relationship $R_1(A, X)$.

5 PLN and AI

While PLN serves as a standalone system, recall that PLN grew out of a desire to build an uncertain reasoning module for use within a more general artificial intelligence framework. In order to completely understand many of the open research problems within PLN, it helps to understand the roles PLN plays within this larger context. To address this issue, we examine PLN from two additional viewpoints. First as PLN relates to intelligent agents, and then we will provide an overview of how PLN fits into the larger OpenCog framework.

5.1 SRAM

Here we very briefly review a simple formal model of intelligent agents called SRAM, for Simple Realistic Agent Model.

Following a theoretical framework developed by Legg and Hutter[8], we consider a class of active agents which observe and explore their environment and also take actions in it, which may affect the environment. The agent sends information to the environment and the environment sends signals to the agent. Agents can also experience rewards.

To this framework, we add a set $\mathcal{M}$ of memory actions which allow agents to maintain memories (of finite size), and at each time step to carry out internal actions on their memories as well as external actions in the environment. We also introduce the notions of *goals* and consider the environment as sending goal-symbols to the agent along with regular observation-symbols. In this extended framework, an interaction sequence looks like

$$m_1 a_1 o_1 g_1 r_1 m_2 a_2 o_2 g_2 r_2 ...$$

where the m_i's represent memory actions, the a_i's represent external actions, the o_i's represent observations, the g_i's represent agent goals, and the r_i's represent rewards. It is assumed that the reward r_i provided to an agent at time i is determined by the goal function g_i. If w is introduced as a single symbol to denote the combination of a memory action and an external action, and y is introduced as a single symbol to denote the combination of an observation, a goal and a reward, we can simplify this interaction sequence as

$$w_1 y_1 w_2 y_2 ...$$

Each goal function maps each finite interaction sequence $I_{g,s,t} = wy_{s:t}$ with g_s corresponding to g, into a value $r_g(I_{g,s,t}) \in [0, 1]$ indicating the value or "raw reward" of achieving the goal during that interaction sequence. The total reward r_t obtained by the agent is the sum of the raw rewards obtained at time t from all goals whose symbols occur in the agent's history before t.

The agent is represented as a function π which takes the current history as input, and produces an action as output. Agents need not be deterministic, an agent may for instance induce a probability distribution over the space of possible actions, conditioned on the current history. In this case we may characterize the agent by a probability distribution $\pi(w_t|wy_{<t})$. Similarly, the environment may be characterized by a probability distribution $\mu(y_k|wy_{<k})$. Taken together, the distributions π and μ define a probability measure over the space of interaction sequences.

Following Legg and Hutter, we will consider the class of environments that are *reward-summable*, meaning that the total amount of reward they return to any agent is bounded by 1. We will also use the term "context" to denote the combination of an environment, a goal function and a reward function. If the agent is acting in environment μ, and is provided with $g_t = g$ for the time-interval $T = t \in \{t_1, ..., t_2\}$, then the *expected goal-achievement* of the agent during the interval is

$$V^{\pi}_{\mu,g,T} \equiv \sum_{t_1}^{t_2} r_i$$

where E is the space of computable, reward-summable environments.

Next, we introduce a second-order probability distribution ν, which is a probability distribution over the space of environments μ. The distribution ν assigns each environment a probability.

What is key in the above formalism is that this second-order probability distribution ties in nicely with the indefinite probabilities framework and allows us to ground a form of possible worlds semantics within experiential semantics. An agent, experiencing a single stream of perceptions, may use this to construct an ensemble of "simulated" possible worlds, which may then be used in various sorts of inferences using a commonplace idea in the field of statistics: subsampling," a form of "bootstrapping."

This notion ties in closely with SRAM, which considers a probability distribution over a space of environments which are themselves probability distributions. What a real agent has is actually a single series of remembered observations. But it can induce a hopeful approximation of this distribution over environments by subsampling its memory and asking: what would it imply about the world if the items in this subsample were the only things I'd seen?

5.2 PLN's Relationship to OpenCog

Now we briefly describe the OCP (OCP) AGI architecture, implemented within the open-source OpenCog AI framework. OCP combines multiple AI paradigms such as uncertain logic, computational linguistics, evolutionary program learning and connectionist attention allocation in a unified architecture. Cognitive processes embodying these different paradigms interoperate together on a common neural-symbolic knowledge store called the Atomspace. The interaction of these processes is designed to encourage the self-organizing emergence of high-level network structures in the Atomspace, including superposed hierarchical and heterarchical knowledge networks, and a self-model network enabling meta-knowledge and meta-learning.

The high-level architecture of OCP involves the use of multiple cognitive processes associated with multiple types of memory to enable an intelligent agent to execute the procedures that it believes have the best probability of working toward its goals in its current context. OCP handles low-level perception and action via an extension called OpenCogBot, which integrates a hierarchical temporal memory system, DeSTIN [1].

OCP's memory types are the declarative, procedural, sensory, and episodic memory types that are widely discussed in cognitive neuroscience [10], plus attentional memory for allocating system resources generically, and intentional memory for allocating system resources in a goal-directed way. Table 1 overviews these memory types, giving key references and indicating the corresponding cognitive processes, and also indicating which of the generic patternist cognitive dynamics each cognitive process corresponds to (pattern creation, association, etc.).

Table 1. Memory Types and Cognitive Processes in OpenCog Prime. The third column indicates the general cognitive function that each specific cognitive process carries out, according to the patternist theory of cognition.

Memory Type	Specific Cognitive Processes	General Cognitive Functions
Declarative	Probabilistic Logic Networks (PLN) [3]; concept blending [2]	pattern creation
Procedural	MOSES (a novel probabilistic evolutionary program learning algorithm) [9]	pattern creation
Episodic	internal simulation engine [4]	association, pattern creation
Attentional	Economic Attention Networks (ECAN) [6]	attention allocation,, association, credit assignment
Intentional	probabilistic goal hierarchy refined by PLN and ECAN, structured according to Psi	credit assignment, pattern creation
Sensory	Supplied by DeSTIN integration	association, attention allocation, pattern creation, credit assignment

6 Conclusion and Future Research Directions

The OpenCog software (with PLN as it core reasoning system) has been used for commercial applications in the area of natural language processing and data mining [5]. A collaboration between Novamente LLC and The Electric Sheep Company demonstrated an OpenCog-controlled virtual dog in a virtual world, that can learn new tricks via imitative and reinforcement learning[4] (see `http://novamente.net/example` for some videos of these virtual dogs in action).

More recently, a new project based at Hong Kong Polytechnic University called M-Lab will explore the creation of generally intelligent humanoid game characters, powered by OpenCog and M-Labs Lucid game engine, with the capability for simple English conversation and realistic human-like emotional dynamics. Once again, as part of this effort, PLN will play a pivotal role, supplying the core planning and inference mechanisms. As these projects proceed, it is clear that new challenges will arise and that PLN will encounter challenges and require alterations and additions. As a standalone system, many challenging problems still remain, most notably forward and backward chaining inference control.

References

1. Arel, I., Rose, D., Coop, R.: Destin: A scalable deep learning architecture with application to high-dimensional robust pattern recognition. In: Proc. AAAI Workshop on Biologically Inspired Cognitive Architectures (2009)
2. Fauconnier, G., Turner, M.: The Way We Think: Conceptual Blending and the Mind's Hidden Complexities. Basic (2002)
3. Goertzel, B., Ikle, M., Goertzel, I., Heljakka, A.: Probabilistic Logic Networks. Springer, Heidelberg (2008)
4. Goertzel, B., Pennachin, C., et al.: An integrative methodology for teaching embodied non-linguistic agents, applied to virtual animals in second life. In: Proc.of the First Conf. on AGI. IOS Press, Amsterdam (2008)
5. Goertzel, B., Pinto, H., Pennachin, C., Goertzel, I.F.: Using dependency parsing and probabilistic inference to extract relationships between genes, proteins and malignancies implicit among multiple biomedical research abstracts. In: Proc. of Bio-NLP 2006 (2006)
6. Goertzel, B., Pitt, J., Ikle, M., Pennachin, C., Liu, R.: Glocal memory: a design principle for artificial brains and minds. Neurocomputing (April 2010)
7. Hart, D., Goertzel, B.: Opencog: A software framework for integrative artificial general intelligence. In: AGI. Frontiers in Artificial Intelligence and Applications, vol. 171, pp. 468–472. IOS Press, Amsterdam (2008), `http://dblp.uni-trier.de/db/conf/agi/agi2008.html#HartG08`
8. Legg, S., Hutter, M.: A formal measure of machine intelligence. In: Proc. of Benelaam 2006 (2007)
9. Looks, M.: Competent Program Evolution. PhD Thesis, Computer Science Department, Washington University (2006)
10. Tulving, E., Craik, R.: The Oxford Handbook of Memory. Oxford U. Press, Oxford (2005)
11. Walley, P.: Statistical Reasoning with Imprecise Probabilities. Chapman-Hall, Boca Raton (1990)
12. Wang, P.: Non-Axiomatic Reasoning System: Exploring the Essence of Intelligence. Ph.D. thesis, Indiana University (1995)

Dynamics of Beliefs

Sébastien Konieczny

CRIL - CNRS
Université d'Artois, Lens, France
konieczny@cril.fr

Abstract. The dynamics of beliefs is one of the major components of any autonomous system, that should be able to incorporate new pieces of information. In this paper we give a quick overview of the main operators for belief change, in particular revision, update, and merging, when the beliefs are represented in propositional logic. And we discuss some works on belief change in more expressive frameworks.

1 Introduction

Every autonomous agent has to use a belief base to model the state of the world. This information is precious since beliefs can be costly to obtain and since they are necessary to carry on reasoning tasks or to take the appropriate decisions. So a first class requirement in order to design intelligent autonomous agents is to try to provide her the means to obtain, and to maintain the most faithful belief base. In particular an agent has to be able to incorporate new pieces of information, and to correct the incorrect beliefs when she detected them. So this dynamics of beliefs is one of the major components of any autonomous agent.

The aim of this paper is to recall the definition of the main belief change operators and the links between them. We focus on the classical case, where the beliefs of the agents are represented using propositional logic, before discussing some extensions in other representational frameworks.

This is a very quick presentation of belief change theory. For a complete introduction the reader should refer to the seminal books on belief revision [35,36,41,64], or the recent special issue of Journal of Philosophical Logic on the 25 Years of AGM Theory [34].

2 Preliminaries

We consider a propositional language $\mathcal{L}$ defined from a finite set of propositional variables $\mathcal{P}$ and the standard connectives.

An interpretation ω is a total function from $\mathcal{P}$ to $\{0, 1\}$. The set of all interpretations is denoted by $\mathcal{W}$. An interpretation ω is a model of a formula $\varphi \in \mathcal{L}$ if and only if it makes it true in the usual truth functional way. $mod(\varphi)$ denotes the set of models of the formula φ, i.e., $mod(\varphi) = \{\omega \in \mathcal{W} \mid \omega \models \varphi\}$. When M is a set of models we denote by φ_M a formula such that $mod(\varphi_M) = M$.

S. Benferhat and J. Grant (Eds.): SUM 2011, LNAI 6929, pp. 61–74, 2011.
© Springer-Verlag Berlin Heidelberg 2011

A *belief base* K is a finite set of propositional formulae. In order to simplify the notations we identify the base K with the formula φ which is the conjunction of the formulae of K[1].

3 Revision

Belief revision aims at changing the status of some beliefs in the base that are contradicted by a more reliable piece of information. Several principles are governing this revision operation:

- First is the primacy of update principle: the new piece of information has to be accepted in the belief base after the revision. This is due to the hypothesis that the new piece of information is more reliable than the current beliefs of the agent[2].
- Second is the principle of coherence: the new belief base after the revision should be a consistent belief base. Asking the beliefs of the agent to be consistent is a natural requirement if one wants the agent to conduct reasoning tasks from her belief base.
- Third is the principle of minimal change: the new belief base after the revision should be as close as possible from the current belief base of the agent. This important principle aims at ensuring that no unnecessary information (noise) is added to the beliefs of the agent during the revision process, and that no unnecessary information is lost during the process: information/beliefs are usually costly to obtain, we do not want to throw them away without any serious reason.

Alchourrón, Gärdenfors and Makinson [2] proposed some postulates in order to formalize these principles for belief revision.

Definition 1 ([48]). *Let φ and μ be two formulas denoting respectively the belief base of the agent, and a new piece of information. Then $\varphi \circ \mu$ is a formula representing the new belief base of the agent. An operator $\circ$ is an AGM belief revision operator if it satisfies the following properties:*

(R1) $\varphi \circ \mu \vdash \mu$
(R2) *If $\varphi \wedge \mu$ is consistent then $\varphi \circ \mu \equiv \varphi \wedge \mu$*
(R3) *If μ is consistent then $\varphi \circ \mu$ is consistent*
(R4) *If $\varphi_1 \equiv \varphi_2$ and $\mu_1 \equiv \mu_2$ then $\varphi_1 \circ \mu_1 \equiv \varphi_2 \circ \mu_2$*
(R5) $(\varphi \circ \mu) \wedge \phi \vdash \varphi \circ (\mu \wedge \phi)$
(R6) *If $(\varphi \circ \mu) \wedge \phi$ is consistent t then $\varphi \circ (\mu \wedge \phi) \vdash (\varphi \circ \mu) \wedge \phi$*

When one works with a finite propositional language the above postulates, proposed by Katsuno and Mendelzon, are equivalent to AGM ones [2,35].

(R1) states that the new piece of information must be believed after the revision. (R2) says that when there is no conflict between the new piece of information and the current

[1] Some approaches are sensitive to syntactical representation. In that case it is important to distinguish between K and the conjunction of its formulae (see e.g. [52]).

[2] If this is not the case one should use a non-prioritized revision operator [42] or a merging operator (see Section 5).

beliefs of the agent, the revision is just the conjunction. (R3) says that revision always lead to a consistent belief base, unless the new piece of information is not consistent. (R4) is an irrelevance of syntax condition, it states that logically equivalent bases must lead to the same result. (R5) and (R6) give conditions on the revision by a conjunction.

Alchourrón, Gärdenfors and Makinson also defined contraction operators, that aim to remove some piece of information from the beliefs of the agent. These contraction operators are closely related to revision operators, since each contraction operator can be used to define a revision operator, through the Levy identity and conversely each revision operator can be used to define a contraction operator through the Harper identity [2,35]. So one can study indifferently revision or contraction operators. So we focus on revision here.

Several representation theorems, that give constructive ways to define AGM revision/contraction operators, have been proposed, such as partial meet contraction/revision [2], epistemic entrenchments [37,35], safe contraction [1], etc. In [48], Katsuno and Mendelzon give a representation theorem, showing that each revision operator corresponds to a faithful assignment, that associates to each base a plausibility preorder on interpretations (this idea can be traced back to Grove systems of spheres [40]).

Definition 2. *A faithful assignment is a function mapping each base φ to a pre-order $\leq_\varphi$ over interpretations such that:*

1. *If $\omega \models \varphi$ and $\omega' \models \varphi$, then $\omega \simeq_\varphi \omega'$*
2. *If $\omega \models \varphi$ and $\omega' \not\models \varphi$, then $\omega <_\varphi \omega'$*
3. *If $\varphi \equiv \varphi'$, then $\leq_\varphi = \leq_{\varphi'}$*

Theorem 1 ([48]). *An operator $\circ$ is an AGM revision operator (ie. it satisfies (R1)-(R6)) if and only if there exists a faithful assignment that maps each base φ to a total pre-order $\leq_\varphi$ such that $mod(\varphi \circ \mu) = \min(mod(\mu), \leq_\varphi)$.*

One of the main problems of this characterization of belief revision is that it does not constrain the operators enough for ensuring a good behavior when we do iteratively several revisions. So one needs to add more postulates and to represent the beliefs of the agent with a more complex structure than a simple belief base. In [26] Darwiche and Pearl proposed a convincing extension of AGM revision. This proposal have been improved by an additional condition in [17,45]. And [55,51] define improvement operators that are a generalization of iterated revision operators.

4 Update

Whereas belief revision should be used to improve the beliefs of the agent by incorporating more reliable pieces of evidence, belief update operators aim at maintaining the belief base of the agent up-to-date, by allowing to modify the base according to a reported change in the world. This distinction between revision and update was made clear in [47,49], where Katsuno and Mendelzon proposed postulates for belief update.

Definition 3 ([47,49]). *An operator $\diamond$ is a (partial) update operator if it satisfies the properties* **(U1)-(U8)**. *It is a total update operator if it satisfies the properties* **(U1)-(U5), (U8), (U9)**.

(U1) $\varphi \diamond \mu \vdash \mu$
(U2) *If* $\varphi \vdash \mu$, *then* $\varphi \diamond \mu \equiv \varphi$
(U3) *If* $\varphi \nvdash \perp$ *and* $\mu \nvdash \perp$ *then* $\varphi \diamond \mu \nvdash \perp$
(U4) *If* $\varphi_1 \equiv \varphi_2$ *and* $\mu_1 \equiv \mu_2$ *then* $\varphi_1 \diamond \mu_1 \equiv \varphi_2 \diamond \mu_2$
(U5) $(\varphi \diamond \mu) \wedge \phi \vdash \varphi \diamond (\mu \wedge \phi)$
(U6) *If* $\varphi \diamond \mu_1 \vdash \mu_2$ *and* $\varphi \diamond \mu_2 \vdash \mu_1$, *then* $\varphi \diamond \mu_1 \equiv \varphi \diamond \mu_2$
(U7) *If* φ *is a complete formula, then* $(\varphi \diamond \mu_1) \wedge (\varphi \diamond \mu_2) \vdash \varphi \diamond (\mu_1 \vee \mu_2)$
(U8) $(\varphi_1 \vee \varphi_2) \diamond \mu \equiv (\varphi_1 \diamond \mu) \vee (\varphi_2 \diamond \mu)$
(U9) *If* φ *is a complete formula and* $(\varphi \diamond \mu) \wedge \phi \nvdash \perp$, *then* $\varphi \diamond (\mu \wedge \phi) \vdash (\varphi \diamond \mu) \wedge \phi$

Most of these postulates are close to the ones of revision. The main differences lie in postulate (U2) that is much weaker than (R2): conversely to revision, even if the new piece of information is consistent with the belief base, the result is generally not simply the conjunction. This illustrates the fact that revision can be seen as a selection process of the most plausible worlds of the current beliefs with respect to the new piece information, whereas update is a transition process: each world of the current beliefs have to be translated to the closest world allowed by the new piece of information. This world-by-world treatment is expressed by postulate (U8).

As for revision, there is a representation theorem in terms of faithful assignment.

Definition 4. *A faithful assignment is a function mapping each interpretation* ω *to a pre-order* $\leq_\omega$ *over interpretations such that if* $\omega \neq \omega'$, *then* $\omega <_\omega \omega'$.

One can easily check that this faithful assignment on interpretations is just a special case of the faithful assignment on bases defined in the previous section on the complete base corresponding to the interpretation.

Katsuno and Mendelzon give two representation theorems for update operators. The first representation theorem, that is the most commonly used, corresponds to partial pre-orders. This use of partial pre-order is one of the differences between belief revision and belief update (note nonetheless that postulates for belief revision can also be adapted to modelize assignements giving partial pre-orders [9]).

Theorem 2 ([47,49]). *An update operator* $\diamond$ *satisfies (U1)-(U8) if and only if there exists a faithful assignment that maps each interpretation* ω *to a partial pre-order* $\leq_\omega$ *such that* $mod(\varphi \diamond \mu) = \bigcup_{\omega \models \varphi} \min(mod(\mu), \leq_\omega)$.

But there is also a second theorem corresponding to total pre-orders.

Theorem 3 ([47,49]). *An update operator* $\diamond$ *satisfies (U1)-(U5), (U8) and (U9) if and only if there exists a faithful assignment that maps each interpretation* ω *to a total pre-order* $\leq_\omega$ *such that* $mod(\varphi \diamond \mu) = \bigcup_{\omega \models \varphi} \min(mod(\mu), \leq_\omega)$.

This characterization of update is quite convincing, but some criticisms can be made that suggest that more elaborate update operators can be studied [43].

5 Merging

Merging operators [4,5,62,58,56] should be used when one wants to combine several belief bases, or wants to take into account several pieces of information of same reliability.

We first need to define a profile of bases, that will represent the set of bases/information one wants to combine:

A *profile* Ψ is a non-empty multi-set (bag) of bases $\Psi = \{\varphi_1, \ldots, \varphi_n\}$ (hence different agents are allowed to exhibit identical bases), and represents a group of n agents. We denote by $\bigwedge \Psi$ the conjunction of bases of $\Psi = \{\varphi_1, \ldots, \varphi_n\}$, i.e., $\bigwedge \Psi = \varphi_1 \wedge \ldots \wedge \varphi_n$. A profile Ψ is said to be consistent if and only if $\bigwedge \Psi$ is consistent. The multi-set union is denoted by $\sqcup$.

Belief merging operators aim at aggregating several bases into a single one. The most basic case is when all the bases have the same strength/importance (see [28] for a discussion on prioritized merging). Often the aggregation has to obey a set of rules, that can be a translation of physical laws or of some knowledge about the result, that form the integrity constraints for the merging. Let us see the postulates for Integrity Constraints merging operators:

Definition 5 ([53]). *Let Ψ be a profile and μ be a formula encoding integrity constraints. Then $\triangle_\mu(\Psi)$ represents the merging of the profile Ψ under the integrity constraints μ. An operator $\triangle$ is an IC merging operator if it satisfies the following properties:*

(IC0) $\triangle_\mu(\Psi) \vdash \mu$
(IC1) *If μ is consistent, then $\triangle_\mu(\Psi)$ is consistent*
(IC2) *If $\bigwedge \Psi$ is consistent with μ, then $\triangle_\mu(\Psi) \equiv \bigwedge \Psi \wedge \mu$*
(IC3) *If $\Psi_1 \equiv \Psi_2$ and $\mu_1 \equiv \mu_2$, then $\triangle_{\mu_1}(\Psi_1) \equiv \triangle_{\mu_2}(\Psi_2)$*
(IC4) *If $\varphi_1 \vdash \mu$ and $\varphi_2 \vdash \mu$, then $\triangle_\mu(\{\varphi_1, \varphi_2\}) \wedge \varphi_1$ is consistent if and only if $\triangle_\mu(\{\varphi_1, \varphi_2\}) \wedge \varphi_2$ is consistent*
(IC5) $\triangle_\mu(\Psi_1) \wedge \triangle_\mu(\Psi_2) \vdash \triangle_\mu(\Psi_1 \sqcup \Psi_2)$
(IC6) *If $\triangle_\mu(\Psi_1) \wedge \triangle_\mu(\Psi_2)$ is consistent, then $\triangle_\mu(\Psi_1 \sqcup \Psi_2) \vdash \triangle_\mu(\Psi_1) \wedge \triangle_\mu(\Psi_2)$*
(IC7) $\triangle_{\mu_1}(\Psi) \wedge \mu_2 \vdash \triangle_{\mu_1 \wedge \mu_2}(\Psi)$
(IC8) *If $\triangle_{\mu_1}(\Psi) \wedge \mu_2$ is consistent, then $\triangle_{\mu_1 \wedge \mu_2}(\Psi) \vdash \triangle_{\mu_1}(\Psi)$*

These postulates are quite close to the ones of revision. The ones that specifically talk about aggregation are (IC4), (IC5) and (IC6). (IC4) is a fairness postulate, that expresses the fact that all the bases have the same importance/weight, so when merging two such bases one can not give more importance to one of them. (IC5) and (IC6) talk about the result of the merging when we join two groups. (IC5) states that all that is common in the merging of the two groups must be selected if we join the two groups. And (IC6) strengthen this condition by asking that the merging obtained when we join the two groups have to be exactly what is commonly chosen by the two groups. These two postulates correspond to well known Pareto conditions (see conditions 5 and 6 of the syncretic assignment).

There is also a representation theorem for merging operators in terms of pre-orders on interpretations [53].

Definition 6. *A* syncretic assignment *is a function mapping each profile Ψ to a total pre-order $\leq_\Psi$ over interpretations such that:*

1. *If $\omega \models \Psi$ and $\omega' \models \Psi$, then $\omega \simeq_\Psi \omega'$*
2. *If $\omega \models \Psi$ and $\omega' \not\models \Psi$, then $\omega <_\Psi \omega'$*

3. *If* $\Psi_1 \equiv \Psi_2$, *then* $\leq_{\Psi_1} = \leq_{\Psi_2}$
4. $\forall \omega \models \varphi\ \exists \omega' \models \varphi'\ \omega' \leq_{\{\varphi\} \sqcup \{\varphi\}'} \omega$
5. *If* $\omega \leq_{\Psi_1} \omega'$ *and* $\omega \leq_{\Psi_2} \omega'$, *then* $\omega \leq_{\Psi_1 \sqcup \Psi_2} \omega'$
6. *If* $\omega <_{\Psi_1} \omega'$ *and* $\omega \leq_{\Psi_2} \omega'$, *then* $\omega <_{\Psi_1 \sqcup \Psi_2} \omega'$

Theorem 4 ([53]). *An operator* $\triangle$ *is an IC merging operator if and only if there exists a syncretic assignment that maps each profile* Ψ *to a total pre-order* $\leq_\Psi$ *such that*

$$mod(\triangle_\mu(\Psi)) = \min(mod(\mu), \leq_\Psi)$$

6 On the Links between Revision, Update and Merging

6.1 Revision vs Update

Intuitively revision operators bring a minimal change to the base by selecting the most plausible models among the models of the new information. Whereas update operators bring a minimal change to each possible world (model) of the base in order to take into account the change described by the new infomation whatever the possible world. So, if we look closely to the representation theorems (theorems 1, 2 and 3), we easily find the following result:

Theorem 5. *If* $\circ$ *is a revision operator (i.e. it satisfies (R1)-(R6)), then the operator* $\diamond$ *defined by* $\varphi \diamond \mu = \bigvee_{\omega \models \varphi} \varphi_{\{\omega\}} \circ \mu$ *is an update operator that satisfies (U1)-(U9).*

So this proposition states that update can be viewed as a kind of pointwise revision.

6.2 Revision vs Merging

Intuitively revision operators select in a formula (the new evidence) the closest information to a ground information (the old base). And, identically, IC merging operators select in a formula (the integrity constraints) the closest information to a ground information (a profile of bases). So following this idea it is easy to make a correspondence between IC merging operators and belief revision operators:

Theorem 6 ([53]). *If* $\triangle$ *is an IC merging operator (it satisfies (IC0-IC8)), then the operator* $\circ$, *defined as* $\varphi \circ \mu = \triangle_\mu(\varphi)$, *is an AGM revision operator (it satisfies (R1-R6)).*

See [53] for more links between belief revision and merging.

7 Other Belief Change Operators

7.1 Confluence Operators

As explained in the previous section, there are close connections between revision, update and merging. Update can be considered as a pointwise revision, and merging as a generalization of revision. So, as illustrated in Figure 1, one can define confluence operators [54] that can be considered as a pointwise merging, and as a generalization of update.

Let us first define p-consistency for profiles:

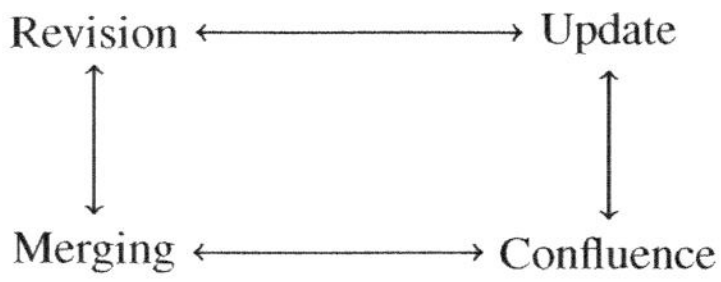

Fig. 1. Revision - Update - Merging - Confluence

Definition 7. *A profile $\Psi = \{\varphi_1, \ldots, \varphi_n\}$ is* p-consistent *if all its bases are consistent, i.e $\forall \varphi_i \in \Psi$, φ_i is consistent.*

Note that p-consistency is much weaker than consistency, the former just asks that all the bases of the profile are consistent, while the later asks that the conjunction of all the bases is consistent.

Definition 8. *An operator $\diamond$ is a confluence operator if it satisfies the following properties:*

(UC0) $\diamond_\mu(\Psi) \vdash \mu$
(UC1) *If μ is consistent and Ψ is p-consistent, then $\diamond_\mu(\Psi)$ is consistent*
(UC2) *If Ψ is complete, Ψ is consistent and $\bigwedge \Psi \vdash \mu$, then $\diamond_\mu(\Psi) \equiv \bigwedge \Psi$*
(UC3) *If $\Psi_1 \equiv \Psi_2$ and $\mu_1 \equiv \mu_2$, then $\diamond_{\mu_1}(\Psi_1) \equiv \diamond_{\mu_2}(\Psi_2)$*
(UC4) *If φ_1 and φ_2 are complete formulae and $\varphi_1 \vdash \mu$, $\varphi_2 \vdash \mu$, then $\diamond_\mu(\{\varphi_1, \varphi_2\}) \wedge \varphi_1$ is consistent if and only $\diamond_\mu(\{\varphi_1, \varphi_2\}) \wedge \varphi_2$ is consistent*
(UC5) $\diamond_\mu(\Psi_1) \wedge \diamond_\mu(\Psi_2) \vdash \diamond_\mu(\Psi_1 \sqcup \Psi_2)$
(UC6) *If Ψ_1 and Ψ_2 are complete profiles and $\diamond_\mu(\Psi_1) \wedge \diamond_\mu(\Psi_2)$ is consistent, then $\diamond_\mu(\Psi_1 \sqcup \Psi_2) \vdash \diamond_\mu(\Psi_1) \wedge \diamond_\mu(\Psi_2)$*
(UC7) $\diamond_{\mu_1}(\Psi) \wedge \mu_2 \vdash \diamond_{\mu_1 \wedge \mu_2}(\Psi)$
(UC8) *If Ψ is a complete profile and if $\diamond_{\mu_1}(\Psi) \wedge \mu_2$ is consistent, then $\diamond_{\mu_1 \wedge \mu_2}(\Psi) \vdash \diamond_{\mu_1}(\Psi) \wedge \mu_2$*
(UC9) $\diamond_\mu(\Psi \sqcup \{\varphi \vee \varphi'\}) \equiv \diamond_\mu(\Psi \sqcup \{\varphi\}) \vee \diamond_\mu(\Psi \sqcup \{\varphi'\})$

See [54] for a representation theorem in terms of assignment for confluence operators. We just give the two results that show how confluence relates with respect to merging and update [54]:

Theorem 7. *If $\diamond$ is a confluence operator (i.e. it satisfies (UC0-UC9)), then the operator $\diamond$, defined as $\varphi \diamond \mu = \diamond_\mu(\varphi)$, is a total update operator (i.e. it satisfies (U1-U9)).*

For relating confluence and merging, we need to use the notion of state:

Definition 9. *A multi-set of interpretations will be called a* state. *We use the letter e, possibly with subscripts, for denoting states. If $\Psi = \{\varphi_1, \ldots, \varphi_n\}$ is a profile and $e = \{\omega_1, \ldots, \omega_n\}$ is a state such that $\forall i\ \omega_i \models \varphi_i$, we say that e is a state of the profile Ψ, or that the state e models the profile Ψ, that will be denoted by $e \models \Psi$.*

If $e = \{\omega_1, \ldots, \omega_n\}$ is a state, we define the profile Ψ_e by putting $\Psi_e = \{\varphi_{\{\omega_1\}}, \ldots, \varphi_{\{\omega_n\}}\}$.

Theorem 8. *If $\triangle$ is an IC merging operator (i.e. it satisfies (IC0-IC8)) then the operator $\diamond$ defined by $\diamond_\mu(\Psi) = \bigvee_{e \models \Psi} \triangle_\mu(\Psi_e)$ is a confluence operator (i.e. it satisfies (UC0-UC9)).*

7.2 Extrapolation and Approaches Based on Sequences of Observations

In [31,32] Dupin and Lang defined extrapolation operators. The idea is, from a sequence of observations at different time points, to try to find the scenarios that best explain the sequence. The principle of minimal change is translated in an inertial assumption, that states that the value of a propositional variable does not change if no change occur. We do not have direct information about the changes, but the observations at different time points inform us on such changes. So, very roughly, these operators can be seen as looking for the most plausible histories compatible with a sequence of observations and minimal change assumptions.

There are others works that deal with sequences of observations such as [57,44] for instance. An interesting operator was proposed by Booth and Nitka [20]. It can be seen as a third-party counterpart of extrapolation. The idea is that an observer observe a sequence of inputs that receives a given agent and a sequence of corresponding outputs (parts of the belief of the agent at that time point). Then the problem is to try to identify the initial beliefs of the agent and her beliefs during the sequence.

7.3 Belief Negotiation

In [16] Booth proposes to aggregate the beliefs of different agents by using a iterative selection-weakening process. The idea is, until the conjunction of the bases is consistent, to select some bases that have to weaken their beliefs. Like belief merging, these belief negotiation operators allow to obtain a consistent belief base from a set of jointly inconsistent bases. But the aim is quite different. In belief merging the aim is to extract as much information as possible from the set of bases, whereas in belief negotiation the aim is to find a potential consensual issue in a (abstract) negotiation process. Several works have tried to use tools from belief change theory in order to modelize abstract negotiation processes [14,15,16,68,59,50,38]. We think that there is still a lot to do in this direction. In particular there is no representation theorem for abstract negotiation.

7.4 Prioritized Merging Operators

In [28] Delgrande, Dubois and Lang propose an interesting discussion on prioritized merging operators. The idea is to merge a set of weighted formulae. The weights are used to stratify the formulae (a formula with a greater weight is more important, even if they are a large number of formula with smaller weights that contradict it).

Delgrande, Dubois and Lang motivate the generality of their approach by showing that classical merging operators (on unweighted formulae) and iterated belief revision operators (à la Darwiche and Pearl [26]) can be considered as two extreme cases of this weighted merging framework.

The main argument is that if one makes the hypothesis that the new pieces of information that come successively in an iterated revision process are about a static world (the usual hypothesis), then there is no reason to give the preference to the last ones. If these information have different reliability, then this can be represented explicitly with the weights of the formulae, in order to take this difference of reliability in the iterated "revision" process if they do not come in the order corresponding to their relative reliability. And the correct way to do that is to make a prioritized merging.

This discussion is interesting since in several papers on iterated revisions, it seems that the authors do not make any distinction between the hypothesis to have more and more recent pieces of information, and the hypothesis to have more and more reliable pieces of information.

The framework of Delgrande, Dubois and Lang identifies the epistemic states as the sequences of formulae that the agent receives. They show that the postulates for iterated belief revision can be obtain as special case of their postulates for weighted merging, and that they can also lead to some postulates of IC merging. This work is interesting since it opens a way for logical characterization of prioritized merging. It could be interesting to try to find a representation theorem in this case, and to look at the generalization of IC merging operators in this prioritized merging framework.

8 Belief Change in Other Representational Frameworks

8.1 Dynamics of Horn Bases

Recently some works have focus on the contraction of Horn bases [27,18,19,29]. This is an interesting case since Horn bases are used for instance for deductive databases and logic programming. Usually works on belief change suppose that the logic is at least as strong as classical propositional logic. But these works on Horn bases show that restrictions of propositional logic exhibit some interesting characteristics. In particular constructions that lead to equivalent classes of operators in the classical case, give rise to different ones for the Horn case.

8.2 Merging of First Order Bases

Lang and Bloch propose to define model-based merging operators using the maximum as aggregation function ($\triangle^{d,\max}$) by using dilation[3] process [12]. One can note that in the original Dalal paper [25], he defines his revision operator with such a dilation function rather than with a distance.

Gorogiannis and Hunter [39] extend this approach in order to define others model-based merging operators using dilations. So, in addition to $\triangle^{d,\max}$, they define $\triangle^{d,\Sigma}$, $\triangle^{d,\textsc{Gmax}}$ and $\triangle^{d,\textsc{Gmin}}$ operators.

The interest of this definition of these operators is that it can be easily extended to first order logic. The usual definition of model-based merging operators is based on the computation of distances between interpretations. So when using logics where the number of interpretations is infinite, this approach is not the more appropriate. The interest of defining these operators with dilations is that they can also be used in this case. This only needs to use the good dilation function. See [39] for a discussion and some examples of dilation functions in the first order logic case.

---

[3] Roughly speaking dilation allows to reach the points/worlds in the neighborhood of a point/world. See [12] to see how to define this formally.

8.3 Merging of Qualitative Constraint Networks

Condotta, Kaci, Marquis and Schwind studied the merging of qualitative constraint networks [22,21]. These methods can be useful for merging constraint networks that represent spatial regions, for instance for Geographical Information Systems it can be necessary to merge spatial databases that come from different sources.

Conflicts that arise in this framework are more subtle that the binary ones in the propositional framework. In this case conflicts can be more or less important. For instance, if we use the Allen algebra, that allows to represent spatial information on segments on a line, namely relations as A BEFORE B, A AFTER[4]. B, A MEET B among others. A conflict between sentences A BEFORE B and A MEET B seems much less important than the one between A BEFORE B and A AFTER B.

This "intensity" that we feel between conflicts allows to define more various merging policies than in the propositional framework.

One can also look at [61,23] to see two examples of merging of spatial regions using logical representations.

8.4 Dynamics of Argumentation Frameworks

There are a lot of works on argumentation as a way to reason about contradictory pieces of information. The basic idea is to use a set of arguments and an attack relation between relations. This is the starting point of Dung abstract argumentation framework [30]. In [24] the problem of merging of argumentation frameworks, where the arguments are distributed among several agents, have been studied. This requires to define a new representation frameworks for argumentation: Partial Argumentation Frameworks, where there are three possible relations between two arguments A and B. Either the agent believes that A attacks B, or he believes that A does not attacks B, or he does not know if A attacks B or not. This last case is necessary to represent the fact that an agent ignores a given argument.

The problem of revision of argumentation systems as been addressed also in several works, such as [33,63,13] for instance.

We think that for both argumentation revision and merging a lot of work is still necessary in order to reach convincing models.

9 Conclusion

We proposed a quick tour of the theory of belief change in classical propositional logic. The core of this theory is quite established now, with a set of important belief change operators that are logically characterized. Still, a lot of developments are possible, for improving existing operators or for defining new classes of change operators.

Another possible way of development is to study the use of these belief change operators in other frameworks than classical logic. As illustrated by the works on horn clauses or on constraint networks, there are some subtleties that appear when one wants to work in these different frameworks.

[4] i.e. B BEFORE A

We focused on purely qualitative approaches here, but there are also a lot of works on belief change (revision, update, merging, etc.) on quantitative frameworks. There are for instance a lot of works on ordinal conditional function [66,67,60], or on change of possibilistic logic bases [6,7,46,8].

Merging is also at work on numerical datas, see for instance [65,3,10] for some examples of numerical data fusion. See [11] for an interesting global overview on (logical and numerical) merging.

References

1. Alchourrón, C., Makinson, D.: On the logic of theory change: Safe contraction. Studia Logica 44, 405–422 (1985)
2. Alchourrón, C.E., Gärdenfors, P., Makinson, D.: On the logic of theory change: Partial meet contraction and revision functions. Journal of Symbolic Logic 50, 510–530 (1985)
3. Ayoun, A., Smets, P.: Data association in multi-target detection using the transferable belief model. International Journal of Intelligent Systems 16(10), 1167–1182 (2001)
4. Baral, C., Kraus, S., Minker, J.: Combining multiple knowledge bases. IEEE Transactions on Knowledge and Data Engineering 3(2), 208–220 (1991)
5. Baral, C., Kraus, S., Minker, J., Subrahmanian, V.S.: Combining knowledge bases consisting of first-order theories. Computational Intelligence 8(1), 45–71 (1992)
6. Benferhat, S., Dubois, D., Lang, J., Prade, H., Saffioti, A., Smets, P.: A general approach for inconsistency handling and merging information in prioritized knowledge bases. In: Proceedings of the Sixth International Conference on Principles of Knowledge Representation and Reasoning (KR 1998), pp. 466–477 (1998)
7. Benferhat, S., Dubois, D., Prade, H.: A computational model for belief change and fusing ordered belief bases. In: Rott, H., Williams, M.A. (eds.) Frontiers in Belief revision. Kluwer, Dordrecht (1999)
8. Benferhat, S., Kaci, S.: Fusion of possibilistic knowledge bases from a postulate point of view. International Journal of Approximate Reasoning 33(3), 255–285 (2003)
9. Benferhat, S., Lagrue, S., Papini, O.: Revision of partially ordered information: Axiomatization, semantics and iteration. In: Proceedings of the Nineteenth International Joint Conference on Artificial Intelligence (IJCAI 2005), pp. 376–381 (2005)
10. Bloch, I., Géraud, T., Maître, H.: Representation and fusion of heterogeneous fuzzy information in the 3d space for model-based structural recognition–application to 3d brain imaging. Artificial Intelligence 148(1-2), 141–175 (2003)
11. Bloch, I., Hunter, A., Appriou, A., Ayoun, A., Benferhat, S., Besnard, P., Cholvy, L., Cooke, R.M., Cuppens, F., Dubois, D., Fargier, H., Grabisch, M., Kruse, R., Lang, J., Moral, S., Prade, H., Saffiotti, A., Smets, P., Sossai, C.: Fusion: General concepts and characteristics. International Journal of Intelligent Systems 16(10), 1107–1134 (2001)
12. Bloch, I., Lang, J.: Towards mathematical morpho-logics. In: Proceedings of the Eigth International Conference on Information Processing and Management of Uncertainty in Knowledge-Based Systems (IPMU 2000), pp. 1405–1412 (2000)
13. Boella, G., da Pereira, C., Tettamanzi, A., van der Torre, L.: Making others believe what they want. In: Artificial Intelligence in Theory and Practice II, pp. 215–224 (2008)
14. Booth, R.: A negotiation-style framework for non-prioritised revision. In: Proceedings of the Eighth Conference on Theoretical Aspects of Rationality and Knowledge (TARK 2001), pp. 137–150 (2001)
15. Booth, R.: Social contraction and belief negotiation. In: Proceedings of the Eighth Conference on Principles of Knowledge Representation and Reasoning (KR 2002), pp. 374–384 (2002)

16. Booth, R.: Social contraction and belief negotiation. Information Fusion 7(1), 19–34 (2006)
17. Booth, R., Meyer, T.: Admissible and restrained revision. Journal of Artificial Intelligence Research 26, 127–151 (2006)
18. Booth, R., Meyer, T., Varzinczak, I.J.: Next steps in propositional horn contraction. In: Proceedings of the Twenty first International Joint Conference on Artificial Intelligence (IJCAI 2009), pp. 702–707 (2009)
19. Booth, R., Meyer, T., Varzinczak, I.J., Wassermann, R.: Horn belief change: A contraction core. In: Proceedings of the Nineteenth European Conference on Artificial Intelligence (ECAI 2010), pp. 1065–1066 (2010)
20. Booth, R., Nittka, A.: Reconstructing an agent's epistemic state from observations. In: Proceedings of the Nineteenth International Joint Conference on Artificial Intelligence (IJCAI 2005), pp. 394–399 (2005)
21. Condotta, J.F., Kaci, S., Marquis, P., Schwind, N.: Merging qualitative constraint networks in a piecewise fashion. In: Proceedings of the Twenty First International Conference on Tools with Artificial Intelligence (ICTAI 2009), pp. 605–608 (2009)
22. Condotta, J.F., Kaci, S., Marquis, P., Schwind, N.: Merging qualitative constraints networks using propositional logic. In: Sossai, C., Chemello, G. (eds.) ECSQARU 2009. LNCS, vol. 5590, pp. 347–358. Springer, Heidelberg (2009)
23. Condotta, J.F., Kaci, S., Marquis, P., Schwind, N.: Majority merging: from boolean spaces to affine spaces. In: Proceedings of the Nineteenth European Conference on Artificial Intelligence (ECAI 2010), pp. 627–632 (2010)
24. Coste-Marquis, S., Devred, C., Konieczny, S., Lagasquie-Schiex, M.C., Marquis, P.: On the merging of dung's argumentation systems. Artificial Intelligence 171, 740–753 (2007)
25. Dalal, M.: Investigations into a theory of knowledge base revision: preliminary report. In: Proceedings of the American National Conference on Artificial Intelligence (AAAI 1988), pp. 475–479 (1988)
26. Darwiche, A., Pearl, J.: On the logic of iterated belief revision. Artificial Intelligence (89), 1–29 (1997)
27. Delgrande, J.P.: Horn clause belief change: Contraction functions. In: Proceedings of the Eleventh International Conference on the Principles of Knowledge Representation and Reasoning (KR 2008), pp. 156–165 (2008)
28. Delgrande, J.P., Dubois, D., Lang, J.: Iterated revision as prioritized merging. In: Proceedings of the Tenth International Conference on Knowledge Representation and Reasoning (KR 2006), pp. 210–220 (2006)
29. Delgrande, J.P., Wassermann, R.: Horn clause contraction functions: Belief set and belief base approaches. In: Proceedings of the Twelfth International Conference on the Principles of Knowledge Representation and Reasoning, KR 2010 (2010)
30. Dung, P.: On the acceptability of arguments and its fundamental role in nonmonotonic reasoning, logic programming and n-person games. Artificial Intelligence 77, 321–357 (1995)
31. Dupin de Saint-Cyr, F., Lang, J.: Belief extrapolation (or how to reason about observations and unpredicted change). In: Proceedings of the Eighth Conference on Principles of Knowledge Representation and Reasoning (KR 2002), pp. 497–508 (2002)
32. Dupin de Saint-Cyr, F., Lang, J.: Belief extrapolation (or how to reason about observations and unpredicted change). Artificial Intelligence 175(2), 760–790 (2011)
33. Falappa, M.A., Kern-Isberner, G., Simari, G.R.: Belief revision and argumentation theory. In: Simari, G., Rahwan, I. (eds.) Argumentation in Artificial Intelligence, pp. 341–360. Springer, US (2009)
34. Fermé, E., Hansson, S.O. (eds.): Journal of Philosophical Logic. Special Issue on 25 Years of AGM Theory, vol. 40(2). Springer, Netherlands (2011)
35. Gärdenfors, P.: Knowledge in flux. MIT Press, Cambridge (1988)

36. Gärdenfors, P. (ed.): Belief Revision. Cambridge University Press, Cambridge (1992)
37. Gärdenfors, P., Makinson, D.: Revisions of knowledge systems using epistemic entrenchment. In: Proceedings of the Second Conference on Theoretical Aspects of Reasoning about Knowledge, pp. 83–95 (1988)
38. Gauwin, O., Konieczny, S., Marquis, P.: Conciliation and consensus in iterated belief merging. In: Godo, L. (ed.) ECSQARU 2005. LNCS (LNAI), vol. 3571, pp. 514–526. Springer, Heidelberg (2005)
39. Gorogiannis, N., Hunter, A.: Merging first-order knowledge using dilation operators. In: Hartmann, S., Kern-Isberner, G. (eds.) FoIKS 2008. LNCS, vol. 4932, pp. 132–150. Springer, Heidelberg (2008)
40. Grove, A.: Two modellings for theory change. Journal of Philosophical Logic 17(157-180) (1988)
41. Hansson, S.O.: A Textbook of Belief Dynamics. Kluwer Academic Publishers, Dordrecht (1997)
42. Hansson, S.O. (ed.): Theoria. Special Issue on non-prioritized belief revision, vol. 63(1-2). Wiley, Chichester (1997)
43. Herzig, A., Rifi, O.: Propositional belief base update and minimal change. Artificial Intelligence 115(1), 107–138 (1999)
44. Hunter, A., Delgrande, J.P.: Belief change in the context of fallible actions and observations. In: Proceedings of the Twenty First National Conference on Artificial Intelligence and the Eighteenth Innovative Applications of Artificial Intelligence Conference, AAAI 2006 (2006)
45. Jin, Y., Thielscher, M.: Iterated belief revision, revised. Artificial Intelligence 171, 1–18 (2007)
46. Kaci, S., Benferhat, S., Dubois, D., Prade, H.: A principled analysis of merging operations in possibilistic logic. In: Proceedings of the Sixteenth Conference in Uncertainty in Artificial Intelligence (UAI 2000), pp. 24–31 (2000)
47. Katsuno, H., Mendelzon, A.O.: On the difference between updating a knowledge base and revising it. In: Proceedings of the Second International Conference on Principles of Knowledge Representation and Reasoning (KR 1991), pp. 387–394 (1991)
48. Katsuno, H., Mendelzon, A.O.: Propositional knowledge base revision and minimal change. Artificial Intelligence 52, 263–294 (1991)
49. Katsuno, H., Mendelzon, A.O.: On the difference between updating a knowledge base and revising it. In: Gärdenfors, P. (ed.) Belief Revision. Cambridge University Press, Cambridge (1992)
50. Konieczny, S.: Belief base merging as a game. Journal of Applied Non-Classical Logics 14(3), 275–294 (2004)
51. Konieczny, S., Medina Grespan, M., Pino Pérez, R.: Taxonomy of improvement operators and the problem of minimal change. In: Proceedings of the Twelfth International Conference on Principles of Knowledge Representation and Reasoning, KR 2010 (2010)
52. Konieczny, S., Lang, J., Marquis, P.: DA^2 merging operators. Artificial Intelligence 157(1-2), 49–79 (2004)
53. Konieczny, S., Pino Pérez, R.: Merging information under constraints: a logical framework. Journal of Logic and Computation 12(5), 773–808 (2002)
54. Konieczny, S., Pino Pérez, R.: Confluence operators. In: Hölldobler, S., Lutz, C., Wansing, H. (eds.) JELIA 2008. LNCS (LNAI), vol. 5293, pp. 272–284. Springer, Heidelberg (2008)
55. Konieczny, S., Pino Pérez, R.: Improvement operators. In: Proceedings of the Eleventh International Conference on Principles of Knowledge Representation and Reasoning (KR 2008), pp. 177–186 (2008)
56. Liberatore, P., Schaerf, M.: Arbitration (or how to merge knowledge bases). IEEE Transactions on Knowledge and Data Engineering 10(1), 76–90 (1998)

57. Liberatore, P., Schaerf, M.: BReLS: A system for the integration of knowledge bases. In: Proceedings of the Seventh Conference on Principles of Knowledge Representation and Reasoning (KR 2000), pp. 145–152 (2000)
58. Lin, J., Mendelzon, A.O.: Merging databases under constraints. International Journal of Cooperative Information System 7(1), 55–76 (1998)
59. Meyer, T., Foo, N., Zhang, D., Kwok, R.: Logical foundations of negotiation: Strategies and preferences. In: Proceedings of the Ninth Conference on Principles of Knowledge Representation and Reasoning (KR 2004), pp. 311–318 (2004)
60. Meyer, T.A.: On the semantics of combination operations. Journal of Applied Non-Classical Logics 11(1-2), 59–84 (2001)
61. Revesz, P.Z.: Model-theoretic minimal chenge operators for constraint databases. In: Afrati, F.N., Kolaitis, P.G. (eds.) ICDT 1997. LNCS, vol. 1186, pp. 447–460. Springer, Heidelberg (1996)
62. Revesz, P.Z.: On the semantics of arbitration. International Journal of Algebra and Computation 7(2), 133–160 (1997)
63. Rotstein, N.D., Moguillansky, M.O., Falappa, M.A., García, A.J., Simari, G.R.: Argument theory change: Revision upon warrant. In: Proceeding of the 2008 Conference on Computational Models of Argument (COMMA 2008), pp. 336–347 (2008)
64. Rott, H.: Change, choice and inference: a study of belief revision and nonmonotonic reasoning. Oxford logic guides. Clarendon, Oxford (2001)
65. Smets, P.: The combination of evidence in the transferable belief model. IEEE Transactions on Pattern Analysis and Machine Intelligence 12(5), 447–458 (1990)
66. Spohn, W.: Ordinal conditional functions: a dynamic theory of epistemic states. In: Harper, W.L., Skyrms, B. (eds.) Causation in Decision, Belief Change, and Statistics, vol. 2, pp. 105–134 (1987)
67. Williams, M.A.: Transmutations of knowledge systems. In: Proceedings of the Fourth International Conference on the Principles of Knowledge Representation and Reasoning (KR 1994), pp. 619–629 (1994)
68. Zhang, D., Foo, N., Meyer, T., Kwok, R.: Negotiation as mutual belief revision. In: Proceedings of the American National Conference on Artificial Intelligence (AAAI 2004), pp. 317–322 (2004)

Fuzzy Classifiers* – Opportunities and Challenges –

Anca Ralescu and Sofia Visa

[1] Machine Learning and Computational Intelligence Laboratory
School of Computing Sciences and Informatics
University of Cincinnati, Cincinnati, OH 45221-0030, USA
Anca.Ralescu@uc.edu
[2] Department of Computer Science
The College of Wooster
Wooster, OH 44691, USA
svisa@wooster.edu

Abstract. Several issues arise when we consider building classifiers in general, and fuzzy classifiers in particular. These issues include but are not limited to attribute/feature selection, adoption of a specific approach/algorithm, evaluate the classifier performance, etc. We consider the opportunities that such classifiers have to offer and contrast them with the challenges they pose.

Keywords: classifiers, fuzzy sets, attribute selection, error models.

1 Introduction

In our daily life we classify: people (as friends, strangers, acquaintance), foods, books, music pieces, images, cities, etc. according to whether we like, dislike, or make no impression, etc. In the process of becoming social beings, we develop our own personal 'classifiers' which take into account our own background and preferences and, of course, the characteristics of the objects of our attention. Classification, and related to it, clustering, have emerged as cornerstones of computer based information processing, regardless of the application domain. Both may be used in image processing [1], in image understanding to extract higher level objects in the image (e.g. differentiate tumors from healthy tissue in medical images, or water from land regions in aerial images), in text processing to gather together documents similar along a certain dimension (e.g. topic); in cyber-security [2]; in medical diagnosis, or in fault detection, classification plays a crucial role and, in general, in fraud detection (e.g. credit card transactions) classification serves to distinguish among millions of valid transactions from those, relatively rare, fraudulent ones [3].

Traditional approaches to classification use statistical tools. In fact, since 1984 the *Journal of Classification* is dedicated to publish work in the area of statistics

* The authors dedicate this paper to Professor Lotfi A. Zadeh on the occasion of his 90th birthday.

S. Benferhat and J. Grant (Eds.): SUM 2011, LNAI 6929, pp. 75–80, 2011.
© Springer-Verlag Berlin Heidelberg 2011

based classification research [4]. With the advent of computer based approaches, machine learning, and data mining, new classification tools have been developed, including new statistical approaches [5].

An alternative, complementary, and often generalizing, approach to classification (and its unsupervised relative clustering) is provided by fuzzy set theory [6]. Since 1965 when he first presented it, Zadeh's concept of fuzzy set, at first glance a modest extension of the classical notion of set, continues have a huge impact on intelligent data analysis. This concept is strongly related to classification, as it introduces the notion of *membership function* and thus *membership degree* of a value to a set, or of a data point to a class. The concept of fuzzy set arose from Zadeh's work in systems theory and found an elegant statement in his Principle of Incompatibility [7].

From this point on this paper is organized as follows. In Section 2 we briefly consider the general issues to be considered when building a classifier. In Section 3 we consider these issues in connection with the fuzzy classifiers and discuss the challenges and opportunities that the adoption of fuzzy sets and fuzzy reasoning entail. We close with a short conclusion section. Since this is a position paper, whose aim is to trigger discussions on the topic of fuzzy classifiers, our presentation is quite general on purpose.

2 Issues to Address When Constructing a Classifier

It is useful to state the classification problem as follows. Let $D = \{\mathbf{x}|x = (x_1, \ldots, x_n)\} \in \Re^n$ be a data set, and $L = \{1, \ldots, m\}$ a set of labels. To construct a classifier is to obtain a rule (a mapping) $f : \Re^n \rightarrow L$, which assigns to each data point $\mathbf{x} \in D$ a label $l \in L$. A special case, referred to as a 2-class classifier, is when the number of labels is $m = 2$. We base our discussion on the 2-class classifier. Often the general, m-class classifier, is obtained from the simpler case. Deriving the rule f is known as the training of the classifier, and that is done based on a training set $T \subset D \times L$. That is, a training tuple is of the form $(\mathbf{x}, l)$. Several choices must be made at this point.

2.1 Selection of the Decision Rule (Class Boundary)

If the training set is viewed as a subset of a high dimensional space ($\leq n$) the decision rule can be viewed as a surface that divides the training set into two subsets each of which corresponding to one class. The usual approach is to parametrize f and to select its parameters by minimizing the overall misclassification errors. If the training data happens to be linearly separable, f is a linear surface. In the case when this is not the case, linear separability can be obtained by using a kernel to achieve implicit mapping into a higher dimensional space where data are linearly separable [8]. Alternatively, neural network based approaches are used to derive the non-linear separating surface. Once trained, a classifier is evaluated on the test data set. In all but the most simple cases when classes are well separated to begin with, errors of classification of the training data are tolerated in order to improve correct classification of test data.

A different view of the decision rule is as a score obtained via a Bayesian argument. Based on the training set the following probabilities are evaluated: probability of the data point x given class C, and probability of the class C, $P(x|C)$, $P(C)$. Then, for a test data point x_0 one computes the posterior probability, $P(C|x_0)$ using Bayes Theorem. Note that x and x_0 refer here a single attribute (dimension) of the data point $\mathbf{x}$ and $\mathbf{x}_0$ respectively.

2.2 Attribute Selection

An important issue concerns the selection of the 'best' attributes. The notion of 'best' very much depends on how the classifier is evaluated. For example, if the classifier evaluation measures the classification error, then the best attributes are those that yield smallest classification error. Another issue to consider here is that of dimensionality: for higher dimensions more data is needed to fill the input space. Often the nature of the problem is that such data is either unavailable or costly to obtain. This leads one to consider attribute selection as an instance of dimensionality reduction.

2.3 Aggregation across Several Attributes

Especially for the Bayes classifiers the issue of aggregating the classification results across various attributes is very important. Although, in theory, the probabilities needed can be defined on the space of the input attributes, in practice, this requires a very large amount of data, and it is often impossible to carry out. Thus the aggregation remains the preferred solutions for those implementing Bayes classifiers. The Naive Bayes rule simply calculates the posterior class probability by multiplying the those determined along each attribute. In effect this rule assumes independence of attributes. Even when this independence is not actually in the data, it is claimed that the Naive Bayes classifier yields good classification results [9]. There are many issues to consider with respect to the Bayesian classifiers, including boosting of small probabilities which, by multiplication, lower the overall Bayes score, and learning from imbalanced data sets, in which the class of interest is very small (has very few elements) compared to the other class. Bayes classifiers can be modified to use a measure of cost or penalty of misclassification so as to penalize differently errors of classification.

2.4 Evaluation of Classifiers

The intuitive approach to evaluate a classifier is based on accuracy, the rate of correct classification. However, in many instances accuracy is not necessarily the best way to evaluate a classifier. For example, in an imbalanced data set where 5% of the data belong to one class, while the remaining 95% belong to another class, a 95% accuracy can be obtained without any training at all, just by classifying all data points in the larger class. Other devices, such as the *confusion matrix* have been introduced to evaluate a classifier with respect to its *precision*, *recall*, or an aggregate meassure, F_α where α quantifies their importance.

3 Fuzzy Classifiers

The use of fuzzy sets in machine learning (classification and clustering), and data mining accounts for a very large body of published research [10]-[14]. Fuzzy techniques are combined with neural networks [16], decision trees probabilistic and statistical approaches, and evolutionary approaches to name just a few. For fuzzy classifiers we seek to obtain a mapping from the attribute space to a set of labels. However, the labels stand for fuzzy sets rather than classical (crisp) sets, the attribute values may also be in fuzzy subsets of the input space, and therefore the decision score, reflects the degree to which a data point belongs to a class. A non-fuzzy decision is usually made based on some rule applied to the decision score (usually the class with highest membership degree is selected as the crisp output of classification).

3.1 Opportunities and Challenges for Fuzzy Classifiers

Taking into consideration the wealth of results obtained integrating fuzzy techniques with existing classifier algorithms we can see that the use of fuzzy sets opens new opportunities for deriving classifiers. Yet, at the same time, it can be argued that these same opportunities pose challenges as well. For example, it can be argued that since the ultimate result is crisp, not fuzzy, and the classifier evaluation is then based on this decision, there is in fact, no need for fuzzy classifiers at all [17]. Here we maintain that this need not be the case, and that instead, we must develop tools specific for the evaluation of fuzzy classifiers. The challenge is to address these tools in a rigorous manner which can be inspected and analyzed.

Deriving the Fuzzy Sets Used in a Fuzzy Classifier. One way to define these fuzzy sets used by a fuzzy classifier is through an initialization-plus-tuning approach, whereby the fuzzy sets in questions are adjusted during the process of training the classifier. The membership functions are selected so as to make the tuning process quite easy. This approach is appealing from an intuitive point of view, and has the advantage that the resulting fuzzy set can be easily expressed by a linguistic label. However, not all classification problems deal with classes whose intuitive meaning is grasped from the beginning. In such cases the fuzzy sets can be obtained directly from the data, either by a clustering procedure [18], or using the mass assignment theory, [13], [14], [19], an approach especially useful for training a classifier from imbalanced data.

Attribute Selection. Often attribute selection for fuzzy classifiers is done in the same way as for traditional classifiers. We find a notable exception in [18] in which a regularity criterion is used in conjunction with a fuzzy model to select the best attributes. However, the criterion does not ensure an overall best subset of attributes, as it stops at a local minimum of the prediction error. Fuzzy techniques offer us the opportunity to obtain a *fuzzy set of attributes* where the membership degree of an attribute reflects meaningfully its importance to classification. The challenge is to derive a formal (as opposed to some ad-hoc weighting of attributes) technique for obtaining such a set.

Aggregation of Classification Results across Attributes. It is perhaps in this respect that fuzzy techniques can bring the most to the problem of classification. The aggregation problem consists on evaluating the classification results obtained across various attributes. In the case of fuzzy classifiers, these results are membership degrees. Unlike the (classical) logic case, where we have only two types of aggregation (conjunction and disjunction), or probabilistic (Bayes) case, where we have only one aggregation (multiplication), many aggregation functions have been defined in the context of fuzzy sets [20]. These functions are often parametrized and therefore subject to training themselves. In addition, fuzzy quantifiers, such as *most, a few, many* can be used to aggregate the results of single attribute classification. Finally, the (fuzzy) result of classification can be qualified by a (fuzzy) probability. All of these contribute to a richer expression of the classification results.

Once the mechanisms for fuzzy classifiers as described above are set in place, we have the opportunity to further analyze the results in addition to producing a crisp classification. For example, a fuzzy classifier supports the notion of graduality and ranking (a data point belongs more to a class than to another, or given two data points, say x_1 and x_2, we can decide which one belongs more to a class C). This analysis is the natural result of using fuzzy techniques, and unlike nonfuzzy cases, it need not be specifically trained for. That is, we do not need to train for ranking in order to obtain it.

3.2 Error Models for Fuzzy Classifiers

The final opportunity and challenge for fuzzy classifiers discussed here is that of new error models. The notions of precision and recall, and the associate F_α measure can be, in theory, easily generalized using fuzzy sets (using Zadeh's extension principle) providing us with tools to further distinguish between classes and classifier results. However, the challenge here is to develop an approach which is at the same time technically correct and computationally efficient.

4 Conclusions

We provided here a brief discussion of fuzzy classifiers, the opportunities and challenges that they open for researchers in machine learning. With each opportunity there comes a challenge and the other way around. A final challenge would be to develop a theory for fuzzy classifiers which, in a any particular approach would be a true generalization of a corresponding nonfuzzy classifier. Perhaps a fuzzy classifier could be a fuzzy set whose level sets relate to crisp classifiers in a manner analogous to the relation between fuzzy and crisp sets as stated in [21].

References

1. Liang, L.R., Looney, C.G.: Competitive fuzzy edge detection. Applied soft computing 3(2), 123–137 (2003)
2. Gomez, J., Dasgupta, D.: Evolving fuzzy classifiers for intrusion detection. In: Proceedings of the 2002 IEEE Workshop on Information Assurance, vol. 6.3, pp. 321–323. IEEE Computer Press, New York (2002)

3. Chan, P., Stolfo, S.: Toward scalable learning with non-uniform class and cost dostributions: A case study in credit card fraud detection. In: Proceedings of Knowledge Discovery and Data Mining, pp. 164–168 (1998)
4. `http://www.springer.com/statistics/statistical+theory+and+methods/journal/357`
5. Vapnik, V.N.: Statistical learning theory. Wiley-Interscience, Hoboken (1998)
6. Zadeh, L.A.: Fuzzy sets. Information Control 8, 338–353 (1965)
7. Zadeh, L.A.: A New Approach to the Analysis of Complex Systems. IEEE Transactions SMC SMC-3, 1 (1973)
8. Cristianini, N., Shawe-Taylor, J.: An introduction to support Vector Machines: and other kernel-based learning methods. Cambridge University Press, Cambridge (2006)
9. Mitchell, T.M.: The discipline of machine learning. Carnegie Mellon University, School of Computer Science, Machine Learning Dept. (2006)
10. Ishibuchi, H., Nakashima, T., Murata, T.: A fuzzy classifier system that generates fuzzy if-then rules for pattern classification problems. In: IEEE International Conference on Evolutionary Computation, vol. 2, pp. 759–764. IEEE, Los Alamitos (1995)
11. Kuncheva, L.I.: How good are fuzzy if-then classifiers? IEEE Transactions on Systems, Man, and Cybernetics, Part B: Cybernetics 30(4), 501–509 (2000)
12. Kuncheva, L.I.: Fuzzy classifier design, vol. 49. Physica, Heidelberg (2000)
13. Visa, S., Ralescu, A.: Learning Imbalanced and Overlapping Classes using Fuzzy Sets. In: Proceedings of the International Conference of Machine Learning, Workshop on Learning from Imbalanced data Sets (II): Learning with Imbalanced Data Sets II, Washington, pp. 97–104 (2003)
14. Visa, S., Ralescu, A.: Fuzzy classifiers for imbalanced, complex classes of varying size. In: Proc. of the IPMU Conference, Perugia, pp. 393–400 (2004)
15. Uebele, V., Abe, S., Lan, M.S.: A neural-network-based fuzzy classifier. IEEE Transactions on Systems, Man and Cybernetics 25(2) (1995)
16. Abe, S.: Pattern classification; neuro-fuzzy methods and their comparison(book). Springer Verlag London, Ltd., London (2001)
17. Huellermeier, E.: Fuzzy-Methods in Machine Learning and Data Mining: Status and Prospects. Fuzzy Sets and Systems 156(3), 387–407 (2005)
18. Sugeno, M., Yasukawa, T.: A Fuzzy-Logic-Based Approach to Qualitative Modeling. IEEE Transactions on fuzzy systems 1(1), 7–31 (1993)
19. Visa, S.: Fuzzy Classifiers for Imbalanced Data Sets. PhD Thesis, Computer Science Department, University of Cincinnati, Cincinnati, Ohio, USA (2007)
20. Ralescu, D.: Cardinality, quantifiers, and the aggregation of fuzzy criteria. Fuzzy Sets and Systems 69, 355–365 (1995)
21. Negoita, C.V., Ralescu, D.: Representation theorems for fuzzy concepts. Kybernetes 4, 169–174 (1975)

A Brief Overview of Research in Argumentation Systems

Guillermo R. Simari

Artificial Intelligence Research & Development Laboratory (LIDIA)
Universidad Nacional del Sur (UNS), Bahía Blanca, Argentina
grs@cs.uns.edu.ar

Abstract. The area of argumentation in Artificial Intelligence has been steadily growing for the last three decades. Many subareas have been delineated within it as the research expanded, giving birth to a field that is exciting, fruitful and rewarding. The challenges are many, and they are met with methods and techniques that have enriched the field of Knowledge Representation and Reasoning. In this paper, a short structured overview of research in the area of Argumentation Systems will be provided in order to lay a foundation for further discussion. This overview will also bring about a personal perspective of the future directions of research and development of the area.

1 Introduction and General Intuitions

Arguing is our natural way of finding a secure footing for our beliefs; it is how we rationally handle conflicting information in order to establish these beliefs. The activity we humans call arguing, and the very nature of an argument, have been the subject of intense inquiry in Philosophy since ancient times (see for instance [15,24]); furthermore, Logic was born from the effort to clarify the presentation and exchange of arguments. More recently in the field of Artificial Intelligence, as the crucible where many disciplines contribute, research on argumentation has expanded and given birth to a field that is exciting, fruitful, and rewarding. The sheer size of the literature precludes a full exploration of the topic, making it clearly out of the scope of any single paper; therefore, we will limit our intent to provide a concise foundation for further discussion, giving a short structured panorama of the research in the area of Argumentation Systems from a computational point of view.

We will set the stage by giving a brief description of the main intuitions that are involved in argumentation. The *argumentation process* reflects a form of reasoning where the conclusion and the way to arrive at it are doubted and effectively challenged, *i.e.*, argumentation is reasoning in a context of disagreement where an audience decides the outcome. This process could be carried out in the privacy of our own minds, as when we try to decide what to believe using our repository of beliefs, or it could be a group activity in which we exchange arguments with other participants sharing the group's repository of beliefs. Consequently, in the first case the process will be regarded as *monological*, or internal,

S. Benferhat and J. Grant (Eds.): SUM 2011, LNAI 6929, pp. 81–95, 2011.
© Springer-Verlag Berlin Heidelberg 2011

and in the second case it will be deemed as *dialogical*, or multi-party. In the first case only one participant is involved, and in the latter at least two agents participate; this distinction is important to understand how the activity progresses. We will start with dialogical argumentation, presenting the general ideas that will be simplified for the monological case.

The dialogical process involves two participants, and begins with the assertion of a statement by one of them that from now on will be referred to as the *proponent*; this initial statement is usually called the *claim* or *thesis*. The other participant, or *opponent*, can accept the claim, ending the process, or could challenge the proponent's claim requesting support for it. The support takes the form of an *argument*, that intuitively is a coherent set of statements leading from a set of premises, or *evidence*, to the conclusion, or claim. The connection between the evidence and the claim is established by some form of *reasoning*; more precisely, an *argument* is a set of statements in which a claim is made, and support is offered for it in an attempt to influence an audience in a context of disagreement. In this case, the disagreement is apparent in the challenge issued by the opponent.

At this point, the opponent can accept the argument or it can decide the argument offered in support of the claim is not effective. Several possibilities for the opponent can arise in this situation: a new argument against the claim can be put forward, the reasoning involved in connecting the evidence and the claim can be challenged with an argument, or part of the evidence can be challenged, becoming new claims subject to the same scrutiny. The first two cases lead to the introduction of an argument from the opponent; these arguments are called counter-arguments because of their role as arguments that work against other arguments. Clearly, this denomination is dependent on the role played and is not inherent to the structure of the argument itself; an argument can be a supporting argument in one case and a counter-argument in a different situation.

The arguing continues with the proponent considering the counter-argument presented. Now, the counter-argument is itself an argument, and therefore can be subject to the same challenges as the original argument. The proponent takes the role of the opponent with respect to the counter-argument, considering the possibility of accepting it or issuing its own challenge. The arguments coming from the proponent are arguments aiming to promote the issue and are mentioned as *pro* arguments, while the arguments produced by the opponent suggest points against it and are referred to as *con* arguments.

From the succinct informal description above we see that in the process of arguing we seek arguments supporting the point of conflict and arguments that try to undermine that support. After finding the pro and con arguments related to the issue, comparison becomes necessary in order to answer the question: which is better? The decision might depend on who is considering the arguments and counter-arguments; this introduces another element in the process: the *audience*. For instance, in a simplified view of a *trial by jury*, the initial claim is the "presumption of innocence"; this presumption is counter-argued by the accusatory part by setting up a case where the accused is shown to be guilty, *i.e.*, not innocent. The evidence is discussed before trial, and sometimes is introduced during the process. The defendant takes the role of challenging the arguments of the prosecutor. In this scenario, the audience that has the power of deciding is the jury with the help of the judge; they will evaluate the merit of the arguments presented.

Turning to the conceptually simpler case of monological argumentation, in which the process is internal to the reasoner, all the roles must be played by the same agent. The source of evidence is its internal belief base, from which all the arguments must be built. In considering the issue, arguments and counter-arguments are introduced and pondered by the agent, who also acts as the deciding audience; *i.e.*, the agent itself introduces and defends the initial thesis, plays the role of opponent, and it is also the arbiter. In this description, it is apparent that the argumentation process has an important role in establishing the beliefs that can be obtained from a repository that could support logically contradictory conclusions.

The task of defining an argumentation system involves several levels [20]. Given a repository of beliefs, or belief base, it can be assumed that some kind of formal language was used to represent the beliefs. Then, the initial level is concerned with the language in which information can be expressed, and with the rules available for the construction of arguments in this language. The next level defines what arguments are, *i.e.*, how pieces of information can be combined to provide support for a claim. The inference mechanism associated with the belief base will provide the reasoning that will link the claim and the premises. The following three levels will address the problem of how arguments interact, defining: (a) when arguments are in conflict, (b) how conflicting arguments can be compared, and (c) which arguments survive the competition between all conflicting arguments. Two extra levels are of interest: the procedural level and the strategy level. In the former, the way in which an actual dispute can be conducted is regulated, *i.e.*, how parties can introduce or challenge new information and state new arguments; here the speech acts that are allowed and the discourse rules governing them are defined. In the latter, rational ways of conducting a dispute within the procedural bounds of the previous layer are provided.

The rest of this work is organized as follows: next we will present Abstract Argumentation Frameworks ; then we will introduce a few examples of systems in which the arguments are constructed from a belief base; after that we will refer to the research challenges facing by researchers; subsequently we will finish with a description of the research context.

2 Abstract Argumentation Frameworks

In a seminal work published in 1995, Phan Minh Dung introduced an abstract theory for argumentation where the central notion is the acceptability of arguments [13]. The main abstraction comes into play disregarding the internal structure of the arguments involved, *i.e.*, arguments are considered to be atomic in nature. In that way, the elements of the theory are reduced to a set of arguments and the consideration of an attack relation defined over that set; a theory thus defined can be visualized as a directed graph where the nodes represent the arguments and the arcs represent an attack from the node (argument) where the arc starts towards the node (argument) where the arc finishes.

Since arguments are involved in an attack relation, in the general case where the attack relation is not empty, not all of them can be thought of as acceptable. It is particularly interesting to decide the status of every argument in the framework; the status of an argument is commonly referred to as its *justification status* [2]. Intuitively, an argument is regarded as justified if its situation is such that it is able to survive the attacks of which it is the target of, and as not justified (or rejected) otherwise. Given an initial set of arguments and an attack relation defined over it, there are several ways to describe properties that a subset of the set of all arguments must satisfy to be accepted or justified together; these are known as *extension-based semantics* or *argumentation semantics.* We will introduce the appropriate definitions below.

An abstract argumentation framework consists of a set of arguments AR and a binary relation *attacks* defined over the arguments in that set. The notation $(A, B) \in attacks$ (or, equivalently, $A\, attacks\, B$) means that there is an attack of A on B. The relation Dung called *attack* in his original paper corresponds to what currently is known as *defeat.* The notion of defeat involves an attack of an argument A on an argument B and establishing the preference of A over B, *i.e.*, defeat is attack plus preference. In Dung's formalism every attack is successful. Formally,

Definition 1. [13] *An argumentation framework is a pair* $AF = \langle AR, attacks\rangle$ *where* AR *is a set of arguments, and* $attacks$ *is a binary relation on* AR*, i.e.,* $attacks \subseteq AR \times AR$.

Example 1. Let $AF = \langle AR, attacks\rangle$ be an abstract argumentation framework as depicted in Figure 2, where the set of arguments is $AR = \{A, B, C, D, E, F, G, H\}$ and $attacks = \{(B, A), (C, B), (D, A), (E, D), (G, H), (H, G)\}$.

An argumentation semantics is the formal definition of the argument evaluation process leading to decide which arguments are able to survive the attacks defined in the framework. These "survivors" will be considered as being able to support their conclusions. Research efforts have produced two different ways of carrying out the evaluation, namely extension-based [2] and labeling-based argumentation semantics [16]. The first one provides a declarative definition, while the second one is procedural in nature. An extension is a subset of arguments contained in the framework, and the extension-based approach specifies how to obtain the subsets that form the set of extensions. Every extension contains a set of arguments that together can be acceptable in the context of the attack relation. The labeling-based approach provides a way of

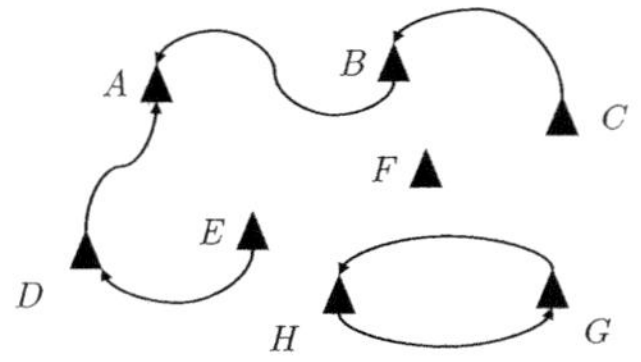

Fig. 1. An abstract AF graph

assigning a label to each argument in the framework, choosing that label from an appropriate set, such as $\{in, out, undecided\}$. The assignment of labels yields a set of labelings that correspond to the extensions found through the declarative method. For an intuitive and didactic introduction to argumentation semantics, see [9].

Dung [13] introduces several argumentation semantics that provide a way of evaluating the status of the arguments in the framework by constructing extensions, *e.g.*, complete, grounded, stable, and preferred semantics. Other semantics have been proposed after the initial definition, *e.g.*, stage, semi-stable, ideal, $CF2$, and prudent semantics.

A set of accepted arguments is characterized in [13] using the concept of *acceptability*, which is a central notion in argumentation, formalized by Dung in the following definition.

Definition 2. [13] *Let* $AF = \langle AR, attacks \rangle$ *be an argumentation framework; an argument* $A \in AR$ *is acceptable with respect to a set of arguments* S *if and only if every argument* B *attacking* A *is attacked by an argument in* S.

If an argument A is acceptable with respect to a set of arguments S then it is also said that S *defends* A. Also, the attackers of the attackers of A are called *defenders* of A. We will use these terms throughout this paper. Acceptability is the main property of Dung's semantic notions, which are summarized in the following definition.

Definition 3. *Let* $AF = \langle AR, attacks \rangle$ *be an argumentation framework; a set of arguments* $S \subseteq AR$ *is said to be*

- *conflict-free if there are no arguments* $A, B \in S$ *such that* A attacks B.
- *admissible if it is conflict-free and defends all its elements.*
- *a preferred extension if* S *is a maximal (for set inclusion) admissible set.*
- *a complete extension if* S *is admissible and it includes every acceptable argument w.r.t.* S.
- *a grounded extension if and only if it is the least (for set inclusion) complete extension.*
- *a stable extension if* S *is conflict-free and it attacks each argument not belonging to* S.

The following example illustrates the concepts introduced above. In the argumentation framework depicted in Figure 2, we have: $\{B, D, H, F\}$ is a conflict free set of arguments, $\{A, C, E\}$ is an admissible set of arguments, $\{A, C, E, F, G\}$ is a preferred extension. It is also a complete extension, $\{A, C, E, F\}$ is the grounded extension,and $\{A, C, E, F, G\}$ and $\{A, C, E, F, H\}$ are stable extensions.

The grounded extension is also the least fixpoint of a simple monotonic *characteristic* function:

$$F_{AF}(S) = \{A : A \text{ is acceptable with respect to } S\}.$$

In [13], results stating conditions of existence and equivalence between these extensions are also introduced. These include that since the empty set is always admissible, there is always one admissible set. Also, that there is always a preferred extension, and that extension is complete. Stable extensions are also preferred. Although it could be empty, the grounded extension is the intersection of all complete extensions. The empty set is not a stable extension, but there are frameworks without stable extensions. When the attack relation presents no cycles, then there is a single extension that is a stable, preferred, complete, and grounded extension.

Several additions to the original definition of abstract frameworks have being presented: Value-base argument systems[3], Argument frameworks with constraints[11], Bipolar argument frameworks[1], Argument frameworks with priorities[17], among many others.

We will now turn to presenting frameworks where the construction of the arguments matters.

3 Arguments with Structure

In an abstract framework, arguments are considered to be atomic entities, *i.e.*, the inner details of their construction plays no role in the formalism. In this section we will be exploring the particular issue of how an argument is built. The possibilities are many, and we will restrict ourselves to a few different systems where the structure of the argument plays an important role in defining the attack relation, a notion that will be extended with the possibility of being unsuccessful. The comparison criterion is another interesting element to be defined. Given an attack, it will be successful only when the argument that attacks is at least as *good* as the argument that receives that attack; here the comparison criterion will decide the matter. Below we will develop the systems in a succinct manner given the space restriction.

3.1 Logical Argumentation

In this section we will give a brief description of the system developed by P. Besnard and A. Hunter, and introduced in [6]; further details can be found in [7,8].[1] This system is based on classical logic, and the details regarding each argument are taken into account. In our introduction we mentioned the three parts that are involved in an argument: the claim, evidence (premises, reasons, support)[2], and the reasoning that connects the evidence with the claim. The claim and the premises are expressed as formulæ in the language of classical logic, and the reasoning or inference method will be limited to deductive inference making the arguments *deductive arguments*. Therefore, the claim is a deductively valid classical consequence of the evidence.

[1] Some of the examples in this section were taken from a Tutorial on Argumentation Systems given by Anthony Hunter in KR'08 in Sydney, Australia.

[2] We will use evidence, premise, reason, and support interchangeably.

In what follows, the existence of a finite set of formulæ Δ is assumed; it is also assumed that every subset of Δ is given an enumeration $\langle \alpha_1; \ldots; \alpha_n \rangle$ of its elements, called its *canonical enumeration.* This enumeration is just a convenient way to indicate the order in which the formulæ in any subset of Δ are assumed to be conjoined to make a formula logically equivalent to that subset. This order has no other meaning, and in particular it does not represent the relative importance of the formulæ in Δ; note that any total order imposed on Δ will satisfy this requirement. The set Δ is regarded as a large information base from which arguments for and against the possible claims are built. No assumption is made about the content of Δ, which can be arbitrarily complex, and possibly inconsistent. In this framework, an argument will be a pair $\langle \Phi, \alpha \rangle$, where Φ is a minimal consistent set of formulæ from which the second element is derived, *i.e.*, (1) $\Phi \vdash \alpha$, (2) $\Phi \nvdash \bot$, and (3) Φ is a minimal subset of Δ satisfying 1.

For instance, from $\Delta = \{\alpha, \alpha \rightarrow \beta, \gamma, \neg\gamma, \neg\gamma \rightarrow \neg\beta\}$, the following are some of the arguments that can be constructed: $\langle\{\alpha, \alpha \rightarrow \beta\}, \beta\rangle$, $\langle\{\alpha, \neg\gamma\}, \alpha \wedge \neg\gamma\rangle$, and $\langle\{\neg\gamma, \neg\gamma \rightarrow \neg\beta, \alpha \rightarrow \beta\}, \neg\alpha\rangle$.

Arguments are not necessarily independent, and it is possible that some encompass others (possibly up to some form of equivalence).

Definition 4. *$\langle \Phi, \alpha \rangle$ is more conservative than $\langle \Psi, \beta \rangle$ iff $\Phi \subseteq \Psi$ and $\beta \vdash \alpha$.*

For instance, $\langle\{\alpha\}, \beta \rightarrow \alpha\rangle$ is more conservative than $\langle\{\alpha, \neg\alpha \vee \neg\beta\}, \neg\beta\rangle$.

The notion of counter-argumentation is introduced through the logical inconsistency of two arguments. Two kinds of counter-argument can be distinguished: *rebutting* and *undercutting* counter-arguments [18]. A rebuttal for $\langle \Phi, \alpha \rangle$ is an argument $\langle \Psi, \beta \rangle$ where $\beta \vdash \neg\alpha$, *i.e.*, their claims are mutually inconsistent. An undercut for $\langle \Phi, \alpha \rangle$ is an argument $\langle \Psi, \neg(\phi_1, \ldots, \phi_n)\rangle$, where $(\phi_1, \ldots, \phi_n) \subseteq \Phi$, *i.e.*, the claim of the counterargument is inconsistent with the support of the attacked argument.

If $\Delta = \{\alpha, \alpha \rightarrow \beta, \gamma, \gamma \rightarrow \neg\alpha\}$, the following arguments and counter-arguments can be constructed: $\langle\{\alpha\}, \alpha\rangle$ rebuts $\langle\{\gamma, \gamma \rightarrow \neg\alpha\}, \neg\alpha\rangle$, the argument $\langle\{\gamma, \gamma \rightarrow \neg\alpha\}, \neg\alpha\rangle$ undercuts $\langle\{\alpha, \alpha \rightarrow \beta\}, \beta\rangle$, and $\langle\{\gamma, \gamma \rightarrow \neg\alpha\}, \neg(\alpha \wedge (\alpha \rightarrow \beta))\rangle$ is a more conservative undercut.

As arguments can be ordered from less conservative to more conservative, there is the notion of *maximally conservative undercuts* for an argument (those that are representative of all undercuts for that argument). A maximally conservative undercut for $\langle \Psi, \beta \rangle$ is an undercut $\langle \Phi, \alpha \rangle$ for $\langle \Psi, \beta \rangle$ such that for all undercuts $\langle \Phi', \alpha' \rangle$ of $\langle \Psi, \beta \rangle$, if $\Phi \subseteq \Phi'$ and $\alpha \vdash \alpha'$ then $\Phi' \subseteq \Phi$ and $\alpha' \vdash \alpha$. Consequently, it can be shown that if $\langle \Psi, \neg(\alpha_1 \wedge \ldots \wedge \alpha_n)\rangle$ is a maximally conservative undercut for $\langle \Phi, \beta)\rangle$, then $\Phi = \{\alpha_1, \ldots, \alpha_n\}$.

A maximally conservative undercut $\langle \Psi, \neg(\alpha_1 \wedge \ldots \wedge \alpha_n)\rangle$ is a canonical undercut for $\langle \Phi, \alpha \rangle$ iff $\alpha_1 \wedge \ldots \wedge \alpha_n$ is the normal form of Φ. For instance, the argument $\langle\{\neg\alpha \vee \neg\beta\}, \neg(\alpha \wedge \beta)\rangle$ is a canonical undercut for $\langle\{\alpha, \beta\}, \alpha \wedge \beta\rangle$.

It can be shown that given two canonical undercuts for the same argument, none is more conservative than the other. Any two canonical undercuts for the same argument have distinct supports, whereas they do have the same consequent.

For each rebuttal of an argument, there is a canonical undercut of the argument that is more conservative than the rebuttal.

Definition 5. *An argument tree for α is a tree where the nodes are arguments such that:*

1. *The root is an argument for α.*
2. *For no node $\langle \Phi, \alpha \rangle$ with ancestor nodes $\langle \Phi_1, \alpha_1 \rangle, \ldots, \langle \Phi_n, \alpha_n \rangle$ is Φ a subset of $\Phi_1 \cup \ldots \cup \Phi_n$.*
3. *Children nodes of a node N are canonical undercuts for N that obey 2.*

A complete argument tree is an argument tree where children nodes of a node N consist of all canonical undercuts for N that obey item 2 above.

There are various ways we can judge individual trees to ascertain whether the root argument is *warranted.* A common definition (*e.g.,* [14]) is *recursive defeat,* which coincides with Dung's definitions for extensions. The marking procedure is the following, where U means undefeated and D means defeated:

1. For any leaf node A_i, $mark(A_i) = \mathsf{U}$.
2. For any non-leaf node A_i, $mark(A_i) = \mathsf{D}$ iff there is a child A_i, s.t. $mark(A_j) = \mathsf{U}$.
3. For any non-leaf node A_i, $mark(A_i) = \mathsf{U}$ iff for all children A_j, $mark(A_j) = \mathsf{D}$.
4. The root argument A_r is warranted iff $mark(A_r) = \mathsf{U}$.

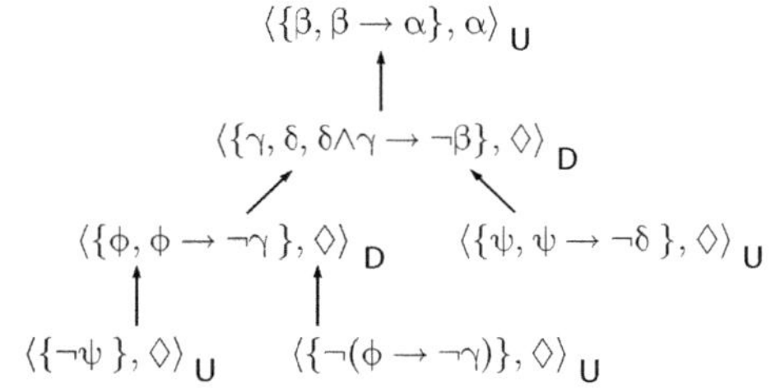

Fig. 2. An Example of a marked Argument Tree

4 Defeasible Logic Programming (DeLP)

Defeasible Logic Programming (DeLP) combines results of Logic Programming and Defeasible Argumentation; the system is fully implemented and available online. A brief explanation is included below (see [14] for full details). It has the declarative capability of representing weak information in the form of *defeasible rules,* and a defeasible argumentation inference mechanism for warranting the entailed conclusions. A DeLP-program $\mathcal{P}$ is a set of facts, strict rules, and defeasible rules defined as follows. Facts are ground literals representing atomic information or the negation of atomic information using strong negation "$\neg$" (*e.g., chicken(little)* or $\neg$*scared(little)*). *Strict Rules* represent non-defeasible information and are denoted $L_0 \leftarrow L_1, \ldots, L_n$, where L_0 is a ground literal and $\{L_i\}_{i>0}$ is a set of ground literals (*e.g., bird* $\leftarrow$ *chicken*) or

$\neg innocent \leftarrow guilty$). *Defeasible Rules* represent tentative information and are denoted $L_0 \prec L_1, \ldots, L_n$, where L_0 is a ground literal and $\{L_i\}_{i>0}$ is a set of ground literals (*e.g.*, $\neg flies \prec chicken$ or $flies \prec chicken, scared$).

When required, $\mathcal{P}$ is denoted (Π, Δ), distinguishing the subset Π of facts and strict rules, and the subset Δ of defeasible rules. *Strong negation* is allowed in the head of rules, and hence may be used to represent contradictory knowledge. From a program (Π, Δ), contradictory literals could be derived; nevertheless, the set Π (which is used to represent non-defeasible information) must possess certain internal coherence. Therefore, no pair of contradictory literals can be derived from Π.

A defeasible rule represents tentative information that may be used if nothing could be posed against it. Observe that strict and defeasible rules are ground; however, schematic rules with variables are used (schematic variables start with an uppercase letter). Consider a DELP-program (Π, Δ) where:

$$\Pi = \{bird(X) \leftarrow chicken(X), chicken(little), chicken(tina), bird(rob), scared(tina)\}$$
$$\Delta = \{flies(X) \prec bird(X), flies(X) \prec chicken(X), scared(X), \neg flies(X) \prec chicken(X)\}$$

This program has three defeasible rules representing tentative information about the flying ability of birds in general, and about regular chickens and scared ones. It also has a strict rule expressing that every *chicken* is a *bird*, and three facts: *tina* and *little* are *chickens*, *rob* is a *bird*, and *tina* is *scared*.

Derivation follows the same mechanism of Logic Programming, not distinguishing between strict and defeasible rules. From a program it is possible to defeasibly derive contradictory literals; *e.g.*, from (Π, Δ) of the program above, it is possible to derive $flies(tina)$ and $\neg flies(tina)$. For the treatment of contradictory knowledge, DELP incorporates a defeasible argumentation formalism. This formalism allows the identification of the pieces of knowledge that are in conflict, and through a *dialectical process* decides which information prevails as warranted. This dialectical process (see below) involves the construction and evaluation of arguments that either support or interfere with the query under analysis, building a *dialectical tree*.

Following [26], an *argument* for a literal L, is a (possibly empty) non-contradictory set of ground defeasible rules $\mathcal{A} \subseteq \Delta$ that, together with the set Π, provide a minimal defeasible proof for L, *i.e.*, (1) L is defeasible derived from $\Pi \cup \mathcal{A}$, (2) $\Pi \cup \mathcal{A}$ in not contradictory, and (3) $\mathcal{A}$ is a minimal subset of Δ satisfying 1, denoted $\langle \mathcal{A}, L \rangle$. The arguments:

$\langle \mathcal{A}_1, flies(t) \rangle = \langle \{flies(t) \prec bird(t)\}, flies(t) \rangle$,
$\langle \mathcal{A}_2, \neg flies(t) \rangle = \langle \{\neg flies(t) \prec chicken(t)\}, \neg flies(t) \rangle$, and
$\langle \mathcal{A}_3, flies(t) \rangle = \langle \{flies(t) \prec chicken(t), scared(t)\}, flies(t) \rangle$

are three arguments built from the program introduced, using t for *tina*.

In DELP, a literal L is *warranted* if there exists a non-defeated argument $\mathcal{A}$ supporting L. To establish if $\langle \mathcal{A}, L \rangle$ is a non-defeated argument, *defeaters* for $\langle \mathcal{A}, L \rangle$ are considered, *i.e.*, counter-arguments that by some criterion are preferred to $\langle \mathcal{A}, L \rangle$. It is important to note that in DeLP the argument comparison criterion is modular, and thus the most appropriate criterion for the domain that is being represented can be selected; *generalized specificity* [27] is the default criterion.

A defeater $\mathcal{D}$ for an argument $\mathcal{A}$ can be *proper* ($\mathcal{D}$ is preferred to $\mathcal{A}$) or *blocking* (unrelated or of the same strength). Since defeaters are arguments, there may exist defeaters for them, and defeaters for these defeaters, and so on. Thus, a sequence of arguments called an *argumentation line* is constructed, where each argument defeats its predecessor. To avoid undesirable sequences, that may represent circular or fallacious argumentation lines, in DeLP an *argumentation line* is *acceptable* if it satisfies certain constraints (see [14]).

The argument $\langle \mathcal{A}_2, \neg flies(t) \rangle$ properly defeats $\langle \mathcal{A}_1, flies(t) \rangle$, $\langle \mathcal{A}_3, flies(t) \rangle$ is a blocking defeater of $\langle \mathcal{A}_2, \neg flies(t) \rangle$, and $[\langle \mathcal{A}_1, flies(t) \rangle, \langle \mathcal{A}_2, \neg flies(t) \rangle, \langle \mathcal{A}_3, flies(t) \rangle]$ is an acceptable argumentation line.

Clearly, there can be more than one defeater for a particular argument $\mathcal{A}$. Therefore, many acceptable argumentation lines could arise from $\mathcal{A}$, leading to a tree structure. We will not introduce the technical definition of dialectical tree in this paper; see [14] for the details. The tree is built from the set of all argumentation lines rooted in the initial argument. In a dialectical tree, every node (except the root) represents a defeater of its parent, and leaves correspond to non-defeated arguments. Each path from the root to a leaf corresponds to a different acceptable argumentation line. A dialectical tree provides a structure for considering all the possible acceptable argumentation lines that can be generated for deciding whether an argument is defeated. We call this tree *dialectical* because it represents an exhaustive dialectical analysis for the argument in its root. Given a literal h and an argument $\langle \mathcal{A}, h \rangle$ to decide whether a literal h is warranted, every node in the dialectical tree $\mathcal{T}(\langle \mathcal{A}, h \rangle)$ is recursively marked as "D" (*defeated*) or "U" (*undefeated*), obtaining a marked dialectical tree $\mathcal{T}^*(\langle \mathcal{A}, h \rangle)$ as follows:

1. All leaves in $\mathcal{T}^*(\langle \mathcal{A}, h \rangle)$ are marked as "U"s, and
2. Let $\langle \mathcal{B}, q \rangle$ be an inner node of $\mathcal{T}^*(\langle \mathcal{A}, h \rangle)$. Then $\langle \mathcal{B}, q \rangle$ will be marked as "U" iff every child of $\langle \mathcal{B}, q \rangle$ is marked as "D". The node $\langle \mathcal{B}, q \rangle$ will be marked as "D" iff it has at least a child marked as "U".

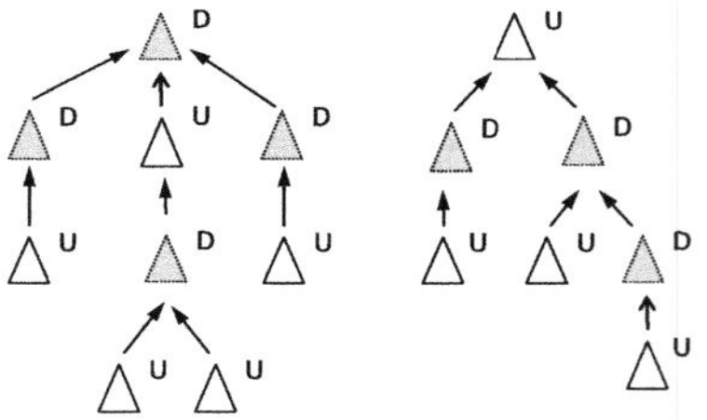

Fig. 3. Two marked dialectical trees

Given an argument $\langle \mathcal{A}, h \rangle$ obtained from $\mathcal{P}$, if the root of $\mathcal{T}^*(\langle \mathcal{A}, h \rangle)$ is marked as "U", then we will say that $\mathcal{T}^*(\langle \mathcal{A}, h \rangle)$ *warrants* h and that h is *warranted* from $\mathcal{P}$. Marked dialectical trees are depicted in Figure 3, where the triangles represent the arguments and the edges denote the defeat relation. At the right of each node, the associated mark ("U" or "D") is shown. Given a query L, three answers are possible: YES, when there is at least one warranted argument $\mathcal{A}$ for L); NO, when there is at least one warranted argument $\mathcal{A}$ for $\neg L$; UNDECIDED, when neither of the previous cases hold.

4.1 Other Approaches

The EU-funded ASPIC project[3] has developed industrial-strength Java components that implement an argumentation system [19] available under an open-source license. These components provide a platform to construct software systems that make use of argumentation as a service to other components.

ArguGRID is another EU-funded program[4], where argumentation technology is used to support rational decision making. In it, a model for building Grid-based applications through the use of multi-agent technologies and argumentation logic to support the formation of dynamic virtual organizations has been developed. Argumentation supports the composition of services for the creation, management, and dynamic evolution of societies of agents. An interesting advantage of the approach is how the interactions between service providers and service consumers is facilitated in a service-oriented environment. The CaSAPI system[5], a part of the ArguGRID effort, is a hybrid argumentation system combining abstract and assumption-based argumentation.

Assumption-Based Argumentation (ABA) is a computational framework conceived to encompass existing approaches to default reasoning in the early 90s. ABA combines Dung's preferred extension semantics for logic programming in argumentation-theoretic terms, and abstract argumentation. Because ABA is an instance of abstract argumentation, all semantic notions for determining the "acceptability" of arguments also apply to arguments in ABA. Moreover, ABA is a general-purpose argumentation framework that can be instantiated to support various applications and specialized frameworks, including: most default reasoning frameworks and problems in legal reasoning, game-theory, practical reasoning, and decision-theory. However, ABA builds actual arguments as deductions supported by assumptions by using inference rules in an underlying logic.

The above systems are examples of technological innovation put to the test of building real world applications; space constraints have prevented us from giving a more extensive review of them. The reader is invited to visit the web pages mentioned in the footnotes. The next two sections will attempt to describe the spectrum of challenges and the different avenues of publication used by the community.

5 Research Challenges

In this section we will give consideration to the research challenges that the argumentation community faces.

In a recently held Perspectives Workshop on the *Theory and Practice of Argumentation Systems* in the Schloss Dagstuhl, several areas where singled out as important and promising in the final report. This work was published as

---

[3] `http://www.argumentation.org/`
[4] `http://www.argugrid.eu/`
[5] `http://www.doc.ic.ac.uk/~ft/CaSAPI/`

Research Challenges for Argumentation [12]; the reader is encouraged to consider that work in order to expand the brief summary included below.

The discussions during the meeting were organized in four areas: Argumentation and the Semantic Web, Argumentation and Decision Support in Application Areas, Argumentation and Multi-Agent Systems, and Argumentation and Social Networks. This arrangement allowed the participants to have separate discussions in a more direct manner; the conclusions of each group were presented and discussed during plenary sessions. In the compact report offered below, only the more important themes will be touched to give an overview of the ideas that were discussed.

An initial remark is pertinent before analyzing each of these four topics. It is important to mention that the argumentation community agrees that research on the general infrastructure for argumentation is required. The research lines dealing with argument construction, argument evaluation, argument visualization, dialogue and protocol managers, and argument presentation and extraction tools, need to be expanded in search of more robust results.

In *Argumentation and the Semantic Web*, three particularly interesting topics were mentioned: Content Integration, Information Acquisition, and Interactive Question Answering. The issues appearing in Content Integration point to the problem of ontology interoperability and the problem of obtaining inferred information from a multitude of sources, where inconsistency and incompleteness could be present. For Information Acquisition, the capabilities of argumentation driven dialogues in a multi-agent environment can power the systems that could help in building ontologies. Interactive Question Answering could be improved by the use of argumentation-based dialogue to support user interaction, providing human-like explanations for the users.

For *Argumentation and Decision Support in Application Areas*, it is valuable to remark that reasoning using argumentation is particularly helpful for the purpose of decision making. As it was described in the previous sections, in argumentation systems the process of reaching a conclusion goes through the consideration of all the reasons for and against that conclusion. This deliberation allows to clearly exhibit the reasons why a decision is made, making the outcome acceptable for a community of agents. In a multi-agent context, virtual agents can explain and justify their decisions to a human, or help humans to reach a decision. Some research lines to pursue include the exploration of how classical decision theory can be extended with the addition of argumentation, how to connect argumentation systems with large data repositories, and how argumentation systems could help in the training of decision makers.

The area of *Argumentation and Multi-Agent Systems* is one of the more promising, showing a number of possible lines of research where argumentation can make a difference; this is reflected in the number of publications and research meetings dedicated to that purpose (see Section 6). An interesting problem to solve, related to the previous topic, is the design of an agent architecture where the deliberative component will be argumentative; implementing such an architecture effectively is another interesting challenge. On the *system* side of a multi-agent system, it is important to advance in the design of a system architecture to aid in the communication, collaboration, and coalition formation. Enabling in software agents the capability of communicating with humans in human-like ways is another goal; although work has been done on this subject, it is necessary to further develop tools able to seamlessly integrate human and artificial cognitive elements.

The Argumentation and Social Networks area presents many opportunities to introduce argumentation. As we have commented in other parts of this paper, argumentation-based reasoning follows the human form of obtaining consequences. By providing argumentation support to structure the exchange of information among humans, virtual social networks will make this interaction more effective. Among the several scenarios that can be mentioned, the sociopolitical debate provides an excellent ground to try new argument-based technologies to structure discourse, as well as both existing and elaborated information.

Clearly, the list could be extended in many ways. Argumentation mirrors our internal reasoning and our external cognitive social behavior. That is the power hidden in the use of the methods and techniques of argumentation: their inner workings are natural to us.

6 Research Context

The work related to Argumentation in Artificial Intelligence has continuously expanded at an increasing rate in the last three decades, producing technical results widely reported in the literature. The top conferences in Artificial Intelligence (IJCAI, AAAI, ECAI, *etc.*) have sessions fully dedicated to Argumentation. Leading scientific journals have published special issues, *e.g.*, Springer's *Journal of Autonomous Agents and Multiagent Systems* (JAAMAS) 2005 [21]; Elsevier's *Artificial Intelligence Journal* (AIJ) 2007 [4]; *IEEE Intelligent Systems* 2007 [22]; Wiley's *International Journal of Intelligent Systems* 2007 [25]. *Argument and Computation* is a journal that started to be published in 2010, aiming at promoting the interaction and cross-fertilization between the fields of argumentation theory and artificial intelligence. Its main focus lies in the research being produced in the fields of artificial intelligence, multi-agent systems, computer science, logic, philosophy, argumentation theory, psychology, cognitive science, game theory, and economics.

A new biannual international conference on *Computational Models of Argument* began in 2006,[6] and a series of well-attended workshops, (*e.g., Argument, Dialogue and Decision* as a special session of the *Non-Monotonic Reasoning Workshop* (NMR) since 2002; *Argumentation in Multi-Agent Systems* (Arg-MAS) held annually with AAMAS since 2004; *Computational Models of Natural Argument* (CMNA) held with IJCAI and ECAI since 2001; and the recently started *International Workshop on the Theory and Applications of Formal Argumentation* (TAFA) with IJCAI-2011.

[6] See `http://www.comma-conf.org`

The existing literature is wide and deep, and to get a gist of the foundations of the area there are a number of possibilities to consider. Two valuable surveys [10,20], although dated, still contain introductory material and a review of the field at the time of their publication, descriptions of the research field, and a wealth of references. A more recent source can be found in [5] where the authors' presentation of the special issue of the *Artificial Intelligence Journal* on Argumentation [4] goes over a wide range of approaches and issues seeking to put them in the context of the historical foundations of argumentation in Artificial Intelligence. They also discuss ideas and themes that have emerged in recent years leading to a significant broadening of the areas in which argumentation based methods are used. Another important characteristic of this paper is its reference section, which contains nearly two hundred valuable items.

Even more recently, two books have been published with the intention of responding to a growing need for in-depth, foundational presentation of the fast-expanding area of Argumentation in Artificial Intelligence. The first one, *The Elements of Argumentation* by P. Besnard and A. Hunter was published in 2008 [7]; it presents the background elements and the necessary techniques for formalizing argumentation in artificial intelligence, covering the emerging formalizations of practical argumentation. The book begins by discussing the nature of argumentation, continues introducing abstract argumentation, logical argumentation, practical argumentation, the comparison of arguments, taking account of the audience, presenting algorithms for argumentation, comparing related approaches and ends with the authors' perspective on the future of the field. The second one, *Argumentation in Artificial Intelligence* by I. Rahwan and G. R. Simari was published in 2009 [23]. The book is an edited collection of chapters written by leading researchers of the field. It begins with an *Introduction to Argumentation Theory* and contains twenty three chapters that have been organized into four parts: *Abstract Argument Systems*, *Arguments with Structure*, *Argumentation in Multi-Agent Systems*, and *Applications*.

References

1. Amgoud, L., Cayrol, C., Lagasquie-Schiex, M.C.: On the bipolarity in argumentation frameworks. In: Delgrande, J.P., Schaub, T. (eds.) NMR, pp. 1–9 (2004)
2. Baroni, P., Giacomin, M.: Semantics of abstract argument systems. In: Rahwan, I., Simari, G.R. (eds.) Argumentation in Artificial Intelligence, pp. 24–44. Springer, Heidelberg (2009)
3. Bench-Capon, T.J.M.: Value-based argumentation frameworks. In: Benferhat, S., Giunchiglia, E. (eds.) NMR, pp. 443–454 (2002)
4. Bench-Capon, T.J.M., Dunne, P.E.: Special Issue on Argumentation in Artificial Intelligence. Artificial Intelligence 171(10-15) (July-October 2005)
5. Bench-Capon, T.J.M., Dunne, P.E.: Argumentation in artificial intelligence. Artificial Intelligence 171(10-15), 619–641 (2007)
6. Besnard, P., Hunter, A.: A Logic-Based Theory of Deductive Arguments. Artif. Intell. 128(1-2), 203–235 (2001)
7. Besnard, P., Hunter, A.: Elements of Argumentation. MIT Press, Cambridge (2008)

8. Besnard, P., Hunter, A.: Argumentation based on classical logic. In: Rahwan, I., Simari, G.R. (eds.) Argumentation in Artificial Intelligence, pp. 133–152. Springer, Heidelberg (2009)
9. Caminada, M.: A Gentle Introduction to Argumentation Semantics (2008), `http://users.numericable.lu/martincaminada/publications/Semantics_Introduction.pdf`
10. Chesñevar, C.I., Maguitman, A.G., Loui, R.P.: Logical models of argument. ACM Computing Surveys 32(4), 337–383 (2000)
11. Coste-Marquis, S., Devred, C., Marquis, P.: Constrained argumentation frameworks. In: Doherty, P., Mylopoulos, J., Welty, C.A. (eds.) KR, pp. 112–122. AAAI Press, Menlo Park (2006)
12. Dix, J., Parsons, S., Prakken, H., Simari, G.R.: Research Challenges for Argumentation. Computer Science - R&D 23(1), 27–34 (2009)
13. Dung, P.M.: On the acceptability of arguments and its fundamental role in nonmonotonic reasoning, logic programming and n-person games. Artificial Intelligence 77(2), 321–358 (1995)
14. García, A.J., Simari, G.R.: Defeasible logic programming: An argumentative approach. Theory and Practice of Logic Programming 4(1-2), 95–138 (2004)
15. Griswold, C.: Plato on rhetoric and poetry. In: Zalta, E.N. (ed.) The Stanford Encyclopedia of Philosophy (Fall 2009 edn.) (2009)
16. Modgil, S., Caminada, M.: Proof theories and algorithms for abstract argumentation frameworks. In: Rahwan, I., Simari, G.R. (eds.) Argumentation in Artificial Intelligence, pp. 105–132. Springer, Heidelberg (2009)
17. Modgil, S.: Reasoning about preferences in argumentation frameworks. Artif. Intell. 173, 901–934 (2009)
18. Pollock, J.L.: Defeasible reasoning. Cognitive Science 11(4), 481–518 (1987)
19. Prakken, H.: An abstract framework for argumentation with structured arguments. Argument and Computation 1, 93–124 (2009)
20. Prakken, H., Vreeswijk, G.: Logics for defeasible argumentation. In: Gabbay, D., Guenthner, F. (eds.) Handbook of Philosophical Logic, vol. 4, pp. 218–319. Kluwer Academic Pub., Dordrecht (2002)
21. Rahwan, I.: Special Issue on Argumentation in Multi-Agent Systems. Journal of Autonomous Agents and Multi-Agent Systems 11(2) (September 2005)
22. Rahwan, I., McBurney, P.: Guest editors' Argumentation Technology. IEEE Intelligent Systems 22(6), 21–23 (2007)
23. Rahwan, I., Simari, G.R.: Argumentation in Artificial Intelligence. Springer, Heidelberg (2009)
24. Rapp, C.: Aristotle's rhetoric. In: Zalta, E.N. (ed.) The Stanford Encyclopedia of Philosophy (Spring 2010 edn.) (2010)
25. Reed, C., Grasso, F.: Recent Advances in Computational Models of Natural Argument. Int. J. Intell. Syst. 22(1), 1–15 (2007)
26. Simari, G.R., Loui, R.P.: A mathematical treatment of defeasible reasoning and its implementation. Artificial Intelligence 53(2-3), 125–157 (1992)
27. Stolzenburg, F., García, A., Chesñevar, C.I., Simari, G.R.: Computing Generalized Specificity. Journal of Non-Classical Logics 13(1), 87–113 (2003)

Maximal Ideal Recursive Semantics for Defeasible Argumentation

Teresa Alsinet[1], Ramón Béjar[1], Lluis Godo[2], and Francesc Guitart[1]

[1] University of Lleida, Jaume II, 69 – 25001 Lleida, Spain
{tracy,ramon,fguitart}@diei.udl.cat
[2] IIIA-CSIC, Campus UAB - 08193 Bellaterra, Spain
godo@iiia.csic.es

Abstract. In a previous work we defined a recursive warrant semantics for Defeasible Logic Programming extended with levels of possibilistic uncertainty for defeasible rules. The resulting argumentation framework, called RP-DeLP, is based on a general notion of collective (non-binary) conflict among arguments allowing to ensure direct and indirect consistency properties with respect to the strict knowledge. An output of an RP-DeLP program is a pair of sets of warranted and blocked conclusions (literals), all of them recursively based on warranted conclusions but, while warranted conclusions do not generate any conflict, blocked conclusions do. An RP-DeLP program may have multiple outputs in case of circular definitions of conflicts among arguments. In this paper we tackle the problem of which output one should consider for an RP-DeLP program with multiple outputs. To this end we define the maximal ideal output of an RP-DeLP program as the set of conclusions which are ultimately warranted and we present an algorithm for computing them in polynomial space and with an upper bound on complexity equal to P^{NP}.

Keywords: defeasible argumentation, recursive warrant semantics, maximal ideal output.

1 Introduction and Motivation

Argumentation frameworks [6,15], can be used as a vehicle for facilitating rationally justifiable decision making when handling incomplete and potentially inconsistent information.

Possibilistic Defeasible Logic Programming (P-DeLP) [1] is a rule-based argumentation framework, extension of (DeLP) [12], where defeasible rules are attached to levels of strength, formalized as degrees of possibilistic necessity. P-DeLP inherits from DeLP the use of dialectical trees as underlying structures for characterizing the semantics for warranted conclusions.

In [1] a new recursive semantics for P-DeLP has been proposed based on a general notion of collective (non-binary) conflict among arguments. In this framework, called *Recursive* P-DeLP (RP-DeLP for short), an output (extension) of a program is a pair of sets of warranted and blocked conclusions, where arguments for warranted and blocked conclusions are recursively based on warranted conclusions but, while warranted conclusions do not generate any conflict with the set of already warranted conclusions and

S. Benferhat and J. Grant (Eds.): SUM 2011, LNAI 6929, pp. 96–109, 2011.
© Springer-Verlag Berlin Heidelberg 2011

the strict part of program, blocked conclusions do.[1] Conclusions that are neither warranted nor blocked correspond to rejected conclusions. The warrant recursive semantics of RP-DeLP ensures the three *rationality postulates* defined by Caminada and Amgoud in [4] without extending the representation of strict rules with transposed rules.

In [14], Pollock proposed a recursive semantics for defeasible argumentation without considering levels of defeasibility, where recursive definitions of defeat between arguments are characterized by means of *inference-graphs*, representing support and (binary) defeat relations between the conclusions of arguments. In RP-DeLP, a program may have multiple outputs (extensions) due to some circular definitions of warranty among arguments that emerge in case of circular definitions of conflicts among arguments. In [1] it was shown that, similar to [14], the latter can be checked by means of *warrant dependency graphs* for a set of arguments. Intuitively, the warrant dependency graph for a set of arguments represents conflict and support relationships among the arguments. In [2] we designed an algorithm which implements a level-wise procedure computing warranted and blocked conclusions until a cycle is found or the unique output is obtained.

In this paper we are interested in the problem of deciding the set of conclusions that can be ultimately warranted in RP-DeLP programs with multiple outputs. The usual skeptical approach would be to adopt the intersection of all possible outputs. However, in addition to the computational limitation, as stated in [14], adopting the intersection of all outputs may lead to an inconsistent output (in the sense of violating the base of the underlying recursive warrant semantics) in case some particular recursive situation among literals of a program occurs. Intuitively, for a conclusion, to be in the intersection does not guarantee the existence of an argument for it that is recursively based on ultimately warranted conclusions.

For instance, consider the following situation involving three conclusions P, Q, and T, where P can be warranted whenever Q is blocked, and vice-versa. Moreover, suppose that T can be warranted when either P or Q are warranted. Then, according to the warrant recursive semantics, we would get two different outputs: one where P and T are warranted and Q is blocked, and the other one where Q and T are warranted and P is blocked. Then, adopting the intersection of both outputs we would get that T would be ultimately warranted, however T should be in fact rejected since neither P nor Q are ultimately warranted conclusions.

According to this example, one could take then as the set of ultimately warranted conclusions of RP-DeLP programs those conclusions in the intersection of all outputs which are recursively based on ultimately warranted conclusions. However, as in RP-DeLP there are levels of defeasibility, this approach could lead to an incomplete solution since we are interested in determining the biggest set of ultimately warranted conclusions with maximum strength.

For instance consider the above example extended with two defeasibility levels as follows. Suppose that P can be warranted with strength α whenever Q is blocked, and

[1] The idea of defining a warrant semantics on the basis of conflicting sets of arguments was proposed in [16] The difference with the collective conflict among arguments in RP-DeLP is that in [16] the conflict is not relative to a set of already warranted conclusions and the strict part of the knowledge base (information we take for granted they hold true).

vice-versa. Moreover, suppose that T can be warranted with strength α whenever P is warranted at least with strength α and that T can be warranted with strength β, with $\beta < \alpha$, independently of the status of conclusions P and Q. Then, again we get two different outputs: one output warrants conclusions P and T with strength α and blocks conclusion Q, and the other one warrants conclusions Q and T with strengths α and β, respectively, and blocks P. Now, adopting conclusions of the intersection which are recursively based on ultimately warranted conclusions, we get that conclusion T is finally rejected, since conclusion T is warranted with a different argument and strength in each output. However, as we are interested in determining the biggest set of warranted conclusions with maximum strength, it seems quite reasonable to reject T at level α but to warrant it at level β.

Therefore, the set of ultimately warranted conclusions we are interested in for RP-DeLP programs has to be characterized by means of a recursive level-wise definition considering at each level the maximum set of conclusions based on warranted information and not involved in neither a conflict nor a circular definition of warranty. In fact, in a different context, this idea corresponds to the *maximal ideal extension* defined by Dung, Mancarella and Toni [9,10] as an alternative skeptical basis for defining collections of justified arguments in the abstract argumentation frameworks promoted by Dung [8] and Bondarenko *et al.* [3].

After this introduction, the rest of the paper is structured as follows. In Section 2 we recall basic definitions from RP-DeLP and then in Section 3 we characterize the maximal ideal output as the set of conclusions which are ultimately warranted for RP-DeLP programs. In Section 4 we present an algorithm for computing the maximal ideal output in polynomial space and with an upper bound on complexity equal to P^{NP}, and in Section 5 we present SAT encodings for the two main queries performed in the algorithm. We end up with some concluding remarks.

2 Preliminaries on RP-DeLP

The *language* of RP-DeLP, denoted $\mathcal{L}_R$, is inherited from the language of logic programming, including the notions of atom, literal, rule and fact. Formulas are built over a finite set of propositional variables $p, q, \ldots$ which is extended with a new (negated) atom "$\sim p$" for each original atom p. Atoms of the form p or $\sim p$ will be referred as literals, and if P is a literal, we will use $\sim P$ to denote $\sim p$ if P is an atom p, and will denote p if P is a negated atom $\sim p$. *Formulas* of $\mathcal{L}_R$ consist of rules of the form $Q \leftarrow P_1 \wedge \ldots \wedge P_k$, where $Q, P_1, \ldots, P_k$ are literals. A fact will be a rule with no premises. We will also use the name *clause* to denote a rule or a fact. The RP-DeLP framework is based on the propositional logic $(\mathcal{L}_R, \vdash_R)$ where the inference operator $\vdash_R$ is defined by instances of the modus ponens rule of the form: $\{Q \leftarrow P_1 \wedge \ldots \wedge P_k, P_1, \ldots, P_k\} \vdash_R Q$. A set of clauses Γ will be deemed as *contradictory*, denoted $\Gamma \vdash_R \bot$, if , for some atom q, $\Gamma \vdash_R q$ and $\Gamma \vdash_R \sim q$.

An RP-DeLP *program* $\mathcal{P}$ is a tuple $\mathcal{P} = (\Pi, \Delta, \preceq)$ over the logic $(\mathcal{L}_R, \vdash_R)$, where $\Pi, \Delta \subseteq \mathcal{L}_R$, and $\Pi \not\vdash_R \bot$. Π is a finite set of clauses representing strict knowledge (information we take for granted they hold true), Δ is another finite set of clauses representing the defeasible knowledge (formulas for which we have reasons to believe they are true). Finally, $\preceq$ is a total pre-order on $\Pi \cup \Delta$ representing levels of defeasibility: $\varphi \prec \psi$ means that φ is more defeasible than ψ. Actually, since formulas in Π are not defeasible, $\preceq$ is such that all formulas in Π are at the top of the ordering. For the sake of a simpler notation we will often refer in the paper to numerical levels for defeasible clauses and arguments rather than to the pre-ordering $\preceq$, so we will assume a mapping $N\colon \Pi \cup \Delta \to [0,1]$ such that $N(\varphi) = 1$ for all $\varphi \in \Pi$ and $N(\varphi) < N(\psi)$ iff $\varphi \prec \psi$.[2]

The notion of *argument* is the usual one. Given an RP-DeLP program $\mathcal{P}$, an argument for a literal (conclusion) Q of $\mathcal{L}_R$ is a pair $\mathcal{A} = \langle A, Q \rangle$, with $A \subseteq \Delta$ such that $\Pi \cup A \not\vdash_R \bot$, and A is minimal (w.r.t. set inclusion) such that $\Pi \cup A \vdash_R Q$. If $A = \emptyset$, then we will call $\mathcal{A}$ a s-argument (s for strict), otherwise it will be a d-argument (d for defeasible). We define the *strength of an argument* $\langle A, Q \rangle$, written $s(\langle A, Q \rangle)$, as follows:

$$s(\langle A, Q \rangle) = 1 \text{ if } A = \emptyset, \text{ and } s(\langle A, Q \rangle) = \min\{N(\psi) \mid \psi \in A\}, \text{ otherwise.}$$

The notion of *subargument* is referred to d-arguments and expresses an incremental proof relationship between arguments which is defined as follows. Let $\langle B, Q \rangle$ and $\langle A, P \rangle$ be two d-arguments such that the minimal sets (w.r.t. set inclusion) $\Pi_Q \subseteq \Pi$ and $\Pi_P \subseteq \Pi$ such that $\Pi_Q \cup B \vdash_R Q$ and $\Pi_P \cup A \vdash_R P$ verify that $\Pi_Q \subseteq \Pi_P$. Then, $\langle B, Q \rangle$ is a *subargument* of $\langle A, P \rangle$, written $\langle B, Q \rangle \sqsubset \langle A, P \rangle$, when either $B \subset A$ (strict inclusion for defeasible knowledge), or $B = A$ and $\Pi_Q \subset \Pi_A$ (strict inclusion for strict knowledge). A literal Q of $\mathcal{L}_R$ is called *justifiable conclusion* w.r.t. $\mathcal{P}$ if there exists an argument for Q, i.e. there exists $A \subseteq \Delta$ such that $\langle A, Q \rangle$ is an argument.

The following notions of acceptable argument, of collective conflict and of warrant dependency graph play a key role to formalize the recursive warrant semantics. If we think of W of a consistent set of already warranted conclusions, an acceptable argument captures the idea of an argument which is based on subarguments already warranted.

Let $\mathcal{P} = (\Pi, \Delta, \preceq)$ be an RP-DeLP program and let W be a set of justifiable conclusions which is consistent w.r.t. Π, i.e. $\Pi \cup W \not\vdash_R \bot$. A d-argument $\langle A, Q \rangle$ is an *acceptable argument* for Q w.r.t. W iff the two following conditions hold:

1. If $\langle B, P \rangle$ is a subargument of $\langle A, Q \rangle$, then $P \in W$.
2. $\Pi \cup W \cup \{Q\} \not\vdash_R \bot$.

The usual notion of attack or defeat relation in an argumentation system is binary. However, in some cases, the conflict relation among arguments is hardly representable as a binary relation when we compare them with the strict part of an RP-DeLP program. Next we formalize the notion of *collective conflict* among acceptable arguments.

Let $\mathcal{A}_1 = \langle A_1, Q_1 \rangle, \ldots, \mathcal{A}_k = \langle A_k, Q_k \rangle$ be acceptable arguments w.r.t. W. We say that the set of arguments $\{\mathcal{A}_1, \ldots, \mathcal{A}_k\}$ generates a conflict w.r.t. W iff the two following conditions hold:

(i) The set of argument conclusions $\{Q_1, \ldots, Q_k\}$ is contradictory w.r.t. $\Pi \cup W$, i.e. $\Pi \cup W \cup \{Q_1, \ldots, Q_k\} \vdash_R \bot$.
(ii) The set of argument conclusions $\{Q_1, \ldots, Q_k\}$ is minimal w.r.t. set inclusion satisfying (i), i.e. if $S \subset \{Q_1, \ldots, Q_k\}$, then $\Pi \cup W \cup S \not\vdash_R \bot$.

---

[2] Actually, a same pre-order $\preceq$ can be represented by many mappings, but we can take any of them to since only the relative ordering is what actually matters.

The warrant dependency graph for a set of arguments represents conflict and support dependencies among them, and will be used later. Let $\mathcal{A}_1 = \langle A_1, Q_1 \rangle, \ldots, \mathcal{A}_k = \langle A_k, Q_k \rangle$ be acceptable arguments w.r.t. W and let $\mathcal{B}_1 = \langle B_1, P_1 \rangle, \ldots, \mathcal{B}_n = \langle B_n, P_n \rangle$ be arguments such that $P_j \notin \{Q_1, \ldots, Q_k\}$ and there exists an argument $\mathcal{S} \in \{\mathcal{A}_1, \ldots, \mathcal{A}_k\}$ with $\mathcal{S} \sqsubset \mathcal{B}_j$, for all $j \in \{1, \ldots, n\}$. The *warrant dependency graph* (V, E) for $\{\mathcal{A}_1, \ldots, \mathcal{A}_k\}$ and $\{\mathcal{B}_1, \ldots, \mathcal{B}_n\}$ w.r.t. W is defined as follows:

1. For every literal $L \in \{Q_1, \ldots, Q_k\} \cup \{P_1, \ldots, P_n\}$, the set of vertices V includes one vertex v_L.
2. For every pair of literals $(L_1, L_2) \in \{P_1, \ldots, P_n\} \times \{Q_1, \ldots, Q_k\}$ such that $L_1 = {\sim} L_2$, the set of directed edges E includes one edge (v_{L_1}, v_{L_2}).
3. For every pair of literals $(L_1, L_2) \in \{Q_1, \ldots, Q_k\} \times \{P_1, \ldots, P_n\}$ such that the argument of L_1 is a subargument of the argument of L_2, the set of directed edges E includes one edge (v_{L_1}, v_{L_2}).
4. For every strict rule $R \leftarrow R_1 \wedge \ldots \wedge R_p \in \Pi$ such that ${\sim} R, R_1, \ldots, R_p \in W \cup \{Q_1, \ldots, Q_k\} \cup \{P_1, \ldots, P_n\}$, the set of directed edges E includes one edge (v_{L_1}, v_{L_2}) for every pair of literals $(L_1, L_2) \in \{P_1, \ldots, P_n\} \times \{Q_1, \ldots, Q_k\}$ such that the argument of L_2 is not a subargument of the argument of L_1, $L_1 \in \{{\sim} R, R_1, \ldots, R_p\}$ and, either $L_2 \in \{R_1, \ldots, R_p\}$ or, L_2 is a subargument of the argument of L_3, for some $L_3 \in \{P_1, \ldots, P_n\}$ such that $L_3 \in \{R_1, \ldots, R_p\}$.
5. Elements of V and E are only obtained by applying the above construction rules.

3 Maximal Ideal Output of an RP-DeLP Program

The maximal ideal output of an RP-DeLP program $\mathcal{P} = (\Pi, \Delta, \preceq)$ is a pair $(\mathit{Warr}, \mathit{Block})$ of warranted and blocked conclusions, respectively, with a maximum strength level such that the arguments of all of them are recursively based on warranted conclusions but, while warranted conclusions do not generate any conflict with the set of already warranted conclusions and any circular definition of warranty, blocked conclusions do.

Since defeasible arguments in an RP-DeLP program may have different levels of strength, the definition of $(\mathit{Warr}, \mathit{Block})$ is done level-wise, starting from the highest level and iteratively going down from one level to next level below. If $1 > \alpha_1 > \ldots > \alpha_p > 0$ are the strengths of d-arguments that can be built within $\mathcal{P}$, we define $\mathit{Warr} = \textit{s-Warr} \cup \textit{d-Warr}$ with $\textit{d-Warr} = \{\textit{d-Warr}(\alpha_1), \ldots, \textit{d-Warr}(\alpha_p)\}$ and $\mathit{Block} = \{\mathit{Block}(\alpha_1), \ldots, \mathit{Block}(\alpha_p)\}$, where *s-Warr* is the set of the warranted conclusions derivable from the strict knowledge Π and $\textit{d-Warr}(\alpha_i)$ and $\mathit{Block}(\alpha_i)$ are respectively the sets of the warranted and blocked conclusions with strength α_i. In the following, we write $\textit{d-Warr}(> \alpha_i)$ to denote $\cup_{\beta > \alpha_i} \textit{d-Warr}(\beta)$, $\textit{d-Warr}(\geq \alpha_i)$ to denote $\cup_{\beta \geq \alpha_i} \textit{d-Warr}(\beta)$ and analogously for $\mathit{Block}(> \alpha_i)$ and $\mathit{Block}(\geq \alpha_i)$, taking $\textit{d-Warr}(> \alpha_1) = \emptyset$ and $\mathit{Block}(> \alpha_1) = \emptyset$.

Before we formalize the maximal ideal warrant recursive semantics for an RP-DeLP program we need to define the following notions of *valid* and *almost valid* arguments with respect to a pair $(\mathit{Warr}, \mathit{Block})$ of warranted and blocked conclusions. A valid argument captures the idea of a non-rejected argument (i.e. a warranted or blocked argument, but not rejected) while an almost valid argument captures the idea of an argument whose rejection is conditional to the warranty of some valid argument.

A d-argument $\langle A, Q\rangle$ of strength α is called *valid* (or not rejected) w.r.t. a pair $(Warr, Block)$ of warranted and blocked conclusions if it satisfies the three following conditions:[3]

(V1) Any subargument $\langle B, P\rangle \sqsubset \langle A, Q\rangle$ of strength α is such that $P \in d\text{-}Warr(\alpha)$.
(V2) $\langle A, Q\rangle$ is acceptable w.r.t. $d\text{-}Warr(> \alpha) \cup \{P \mid \langle B, P\rangle \sqsubset \langle A, Q\rangle$ *and* $s(\langle B, P\rangle) = \alpha\}$.
(V3) $Q \not\in d\text{-}Warr(> \alpha) \cup Block(> \alpha)$, and $\sim Q \not\in Block(> \alpha)$.

The intuition underlying this definition is as follows. A d-argument $\langle A, Q\rangle$ of strength α is considered valid whenever the conclusions of all its subarguments are warranted, the argument is acceptable w.r.t. $d\text{-}Warr(> \alpha)$ and the conclusions of its subarguments, and there does not exist a valid argument neither for Q nor for $\sim Q$ of strength greater than α.

Let $\mathcal{A}$ be a set of valid arguments w.r.t. $(Warr, Block)$ of strength α. A d-argument $\langle B, P\rangle$ of strength α is called *almost valid* w.r.t. $\mathcal{A}$ and $(Warr, Block)$ if it satisfies the following six conditions:

(AV1) There does not exist an argument for P of strength α that is valid w.r.t. $(Warr, Block)$.
(AV2) Any subargument $\langle C, R\rangle \sqsubset \langle B, P\rangle$ of strength $\beta > \alpha$ is such that $R \in d\text{-}Warr(\beta)$.
(AV3) $\Pi \cup d\text{-}Warr(> \alpha) \cup \{R \mid \langle C, R\rangle \sqsubset \langle B, P\rangle$ *and* $s(\langle C, R\rangle) = \alpha\} \cup \{P\} \not\vdash_R \bot$.
(AV4) $P \not\in d\text{-}Warr(> \alpha) \cup Block(> \alpha)$, and $\sim P \not\in Block(> \alpha)$.
(AV5) For every subargument $\langle C, R\rangle \sqsubset \langle B, P\rangle$ of strength α such that $R \not\in Warr(\alpha)$, it holds that either $\langle C, R\rangle \in \mathcal{A}$, or $R, \sim R \not\in Block(\geq \alpha)$.
(AV6) There exists at least an argument $A \in \mathcal{A}$ such that $A \sqsubset \langle B, P\rangle$.

An almost valid argument captures the idea of an argument based on valid arguments of $\mathcal{A}$ and which status is valid (not rejected) whenever these arguments are warranted, and rejected, otherwise.

Definition 1. *The* maximal ideal output of an RP-DeLP program $\mathcal{P} = (\Pi, \Delta, \preceq)$ *is a pair* $(Warr, Block)$*, such that d-Warr and Block are required to satisfy the following recursive constraint: for every valid d-argument* $\langle A, Q\rangle$ *of strength* α *it holds that:*

- $Q \in Block(\alpha)$ *whenever one of the two following conditions holds:*
  (B1) There exists a set $\mathcal{G}$ *of valid arguments of strength* α *with* $\langle A, Q\rangle \not\in \mathcal{G}$ *such that the two following conditions hold:*
    (G1) $\langle A, Q\rangle \not\sqsubset \mathcal{C}$*, for all* $\mathcal{C} \in \mathcal{G}$*, and*
    (G2) $\mathcal{G} \cup \{\langle A, Q\rangle\}$ *generates a conflict w.r.t.* $d\text{-}Warr(> \alpha) \cup \{P \mid$ *there exists* $\langle B, P\rangle \sqsubset \mathcal{C}$ *for some* $\mathcal{C} \in \mathcal{G} \cup \{\langle A, Q\rangle\}\}$.
  (B2) There exists a set $\mathcal{A}$ *of valid arguments of strength* α *with* $\langle A, Q\rangle \in \mathcal{A}$ *such that the three following conditions hold:*
    (A1) $\langle A, Q\rangle \not\sqsubset \mathcal{C}$*, for all* $\mathcal{C} \in \mathcal{A}$.

[3] Notice that if $\langle A, Q\rangle$ is an acceptable argument w.r.t. $d\text{-}Warr(> \alpha)$, then $\langle A, Q\rangle$ is valid whenever condition (V3) holds.

(A2) There exists a set $\mathcal{B}$ of almost valid arguments w.r.t. $\mathcal{A}$ of strength α such that there is a cycle in the warrant dependency graph for $\mathcal{A}$ and $\mathcal{B}$, and any argument $\mathcal{C} \in \mathcal{A}$ is such that the conclusion of $\mathcal{C}$ is either a vertex of the cycle or $\mathcal{C}$ does not satisfy condition (B1).

(A3) For some vertex $v \in V$ of the cycle either v is the vertex of conclusion Q or v is the vertex of some other conclusion in $\mathcal{A}$ and there exists a path from v to the the vertex of conclusion Q.

– *Otherwise, $Q \in$ d-Warr(α).*

The intuition underlying the maximal ideal output definition is as follows. The conclusion of every valid or not rejected d-argument $\langle A, Q \rangle$ of strength α is either warranted or blocked. Then, it is eventually blocked if either (B1) it is involved in some conflict w.r.t. *d-Warr*$(> \alpha)$ and a set $\mathcal{G}$ of valid arguments of strength α whose supports do not depend on $\langle A, Q \rangle$ (conditions (G1) and (G2)), or (B2) the warranty of $\langle A, Q \rangle$ depends on some circular definition of conflict between arguments; otherwise, it is warranted.

Conditions (A1)-(A3) check whether the warranty of $\langle A, Q \rangle$ depends on some circular definition of conflict between a set $\mathcal{A}$ of valid arguments of strength α whose supports do not depend on $\langle A, Q \rangle$ and a set $\mathcal{B}$ of almost valid arguments whose supports depend on some argument in $\mathcal{A}$. In fact, the idea here is that if the warranty of $\langle A, Q \rangle$ depends on some circular definition of conflict between the arguments of $\mathcal{A}$ and $\mathcal{B}$, one could consider two different extensions (status) for conclusion Q: one with Q warranted and another one with Q blocked. Therefore, conclusion Q is blocked for the maximal ideal output. In general, the arguments of $\mathcal{A}$ and $\mathcal{B}$ involved in a cycle are respectively blocked and rejected for the maximal ideal output.

The following example shows a circular definition of conflict among arguments involving strict knowledge. Consider the RP-DeLP program $\mathcal{P} = (\Pi, \Delta, \preceq)$ with

$$\Pi = \{y, \sim y \leftarrow p \wedge r, \sim y \leftarrow q \wedge s\},$$
$$\Delta = \{p, q, t, r \leftarrow q, s \leftarrow p, t \leftarrow p, t \leftarrow q\} \text{ and}$$

two defeasibility levels for Δ as follows: $\{t\} \prec \{p, q, r \leftarrow q, s \leftarrow p, t \leftarrow p, t \leftarrow q\}$. Assume α_1 is the level of $\{p, q, r \leftarrow q, s \leftarrow p, t \leftarrow p, t \leftarrow q\}$ and α_2 is the level of $\{t\}$, with $1 > \alpha_1 > \alpha_2 > 0$. Obviously, *s-Warr* $= \{y\}$ and, at level α_1, arguments $\mathcal{A}_1 = \langle \{p\}, p \rangle$ and $\mathcal{A}_2 = \langle \{q\}, q \rangle$ are valid, and thus, conclusions p and q may be warranted or blocked but not rejected. Moreover, arguments $\mathcal{B}_1 = \langle \{q, r \leftarrow q\}, r \rangle$, $\mathcal{B}_2 = \langle \{p, s \leftarrow p\}, s \rangle$, $\mathcal{B}_3 = \langle \{q, t \leftarrow q\}, t \rangle$ and $\mathcal{B}_4 = \langle \{p, t \leftarrow p\}, t \rangle$ are almost valid w.r.t. $\mathcal{A}_1$ and $\mathcal{A}_2$. Figure 1 shows the warrant dependency graph for $\{\mathcal{A}_1, \mathcal{A}_2\}$ and $\{\mathcal{B}_1, \mathcal{B}_2, \mathcal{B}_3, \mathcal{B}_4\}$. Conflict and support dependencies among arguments are represented as dashed and solid arrows, respectively. The cycle of the graph expresses that (1) the warranty of p depends on a (possible) conflict with r; (2) the support of r depends on q (i.e., r is valid whenever q is warranted); (3) the warranty of q depends on a (possible) conflict with s; and (4) the support of s depends on p (i.e., s is valid whenever p is warranted). Then, conclusions p and q are blocked, and conclusions r and s are rejected. Remark that conclusion t is also rejected at level α_1 since (5) the support of $\mathcal{B}_3$ depends on p, (6) the support of $\mathcal{B}_4$ depends on q, and p and q are blocked. Therefore,

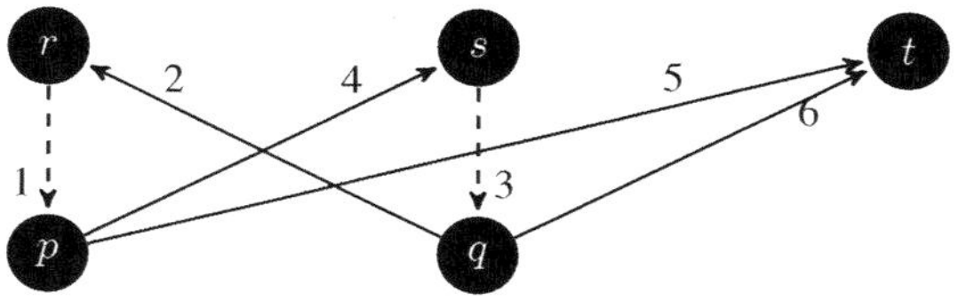

Fig. 1. Warrant dependency graph for $\mathcal{P}$

d-Warr$(\alpha_1) = \emptyset$ and *Block*$(\alpha_1) = \{p, q\}$. Finally, at level α_2, $\langle\{t\}, t\rangle$ is the unique valid argument and therefore conclusion t is warranted. Hence, *d-Warr*$(\alpha_2) = \{t\}$ and *Block*$(\alpha_2) = \emptyset$.

Next propositions[4] state that the maximal ideal output of an RP-DeLP program is unique and satisfies the indirect consistency and closure properties defined by Caminada and Amgoud[4] with respect to the strict knowledge. Moreover, it could be proved that the maximal ideal output of an RP-DeLP program contains all conclusions in the intersection of all outputs whose arguments are recursively based on ultimately warranted conclusions.

Proposition 1 (Unicity of the maximal ideal output). *Let $\mathcal{P} = (\Pi, \Delta, \preceq)$ be an RP-DeLP program. The pair (Warr, Block) of warranted and blocked conclusions that satisfies the maximal ideal output characterization for $\mathcal{P}$ is unique.*

Proposition 2 (Indirect consistency and closure). *Let $\mathcal{P} = (\Pi, \Delta, \preceq)$ be an RP-DeLP program with defeasibility levels $1 > \alpha_1 > \ldots > \alpha_p > 0$ and let (Warr, Block) be the maximal ideal output for $\mathcal{P}$. Then,*

(i) $\Pi \cup Warr \not\vdash_R \bot$, *and*
(ii) if $\Pi \cup$ *d-Warr*$(\geq \alpha_i) \vdash_R Q$ *and* $\Pi \cup$ *d-Warr*$(> \alpha_i) \not\vdash_R Q$, *then either* $Q \in$ *d-Warr*(α_i), *or* $Q \in$ *Block*$(> \alpha_i)$, *or* $\sim Q \in$ *Block*$(> \alpha_i)$.

Notice that for the particular case of considering just one defeasibility level for Δ, the closure property reads as follows : if $\Pi \cup Warr \vdash_R Q$, then $Q \in Warr$.

4 On the Computation of the Maximal Ideal Output

From a computational point of view, the maximal ideal output of an RP-DeLP program can be computed by means of a level-wise procedure, starting from the highest level and iteratively going down from one level to next level below. Then, at every level it is necessary to determine the status (warranted or blocked) of each valid argument. Next we design an algorithm which implements this level-wise procedure computing warranted and blocked conclusions by checking the existence of conflicts between arguments and cycles at some warrant dependency graph. In the following we use the notation $W(1)$ for *s-Warr*, $W(\alpha)$ and $W(\geq \alpha)$ for *d-Warr*(α) and *d-Warr*$(\geq \alpha)$, respectively, and $B(\alpha)$ and $B(\geq \alpha)$ for *Block*(α) and *Block*$(\geq \alpha)$, respectively.

[4] Proofs can be found in a forthcoming extended version of this paper.

Algorithm `Computing warranted conclusions`
Input $\mathcal{P} = (\Pi, \Delta, \preceq)$: An RP-DeLP program
Output (W, B): maximal ideal output for $\mathcal{P}$

$W(1) := \{Q \mid \Pi \vdash_R Q\}$
$B := \emptyset$
$\alpha :=$ `maximum_level`(Δ)
while $(\alpha > 0)$ **do**
  `level_computing`(α, W, B)
  $\alpha :=$ `next_level`(Δ)
end while

The algorithm `Computing warranted conclusions` first computes the set of warranted conclusions $W(1)$ form the set of strict clauses Π. Then, for each defeasibility level $1 > \alpha > 0$, the procedure `level_computing` determines all warranted and blocked conclusions with strength α. Remark that for every level α, the procedure `level_computing` receives $W(> \alpha)$ and $B(> \alpha)$ as input and produces $W(\geq \alpha)$ and $B(\geq \alpha)$ as output.

Procedure `level_computing` (**in** α; **in_out** W, B)
  $VA := \{\langle A, Q\rangle$ with strength $\alpha \mid \langle A, Q\rangle$ is valid w.r.t. W and $B\}$
  while $(VA \neq \emptyset)$
    while $(\exists \langle A, Q\rangle \in VA \mid$
      $\neg$ `conflict`$(\alpha, \langle A, Q\rangle, VA, W,$ `not_dependent`$(\alpha, \langle A, Q\rangle, VA, W, B))$
      and $\neg$ `cycle`$(\alpha, \langle A, Q\rangle, VA, W,$ `almost_valid`$(\alpha, VA, W, B))$ **do**
        $W(\alpha) := W(\alpha) \cup \{Q\}$
        $VA := VA \backslash \{\langle A, Q\rangle\} \cup \{\langle C, P\rangle$ with strength $\alpha \mid$
          $\langle C, P\rangle$ is valid w.r.t. W and $B\}$
    end while
    $I := \{\langle A, Q\rangle \in VA \mid$ `conflict`$(\alpha, \langle A, Q\rangle, VA, W, \emptyset)$
      or `cycle`$(\alpha, \langle A, Q\rangle, VA, W,$ `almost_valid`$(\alpha, VA, W, B))\}$
    $B(\alpha) := B(\alpha) \cup \{Q \mid \langle A, Q\rangle \in I\}$
    $VA := VA \backslash I$
  end while

For every level α the procedure `level_computing` first computes the set VA of valid arguments[5] w.r.t. $W(> \alpha)$ and $B(> \alpha)$. Then, the set VA of valid arguments is dynamically updated depending on new warranted and blocked conclusions with strength α. The procedure `level_computing` finishes when the status for every valid argument is computed. The status of a valid argument is computed by means of the four following auxiliary functions.

Function `almost_valid`(**in** α, VA, W, B) **return** AV: set of arguments
  $AV := \{\langle C, P\rangle$ with strength $\alpha \mid \langle C, P\rangle$ satisfies conditions (AV1)-(AV6) w.r.t. $VA\}$
  return(AV)
Function `not_dependent`(**in** $\alpha, \langle A, Q\rangle, VA, W, B$) **return** ND: set of arguments
  $AV :=$ `almost_valid`(α, VA, W, B)
  $ND := \{\langle C, P\rangle \in AV \mid \langle A, Q\rangle \not\sqsubset \langle C, P\rangle\}$
  return(ND)

[5] Notice that an argument $\langle A, Q\rangle$ with strength α is *valid* w.r.t. $W(> \alpha)$ and $B(> \alpha)$ if $\langle A, Q\rangle$ is acceptable w.r.t. $W(> \alpha)$ and satisfies condition (V3).

Function `conflict`(**in** α, $\langle A, Q\rangle$, *VA*, W, *ND*) : **return Boolean**
  return ($\exists\ S \subseteq VA\backslash\{\langle A, Q\rangle\} \cup ND$ such that $\Pi \cup W(\geq \alpha) \cup \{P \mid \langle B, P\rangle \in S\} \not\vdash_R \bot$
    and $\Pi \cup W(\geq \alpha) \cup \{P \mid \langle B, P\rangle \in S\} \cup \{Q\} \vdash_R \bot$)
Function `cycle`(**in** α, $\langle A, Q\rangle$, *VA*, W, *AV*) : **return Boolean**
  return (there is a cycle in the warrant dependency graph for *VA* and *AV* **and** the vertex of $\langle A, Q\rangle$ is a vertex of the cycle **or** there exists a path from a vertex in *VA* of the cycle to the the vertex of $\langle A, Q\rangle$)

The function `conflict` checks (possible) conflicts among the argument $\langle A, Q\rangle$ and the set *VA* of valid arguments extended with the set *ND* of arguments. The set *ND* of arguments takes two different values: the empty set and the set of almost valid arguments whose supports depend on some argument in $VA\backslash\{\langle A, Q\rangle\}$. The empty set value is used to detect conflicts between the argument $\langle A, Q\rangle$ and the arguments in *VA*, and thus, every valid argument involved in a conflict is blocked. On the other hand, the value set of almost valid arguments which do not depend on argument $\langle A, Q\rangle$ is used to detect possible conflicts between the argument $\langle A, Q\rangle$ and the arguments in $VA \cup ND$, and thus, every valid argument involved in a possible conflict remains as valid. In fact, the function `almost_valid` computes the set of almost valid arguments that satisfy conditions (AV1)-(AV6) w.r.t. the current set of valid arguments. The function `not_dependent` considers almost valid arguments w.r.t. the current set of valid arguments which do not depend on $\langle A, Q\rangle$. Finally, the function `cycle` checks the existence of a cycle in the warrant dependency graph for the current set of valid arguments and its set of almost valid arguments, and verifies whether the vertex of argument $\langle A, Q\rangle$ is in the cycle or there exists a path from a vertex of the cycle to it.

One of the main advantages of the maximal ideal warrant recursive semantics for RP-DeLP is from the implementation point of view. Warrant semantics based on dialectical trees, like DeLP [5,7], might consider an exponential number of arguments with respect to the number of rules of a given program. The previous algorithm can be implemented to work in polynomial space[6], with a complexity upper bound equal to P^{NP}. This can be achieved because it is not actually necessary to find all the valid arguments for a given literal Q, but only one witnessing a valid argument for Q is enough. Analogously, function `not_dependent` can be implemented to generate at most one almost valid argument, not dependent on $\langle A, Q\rangle$, for a given literal. The only function that in the worst case can need an exponential number of arguments is `cycle`, but it can be shown that whenever `cycle` returns true for $\langle A, Q\rangle$, then a conflict will be detected with the almost valid arguments not dependent on $\langle A, Q\rangle$, so warranted literals can be detected without function `cycle`. Also, blocked literals detected by function `cycle` can also be detected by checking the stability of the set of valid arguments after two consecutive iterations, so it is not necessary to explicitly compute dependency graphs. Next, observe that the following queries can be implemented with NP algorithms:

1. Whether a literal P is a conclusion of some argument returned by `not_dependent`(α, $\langle A, Q\rangle$, *VA*, W, B). To check the existence of an almost valid argument $\langle C, P\rangle$ not dependent on $\langle A, Q\rangle$, we can non-deterministically guess a subset of rules, and check in polynomial time whether they actually generate the desired argument for P, as all the conditions for an almost valid argument can be

[6] Details can be found in a forthcoming extended version of this paper.

checked in polynomial time and also the condition of not being dependent on the literal Q.

2. Whether the function conflict(**in** α, $\langle A, Q\rangle$, *VA*, W, *ND*) returns true. To check the existence of a conflict, we can non-deterministically guess a subset of literals S from $\{P \mid \langle B, P\rangle \in \mathit{VA} \setminus \{\langle A, Q\rangle\} \cup \mathit{ND}\}$ and check in polynomial time whether i) $\Pi \cup W(\geq \alpha) \cup S \not\vdash \bot$ and ii) $\Pi \cup W(\geq \alpha) \cup S \cup \{Q\} \vdash \bot$.

Then, as the maximum number of times that these queries need to be executed before the set of conclusions associated with *VA* becomes stable is polynomial in the size of the input program, the P^{NP} upper bound follows.

5 SAT Encodings for Finding Warranted Literals

The previous algorithm to find warranted literals needs to compute two main queries during its execution: i) whether an argument $\langle C, P\rangle$ with strength α is almost valid (computed in function almost_valid) and ii) whether there is a conflict for a valid argument $\langle A, Q\rangle$. Remember that the explicit computation of cycles in warrant dependency graphs can be avoided. We present here SAT encodings for resolving both queries with a SAT solver. In a forthcoming extended version of this paper[7], we show empirical results obtained with an implementation of our algorithm that uses these SAT encodings. The results show that, at least on randomly generated instances, the practical complexity is strongly dependent on the size of the strict part of the program, as for the same number of variables RP-DeLP programs with different size for their strict part can range from trivially solvable to exceptionally hard.

5.1 Looking for Almost Valid Arguments

The idea for encoding the problem of searching almost valid arguments is based on the same idea behind successful SAT encodings for solving STRIPS planning problems [13]. In a STRIPS planning problem, given an initial state, described with a set of predicates, the goal is to decide whether a desired goal state can be achieved by means of the application of a suitable sequence of actions. Each action has a set preconditions, when they hold true the action can be executed and as a result certain facts become true and some others become false (its effects). Hence executing an action changes the current state, and the application of a sequence of actions creates a sequence of states. The planning problem is to find a sequence of actions such that, when executed, the obtained final state satisfies the goal state.

In our case, the search for an almost valid argument $\langle C, P\rangle$ can be seen as the search for a *plan* for producing P, taking as the initial set of facts some subset of a set of literals in which we already trust. We call such initial set the base set of literals[8], and we say that they are true at the first step of the argument. Given the set of possible rules for almost valid arguments with the current strength α and a given state[9], if we execute

[7] In preparation.

[8] For an almost valid argument, the base set can contain only warranted and valid literals.

[9] These are rules R satisfying: i) either $N(R) > \alpha$ and $\mathit{Body}(R) \setminus W(> \alpha) \neq \emptyset$, or $N(R) = \alpha$; ii) $\mathit{Body}(R) \cap B(\geq \alpha) = \emptyset$; and iii) $\mathit{Head}(R), {\sim}\mathit{Head}(R) \not\in W(\geq \alpha) \cup B(\geq \alpha)$.

all the rules that have their precondition satisfied we obtain a new state, that contains all the previous literals plus the possible new ones obtained. This process can be repeated iteratively, obtaining a sequence of states $\mathcal{S} = \{S_0, S_1, \ldots, S_t\}$ and a sequence of sets of executed rules $\mathcal{R} = \{R_0, R_1, \ldots, R_{t-1}\}$, until we reach a final state S_t in which the execution of any possible rule does not increase the set of literals already in S_t. If starting from an initial set S_0 that contains all the current valid and warranted literals the final state S_t contains P, that means that an almost valid argument for P could be obtained from theses sequences, if we could find selected subsets such that their literals and rules satisfy the conditions for an almost valid argument for P. Observe that in this forward reasoning process some of the conditions for almost valid arguments have already been satisfied, but the existence of an argument that satisfies the consistency conditions is not secured with that process.

So, we consider encoding as a SAT instance the search for an almost valid argument $\langle C, P\rangle$ from the sequences $\mathcal{S}$ and $\mathcal{R}$ defined above. That is, a SAT instance with variables to represent all the possible literals we can select from each set S_i: $\{v^i_L \mid L \in S_i, 0 \leq i \leq t\}$, plus variables to represent all the possible rules R we can select from each set R_i: $\{v^i_R \mid R \in R_i, 0 \leq i < t\}$. In order to check that the variables set to true represent a well formed argument that is almost valid, we add clauses for ensuring that:

1. If variable v^i_L is true, then either v^{i-1}_L is true or one of the variables v^{i-1}_R, with $\mathit{Head}(R) = L$, is true.
2. If a variable v^i_R is true, then for all the literals L in its body v^i_L must be true.
3. If variable v^i_L is true, then v^{i+1}_L is also true.
4. The variable v^t_P must be true.
5. No two contradictory variables v^t_L and $v^t_{\sim L}$ can be both true.

In addition, in order to satisfy the consistency of the literals of the argument with respect to the closure of the strict knowledge Π, we create also an additional set of variables V_Π and set of clauses R_Π. The set of variables V_Π contains a variable v^Π_L for each literal that appears in the logical closure of the set $S_t \cup W$ with respect to the strict rules.

Then, we add the following clauses to check the consistency with Π:

1. If a literal is selected for the argument (v^t_L set to true) then v^Π_L must also be true.
2. For any $L \in W$, v^Π_L must be true.
3. For any rule $R \in \Pi$ that was executed when computing the logical closure, if for all the literals L in its body v^Π_L is true, then $v^\Pi_{\mathit{Head}(R)}$ must be true.
4. No two contradictory variables v^Π_L and $v^\Pi_{\sim L}$ can be both true.

5.2 Looking for Collective Conflicts

We reduce the query computed by function `conflict`, to a query where we consider finding the set of conflict literals that are the conclusions of the corresponding conflict set of arguments. Basically, for finding this conflict set of literals S for a valid argument $\langle A, Q\rangle$ from the base set of literals considered in function `conflict`, i.e. the set $G = \{P \mid \langle B, P\rangle \in \mathit{VA} \backslash \{\langle A, Q\rangle\} \cup \mathit{ND}\}$, we have to find two arguments $\langle A_1, L\rangle$, $\langle A_2, {\sim}L\rangle$

using only rules from Π, literals $W \cup \{Q\}$ and a subset S from G, but such that when Q is not used, no conflict (generation of L and $\sim L$ for any L with strict rules) is produced with such set S. So, this can be seen as a simple extension of the previous query, where now we have to look for two arguments, instead of only one, although both arguments must be for two contradictory literals. That is, the SAT formula contains variables for encoding arguments that use as base literals $W \cup G \cup \{Q\}$ and rules from Π (with the same scheme of the previous SAT encoding for almost valid arguments), with an additional set of conflict variables to encode the set of possible conflicts that can be, potentially, generated from $W \cup G \cup \{Q\}$ using rules from Π, in order to be able to force the existence of at least one conflict. There is also an additional set of variables and clauses for encoding the subproblem of checking that S, when Q is not used, does not generate any conflict.

So, the SAT formula contains two different parts. A first part is devoted to checking that the selected set of literals S plus $\{Q\}$ is a conflict set (i.e. if $\Pi \cup W(\geq \alpha) \cup S \cup \{Q\} \vdash_R \bot$). This set of variables and clauses is similar to the previous one for finding almost valid arguments, but in this case for finding two arguments starting from a subset of $W \cup G$ and forcing the inclusion of $\{Q\}$. That is, the clauses of this first part are:

1. A clause that states that the literal Q must be true at the first step.
2. A clause that states that at least one conflict variable c_L must be true.
3. For every conflict variable c_L, a clause that states that if c_L is true then literals L and $\sim L$ must be true at the final step of the argument.
4. The rest of clauses are the same ones described in the first part of the previous encoding, except the clauses of the item 5 that are not included, but now considering as possible literals and rules at every step the ones computed from the base set $W \cup G \cup \{Q\}$ and using only strict rules.

The process for computing the possible literals and rules that can be potentially applied in every step of the argument is the same forward reasoning process presented for the previous encoding. This same process is used for discovering the set of conflict variables c_L that need to be considered, because we can potentially force the conflict c_L if at the end of this process both L and $\sim L$ appear as reachable literals.

A second part is devoted to checking that the selected set S, without using Q, does not cause any conflict with the strict rules. This is a set of variables and clauses that ensures that the selected set of literals (minus Q) at the first step of the argument encoded in the first part of the formula, when we consider the closure with respect to the strict clauses, does not generate a conflict. So this second part of the formula contains a variable for any literal that appears in the logical closure of $G \cup W$ with respect to the strict rules. Actually, this second part of the formula is analogous to the second part of the formula for the previous encoding.

6 Conclusions and Future Work

In this paper we have tackled the problem of deciding which set of ultimately warranted conclusions should be considered for RP-DeLP programs with multiple outputs according to a recursive warrant semantics. A natural solution to this problem could be to adopt the intersection of all possible outputs, however, as it has been shown, this can lead to an inconsistent output when some recursive situation occurs between the arguments of a program. So we have defined the maximal ideal output for RP-DeLP programs as the set of ultimately warranted conclusions characterized by means of a recursive level-wise definition considering at each defeasibility level the maximum set of conclusions based on warranted information and not involved in neither a conflict nor a circular definition of warranty. We have also designed and implemented an algorithm with an upper bound on complexity equal to P^{NP} for computing the warranty status of arguments according to the new maximal ideal recursive semantics.

Acknowledgments. Authors are thankful to the anonymous reviewers for their helpful comments. Research partially funded by the Spanish MICINN projects ARINF (TIN2009-14704-C03-01/03) and TASSAT (TIN2010-20967-C04-01/03), by CONSOLIDER (CSA*V*2007-0022), by ESF Eurocores-LogICCC/MICINN (FFI2008-03126-E/FILO) and by University of Lleida pre-doctoral program.

References

1. Alsinet, T., Béjar, R., Godo, L.: A characterization of collective conflict for defeasible argumentation. In: COMMA, pp. 27–38 (2010)
2. Alsinet, T., Béjar, R., Godo, L.: A computational method for defeasible argumentation based on a recursive warrant semantics. In: Kuri-Morales, A., Simari, G.R. (eds.) IBERAMIA 2010. LNCS, vol. 6433, pp. 40–49. Springer, Heidelberg (2010)
3. Bondarenko, A., Dung, P.M., Kowalski, R.A., Toni, F.: An abstract, argumentation-theoretic approach to default reasoning. Artif. Intell. 93, 63–101 (1997)
4. Caminada, M., Amgoud, L.: On the evaluation of argumentation formalisms. Artif. Intell. 171(5-6), 286–310 (2007)
5. Cecchi, L., Fillottrani, P., Simari, G.: On the complexity of delp through game semantics. In: NMR, pp. 386–394 (2006)
6. Chesñevar, C., Maguitman, A., Loui, R.: Logical Models of Argument. ACM Computing Surveys 32(4), 337–383 (2000)
7. Chesñevar, C., Simari, G., Godo, L.: Computing dialectical trees efficiently in possibilistic defeasible logic programming. In: Baral, C., Greco, G., Leone, N., Terracina, G. (eds.) LPNMR 2005. LNCS (LNAI), vol. 3662, pp. 158–171. Springer, Heidelberg (2005)
8. Dung, P.M.: On the acceptability of arguments and its fundamental role in nonmonotonic reasoning, logic programming and n-person games. Artif. Intell. 77(2), 321–358 (1995)
9. Dung, P.M., Mancarella, P., Toni, F.: A dialectic procedure for sceptical, assumption-based argumentation. In: COMMA, pp. 145–156 (2006)
10. Dung, P.M., Mancarella, P., Toni, F.: Computing ideal sceptical argumentation. Artif. Intell. 171(10-15), 642–674 (2007)
11. Een, N., Sorensson, N.: An extensible SAT-solver. In: Giunchiglia, E., Tacchella, A. (eds.) SAT 2003. LNCS, vol. 2919, pp. 502–518. Springer, Heidelberg (2004)
12. García, A., Simari, G.: Defeasible Logic Programming: An Argumentative Approach. Theory and Practice of Logic Programming 4(1), 95–138 (2004)
13. Kautz, H.A., Selman, B.: Unifying sat-based and graph-based planning. In: IJCAI, pp. 318–325 (1999)
14. Pollock, J.L.: A recursive semantics for defeasible reasoning. In: Rahwan, I., Simari, G.R. (eds.) Argumentation in Artificial Intelligence. ch.9, pp. 173–198. Springer, Heidelberg (2009)
15. Prakken, H., Vreeswijk, G.: Logical Systems for Defeasible Argumentation. In: Gabbay, D., Guenther, F. (eds.) Handbook of Phil. Logic, pp. 219–318. Kluwer, Dordrecht (2002)
16. Vreeswijk, G.: Abstract argumentation systems. Artif. Intell. 90(1-2), 225–279 (1997)

Argumentation Frameworks as Constraint Satisfaction Problems

Leila Amgoud[1] and Caroline Devred[2]

[1] IRIT, University of Toulouse, France
Leila.Amgoud@irit.fr
[2] LERIA, University of Angers, France
devred@info.univ-angers.fr

Abstract. This paper studies how to encode the problem of computing the extensions of an argumentation framework (under a given semantics) as a constraint satisfaction problem (CSP). Such encoding is of great importance since it makes it possible to use the very efficient solvers (developed by the CSP community) for computing the extensions. We focus on three families of frameworks: Dung's abstract framework, its constrained version and preference-based argumentation frameworks.

Keywords: Default Reasoning, Argumentation, CSP.

1 Introduction

Argumentation is a reasoning model based on the construction and evaluation of interacting arguments. An argument is a reason for believing in a statement, doing an action, pursuing a goal, etc.

Argumentation theory is gaining an increasing interest in Artificial Intelligence, namely for reasoning about defeasible/uncertain information, making decisions under uncertainty, learning concepts, and modeling agents' interactions (see [1]).

The most abstract argumentation framework has been proposed in the seminal paper [15] by Dung. It consists of a set of *arguments*, a *binary relation* representing *attacks* among arguments, and *semantics* for evaluating the arguments. A semantics describes when a set of arguments, called *extension*, is acceptable without bothering on how to compute that set. This framework has been extended in different ways in the literature. In [2,3], arguments are assumed to not have the same strength while in [9] an additional constraint on arguments may be available. In both works, Dung's semantics are used to evaluate arguments, thus to compute the extensions.

In [9,12,13,17], different decision problems related to the implementation of those semantics have been identified and the computational complexity of each problem studied. The results are a bit disappointing since they show that the most important decision problems (like for instance testing whether a framework has a stable set of arguments) are costly. Some algorithms that compute extensions under some semantics have been developed, for instance in [8,11,18].

S. Benferhat and J. Grant (Eds.): SUM 2011, LNAI 6929, pp. 110–122, 2011.
© Springer-Verlag Berlin Heidelberg 2011

However, the efficiency of those algorithms was not proved. They are neither tested on benchmarks nor compared to other algorithms developed for the same purpose.

Besides, there is a huge literature on *Constraints Satisfaction Problems* (CSP) since many real-world problems can be described as CSPs. A CSP consists of a set of *variables*, a (generally finite) *domain* for each variable and a set of *constraints*. Each constraint is defined over a subset of variables and limits the combination of values that the variables in this subset can take. The goal is to find an assignment to the variables which satisfies all the constraints. In some problems, the goal is to find all such assignments. Solving a constraint satisfaction problem on a finite domain is an NP-complete problem in general. In order to be solved in a reasonable time, different *solvers* have been developed. They use a form of search based on variants of backtracking, constraint propagation and local search [19].

Our aim is to be able to use those powerful solvers for computing the extensions of an argumentation framework. For that purpose, we study in this paper how to encode an argumentation framework as a CSP. We particularly focus on three families of frameworks: Dung's framework, constrained argumentation framework and preference-based argumentation framework (where arguments may have different strengths). For each family, we propose different CSPs which compute the extensions of the framework under different acceptability semantics. In each CSP, arguments play the role of variables and the attacks represent mainly the constraints.

This paper is organized as follows: Section 2 recalls the basic concepts of a CSP. Section 3 recalls Dung's argumentation framework and shows how it is encoded as a CSP. Section 4 recalls the constrained version of Dung's framework and presents its encoding as a CSP. Section 5 presents preference-based argumentation frameworks as well as their encoding as CSPs. In Section 6, we compare our approach to existing works on the topic. The last section is devoted to concluding remarks and perspectives. Due to space limitation, the proofs are not included in the paper.

2 Constraint Satisfaction Problems (CSPs)

Formally speaking, a *constraint satisfaction problem* (or CSP) is defined by a set of *variables*, $x_1, \ldots, x_n$, and a set of *constraints* $c_1, \ldots, c_m$. Each variable x_i takes its values from a *finite domain* $\mathcal{D}_i$, and each constraint c_i involves some subset of the variables and specifies the allowable combinations of values for that subset.

Definition 1 (CSP). *A* CSP *instance is a triple* $(\mathcal{X}, \mathcal{D}, \mathcal{C})$ *where:*

- $\mathcal{X} = \{x_1, \ldots, x_n\}$ *is a set of variables,*
- $\mathcal{D} = \{\mathcal{D}_1, \ldots, \mathcal{D}_n\}$ *is a set of finite domains for the variables, and*
- $\mathcal{C} = \{c_1, \ldots, c_m\}$ *is a set of constraints.*

Each constraint c_i is a pair (h_i, H_i) where

- $h_i = (x_{i1}, \ldots, x_{ik})$ *is a k-tuple of variables*
- H_i *is a k-ary relation over* $\mathcal{D}$, *i.e.* H_i *is a subset of all possible variable values representing the allowed combinations of simultaneous values for the variables in* h_i.

A state of the problem is defined by an *assignment* of values to some or all of the variables.

Definition 2 (Assignment). *An* assignment v *for a CSP instance* $(\mathcal{X}, \mathcal{D}, \mathcal{C})$ *is a mapping that assigns to every variable* $x_i \in \mathcal{X}$ *an element* $v(x_i) \in \mathcal{D}_i$. *An assignment* v *satisfies a constraint* $((x_{i1}, \ldots, x_{ik}), H_i) \in \mathcal{C}$ *iff* $(v(x_{i1}), \ldots, v(x_{ik})) \in H_i$.

Finally, a solution of a CSP is defined as follows:

Definition 3 (Solution). *A* solution *of a CSP instance* $(\mathcal{X}, \mathcal{D}, \mathcal{C})$ *is an assignment* v *that satisfies all the constraints in* $\mathcal{C}$ *and in which all the variables of* $\mathcal{X}$ *are assigned a value. We write* $(v(x_1), \ldots, v(x_n))$ *to denote the solution.*

3 Abstract Frameworks

This section recalls Dung's argumentation framework and presents the different corresponding CSPs which return its extensions under various semantics.

3.1 Dung's Framework

In [15], Dung has developed the most abstract argumentation framework in the literature. It consists of a set of arguments and an attack relation between them.

Definition 4 (Argumentation framework). *An* argumentation framework *(AF) is a pair* $\mathcal{F} = (\mathcal{A}, \mathcal{R})$ *where* $\mathcal{A}$ *is a set of arguments and* $\mathcal{R}$ *is an attack relation* ($\mathcal{R} \subseteq \mathcal{A} \times \mathcal{A}$). *The notations* $a\mathcal{R}b$ *or* $(a, b) \in \mathcal{R}$ *mean that the argument* a attacks *the argument* b.

Different *acceptability semantics* for evaluating arguments have been proposed in the same paper [15]. Each semantics amounts to define sets of acceptable arguments, called *extensions*. Before recalling those semantics, let us first introduce the two basic properties underlying them, namely *conflict-freeness* and *defence*.

Definition 5 (Conflict-free, Defence). *Let* $\mathcal{F} = (\mathcal{A}, \mathcal{R})$ *be an AF and* $\mathcal{B} \subseteq \mathcal{A}$.

- $\mathcal{B}$ *is* conflict-free *iff* $\nexists\, a, b \in \mathcal{B}$ *s.t.* $a\mathcal{R}b$.
- $\mathcal{B}$ defends *an argument* a *iff for all* $b \in \mathcal{A}$ *s.t.* $b\mathcal{R}a$, *there exists* $c \in \mathcal{B}$ *s.t.* $c\mathcal{R}b$.

The following definition recalls the acceptability semantics proposed in [15].

Definition 6 (Acceptability semantics). *Let $\mathcal{F} = (\mathcal{A}, \mathcal{R})$ be an AF and $\mathcal{B} \subseteq \mathcal{A}$.*

- $\mathcal{B}$ *is an* admissible *set iff it is conflict-free and defends its elements.*
- $\mathcal{B}$ *is a* preferred *extension iff it is a maximal (for set $\subseteq$) admissible set.*
- $\mathcal{B}$ *is a* stable *extension iff it is a preferred extension that attacks any argument in $\mathcal{A} \setminus \mathcal{B}$.*
- $\mathcal{B}$ *is a* complete *extension iff it is conflict-free and it contains all the arguments it defends.*
- $\mathcal{B}$ *is a* grounded *extension iff it is a minimal (for set $\subseteq$) complete extension.*

Example 1. Let us consider the framework $\mathcal{F}_1 = (\mathcal{A}_1, \mathcal{R}_1)$ where $\mathcal{A}_1 = \{a, b, c, d\}$ and $\mathcal{R}_1 = \{(a, b), (b, c), (c, d), (d, a)\}$. $\mathcal{F}_1$ has two preferred and stable extensions: $\mathcal{B}_1 = \{a, c\}$ and $\mathcal{B}_2 = \{b, d\}$ while its grounded extension is the empty set.

3.2 Computing Dung's Semantics by CSPs

In this section, we propose four mappings of Dung's argumentation framework into CSP instances. The idea is: starting from an argumentation framework, we define a CSP instance whose solutions are the extensions of the framework under a given acceptability semantics. In the four instances, arguments play the role of variables that is, a variable is associated to each argument. Each variable may take two values 0 or 1 meaning that the corresponding argument is rejected or accepted. Thus, the domains of the variables are all *binary*. Things are different with the constraints. We show that according to the semantics that is studied, the definition of a constraint changes.

Let us start with a CSP instance that computes the conflict-free sets of arguments. In this case, each attack $(a, b) \in \mathcal{R}$ gives birth to a constraint which says that the two variables a and b cannot take value 1 at the same time. This means that the two corresponding arguments cannot belong to the same set. This constraint has the following form: $((a, b), ((0, 0), (0, 1), (1, 0)))$. Note that this is equivalent to the cases where the propositional formula $a \Rightarrow \neg b$ is true (i.e. gets value 1). For simplicity reasons, throughout the paper we will use propositional formulas for encoding constraints. Solving a CSP amounts thus to finding the models of the set of constraints.

Definition 7 (Free CSP). *Let $\mathcal{F} = (\mathcal{A}, \mathcal{R})$ be an argumentation framework. A* free CSP *associated with $\mathcal{F}$ is a tuple $(\mathcal{X}, \mathcal{D}, \mathcal{C})$ where $\mathcal{X} = \mathcal{A}$, for each $a_i \in \mathcal{X}$, $\mathcal{D}_i = \{0, 1\}$ and $\mathcal{C} = \{a \Rightarrow \neg b \mid (b, a) \in \mathcal{R}\}$.*

It can be checked that $|\mathcal{C}| = |\mathcal{R}|$. The following result shows that the solutions of this CSP are the conflict-free sets of arguments of the corresponding AF.

Theorem 1. *Let $(\mathcal{X}, \mathcal{D}, \mathcal{C})$ be the CSP instance associated with the AF $\mathcal{F} = (\mathcal{A}, \mathcal{R})$. The tuple $(v(x_1), \ldots, v(x_n))$ is a solution of the CSP iff the set $\{x_j, \ldots, x_k\}$ s.t. $v(x_i) = 1$ is conflict-free (with $i = j \ldots, k$).*

Let us consider the argumentation framework $\mathcal{F}_1$ defined in Example 1.

Example 1 (Cont): The CSP corresponding to $\mathcal{F}_1$ is $(\mathcal{X}, \mathcal{D}, \mathcal{C})$ s.t. $\mathcal{X} = \{a, b, c, d\}$, $\mathcal{D} = \{\{0,1\}, \{0,1\}, \{0,1\}, \{0,1\}\}$, $\mathcal{C} = \{a \Rightarrow \neg d, b \Rightarrow \neg a, c \Rightarrow \neg b, d \Rightarrow \neg c\}$. This CSP has the following solutions: $(0,0,0,0)$, $(1,0,0,0)$, $(0,1,0,0)$, $(0,0,1,0)$, $(0,0,0,1)$, $(1,0,1,0)$ and $(0,1,0,1)$. Thus, the sets $\{\}$, $\{a\}$, $\{b\}$, $\{c\}$, $\{d\}$, $\{a,c\}$ and $\{b,d\}$ are conflict-free.

Let us now study the case of stable semantics. Stable extensions are computed by a CSP which considers that an argument and its attackers cannot have the same value.

Definition 8 (Stable CSP). *Let $\mathcal{F} = (\mathcal{A}, \mathcal{R})$ be an argumentation framework. A* stable CSP *associated with $\mathcal{F}$ is a tuple $(\mathcal{X}, \mathcal{D}, \mathcal{C})$ where $\mathcal{X} = \mathcal{A}$, $\forall a_i \in \mathcal{X}$, $\mathcal{D}_i = \{0,1\}$ and $\mathcal{C} = \{a \Leftrightarrow \bigwedge_{b:(b,a) \in \mathcal{R}} \neg b \mid a \in \mathcal{A}\}$.*

It is worth mentioning that the previous definition is inspired from [10].

Theorem 2. *Let $(\mathcal{X}, \mathcal{D}, \mathcal{C})$ be a stable CSP associated with $\mathcal{F} = (\mathcal{A}, \mathcal{R})$. The tuple $(v(x_1), \ldots, v(x_n))$ is a solution of the CSP iff the set $\{x_j, \ldots, x_k\}$ s.t. $v(x_i) = 1$ is a stable extension of $\mathcal{F}$.*

Let us illustrate this result on the following example.

Example 1 (Cont): The stable CSP corresponding to $\mathcal{F}_1$ is $(\mathcal{X}, \mathcal{D}, \mathcal{C})$ s.t. $\mathcal{X} = \{a, b, c, d\}$, $\mathcal{D} = \{\{0,1\}, \{0,1\}, \{0,1\}, \{0,1\}\}$, and $\mathcal{C} = \{a \Leftrightarrow \neg d, b \Leftrightarrow \neg a, c \Leftrightarrow \neg b, d \Leftrightarrow \neg c\}$. This CSP has two solutions: $(1,0,1,0)$ and $(0,1,0,1)$. The sets $\{a,c\}$ and $\{b,d\}$ are the two stable extensions of $\mathcal{F}_1$.

The two previous CSPs are simple since attacks are directly transformed into constraints. The notion of defence is not needed in both cases. However, things are not so obvious with admissible semantics. The following definition shows that a CSP which computes the admissible sets of an AF should consider both the attacks and the defence in its constraints.

Definition 9 (Admissible CSP). *Let $\mathcal{F} = (\mathcal{A}, \mathcal{R})$ be an argumentation framework. An* admissible CSP *associated with $\mathcal{F}$ is a tuple $(\mathcal{X}, \mathcal{D}, \mathcal{C})$ where $\mathcal{X} = \mathcal{A}$, for each $a_i \in \mathcal{X}$, $\mathcal{D}_i = \{0,1\}$ and $\mathcal{C} = \{(a \Rightarrow \bigwedge_{b:(b,a) \in \mathcal{R}} \neg b) \wedge (a \Rightarrow \bigwedge_{b:(b,a) \in \mathcal{R}} (\bigvee_{c:(c,b) \in \mathcal{R}} c)) \mid a \in \mathcal{A}\}$.*

The following result shows that the solutions of an admissible CSP provide the admissible extensions of the corresponding argumentation framework.

Theorem 3. *Let $(\mathcal{X}, \mathcal{D}, \mathcal{C})$ be an admissible CSP associated with an AF $\mathcal{F}$. $(v(x_1), \ldots, v(x_n))$ is a solution of the CSP iff the set $\{x_j, \ldots, x_k\}$ s.t. $v(x_i) = 1$ is an admissible set of $\mathcal{F}$.*

Let us illustrate the notion of admissible CSP with a simple example.

Example 2. Let us consider the framework $\mathcal{F}_2 = (\mathcal{A}_2, \mathcal{R}_2)$ where $\mathcal{A}_2 = \{a, b, c, d\}$ and $\mathcal{R}_2 = \{(c, b), (d, b), (b, a)\}$. The admissible CSP associated with $\mathcal{F}_2$ is $(\mathcal{X}, \mathcal{D}, \mathcal{C})$ where: $\mathcal{X} = \mathcal{A}_2$, $\mathcal{D} = \{\{0,1\}, \{0,1\}, \{0,1\}, \{0,1\}\}$ and $\mathcal{C} = \{d \Rightarrow \top$, $c \Rightarrow \top$, $b \Rightarrow \neg c \wedge \neg d$, $a \Rightarrow \neg b, a \Rightarrow c \vee d\}$. This CSP has the following solutions: $(0,0,0,0)$, $(0,0,1,0)$, $(0,0,0,1)$, $(0,0,1,1)$ $(1,0,1,0)$, $(1,0,0,1)$, $(1,0,1,1)$. These solutions return the admissible sets of $\mathcal{F}_2$, that is: $\{\}$, $\{c\}$, $\{d\}$, $\{c, d\}$, $\{a, c\}$, $\{a, d\}$ and $\{a, c, d\}$.

As preferred extensions are maximal (for set inclusion) admissible sets, then they are computed by an admissible CSP.

Theorem 4. *Let $(\mathcal{X}, \mathcal{D}, \mathcal{C})$ be an admissible CSP associated with an AF $\mathcal{F}$. Each maximal (for set inclusion) set $\{x_j, \ldots, x_k\}$, s.t. $v(x_i) = 1$ and $(v(x_1), \ldots, v(x_n))$ is a solution of the CSP, is a preferred extension of $\mathcal{F}$.*

Let us come back to Example 2.

Example 2 (Cont): It is clear that the last solution $(1,0,1,1)$ is the one which returns the only preferred extension of $\mathcal{F}_2$, i.e. $\{a, c, d\}$.

Complete extensions are also computed by a CSP which takes into account the notion of defence in the constraints.

Definition 10 (Complete CSP). *Let $\mathcal{F} = (\mathcal{A}, \mathcal{R})$ be an argumentation framework. A* complete CSP *associated with $\mathcal{F}$ is a tuple $(\mathcal{X}, \mathcal{D}, \mathcal{C})$ where $\mathcal{X} = \mathcal{A}$, for each $a_i \in \mathcal{X}$, $\mathcal{D}_i = \{0, 1\}$ and $\mathcal{C} = \{(a \Rightarrow \bigwedge_{b:(b,a)\in\mathcal{R}} \neg b) \wedge (a \Leftrightarrow \bigwedge_{b:(b,a)\in\mathcal{R}} (\bigvee_{c:(c,b)\in\mathcal{R}} c)) \mid a \in \mathcal{A}\}$.*

Note that there is a slight difference between the constraints of an admissible CSP and those of a complete CSP. Since a complete extension should contain all the arguments it defends, then an argument and all its defenders should be in the same set. However, the only requirement on an admissible set is that it defends its arguments. This is encoded by a simple logical implication.

Theorem 5. *Let $(\mathcal{X}, \mathcal{D}, \mathcal{C})$ be a complete CSP associated with an AF $\mathcal{F}$. The tuple $(v(x_1), \ldots, v(x_n))$ is a solution of the CSP iff the set $\{x_j, \ldots, x_k\}$, s.t. $v(x_i) = 1$ is a complete extension of $\mathcal{F}$.*

Example 2 (Cont): The complete CSP associated with $\mathcal{F}_2$ is $(\mathcal{X}, \mathcal{D}, \mathcal{C})$ where: $\mathcal{X} = \mathcal{A}_2, \mathcal{D} = \{\{0,1\}, \{0,1\}, \{0,1\}, \{0,1\}\}$ and $\mathcal{C} = \{d \Rightarrow \top, c \Rightarrow \top, b \Rightarrow \neg c \wedge \neg d$, $a \Rightarrow \neg b, a \Leftrightarrow c \vee d\}$. This CSP has one solution which is $(1,0,1,1)$. Thus, $\mathcal{F}_2$ has the set $\{a, c, d\}$ as its unique complete extension.

Since grounded extension is a minimal (for set inclusion) complete extension, then it is computed by a complete CSP as follows.

Theorem 6. *Let $(\mathcal{X}, \mathcal{D}, \mathcal{C})$ be a complete CSP associated with an AF $\mathcal{F}$. The grounded extension of $\mathcal{F}$ is the minimal (for set inclusion) set $\{x_j, \ldots, x_k\}$ s.t. $v(x_i) = 1$ and $(v(x_1), \ldots, v(x_n))$ is a solution of the CSP.*

Example 2 (Cont): The grounded extension of $\mathcal{F}_2$ is $\{a, c, d\}$ which is returned by the unique solution of the complete CSP corresponding to $\mathcal{F}_2$.

4 Constrained Frameworks

This section recalls the constrained version of Dung's argumentation framework and proposes its mappings to CSPs.

4.1 Basic Definitions

The basic argumentation framework of Dung has been extended in [9] by adding a *constraint* on arguments. This constraint should be satisfied by Dung's extensions (under a given semantics). For instance, in Example 1, one may imagine a constraint which requires that the two arguments a and c belong to the same extension. Note that this constraint is satisfied by $\mathcal{B}_1$ but not by $\mathcal{B}_2$. Thus, $\mathcal{B}_1$ would be the only extension of the framework.

The constraint is a formula of a propositional language $\mathcal{L}_\mathcal{A}$ whose alphabet is exactly the set $\mathcal{A}$ of arguments. Thus, each argument in $\mathcal{A}$ is a literal of $\mathcal{L}_\mathcal{A}$. $\mathcal{L}_\mathcal{A}$ contains all the formulas that can be built using the usual logical operators ($\wedge$, $\vee$, $\Rightarrow$, $\neg$, $\Leftrightarrow$) and the constant symbols ($\top$ and $\bot$).

Definition 11 (Constraint, Completion). *Let $\mathcal{A}$ be a set of arguments and $\mathcal{L}_\mathcal{A}$ its corresponding propositional language.*

- *$\mathscr{C}$ is a* constraint *on $\mathcal{A}$ iff $\mathscr{C}$ is a formula of $\mathcal{L}_\mathcal{A}$.*
- *The* completion *of a set $\mathcal{B} \subseteq \mathcal{A}$ is $\widehat{\mathcal{B}} = \{a \mid a \in \mathcal{B}\} \cup \{\neg a \mid a \in \mathcal{A} \setminus \mathcal{B}\}$.*
- *A set $\mathcal{B} \subseteq \mathcal{A}$* satisfies *$\mathscr{C}$ iff $\widehat{\mathcal{B}}$ is a model of $\mathscr{C}$ ($\widehat{\mathcal{B}} \models \mathscr{C}$).*

The completion of a set $\mathcal{B}$ of arguments is a set in which each argument of $\mathcal{A}$ appears either as a positive literal if the argument belongs to $\mathcal{B}$ or as a negative one otherwise. Thus, $|\widehat{\mathcal{B}}| = |\mathcal{A}|$.

A constrained argumentation framework (CAF) is defined as follows:

Definition 12 (CAF). *A* constrained argumentation framework *(CAF) is a triple $\mathcal{F} = (\mathcal{A}, \mathcal{R}, \mathscr{C})$ where $\mathcal{A}$ is a set of arguments, $\mathcal{R} \subseteq \mathcal{A} \times \mathcal{A}$ is an attack relation and $\mathscr{C}$ is a constraint on the set $\mathcal{A}$.*

Dung's semantics are extended to the case of CAFs. The idea is to compute Dung's extensions (under a given semantics), and to keep among those extensions only the ones that satisfy the constraint $\mathscr{C}$.

Definition 13 (C-admissible set). *Let $\mathcal{F} = (\mathcal{A}, \mathcal{R}, \mathscr{C})$ be a CAF and $\mathcal{B} \subseteq \mathcal{A}$. The set $\mathcal{B}$ is $\mathscr{C}$-*admissible *in $\mathcal{F}$ iff:*

1. *$\mathcal{B}$ is admissible,*
2. *$\mathcal{B}$ satisfies the constraint $\mathscr{C}$.*

In [15], it has been shown that the empty set is always admissible. However, it is not always $\mathscr{C}$-admissible since the set $\widehat{\varnothing}$ does not always imply $\mathscr{C}$.

Definition 14 (C-preferred, C-stable extension). *Let $\mathcal{F} = (\mathcal{A}, \mathcal{R}, \mathscr{C})$ be a CAF and $\mathcal{B} \subseteq \mathcal{A}$.*

- *$\mathcal{B}$ is a $\mathscr{C}$-*preferred extension *of $\mathcal{F}$ iff $\mathcal{B}$ is maximal for set-inclusion among the $\mathscr{C}$-admissible sets.*
- *$\mathcal{B}$ is a $\mathscr{C}$-*stable extension *of $\mathcal{F}$ iff $\mathcal{B}$ is a $\mathscr{C}$-preferred extension that attacks all arguments in $\mathcal{A} \backslash \mathcal{B}$.*

The following result summarizes the links between the extensions of a CAF $\mathcal{F} = (\mathcal{A}, \mathcal{R}, \mathscr{C})$ and those of its basic version $\mathcal{F}' = (\mathcal{A}, \mathcal{R})$.

Theorem 7. *[9] Let $\mathcal{F} = (\mathcal{A}, \mathcal{R}, \mathscr{C})$ be a CAF and $\mathcal{F}' = (\mathcal{A}, \mathcal{R})$ be its basic version.*

- *For each $\mathscr{C}$-preferred extension $\mathcal{B}$ of $\mathcal{F}$, there exists a preferred extension $\mathcal{B}'$ of $\mathcal{F}'$ s.t. $\mathcal{B} \subseteq \mathcal{B}'$.*
- *Every $\mathscr{C}$-stable extension of $\mathcal{F}$ is a stable extension of $\mathcal{F}'$. The converse does not hold.*

It is worth noticing that when the constraint of a CAF is a tautology, then the extensions of this CAF coincide with those of its basic version (i.e. the argumentation framework without the constraint).

Let us illustrate this notion of CAFs through a simple example.

Example 1 (Cont): Assume an extended version of the argumentation framework $\mathcal{F}_1$ where we would like to accept the two arguments a and c. This is encoded by a constraint $\mathscr{C} : a \wedge c$. It can be checked that the CAF $(\mathcal{A}_1, \mathcal{R}_1, \mathscr{C})$ has one $\mathscr{C}$-stable extension which is $\mathcal{B}_1 = \{a, c\}$. Note that $\mathcal{B}_2 = \{b, d\}$ is a stable extension of $\mathcal{F}_1$ but not a $\mathscr{C}$-stable extension of its constrained version.

4.2 Mappings into CSPs

Let $\mathcal{F} = (\mathcal{A}, \mathcal{R}, \mathscr{C})$ be a given CAF. In order to compute its $\mathscr{C}$-extensions under different semantics, we follow the same line of research as in the previous section. The only difference is that in addition to the constraints defined in Section 3.2, there is an additional constraint which is $\mathscr{C}$.

Let us start with $\mathscr{C}$-stable extensions. They are computed by the stable CSP given in Def. 8 augmented by the constraint $\mathscr{C}$ in its set $\mathcal{C}$.

Definition 15 ($\mathscr{C}$-stable CSP). *Let $\mathcal{F} = (\mathcal{A}, \mathcal{R}, \mathscr{C})$ be a constrained argumentation framework. A $\mathscr{C}-$*stable CSP *associated with $\mathcal{F}$ is a tuple $(\mathcal{X}, \mathcal{D}, \mathcal{C})$ where $\mathcal{X} = \mathcal{A}$, for each $a_i \in \mathcal{X}$, $\mathcal{D}_i = \{0, 1\}$ and $\mathcal{C} = \{\mathscr{C}\} \cup \{a \Leftrightarrow \bigwedge_{b:(b,a)\in\mathcal{R}} \neg b \mid a \in \mathcal{A}\}$.*

Note that the constraints in $\mathcal{C}$ are all propositional formulas built over a language $\mathcal{L}_{\mathcal{A}}$ whose alphabet is the set $\mathcal{A}$ of arguments. We show next that the solutions of a $\mathscr{C}$-stable CSP return all the $\mathscr{C}$-stable extensions of the corresponding CAF.

Theorem 8. *Let $(\mathcal{X}, \mathcal{D}, \mathcal{C})$ be a $\mathscr{C}$-stable CSP associated with a CAF $\mathcal{F} = (\mathcal{A}, \mathcal{R}, \mathscr{C})$. The tuple $(v(x_1), \ldots, v(x_n))$ is a solution of the CSP iff the set $\{x_j, \ldots, x_k\}$ such that $v(x_i) = 1$ is a $\mathscr{C}$-stable extension of $\mathcal{F}$.*

Example 1 (Cont): The $\mathscr{C}$-stable CSP associated with the CAF extending $\mathcal{F}_1$ with the constraint $\mathscr{C} : a \wedge c$ is $(\mathcal{X}, \mathcal{D}, \mathcal{C})$ s.t. $\mathcal{X} = \{a, b, c, d\}$, $\mathcal{D} = \{\{0, 1\}, \{0, 1\}, \{0, 1\}, \{0, 1\}\}$, and $\mathcal{C} = \{a \wedge c, a \Leftrightarrow \neg d, b \Leftrightarrow \neg a, c \Leftrightarrow \neg b, d \Leftrightarrow \neg c\}$. This CSP has one solution which is $(1, 0, 1, 0)$. It returns the $\mathscr{C}$-stable extension $\{a, c\}$ of the CAF.

A CSP which computes the $\mathscr{C}$-admissible sets of a CAF is grounded on the admissible CSP introduced in Definition 9.

Definition 16 ($\mathscr{C}$-admissible CSP). *Let $\mathcal{F} = (\mathcal{A}, \mathcal{R}, \mathscr{C})$ be a constrained argumentation framework. A $\mathscr{C}$-*admissible CSP *associated with $\mathcal{F}$ is a tuple $(\mathcal{X}, \mathcal{D}, \mathcal{C})$ where $\mathcal{X} = \mathcal{A}$, for each $a_i \in \mathcal{X}$, $\mathcal{D}_i = \{0, 1\}$ and $\mathcal{C} = \{\mathscr{C}\} \cup \{(a \Rightarrow \bigwedge_{b:(b,a)\in\mathcal{R}} \neg b) \wedge (a \Rightarrow \bigwedge_{b:(b,a)\in\mathcal{R}} (\bigvee_{c:(c,b)\in\mathcal{R}} c)) \mid a \in \mathcal{A}\}$.*

We show that the solutions of this CSP are $\mathscr{C}$-admissible extensions of the corresponding CAF.

Theorem 9. *Let $(\mathcal{X}, \mathcal{D}, \mathcal{C})$ be a $\mathscr{C}$-admissible CSP associated with a CAF $\mathcal{F} = (\mathcal{A}, \mathcal{R}, \mathscr{C})$. The tuple $(v(x_1), \ldots, v(x_n))$ is a solution of the CSP iff the set $\{x_j, \ldots, x_k\}$ s.t. $v(x_i) = 1$ is a $\mathscr{C}$-admissible set of the CAF $\mathcal{F}$.*

$\mathscr{C}$-preferred extensions are maximal (for set inclusion) admissible sets, then the following result follows from the previous one.

Theorem 10. *Let $(\mathcal{X}, \mathcal{D}, \mathcal{C})$ be a $\mathscr{C}$-admissible CSP associated with a CAF $\mathcal{F} = (\mathcal{A}, \mathcal{R}, \mathscr{C})$. Each maximal (for set inclusion) set $\{x_j, \ldots, x_k\}$, s.t. $v(x_i) = 1$ and $(v(x_1), \ldots, v(x_n))$ is a solution of the CSP, is a $\mathscr{C}$-preferred extension of $\mathcal{F}$.*

5 Preference-Based Frameworks

Is is well acknowledged in argumentation literature that arguments may not have the same strength. For instance, arguments built from certain information are stronger than arguments built from uncertain information. Consequently, in [2] Dung's framework has been extended in such a way to take into account the strengths of arguments when evaluating them. The idea is to consider in addition to the attack relation, another binary relation $\succeq$ which represents preferences between arguments. This relation can be instantiated in different ways. Writing $a \succeq b$ means that a is at least as good as b. Let $\succ$ be the strict relation associated with $\succeq$. It is defined as follows: $a \succ b$ iff $a \succeq b$ and not $b \succeq a$. In Dung's framework, an attack always succeeds (if the attacked argument is not defended). In preference-based frameworks, an attack may fail if the attacked argument is stronger than its attacker.

Definition 17 (PAF). *A preference-based argumentation framework (PAF) is a tuple $\mathcal{F} = (\mathcal{A}, \mathcal{R}, \succeq)$ where $\mathcal{A}$ is a set of arguments, $\mathcal{R} \subseteq \mathcal{A} \times \mathcal{A}$ is an attack relation and $\succeq$ is (partial or total) preorder on $\mathcal{A}$ ($\succeq \subseteq \mathcal{A} \times \mathcal{A}$).*
The extensions of $\mathcal{F}$ (under any semantics) are those of the AF $(\mathcal{A}, \texttt{Def})$ where $(a, b) \in \texttt{Def}$ iff $(a, b) \in \mathcal{R}$ and not$(b \succ a)$.

Let us now show how to compute the extensions of a PAF with a CSP. The following CSP computes the conflict-free sets of arguments in a PAF.

Definition 18. *Let $\mathcal{F} = (\mathcal{A}, \mathcal{R}, \succeq)$ be a PAF. A CSP associated with $\mathcal{F}$ is a tuple $(\mathcal{X}, \mathcal{D}, \mathcal{C})$ where $\mathcal{X} = \mathcal{A}$, for each $a_i \in \mathcal{X}$, $\mathcal{D}_i = \{0, 1\}$ and $\mathcal{C} = \{a \Rightarrow \neg b$ s.t. $(a, b) \in \mathcal{R}$ and not$(b \succ a)\}$.*

Theorem 11. *Let $(\mathcal{X}, \mathcal{D}, \mathcal{C})$ be a CSP instance associated with a PAF $\mathcal{F} = (\mathcal{A}, \mathcal{R}, \succeq)$. The tuple $(v(x_1), \ldots, v(x_n))$ is a solution of the CSP iff the set$\{x_j, \ldots, x_k\}$ s.t. $v(x_i) = 1$ is conflict-free in PAF $\mathcal{F}$.*

Stable extensions of a PAF are computed by a slightly modified version of stable CSP.

Definition 19 (Pref-stable CSP). *Let $\mathcal{F} = (\mathcal{A}, \mathcal{R}, \succeq)$ be a PAF. A* pref stable CSP *associated with $\mathcal{F}$ is a tuple $(\mathcal{X}, \mathcal{D}, \mathcal{C})$ where $\mathcal{X} = \mathcal{A}$, for each $a_i \in \mathcal{X}$, $\mathcal{D}_i = \{0, 1\}$ and $\mathcal{C} = \{a \Leftrightarrow \bigwedge_{b:(b,a) \in \mathcal{R} \text{ and } not(a \succ b)} \neg b \mid a \in \mathcal{A}\}$.*

Theorem 12. *Let $(\mathcal{X}, \mathcal{D}, \mathcal{C})$ be a pref stable CSP associated with a PAF $\mathcal{F} = (\mathcal{A}, \mathcal{R}, \succeq)$. The tuple $(v(x_1), \ldots, v(x_n))$ is a solution of this CSP iff the set $\{x_j, \ldots, x_k\}$ s.t. $v(x_i) = 1$ is a stable extension of $\mathcal{F}$.*

Example 1 (Cont): Assume a PAF with $\mathcal{A}_1$ as its set of arguments, $\mathcal{R}_1$ its attack relation and that $b \succ a$ and $d \succ c$. Its corresponding pref stable CSP is $(\mathcal{X}, \mathcal{D}, \mathcal{C})$ s.t. $\mathcal{X} = \{a, b, c, d\}$, $\mathcal{D} = \{\{0, 1\}, \{0, 1\}, \{0, 1\}, \{0, 1\}\}$, and $\mathcal{C} = \{a \Leftrightarrow \neg d, b \Leftrightarrow \top, c \Leftrightarrow \neg b, d \Leftrightarrow \top\}$. This CSP has one solution: $(0, 1, 0, 1)$. Thus, the set $\{b, d\}$ is the unique stable extensions of this PAF.

A CSP which computes the admissible sets of a PAF is an extended version of admissible CSP.

Definition 20 (Pref-admissible CSP). *Let* $\mathcal{F} = (\mathcal{A}, \mathcal{R}, \succeq)$ *be a PAF. A* pref-admissible CSP *associated with* $\mathcal{F}$ *is a tuple* $(\mathcal{X}, \mathcal{D}, \mathcal{C})$ *where* $\mathcal{X} = \mathcal{A}$*, for each* $a_i \in \mathcal{X}$*,* $\mathcal{D}_i = \{0, 1\}$ *and* $\mathcal{C} = \{(a \Rightarrow \bigwedge_{b:(b,a)\in\mathcal{R} \text{ and } not(a\succ b)} \neg b) \wedge (a \Rightarrow \bigwedge_{b:(b,a)\in\mathcal{R} \text{ and } not(a\succ b)} (\bigvee_{c:(c,b)\in\mathcal{R} \text{ and } not(b\succ c)} c)) \mid a \in \mathcal{A}\}$.

We show that the solutions of this CSP are admissible extensions of the corresponding PAF.

Theorem 13. *Let* $(\mathcal{X}, \mathcal{D}, \mathcal{C})$ *be a pref-admissible CSP associated with a PAF* $\mathcal{F}$*. The tuple* $(v(x_1), \ldots, v(x_n))$ *is a solution of this CSP iff the set* $\{x_j, \ldots, x_k\}$ *s.t.* $v(x_i) = 1$ *is an admissible set of* $\mathcal{F}$*.*

As preferred extensions are maximal (for set inclusion) admissible sets, then the following result follows from the previous one.

Theorem 14. *Let* $(\mathcal{X}, \mathcal{D}, \mathcal{C})$ *be a pref-admissible CSP associated with a PAF* $\mathcal{F}$*. Each maximal (for set inclusion) set* $\{x_j, \ldots, x_k\}$*, s.t.* $v(x_i) = 1$ *and* $(v(x_1), \ldots, v(x_n))$ *is a solution of the CSP, is a preferred extension of* $\mathcal{F}$*.*

Complete extensions are computed by a revised version of complete CSP.

Definition 21 (Pref-complete CSP). *Let* $\mathcal{F} = (\mathcal{A}, \mathcal{R}, \succeq)$ *be a PAF. A* pref-complete CSP *associated with* $\mathcal{F}$ *is a tuple* $(\mathcal{X}, \mathcal{D}, \mathcal{C})$ *where* $\mathcal{X} = \mathcal{A}$*, for each* $a_i \in \mathcal{X}$*,* $\mathcal{D}_i = \{0, 1\}$ *and* $\mathcal{C} = \{(a \Rightarrow \bigwedge_{b:(b,a)\in\mathcal{R} \text{ and } not(a\succ b)} \neg b) \wedge (a \Leftrightarrow \bigwedge_{b:(b,a)\in\mathcal{R} \text{ and } not(a\succ b)} (\bigvee_{c:(c,b)\in\mathcal{R} \text{ and } not(b\succ c)} c)) \mid a \in \mathcal{A}\}$.

Theorem 15. *Let* $(\mathcal{X}, \mathcal{D}, \mathcal{C})$ *be a pref-complete CSP associated with a PAF* $\mathcal{F}$*. The tuple* $(v(x_1), \ldots, v(x_n))$ *is a solution of the CSP iff the set* $\{x_j, \ldots, x_k\}$*, s.t.* $v(x_i) = 1$ *is a complete extension of* $\mathcal{F}$*.*

The grounded extension of a PAF is computed by the pref-complete CSP as follows.

Theorem 16. *Let* $(\mathcal{X}, \mathcal{D}, \mathcal{C})$ *be a pref-complete CSP associated with a PAF* $\mathcal{F}$*. The grounded extension of* $\mathcal{F}$ *is the minimal (for set inclusion) set* $\{x_j, \ldots, x_k\}$ *s.t.* $v(x_i) = 1$ *and* $(v(x_1), \ldots, v(x_n))$ *is a solution of the CSP.*

6 Related Work

There are very few attempts in the literature for modeling argumentation frameworks as a CSP. To the best of our knowledge, the only works on the topic are [4,5].

In [5], the authors have studied the problem of encoding *weighted* argumentation frameworks by semirings. In a weighted framework, attacks do not necessarily have the same weights. Thus, a weight (i.e. a value between 0 and 1) is associated with each attack between two arguments. When all the attacks have weight 1, the corresponding framework collapses with Dung's abstract framework recalled in Section 3.

In [5], it has been shown how to compute stable and complete extensions by semirings. In our paper, we have proposed an alternative approach for computing those semantics and other semantics (like preferred and grounded semantics). The approach is simpler and more natural. While in [5], the authors have used soft CSP, in our paper we have used simple CSP. Moreover, we have studied more semantics and two extended versions of Dung's framework: the constrained version proposed in [9] and the preferred version proposed in [2].

The works presented in [4,9] are closer to our. In these papers, the authors have encoded Dung's framework as a satisfiability problem (SAT). In [7], it has been shown that SAT is a particular case of CSPs and a mapping from SAT to CSP has been given. In our paper, we took advantage of that mapping and we presented different CSPs which encode Dung's semantics not only for Dung's framework, but also for constrained frameworks and preference-based frameworks.

In [18] an implementation of Dung's semantics using answer set programming (ASP) has been provided. Thus, it is complementary to our work. Moreover, the ASP literature has shown that there are links between ASP and CSP.

7 Conclusion

In this paper, we have expressed the problem of computing the extensions of an argumentation framework under a given semantics as a CSP. We have investigated three types of frameworks: Dung's argumentation framework [15], its constrained version proposed in [9], and its extension with preferences [2]. For each of these frameworks, we have proposed different CSPs which compute their extensions under various semantics, namely admissible, preferred, stable, complete and grounded.

Such mappings are of great importance since they allow the use of the *efficient solvers* that have been developed by CSP community. Thus, the efficiency of our different CSPs depend on that of the solver that is chosen to solve them. Note also that the CSP version of Dung's argumentation framework is as simple as this latter since a CSP can be represented as a graph.

It is worth mentioning that in the particular case of grounded semantics, there is an additional test of minimality that is required after computing the solutions of the corresponding CSP. This increases thus the complexity of computing the grounded extension of an argumentation framework. Consequently, this particular extension should be computed using existing algorithms in argumentation literature [1] and not by a CSP.

There are a number of ways to extend this work. One future direction consists of proposing the CSPs that return other semantics like semi-stable [6] and ideal [14]. Another idea consists of encoding weighted argumentation frameworks [16] as CSPs. In a weighted framework, attacks may not have the same importance. Such framework can be encoded by valued CSP in which constraints are associated with weights.

References

1. Rahwan, I., Simari, G. (eds.): Argumentation in Artificial Intelligence. Springer, Heidelberg (2009)
2. Amgoud, L., Cayrol, C.: A reasoning model based on the production of acceptable arguments. Annals of Mathematics and Artificial Intelligence 34, 197–216 (2002)
3. Bench-Capon, T.J.M.: Persuasion in practical argument using value-based argumentation frameworks. Journal of Logic and Computation 13(3), 429–448 (2003)
4. Besnard, P., Doutre, S.: Checking the acceptability of a set of arguments. In: NMR, pp. 59–64 (2004)
5. Bistarelli, S., Santini, F.: A common computational framework for semiring-based argumentation systems. In: ECAI, pp. 131–136 (2010)
6. Caminada, M.: Semi-stable semantics. In: Proceedings of the 1st International Conference on Computational Models of Argument (COMMA 2006), pp. 121–130 (2006)
7. Castell, T., Fargier, H.: Propositional satisfaction problems and clausal csps. In: ECAI, pp. 214–218 (1998)
8. Cayrol, C., Doutre, S., Mengin, J.: On decision problems related to the preferred semantics for argumentation frameworks. Journal of Logic and Computation 13(3), 377–403 (2003)
9. Coste-Marquis, S., Devred, C., Marquis, P.: Constrained argumentation frameworks. In: KR, pp. 112–122 (2006)
10. Creignou, N.: The class of problems that are linearly equivalent to satisfiability or a uniform method for proving np-completeness. Theor. Comput. Sci. 145(1&2), 111–145 (1995)
11. Devred, C., Doutre, S., Lefèvre, C., Nicolas, P.: Dialectical proofs for constrained argumentation. In: COMMA, pp. 159–170 (2010)
12. Dimopoulos, Y., Nebel, B., Toni, F.: Preferred arguments are harder to compute than stable extensions. In: IJCAI 1999, pp. 36–43 (1999)
13. Dimopoulos, Y., Nebel, B., Toni, F.: Finding admissible and preferred arguments can be very hard. In: KR 2000, pp. 53–61 (2000)
14. Dung, P.M., Mancarella, P., Toni, F.: Computing ideal skeptical argumentation. Artificial Intelligence Journal 171, 642–674 (2007)
15. Dung, P.M.: On the acceptability of arguments and its fundamental role in non-monotonic reasoning, logic programming and n-person games. Artificial Intelligence Journal 77, 321–357 (1995)
16. Dunne, P., Hunter, A., McBurney, P., Parsons, S., Wooldridge, M.: Inconsistency tolerance in weighted argument systems. In: AAMAS, pp. 851–858 (2009)
17. Dunne, P., Wooldridge, M.: Complexity of abstract argumentation. In: Rahwan, I., Simari, G. (eds.) Argumentation in Artificial Intelligence. ch.4, pp. 85–104 (2009)
18. Egly, U., Gaggl, S., Woltran, S.: Answer-set programming encodings for argumentation frameworks. In: Technical report DBAI-TR-2008-62, Technische Universitat Wien (2008)
19. Kumar, V.: Depth-first search. Encyclopaedia of Artificial Intelligence 2, 1004–1005 (1987)

On the Equivalence of Logic-Based Argumentation Systems

Leila Amgoud and Srdjan Vesic

IRIT – CNRS, 118 route de Narbonne
31062 Toulouse Cedex 9, France
{amgoud,vesic}@irit.fr

Abstract. Equivalence between two argumentation systems means mainly that the two systems return the same outputs. It can be used for different purposes, namely in order to show whether two systems that are built over the same knowledge base but with distinct attack relations return the same outputs, and more importantly to check whether an infinite system can be reduced into a finite one.

Recently, the equivalence between abstract argumentation systems was investigated. Two categories of equivalence criteria were particularly proposed. The first category compares directly the outputs of the two systems (e.g. their extensions) while the second compares the outputs of their extended versions (i.e. the systems augmented by the same set of arguments). It was shown that only identical systems are equivalent w.r.t. those criteria.

In this paper, we study when two logic-based argumentation systems are equivalent. We refine existing criteria by considering the internal structure of arguments and propose new ones. Then, we identify cases where two systems are equivalent. In particular, we show that under some reasonable conditions on the logic underlying an argumentation system, the latter has an equivalent finite subsystem. This subsystem constitutes a threshold under which arguments of the system have not yet attained their final status and consequently adding a new argument may result in status change. From that threshold, the statuses of all arguments become stable.

1 Introduction

One of the most abstract argumentation systems was proposed by Dung [6]. It consists of a *set of arguments* and a binary relation representing *conflicts* among arguments. Those conflicts are then solved using a *semantics* which amounts to define acceptable sets of arguments, called *extensions*. From the extensions, a *status* is assigned to each argument. An argument is *skeptically* accepted if it appears in each extension, it is *credulously* accepted if it belongs to at least one extension, and finally it is *rejected* if it is not in any extension.

Several works were done on this system. Some of them extended it with new features like preferences between arguments (e.g. [2,4]) or weights on attacks (e.g. [7]), others defined new semantics that solve some problems encountered with Dung's ones (e.g. [3,5]) and another category of works instantiated the system for application purposes. More recently, the question of *equivalence* between two abstract argumentation systems

S. Benferhat and J. Grant (Eds.): SUM 2011, LNAI 6929, pp. 123–136, 2011.
© Springer-Verlag Berlin Heidelberg 2011

was tackled by Oikarinen and Woltran [9]. To the best of our knowledge this is the only work on this issue. The authors proposed two kinds of equivalence: *basic equivalence* and *strong equivalence*. According to basic equivalence, two systems are equivalent if they have the same extensions (resp. the same sets of skeptically/credulously accepted arguments). However, these criteria were not studied by Oikarinen and Woltran. Instead, they concentrated on strong equivalence. Two systems are strongly equivalent if they have the same extensions (resp. the same sets of skeptically/credulously accepted arguments) even after extending both systems by any set of arguments. The authors investigated under which conditions two systems are strongly equivalent. They have shown that when there are no self-attacking arguments, which is the case in most argumentation systems, and particularly in most logic-based argumentation systems as shown by Amgoud and Besnard [1], then two systems are strongly equivalent if and only if they coincide, i.e. they are the same. This makes the notion of strong equivalence a nice theoretical property, but without any practical applications.

In this paper, we study when two logic-based argumentation systems are equivalent. We refine existing criteria by considering the internal structure of arguments and propose new ones. We identify interesting cases where two systems are equivalent. In particular, we show that under some reasonable conditions on the logic underlying an argumentation system, the latter has an equivalent finite subsystem, which constitutes a threshold under which arguments of the system have not yet attained their final status and consequently any new argument may result in status change. From that threshold, the statuses of all arguments become stable.

The paper is structured as follows: in Section 2, we recall the logic-based argumentation systems we are interested in. In Section 3, we propose three equivalence criteria that refine the basic ones and study when two systems are equivalent w.r.t. each criterion. In Section 4, we refine the three criteria of strong equivalence and give the conditions under which they hold. Section 5 studies when the status of an argument may change when a new argument is received or removed from a system. The last section is devoted to some concluding remarks and perspectives. All the proofs are put in an appendix.

2 Logic-Based Argumentation Systems

This section describes the logical instantiations of Dung's argumentation system we are interested in. They are built around *any* monotonic logic whose consequence operator satisfies the five postulates proposed by Tarski [10]. Indeed, according to those postulates, a *monotonic logic* is a pair $(\mathcal{L}, \mathtt{CN})$ where $\mathcal{L}$ is any set of *well-formed formulae* and $\mathtt{CN}$ is a *consequence operator*, i.e. a function from $2^{\mathcal{L}}$ to $2^{\mathcal{L}}$ that satisfies the following five postulates:

- $X \subseteq \mathtt{CN}(X)$ **(Expansion)**
- $\mathtt{CN}(\mathtt{CN}(X)) = \mathtt{CN}(X)$ **(Idempotence)**
- $\mathtt{CN}(X) = \bigcup_{Y \subseteq_f X} \mathtt{CN}(Y)$[1] **(Finiteness)**
- $\mathtt{CN}(\{x\}) = \mathcal{L}$ for some $x \in \mathcal{L}$ **(Absurdity)**
- $\mathtt{CN}(\emptyset) \neq \mathcal{L}$ **(Coherence)**

[1] The notation $Y \subseteq_f X$ means that Y is a finite subset of X.

Intuitively, $\text{CN}(X)$ returns the set of formulae that are logical consequences of X according to the logic at hand. Almost all well-known logics (classical logic, intuitionistic logic, modal logics, ...) are special cases of Tarski's notion of monotonic logic. In such a logic, a set X of formulae is *consistent* iff its set of consequences is not the set $\mathcal{L}$. For two formulae $x, y \in \mathcal{L}$, we say that x and y are equivalent, denoted by $x \equiv y$, iff $\text{CN}(\{x\}) = \text{CN}(\{y\})$. Arguments are built from a *knowledge base* Σ which is a finite subset of $\mathcal{L}$.

Definition 1 (Argument). *Let* $(\mathcal{L}, \text{CN})$ *be a Tarskian logic and* $\Sigma \subseteq \mathcal{L}$. *An* argument *built from* Σ *is a pair* (X, x) *s.t.*

- $X \subseteq \Sigma$,
- X *is consistent,*
- $x \in \text{CN}(X)$,
- $\nexists X' \subset X$ *s.t.* $x \in \text{CN}(X')$.

X *is the* support *of the argument and* x *its* conclusion.

Notations: For an argument $a = (X, x)$, $\text{Conc}(a) = x$ and $\text{Supp}(a) = X$. For a set $\mathcal{S} \subseteq \mathcal{L}$, $\text{Arg}(\mathcal{S}) = \{a \mid a$ is an argument (in the sense of Definition 1) and $\text{Supp}(a) \subseteq \mathcal{S}\}$. The set of all arguments that can be built from the language $\mathcal{L}$ will be denoted by $\text{Arg}(\mathcal{L})$. For any $\mathcal{E} \subseteq \text{Arg}(\mathcal{L})$, $\text{Base}(\mathcal{E}) = \bigcup_{a \in \mathcal{E}} \text{Supp}(a)$.

The previous definition specified what we accept as an argument. An attack relation $\mathcal{R}$ is defined on a given set $\mathcal{A}$ of arguments, i.e. $\mathcal{R} \subseteq \mathcal{A} \times \mathcal{A}$. The writing $a\mathcal{R}b$ or $(a, b) \in \mathcal{R}$ means that argument a *attacks* argument b. A study on how to choose an appropriate attack relation was recently carried out by Amgoud and Besnard [1]. Some basic properties of an attack relation were also discussed by Gorogiannis and Hunter [8]. Examples of such properties are recalled below.

C1 $\forall a, b, c \in \mathcal{A}$, if $\text{Conc}(a) \equiv \text{Conc}(b)$ then $a\mathcal{R}c$ iff $b\mathcal{R}c$
C2 $\forall a, b, c \in \mathcal{A}$, if $\text{Supp}(a) = \text{Supp}(b)$ then $c\mathcal{R}a$ iff $c\mathcal{R}b$

The first property says that two arguments having equivalent conclusions attack exactly the same arguments. The second property says that arguments having the same supports are attacked by the same arguments. In this paper, we study attack relations verifying these two properties. That is, from now on, we suppose that an attack relation verifies $C1$ and $C2$.

An argumentation system is defined as follows.

Definition 2 (Argumentation system). *An* argumentation system *(AS) built from a knowledge base* Σ *is a pair* $\mathcal{F} = (\mathcal{A}, \mathcal{R})$ *where* $\mathcal{A} \subseteq \text{Arg}(\Sigma)$ *and* $\mathcal{R} \subseteq \mathcal{A} \times \mathcal{A}$ *is an attack relation which verifies* $C1$ *and* $C2$.

In the rest of the paper, we do *not* implicitly suppose that two arbitrary AS are built from the *same* knowledge base. We also assume that arguments are evaluated using stable semantics. Note that this is not a substantial limitation since the main purpose of this paper is to explore equivalence and strong equivalence in logic-based argumentation and not to study the subtleties of different semantics. For all the main results of this paper, similar ones can be proved for all well-known semantics.

Definition 3 (Stable semantics). *Let $\mathcal{F} = (\mathcal{A}, \mathcal{R})$ be an AS and $\mathcal{E} \subseteq \mathcal{A}$.*

- *$\mathcal{E}$ is* conflict-free *iff $\nexists a, b \in \mathcal{E}$ s.t. $a\mathcal{R}b$.*
- *$\mathcal{E}$ is a* stable extension *iff $\mathcal{E}$ is conflict-free and attacks any argument in $\mathcal{A} \setminus \mathcal{E}$.*

Let $\mathtt{Ext}(\mathcal{F})$ denote the set of all the stable extensions of $\mathcal{F}$.

A status is assigned to each argument as follows.

Definition 4 (Status of arguments). *Let $\mathcal{F} = (\mathcal{A}, \mathcal{R})$ be an AS and $a \in \mathcal{A}$.*

- *a is* skeptically accepted *iff $\mathtt{Ext}(\mathcal{F}) \neq \emptyset$ and $\forall \mathcal{E} \in \mathtt{Ext}(\mathcal{F})$, $a \in \mathcal{E}$*
- *a is* credulously accepted *iff $\exists\, \mathcal{E} \in \mathtt{Ext}(\mathcal{F})$ s.t. $a \in \mathcal{E}$*
- *a is* rejected *iff $\nexists \mathcal{E} \in \mathtt{Ext}(\mathcal{F})$ s.t. $a \in \mathcal{E}$*

Note that there are three possible statuses of an argument. An argument is either 1) skeptically and credulously accepted, or 2) only credulously accepted, or 3) rejected. Let $\mathtt{Status}(a, \mathcal{F})$ be a function which returns the status of an argument a in an AS $\mathcal{F}$. We assume that this function returns three different values corresponding to the three possible situations.

Property 1. Let $\mathcal{F} = (\mathcal{A}, \mathcal{R})$ be an argumentation system and $a, a' \in \mathcal{A}$. If $\mathtt{Supp}(a) = \mathtt{Supp}(a')$, then $\mathtt{Status}(a, \mathcal{F}) = \mathtt{Status}(a', \mathcal{F})$.

In addition to extensions and the status of arguments, other outputs are returned by an AS. These are summarized in the next definition.

Definition 5 (Outputs of an AS). *Let $\mathcal{F} = (\mathcal{A}, \mathcal{R})$ be an AS built over a knowledge base Σ.*

- $\mathtt{Sc}(\mathcal{F}) = \{a \in \mathcal{A} \mid a \textit{ is skeptically accepted}\}$
- $\mathtt{Cr}(\mathcal{F}) = \{a \in \mathcal{A} \mid a \textit{ is credulously accepted}\}$
- $\mathtt{Output}_{sc}(\mathcal{F}) = \{\mathtt{Conc}(a) \mid a \textit{ is skeptically accepted}\}$
- $\mathtt{Output}_{cr}(\mathcal{F}) = \{\mathtt{Conc}(a) \mid a \textit{ is credulously accepted}\}$
- $\mathtt{Bases}(\mathcal{F}) = \{\mathtt{Base}(\mathcal{E}) \mid \mathcal{E} \in \mathtt{Ext}(\mathcal{F})\}$

3 Basic Equivalence of Argumentation Systems

Three criteria for the notion of basic equivalence were proposed [9]. They compare the outputs of systems as follows. Let $\mathcal{F} = (\mathcal{A}, \mathcal{R})$ and $\mathcal{F}' = (\mathcal{A}', \mathcal{R}')$ be two argumentation systems. The following three criteria are used:

- $\mathtt{Ext}(\mathcal{F}) = \mathtt{Ext}(\mathcal{F}')$
- $\mathtt{Sc}(\mathcal{F}) = \mathtt{Sc}(\mathcal{F}')$
- $\mathtt{Cr}(\mathcal{F}) = \mathtt{Cr}(\mathcal{F}')$

While these criteria are meaningful, they are too rigid. Let us consider two argumentation systems grounded on propositional logic. Assume that the first system has one stable extension which is $\{(\{x\}, x)\}$ while the second system has $\{(\{x\}, x \wedge x)\}$ as its unique stable extension. According to the three previous criteria, the two systems are not equivalent. In what follows, we refine the three criteria by taking into account the internal structure of arguments via a notion of equivalent arguments.

Definition 6 (Equivalent arguments). *For two arguments $a, a' \in \texttt{Arg}(\mathcal{L})$, a is* equivalent *to a', denoted by $a \approx a'$, iff* $\texttt{Supp}(a) = \texttt{Supp}(a')$ *and* $\texttt{Conc}(a) \equiv \texttt{Conc}(a')$.

Note that this relation of equivalence was also used by Gorogiannis and Hunter [8].

The following property shows that equivalent arguments w.r.t. relation $\approx$ behave in the same way w.r.t. attacks.

Property 2. Let $\mathcal{F} = (\mathcal{A}, \mathcal{R})$ be an argumentation system. For all $a, a', b, b' \in \mathcal{A}$, if $a \approx a'$ and $b \approx b'$, then $a\mathcal{R}b$ iff $a'\mathcal{R}b'$.

Note that relation $\approx$ is an *equivalence relation* (i.e. reflexive, symmetric and transitive). The equivalence between two arguments is extended to equivalence between sets of arguments as follows.

Definition 7 (Equivalent sets of arguments). *Let $\mathcal{E}, \mathcal{E}' \subseteq \texttt{Arg}(\mathcal{L})$. $\mathcal{E}$ is* equivalent *to $\mathcal{E}'$, denoted by $\mathcal{E} \sim \mathcal{E}'$, iff $\forall a \in \mathcal{E}, \exists a' \in \mathcal{E}'$ s.t. $a \approx a'$ and $\forall a' \in \mathcal{E}', \exists a \in \mathcal{E}$ s.t. $a \approx a'$.*

We can now define a flexible notion of equivalence between argumentation systems.

Definition 8 (Equivalence between two AS). *Let $\mathcal{F} = (\mathcal{A}, \mathcal{R})$ and $\mathcal{F}' = (\mathcal{A}', \mathcal{R}')$ be two argumentation systems grounded on the same logic $(\mathcal{L}, \texttt{CN})$. The two systems $\mathcal{F}$ and $\mathcal{F}'$ are* EQi-equivalent *iff criterion EQi below holds:*

EQ1 $\exists f : \texttt{Ext}(\mathcal{F}) \to \texttt{Ext}(\mathcal{F}')$ *s.t. f is a bijection and* $\forall \mathcal{E} \in \texttt{Ext}(\mathcal{F}), \mathcal{E} \sim f(\mathcal{E})$
EQ2 $\texttt{Sc}(\mathcal{F}) \sim \texttt{Sc}(\mathcal{F}')$
EQ3 $\texttt{Cr}(\mathcal{F}) \sim \texttt{Cr}(\mathcal{F}')$

For two equivalent argumentation systems $\mathcal{F}$ and $\mathcal{F}'$, we will write $\mathcal{F} \equiv_{EQX} \mathcal{F}'$, with $X \in \{1, 2, 3\}$.

It is easy to show that each criterion EQi refines one criterion among those proposed by Oikarinen and Woltran.

Property 3. Let $\mathcal{F}$ and $\mathcal{F}'$ be two argumentation systems grounded on the same logic $(\mathcal{L}, \texttt{CN})$.

- If $\texttt{Ext}(\mathcal{F}) = \texttt{Ext}(\mathcal{F}')$, then $\mathcal{F} \equiv_{EQ1} \mathcal{F}'$.
- If $\texttt{Sc}(\mathcal{F}) = \texttt{Sc}(\mathcal{F}')$, then $\mathcal{F} \equiv_{EQ2} \mathcal{F}'$.
- If $\texttt{Cr}(\mathcal{F}) = \texttt{Cr}(\mathcal{F}')$, then $\mathcal{F} \equiv_{EQ3} \mathcal{F}'$.

Note that the converses are not always true. We show also that when two systems are equivalent w.r.t. EQ1, then they are also equivalent w.r.t. EQ2 and EQ3. This means that criterion EQ1 is more general than the others.

Theorem 1. *Let $\mathcal{F}$ and $\mathcal{F}$ be two argumentation systems. If $\mathcal{F} \equiv_{EQ1} \mathcal{F}'$, then $\mathcal{F} \equiv_{EQ2} \mathcal{F}'$ and $\mathcal{F} \equiv_{EQ3} \mathcal{F}'$.*

It can also be checked that equivalent arguments from equivalent systems have the same status.

Theorem 2. *Let $\mathcal{F} = (\mathcal{A}, \mathcal{R}), \mathcal{F}' = (\mathcal{A}', \mathcal{R}')$ be two argumentation systems. If $\mathcal{F} \equiv_{EQ1} \mathcal{F}'$, then for all $a \in \mathcal{A}$ and for all $a' \in \mathcal{A}'$, if $a \approx a'$ then* $\texttt{Status}(a, \mathcal{F}) = \texttt{Status}(a', \mathcal{F}')$.

In order to show that outputs of equivalent systems are equivalent as well, we need the following notion.

Definition 9 (Equivalent sets of formulae). *Let $X, Y \subseteq \mathcal{L}$. We say that X and Y are equivalent, denoted by $X \cong Y$, iff $\forall x \in X, \exists y \in Y$ s.t. $x \equiv y$ and $\forall y \in Y, \exists x \in X$ s.t. $x \equiv y$.*

For example, in case of the propositional logic, this allows to say that the two sets $\{x, \neg\neg y\}$ and $\{x, y\}$ are equivalent. Note that if $X \cong Y$, then $\mathtt{CN}(X) = \mathtt{CN}(Y)$. However, the converse is not true. For instance, $\mathtt{CN}(\{x \wedge y\}) = \mathtt{CN}(\{x, y\})$ while $\{x \wedge y\} \ncong \{x, y\}$. One may ask why not to use the equality of $\mathtt{CN}(X)$ and $\mathtt{CN}(Y)$ in order to say that X and Y are equivalent? The answer is given by the following example of two AS whose credulous conclusions are respectively $\{x, \neg x\}$ and $\{y, \neg y\}$. It is clear that $\mathtt{CN}(\{x, \neg x\}) = \mathtt{CN}(\{y, \neg y\})$ while the two sets are different.

The next result shows that if two argumentation systems are equivalent w.r.t. $EQ1$, then their sets of skeptical (credulous) conclusions are equivalent, and the bases of their extensions coincide (i.e. are the same).

Theorem 3. *Let $\mathcal{F} = (\mathcal{A}, \mathcal{R})$ and $\mathcal{F}' = (\mathcal{A}', \mathcal{R}')$ be two AS. If $\mathcal{F} \equiv_{EQ1} \mathcal{F}'$, then:*

- $\mathtt{Output}_{sc}(\mathcal{F}) \cong \mathtt{Output}_{sc}(\mathcal{F}')$
- $\mathtt{Output}_{cr}(\mathcal{F}) \cong \mathtt{Output}_{cr}(\mathcal{F}')$
- $\mathtt{Bases}(\mathcal{F}) = \mathtt{Bases}(\mathcal{F}')$

Since equivalent systems preserve all their important outputs, then we can exchange a given system with an equivalent one. In what follows, we show how we can take advantage of this notion of equivalence in order to reduce the number of arguments in an AS. The idea is to take exactly one argument from each equivalence class of $\mathcal{A}/\approx$. A resulting system is called *core*. Let X be a given set and $\sim$ an equivalence relation on it. For all $x \in X$, we write $[x] = \{x' \in X \mid x' \sim x\}$ and $X/\sim\, = \{[x] \mid x \in X\}$.

Definition 10 (Core). *Let $\mathcal{F} = (\mathcal{A}, \mathcal{R})$ be an argumentation system. An argumentation system $\mathcal{F}' = (\mathcal{A}', \mathcal{R}')$ is a* core *of $\mathcal{F}$ iff:*

- $\mathcal{A}' \subseteq \mathcal{A}$
- $\forall C \in \mathcal{A}/\approx,\ |C \cap \mathcal{A}'| = 1$
- $\mathcal{R}' = \mathcal{R}_{|\mathcal{A}'}$, *where* $\mathcal{R}_{|\mathcal{A}'} = \{(a, b) \mid (a, b) \in \mathcal{R}$ *and* $a, b \in \mathcal{A}'\}$, *i.e. the* restriction *of $\mathcal{R}$ on $\mathcal{A}'$.*

The fact that at least one representative of each equivalence class is included in a core allows us to show that any core of an AS is equivalent with the latter.

Theorem 4. *If $\mathcal{F}'$ is a core of an argumentation system $\mathcal{F}$, then $\mathcal{F} \equiv_{EQ1} \mathcal{F}'$.*

We now provide a condition which guarantees that any core of any argumentation system built from a finite knowledge base is finite. This is the case for logics in which any consistent finite set of formulae has finitely many logically non-equivalent consequences. To formalize this, we use the following notation for a set of logical consequences made from consistent subsets of a given set: For any $X \subseteq \mathcal{L}$, $\mathtt{Cncs}(X) = \{x \in \mathcal{L} \mid \exists Y \subseteq X$ s.t. $\mathtt{CN}(Y) \neq \mathcal{L}$ and $x \in \mathtt{CN}(Y)\}$. We show that if $\mathtt{Cncs}(\Sigma)$ has a finite number of equivalence classes, then any core of $\mathcal{F}$ is finite (i.e. with a finite set of arguments).

Theorem 5. *Let $\mathcal{F} = (\mathcal{A}, \mathcal{R})$ be an argumentation system built from a finite knowledge base Σ. If $\mathtt{Cncs}(\Sigma)/\equiv$ is finite, then any core of $\mathcal{F}$ is finite.*

This result is of great importance since it shows that instead of working with an infinite argumentation system which is costly, one can focus only on its core which is finite. Recall that generally, logic-based argumentation systems are infinite. This is for instance the case of systems that are grounded on propositional logic.

4 Strong Equivalence of Argumentation Systems

In this section, we study strong equivalence between logic-based argumentation systems. As mentioned before, two argumentation systems are strongly equivalent iff after adding the same set of arguments to both systems, the new systems are equivalent w.r.t. any of the basic criteria given in Definition 8.

Recall that $\mathtt{Arg}(\mathcal{L})$ is the set of all arguments that can be built from a logical language $(\mathcal{L}, \mathtt{CN})$. Let $\mathcal{R}(\mathcal{L})$ be an attack relation on the set $\mathtt{Arg}(\mathcal{L})$, i.e. $\mathcal{R}(\mathcal{L}) \subseteq \mathtt{Arg}(\mathcal{L}) \times \mathtt{Arg}(\mathcal{L})$. As in the first part of the paper, we assume that $\mathcal{R}(\mathcal{L})$ verifies properties $C1$ and $C2$.

Let $\mathcal{F} = (\mathcal{A}, \mathcal{R})$ be an argumentation system where $\mathcal{A} \subseteq \mathtt{Arg}(\mathcal{L})$ and $\mathcal{R} = \mathcal{R}(\mathcal{L})_{|\mathcal{A}}$. Augmenting $\mathcal{F}$ by an arbitrary set $\mathcal{B}$ of arguments ($\mathcal{B} \subseteq \mathtt{Arg}(\mathcal{L})$) results in a new system denoted by $\mathcal{F} \oplus \mathcal{B}$, where $\mathcal{F} \oplus \mathcal{B} = (\mathcal{A}_b, \mathcal{R}_b)$ with $\mathcal{A}_b = \mathcal{A} \cup \mathcal{B}$ and $\mathcal{R}_b = \mathcal{R}(\mathcal{L})_{|\mathcal{A}_b}$.

Definition 11 (Strong equivalence between two AS). *Let $\mathcal{F} = (\mathcal{A}, \mathcal{R})$ and $\mathcal{F}' = (\mathcal{A}', \mathcal{R}')$ be two argumentation systems built using the same logic $(\mathcal{L}, \mathtt{CN})$. The two systems $\mathcal{F}$ and $\mathcal{F}'$ are* EQi-strongly equivalent *iff criterion EQiS below holds:*

EQ1S $\forall \mathcal{B} \subseteq \mathtt{Arg}(\mathcal{L})$, $\mathcal{F} \oplus \mathcal{B} \equiv_{EQ1} \mathcal{F}' \oplus \mathcal{B}$
EQ2S $\forall \mathcal{B} \subseteq \mathtt{Arg}(\mathcal{L})$, $\mathcal{F} \oplus \mathcal{B} \equiv_{EQ2} \mathcal{F}' \oplus \mathcal{B}$
EQ3S $\forall \mathcal{B} \subseteq \mathtt{Arg}(\mathcal{L})$, $\mathcal{F} \oplus \mathcal{B} \equiv_{EQ3} \mathcal{F}' \oplus \mathcal{B}$.

In the remainder of the paper, we will use the terms 'strongly equivalent w.r.t. EQi' and 'equivalent w.r.t. EQiS' to denote the same thing (where $i \in \{1, 2, 3\}$).

Property 4. If two argumentation systems are strongly equivalent w.r.t. EQ1S (resp. EQ2S, EQ3S), then they are equivalent w.r.t. EQ1 (resp. EQ2, EQ3).

The following property establishes the links between the three criteria of strong equivalence.

Property 5. Let $\mathcal{F}$ and $\mathcal{F}'$ be two argumentation systems. If $\mathcal{F} \equiv_{EQ1S} \mathcal{F}'$, then $\mathcal{F} \equiv_{EQ2S} \mathcal{F}'$ and $\mathcal{F} \equiv_{EQ3S} \mathcal{F}'$.

We have already pointed out that in logic-based argumentation, there are no self-attacking arguments [1]. Formally, $\nexists a \in \mathtt{Arg}(\mathcal{L})$ such that $(a, a) \in \mathcal{R}(\mathcal{L})$. Furthermore, it was proved that if there are no self-attacking arguments, then any two argumentation systems are strongly equivalent (w.r.t. any of the three criteria used by Oikarinen and Woltran) if and only if they coincide [9]. In what follows, we show that if the structure of arguments is taken into account and if criteria are relaxed as we proposed in Definition 11, then there are cases where different systems are strongly equivalent. More precisely, we show that if $\mathcal{F} = (\mathcal{A}, \mathcal{R})$ and $\mathcal{F}' = (\mathcal{A}', \mathcal{R}')$ where $\mathcal{R}$ and $\mathcal{R}'$ are restrictions of $\mathcal{R}(\mathcal{L})$ on $\mathcal{A}$ and $\mathcal{A}'$, and if $\mathcal{A} \sim \mathcal{A}'$ then $\mathcal{F} \equiv_{EQ1S} \mathcal{F}'$.

Theorem 6. *Let $\mathcal{F} = (\mathcal{A}, \mathcal{R})$ and $\mathcal{F}' = (\mathcal{A}', \mathcal{R}')$ be two argumentation systems. If $\mathcal{A} \sim \mathcal{A}'$ then $\mathcal{F} \equiv_{EQ1S} \mathcal{F}'$.*

From the previous theorem, we conclude that if the sets of arguments of two systems are equivalent w.r.t. $\sim$, then they are also strongly equivalent w.r.t. EQ2 and EQ3.

Corollary 1. *Let $\mathcal{F} = (\mathcal{A}, \mathcal{R})$ and $\mathcal{F}' = (\mathcal{A}', \mathcal{R}')$ be two argumentation systems. If $\mathcal{A} \sim \mathcal{A}'$ then $\mathcal{F} \equiv_{EQ2S} \mathcal{F}'$ and $\mathcal{F} \equiv_{EQ3S} \mathcal{F}'$.*

As in the basic case, strong equivalence can be used in order to reduce the computational cost of an argumentation system by removing unnecessary arguments. We provide a condition under which a given argumentation system has a *finite strongly equivalent system*.

Theorem 7. *Let $\mathcal{F} = (\mathcal{A}, \mathcal{R})$ be an argumentation system built over a knowledge base Σ. If $\texttt{Cncs}(\Sigma)/\equiv$ is finite, then there exists an argumentation system $\mathcal{F}' = (\mathcal{A}', \mathcal{R}')$ such that $\mathcal{F} \equiv_{EQ1S} \mathcal{F}'$ and $\mathcal{A}'$ is finite.*

The following corollary follows directly.

Corollary 2. *Let $\mathcal{F} = (\mathcal{A}, \mathcal{R})$ be an argumentation system built over a knowledge base Σ. If $\texttt{Cncs}(\Sigma)/\equiv$ is finite, then there exists an argumentation system $\mathcal{F}' = (\mathcal{A}', \mathcal{R}')$ s.t. $\mathcal{F} \equiv_{EQ2S} \mathcal{F}'$ and $\mathcal{F} \equiv_{EQ3S} \mathcal{F}'$ and $\mathcal{A}'$ is finite.*

This result is of great importance. It shows that our criteria are useful since on the one hand, there are situations when *different* systems are equivalent, and on the other hand, our criteria allow to reduce an infinite system to a finite one.

5 Dynamics of Argument Status

Let us now show when the previous results may be used when studying dynamics of argumentation systems. The problem we are interested in is defined as follows: Given an argumentation system $\mathcal{F} = (\mathcal{A}, \mathcal{R})$ where $\mathcal{A} \subseteq \texttt{Arg}(\mathcal{L})$ and $\mathcal{R} = \mathcal{R}(\mathcal{L})_{|\mathcal{A}}$, when the status of any argument $a \in \mathcal{A}$ may evolve if a new argument $e \in \texttt{Arg}(\mathcal{L})$ is received or if an argument $e \in \mathcal{A}$ is removed. When $\mathcal{F}$ is extended by e, the resulting system is denoted by $\mathcal{F} \oplus \{e\}$. When an argument e is removed from $\mathcal{F}$, the new system is denoted by $\mathcal{F} \ominus \{e\} = (\mathcal{A}', \mathcal{R}')$ is defined as $\mathcal{A}' = \mathcal{A} \setminus \{e\}$ and $\mathcal{R}' = \mathcal{R}(\mathcal{L})_{|\mathcal{A}'}$.

5.1 Extending an AS by New Argument(s)

Let $\mathcal{F} = (\mathcal{A}, \mathcal{R})$ be an argumentation system and $\texttt{Base}(\mathcal{A}) = \Sigma$. By definition of $\mathcal{F}$, the set $\mathcal{A}$ is a subset of $\texttt{Arg}(\Sigma)$ (the set of all arguments that may be built from Σ). Let $\mathcal{F}_c = (\texttt{Arg}(\Sigma), \mathcal{R}(\mathcal{L})_{|\texttt{Arg}(\Sigma)})$ denote the *complete* version of $\mathcal{F}$. We also say that $\mathcal{F}$ is *incomplete* iff $\mathcal{A} \subset \texttt{Arg}(\Sigma)$. Note that, generally for reasoning over a knowledge base, a complete system is considered. However, in dialogues the exchanged arguments do not necessarily constitute a complete system. To say it differently, it may be the case that other arguments may be built using the exchanged information (the formulas of the supports of exchanged arguments).

In what follows, we show that the statuses of arguments in an incomplete system are floating in case that system does not contain a core of the complete system. However, as soon as an incomplete system is a core or contains a core of the complete system, then the status of each argument becomes fixed and will never change when a new argument from $\texttt{Arg}(\Sigma)$ is received.

Theorem 8. *Let* $\mathcal{F} = (\mathcal{A}, \mathcal{R})$ *and* $\mathcal{F}_c = (\texttt{Arg}(\texttt{Base}(\mathcal{A})), \mathcal{R}(\mathcal{L})_{|\texttt{Arg}(\texttt{Base}(\mathcal{A}))})$ *be two argumentation systems. If there exists a core* $(\mathcal{A}', \mathcal{R}')$ *of* $\mathcal{F}_c$ *s.t.* $\mathcal{A}' \subseteq \mathcal{A}$*, then* $\forall e \in \texttt{Arg}(\texttt{Base}(\mathcal{A}))$ *the following hold:*

- $\mathcal{F} \equiv_{EQ1} \mathcal{F} \oplus \{e\}$
- $\forall a \in \mathcal{A}$, $\texttt{Status}(a, \mathcal{F}) = \texttt{Status}(a, \mathcal{F} \oplus \{e\})$
- $\texttt{Status}(e, \mathcal{F} \oplus \{e\}) = \texttt{Status}(a, \mathcal{F})$*, where* $a \in \mathcal{A}$ *is any argument s.t.* $\texttt{Supp}(a) = \texttt{Supp}(e)$.

We now show that when a system does not contain a core of the system built over its base, new arguments may change the status of the existing ones.

Example 1. Let $(\mathcal{L}, \texttt{CN})$ be propositional logic and let us consider the attack relation defined as follows: $\forall a, b \in \texttt{Arg}(\mathcal{L})$, $a\mathcal{R}b$ iff $\exists h \in \texttt{Supp}(b)$ s.t. $\texttt{Conc}(a) \equiv \neg h$. Let $\mathcal{F} = (\mathcal{A}, \mathcal{R})$ with $\mathcal{A} = \{a_1, a_2\}$ s.t. $a_1 = (\{x, x \rightarrow y\}, y)$ and $a_2 = (\{\neg x\}, \neg x)$. It can be checked that $a_2\mathcal{R}a_1$. Thus, a_2 is skeptically accepted while a_1 is rejected. Note that $\texttt{Base}(\mathcal{A}) = \{x, \neg x, x \rightarrow y\}$, thus $e = (\{x\}, x) \in \texttt{Arg}(\texttt{Base}(\mathcal{A}))$. In the new system $\mathcal{F} \oplus \{e\}$, the two arguments both change their statuses.

The previous example illustrates a situation where an argumentation system does not contain a core of the system constructed from its base. This means that not all crucial arguments are considered in $\mathcal{F}$; thus, it is not surprising that it is possible to revise arguments' statuses.

5.2 Removing Argument(s) from an AS

We have already seen that extracting a core of an argumentation system is a compact way to represent the original system. In that process, redundant arguments are deleted from the original system. In this subsection, we show under which conditions deleting argument(s) does not influence the status of other arguments.

As one may expect, if an argument e is deleted from an argumentation system $\mathcal{F} = (\mathcal{A}, \mathcal{R})$ and if the resulting system $\mathcal{F} \ominus \{e\}$ is a core or contains a core of the complete version of $\mathcal{F}$, then all arguments in $\mathcal{A}$ keep their original status.

Theorem 9. *Let* $\mathcal{F} = (\mathcal{A}, \mathcal{R})$ *be an argumentation system,* $\mathcal{F}_c = (\texttt{Arg}(\texttt{Base}(\mathcal{A})), \mathcal{R}(\mathcal{L})_{|\texttt{Arg}(\texttt{Base}(\mathcal{A}))})$ *its complete version. Let* $e \in \mathcal{A}$*. If* $\mathcal{F} \ominus \{e\}$ *contains a core of* $\mathcal{F}_c$*, then the following hold:*

- $\mathcal{F} \equiv_{EQ1} \mathcal{F} \ominus \{e\}$
- $\forall a \in \mathcal{A} \setminus \{e\}, \texttt{Status}(a, \mathcal{F}) = \texttt{Status}(a, \mathcal{F} \ominus \{e\})$.

It can be shown that Theorem 8 (resp. Theorem 9) is true even if a (finite or infinite) set of arguments is added (resp. deleted) to $\mathcal{F}$. In order to simplify the presentation, only results when one argument is added (deleted) were presented. The general result (when an arbitrary set of arguments is added/removed) is proved in Lemma 2 in the Appendix.

6 Conclusion

In this paper, we studied the problem of equivalence and strong equivalence between logic-based argumentation systems. While there are no works on equivalence in argumentation, previous works on strong equivalence are disappointing, since according to the proposed criteria [9] no different systems may be equivalent, the only exception being a case when systems contain self-attacking arguments, which is never a case in logical based argumentation [1]. Thus, this notion has no practical application since two different systems are never strongly equivalent.

In this paper, we have refined existing criteria and defined new ones by taking into account the structure of arguments. Since almost all applications of Dung's abstract argumentation system are obtained by constructing arguments from a given knowledge base, using a given logic, we studied the most general case in logic-based argumentation: we conducted our study for *any* logic which satisfies five basic properties proposed by Tarski [10]. We proposed flexible equivalence criteria and we showed when two systems are equivalent and strongly equivalent w.r.t. those criteria. The results show that for almost all well-known logics, even for an infinite argumentation system, it can be possible to find a finite system which is strongly equivalent to it.

References

1. Amgoud, L., Besnard, P.: Bridging the gap between abstract argumentation systems and logic. In: Godo, L., Pugliese, A. (eds.) SUM 2009. LNCS, vol. 5785, pp. 12–27. Springer, Heidelberg (2009)
2. Amgoud, L., Cayrol, C.: A reasoning model based on the production of acceptable arguments. Annals of Mathematics and Artificial Intelligence 34, 197–216 (2002)
3. Baroni, P., Giacomin, M., Guida, G.: Scc-recursiveness: a general schema for argumentation semantics. Artificial Intelligence Journal 168, 162–210 (2005)
4. Bench-Capon, T.J.M.: Persuasion in practical argument using value-based argumentation frameworks. Journal of Logic and Computation 13(3), 429–448 (2003)
5. Caminada, M.: Semi-stable semantics. In: Proceedings of the 1st International Conference on Computational Models of Argument (COMMA 2006), pp. 121–130 (2006)
6. Dung, P.M.: On the acceptability of arguments and its fundamental role in nonmonotonic reasoning, logic programming and n-person games. Artificial Intelligence Journal 77, 321–357 (1995)
7. Dunne, P., Hunter, A., McBurney, P., Parsons, S., Wooldridge, M.: Inconsistency tolerance in weighted argument systems. In: AAMAS, pp. 851–858 (2009)
8. Gorogiannis, N., Hunter, A.: Instantiating abstract argumentation with classical logic arguments: Postulates and properties. Artificial Intelligence Journal (in press, 2011)

9. Oikarinen, E., Woltran, S.: Characterizing strong equivalence for argumentation frameworks. In: Proceedings of KR 2010 (2010)
10. Tarski, A.: On Some Fundamental Concepts of Metamathematics. In: Woodger, J.H. (ed.) Logic, Semantics, Metamathematic. Oxford Uni. Press, Oxford (1956)

Appendix

Proof. of Property 1. Let $\mathcal{F} = (\mathcal{A}, \mathcal{R})$ be an AS and $a, a' \in \mathcal{A}$ such that $\texttt{Supp}(a) = \texttt{Supp}(a')$. We prove that for every stable extension $\mathcal{E}$, we have $a \in \mathcal{E}$ iff $a' \in \mathcal{E}$. Let us assume that $a \in \mathcal{E}$ and $a' \notin \mathcal{E}$. Since $\mathcal{E}$ is a stable extension, then $\exists b \in \mathcal{E}$ s.t. $b\mathcal{R}a'$. Since $\mathcal{R}$ satisfies property $C2$, then $b\mathcal{R}a$ which contradicts the fact that $\mathcal{E}$ is a stable extension. The case $a \notin \mathcal{E}$ and $a' \in \mathcal{E}$ is symmetric. This means that each extension of $\mathcal{F}$ either contains both a and a' or does not contain any of those two arguments. Consequently, the statuses of those arguments must coincide.

Proof. of Property 2. Let $\mathcal{F} = (\mathcal{A}, \mathcal{R})$ be an AS and $a, a', b, b' \in \mathcal{A}$ such that $a \approx a'$ and $b \approx b'$. Assume that $a\mathcal{R}b$. Since $\texttt{Supp}(b) = \texttt{Supp}(b')$ then from $C2$, it follows that $a\mathcal{R}b'$. From $C1$ and the fact that $\texttt{Conc}(a) \equiv \texttt{Conc}(a')$, we get $a'\mathcal{R}b'$. To show that $a'\mathcal{R}b'$ implies $a\mathcal{R}b$ is similar.

Proof. of Theorem 1. Let $\mathcal{F} = (\mathcal{A}, \mathcal{R})$, $\mathcal{F}' = (\mathcal{A}', \mathcal{R}')$ be two AS such that $\mathcal{F} \equiv_{EQ1} \mathcal{F}'$.

- Let us prove that $\texttt{Sc}(\mathcal{F}) \sim \texttt{Sc}(\mathcal{F}')$. If $\texttt{Ext}(\mathcal{F}) = \emptyset$, then from $\mathcal{F} \equiv_{EQ1} \mathcal{F}'$, $\texttt{Ext}(\mathcal{F}') = \emptyset$. In this case, $\texttt{Sc}(\mathcal{F}) \sim \texttt{Sc}(\mathcal{F}')$ holds trivially, since $\texttt{Sc}(\mathcal{F}) = \texttt{Sc}(\mathcal{F}') = \emptyset$. Assume now that $\texttt{Ext}(\mathcal{F}) \neq \emptyset$.
  Let $\texttt{Sc}(\mathcal{F}) = \emptyset$. We will prove that $\texttt{Sc}(\mathcal{F}') = \emptyset$. Suppose the contrary and let $a' \in \texttt{Sc}(\mathcal{F}')$. Let $\mathcal{E}' \in \texttt{Ext}(\mathcal{F}')$. Argument a' is skeptically accepted, thus $a' \in \mathcal{E}'$. Let f be a bijection from $\mathcal{F} \equiv_{EQ1} \mathcal{F}'$, and let us denote $\mathcal{E} = f^{-1}(\mathcal{E}')$. From $\mathcal{F} \equiv_{EQ1} \mathcal{F}'$, we obtain $\mathcal{E} \in \texttt{Ext}(\mathcal{F})$. Furthermore, $\mathcal{E} \sim \mathcal{E}'$, and, consequently, $\exists a \in \mathcal{E}$ s.t. $a \approx a'$. Theorem 2 implies that a is skeptically accepted in $\mathcal{F}$, contradiction.
  Let $\texttt{Sc}(\mathcal{F}) \neq \emptyset$ and let $a \in \texttt{Sc}(\mathcal{F})$. Since $\mathcal{F} \equiv_{EQ1} \mathcal{F}'$, and a is in at least one extension, then $\exists a' \in \mathcal{A}'$ s.t. $a' \approx a$. From $\mathcal{F} \equiv_{EQ1} \mathcal{F}'$ and from Theorem 2, a' is skeptically accepted in $\mathcal{F}'$. Thus $\forall a \in \texttt{Sc}(\mathcal{F})$, $\exists a' \in \texttt{Sc}(\mathcal{F}')$ s.t. $a' \approx a$. To prove that $\forall a' \in \texttt{Sc}(\mathcal{F}')$, $\exists a \in \texttt{Sc}(\mathcal{F})$ s.t. $a \approx a'$ is similar.

- We can easily see that $\texttt{Ext}(\mathcal{F}) = \emptyset$ iff $\texttt{Ext}(\mathcal{F}') = \emptyset$ and that $\texttt{Ext}(\mathcal{F}) = \{\emptyset\}$ iff $\texttt{Ext}(\mathcal{F}') = \{\emptyset\}$. Let $a \in \texttt{Cr}(\mathcal{F})$. We prove that $\exists a' \in \texttt{Cr}(\mathcal{F}')$ s.t. $a \approx a'$. Since $a \in \texttt{Cr}(\mathcal{F})$ then $\exists \mathcal{E} \in \texttt{Ext}(\mathcal{F})$ s.t. $a \in \mathcal{E}$. Let f be a bijection between from $\mathcal{F} \equiv_{EQ1} \mathcal{F}'$ and let $\mathcal{E}' = f(\mathcal{E})$. From $\mathcal{F} \equiv_{EQ1} \mathcal{F}'$, we obtain that $\mathcal{E} \sim \mathcal{E}'$, thus $\exists a' \in \mathcal{E}'$ s.t. $a \approx a'$. This means that $\forall x \in \texttt{Cr}(\mathcal{F})$, $\exists x' \in \texttt{Cr}(\mathcal{F}')$ such that $x \approx x'$. To prove that $\forall a' \in \texttt{Cr}(\mathcal{F}')$, $\exists a \in \texttt{Cr}(\mathcal{F})$ such that $a \approx a'$ is similar. Thus, $\texttt{Cr}(\mathcal{F}) \sim \texttt{Cr}(\mathcal{F}')$.

Proof. of Theorem 2. If $\mathcal{F}$ has no extensions, then all arguments in $\mathcal{F}$ and $\mathcal{F}'$ are rejected. Thus, in the rest of the proof, we study the case when $\texttt{Ext}(\mathcal{F}) \neq \emptyset$. We will prove that for any extension $\mathcal{E}$ of $\mathcal{F}$, $a \in \mathcal{E}$ iff $a' \in f(\mathcal{E})$, where $f : \texttt{Ext}(\mathcal{F}) \rightarrow \texttt{Ext}(\mathcal{F}')$ is a bijection s.t. $\forall \mathcal{E} \in \texttt{Ext}(\mathcal{F})$, $\mathcal{E} \sim f(\mathcal{E})$. Let $\mathcal{E} \in \texttt{Ext}(\mathcal{F})$, let $a \in \mathcal{E}$ and let $a' \in \mathcal{A}'$

with $a \approx a'$. Let $\mathcal{E}' = f(\mathcal{E})$; we will prove that $a' \in \mathcal{E}'$. From $\mathcal{F} \equiv_{EQ1} \mathcal{F}'$, one obtains $\exists a'' \in \mathcal{E}'$ s.t. $a \approx a''$. (Note that we do not know whether $a' = a''$ or not.) We will prove that $\{a'\} \cup \mathcal{E}'$ is conflict-free. Let us suppose the contrary. This means that $\exists x \in \mathcal{E}'$ s.t. $x\mathcal{R}'a'$ or $a'\mathcal{R}'x$. From $(x\mathcal{R}'a' \vee a'\mathcal{R}'a')$, we have $(x\mathcal{R}'a'' \vee a''\mathcal{R}'x)$, which contradicts the fact that $\mathcal{E}'$ is a stable extension. We conclude that $\{a'\} \cup \mathcal{E}'$ is conflict-free. Since $\mathcal{E}'$ is a stable extension, it attacks any argument $y \notin \mathcal{E}'$. Since $\mathcal{E}'$ does not attack a', then $a' \in \mathcal{E}'$.

This means that we showed that for any $\mathcal{E} \in \mathtt{Ext}(\mathcal{F})$, if $a \in \mathcal{E}$ then $a' \in f(\mathcal{E})$. Let $a \notin \mathcal{E}$ and let us prove that $a' \notin f(\mathcal{E})$. Suppose the contrary, i.e. suppose that $a' \in f(\mathcal{E})$. Since we made exactly the same hypothesis on $\mathcal{F}$ and $\mathcal{F}'$, by using the same reasoning as in the first part of the proof, we can prove that $a \in \mathcal{E}$, contradiction. This means that $a' \notin f(\mathcal{E})$. So, we proved that for any extension $\mathcal{E} \in \mathtt{Ext}(\mathcal{F})$, we have $a \in \mathcal{E}$ iff $a' \in f(\mathcal{E})$.

If a is skeptically accepted, then for any $\mathcal{E} \in \mathtt{Ext}(\mathcal{F})$, $a \in \mathcal{E}$. Let $\mathcal{E}' \in \mathtt{Ext}(\mathcal{F}')$. Then, from $\mathcal{F} \equiv_{EQ1} \mathcal{F}'$, there exists $\mathcal{E} \in \mathtt{Ext}(\mathcal{F})$ s.t. $\mathcal{E}' = f(\mathcal{E})$. Since $a \in \mathcal{E}$, then $a' \in \mathcal{E}'$. If a is not skeptically accepted, then $\exists \mathcal{E} \in \mathtt{Ext}(\mathcal{F})$ s.t. $a \notin \mathcal{E}$. It is clear that $\mathcal{E}' = f(\mathcal{E})$ is an extension of $\mathcal{F}'$ and that $a' \notin \mathcal{E}'$. Thus, in this case a' is not skeptically accepted in $\mathcal{F}'$.

Let a be credulously accepted in $\mathcal{F}$ and let $\mathcal{E} \in \mathtt{Ext}(\mathcal{F})$ be an extension s.t. $a \in \mathcal{E}$. Then, $a' \in f(\mathcal{E})$, thus a' is credulously accepted in $\mathcal{F}'$. It is easy to see that the case when a is not credulously accepted in $\mathcal{F}$ and a' is credulously accepted in $\mathcal{F}'$ is not possible.

If a is rejected in $\mathcal{F}$, then a is not credulously accepted, thus a' is not credulously accepted which means that it is rejected.

Proof. of Theorem 3. Let $\mathcal{F} \equiv_{EQ1} \mathcal{F}'$. From Theorem 1, we obtain $\mathcal{F} \equiv_{EQ2} \mathcal{F}'$ and $\mathcal{F} \equiv_{EQ3} \mathcal{F}'$. It is easy to see that this implies $\mathtt{Output}_{sc}(\mathcal{F}) \cong \mathtt{Output}_{sc}(\mathcal{F}')$ and $\mathtt{Output}_{cr}(\mathcal{F}) \cong \mathtt{Output}_{cr}(\mathcal{F}')$. Considering the third part of the theorem, let f be a bijection from $\mathcal{F} \equiv \mathcal{F}'$, let $\mathcal{E} \in \mathtt{Ext}(\mathcal{F})$ and $\mathcal{E}' = f(\mathcal{E})$. One can easily check that $\mathtt{Base}(\mathcal{E}) = \mathtt{Base}(\mathcal{E}')$. This means that $\forall \mathcal{E} \in \mathtt{Ext}(\mathcal{F})$, $\exists \mathcal{E}' \in \mathtt{Ext}(\mathcal{F}')$ s.t. $\mathtt{Base}(\mathcal{E}) = \mathtt{Base}(\mathcal{E}')$. To see that $\forall \mathcal{E}' \in \mathtt{Ext}(\mathcal{F}')$, $\exists \mathcal{E} \in \mathtt{Ext}(\mathcal{F})$ s.t. $\mathtt{Base}(\mathcal{E}) = \mathtt{Base}(\mathcal{E}')$ is similar. Consequently, $\mathtt{Bases}(\mathcal{F}) = \mathtt{Bases}(\mathcal{F}')$.

Lemma 1. *Let $\mathcal{R}(\mathcal{L}) \subseteq \mathtt{Arg}(\mathcal{L}) \times \mathtt{Arg}(\mathcal{L})$ be an attack relation on the set of all arguments built from $\mathcal{L}$. Let $\mathcal{F} = (\mathcal{A}, \mathcal{R})$ and $\mathcal{F}' = (\mathcal{A}', \mathcal{R}')$ be two AS such that $\mathcal{A}, \mathcal{A}' \subseteq \mathtt{Arg}(\mathcal{L})$ and $\mathcal{R} = \mathcal{R}(\mathcal{L})_{|\mathcal{A}}$, $\mathcal{R}' = \mathcal{R}(\mathcal{L})_{|\mathcal{A}'}$. If $\mathcal{A} \sim \mathcal{A}'$, then $\mathcal{F} \equiv_{EQ1} \mathcal{F}'$.*

Proof. Let us first suppose that $\mathtt{Ext}(\mathcal{F}) \neq \emptyset$ and let us define the function $f' : 2^{\mathcal{A}} \to 2^{\mathcal{A}'}$ as follows: $f'(B) = \{a' \in \mathcal{A}' \mid \exists a \in B \text{ s.t. } a' \approx a\}$.

Let f be the restriction of f' to $\mathtt{Ext}(\mathcal{F})$. We will prove that the image of this function is $\mathtt{Ext}(\mathcal{F}')$ and that f is a bijection between $\mathtt{Ext}(\mathcal{F})$ and $\mathtt{Ext}(\mathcal{F}')$ which verifies EQ1.

- First, we will prove that for any $\mathcal{E} \in \mathtt{Ext}(\mathcal{F})$, $f(\mathcal{E}) \in \mathtt{Ext}(\mathcal{F}')$. Let $\mathcal{E} \in \mathtt{Ext}(\mathcal{F})$ and let $\mathcal{E}' = f(\mathcal{E})$. We will prove that $\mathcal{E}'$ is conflict-free. Let $a', b' \in \mathcal{E}'$. There must exist $a, b \in \mathcal{E}$ s.t. $a \approx a'$ and $b \approx b'$. Since $\mathcal{E}$ is an extension, $\neg(a\mathcal{R}b)$ and $\neg(b\mathcal{R}a)$. By applying Property 2 on $(\mathtt{Arg}(\mathcal{L}), \mathcal{R}(\mathcal{L}))$, we have that $\neg(a'\mathcal{R}'b')$ and $\neg(b'\mathcal{R}'a')$. Let $x' \in \mathcal{A}' \setminus \mathcal{E}'$. Then $\exists x \in \mathcal{A}$ s.t. $x \approx x'$. Note also that it must be that

$x \notin \mathcal{E}$. Since $\mathcal{E} \in \texttt{Ext}(\mathcal{F})$, then $\exists y \in \mathcal{E}$ s.t. $y\mathcal{R}x$. Note that $\exists y' \in \mathcal{E}'$ s.t. $y' \approx y$. From Property 2, $y'\mathcal{R}'x'$.

- We have shown that the image of f is the set $\texttt{Ext}(\mathcal{F}')$. We will now prove that $f : \texttt{Ext}(\mathcal{F}) \to \texttt{Ext}(\mathcal{F}')$ is injective. Let $\mathcal{E}_1, \mathcal{E}_2 \in \texttt{Ext}(\mathcal{F})$ with $\mathcal{E}_1 \neq \mathcal{E}_2$ and $\mathcal{E}' = f(\mathcal{E}_1) = f(\mathcal{E}_2)$. We will show that if $\mathcal{E}_1 \sim \mathcal{E}_2$ then $\mathcal{E}_1 = \mathcal{E}_2$. Without loss of generality, let $\exists x \in \mathcal{E}_1 \setminus \mathcal{E}_2$. Then, from $\mathcal{E}_1 \sim \mathcal{E}_2$, $\exists x' \in \mathcal{E}_2$, s.t. $x' \approx x$. Then, since $x \in \mathcal{E}_1$ and $x \notin \mathcal{E}_2$, from the proof of Property 1 we obtain that $x' \in \mathcal{E}_1$ and $x' \notin \mathcal{E}_2$. Contradiction with $x' \in \mathcal{E}_2$. This means that $\neg(\mathcal{E}_1 \sim \mathcal{E}_2)$. Without loss of generality, $\exists a_1 \in \mathcal{E}_1 \setminus \mathcal{E}_2$ s.t. $\nexists a_2 \in \mathcal{E}_2$ s.t. $a_1 \approx a_2$. Let $a' \in \mathcal{A}'$ s.t. $a' \approx a_1$. Recall that $\mathcal{E}' = f(\mathcal{E}_2)$. Thus, $\exists a_2 \in \mathcal{E}_2$ s.t. $a_2 \approx a'$. Contradiction.
- We show that $f : \texttt{Ext}(\mathcal{F}) \to \texttt{Ext}(\mathcal{F}')$ is surjective. Let $\mathcal{E}' \in \texttt{Ext}(\mathcal{F}')$, and let us show that $\exists \mathcal{E} \in \texttt{Ext}(\mathcal{F})$ s.t. $\mathcal{E}' = f(\mathcal{E})$. Let $\mathcal{E} = \{a \in \mathcal{A} \mid \exists a' \in \mathcal{E}'$ s.t. $a \approx a'\}$. From Property 2 we see that $\mathcal{E}$ is conflict-free. For any $b \in \mathcal{A} \setminus \mathcal{E}$, $\exists b' \in \mathcal{A}' \setminus \mathcal{E}'$ s.t. $b \approx b'$. Since $\mathcal{E}' \in \texttt{Ext}(\mathcal{F}')$, then $\exists a' \in \mathcal{E}'$ s.t. $a'\mathcal{R}'b'$. Now, $\exists a \in \mathcal{E}$ s.t. $a \approx a'$; from Property 2, $a\mathcal{R}b$. Thus, $\mathcal{E}$ is a stable extension in $\mathcal{F}$.
- We will now show that $f : \texttt{Ext}(\mathcal{F}) \to \texttt{Ext}(\mathcal{F}')$ verifies the condition of EQ1. Let $\mathcal{E} \in \texttt{Ext}(\mathcal{F})$ and $\mathcal{E}' = f(\mathcal{E})$. Let $a \in \mathcal{E}$. Then, $\exists a' \in \mathcal{A}'$ s.t. $a' \approx a$. From the definition of f, it must be that $a' \in \mathcal{E}'$. Similarly, if $a' \in \mathcal{E}'$, then must be an argument $a \in \mathcal{A}$ s.t. $a \approx a'$, and again from the definition of the function f, we conclude that $a \in \mathcal{E}$.

From all above, we conclude that $\mathcal{F} \equiv_{EQ1} \mathcal{F}'$. Let us take a look at the case when $\texttt{Ext}(\mathcal{F}) = \emptyset$. We will show that $\texttt{Ext}(\mathcal{F}') = \emptyset$. Suppose not and let $\mathcal{E}' \in \texttt{Ext}(\mathcal{F}')$. Let us define $\mathcal{E} = \{a \in \mathcal{A} \mid \exists a' \in \mathcal{E}'$ s.t. $a \approx a'\}$. From Property 2, $\mathcal{E}$ must be conflict-free. The same property shows that for any $b \in \mathcal{A} \setminus \mathcal{E}$, $\exists a \in \mathcal{E}$ s.t. $a\mathcal{R}b$. Thus, $\mathcal{E}$ is a stable extension in $\mathcal{F}$. Contradiction with the hypothesis that $\texttt{Ext}(\mathcal{F}) = \emptyset$.

Proof. of Theorem 4. The result is obtained by applying Lemma 1 on $\mathcal{F}$ and $\mathcal{F}'$.

Proof. of Theorem 5. Let $\mathcal{F}' = (\mathcal{A}', \mathcal{R}')$ be a core of $\mathcal{F}$ and let us prove that $\mathcal{F}'$ is finite. Since Σ is finite, then $\{\texttt{Supp}(a) \mid a \in \mathcal{A}'\}$ must be finite. If for all $H \in \{\texttt{Supp}(a) \mid a \in \mathcal{A}'\}$, the set $\{a \in \mathcal{A}' \mid \texttt{Supp}(a) = H\}$, is finite, then the set $\mathcal{A}'$ is clearly finite. Else, there exists $H_0 \in \{\texttt{Supp}(a) \mid a \in \mathcal{A}'\}$, s.t. the set $\mathcal{A}_{H_0} = \{a \in \mathcal{A}' \mid \texttt{Supp}(a) = H_0\}$ is infinite. By the definition of $\mathcal{A}'$, one obtains that $\forall a, b \in \mathcal{A}_{H_0}$, $\texttt{Conc}(a) \not\equiv \texttt{Conc}(b)$. It is clear that $\forall a \in \mathcal{A}_{H_0}$, $\texttt{Conc}(a) \in \texttt{Cncs}(\Sigma)$. This implies that there are infinitely many different formulae having logically non-equivalent conclusions in $\texttt{Cncs}(\Sigma)$, formally, set $\texttt{Cncs}(\Sigma)/\equiv$ is infinite, contradiction.

Proof. of Theorem 6. Let $\mathcal{B} \subseteq \texttt{Arg}(\mathcal{L})$. Since $\mathcal{A} \sim \mathcal{A}'$ then clearly $\mathcal{A} \cup \mathcal{B} \sim \mathcal{A}' \cup \mathcal{B}$. From Lemma 1, we obtain that $\mathcal{F} \oplus \mathcal{B} \equiv_{EQ1} \mathcal{F}' \oplus \mathcal{B}$. Thus, $\mathcal{F} \equiv_{EQ1S} \mathcal{F}'$.

Proof. of Theorem 7. Let $\mathcal{A}' \subseteq \mathcal{A}$ be a set defined as follows: $\forall a \in \mathcal{A}\ \exists! a' \in \mathcal{A}'$ s.t. $a' \approx a$. It is clear that $\mathcal{F}' = (\mathcal{A}', \mathcal{R}' = \mathcal{R}_{|\mathcal{A}'})$ is a core of $\mathcal{F}$. Since $\mathcal{A} \sim \mathcal{A}'$, then from Theorem 6, $\mathcal{F} \equiv_{EQ1S} \mathcal{F}'$. From Theorem 5, $\mathcal{F}'$ is finite.

Lemma 2. *Let $\mathcal{F} = (\mathcal{A}, \mathcal{R})$ be an AS built from Σ which contains a core of $\mathcal{G} = (\mathcal{A}_g = \texttt{Arg}(\Sigma), \mathcal{R}_g = \mathcal{R}(\mathcal{L})|_{\mathcal{A}_g})$ and let $\mathcal{E} \subseteq \texttt{Arg}(\Sigma)$. Then:*

- $\mathcal{F} \equiv_{EQ1} \mathcal{F} \oplus \mathcal{E}$
- $\forall a \in \mathcal{A}, \mathtt{Status}(a, \mathcal{F}) = \mathtt{Status}(a, \mathcal{F} \oplus \mathcal{E})$
- $\forall e \in \mathcal{E} \setminus \mathcal{A}, \mathtt{Status}(e, \mathcal{F} \oplus \mathcal{E}) = \mathtt{Status}(a, \mathcal{F})$, *where* $a \in \mathcal{A}$ *is any argument s.t.* $\mathtt{Supp}(a) = \mathtt{Supp}(e)$.

Proof. Let $\mathcal{F}' = \mathcal{F} \oplus \mathcal{E}$ with $\mathcal{F}' = (\mathcal{A}', \mathcal{R}')$ and let $\mathcal{H} = (\mathcal{A}_h, \mathcal{R}_h)$ be a core of $\mathcal{G}$ s.t. $\mathcal{A}_h \subseteq \mathcal{A}$. We will first show that $\mathcal{H}$ is a core of both $\mathcal{F}$ and $\mathcal{F}'$. Let us first show that $\mathcal{H}$ is a core of $\mathcal{F}$. We will show that all conditions of Definition 10 are verified.

- From what we supposed, we have that $\mathcal{A}_h \subseteq \mathcal{A}$.
- We will show that $\forall a \in \mathcal{A}\ \exists! a' \in \mathcal{A}_h$ s.t. $a' \approx a$. Let $a \in \mathcal{A}$. Since $a \in \mathcal{A}_g$ and $\mathcal{H}$ is a core of $\mathcal{G}$, then $\exists! a' \in \mathcal{A}_h$ s.t. $a' \approx a$.
- Since $\mathcal{R} = \mathcal{R}(\mathcal{L})|_{\mathcal{A}}$ and $\mathcal{R}_h = \mathcal{R}(\mathcal{L})|_{\mathcal{A}_h}$ then from $\mathcal{A}_h \subseteq \mathcal{A}$ we obtain that $\mathcal{R}_h = \mathcal{R}|_{\mathcal{A}_h}$.

Thus, $\mathcal{H}$ is a core of $\mathcal{F}$. Let us now show that $\mathcal{H}$ is also a core of $\mathcal{F}'$:

- Since $\mathcal{A}_h \subseteq \mathcal{A}$ and $\mathcal{A} \subseteq \mathcal{A}'$ then $\mathcal{A}_h \subseteq \mathcal{A}'$.
- Let $a \in \mathcal{A}'$. Since $a \in \mathcal{A}_g$ and $\mathcal{H}$ is a core of system $\mathcal{G}$, then $\exists! a' \in \mathcal{A}_h$ s.t. $a' \approx a$.
- Since $\mathcal{R}' = \mathcal{R}(\mathcal{L})|_{\mathcal{A}'}$, $\mathcal{R}_h = \mathcal{R}(\mathcal{L})|_{\mathcal{A}_h}$ and $\mathcal{A}_h \subseteq \mathcal{A}'$, then we obtain that $\mathcal{R}_h = \mathcal{R}'|_{\mathcal{A}_h}$.

We have shown that $\mathcal{H}$ is a core of $\mathcal{F}$ and of $\mathcal{F}'$. From Theorem 4, $\mathcal{F} \equiv_{EQ1} \mathcal{H}$ and $\mathcal{F}' \equiv_{EQ1} \mathcal{H}$. Since $\equiv$ is an equivalence relation, then $\mathcal{F} \equiv_{EQ1} \mathcal{F}'$. Let $a \in \mathcal{A}$. From Theorem 2, $\mathtt{Status}(a, \mathcal{F}) = \mathtt{Status}(a, \mathcal{F}')$.

Let $e \in \mathcal{A}' \setminus \mathcal{A}$ and let $a \in \mathcal{A}$ be an argument such that $\mathtt{Supp}(a) = \mathtt{Supp}(e)$. From Property 1, we obtain $\mathtt{Status}(e, \mathcal{F}') = \mathtt{Status}(a, \mathcal{F}')$. Since we have just seen that $\mathtt{Status}(a, \mathcal{F}') = \mathtt{Status}(a, \mathcal{F})$, then $\mathtt{Status}(e, \mathcal{F}') = \mathtt{Status}(a, \mathcal{F})$.

Proof. of Theorem 8. This result is a consequence of Lemma 2.

Proof. of Theorem 9. This result is a consequence of Lemma 2.

Bipolarity in Argumentation Graphs: Towards a Better Understanding

Claudette Cayrol and Marie-Christine Lagasquie-Schiex

IRIT-UPS, Université de Toulouse
{ccayrol,lagasq}@irit.fr

Abstract. Different abstract argumentation frameworks have been used for various applications within multi-agents systems. Among them, bipolar frameworks make use of both attack and support relations between arguments. However, there is no single interpretation of the support, and the handling of bipolarity cannot avoid a deeper analysis of the notion of support. In this paper we consider three recent proposals for specializing the support relation in abstract argumentation : the deductive support, the necessary support and the evidential support. These proposals have been developed independently within different frameworks. We restate these proposals in a common setting, which enables us to undertake a comparative study of the modellings obtained for the three variants of the support. We highlight relationships and differences between these variants, namely a kind of duality between the deductive and the necessary interpretations of the support.

1 Introduction

Formal model of argumentation have recently received considerable interest across different AI communities, like defeasible reasoning and multi-agent systems. Typical applications such as for instance negotiation and practical reasoning represent pieces of knowledge and opinions as arguments and reach some conclusion or decision on the basis of interacting arguments.

Abstract argumentation frameworks model arguments as atomic entities, ignoring their internal structure and focusing on the interactions between arguments, or sets of arguments. Several semantics can be defined that formalize different intuitions about which arguments to accept from a given framework.

The first abstract framework introduced by [5] limits the interactions to conflicts between arguments which the so-called attack binary relation. Several specialized or extended versions of Dung's framework have been proposed, namely the bipolar framework [3] which is capable of modelling a kind of positive interaction expressed by a support relation. Positive interaction between arguments has been first introduced by [6,10]. In [3], the support relation is left general so that the bipolar framework keeps a high level of abstraction. The associated semantics are based on the combination of the attack relation with the support relation which results in new complex attack relations. However, introducing the notion of support between arguments within abstract frameworks has been

S. Benferhat and J. Grant (Eds.): SUM 2011, LNAI 6929, pp. 137–148, 2011.
© Springer-Verlag Berlin Heidelberg 2011

a controversial issue and some counterintuitive results have been obtained, showing that the combination of both interactions cannot avoid a deeper analysis of the notion of support.

Moreover, there is no single interpretation of the support. Indeed, recently, a number of researchers proposed specialized variants of the support relation. Each specialization can be associated with an appropriate modelling using an appropriate complex attack. However, these proposals have been developed quite independently, based on different intuitions and with different formalizations. In this paper we do not want to discuss all the criticisms which have been advanced, but our purpose is to show that bipolar abstract frameworks provide a convenient way to model and discuss various kinds of support. In particular, we address a comparative study of these proposals, in a common setting.

Section 2 presents a brief review of the classical and bipolar abstract argumentation frameworks. In Section 3 we discuss three specializations of the notion of support and propose an appropriate modelling for each of them in the bipolar framework. In Section 4 we conclude and give some perspectives for future work.

2 Background on Abstract Argumentation Frameworks

2.1 Dung Argumentation Framework

Dung's seminal abstract framework consists of a set of arguments and one type of interaction between them, namely attack. What really means is the way arguments are in conflict.

Definition 1 (Dung AF). *A Dung's argumentation framework (*AF*, for short) is a pair $\langle \mathcal{A}, \mathcal{R} \rangle$ where $\mathcal{A}$ is a finite and non-empty set of arguments and $\mathcal{R}$ is a binary relation over $\mathcal{A}$ (a subset of $\mathcal{A} \times \mathcal{A}$), called the* attack relation.

An argumentation framework can be represented by a directed graph, called the *interaction graph*, in which the nodes represent arguments and the edges are defined by the attack relation: $\forall a, b \in \mathcal{A}$, $a\mathcal{R}b$ is represented by $a \not\rightarrow b$.

Definition 2 (Admissibility in AF). *Given $\langle \mathcal{A}, \mathcal{R} \rangle$ and $S \subseteq \mathcal{A}$,*

- *S is* conflict-free *iff there are no arguments $a, b \in S$, such that $a\mathcal{R}b$.*
- *$a \in \mathcal{A}$ is* acceptable *with respect to S iff $\forall b \in \mathcal{A}$ such that $b\mathcal{R}a$, $\exists c \in S$ such that $c\mathcal{R}b$.*
- *S is* admissible *iff S is conflict-free and each argument in S is acceptable with respect to S.*

Standard semantics introduced by Dung (preferred, stable, grounded) enable to characterize admissible sets of arguments that satisfy some form of optimality.

Definition 3 (Extensions). *Given $\langle \mathcal{A}, \mathcal{R} \rangle$ and $S \subseteq \mathcal{A}$,*

- *S is* a preferred extension *of $\langle \mathcal{A}, \mathcal{R} \rangle$ iff it is a maximal (with respect to $\subseteq$) admissible set.*

- S *is* a stable extension *of* $\langle \mathcal{A}, \mathcal{R} \rangle$ *iff it is conflict-free and for each* $a \notin S$*, there is* $b \in S$ *such that* $b\mathcal{R}a$*.*
- S *is* the grounded extension *of* $\langle \mathcal{A}, \mathcal{R} \rangle$ *iff it is the least (with respect to* $\subseteq$*) admissible set* X *such that each argument acceptable with respect to* X *belongs to* X*.*

Example 1. *Let* AF *be defined by* $\mathcal{A} = \{a, b, c, d, e\}$ *and* $\mathcal{R}_{\text{att}} = \{(a, b), (b, a), (b, c), (c, d), (d, e), (e, c)\}$ *and represented by the following graph. There are two preferred extensions (*$\{a\}$ *and* $\{b, d\}$*), one stable extension (*$\{b, d\}$*) and the grounded extension is the empty set.*

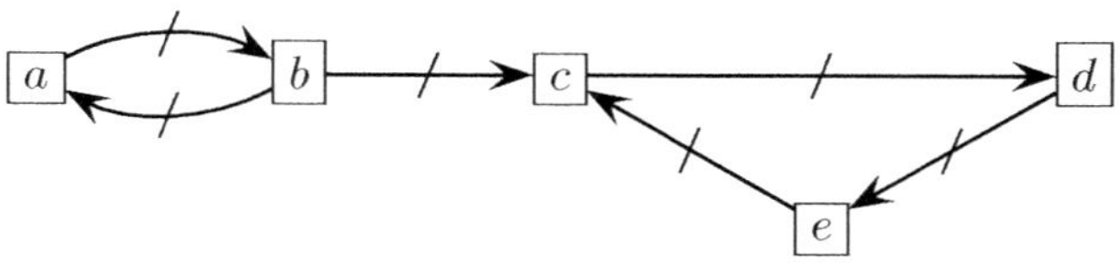

2.2 Bipolar Argumentation Framework

The abstract bipolar argumentation framework [3] [4] extends Dung's framework in order to take into account both negative interactions expressed by the attack relation and positive interactions expressed by a support relation.

Definition 4 (BAF). *A bipolar argumentation framework (*BAF*, for short) is a tuple* $\langle \mathcal{A}, \mathcal{R}_{\text{att}}, \mathcal{R}_{\text{sup}} \rangle$ *where* $\mathcal{A}$ *is a finite and non-empty set of arguments,* $\mathcal{R}_{\text{att}}$ *(resp.* $\mathcal{R}_{\text{sup}}$*) is a binary relation over* $\mathcal{A}$ *called the* attack relation *(resp. the* support relation*).*

A BAF can still be represented by a directed graph $\mathcal{G}_b$ called the *bipolar interaction graph*, with two kinds of edges. Let a_i and $a_j \in \mathcal{A}$, $a_i \mathcal{R}_{\text{att}} a_j$ (resp. $a_i \mathcal{R}_{\text{sup}} a_j$) means that a_i attacks a_j (resp. a_i supports a_j) and it is represented by $a \not\rightarrow b$ (resp. by $a \rightarrow b$).

Example 2. *For instance, in the following graph representing a* BAF*, there is a support from* g *to* d *and an attack from* b *to* a

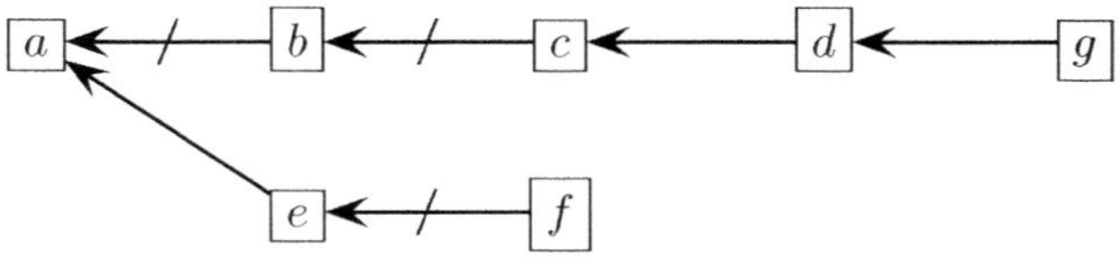

New kinds of attack emerge from the interaction between the direct attacks and the supports.

For instance, the supported attack and the secondary attack have been introduced in [4] (and previously in [3] with a different terminology).

Definition 5 ([4] Complex attacks in BAF)

- *There is a* supported attack *from a to b iff there is a sequence $a_1\mathcal{R}_1 \dots \mathcal{R}_{n-1}a_n$, $n \geq 3$, with $a_1 = a$, $a_n = b$, $\forall i = 1 \dots n-2$, $\mathcal{R}_i = \mathcal{R}_{\text{sup}}$ and $\mathcal{R}_{n-1} = \mathcal{R}_{\text{att}}$.*
- *There is a* secondary attack *from a to b iff there is a sequence $a_1\mathcal{R}_1 \dots \mathcal{R}_{n-1}a_n$, $n \geq 3$, with $a_1 = a$, $a_n = b$, $\mathcal{R}_1 = \mathcal{R}_{\text{att}}$ and $\forall i = 2 \dots n-1$, $\mathcal{R}_i = \mathcal{R}_{\text{sup}}$.*

Note that the above definitions combine a direct attack with a sequence of direct supports, that is a direct or indirect support. In the following, *a supports b* means that there is a sequence of direct supports from a to b.

Example 2 (cont'd) *In this example, there is a supported attack from g (or d) to b and a secondary attack from f to a.*

From these complex attacks, new notions of conflict-freeness can be obtained. Moreover, the notion of coherence of a set of arguments can be still enforced with the notion of safety [3].

Definition 6 ([3] Safety in BAF). *Given $S \subseteq \mathcal{A}$, S is* safe *iff there are no arguments $a, b \in S$, and $c \in \mathcal{A}$ such that*

- *b* supports *c or $c \in S$ and*
- *there is a* supported attack *or a* direct attack *from a to c.*

Admissibility in a bipolar argumentation framework can be defined as in Dung's framework by combining acceptability and conflict-freeness. Different definitions can be proposed depending on the notion of attack (direct, supported, secondary, ...) and on the notion of coherence which are used.

3 Modelling Various Kinds of Support

3.1 A Need for Specialization of Support

Handling support and attack at an abstract level has the advantage to keep genericity. An abstract bipolar framework is useful as an analytic tool for studying different notions of complex attacks, complex conflicts, and new semantics taking into account both kinds of interactions between arguments. However, the drawback is the lack of guidelines for choosing the appropriate definitions and semantics depending on the application. For instance, in Dung's framework, whatever the semantics, the acceptance of an argument which is not attacked is guaranteed. Is it always desirable in a bipolar framework? Two related questions are: Can arguments stand in an extension without being supported? Can arguments be used as attackers without being supported? It may depend on the interpretation of the support, as shown below.

In the following, we discuss three specialized variants of the support relation, which have been proposed recently: the deductive support, the evidential support and the necessary support. Let us first briefly give the underlying intuition, then some illustrative examples.

Deductive support [1] is intended to capture the following intuition: If $a\mathcal{R}_{\text{sup}}b$ then the acceptance of a implies the acceptance of b, and as a consequence the non-acceptance of b implies the non-acceptance of a.

Evidential support [8,9] enables to distinguish between *prima-facie* and standard arguments. *Prima-facie* arguments do not require any support from other arguments to stand, while standard arguments must be supported by at least one *prima-facie* argument.

Necessary support [7], is intended to capture the following intuition: If $a\mathcal{R}_{\text{sup}}b$ then the acceptance of a is necessary to get the acceptance of b, or equivalently the acceptance of b implies the acceptance of a.

The following example shows that different interpretations of the support can be given, and that, according to the considered interpretation, some complex attacks need to be considered, while others are counterintuitive.

Example 3. *This example has been inspired from [7] (and also from a variant in [1]). Let us consider the following knowledge: Obtaining a Bachelor's degree with honors (BH) supports obtaining a scholarship (S) and suppose that having at least one bad mark (BM) does not allow to obtain the honors (even if the average of marks normally allows it). One possible interpretation of the support is: obtaining a bachelor's degree is necessary for obtaining a scholarship. So, if we don't have a BH then we are sure that we don't have S.*

Now let us suppose that obtaining S may be also fulfilled if the student justifies modest incomes (MI). A more appropriate interpretation of the support is a deductive one. In that case, a secondary attack from BM to S would be counterintuitive. Moreover, it is known that making a blank copy (BC) supports having a very bad mark. With a deductive interpretation of that support, it makes sense to add a supported attack from BC to BH. Finally, we add the knowledge: having a very good mark for each test of the examination (VG) supports obtaining a Bachelor's degree with honors.

The whole example can be formalized in a BAF *represented by the following graph:*

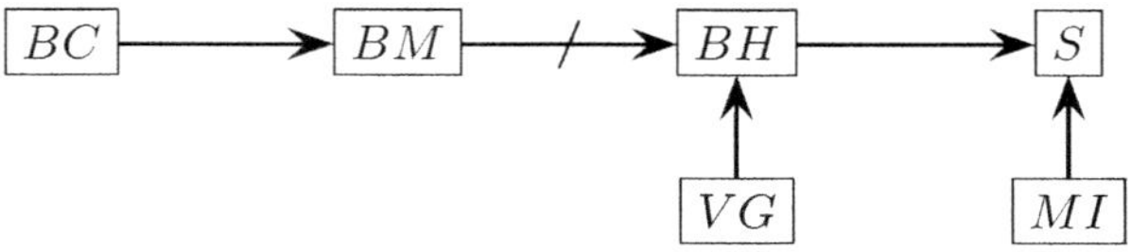

Example 4. *(Example justifying a secondary attack) Let us consider the following dialogue between three agents:*

- *Agent 1: The room is dark, so I will light up the lamp.*
- *Agent 2: But the electric meter does not work.*
- *Agent 1: Are you sure?*
- *Agent 3: The electrician has detected a failure (F)*

This dialogue shows interactions between the positions RD (the room is dark), LL (the lamp will light up), EW (the electric meter works), and F (there is a failure in the electric meter). These interactions can be formalized in a BAF *represented by the following graph:*

The intuitive interpretation of the support is a necessary one since the lamp cannot light up when the electric meter does not work. In that case, it makes sense to add a secondary attack from F to LL .

Example 5. *(Example for an evidential support) Let us consider the* BAF *represented by the graph:*

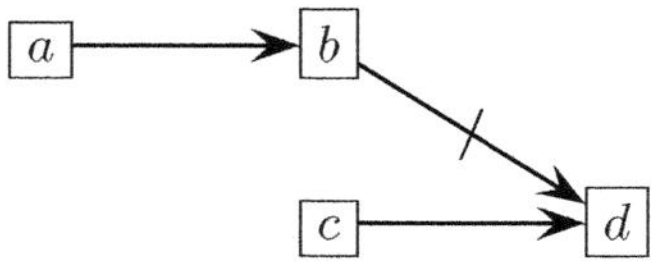

Assume first that the only prima-facie *argument is c. So, d may stand, but neither a nor b is grounded in* prima-facie *arguments. As a consequence, the attack on d cannot be taken into account. So, c and d will be accepted. Assume now that the* prima-facie *arguments are a and c. So, b and d may stand and the attack on d must be considered. In that case the accepted arguments are a, b and c. In order to reinstate d, an attack could be added either from c to b either from c to a. Indeed, an attack from c to a invalidates the attack on d by rendering b unsupported. Finally, assume that the* prima-facie *arguments are a, b and c. The attack from b to d holds without the support by a. So an attack from c to a does not enable to reinstate d. There must be an attack from c to b.*

3.2 A Framework for a Comparative Study of Various Supports

We propose to restate various notions of support in the BAF framework. We will show that each specialized variant of the support can be associated with appropriate complex attacks. Then, we will be able to highlight links between these various notions of support.

Deductive and necessary supports. We first discuss the deductive and necessary support, and prove that these two specializations of the support are indeed dual. As a consequence, these two kinds of support can be handled simultaneously in a bipolar framework. Moreover, defining admissibility in a bipolar framework with deductive and necessary support can be done exactly as in Dung's AF, as follows: For each specialized support, the combination of the direct attacks and the supports results in the addition of appropriate complex attacks. Then Definition 2 can be applied, where $\mathcal{R}$ represents a direct or a complex attack. So, once the complex attacks have been added, we recover a classical Dung AF.

As explained above, a deductive support is intended to enforce the following constraint: If $a\mathcal{R}_{\text{sup}}b$ then the acceptance of a implies the acceptance of b, and as a consequence the non-acceptance of b implies the non-acceptance of a. Suppose now that $c\mathcal{R}_{\text{att}}b$. The acceptance of c implies the non-acceptance of b and so the non-acceptance of a. This strong constraint can be taken into account by introducing a new attack, called mediated attack.

Definition 7 ([1] Mediated attack). *There is a* mediated attack *from a to b iff there is a sequence $a_1\mathcal{R}_{\text{sup}} \dots \mathcal{R}_{\text{sup}}a_{n-1}$, and $a_n\mathcal{R}_{\text{att}}a_{n-1}$, $n \geq 3$, with $a_1 = b$, $a_n = a$.*

Example 3 (cont'd) . *From $VG\mathcal{R}_{\text{sup}}BH$ and $BM\mathcal{R}_{\text{att}}BH$, the mediated attack $BM\mathcal{R}_{\text{att}}VG$ will be added.*

Moreover, the deductive interpretation of the support justifies the introduction of supported attacks (cf Definition 5). If $a\mathcal{R}_{\text{sup}}b$ and $b\mathcal{R}_{\text{att}}c$, the acceptance of a implies the acceptance of b and the acceptance of b implies the non-acceptance of c. So, the acceptance of a implies the non-acceptance of c.

In the following, deductive support will be called *d-support.*

Definition 8 (Modelling deductive support). *The combination of the direct attacks and the d-supports results in the addition of supported attacks and mediated attacks.*

Example 3 (cont'd) . *The following complex attacks are added: a supported attack from from BC to BH and a mediated attack from BM to VG. Then supports can be ignored, and we obtain the following AF:*

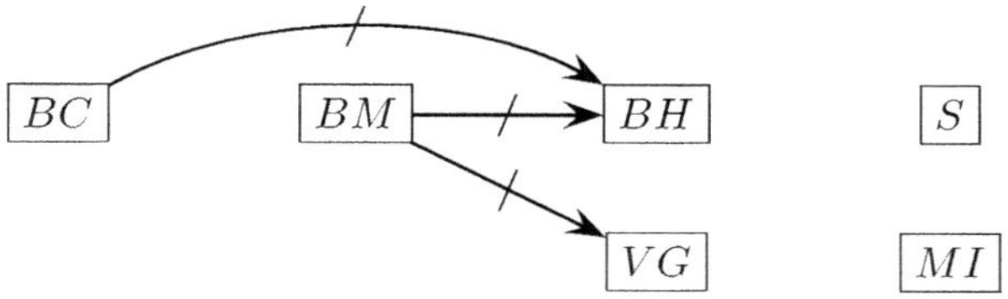

This AF has one preferred (and also stable and grounded) extension $\{BC, BM, S, MI\}$.

As explained above, modelling deductive support in a BAF can be done in considering the associated Dung AF consisting of the same arguments and of the relation built from the direct attacks, the supported attacks and the mediated attacks. Note that the notion of conflict-free in this new AF exactly corresponds to the notion of safety (issued from [3] and recalled in Definition 6) which has been proposed in a BAF for enforcing the notion of coherence.

Necessary support corresponds to the following interpretation: If $a\mathcal{R}_{\text{sup}}b$ then the acceptance of a is necessary to get the acceptance of b, or equivalently the acceptance of b implies the acceptance of a. Suppose now that $c\mathcal{R}_{\text{att}}a$. The acceptance of c implies the non-acceptance of a and so the non-acceptance of b. This constraint can be taken into account by introducing a new attack, called extended attack in [7]. Indeed, it is exactly the secondary attack presented above [4]. Moreover, another kind of complex attack can be justified: If $a\mathcal{R}_{\text{sup}}b$ and $a\mathcal{R}_{\text{att}}c$, the acceptance of b implies the acceptance of a and the acceptance of a implies the non-acceptance of c. So, the acceptance of b implies the non-acceptance of c. This constraint relating b and c should be enforced by adding a new complex attack from b to c. However, this complex attack has not been considered in [7]. In the following, necessary support will be called *n-support.*

Deductive support and necessary support have been introduced independently. However, they correspond to dual interpretations of the support in the following sense: *a n-supports b* is equivalent to *b d-supports a*. Besides, it is easy to see that the constructions of mediated attack and secondary attack are dual in the following sense: the mediated attacks obtained by combining the attack relation $\mathcal{R}_{\text{att}}$ and the support relation $\mathcal{R}_{\text{sup}}$ are exactly the secondary attacks obtained by combining the attack relation $\mathcal{R}_{\text{att}}$ and the support relation which is the inverse relation of $\mathcal{R}_{\text{sup}}$. Moreover, the complex attacks which are missing in [7] as evoked previously can be recovered by considering the supported attacks built from $\mathcal{R}_{\text{att}}$ and the inverse of $\mathcal{R}_{\text{sup}}$.

Consequently, the modelling by the addition of appropriate complex attacks satisfies this duality.

Definition 9 (Modelling necessary support). *The combination of the direct attacks and the n-supports can be handled by turning the n-supports into the dual d-supports and then adding the supported attacks and mediated attacks.*

Example 6. *We complete Ex 5 by adding an attack from c to a:*

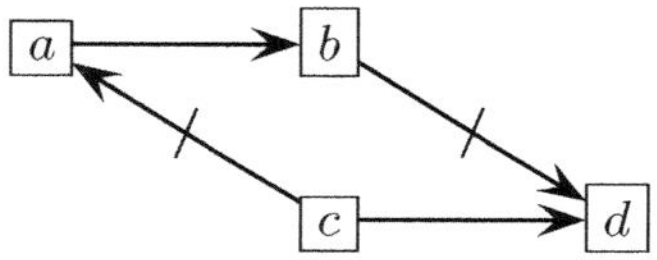

Assume that the support relation has been given a necessary interpretation. That is a is necessary for b and c is necessary for d. It is equivalent to consider that there is a deductive support from b to a and also from d to c. Then, we add a supported attack from d to a and a mediated attack from c to b. The resulting AF is represented by:

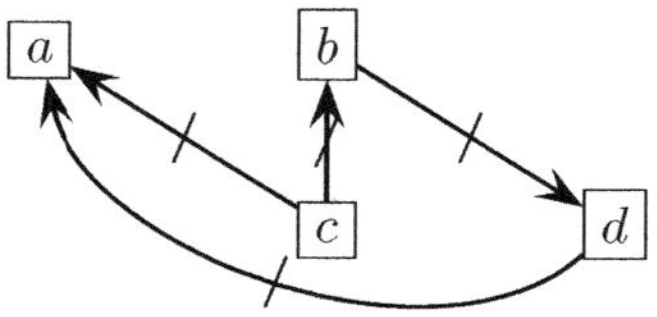

It follows that $\{c, d\}$ *is the only preferred (and also stable and grounded) extension.*

Evidential support. Evidential support [8,9] is intended to capture the notion of *support by evidence*: an argument cannot be accepted unless it is supported by evidence. Evidence is represented by a special argument, and the arguments which are directly supported by this special argument are called *prima-facie* arguments. Arguments can be accepted only if they are supported (directly or indirectly) by *prima-facie* arguments. Besides, only supported arguments can be used to attack other arguments.

In Oren's evidential argument framework, attacks and supports may be carried out by a set of arguments (and not only by a single argument). However, for the purpose of comparing different specializations of the notion of support, we will restrict the presentation of evidential support to the case where attacks and supports are carried out by single arguments. All the definitions that we give in the following are inspired by those given in [8,9].

Given a BAF $\langle \mathcal{A}, \mathcal{R}_{\text{att}}, \mathcal{R}_{\text{sup}} \rangle$, we distinguish a subset $\mathcal{A}_e \subseteq \mathcal{A}$ of arguments which do not require any support to stand. These arguments will be called self-supported and correspond to the *prima-facie* arguments. We recall that in a BAF, a *supports* b means that there is a sequence of direct supports from a to b.

So, evidential support (or *e-support* for short) can be defined as a particular case of this notion of (direct or indirect) support.

Definition 10 (e-support)

- a *is* e-supported *iff either* $a \in \mathcal{A}_e$ *or there exists* b *such that* b *is e-supported and* $b\mathcal{R}_{\text{sup}}a$.
- a *is* e-supported by S *(or* S e-supports a*) iff either* $a \in \mathcal{A}_e$ *or there is an elementary sequence* $b_1\mathcal{R}_{\text{sup}} \dots \mathcal{R}_{\text{sup}}b_n\mathcal{R}_{\text{sup}}a$ *such that* $\{b_1 \dots b_n\} \subseteq S$ *and* $b_1 \in \mathcal{A}_e$.
- S *is* self-supporting *iff* S *e-supports each of its elements.*

Example 6 (cont'd) . *Assume that* $\mathcal{A}_e = \{a, c\}$*. Then* b *is e-supported by* $\{a\}$*,* d *is e-supported by* $\{c\}$*. The set* $\{c, d\}$ *is self-supporting.*

The combination of the direct attacks and the evidential support results in restrictions on the notion of attack and also on the notion of acceptability. The first idea is that only e-supported arguments may be used to make a direct attack on other arguments. This is formalized by the notion of e-supported attack.

Definition 11 (e-supported attack). S *carries out an* e-supported attack *on* a *iff there exists* $b \in S$ *such that* $b\mathcal{R}_{\text{att}}a$ *and* b *is e-supported by* S.

The second idea concerns reinstatement: If a is attacked by b, which is e-supported, a can be reinstated either by a direct attack on b or by an attack on c such that without c, b would be no longer supported. In order to enforce this idea, minimal (for set-inclusion) e-supported attacks have to be considered. It is easy to prove that:

Proposition 1. X *is a* minimal e-supported attack *on* a *iff* X *is the set of arguments appearing in a minimal length elementary sequence* $b_1\mathcal{R}_{\text{sup}} \dots \mathcal{R}_{\text{sup}}b_n$ *such that* $b_1 \in \mathcal{A}_e$ *and* $b_n\mathcal{R}_{\text{att}}a$.

Note that a minimal e-supported attack corresponds to a particular case of a supported attack as defined in Definition 5. In the case when $b_1\mathcal{R}_{\text{sup}} \dots \mathcal{R}_{\text{sup}}b_n$ with $b_1 \in \mathcal{A}_e$ and $b_n\mathcal{R}_{\text{att}}a$, each b_i carries out a supported attack on a.

Now, following Oren's evidential argument framework, we propose a new definition for acceptability. There are two conditions on S, for a being acceptable wrt S. The first one is classical and concerns defence or reinstatement: S must invalidate each minimal e-supported attack on a (either by attacking the attacker of a or by rendering this attacker unsupported). The second condition requires that S e-supports a.

Definition 12 (e-acceptability). *a is* e-acceptable *wrt S iff*

- *For each minimal e-supported attack X on a, there exists $b \in S$ and $x \in X$ such that $b\mathcal{R}_{\text{att}}x$ and*
- *a is e-supported by S.*

Definition 13 (e-admissibility). *S is* e-admissible *iff*

- *Each element of S is e-acceptable wrt S and*
- *there are no arguments $a, b \in S$, such that $a\mathcal{R}_{\text{att}}b$.*

Example 6 (cont'd) . *Assume that $\mathcal{A}_e = \{a, c\}$. There is only one minimal e-supported attack on d: $\{a, b\}$. As $c\mathcal{R}_{\text{att}}a$ and d is e-supported by $\{c\}$, we have that d is e-acceptable wrt $\{c\}$. Then, $\{c, d\}$ is e-admissible. Note that there is no e-supported attack on b. However, b does not belong to any e-admissible set, because no e-admissible set e-supports b. Assume now that $\mathcal{A}_e = \{a, b, c\}$. $\{b\}$ is the only minimal e-supported attack on d. As no argument attacks b, no e-admissible set contains d. The only e-admissible set is $\{c, b\}$.*

The above example enables us to highlight the relationship between the notion of evidential support and the notion of necessary support. It seems that evidential support can be viewed as a kind of weak necessary support, in the following sense: Assume that b is supported by a and c; with the necessary support interpretation, the acceptance of b implies the acceptance of a *and* the acceptance of c; with the evidential interpretation, the acceptance of b implies the acceptance of a *or* the acceptance of c, if b is not self-supported and, if b is self-supported, the acceptance of b implies no constraint on a and c.

The above comment suggests to consider the particular case when each argument is self-supported, that is $\mathcal{A}_e = \mathcal{A}$. In that case, X is a *minimal e-supported attack* on a iff X is reduced to one argument which directly attacks a. So, classical acceptability is recovered: a is e-acceptable wrt S iff a is acceptable wrt S in Dung's sense. And as each argument is self-supported, we also recover classical admissibility. That is to say that the support relation is ignored.

Another interesting case occurs when self-supported arguments are exactly those which do not have any support, that is $\mathcal{A}_e = \{a \in \mathcal{A}$ / there does not exist b such that $b\mathcal{R}_{\text{sup}}a\}$. However, even in that particular case, evidential support cannot be modelled with necessary support, as shown by the following example.

Example 7. *We complete Ex 6 by adding an argument e and a support from e to b:*

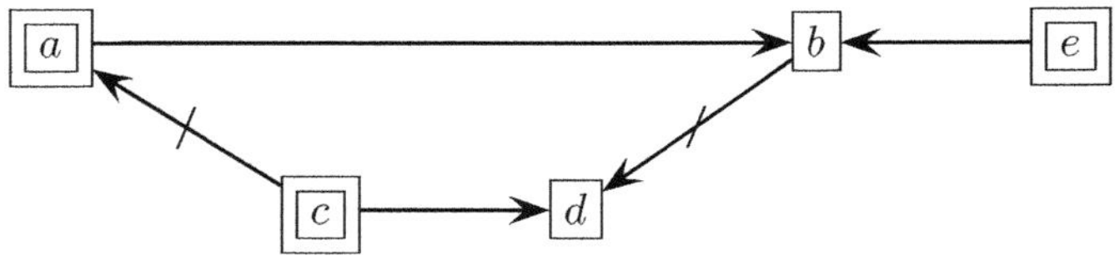

Assume that $\mathcal{A}_e = \{a, c, e\}$ (this is represented by a double box around the elements of $\mathcal{A}_e$). The only $\subseteq$-maximal e-admissible set is $\{c, e, b\}$. Indeed, d is not e-acceptable wrt $\{c\}$ since $\{e, b\}$ is a minimal e-supported attack on d and neither

b nor e is attacked. Now, if we handle the same graph with necessary supports, we first take the inverse of $\mathcal{R}_{\text{sup}}$ *and then add supported and mediated attacks. This results in adding an attack from d to a and an attack from c to b:*

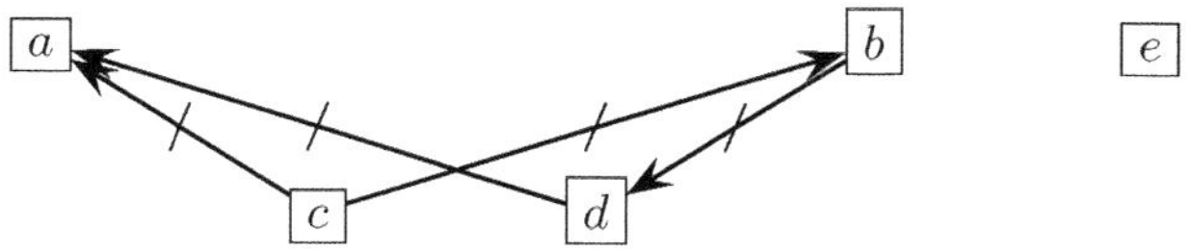

Taking into account these new attacks, the set is $\{c, b, e\}$ *is no longer admissible (there is a conflict between c and b) and* $\{c, d\}$ *becomes admissible.*

The above example shows that the notion of evidential support cannot be reduced to strict necessary support (nor to deductive support). So, it is not possible to handle together in the same bipolar framework evidential support and necessary / deductive support.

4 Conclusions and Future Works

In this paper, we have considered three recent proposals for specializing the support relation in abstract argumentation : the deductive support, the necessary support and the evidential support. These proposals have been developed independently within different frameworks and with appropriate modellings, based on different intuitions.

We have restated these proposals in a common setting, the bipolar argumentation framework. Basically, the idea is to keep the original arguments, to add complex attacks defined by the combination of the original attack and the support, and to modify the classical notions of acceptability. We have proposed a comparative study of the modellings obtained for the considered variants of the support, which has enabled us to highlight relationships and differences between these variants. Namely, we have shown a kind of duality between the deductive and the necessary interpretations of support, which results in a duality in the modelling by complex attacks. In contrast, the evidential interpretation is quite different and cannot be captured with deductive or necessary supports.

This work is a first step towards a better understanding of the notion of support in argumentation.

For future works, we first intend to go deeply into the comparative study, by exploiting another interesting idea : a bipolar framework can be turned into a meta-argumentation framework, which instantiates Dung's framework with meta-arguments. This meta-argumentation framework enables to reuse Dung's principles and properties. Meta-arguments may represent groups of arguments, or may just be auxiliary arguments representing pairs of interacting arguments. So, the set of arguments and the interactions are completely different from the original ones. Some promising work has been done by [9] which proposes a mapping from the evidential argumentation framework to meta-argumentation framework. This mapping is based on the use of maximal self-supporting paths and preserves semantics. We plan to follow that research direction for other variants of the support.

Another interesting topic for further research is to restate the study of the support in the more general setting of Abstract Dialectical Frameworks (ADF for short) [2]. Such frameworks also enable to express acceptance conditions over interacting nodes of a graph. These conditions are much more flexible than the conditions described for deductive, necessary or evidential support. For instance, if c depends on a and b, the following constraint can be taken into account: The acceptance of b and the non-acceptance of a imply the acceptance of c. However, the status of a node in the graph only depends on the status of its parents in the graph. So, the strict deductive interpretation of the support dependency cannot be taken into account. For instance,if x d-supports s and $y\mathcal{R}_{\text{att}}s$, no condition in ADF will ensure that x and y cannot be accepted together, since x has no parent.

Acknowledgements. To the French National Research Agency (ANR).

References

1. Boella, G., Gabbay, D.M., van der Torre, L., Villata, S.: Support in abstract argumentation. In: Proc. of COMMA, pp. 111–122. IOS Press, Amsterdam (2010)
2. Brewka, G., Woltran, S.: Abstract dialectical frameworks. In: Proc. of KR, pp. 102–111 (2010)
3. Cayrol, C., Lagasquie-Schiex, M.-C.: On the acceptability of arguments in bipolar argumentation frameworks. In: Godo, L. (ed.) ECSQARU 2005. LNCS (LNAI), vol. 3571, pp. 378–389. Springer, Heidelberg (2005)
4. Cayrol, C., Lagasquie-Schiex, M.-C.: Coalitions of arguments: a tool for handling bipolar argumentation frameworks. International Journal of Intelligent Systems 25, 83–109 (2010)
5. Dung, P.M.: On the acceptability of arguments and its fundamental role in nonmonotonic reasoning, logic programming and n-person games. Artificial Intelligence 77, 321–357 (1995)
6. Karacapilidis, N., Papadias, D.: Computer supported argumentation and collaborative decision making: the HERMES system. Information systems 26(4), 259–277 (2001)
7. Nouioua, F., Risch, V.: Bipolar argumentation frameworks with specialized supports. In: Proc. of ICTAI, pp. 215–218. IEEE Computer Society, Los Alamitos (2010)
8. Oren, N., Norman, T.J.: Semantics for evidence-based argumentation. In: Proc. of COMMA, pp. 276–284 (2008)
9. Oren, N., Reed, C., Luck, M.: Moving between argumentation frameworks. In: Proc. of COMMA, pp. 379–390. IOS Press, Amsterdam (2010)
10. Verheij, B.: Deflog: on the logical interpretation of prima facie justified assumptions. Journal of Logic in Computation 13, 319–346 (2003)

Handling Enthymemes in Time-Limited Persuasion Dialogs*

Florence Dupin de Saint-Cyr

IRIT, Toulouse, France

Abstract. This paper is a first attempt to define a framework to handle enthymeme in a time-limited persuasion dialog. The notion of incomplete argument is explicited and a protocol is proposed to regulate the utterances of a persuasion dialog with respect to the three criteria of consistency, non-redundancy and listening. This protocol allows the use of enthymemes concerning the support or conclusion of the argument, enables the agent to retract or re-specify an argument. The system is illustrated on a small example and some of its properties are outlined.

1 Introduction

Many persuasion debates have marked human history: Herodotus debate on the three government types, Valiadolid debate, the Bohr-Einstein debate about quantum mechanics, presidential TV debates... The "winner" is often considered as very clever and skilled. Indeed oratory featured in the original Olympics and there exist teaching lessons for being a good orator (e.g. [19]). A good orator is someone who is able to make his point of view adopted by the public whatever the truth is and whatever his adversary may say. This skillness and cleverness is a big challenge for human being in everyday life as well as in History since debates are both very common and very influential. This is why it is important that artificial intelligence focuses on this field of research. This implies to develop at least two features: representing and handling persuasion dialogs, designing good artificial orators (able to find strategies to win a debate).

The first feature has already been widely developed in the literature (see e.g. [3,5,8,10,11,17]) but as far as we know the dialog persuasion systems that have been developed either do not define what is an argument or always assume that an argument is a "perfect" minimal proof of a formula. Our purpose is to develop a system in which it is possible to express an argument, called "approximate argument" by [13], that takes into account implicit information. Indeed, it is generally admitted that an argument is composed of two parts: a support and a claim, such that the support is a logical minimal proof of the claim, this kind of argument is called "logical argument" by [13]. In everyday life, there is nearly no "logical argument", since it is not useful and maybe tiring to completely justify a given claim, we often give an argument without mentioning implicit common knowledge. Otherwise argumentation would be very long to express and boring to listen (and could be recursively infinite when each part of the support

* This work was funded by the ANR project LELIE on risk analysis and prevention (`http://www.irit.fr/recherches/ILPL/lelie/accueil.html`).

S. Benferhat and J. Grant (Eds.): SUM 2011, LNAI 6929, pp. 149–162, 2011.
© Springer-Verlag Berlin Heidelberg 2011

of a claim should in turn be completely explained). Shortly speaking a logical argument is not into line with Gricean maxims. Approximate arguments, called enthymeme by Aristotle, is a syllogism keeping at least one of the premises or conclusion unsaid. Enthymemes have already been studied in the literature [25,23,7,14], but no formal persuasion dialog system able to handle enthymeme has yet been defined.

Handling enthymeme in persuasion dialog has two advantages: first it allows to deal with more concrete cases where agents want to shorten their arguments, second it may involve strategic matter, namely droping a premise may remove a possible attack or may enable to cheat by pretending that implicit knowledge can help to prove a claim while it is not the case... The problem of implicit knowledge was one of the motivation for non-monotonic reasoning which aims at reasonning despite a lack of information. In enthymeme handling, this is not the only aim, it may also be interesting to focus on what is missing. The following example is given by Schopenhauer [19] to exemplify the "extension stratagem" which is the first among the 38 stratagems he designed for taking victory in a dispute (without worrying about the objective truth). Here, the argument of Schopenhauer's opponent is an enthymeme.

Example 1 ([19]). "I asserted that the English were supreme in drama. My opponent attempted to give an instance to the contrary, and replied that it was a well-known fact that in music, and consequently in opera, they could do nothing at all. I repelled the attack by reminding him that music was not included in dramatic art, which covered tragedy and comedy alone. This he knew very well. What he had done was to try to generalize my proposition, so that it would apply to all theatrical representations, and, consequently, to opera and then to music, in order to make certain of defeating me."

In the following, we first reintroduce enthymemes in a logical framework, then we present the utterances that can be done in a persuasion dialog with enthymemes. For instance we propose a new speech act, Complete, that allows to precise an already expressed approximate argument. We then develop the protocol that governs a persuasion dialog allowing enthymemes (first defined in [9]). While in concrete dialogs it is often the case that people do not listen to each other or are inconsistent, here the protocol enforces consistency, non redundancy and listening. A novelty of our proposal is the representation of time-limited persuasion dialogs in which each speaker is given a fixed speech-time, this ensures that every dialog has an end and enforces agents to take time into account when uttering their arguments. Although the example we provide is very short and simple, its size allows us to use it along the paper even if it does not show all the strategical aspects of enthymemes (left for further research).

2 Arguments and Enthymemes

We consider a logical language $\mathscr{L}$, where Greek letters (e.g. φ, ψ) denote formulas, $\vdash$ the logical inference, $\bot$ the contradiction. Let us recall the definition of [13]:

Definition 1 (logical and approximate arguments [6,13])

A logical argument *is a pair* $\langle S, \varphi \rangle$ *such that:*
$$\begin{cases} (1) & S \subseteq \mathscr{L}, \varphi \in \mathscr{L} \\ (2) & S \nvdash \bot, \\ (3) & S \vdash \varphi, \\ (4) & \nexists S' \subset S \ s.t.\ S' \vdash \varphi \end{cases}$$

An approximate argument *is a pair* $\langle S, \varphi \rangle$ *where the* support $S \subseteq \mathscr{L}$ *is a set of propositional formulas, and the* claim *is* $\varphi \in \mathscr{L}$.
Arg_{Σ} *(resp.* AArg_{Σ}*) denotes the set of logical (resp. approximate) arguments that can be built from a set of formulas* Σ*, lower-case Latin letters (e.g.* a, b*) denote arguments.*

Notation: if A *is a set of arguments, the set of formulas pertaining to the supports and claims of these arguments is denoted by* $\mathsf{form}(A)$, $\mathsf{form}(A) = \bigcup_{\langle S,\varphi\rangle \in A} S \cup \{\varphi\}$.

In other words, an approximate argument is simply a pair (support,claim) and when the support is a minimal proof of the claim this argument is called a logical argument. Note that an approximate argument does not need to have a consistent support S and it is not required that its conclusion φ is a logical consequence of S. In order to be able to deal with arguments that have incomplete support or incompletely developed conclusion we first define an incomplete argument and then extend the enthymeme formalization proposed in [7].

Definition 2 (incomplete argument). *An* incomplete argument *is a pair* $\langle S, \varphi \rangle$ *where* $S \subseteq \mathscr{L}$ *and* $\varphi \in \mathscr{L}$ *(i.e.,* $\langle S, \varphi \rangle$ *is an approximate argument) such that:*

$$\begin{cases} (1) & S \nvdash \varphi \\ (2) & \exists \psi \in \mathscr{L} \text{ s.t. } \langle S \cup \{\psi\}, \varphi \rangle \text{ is a logical argument} \end{cases}$$

IArg_{Σ} *denotes the set of incomplete arguments that can be built on a set of formulas* Σ.

In this definition, the first condition expresses the fact that the argument is strictly incomplete, *i.e.*, the support is not sufficient to infer the conclusion. The second one imposes that it is possible to complete it in order to obtain a logical argument. Logical or incomplete arguments are particular distinct cases of approximate arguments:

Proposition 1. $\text{IArg}_{\Sigma} \cap \text{Arg}_{\Sigma} = \varnothing$ *and* $\text{IArg}_{\Sigma} \cup \text{Arg}_{\Sigma} \subseteq \text{AArg}_{\Sigma}$.

Note that the support of an incomplete argument should be consistent or else adding any formula to it would still give an inconsistent support (hence violate condition (2) for logical arguments). Moreover S should be consistent with φ, more formally:

Remark 1. If $\langle S, \varphi \rangle$ is an incomplete argument then $S \nvdash \bot$ and $S \nvdash \neg\varphi$.

This remark shows that this notion is a slight variation of Hunter's concept of *precursor*, which he defines as an approximate argument $\langle S, \varphi \rangle$ such that $S \nvdash \varphi$ and $S \nvdash \neg\varphi$. Hence an "incomplete argument" is a "precursor" but the converse is false. The small difference lays in the fact that a completed precursor may not be minimal, for instance $\langle \{a, b, a \wedge b\}, c \rangle$ is a "precursor" and not an "incomplete argument" since any completion would have a non minimal support (i.e., in Definition 1, (4) will not hold).

Example 1 (continued): *The argument proposed by Schopenhauer's opponent is an incomplete argument. Indeed, "in music* (m)*, and consequently in opera* (o)*, English are not supreme* ($\neg s$)*" maybe transcribed into the following approximate argument:* $a_1 = \langle \{m \rightarrow \neg s\}, o \rightarrow \neg s \rangle$. *And by adding the formula* $o \rightarrow m$ *to its support we obtain the following logical argument:* $b_1 = \langle \{m \rightarrow \neg s, o \rightarrow m\}, o \rightarrow \neg s \rangle$.

Definition 3 (enthymeme). *Let* $a = \langle S, \varphi \rangle$ *and* $a' = \langle S', \varphi' \rangle$ *being approximate arguments,* $\langle S', \varphi' \rangle$ completes $\langle S, \varphi \rangle$ *iff*[1] $\begin{cases} (1)\ S \subset S' \text{ and } \varphi = \varphi' \text{ or} \\ (2)\ S \subseteq S' \text{ and } \{\varphi'\} \cup S' \vdash \varphi \text{ and } \varphi \neq \varphi' \end{cases}$

a is an enthymeme *for a' iff a' is a logical argument and a' completes a.*

In other words, there are two ways to "complete" an argument: either by adding premises, then the support should be strictly included in the completed support or by specifying the conclusion, then it should be inferred by the union of the completed conclusion and support but should differ from the previous conclusion. Our definition extends the definition of [7] in the sense that it allows to cover arguments whose conclusion is an implicit claim requiring implicit support (the following example would not be considered as an enthymeme by [7]).

Example 1 (continued): *We may build an infinity of logical arguments decoding an incomplete argument. For instance, a_1 is an enthymeme for the logical argument b_1 but also for the logical argument:* $\gamma_1 = \langle \{m \to \neg s, o \to m, o, o \to d\}, \neg(d \to s) \rangle$.

Note that completion does not necessary give a logical argument since the initial approximate argument may have an inconsistent or redundant support, or the completion may be too weak for being a logical proof of the claim. Moreover, even logical arguments may be completed, since the completion may concern the conclusion.

Remark 2. When a is an enthymeme for b, it is not necessarily the case that $b \in \text{IArg}$.

The following function gives the set of logical arguments that can be built from a knowledge base Σ and that are enthymemes for a given argument.

Definition 4 (Decode). *Let* $\Sigma \subseteq \mathcal{L}$ *and* $\langle S, \varphi \rangle \in \text{AArg}$, $\mathsf{Decode}_\Sigma(\langle S, \varphi \rangle) = \{\langle S', \varphi' \rangle \in \text{Arg}$ *such that* $S' \setminus S \subseteq \Sigma$, $\varphi' \in \Sigma$ *and* $\langle S, \varphi \rangle$ *is an enthymeme for* $\langle S', \varphi' \rangle\}$.

In the previous example, it holds that $b_1, \gamma_1 \in \mathsf{Decode}_{\mathcal{L}}(a_1)$. It is easy to see that incomplete arguments are particular enthymemes for logical arguments that have the same conclusion ($\varphi = \varphi'$). In other words,

Proposition 2. *If* $\langle S, \varphi \rangle$ *is an incomplete argument then* $\mathsf{Decode}_{\mathcal{L}}(\langle S, \varphi \rangle) \neq \varnothing$.

Proof. If $a = \langle S, \varphi \rangle$ is an incomplete argument then there exists $\psi \in \mathcal{L}$ such that $\langle S \cup \{\psi\}, \varphi \rangle$ is a logical argument. Besides $\psi \notin S$ since $S \nvdash \varphi$ while $S \cup \{\psi\} \vdash \varphi$, hence $S \subset S \cup \{\psi\}$, in other words, there exists $a' \in \text{Arg}$ s.t. a' completes a.

3 A Protocol for Persuasion Dialogs

In persuasion dialogs [12,24], two or more participants are arguing about a claim, and each party is trying to persuade the other participants to adopt its point of view (i.e. to agree with the "right" claim). The set of symbols representing agents is denoted AG but in the following we are going to focus on a dialog between only two agents (named

[1] "iff" stands for "if and only if", $\subset$ denotes strict set-inclusion.

1 and 2, this assumption is convenient but a generalization for more agents can easily be done).The current agent will often be called x in that case the other agent will be denoted by $\overline{x}$. A communicative act, called move, is defined below:

Definition 5 (Moves). *A* move *is a triplet* $(sender, act, content)$, *where* $sender \in AG$, $act \in$ {Accept, Agree, Argue, Assert, Challenge, Close, Dismantle, Quiz, Quizlink, Complete, Retract} *and* ($content = \emptyset$ *or* $content \in \mathscr{L}$ *or* $content \in$ AArg *or* $content \in$ AArg $\times$ AArg*).*
For a given move, the functions Sender, Act *and* Content *are returning respectively, the first, second and third element of the triplet. When there is no ambiguity about the sender, the move is denoted by* $(act\ content)$.
A move m *is* well-formed *if* (Act(m) Content(m)) *matches a utterance given in Table 1. Let* $\mathscr{M}$ *be the set of all well-formed moves.*

We consider a set of eleven speech acts that are described in Table 1 with their three associated effects (locutionary, illocutionary and perlucotionary [4]). Namely, the six usual speech acts used in persuasion dialog are augmented with Quiz and Agree proposed by [7] enabling to handle incomplete arguments, to which we add Quizlink enabling to ask for a completion concerning the conclusion of an argument, Complete allowing to precise an argument and Dismantle for retracting an argument. Although some speech acts are "assertive" (according to Searle [20]) namely Assert and Argue, we claim that they are "commissives" in the sense that they commit the utterer to both be able to explain them when challenged and avoid to contradict them. Moreover the

Table 1. Speech acts for enthymeme persuasion dialogs

Utterance	Meaning	Speaker's intention	Effects on the audience
Accept φ	acceptance of φ	to announce that he accepts φ	the speaker is associated with φ
Agree $\langle S, \varphi\rangle$	acceptance of $\langle S, \varphi\rangle$	to announce that he accepts $\langle S, \varphi\rangle$ as an enthymeme	the speaker is associated with formulas of $S \cup \{\varphi\}$ and knows a logical argument that decodes $\langle S, \varphi\rangle$
Argue $\langle S, \varphi\rangle$	providing a set of formulas S which may support φ if completed	to prove that φ is justified	the speaker is associated with formulas of $S \cup \{\varphi\}$ and knows a logical argument that decodes $\langle S, \varphi\rangle$
Assert φ	statement of assertion φ	to make the hearers believe φ	the speaker is associated with φ
Challenge φ	seeking for arguments supporting φ	to obtain arguments for φ	the receiver must justify φ
Close	closing the dialog	to announce that he has nothing to add	the speaker can no longer participate to the dialog.
Complete (a, b)	providing a new argument b completing a previous one a	to explicit an incomplete argument	the speaker is associated with the support and claim of b and knows a logical argument that decodes b
Dismantle $\langle S, \varphi\rangle$	withdrawal of argument $\langle S, \varphi\rangle$	to renounce to the fact that S is a proof of φ	the speaker is no more associated with $\langle S, \varphi\rangle$
Quiz $\langle S, \varphi\rangle$	seeking for a completion of argument $\langle S, \varphi\rangle$	to obtain a more detailed argument for φ	the receiver must complete $\langle S, \varphi\rangle$
Quizlink $\langle S, \varphi\rangle$	seeking for a link between $\langle S, \varphi\rangle$ and the dialog	to obtain a completion in which the implicit conclusion is disclosed	the receiver must complete at least the conclusion of $\langle S, \varphi\rangle$
Retract φ	withdrawal of assertion φ	to restore consistency or to renounce to prove φ	the speaker is no more associated with φ

"directive" speech acts such as Challenge or Quiz induce commitments for the hearer to answer to these questions. While Close is clearly a "declarative" speech act, it is less obvious for Retract, Dismantle, Accept and Agree since they not only are "declarative" but also "assertive" (because the accepted or agreed formulas are known as if the utterer had asserted them, the retracted formulas or dismantled arguments correspond to assertion of the form "I assert neither φ nor $\neg\varphi$" "I assert neither that S is a valid proof for φ nor that it is not") and "commissives"(since they are assertive).

These commitments are stored in a base called "commitment store" (first introduced by [12] in the dialog game DC). The protocol is a boolean function that uses it in order to check if a move is acceptable at a given stage of the dialog as follows:

Definition 6 (Persuasion dialog). *Let $p \in \mathbb{N} \cup \{\infty\}$, let (F, A) be a common knowledge base s.t. $F \subseteq \mathscr{L}$ and $A \subseteq$ AArg and $F \cup \mathsf{form}(A)$ is consistent.*

A sequence $(m_n)_{n \in [\![1,p]\!]}$ is a persuasion dialog *of* length p *based on (F, A) iff*

$$\begin{cases} \textsf{Act}(m_1) = \textsf{Assert} \text{ and } \forall n \in [\![1,p]\!], m_n \text{ is a well formed move} \\ \text{and there is a sequence } (CS_n)_{n \in [\![1,p+1]\!]} \text{ such that:} \forall n \in [\![1,p+1]\!], CS_n \text{ is a tuple} \\ (F_1, A_1, R_1, F^\circ, A^\circ, F_2, A_2, R_2) \in 2^{\mathscr{L}} \times 2^{\mathrm{AArg}_{\mathscr{L}}} \times 2^{\mathscr{M}} \times 2^{\mathscr{L}} \times 2^{\mathrm{AArg}_{\mathscr{L}}} \times 2^{\mathscr{L}} \\ \times 2^{\mathrm{AArg}_{\mathscr{L}}} \times 2^{\mathscr{M}} \text{ called commitment store at stage } n \text{ satisfying:} \\ \text{-Starting condition: } CS_1 = (\varnothing, \varnothing, \varnothing, F, A, \varnothing, \varnothing, \varnothing) \\ \text{-Current conditions: } \forall n \in [\![1,p-1]\!], \text{ let } x = \textsf{Sender}(m_n), \\ \quad \textsf{Close} \notin R_x \text{ and } \mathtt{precond}(m_n) \text{ is true in } CS_n \text{ and} \\ \quad CS_{n+1} = (F_1', A_1', R_1', F^{\circ\prime}, A^{\circ\prime}, F_2', A_2', R_2') \text{ defined by } \mathtt{effect}(m_n, CS_n) \\ \text{-Ending conditions: let } x = \textsf{Sender}(m_p), \\ \quad \mathtt{precond}(m_p) \text{ is true in } CS_p \text{ and } \textsf{Act}(m_p) = \textsf{Close} \text{ and } \textsf{Close} \in R_{\overline{x}} \end{cases}$$

where `precond` *and* `effect` *are given in Table 2* [2].

The commitment store used in this definition is made of three knowledge bases:

- a base containing *common knowledge* divided into two parts: the common formulas, denoted by F°, and the common arguments, denoted by A°,
- and the two commitment stores of *each agents*, each one divided into three parts[3]:
  - the first two parts contain the *assertive commitments* of agent x separated into a set F_x of propositional formulas, and a set A_x of approximate arguments
  - the third one contains the *commitments towards the other agent*, i.e., the requests to which x should answer, denoted by R_x.

In this definition, the "starting condition" is simply an initialisation of the commitment store. The "current conditions" require that, in order to make a move, the sender should not have already closed his participation to the dialog, and ensure that every move is done under the adequate preconditions, each move has an effect on the commitment store (preconditions and effects are described in Table 2). Note that the deterministic definition of the effects of the moves induces that the sequence of commitment store stages associated to a persuasion dialog is *unique*. The "ending conditions" ensure that the last move of the dialog is a Close move and that the other agent has already closed

[2] For shortness, in the "effect" column, the sets that remain unchanged are not mentioned, and K_x denotes the set of formulas $F_x \cup \mathsf{form}(A_x) \cup F^\circ \cup \mathsf{form}(A^\circ)$.

[3] The distinction between internal and external commitment is inspired from [1].

Table 2. Effects and conditions of a move from x towards $\overline{x}$

m	precond(m)[3]	effect(m, $(F_x, A_x, R_x, F^\circ, A^\circ, F_{\overline{x}}, A_{\overline{x}}, R_{\overline{x}})$)
Accept φ	$\varphi \in F_{\overline{x}}$ and $\{\varphi\} \cup K_x$ consistent	$R'_x = R_x \backslash \{(\mathsf{Accept}\,\varphi)\}, F^{\circ\prime} = F^\circ \cup \{\varphi\}$ $F'_{\overline{x}} = F_{\overline{x}} \backslash \{\varphi\}$ $R'_{\overline{x}} = R_{\overline{x}} \backslash \left\{ \begin{array}{l} (\mathsf{Challenge}\,\varphi), (\mathsf{Quiz}\,\langle S,\varphi\rangle) \\ (\mathsf{Quizlink}\,\langle S,\varphi\rangle) \end{array} \middle\vert\, S \subseteq \mathscr{L} \right\}$
Agree $\langle S,\varphi\rangle$	$\langle S,\varphi\rangle \in A_{\overline{x}}$ and $S \cup \{\varphi\} \cup K_x$ consistent	$R'_x = R_x \backslash \{(\mathsf{Agree}\,\langle S,\varphi\rangle)\}, F^{\circ\prime} = F^\circ \cup \{\varphi\} \cup S,$ $A^{\circ\prime} = A^\circ \cup \{\langle S,\varphi\rangle\}, A'_{\overline{x}} = A_{\overline{x}} \backslash \{\langle S,\varphi\rangle\}$ $R'_{\overline{x}} = R_{\overline{x}} \backslash \{(\mathsf{Challenge}\,\varphi), (\mathsf{Quiz}\,\langle S,\varphi\rangle), (\mathsf{Quizlink}\,\langle S,\varphi\rangle)\}$
Argue $\langle S,\varphi\rangle$	$\langle S,\varphi\rangle \notin A^\circ \cup A_x \cup A_{\overline{x}}$ and $S \cup \{\varphi\} \cup K_x$ consistent	$A'_x = A_x \cup \{\langle S,\varphi\rangle\}, R'_{\overline{x}} = R_{\overline{x}} \cup \{(\mathsf{Agree}\,\langle S,\varphi\rangle)\}$
Assert φ	$\varphi \notin K_x \cup F_{\overline{x}}$ and $\{\varphi\} \cup K_x$ consistent	$F'_x = F_x \cup \{\varphi\},$ $R'_{\overline{x}} = R_{\overline{x}} \cup \{(\mathsf{Accept}\,\varphi)\}$
Challenge φ	$\varphi \in F_{\overline{x}}$ and$(\mathsf{Challenge}\,\varphi) \notin R_{\overline{x}}$ and $\nexists \langle S,\varphi\rangle \in A_x \cup A_{\overline{x}} \cup A^\circ$	$R'_{\overline{x}} = R_{\overline{x}} \cup \{(\mathsf{Challenge}\,\varphi)\}$
Close	$R_x = \varnothing$	$R'_x = \{\mathsf{Close}\}$
Complete $(a = \langle S,\varphi\rangle,$ $b = \langle S',\varphi'\rangle)$	$a \in A_x$ and (b completes a) and $b \notin A^\circ \cup A_x \cup A_{\overline{x}}$ and $S' \cup \{\varphi'\} \cup K_x$ consistent	$A'_x = A_x \backslash \{a\} \cup \{b\}$ $R'_{\overline{x}} = R_{\overline{x}} \backslash \{(\mathsf{Agree}\,a)\} \cup \{(\mathsf{Agree}\,b)\}$ $R'_x = R_x \backslash \{(\mathsf{Quiz}\,a), (\mathsf{Quizlink}\,a)\}$ $\cup\{(\mathsf{Quizlink}\,b)$, if $(\mathsf{Quizlink}\,a) \in R_x$ and $\varphi = \varphi'\}$
Dismantle a	$a \in A_x$	$A'_x = A_x \backslash \{a\}$ $R'_x = R_x \backslash \{(\mathsf{Quiz}\,a), (\mathsf{Quizlink}\,a)\},$ $R'_{\overline{x}} = R_{\overline{x}} \backslash \{(\mathsf{Agree}\,a)\}$
Quiz a	$a \in A_{\overline{x}}$ and $(\mathsf{Quiz}\,a) \notin R_{\overline{x}}$ and $\mathsf{Decode}_{K_x}(a) = \varnothing$	$R'_{\overline{x}} = R_{\overline{x}} \cup \{(\mathsf{Quiz}\,a)\}$
Quizlink a	$a \in A_{\overline{x}}$ and $(\mathsf{Quizlink}\,a) \notin R_{\overline{x}}$ and $\nexists \langle S',\varphi'\rangle \in \mathsf{Decode}_{K_x}(a)$ s.t. $\left\{ \begin{array}{l} \{\varphi', \neg\varphi'\} \cap (F_x \cup \mathsf{form}(A_x)) \neq \varnothing \\ \text{or } \varphi' \in F_{\overline{x}} \cup \mathsf{form}(A_{\overline{x}}) \end{array} \right.$	$R'_{\overline{x}} = R_{\overline{x}} \cup \{(\mathsf{Quizlink}\,a)\}$
Retract φ	$\varphi \in F_x$ [3]with $K_x = F_x \cup \mathsf{form}(A_x) \cup F^\circ \cup \mathsf{form}(A^\circ)$	$F'_x = F_x \backslash \{\varphi\}$ $A'_x = A_x \backslash \{\langle S,\psi\rangle \mid S \subseteq \mathscr{L}, (\varphi \in S \text{ or } \psi = \varphi)\},$ $R'_x = R_x \backslash \left\{ \begin{array}{l} (\mathsf{Challenge}\,\varphi), \\ (\mathsf{Quiz}\,\langle S,\psi\rangle) \\ (\mathsf{Quizlink}\,\langle S,\psi\rangle) \end{array} \middle\vert \begin{array}{l} S \subseteq \mathscr{L}, \\ \varphi \in S \text{ or } \psi = \varphi \end{array} \right\}$ $R'_{\overline{x}} = R_{\overline{x}} \backslash \left\{ \begin{array}{l} (\mathsf{Accept}\,\varphi), \\ (\mathsf{Agree}\,\langle S,\psi\rangle) \end{array} \middle\vert \begin{array}{l} S \subseteq \mathscr{L}, \\ \varphi \in S \text{ or } \psi = \varphi \end{array} \right\}$

his participation to the dialog. If these "ending conditions" are not possible then the dialog has no end. This may seem not realistic, but it is often the case that the physical end of a debate is not really a true ending of the subject since the participants may still not be convinced by the arguments of their adversary, and the dialog will continue at another occasion. Let us describe more precisely how each move is taken into account according to Table 2:

An agent x may "accept" a formula φ (respectively "agree" about an argument $\langle S, \varphi\rangle$) only if this formula (resp. argument) has been uttered by the other agent $\overline{x}$ and is consistent with both the common knowledge and all the formulas that x has already uttered. After the Accept move the formula becomes common knowledge and is no more considered as $\overline{x}$ own utterance. Similarly, an Agree move introduces the formulas used in the argument as well as the argument itself into the common knowledge and the argument is removed from $\overline{x}$ own arguments. When these moves are uttered the commitments of the speaker to accept (or agree) this formula (or argument) are fulfilled hence removed from the commitment store, moreover the requests he may have made about this formula (or argument) are nomore committing his adversary.

The protocol is designed for obtaining "rational" dialogs, in terms of *non redundancy, self consistency* and *listening*. Hence, in order to Assert a formula (or Argue an argument) this formula (argument) should not have already been asserted and should be consistent with common knowledge and with what the utterer has already said. Each Assert or Argue move commits the receiver to accept or agree with it or to make the sender retract or dismantle it. This commitment is dropped when the move is retracted (or dismantled or completed) by his utterer or accepted (or agreed) by the receiver.

The Challenge is authorized only if the formula has been uttered by the other agent (and if it is not common knowledge) but not already been proved nor challenged. After this move the other agent is committed to give an argument for the challenged formula or to "retract" it, this is translated by adding (Challenge φ) to his requests commitment store. This request will be removed when an argument whose claim is φ will be agreed or if the formula φ is directly accepted.

A Close move requires that all the commitments of the agent are fulfilled. After this move the agent is not allowed to speak anymore: for this purpose an artificial commitment Close is added to his request commitment store, and, as described in the "current condition", no move can be done by agent x if Close is present in R_x.

"Completing" an argument $a = \langle S, \varphi\rangle$ by $b = \langle S', \varphi'\rangle$ should be done by giving a more precise argument b than a (uttered by the current speaker, but not yet agreed by the hearer) *i.e.*, a logical or incomplete argument that completes it. The new argument b should be consistent and not already present. After the utterance, b replaces a in the set of uttered arguments, some commitments of the utterer may nomore be appropriate: namely, Quiz a , Quizlink a. But a request commitment maybe inherited, namely if there was a request Quizlink a and $\varphi = \varphi'$, then it becomes Quizlink b.

Dismantle (or Retract) allows to remove an argument (respectively a formula) from the utterances of agent x. The commitments concerning this argument or formula are also removed from the request commitment stores of the sender and the receiver.

A move (Quiz $\langle S, \varphi\rangle$) can be done by an agent x only if there is no logical argument completing $\langle S, \varphi\rangle$ that can be built from the common knowledge and the formulas already asserted by x (this set is called K_x for short). A (Quizlink a) move requires that there is no obvious link between a and something said by x (positively or negatively) neither with a previous assertion of $\overline{x}$.

Example 1 (continued): *Let us consider the following persuasion sub-dialog:*

$$D = \begin{pmatrix} (Shopenhauer, & \textsf{Assert}, & d \rightarrow s &), \\ (Adversary, & \textsf{Argue}, & \langle\{m \rightarrow \neg s\}, o \rightarrow \neg s\rangle &), \\ (Shopenhauer, & \textsf{Argue}, & \langle\{d \leftrightarrow t \vee s\}, m \rightarrow \neg d\rangle &), \\ (Adversary, & \textsf{Agree}, & \langle\{d \leftrightarrow t \vee s\}, m \rightarrow \neg d\rangle &) \end{pmatrix}$$

Suppose that common knowledge is: $F^\circ = \{o \rightarrow m, o\}$ *meaning that "opera is music" and that "opera exists". Table 3 describes the commitment stores of each participant, with* a_1, a_2 *denoting respectively* $\langle\{m \rightarrow \neg s\}, o \rightarrow \neg s\rangle$ *and* $\langle\{d \leftrightarrow t \vee s\}, m \rightarrow \neg d\rangle$.

After these moves the dialog is not finished since two requests are not yet answered. Schopenhauer has two options either (1) he agrees with his adversary's argument a_1 *(since it is consistent with common knowledge) then he would have no more* *commitments and his adversary will be obliged either to accept the first claim or to provide another argument against it, or (2) he may ask his adversary to precise the link that argument a_1 has with the formulas already asserted. In that case the adversary would not be able to* Complete *his argument since the logical argument that completes a_1 and that has a link with the subject is γ_1 ($\langle\{m \to \neg s, o \to m, o, o \to d\}, \neg(d \to s)\rangle$) whose support is now inconsistent with the common knowledge (see Table 3).*

Table 3. Commitments stores of Schopenhauer and his Adversary

Schopenhauer			Common knowledge		Adversary		
After the third move							
Formulas	Arguments	Requests	Formulas (F°)	Arguments (A°)	Formulas	Arguments	Requests
$d \to s$	a_2	(Agree a_1)	$o \to m$ o			a_1	(Accept $d \to s$) (Agree a_2)
After the fourth move							
$d \to s$		(Agree a_1)	$o \to m$ o $d \leftrightarrow t \vee s$ $m \to \neg d$	a_2		a_1	(Accept $d \to s$)
If the move ($Shopenhauer$, Quizlink, a_1) is done Then the move ($Adversary$, Dismantle, a_1) should be done, leading to:							
$d \to s$			$o \to m$ o $d \leftrightarrow t \vee s$ $m \to \neg d$	a_2			(Accept $d \to s$)
If the Adversary has no other argument linked with the subject, then he is forced to do the move ($Adversary$, Accept, $d \to s$) in order to be authorized to close the dialog:							
			$o \to m$ o $d \leftrightarrow t \vee s$ $m \to \neg d$ $d \to s$	a_2			
Schopenhauer			Common knowledge		Adversary		

Proposition 3. *Two* Close *moves belonging to a persuasion dialog have distinct senders.*

Proof. Due to the definition of the precondition of Close, in order to be able to do it there should not remain any commitment unfulfilled, however a Close move commits the sender by adding Close to its requirement commitment store R_x.

Remark 3. Even if a persuasion dialog has a finite length p, it may be the case that, in CS_{p+1}, neither Content$(m_1) \in F^\circ$ nor $\neg$Content$(m_1) \in F^\circ$

Proof. Consider the dialog ((1, Assert, φ), (1, Retract, φ)).

Definition 7 (Output). *Let D be a persuasion dialog of length p, with $(CS_n)_{n\in[\![1,p+1]\!]}$ its sequence of commitment stages, the output of the dialog,* Output(D)*, is:*

- Undecided *if $p = \infty$*
- *otherwise*
  - Public agreement that φ holds, *if in CS_{p+1} it holds that $\varphi \in F^\circ$*
  - Public agreement that $\neg\varphi$ holds, *if in CS_{p+1} it holds that $\neg\varphi \in F^\circ$*
  - No public agreement on φ *otherwise*

4 A Protocol for Time-Limited Persuasion Dialogs

Since a persuasion dialog may be infinite, we introduce particular persuasion dialogs where the speaking time is restricted, indeed it is often the case that the speakers of a public debate have to keep strictly to a given speaking time. This notion requires to define first the duration associated to a move.

Definition 8 (Move duration). *We assume a function* $size : \mathscr{L} \rightarrow \mathbb{N}^\star$ *that associates an integer to each formula (e.g., the size of a binary encoding of this formula).*
The duration $d(m)$ *of a move* m *is equal to:*

$$d(m) = \begin{cases} 1 + size(\varphi) & \text{if } m = (\textsf{Assert}\,\varphi) \\ 1 + \sum_{\psi \in S} size(\psi) + size(\varphi) & \text{if } m = (\textsf{Argue}\,\langle S, \varphi \rangle) \\ 1 + \sum_{\psi \in S'} size(\psi) + size(\varphi') & \text{if } m = (\textsf{Complete}\,(\langle S, \varphi \rangle, \langle S' \varphi' \rangle)) \\ 1 & \text{if } \textsf{Act}(m) \notin \{\textsf{Assert}, \textsf{Argue}, \textsf{Complete}\}, \end{cases}$$

There is a link between taking duration into account and allowing enthymemes, since the usual reason to use an enthymeme is for sake of shortness (even if sometimes it is more a strategical choice). In the above definition, the duration of every move is one except for Assert, Argue and Complete moves where it is strictly greater than 1. This may seem artificial but it is based on the fact that those three moves are introducing new formulas (hence requiring time to express them) while other moves refer to already expressed formulas (hence could be shortly expressed by using only a reference). Now, we introduce the time-limited persuasion dialog which is a variant of a persuasion dialog:

Definition 9 (Time-limited persuasion dialog). *Let* $p \in \mathbb{N} \cup \{\infty\}$ *and* $T \in \mathbb{N}$, *let* $(F, A) \in 2^{\mathscr{L}} \times 2^{\text{AArg}}$ *be a common knowledge base such that* $F \cup \textsf{form}(A)$ *is consistent.*

A sequence of moves $(m_n)_{n \in [\![1,p]\!]}$, *is a* T-limited persuasion dialog *of* length p *based on* (F, A) *iff:*

$$\begin{cases} \textsf{Act}(m_1) = \textsf{Assert} \text{ and } \forall n \in [\![1,p]\!], m_n \text{ is a well formed move and there is a} \\ \text{sequence } (CS_n)_{n \in [\![1,p+1]\!]} \text{ where } \forall n \in [\![1,p+1]\!], CS_n \text{ is a tuple } (F_1, A_1, R_1, \\ dur_1, F^\circ, A^\circ, F_2, A_2, R_2, dur_2) \text{ called commitment store at stage } n \text{ with } dur_1, \\ dur_2 \in \mathbb{N} \text{ representing the agents remaining speaking time and } F^\circ, F_1, F_2 \subseteq \mathscr{L}, \\ A^\circ, A_1, A_2 \subseteq \text{AArg}_{\mathscr{L}}, R_1, R_2 \subseteq \mathscr{M}, \text{ satisfying:} \\ \text{-Starting condition: } CS_1 = (\varnothing, \varnothing, \varnothing, T, F, A, \varnothing, \varnothing, \varnothing, T) \\ \text{-Current conditions: } \forall n \in [\![1,p-1]\!], \text{ let } x = \textsf{Sender}(m_n), \\ \quad \textsf{Close} \notin R_x \text{ and } \texttt{precond}(m_n) \text{ is true in } CS_n \text{ and } dur_x \geq d(m_n) \text{ and} \\ \quad CS_{n+1} = (F_1', A_1', R_1', dur_1', F^{\circ\prime}, A^{\circ\prime}, F_2', A_2', R_2', dur_2') \\ \quad \text{defined by } \texttt{effect}(m_n, CS_n) \text{ with } dur_x' = dur_x - d(m_n), dur_{\overline{x}}' = dur_{\overline{x}} \\ \text{-Ending conditions: let } x = \textsf{Sender}(m_p), \\ \quad \textsf{Close} \notin R_x \text{ and } \texttt{precond}(m_p) \text{ is true in } CS_p \text{ and } \textsf{Act}(m_p) = \textsf{Close} \\ \text{and } dur_x \geq d(m_p) \text{ and } \begin{cases} \text{- either } \textsf{Act}(m_p) = \textsf{Close} \text{ and } \textsf{Close} \in R_{\overline{x}} \\ \text{- or } dur_x - d(m_p) = 0 \text{ and } dur_{\overline{x}} = 0 \end{cases} \end{cases}$$

where `precond` *and* `effect` *are given in Table 2.*

The last condition expresses the termination condition for the dialog: the dialog may finish either because the two agents agree to close it or because they have no more speaking time.

Proposition 4. *A T-limited persuasion dialog of length p is such that:*

$$\forall x \in AG, \forall k \in [\![1,p]\!] \quad dur_x(k) \geq 0$$

$$\forall x \in AG, \quad \sum_{\{m \in D | \mathsf{Sender}(m)=x\}} d(m) \leq T$$

Proposition 5. *A T-limited persuasion dialog is finite.*

This last property is important since by allowing enthymemes and requests about them, it is possible that a dialog may never end. Indeed an argument may be "Quizlinked" eternally (when there is no common knowledge) since explaining a concrete fact could require to go back to reasons involving the Big-Bang theory.

Definition 10 (Output of a time limited persuasion dialog). *Let D be a T-limited persuasion dialog of length p, with $(CS_n)_{n \in [\![1,p+1]\!]}$ its associated sequence of commitment stages, the output of the dialog, denoted by* Output*(D), is:*
– Public agreement that *φ holds,*   *if, in CS_{p+1}, $\varphi \in F^{\circ}$*
– Public agreement that *$\neg\varphi$ holds,*   *if, in CS_{p+1}, $\neg\varphi \in F^{\circ}$*
– No public agreement on *φ,*   *else*

The following proposition shows that when a dialog is closed by the two participants then they have fulfilled all their commitments (hence their commitment store contains only the Close commitment).

Proposition 6. *If* Close *appears twice in a time-limited persuasion dialog such that $(CS_n)_{n \in [\![1,p+1]\!]}$ is its associated sequence of commitment stages then in CS_{p+1} it holds that $R_x = R_{\overline{x}} = \{Close\}$ and $F_x = F_{\overline{x}} = \varnothing$ and $A_x = A_{\overline{x}} = \varnothing$.*

Proof. Due to Proposition 3, the two Close moves have been done by two distinct agents. Moreover, in order to do a Close move the commitments toward the other agent should be empty, i.e, $R_x = R_{\overline{x}} = \varnothing$. Furthermore, if all their commitments are fulfilled then they either have agreed or accepted all adversary's arguments or he has retracted them.

Proposition 7. *If D is a time-limited persuasion dialog of length p and $(CS_n)_{n \in [\![1,p+1]\!]}$ its associated sequence of commitment stages, then $\forall n \in [\![1, p+1]\!]$, $\forall x \in AG$,*

- *$F_x \cup \mathsf{form}(A_x) \cup F^{\circ} \cup \mathsf{form}(A^{\circ})$ consistent and*
- *$F_x \cap F^{\circ} = \varnothing$ and*
- *$A_x \cap A^{\circ} = \varnothing$*

Corollary 1. *If D is a time-limited persuasion dialog of length p and $(CS_n)_{n \in [\![1,p+1]\!]}$ is its associated sequence of commitment stages, then,*

$$\text{if } \mathsf{Output}(D) = \varphi \text{ then } \begin{cases} \varphi \cup F_x \cup F_{\overline{x}} \cup F^{\circ} \text{ is consistent and} \\ \nexists \langle S, \psi \rangle \in A_1 \cup A_2 \cup A^{\circ} \text{ s.t. } \psi \vdash \neg\varphi \end{cases}$$

The following propositon shows that common knowledge may only increase with the persuasion dialog.

Proposition 8. *If D is a time-limited persuasion dialog of length p based on a common knowledge base (F, A), and if $(CS_n)_{n \in [\![1,p+1]\!]}$ is its associated sequence of commitment stages, then $\forall n \in [\![1, p+1]\!]$, in CS_n, it holds that $F \subseteq F^{\circ}$ and $A \subseteq A^{\circ}$*

5 Concluding Remarks

This work is a preliminary study on handling enthymemes in persuasion dialogs. The ambition was to deal with incomplete information both in the premises and in the claim of an argument. The latter is more difficult to handle and has required to introduce a new speech act Quizlink allowing to ask for an insight about what is hiding behind the claim. In some cases, one may not disagree with an argument that is not related with the subject but when he understands the underlying implication he wants to reject it. This is why it is necessary to allow the agent to "dismantle" an argument even if this argument is not attacked.

In this work we only represent what is publicly uttered, since we consider that we do not have access to the agent's mind. This way to apprehend the public statements is also done for instance by [10], where a public utterance is called "grounded". Their proposal is a framework to represent and reason about the public knowledge during a persuasion dialog, their approach allows to deal with inconsistent assertions (which is not allowed in our framework) like in Walton & Krabbe's system PPD_0. This feature seems more realistic since it is up to the other agent to detect and denounce inconsistency by asking to its adversary to "resolve" it. Dealing with possible inconsistent assertions is a challenge for further developments of our approach, however we could argue that a well designed protocol should enforce that what is public is consistent (in order to obtain a debate of hight quality that is civilised and respectful of the audience). Note that [10] do not deal with "dark side commitments" [24] (implicit assertions that are difficult to concede explicitely in front of a public) since they do not want to take into account the agent's mind but rather focus on what is observable and objective. In our approach a "dark side commitment" can be encountered when decoding an enthymeme, it may be the necessary piece to add in order to obtain a logical argument and may be revealed by means of a Quizlink or a Quiz move.

Our definition of incomplete argument maybe compared to the notion of "partial argument" given in [22], which is a set of default rules, and conclusion such that there exists a minimal set of strict rules and facts coming from a given knowledge base that allows to defeasibly derive the conclusion. Beyond the fact that we do not consider default reasoning, our incomplete argument, and particularly the ones that are enthymemes are not lacking information because it is not available as in the work of [22] but rather because the lacking information is considered as obvious or worthwile to conceal. The use of enthymemes is not necessarily a proof of weakness but rather "a highly adaptative argumentation strategy, given the need of everyday reasoners to optimize their cognitive resources" as it is claimed in [15,16]. The computation of the completion of enthymeme, namely how to define our function Decode, is out of the scope of the paper but as already been studied by several authors (see e.g. [23,14] in which several examples are analysed in order to understand how implicit premises can be discovered, and which provides a set of argumentation schemes that can be used as a guide for finding them). Moreover, it seems important to take into account that in enthymemes, the link between the premises and the claim is not necessarily classical logic inference, this is why [14] proposes to view it as a presumptive type of argumentation scheme. The notion of non-classical inference is also suggested by [18] that defines an argument as a pair where the premises provide backing for the claim but do not necessarily infere it. Indeed in this work, the argument is composed by a set of litterals (for the premises) and a literal (for the claim), with the only constraint that no premise is equal to the claim or its negation. This work is related to our own on another aspect since the argument is evaluated with respect to a knowledge base called "evidence", this base plays a similar role than our "common knowledge base" and represents the context in which arguments are to take into account. The coherence and redundancy notions of an argument with respect to the "evidence" are also introduced, this slightly differs from our approach in which the protocol ensures "self-coherence" of an agent (hence it implies both the common knowledge and the agent utterances), non-redundancy (based on common knowledge but also on agents utterances) and listening (this last is not related to [18] since they do not deal with dialog systems). In [18] the use of evidence it not done in order to complete arguments as we do with the common knowledge but to decide about their status. Besides a very appealing aspect of this approach is that evidence may evolve hence may imply changes in the arguments status while in our proposal, common knowledge may only increase consistently. Again the non-monotonic aspect seems to be interesting to consider in future studies.

During dialogs, the public utterances are stored and may evolve when arguments are retracted or replaced. Since enthymemes are possible and based on implicit information that is often common knowledge, we use a common knowledge base that is public. An advantage of our proposal is that at the end of the dialog all agreed assertions are kept in this common knowledge base that can hence only increase during the dialog, and that can be used for future dialogs.

A very appealing development of this framework concerns the strategical part, we plan to translate our protocol rules into the Game player project language GDL2 [21], indeed in GDL2 it is possible to handle games with imperfect information. With this translation, strategies coming from game theory and strategies dedicated to dialog games (e.g. [2]) coud be compared. Moreover, the move duration would have to be taken into account for the strategical aspect.

Acknowledgments. The author thanks Pierre Bisquert for the useful reference to Schopenhauer. Some words of thanks also go to the reviewers for valuable suggestions of improvements and very appropriate bibliographic references.

References

1. Amgoud, L., Dupin de Saint Cyr, F.: Towards ACL semantics based on commitments and penalties. In: European Conf. on Artif. Intelligence (ECAI), pp. 235–239. IOS Press, Amsterdam (2006)
2. Amgoud, L., Maudet, N.: Strategical considerations for argumentative agents (preliminary report). In: Proceedings of the 9th International Workshop on Non-Monotonic Reasoning (NMR), pp. 409–417 (2002), Special session on Argument, Dialogue, Decision
3. Amgoud, L., Maudet, N., Parsons, S.: Modelling dialogues using argumentation. In: Proc. of the International Conference on Multi-Agent Systems, Boston, MA, pp. 31–38 (2000)
4. Austin, J.: How to Do Things With Words, Cambridge (Mass.), 1962, 2nd edn. Harvard University Press, Paperback (2005)
5. Bench-Capon, T.: Persuasion in practical argument using value-based argumentation frameworks. J. of Logic and Computation 13(3), 429–448 (2003)

6. Besnard, P., Hunter, A.: A logic-based theory of deductive arguments. Artificial Intelligence 128(1-2), 203–235 (2001)
7. Black, E., Hunter, A.: Using enthymemes in an inquiry dialogue system. In: Proc of the 7th Int. Conf. on Autonomous Agents and Multiag. Syst. (AAMAS 2008), pp. 437–444 (2008)
8. Dunne, P., Bench-Capon, T.: Two party immediate response disputes: Properties and efficiency. Artificial Intelligence 149, 221–250 (2003)
9. Dupin de Saint-Cyr, F.: A first attempt to allow enthymemes in persuasion dialogs. In: DEXA International Workshop: Data, Logic and Inconsistency, DALI (2011)
10. Gaudou, B., Herzig, A., Longin, D.: A Logical Framework for Grounding-based Dialogue Analysis. Electronic Notes in Theoretical Computer Science 157(4), 117–137 (2006)
11. Gordon, T.: The pleadings game. Artificial Intelligence and Law 2, 239–292 (1993)
12. Hamblin, C.: Fallacies. Methuen, London (1970)
13. Hunter, A.: Real arguments are approximate arguments. In: Proceedings of the 22nd AAAI Conference on Artificial Intelligence (AAAI 2007), pp. 66–71. MIT Press, Cambridge (2007)
14. Macagno, F., Walton, D.: Enthymemes, argumentation schemes, and topics. Logique et Analyse 205, 39–56 (2009)
15. Paglieri, F.: No more charity, please! enthymematic parsimony and the pitfall of benevolence. In: Dissensus and the search for common ground: Proc. of OSSA 2007, pp. 1–26 (2007)
16. Paglieri, F., Woods, J.: Enthymematic parsimony. Synthese 178, 461–501 (2011)
17. Parsons, S., McBurney, P.: Games that agents play: A formal framework for dialogues between autonomous agents. J. of Logic, Language and Information 11(3), 315–334 (2002)
18. Rotstein, N., Moguillansky, M., García, A., Simari, G.: A dynamic argumentation framework. In: COMMA, pp. 427–438 (2010)
19. Schopenhauer, A.: The Art of Always Being Right: 38 Ways to Win an Argument (1831), `http://en.wikisource.org/wiki/The_Art_of_Always_Being_Right`, Orig. title: Die Kunst, Recht zu behalten (Transl. by T. Saunders in 1896)
20. Searle, J.: Speech acts: An essay in the philosophy of language. Cambridge U. Press, Cambridge (1969)
21. Thielscher, M.: A general game description language for incomplete information games. In: Proceedings of AAAI, pp. 994–999 (2010)
22. Thimm, M., Garcia, A., Kern-Isberner, G., Simari, G.: Using collaborations for distributed argumentation with defeasible logic programming. In: Proceedings of the 12th Int. Workshop on Non-Monotonic Reasoning (NMR 2008), pp. 179–188 (2008)
23. Walton, D.: The three bases for the enthymeme: A dialogical theory. Journal of Applied Logic 6, 361–379 (2008)
24. Walton, D., Krabbe, E.: Commitment in Dialogue: Basic Concepts of Interpersonal Reasoning. State University of New York Press, Albany (1995)
25. Walton, D., Reed, C.: Argumentation schemes and enthymemes. Synthese 145, 339–370 (2005)

Argumentation Frameworks with Necessities

Farid Nouioua and Vincent Risch

LSIS - UMR CNRS 6168,
Avenue Escadrille Normandie Niemen,
13397, Marseille Cedex 20, France
{farid.nouioua,vincent.risch}@lsis.org

Abstract. In this paper, we introduce argumentation frameworks with necessities (AFNs), an extension of Dung's argumentation frameworks (AFs) taking into account a necessity relation as a kind of support relation between arguments (an argument is necessary for another). We redefine the acceptability semantics for these extended frameworks and we show how the necessity relation allows a direct and easy correspondence between a fragment of logic programs (LPs) and AFNs. We introduce then a further generalization of AFNs that extends the necessity relation to deal with sets of arguments. We give a natural adaptation of the acceptability semantics to this new context and show that the generalized frameworks allow to encode arbitrary logic programs.

Keywords: abstract argumentation, necessity relation, acceptability semantics, bipolarity, logic programming.

1 Motivation

Abstract argumentation frameworks (AFs) initiated by P.M.Dung [10] have recently received considerable interest. An AF consists simply of a set of abstract objects, the arguments, and a binary attack relation. Despite their simplicity, AFs provide a powerful tool to capture various non-monotonic reasoning approaches. Many extensions have been proposed to initial AFs in order to enrich them by further features. Since arguing often involves the exchange of arguments for or against a given position, one of the relevant features in argumentation is that of support on which we focus in this paper.

Roughly speaking, we can distinguish two approaches in treating the idea of support in abstract argumentation. In the first approach support is taken in the sense of logical inference. For example [1] considers the issue of building an AF from a logical knowledge base. Following [2], the used arguments are constructed from the knowledge base as couples of the form $(support, conclusion)$ where $support$ is a minimal set of formulas that infers the $conclusion$. Here, the support is an internal mechanism to the argument itself. Another work proposes constrained argumentation frameworks [9] that add to AFs propositional constraints manipulating arguments as propositional atoms. The support as a logical inference operates here at a logical level which is complementary to the abstract argumentative level. The second approach, into which our work fits, consists in adding to AFs an explicit support relation between arguments. A well known work in this direction are the bipolar argumentation frameworks (BAFs) [7] [8]. Its main drawback lies in that supported and indirect attacks proposed in this model

S. Benferhat and J. Grant (Eds.): SUM 2011, LNAI 6929, pp. 163–176, 2011.
© Springer-Verlag Berlin Heidelberg 2011

may lead to counter-intuitive results if the support relation has a precise meaning like that of necessity. For example, consider this dialogue :

Agent A: I will light up this room; **Agent B:** It is not true, the lamp will not light up;
Agent A: I will open the switch; **Agent B:** But the meter will not work;
Agent A: Really ?; **Agent B:** Yes, the electrician has detected a failure in the meter.

Let us consider the following arguments: The room will be dark (RD); The lamp will light up (LL); the switch will be open (SO); the meter works (MW); there is a failure in the meter (F). In the resulting BAF we have the relations : F $attacks$ MW, LL $attacks$ RD, MW $Supports$ LL and SO $Supports$ LL. According to [7] the unique stable extension is $E = \{F, SO\}$, i.e., there is a failure in the meter and the switch is open. We naturally expect that the extension contains also RD (the room will stay dark) but this is not the case because of the supported attack of RD by SO. However, this attack is explained only by the fact that SO supports LL which attacks RD, but, since $LL \notin E$, there is no reason to consider as successful the attack of RD by SO. This problem persists with all kinds of extensions presented in [7] but it can be fixed by remarking that the support relation used in the example is a necessity. The reader can easily check that the only stable extension obtained by applying the approach described in this paper is $\{F, SO, RD\}$ which corresponds to our expectations.

Subsequent works that are interested in the notion of support include the evidence-based argumentation approach [18] and more recently the work in [3] about deductive and defeasible supports as well as the abstract dialectical frameworks [6].

The support relation may have different meanings that may require completely different treatments. Our concern here is not to provide an exhaustive enumeration of these possible meanings. Instead we will focus on two of them that are very intuitive : the necessity and the sufficiency relations. In AFs, "a attacks b" is interpreted by : "if a is accepted then b is not accepted". Similarly, we will interpret "a is necessary for b" exactly as "b is sufficient for a" by : "if b is accepted then a is accepted". Thus, without loss of generality we can focus on only one of these two relations. We chose necessity because of its adequacy for LPs (see sections 4. and 5.). The motivation of this work is to propose new direct and very simple methods to link argumentation frameworks and LPs. This kind of cheap translations is very beneficial : among other thinks, it allows to reuse in one formalism the methods and algorithms developed in the other formalism and also to take benefit of the important developments in LPs domain to implement efficiently the new AFs. To do so, we propose in this paper an extension of Dung's AFs with the necessity relation and we consider the following points : (1) we highlight the kind of interactions that result between necessities and attacks; (2) in light of these interactions, we consider the acceptability semantics in the new context. Since necessities are considered at the same level of attacks, instead of proposing new alternatives, we integrate the necessity relation in redefining the main existing semantics. This integration is not performed without additional problems for which we provide suitable solutions; (3) we consider then the issue of relating argumentation systems with LPs. This issue initiated in [10] continues to be an interesting research topic (see for example [11] [16] [20]). In our context, we show that adding the necessity relation not only doesn't alter the existing connection between AFs and LPs but also simplifies it.

First, we recall some basics about answer set programming and Dung's abstract AFs. Then, we introduce the AFNs, we generalize their acceptability semantics and we show how to represent them as classical AFs. After that, we discuss the correspondence between AFNs and a fragment of LPs. Then, we introduce a further extension of AFNs that generalizes the necessity relation to sets of arguments. After the adaptation of the key notions and acceptability semantics to this new context, we show how this generalized AFNs cope properly with arbitrary logic programs. Finally, a last section includes a discussion of related works as well as some perspectives of future work.

2 Preliminaries

2.1 Answer Sets and ι–Answer Sets for LPs

A *normal LP* is a finite set of rules of the form :

$$a_0 \leftarrow a_1, ..., a_m, not\ a_{m+1}, ..., not\ a_n \tag{1}$$

where $0 \leq m \leq n$ and a_i $(0 \leq i \leq n)$ are atoms. For a rule r of the form (1) we define : $head(r) = a_0$, $body^+(r) = \{a_1, ..., a_m\}$ and $body^-(r) = \{a_{m+1}, ..., a_n\}$. The previous definition is generalized to any LP Π as follows : $head(\Pi) = \{head(r) | r \in \Pi\}$, $body^+(\Pi) = \bigcup_{r\in\Pi} body^+(r)$ and $body^-(\Pi) = \bigcup_{r\in\Pi} body^-(r)$. A LP Π is *basic* if $body^-(\Pi) = \emptyset$. A set of atoms X is closed under a basic LP Π if, for any $r \in \Pi$, $head(r) \in X$ whenever $body^+(r) \subseteq X$. $Cn(\Pi)$ denotes the smallest set of atoms closed under Π. The *reduct* of a program Π relative to a set of atoms X is the basic program Π^X given by : $\Pi^X = \{head(r) \leftarrow body^+(r) | r \in \Pi, body^-(r) \cap X = \emptyset\}$. A set X of atoms is an *answer set* of Π iff $X = Cn(\Pi^X)$ [13]. The *generating rules* and the *applicable rules* of Π relative to X are given respectively by : $G_\Pi(X) = \{r \in \Pi | body^+(r) \subseteq X, body^-(r) \cap X = \emptyset\}$ and $Ap_\Pi(X) = \{r \in \Pi | body^+(r) \subseteq X, body^-(r) \cap X = \emptyset, head(r) \in X\}$.

It is easy to see that we always have: $Ap_\Pi(X) \subseteq G_\Pi(X)$. *$\iota$-Answer sets* [12] are proposed as the counterpart of Lukaszewicz justified extensions [14] in the domain of default logics [19] to overcome the problem of non-modularity in the construction of an answer set which needs to inspect all the rules at once. Its advantage is that, unlike answer sets, it is possible to incrementally construct a ι-answer set or to locally validate a (partial) construction of it (for more details, see [14], [12]).

Let $Cn^+(\Pi) = Cn(\Pi^\emptyset)$.A set X of atoms is a ι-answer set of Π if $X = Cn^+(\Pi')$ for some $\subseteq$-maximal $\Pi' \subseteq \Pi$ satisfying the following two conditions:

$$(C1) : body^+(\Pi') \subseteq Cn^+(\Pi'), \qquad (C2) : body^-(\Pi') \cap Cn^+(\Pi') = \emptyset.$$

For a LP Π and a set of atoms X, we have these results : (1) *Incremental construction*: if X verifies $X = Cn^+(Ap(X))$ then there is a ι-answer set X' such that $X \subseteq X'$; (2) *Existence*: for every LP Π, there is at least one ι-answer set. This property does not hold in general for answer sets; (3) *Answer sets are ι-answer sets*: if X is an answer set of Π then X is a ι-answer set of Π, but not *vice versa*. Moreover, if X is a ι-answer set of Π then X is an answer set of Π iff $Ap_\Pi(X) = G_\Pi(X)$[1].

[1] In the rest of the paper we assume that any LP Π we use, verifies the property : $body^+(\Pi) \subseteq head(\Pi)$. In fact, any rule r such that $body^+(r) \nsubseteq head(\Pi)$ is never applied and it is easy to show that by removing it from Π, we do not alter neither its ι-answer sets nor its answer sets.

2.2 Dung's Argumentation Frameworks

Dung's AFs [10] are defined by a pair $F = \langle A, R \rangle$ where A is a set of arguments and R is a binary attack relation over A. A set $S \subseteq A$ attacks an argument b iff there is $a \in S$ such that $a \; R \; b$. S is *conflict-free* iff there is no $a, b \in S$ such that $a \; R \; b$. The $\subseteq$-maximal conflict-free subsets of A are called *naive extensions* [4] and represent a first manner to construct sets of acceptable arguments. Many other acceptability semantics have been proposed in [10]. Among them, we will focus in this paper on the following semantics widely used in practice : (1) S is an *admissible set* iff S is conflict-free and $\forall a \in A \backslash S$, if $a \; R \; b$ for some $b \in S$ then $S \; R \; a$; (2) a *preferred extension* is a $\subseteq$-maximal admissible set; (3) S is a *stable extension* iff S is conflict-free and $\forall a \in A \backslash S$, $S \; R \; a$.

Dung proposed in [10] methods to translate any AF into a LP and *vice versa*. If the passage from an AF to a LP is very simple, the passage in the opposite side is less direct and requires the computation of a number of arguments which may be, in general, exponential relative to the number of atoms in the program [2].

3 Bipolar AFs with Necessities

Let us first give some basic definitions. We start by the formal definition of an AFN which adds to the classical Dung's AF a new (positive) relation between arguments

Definition 1. An AFN is defined by $\langle A, R, N \rangle$ where A is a set of arguments, $R \subseteq A \times A$ (resp. $N \subseteq A \times A$) is a binary *attack* (resp. *necessity*) relation over A: for $a, b \in A$, $a \; R \; b$ (resp. $a \; N \; b$) means that a attacks (resp. is necessary for) b.

New attacks and necessities are deduced from the direct ones. On the one hand, the transitive closure of the necessity relation is interpreted as a necessity (if a is necessary for b and b is necessary for c then a is necessary for c). On the other hand, new attacks emerge in two cases : if a attacks c and c is necessary for b then a attacks b, and if c attacks b and c is necessary for a then a attacks b (accepting a means that c is also accepted which excludes b). Comparing with [7], the first case corresponds to the indirect attack but the second case does not correspond to the supported attack.

Definition 2. Let $\Delta = \langle A, R, N \rangle$ be an AFN. There is an extended necessity from a to b ($a \; N^+ \; b$) iff there is a sequence: $a_1 \; N \; ... \; N \; a_n \;\; (n \geq 2)$ where $a_1 = a$, $a_n = b$. There is an extended attack of b by a ($a \; R^+ \; b$) iff either we have : $a \; R \; b$ or there is $c \in A$ such that $a \; R \; c \; N^+ \; b$ or $c \; R \; b$ and $c \; N^+ \; a$.

Clearly a classical AF is a particular case of AFN where $N = \emptyset$. When $N \neq \emptyset$, a main requirement in defining acceptability is to avoid cycles of necessities that reflect a kind of deadlock which can be seen as a form of the fallacy "*begging the question*".

[2] From a given LP Π, the method in [10] constructs all the arguments of the form $(\{\neg b_1, \cdots \neg b_m\}, k)$ where $K = \{\neg b_1, \cdots \neg b_m\}$ is a set of negative literals and k is an atom which is a defeasible consequence of $\{\neg b_1, \cdots \neg b_m\}$, i.e., there is a sequence $(e_0, \cdots e_n)$ with $e_n = k$ and for each e_i either $e_i \leftarrow$ is a rule of Π or e_i is the head of a rule $e_i \leftarrow a_1, \cdots a_t, not \; a_{t+1}, \cdots not \; a_{t+r}$ of Π such that $a_1, \cdots, a_t$ belong to the preceding members of the sequence and $\neg a_{t+1}, \cdots, \neg a_{t+r}$ boelong to K. K is called the support of k.

Definition 3. Let $\Delta = \langle A, R, N \rangle$ be an AFN and $a \in A$. a is necessity-cycle free (N-Cycle-Free) iff it is not the case that $a\ N^+\ a$ or that $b\ N^+\ a$ and $b\ N^+\ b$. A set $S \subseteq A$ is N-Cycle-Free iff all their elements are N-Cycle-Free.

Now, we are ready to introduce the key notions of *coherent* and *strongly coherent sets*. The latter will generalize the notion of conflict-freeness in original Dung's AFs.

Definition 4. Let $\Delta = \langle A, R, N \rangle$ be an AFN and $S \subseteq A$. S is said to be *coherent* iff S is N-Cycle-Free and closed under N^{-1} (if $a \in S$ then $b \in S$ for each $b\ N\ a$). S is *strongly coherent* iff it is coherent and conflict-free w.r.t R.

The coherence of a set S excludes the risk of having a necessity-cycle S and ensures that S provides to each of its arguments all its necessary arguments. Notice that in Dung's AFs ($N = \emptyset$), the strong coherence is reduced to classical conflict-freeness.

3.1 Adaptation of Acceptability Semantics to AFNs

The acceptability semantics for AFNs follow roughly the same principles of that for AFs and uses the notion of strong coherence instead of conflict-freeness. The naive and stable extension are defined as follows :

Definition 5. Let $\Delta = \langle A, R, N \rangle$ be an AFN and $S \subseteq A$. (1) S is a naive extension of Δ iff S is a $\subseteq$-maximal strongly coherent subset of A; (2) S is a stable extension of Δ iff S is a strongly coherent subset of A such that for each $a \in A \setminus S$ either $S\ R\ a$ or $b\ N\ a$ for some $b \in A \setminus S$.

Proposition 1 characterizes the arguments inside and outside a stable extension.

Proposition 1. Let $\Delta = \langle A, R, N \rangle$ be an AFN and $S \subseteq A$. S is a stable extension of Δ iff : (1) S is N-Cycle-Free and (2) ($a \in S$) iff ($\forall b \in A$, if $b\ R\ a$ then $b \notin S$ and if $b\ N\ a$ then $b \in S$).

The ($\Rightarrow$) part of condition (2) means that if an argument a is in S, then each argument that attacks it is outside S and each argument which is necessary for it is inside S. The ($\Leftarrow$) part states that if an argument a is not in S, then either it is attacked by S or there is an argument outside S which is necessary for a. The relationship between naive and stable extensions is given by the following proposition.

Proposition 2. Each stable extension is a naive extension but not vice versa.
Let us now turn to the definition of admissible sets and preferred semantics for AFNs by using the notions of coherence and strong coherence.

Definition 6. Let $\Delta = \langle A, R, N \rangle$ be an AFN and $S \subseteq A$. S is an admissible set of Δ iff S is strongly coherent and if $b\ R\ S$ then for each coherent subset $S' \subseteq A \setminus S$ such that $b \in S'$ we have $S\ R\ S'$. A preferred extension is a $\subseteq$-maximal admissible set.

Definition 6 may be seen as an extension of Dung's definition of admissibility in AFs where conflict-freeness is replaced by strong coherence and self-defense concerns extended attacks (and not only the direct ones) and is required only against arguments that are N-Cycle-Free. This is the claim of the following proposition 3.

Proposition 3. Let $\Delta = \langle A, R, N \rangle$ be an AFN and $S \subseteq A$. S is an admissible set of Δ iff S is coherent and for each $b \in A \setminus S$, if b is N-Cycle-Free and $b\ R\ S$ then, $S\ R^+\ b$.

It is worth noticing that stable and preferred semantics for AFNs preserve the properties of stable and preferred extensions for Dung's AFs. This is the case for the existence of preferred extensions and the fact that stable extensions are also preferred.

Proposition 4. Let $\Delta = \langle A, R, N \rangle$ be an AFN. We have : (1) there is always at least one preferred extension for Δ; (2) each stable extension of Δ is a preferred extension of Δ, but not vice versa.

Example 1. Figures 1-(1) and 1-(2) depict two examples of AFNs where continuous edges represent attacks and dashed edges represent necessities. The system of figure 1-(1) has three naive extensions, $\{r_2\}$, $\{r_1, r_3\}$ and $\{r_4, r_5\}$. Two of them are stable extensions, $\{r_2\}$ and $\{r_4, r_5\}$. Let us check for instance that $S = \{r_4, r_5\}$ is a stable extension. Only r_4 has attackers (r_1 and r_2) that are outside S and only r_5 has a necessary argument (r_4) which is inside S. Moreover, each argument outside S is either attacked by S (the case of r_1, r_2 and r_3) or has a necessary argument outside S (the case of r_1). The admissible sets of this system are $\{r_2\}$ and $\{r_4, r_5\}$ that are also the preferred and the stable extensions. Let us check for example that $\{r_2\}$ is admissible but $\{r_1, r_3\}$ is not. r_2 is attacked by r_5 and all coherent sets containing r_5 must contain r_4 which is attacked by r_2. For $\{r_1, r_3\}$ we have : r_3 is attacked by r_2 but $\{r_2\}$ which is a coherent set containing r_2 is not attacked by $\{r_1, r_3\}$.

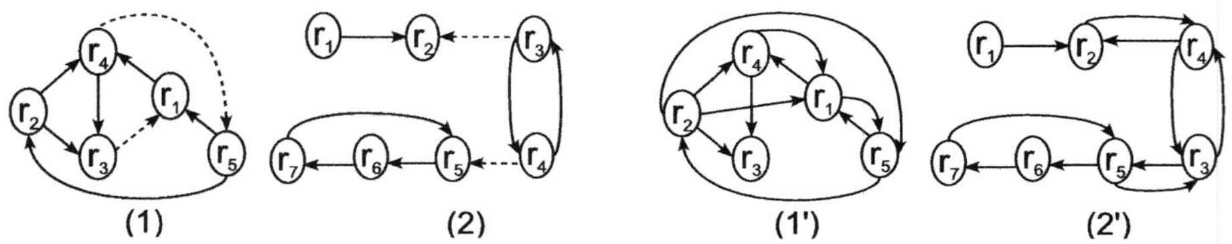

Fig. 1. Examples of two AFNs and two AFs

The system depicted in figure 1-(2) has eight naive extensions and only one of them is stable : $\{r_1, r_3, r_6\}$. It has two preferred extensions : $\{r_1, r_3, r_6\}$ and $\{r_1, r_4\}$.

It is easy to check that for $N = \emptyset$, naive, stable and preferred extensions of the AFN $\Delta = \langle A, R, \emptyset \rangle$ coincide with the corresponding extensions of the AF $F = \langle A, R \rangle$.

3.2 AFN as a Classical AF

We have seen above that an AF can be considered as an AFN whose necessity relation is empty. In this section we consider the opposite question and show how to represent any AFN by an AF having at most the same number of arguments. The idea is to keep only arguments that are N-Cycle-Free and use the extended attacks as attack relation.

Definition 7. Let $\Delta = \langle A, R, N \rangle$ be an AFN. Δ can be represented by an AF $F_\Delta = \langle A_\Delta, R_\Delta \rangle$ such that : $A_\Delta = \{a \in A|$ a is N-Cycle-Free$\}$ and for each $x, y \in A_\Delta$, $x\ R_\Delta\ y$ iff $x\ R^+\ y$ (R_Δ is the restriction of R^+ on A_Δ).

It turns out that strong coherence in the AFN is stronger than conflict-freeness in the corresponding AF. However naive, stable and preferred extensions are all preserved.

Proposition 5. Let Δ be an AFN, F_Δ its corresponding AF and $S \subseteq A$. We have : (1) if S is strongly coherent in Δ then S is conflict-free in F_Δ; (2) S is a naive (resp. stable, preferred) extension in Δ iff S is a naive (resp. stable, preferred) extension in F_Δ.

Example 1 (cont.). Let Δ_1 and Δ_2 be the AFNs depicted in figures 1-(1) and 1-(2) respectively. The corresponding AFs F_{Δ_1} and F_{Δ_2} are represented in figures 1-(1') and 1-(2'). $\{r_2\}$, $\{r_1, r_3\}$, $\{r_4, r_5\}$ are the naive extensions and $\{r_4, r_5\}$, $\{r_2\}$ are the stable and the preferred extensions of both Δ_1 and F_{Δ_1}. Similarly, we can check that and Δ_2 and F_{Δ_1} share the same naive, stable and preferred extensions.

4 AFN and Logic Programs

The aim of this section is to take advantage of the necessity relation to propose a new method to represent a LP as an AFN. The key idea of this method is to consider a rule itself as an argument. The advantage of this method is to provide an immediate translation where the number of resulting arguments is no more exponential w.r.t the number of atoms but equals the number of rules in the program. However, following this method, AFNs captures only a fragment of LPs. The fragment considered here is denoted $Frag$ and it contains LPs where each atom is given by only one rule of the LP. Formally, a LP Π is in $Frag$ iff $\forall r_1, r_2 \in \Pi$, if $r_1 \neq r_2$ then $head(r_1) \neq head(r_2)$.

Let Π be a LP of $Frag$ containing n rules, $\Pi = \{r_1, ..., r_n\}$. The corresponding AFN is $\Delta_\Pi = \langle \Pi, R_\Pi, N_\Pi \rangle$ such that : its arguments are the rules of Π, $\forall r_1, r_2 \in \Pi, r_1\ R_\Pi\ r_2$ iff $head(r_1) \in body^-(r_2)$ and $\forall r_1, r_2 \in \Pi, r_1\ N_\Pi\ r_2$ iff $head(r_1) \in body^+(r_2)$. Then, there is a correspondence between ι-answer sets (resp. answer sets) of a LP of $Frag$ and naive (resp. stable) extensions in the corresponding AFN.

Proposition 6. Let Π be a LP of $Frag$ and $\Delta_\Pi = \langle \Pi, R_\Pi, N_\Pi \rangle$ be the corresponding AFN. A set X is a ι-answer set (resp. answer set) of Π, with Π' as the corresponding maximal subset of Π iff Π' is a naive (resp. stable) extension of Δ_Π.

Example 2. Consider the program Π_1 :

$$r_1 : a \leftarrow c,\ not\ e$$
$$r_2 : b \leftarrow not\ e$$
$$r_3 : c \leftarrow not\ d,\ not\ b$$
$$r_4 : d \leftarrow not\ a,\ not\ b$$
$$r_5 : e \leftarrow d$$

The ι-answer sets of Π are $\{d,e\}$, $\{b\}$ and $\{a,c\}$. For $\{d,e\}$, the corresponding maximal subset of Π_1 is $\Pi'_1 = \{r_4, r_5\}$. Since $body^+(\Pi'_1) = \{d\} \subseteq \{d,e\}$ and $body^-(\Pi'_1) \cap Cn^+(\Pi'_1) = \{a,b\} \cap \{d,e\} = \emptyset$, conditions (C1) and (C2) hold for Π'_1 and we can check that Π'_1 is maximal. Moreover, $G_{\Pi_1}(\{d,e\}) = Ap_{\Pi_1}(\{d,e\}) = \{r_4, r_5\}$, so $\{d,e\}$ is also an answer set. Similarly we can check that $\{b\}$ is a ι-answer set and an answer set and that $\{a,c\}$ is a ι-answer set but not an answer set because $G_{\Pi_1}(\{a,c\}) \neq Ap_{\Pi_1}(\{a,c\})$. The AFN $\Delta_{\Pi_1} = \langle \Pi_1, R_{\Pi_1}, N_{\Pi_1} \rangle$ corresponding to Π_1 is that depicted in figure 1-(1). The three ι-answer sets of Π_1, $\{d,e\}$, $\{b\}$ and $\{a,c\}$ have, respectively, $\{r_4, r_5\}$, $\{r_2\}$ and $\{r_1, r_3\}$ as maximal subset Π'_1 of Π_1. These sets are exactly the naive extensions of Δ_{Π_1}. Among them only $\{r_4, r_5\}$ and $\{r_2\}$ are stable extensions of Δ_{Π_1} and they correspond to the two answer sets of Π_1, $\{d,e\}$ and $\{b\}$.

Remark 1. Conversely, any AFN $\Delta = \langle A, R, N \rangle$ can be translated into a LP Π_Δ of the fragment $Frag$. Each $r \in A$ gives rise to an atom h_r and a rule r with $head(r) = h_r$, $body^+(r) = \{h_s \mid s\ N\ r\}$, $body^-(r) = \{h_s \mid s\ R\ r\}$. To prove that this translation, preserves the mapping between naive (resp. stable) extensions of Δ and ι-answer (resp. answer) sets of Π_Δ, just remark that translating Π_Δ by using the method described in this section, gives exactly Δ. The wished results follow then from propositions 6.

5 Generalized Argumentation Frameworks with Necessities

Now, we extend the AFNs so that the necessity relation expresses the fact that a given argument requires at least one element among a set of arguments. The resulting frameworks are called : the Generalized Argumentation Frameworks with Necessities (GAFN).

Definition 8. A GAFN is defined by $\Delta = \langle A, R, DN \rangle$ where A is a set of arguments, R is a binary attack relation over A and $DN \subseteq ((2^A \setminus \emptyset) \times A)$ is a necessity relation. $E\ DN\ b$ means that b requires at least one of the arguments of E.

We adapt slightly the graphical representation to the case of sets of arguments (see figure 2). Closedness under DN^{-1} is adapted to the case of GAFNs as follows :

Definition 9. A set $S \subseteq A$ is said to be closed under DN^{-1} iff for each $a \in S$ and $E \subseteq A$ such that $E\ DN\ a$, we have $E \cap S \neq \emptyset$.

Fig. 2. Adapted representation of the necessity relation for the case of : $\{a_1, a_2\}\ DN\ b$

Definition 10 adapts the notion of N-Cycle-Freeness to the case of GAFNs. The notions of coherence and strong coherence are given in definition 11.

Definition 10. Let $\Delta = \langle A, R, DN \rangle$ be a GAFN, $S \subseteq A$ and $a \in S$. a is N-Cycle-Free in S iff for each $E \subseteq A$ s.t. $E\ DN\ a$, either $E \cap S = \emptyset$, or $\exists b \in E \cap S$ such that b is N-Cycle-Free in S. S is N-Cycle-Free iff each $a \in S$ is N-Cycle-Free in S.

Definition 11. Let $\Delta = \langle A, R, DN \rangle$ be a GAFN and $S \subseteq A$. S is coherent iff S is N-Cycle-Free and closed under N^{-1}. S is strongly coherent iff S is coherent and conflict-free w.r.t R.

Now, we can generalize the acceptability semantics in a very similar way as above :

Definition 12. Let $\Delta = \langle A, R, DN \rangle$ be a $GAFN$ and $S \subseteq A$. (1) S is a naive extension of Δ iff S is a $\subseteq$-maximal strongly coherent set; (2) S is a stable extension of Δ iff S is a strongly coherent set and for each $a \in A \setminus S$ either $S\ R\ a$ or $\exists E \subseteq A$ such that $E\ DN\ a$ and $E \cap S = \emptyset$; (3) S is an admissible set of Δ iff S is strongly coherent and if $b\ R\ S$ then for each coherent subset $S' \subseteq A \setminus S$ such that $b \in S'$ we have $S\ R\ S'$; (4) S is a preferred extension of Δ iff a $\subseteq$-maximal admissible set of Δ.

The main results about acceptability semantics discussed above, remain true in the context of GAFNs. These results are summarized in the following proposition :

Proposition 7. Let $\Delta = \langle A, R, DN \rangle$ be a $GAFN$ and $S \subseteq A$. We have : (1) S is a stable extension of Δ iff : (a) S is N-Cycle-Free and (b) ($a \in S$) iff ($\forall b \in A$, if $b\ R\ a$ then $b \notin S$ and $\forall E \subseteq A$, if $E\ DN\ a$ then $E \cap S \neq \emptyset$); (2) each stable extension of Δ is a naive extension of Δ but not vice versa; (3) there is at least one preferred extension of Δ; (4) each stable extension of Δ is a preferred extension of Δ but not *vice versa*.

It is not difficult to check that AFNs are a particular case of GAFNs where the sets of necessary arguments are reduced to single arguments.

Example 3. Consider the GAFNs represented in figure 3.

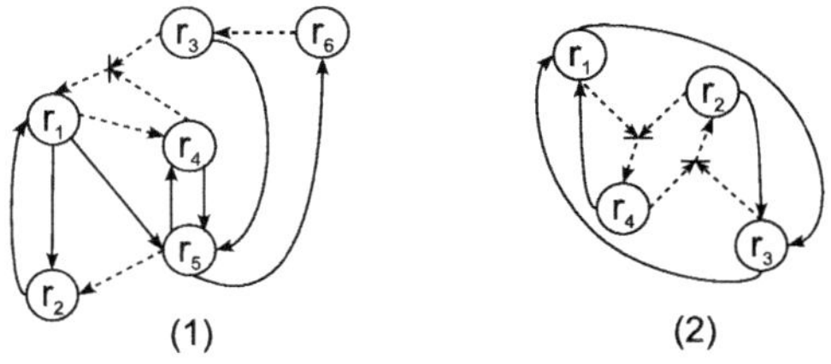

Fig. 3. Two examples of GAFNs

The system of figure 3-(1) is a GAFN because of the relation $\{r_3, r_4\}$ DN r_1. $X_1 = \{r_1, r_4\}$ is closed under DN^{-1} since the only set necessary for r_4 (resp. r_1) is $\{r_1\}$ (resp. $\{r_3, r_4\}$) and $r_1 \in X_1$ (resp. $r_4 \in X_1$) but X_1 is not N-Cycle-Free because neither r_1 nor r_4 is N-Cycle-Free in X_1. $X_2 = \{r_1, r_3\}$ is N-Cycle-Free but not closed under DN^{-1} : we have $\{r_6\}$ DN r_3 but $\{r_6\} \cap X_2 = \emptyset$. Finally, $X_3 = \{r_1, r_3, r_6\}$ is both N-Cycle-Free and closed under DN^{-1}. It is then coherent. The GAFN of figure 3-(1) has two stable extensions : $S_1 = \{r_1, r_3, r_4, r_6\}$ and $S_2 = \{r_2, r_5\}$. In fact, r_6 is N-Cycle-Free (there is no necessary argument for r_6 in S_1) which makes r_3 N-Cycle-Free, which makes r_1 N-Cycle-Free and eventually this makes r_4 also N-Cycle-Free. S_1 is then N-Cycle-Free. All the attackers of S_1 are outside S_1 and for each $a \in S_1$ and each set E DN a, we have $E \cap S \neq \emptyset$. On the other hand, r_2 and r_5 are both attacked by S_1. By the same reasoning, we can easily check that S_2 is a stable extension too. Notice that r_3 is not accepted because : $\{r_6\}$ DN r_3 and $\{r_6\} \cap S_2 = \emptyset$. We can verify that the preferred extensions of this GAFN coincide with its stable extensions.

The GAFN of figure 3-(2) has no stable extension. Indeed, it is easy to see that the only strongly coherent subsets here are $\{r_1\}$, $\{r_3\}$ and $\emptyset$. None of them is stable and only $\emptyset$ is admissible and represent the only preferred extension. Let us take for instance $\{r_1\}$. It is attacked by r_3 and r_4 but $\{r_1, r_2, r_4\}$ is a coherent set containing r_4 and not attacked by $\{r_1\}$. $\{r_1\}$ is then not admissible and thus neither preferred nor stable.

Now, let us show how GAFNs allow to encode in a simple way any LP. Let Π be an arbitrary LP containing n rules, $\Pi = \{r_1, ..., r_n\}$. The AFN that corresponds to Π is $\Delta_\Pi = \langle \Pi, R_\Pi, DN_\Pi \rangle$ such that : its arguments are the rules of Π, its attack relation R_Π is defined by : $\forall r_1, r_2 \in \Pi, r_1\ R_\Pi\ r_2$ iff $head(r_1) \in body^-(r_2)$ and its necessity relation DN_Π is defined as follows : let E be a set of rules sharing the same head that we denote by head(E) and r be a rule of Π. We let $E\ DN_\Pi\ r$ iff $head(E) \in body^+(r)$. Then, we have a one to one correspondence between ι-answer sets (resp. answer sets) of Π and naive (resp. stable) extensions of Δ_Π.

Proposition 8. Let Π be an arbitrary LP and $\Delta_\Pi = \langle \Pi, R_\Pi, DN_\Pi \rangle$ be the corresponding GAFN. A set X is a ι-answer set (resp. answer set) of Π, with Π' as the corresponding maximal subset of Π iff Π' is naive (resp. stable) extension of Δ_Π.

Example 4. Consider the following programs Π_2 and Π_3 :

(Π_2)	$r_1 : a \leftarrow c,\ not\ b$	(Π_3) $r_1 : p \leftarrow not\ q$
	$r_2 : b \leftarrow d,\ not\ a$	$r_2 : p \leftarrow q$
	$r_3 : c \leftarrow e$	$r_3 : q \leftarrow not\ p$
	$r_4 : c \leftarrow a,\ not\ d$	$r_4 : q \leftarrow p$
	$r_5 : d \leftarrow not\ c$	
	$r_6 : e \leftarrow not\ d$	

The GAFN $\Delta_{\Pi_2} = \langle A_{\Pi_2}, R_{\Pi_2}, DN_{\Pi_2} \rangle$ is represented in figure 3-(1) and the GAFN $\Delta_{\Pi_3} = \langle A_{\Pi_3}, R_{\Pi_3}, DN_{\Pi_3} \rangle$ in figure 3-(2). The program Π_2 has two ι-answer sets : $X_1 = \{a, c, e\}$ and $X_2 = \{b, d\}$ that are also its answer sets. Their corresponding maximal subsets are $S_1 = \{r_1, r_3, r_4, r_6\}$ and $S_2 = \{r_2, r_5\}$ respectively, that have been shown in example 3 to be the stable extensions of Δ_{Π_2}. Moreover, as all ι-answer sets of Π_2 are also answer sets, no subset $S \subseteq A_{\Pi_2}$ other than S_1 and S_2 is a maximal strongly coherent subset. The program Π_3 has no ι-answer set and so no answer set too. Indeed, there is no subset Π'_4 of Π_3 that verifies conditions (C1) and (C2).[3].

Remark 2. It is the head of a rule which determines whether it attacks and/or is necessary for other rules. Hence, for a set E of rules sharing the same head, if a rule of E attacks a rule r then all the rules of E must attack r, and if $E' \subseteq E$ is necessary for a rule r, then the whole E is necessary for r. Thus, to use a direct method as in Remark 1 to translate a GAFN $\Delta = \langle A, R, DN \rangle$ into a LP Π_Δ, the following conditions must hold : (1) if $E_1\ DN\ r_1$, $E_2\ DN\ r_2$ and $E_1 \neq E_2$ then $E_1 \cap E_2 = \emptyset$; (2) if $E\ DN\ r$ for some $r \in A$ and $r'\ R\ s$ for some $r' \in E$ and $s \in A$ then $r''\ R\ s$ for all $r'' \in E$.

Let $\Delta = \langle A, R, DN \rangle$ be a GAFN satisfying the previous conditions, let $E_1, ..., E_k$ be all the subsets of A such that $E_i\ DN\ r$ for some $r \in A$ and let $P = \{r_{k+1}, ..., r_n\}$ be the (possibly) other arguments of A that are not involved in any E_i. We associate to each E_i an atom h_i $(1 \leq i \leq k)$ and to each r_j (if any) of P an atom h_j $(k+1 \leq j \leq n)$. Then, each argument $r \in A$ gives rise to a rule r in Π_Δ such that : $head(r) = h_i$ $(1 \leq i \leq k)$ if $r \in E_i$ for some E_i and $head(r) = h_j$ $(k+1 \leq j \leq n)$ if $r = r_j$ for some $r_j \in P$; $body^+(r) = \{h_m \mid E_m\ DN\ r\}$ and $body^-(r) = \{h_m \mid s\ R\ r\ and\ (s \in E_m\ or\ s = r_m \in P)\}$. To show that this translation, preserves a one to one correspondence between naive (resp. stable) extensions of Δ and the ι-answer sets (resp. answer sets) of Π_Δ, it suffices to show that translating Π_Δ by using the method described in this section, gives exactly Δ. The wished results follow then from proposition 8.

6 Discussion and Related Work

In this paper we introduced argumentation frameworks with necessities that generalize Dung's AFs with a new support relation having the meaning of necessity. We redefined acceptability semantics including naive, stable, admissible and preferred extensions and showed that the main properties of these semantics are kept in this new context. We also proposed a new method that takes advantage of the necessity relation to relate a fragment of logic programs with AFNs in a very simple way. Then we further generalized the previous framework to let necessity relation deal with sets of arguments and redefined the acceptability semantics within this generalized frameworks that allow to cope with arbitrary logic programs in the same simple way as with AFNs.

We have already discussed in section 1. the main drawback of the approach of BAFs presened in [7] [8]. For more technical differences, we can state that [7] considers only acyclic systems while our work handles the general case where both cycles of necessities and of attacks are allowed. Moreover, out work generalizes the support relation to deal with sets of arguments and relates both AFNs and GAFNs to LPs.

The work presented in [3] started from a criticism of BAFs on two points, namely, the loss of admissibility in the sense of Dung of the extension obtained from the metamodel using coalitions [8], and the handling of attacks in the context of support relations. The proposed approach develops the so-called deductive support and introduces

[3] Although we have $Ap_{\Pi_3}(\{p,q\}) = G_{\Pi_3}(\{p,q\}) = \{r_2, r_4\}$, $\{p,q\}$ is not a ι-answer set because for the corresponding $\Pi'_4 = \{r_2, r_4\}$ the condition (C1) does not hold : $body^+(\Pi'_4) = \{p,q\} \not\subseteq Cn^+(\Pi'_4) = \emptyset$. Consequently, $\{p,q\}$ is not an answer set either. This is explained by the fact that in Δ_{Π_3}, $\{r_2, r_4\}$ is not strongly coherent because it is not N-Cycle-Free.

mediated attacks instead of indirect attacks. The authors show that the admissibility of extensions is then restored. It turns out that the deductive support is nothing but the sufficiency relation. As discussed in section 1, this relation corresponds merely to the inverse of the necessity relation. Thus, by inversing the direction of deductive support in [3] (which gives necessity relations), the mediated attack and the supported attacks correspond respectively to the first and the second cases of our extended attacks (see definition 2). However, instead of imposing the use of only one type of support relation, our approach can start from a system where the two types are freely expressed and then reduce in a preliminary stage all the relations to one type. Notice that all the results of our paper hold for a sufficiency relation (a deductive support) by simply replacing the necessity relation N by a deductive relation, say D, and using closedness under D instead of closedness under N^{-1}. The subsequent development in [3] is different from ours: while they define a meta-argumentation model to handle supports and introduce defeasible supports, in our work we generalize our frameworks to represent arguments supported by sets of arguments, we redefine Dung's acceptability semantics in the resulting frameworks and we show how these frameworks capture arbitrary LPs.

A recent work developed in [6] proposes abstract dialectical frameworks (ADF), a powerful generalization of Dung's AFs that formalizes the idea of proof standards, widely studied in legal reasoning domain. This idea is captured in ADFs by linking each argument to a set of arguments (its parents) and introducing the notion of acceptance conditions that determine whether an argument is accepted or not according to the acceptance status of its parents. The main results of the paper concerns however a sub-class of ADFs called bipolar ADFs (BADFs), where the relation between an argument and a parent plays always one role : either an attack or a support. It is not difficult to see that our AFNs are a kind of BAFs in the sense of [6]. GAFNs does not fit directly into ADFs since an argument in a GAFN may have as a parent a set of arguments (the set of parents is a set of sets) while a parent in an ADF is a single argument. But by adding further arguments one can translate a GAFN into a BADF. A main difference between our work and [6] lies on the method used to generalize stable and admissible semantics. [6] adapts techniques from logic programming, namely Gelfond/Lifschitz reduct, to avoid necessity cycles. In our work, thanks to the notions of coherence and strong coherence used instead of conflict-freeness, we keep our definitions similar to that in Dung's original AFs. Another point is that in the method we use to encode a LP as a BAFN, each rule is represented by an argument which gives an homogeneous view of the meaning of an argument. In [6], a similar homogenous representation using atoms as arguments is proposed but as pointed out in the paper, it leads in general to an ADF which may not be bipolar. To obtain a BADF, one must introduce new arguments designating rules. The resulting representation is then heterogeneous in the sense that arguments may refer to rules or to atoms. Finally, the opposite question, i.e., the representation of ADFs as LPs is not explicitly considered in [6], but our work, gives the constraints that a GAFN must satisfy to ensure its direct encoding as a LP.

Another approach that shares some features with ours is the evidence based argumentation [18]. This approach considers that only arguments that have some evidential support can attack other arguments. The evidential support comes either directly from the environment to some particular arguments or from a chain of supports that originates in such particular arguments. A very similar idea is present in our work. Indeed, to ensure admissibility of a set, we must guarantee just the response to attacks coming from arguments that are N-Cycle-Free, i.e., those that have no need for a support or that are ultimately supported by arguments that have no need for a support. An interesting perspective is to see how to use our model in the context of evidential reasoning.

Notice that AFNs cannot be reduced to constrained AFs [9] where $a\ N\ b$ is replaced by the implication $b \Rightarrow a$. An extension of a contrained AF is an extension of the corresponding AF that verifies the given constraints and this is not true for AFNs. For example, the AFN $\langle A = \{a, b, c\}, R = \{a, b\}, N = \{b, c\}\rangle$ has one stable extension : $\{a\}$, but the constrained framework $\langle A, R, C = c \Rightarrow b\rangle$ has no stable extension (the only stable extension of $\langle A, R\rangle$ is $\{a, c\}$ which does not verify the constraint C).

Many recent works was interested in handling preferences among arguments in AFs (see for example [5] for a brief synthesis). The main challenge in these works is to deal with the possible conflict between preferences and attacks. We want to study the impact of adding preferences to (G)AFNs and to exploit the obtained results in the context of LPs with preferences about its rules and/or atoms. Another perspective is to elaborate adapted forms of dialectical proof procedures and labelling algorithms (see [15]) for (G)AFNs. The link with LPs allows then to use these new algorithms in various applications. For example, a dialectical proof procedure may be more suitable than existing ASP solvers if we want to formulate targeted queries to a knowledge base represented by a LP (for such an application, see for example [17] which proposes a non-monotonic reasoning to detect causes of accidents from their textual descriptions). Finally, we plan also to extend our framework to cope with disjunctive LPs.

References

1. Amgoud, L., Besnard, P.: Bridging the gap between abstract argumentation systems and logic. In: Godo, L., Pugliese, A. (eds.) SUM 2009. LNCS, vol. 5785, pp. 12–27. Springer, Heidelberg (2009)
2. Besnard, P., Hunter, A.: Elements of Argumentation. MIT Press, Cambridge (2008)
3. Boella, G., Gabbay, D.M., Van Der Torre, L., Villata, S.: Support in Abstract Argumentation. In: COMMA 2010, pp. 40–51. IOS Press, Amsterdam (2010)
4. Bondarenko, A., Dung, P.M., Kowalski, R., Toni, F.: An abstract, argumentation-theoretic approach to default reasoning. Artificial Intelligence 93, 63–101 (1997)
5. Bourguet, J., Amgoud, L., Thomapoulos, R.: Towards a unified model of preference-based argumentation. In: Link, S., Prade, H. (eds.) FoIKS 2010. LNCS, vol. 5956, pp. 326–344. Springer, Heidelberg (2010)
6. Brewka, G., Woltran, S.: Abstract Dialectical Frameworks. In: International Conference on the Principles of Knowledge Representation and Reasoning (KR 2010), pp. 102–111 (2010)
7. Cayrol, C., Lagasquie-Schiex, M.C.: On the acceptability of arguments in bipolar argumentation frameworks. In: Godo, L. (ed.) ECSQARU 2005. LNCS (LNAI), vol. 3571, pp. 378–389. Springer, Heidelberg (2005)
8. Cayrol, C., Lagasquie-Schiex, M.-C.: Coalitions of arguments: A tool for handling bipolar argumentation frameworks. Int. J. Intell. Syst. 25(1), 83–109 (2010)
9. Coste-Marquis, S., Devred, C., Marquis, P.: Constrained Argumentation Frameworks. In: International Conference on Principles of Knowledge Representation and Reasoning (KR 2006), pp. 112–122. AAAI Press, Lake District (2006)

10. Dung, P.M.: On the acceptability of arguments and its fundamental role in nonmonotonic reasoning, logic programming and n-person games. Artificial Intelligence 77(2), 321–357 (1995)
11. Egly, U., Gaggl, S.A., Woltran, S.: Answer-set programming encodings for argumentation frameworks. Argument and Computation 1(2), 147–177 (2010)
12. Gebser, M., Gharib, M., Mercer, R., Schaub, T.: Monotonic Answer Set Programming. Journal of Logic and Computation 19(4), 539–564 (2009)
13. Gelfond, M., Lifschitz, V.: Classical negation in logic programs and disjunctive databases. New Generation Computing 9, 365–385 (1991)
14. Łukaszewicz, W.: Considerations on Default Logic: An Alternative Approach. Computational Intelligence 4, 1–16 (1988)
15. Modgil, S., Caminada, M.: Proof theories and algorithms for abstract argumentation frameworks. In: Rahwan, I., Simari, G. (eds.) Argumentation in Artificial Intelligence, pp. 105–129. Springer, Heidelberg (2009)
16. Nieves, J.C., Cortes, U., Osorio, M.: Preferred extensions as stable models. TPLP 8(4), 527–543 (2008)
17. Kayser, D., Nouioua, F.: From the description of an accident to its causes. Artificial Intelligence 173(12-13), 1154–1193 (2010)
18. Oren, N., Norman, T.J.: Semantics for Evidence-Based Argumentation, in Computational Models of Argument. In: COMMA 2008, Frontiers in Artificial Intelligence and Applications, pp. 276–284 (2008)
19. Reiter, R.: A Logic for Default reasoning. Artificial Intelligence 13(1-2), 81–132 (1980)
20. Wu, Y., Caminada, M., Gabbay, D.: Complete Extensions in Argumentation Coincide with 3-Valued Stable Models in Logic Programming. Studia logica 93(2-3), 383–403 (2009)

A Heuristics-Based Pruning Technique for Argumentation Trees

Nicolás D. Rotstein*, Sebastian Gottifredi,
Alejandro J. García, and Guillermo R. Simari

National Council of Scientific and Technical Research (CONICET)
Artificial Intelligence Research & Development Laboratory (LIDIA)
Universidad Nacional del Sur (UNS), Bahía Blanca, Argentina
nico.rotstein@abdn.ac.uk, {sg,ajg,grs}@cs.uns.edu.ar

Abstract. Argumentation in AI provides an inconsistency-tolerant formalism capable of establishing those pieces of knowledge that can be warranted despite having information in contradiction. Computation of warrant tends to be expensive; in order to alleviate this issue, we propose a heuristics-based pruning technique over dialectical trees. Empirical testing shows that in most cases our approach answers queries much faster than the usual techniques, which prune with no guide.

1 Introduction and Motivation

The theory on computational argumentation is usually focused on bringing new theoretical elements to augment the expressive capability of the formalism. Other extensions are also devoted to handle different aspects of the argumentation process, from its dynamics [15] to its capability to represent dialogues, negotiations, and other features [4]. Complementarily, some approaches study the suitability of argumentation within different application contexts, such as Multi-Agent Systems and the Semantic Web. However, many complications lying on the practical side of argumentation have not been completely addressed. Implementations have not yet achieved maturity; the few systems available are still at an experimental stage and have never been tested against large amounts of data. This is understandable for a rather young discipline like argumentation in AI.

Nonetheless, as the theoretical foundations become stronger, the community is starting to pay attention to the computational tractability of argumentation [2,6]. In this article, we take on this concern and put our focus on the computation of warrant through dialectical trees, *i.e.,* focusing on a single argument at a time. When this is performed in massive argumentation [14], two difficulties can be envisioned: either dialectical trees are small and too many, or they are few but large. In this paper we address the latter issue, attempting to build smaller trees via a *pruning technique.* Thus, a smaller amount of arguments would be required to determine the status of the root argument. In order

* Employed by the University of Aberdeen for dot.rural Digital Economy Research, under the RCUK Digital Economy Programme, http://www.dotrural.ac.uk

S. Benferhat and J. Grant (Eds.): SUM 2011, LNAI 6929, pp. 177–190, 2011.
© Springer-Verlag Berlin Heidelberg 2011

to confirm this improvement, experimental tests were performed over large argumentation frameworks, randomly generated, containing up to 500 arguments.

In this article we apply the pruning technique over a variation of the original framework for argumentation that considers a *universal* set of arguments along with a subset of currently *active* ones. However, the interest on such a pruning technique is not constrained to this framework only. For instance, think of a rule-based argumentation formalism. Provided that there are no functional letters, we could build the universal set of arguments as a subset of the (finite) *Herbrand Base*. Then, as the state of the world changes, facts would be asserted and retracted accordingly, therefore activating and deactivating arguments. Here we will show how to compute a heuristics from the universal set of arguments in order for it to be useful at any given state of the world. We contend that many implementations for argument frameworks would benefit from these results.

The approach used for pruning will be based on the abstract notion of *argument strength*, which indicates the likelihood of an argument to be ultimately defeated. We propose a concrete formula showing the desired behaviour. Arguments' strength is used as a heuristic value to sort the attackers of an inner node during the construction of dialectical trees. We will show that pruning opportunities are likely to appear, and entire subtrees can be omitted without affecting the calculation of the root's warrant. We introduce a strategy to build these *dialectical bonsai*[1], based on our proposal of argument strength. Empirical testing has shown that strength calculation allows for virtually instantaneous response to queries. Strength calculation also allows for storing arguments and attacks, which is a clear advantage. This analysis shows that strength calculation does not undermine the gain obtained by bonsai, in terms of time. Although the goal of this paper is mainly practical, in this first approach we study the suitability of the approach applied over a particular flavour of Dung's framework [10].

2 Theoretical Basis

Nowadays, Dung's abstract argumentation framework has become the standard to analyse and apply new ideas to any argumentation-based setting. In this article, we will use a slightly different version of the classic framework that allows for the representation of both *active* and *inactive* arguments. This *dynamic abstract argumentation framework* (DAF) is a simpler version than the one presented in [15], getting rid of all the representational intricacies that are not relevant for this line of research. This version of the DAF consists of a universal set of arguments holding every conceivable argument along with a subset of it containing only the active ones. These active arguments represent the current state of the world and are the only ones that can be used by the argumentation machinery to make inferences and calculate warrant. Therefore, at a given moment, those arguments from the universal set that are not active represent reasons that, though valid, cannot be taken into consideration due to the current

[1] The pruning technique attempts to keep trees small while retaining the properties of the entire tree, like a bonsai.

context. At some point, active arguments could become inactive, thus no longer used for inference, or *vice versa*. As in [10], we have a set containing attacks between pairs of arguments.

Definition 1 (Dynamic Argumentation Framework). *A* ***dynamic argumentation framework****, or DAF, is a triple* $(\mathbb{U}, \hookrightarrow)[\mathbb{A}]$*, where* $\mathbb{U}$ *is the universal set of arguments,* $\hookrightarrow \subseteq \mathbb{U}\times\mathbb{U}$ *is the attack relation between arguments, and* $\mathbb{A} \subseteq \mathbb{U}$ *is the subset of active arguments.*

The DAF yields a graph of arguments connected by the attack relation. An "active subgraph" could be considered, containing only active arguments. In argumentation, the challenge consists in finding out which active arguments prevail after all things considered, *i.e.*, those arguments that are warranted. To this end, the notion of *argumentation semantics* has been extensively studied [3]. In this article, warrant of arguments will be determined on top of the dialectical tree for each one of them, assuming a particular marking criterion (see Assumption 1.)

A dialectical tree is conformed by a set of argumentation lines; each of which is a non-empty sequence λ of arguments from a DAF, where each argument in λ attacks its predecessor in the line. An argumentation line should be non-circular (an argument should not occur twice in the same argumentation line) in order to avoid infinite lines, and it also should be exhaustive (no more arguments can be added to it.)

Definition 2 (Argumentation Line). *Given a DAF* $\tau = (\mathbb{U}, \hookrightarrow)[\mathbb{A}]$*, and* $\mathsf{B}_1, \ldots, \mathsf{B}_n \in \mathbb{U}$*, an* ***argumentation line*** λ *in* τ *is a (non-empty) finite sequence of arguments* $[\mathsf{B}_1, \ldots, \mathsf{B}_n]$ *such that,* $\forall \mathsf{B}_i, \mathsf{B}_j$ *with* $i \neq j$ *and* $1 < i, j \leq n$*,* $\mathsf{B}_i \hookrightarrow \mathsf{B}_{i-1}$*,* $\mathsf{B}_i \neq \mathsf{B}_j$ *and* $\nexists \mathsf{C} \in \mathbb{U}$ *such that* $\mathsf{C} \hookrightarrow \mathsf{B}_n$*. The set of all argumentation lines in* τ *is noted as* $\mathfrak{Lines}_\tau$*.*

The first argument in an argumentation line is called the *root* whereas the last one is the *leaf* of λ. Arguments in these lines are classified according to their role wrt. the root argument: a *pro argument* (respectively, *con*) in an argumentation line is placed at an odd (respectively, even) position.

Note that the definition for an argumentation line takes into account every argument in the universal set. A restricted version of argumentation line could be considered, setting its domain within the set of active arguments. This variant is called *active argumentation line*. Since regular argumentation lines as defined above include both active and inactive arguments, we will refer to them as *potential argumentation lines*, to emphasise their meaning.

As said before, the warrant status of an argument will be determined by analysing the dialectical tree rooted in it. A dialectical tree rooted in an argument A will be built from a set of argumentation lines rooted in A.

Definition 3 (Dialectical Tree). *Given a DAF* $\tau = (\mathbb{U}, \hookrightarrow)[\mathbb{A}]$ *and an argument* $\mathsf{A} \in \mathbb{U}$*, the* ***dialectical tree*** $\mathcal{T}(\mathsf{A})$ *rooted in* A *from* τ *is built from a set* $X \subseteq \mathfrak{Lines}_\tau$ *of lines rooted in* A*, such that an argument* C *in* $\mathcal{T}(\mathsf{A})$ *is:*

– *a* ***node*** *iff* $\mathsf{C} \in \lambda \in X$
– *a* ***child*** *of a node* B *in* $\mathcal{T}(\mathsf{A})$ *in* λ *iff* $\lambda = [\ldots, \mathsf{B}, \mathsf{C}, \ldots]$, $\lambda \in X$.
– *a* ***leaf*** *of* $\mathcal{T}(\mathsf{A})$ *in* λ *iff* C *is a leaf in* $\lambda \in X$.

The set of all dialectical trees in τ *is noted as* $\mathfrak{Trees}_\tau$.

Dialectical trees are built from argumentation lines and can be classified in a similar way. Therefore, *potential dialectical trees* are built from potential argumentation lines: those containing arguments from the universal set. Similarly, *active dialectical trees* are built from active argumentation lines, which include only active arguments.

The computation of warrant through dialectical trees usually relies on a marking criterion that could be defined according to any conceivable policy; its main objective is to assign a status to each argument in the dialectical tree. The status of the root argument would tell whether it is warranted. An abstract specification for a sensible marking criterion was given in [16]. Here we present a particular version of the *marking function* that assigns either "**D**" (defeated) or "**U**" (undefeated) to each argument.

Definition 4 (Marking Function). *Given a DAF* τ*, a* ***marking function*** *over arguments in* τ *is any function* $\mathfrak{m} : \mathbb{U} \times \mathfrak{Lines}_\tau \times \mathfrak{Trees}_\tau \rightarrow \{\mathbf{D}, \mathbf{U}\}$.

Since the same argument can appear in different lines of the same tree, the marking function needs to address it through line and tree, *e.g.,* an argument could be marked as **D** in some lines and **U** in others. Although Definition 4 indicates that $\mathfrak{m}$ is defined for the whole cartesian product of acceptable lines, trees and arguments in a DAF, an implementation would probably define the function only for arguments within a given acceptable line within a given acceptable tree. There is a case in which an argument B can be associated to several lines, interchangeably: when the path from the root to B coincides in these lines. We will not address this issue with any particular convention, since it will not be problematic: the mark of B in all these lines will be just the same (see Example 1). In this article we assume the marking criterion given in DeLP [11].

Assumption 1. *An argument is marked as* **D** *iff it has an attacker marked as* **U**. *Otherwise it is marked as* **U**.

Example 1. Consider the dialectical tree $\mathcal{T}(\mathsf{A})$ below, depicted along with its two argumentation lines λ_1 and λ_2. White (black) triangles are used to denote arguments marked as undefeated (defeated).

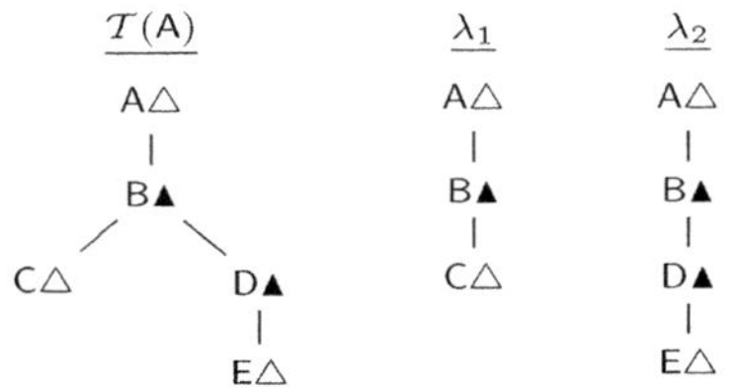

Leaves are undefeated and B is defeated due to C being undefeated. Also note that A and B belong to both lines, thus receiving the same mark.

The construction of *dialectical bonsai* depends on the capability of dialectical trees to be pruned. Given a tree $\mathcal{T}$, a bonsai of $\mathcal{T}$ will be a pruned version of it such that the marking applied to the bonsai retains, at least, the same marking for the root than the one in $\mathcal{T}$. That is, during the construction of dialectical trees, it should be possible to determine when the construction procedure has collected enough information to compute the same mark for the root as the one in the non-pruned tree. A simple example is a scenario in which there is an undefeated attacker for the root; when such an argument is found, the dialectical analysis should be stopped and the root marked as defeated. To simplify things, we will assume the *and-or* pruning technique explained in [9], which we call *1U pruning* (formally introduced in Definition 7): *whenever an* ***attacker*** *for an inner argument* A *is found as* ***undefeated****,* A *can be directly marked as* ***defeated*** *and the rest of the attackers for* A *can be ignored.*

Although this technique is not too restrictive and could even be used often, finding all the opportunities for pruning (thus obtaining the smallest possible trees) requires just plain luck, since undefeated attackers must be found first (see Example 2), and undefeatedness is not predictable –indeed, its discovery is the reason to build dialectical trees.

Example 2. Consider the dialectical tree $\mathcal{T}(\mathsf{A})$ depicted below, and three possible prunings: $\mathcal{P}^i_{\mathcal{T}}(\mathsf{A}), 1 \leq i \leq 3$. White (respectively, black) triangles are used to denote arguments marked as undefeated (respectively, defeated). Depending on the order in which attackers for A are selected, different prunings come up.

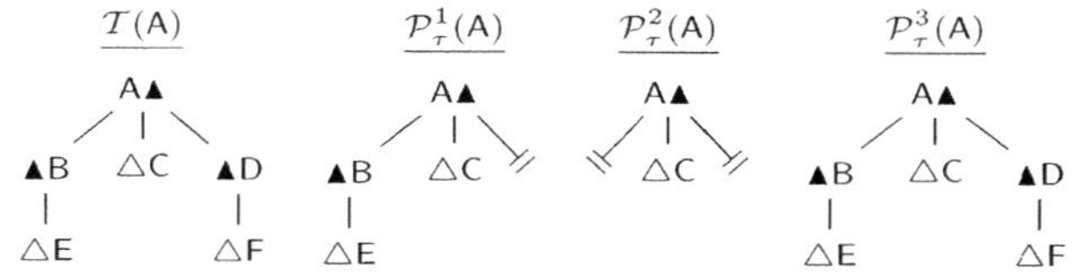

– $\mathcal{P}^1_{\mathcal{T}}(\mathsf{A})$: C is selected after B but before D, so D is cut off;
– $\mathcal{P}^2_{\mathcal{T}}(\mathsf{A})$: C is selected first, both B and D are cut off;
– $\mathcal{P}^3_{\mathcal{T}}(\mathsf{A})$: C is selected last, no pruning occurs.

Note that pruning argument C would lead to a faulty pruning, marking the root as undefeated. Such a pruning would not qualify as a bonsai.

Generally in the literature, when using a pruning technique while building trees the expansion of attackers (*i.e.,* children) follows no criterion, and this happens both in theoretical approaches as well as in implementations. The former usually present a set of arguments to choose from, whereas the latter are often rule-based, and rules are placed rather arbitrarily, and looked up in a top-to-bottom fashion. To sum up, when building dialectical trees there is no available information to know beforehand how to augment the possibilities of pruning. An external mechanism should provide such knowledge. Next we introduce the concept of argument strength.

3 An Approach to Argument Strength

Our approach to the notion of *argument strength* is similar to that in [12,5], and it is based on the following statement: *"an argument is as strong as weak are its attackers"*. In this way, the number of attackers is not the only parameter that affects an argument's strength. This is particularly interesting, since a strategy would be flawed if based solely on that number. For instance, given a pro argument K within any dialectical tree, a subtree rooted in K packed with pro arguments should give K a high strength value. The strength of an argument in a tree should somehow codify how likely is this argument to be ultimately un/defeated. An argument with great strength could even be considered as a leaf, which would improve pruning at the risk of losing soundness. Strength values will be used by a heuristic method to prune the tree, as will be explained later. The idea behind this strength measure is to codify the likeliness of an argument to be ultimately defeated.

Next we propose a formula to calculate strength [5] for an argument A that works over the potential tree rooted in A. In order to get A's strength, the method relies on the strength of A's immediate attackers, which in turn rely on the strength of their own immediate attackers, and so on, thus leading to the consideration of the entire tree. The definition of a set of rationality postulates modelling the intuitions given above is underway but, however, out of the scope of this article. Similarly, other formulas showing the desired behaviour could be proposed.

Definition 5 (Argument Strength). *The* ***strength*** *of an argument* B *in a line* λ *in a potential tree* $\mathcal{T}(\mathsf{B})$ *is calculated as:*

$$\mu(\mathsf{B}, \lambda_i, \mathcal{T}(\mathsf{B})) = \frac{1}{1 + \sum_i (\mu(\mathsf{C}_i, \lambda_i, \mathcal{T}(\mathsf{B})))},$$

where C_i *is a child of* B *in a line* λ_i *within* $\mathcal{T}(\mathsf{B})$.

Again, as arguments might appear in different lines of the same tree, they have to be individualised through these three parameters. The strength of an argument A in the context of a DAF τ is calculated as $\mu(\mathsf{A}, \cdot, \mathcal{T}(\mathsf{A}))$, and the shortcut is $\mu(\mathsf{A})$. Note that the proposed formula captures the intuition that an argument's strength has an inverse correlation with the strength of its attackers.

Each argument's strength is calculated by building the potential dialectical tree rooted in it. Although each non-root argument in this potential tree has an associated strength, this measure is *local*. That is, this is not the *actual* strength value they would have in the potential tree rooted in them, but just a partial measure towards the calculation of the root's strength. Moreover, an argument appearing twice in a tree would probably have two different local strength measures. Nonetheless, the strength value associated to an argument will be its actual strength value.

Example 3. Suppose a DAF $(\{\mathsf{A},\mathsf{B},\mathsf{C},\mathsf{D}\},\{(\mathsf{B},\mathsf{A}),(\mathsf{C},\mathsf{A}),(\mathsf{A},\mathsf{C}),(\mathsf{D},\mathsf{C})\})[\{\mathsf{A},\mathsf{B},\mathsf{C},\mathsf{D}\}]$. The potential trees for $\mathcal{T}(\mathsf{A})$ and $\mathcal{T}(\mathsf{C})$ are shown below. In the former tree, arguments B and D receive no attacks, thus $\mu(\mathsf{B},\lambda_1,\mathcal{T}(\mathsf{A})) = 1 = \mu(\mathsf{D},\lambda_2,\mathcal{T}(\mathsf{A}))$. Argument C is attacked just by D, then $\mu(\mathsf{C},\lambda_2,\mathcal{T}(\mathsf{A})) = 1/(1+1) = 0.5$. Finally, A is attacked by B and C, which sum 1.5, and therefore $\mu(\mathsf{A}) = 1/(1+1.5) = 0.4$. The rest of the strength values are local. The actual strength of C is $\mu(\mathsf{C}) = 1/(1+1.5) = 0.4$, since in its potential tree $\mathcal{T}(\mathsf{C})$ has D as undefeated attacker with $\mu(\mathsf{D},\lambda_1,\mathcal{T}(\mathsf{C})) = 1$ and A as defeated attacker with $\mu(\mathsf{A},\lambda_2,\mathcal{T}(\mathsf{C})) = 1/(1+1) = 0.5$. Note that the local strength value 0.5 for C in the potential tree for A differs from C's actual strength value 0.4, computed from its potential tree.

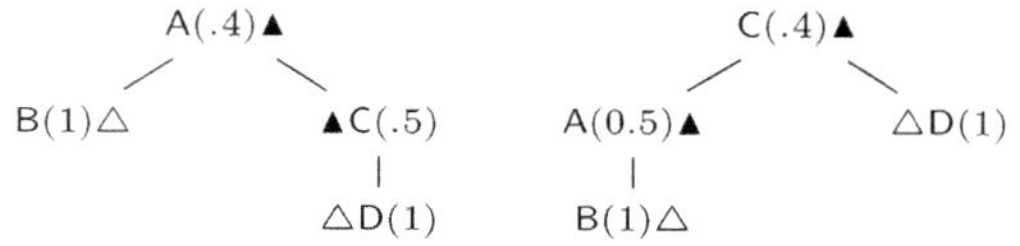

4 Building Dialectical Bonsai

In this article, we will use strength values to devise a heuristics-based method to construct *dialectical bonsai*. Remember that the strength of an argument is computed on top of the potential dialectical tree associated to it. Therefore, these values will be an indication of how likely to be defeated an argument is, but given a specific scenario the real strength of an argument (corresponding to the active tree rooted in it) could greatly differ from the one calculated from the potential tree. That is, if strength values were to be kept up to date, they should be recalculated every time the situation changes, which is rather undesirable. In this article we propose a more pragmatic approach, in which each strength value is computed just once from the corresponding potential dialectical tree. Thus, when faced to particular situations, strength values end up being approximated; however they are still useful as a heuristics, as demonstrated by experimental testing.

Example 4. Consider Example 3, where $\mu(\mathsf{A}) = 0.4, \mu(\mathsf{B}) = \mu(\mathsf{D}) = 1, \mu(\mathsf{C}) = 0.4$. Let us assume a world in which A and D are inactive. In this case, C has no defeaters, but its strength value is still 0.4, as it would have been calculated beforehand, from its potential tree.

From precalculated strength values we seek to build smaller trees via pruning. Next, we formalise the generalised notion of pruning, according to the intuitions previously presented.

Definition 6 (Pruning). *Let $\mathcal{T}(\mathsf{A})$ be a dialectical tree in the context of a DAF $\tau = (\mathbb{U},\hookrightarrow)[\mathbb{A}]$, $\mathsf{S_t} \subseteq \mathbb{U}$, the set of arguments in $\mathcal{T}(\mathsf{A})$, and $\mathsf{E_t} \subseteq \hookrightarrow$, the set of edges in $\mathcal{T}(\mathsf{A})$. A* ***pruning*** *$\mathcal{P}(\mathsf{A})$ for $\mathcal{T}(\mathsf{A})$ is a tree rooted in A with a set of arguments $\mathsf{S_p} \subseteq \mathsf{S_t}$ and a set of edges $\mathsf{E_p} \subseteq \mathsf{E_t}$. An* ***active pruning*** *is a pruning for an active dialectical tree.*

The above definition gives a general notion of what we consider as a pruning. However, not every pruning of a tree qualifies as a dialectical bonsai. As stated before, the requirement for a pruning to be a bonsai is to yield the same information than the complete tree about the warrant status of the root argument. Next, we introduce the particular kind of pruning used in this paper, and then we formally define the concept of dialectical bonsai. The reader should know that this is not a new concept, but used in existing argumentation systems, such as DeLP [11].

Definition 7 (1U Pruning). *Given a DAF* τ *and a pruning* $\mathcal{P}(\mathsf{A})$ *for a dialectical tree* $\mathcal{T}(\mathsf{A})$*, let* B *be an inner node in* $\mathcal{P}(\mathsf{A})$ *with a set of attackers* Γ *in* $\mathcal{T}(\mathsf{A})$ *such that the subset of attackers in* $\mathcal{P}(\mathsf{A})$ *is* $\Gamma' \subseteq \Gamma$*. The pruning* $\mathcal{P}(\mathsf{A})$ *is a* ***1U pruning*** *for* $\mathcal{T}(\mathsf{A})$ *iff* $\exists \mathsf{B}_i \in \Gamma, \mathfrak{m}(\mathsf{B}_i, \lambda_i, \mathcal{P}(\mathsf{A})) = \mathbf{U}$ *implies that there is exactly one argument* $\mathsf{B}_k \in \Gamma', \mathfrak{m}(\mathsf{B}_k, \lambda_k, \mathcal{P}(\mathsf{A})) = \mathbf{U}$.

In words, a 1U pruning is a pruning $\mathcal{P}(\mathsf{A})$ such that for any set of attackers with at least one attacker marked as **U** in the original tree $\mathcal{T}(\mathsf{A})$, the subset of attackers that stays in $\mathcal{P}(\mathsf{A})$ has exactly one undefeated attacker. The definition does not specify how to treat a set Γ with all defeated arguments. Note that any subset would work, even the empty set. However, an implementation would most likely check all the defeaters to find out that they are defeated.

Definition 8 (Dialectical Bonsai). *Let* $\mathcal{T}(\mathsf{A})$ *be an active dialectical tree in the context of a DAF* τ*, a* ***dialectical bonsai*** $\mathcal{B}(\mathsf{A})$ *for* $\mathcal{T}(\mathsf{A})$ *is a pruning of* $\mathcal{T}(\mathsf{A})$ *with a set of arguments* $\mathsf{S_b}$ *verifying:*

$$\mathfrak{m}(\mathsf{B}_i, \lambda_k, \mathcal{T}(\mathsf{A})) = \mathfrak{m}(\mathsf{B}_i, \lambda_i, \mathcal{B}(\mathsf{A})), \textit{ for every } \mathsf{B}_i \in \mathsf{S_b},$$

where paths from A *to* B_i *in* $\mathcal{B}(\mathsf{A})$ *and* $\mathcal{T}(\mathsf{A})$ *are equal.*

Proposition 1. *A 1U pruning for an active dialectical tree* $\mathcal{T}$ *is a dialectical bonsai for* $\mathcal{T}$*.*

Proof. *Let* $\mathcal{P}(\mathsf{A})$ *be a 1U pruning of a dialectical tree* $\mathcal{T}(\mathsf{A})$*. Since for any set of attackers for a given argument* B *in the original tree* $\mathcal{T}(\mathsf{A})$ *with at least one undefeated attacker, exactly one of these remains in the 1U pruning* $\mathcal{P}(\mathsf{A})$ *by Definition 7, then* B *is defeated both in* $\mathcal{T}(\mathsf{A})$ *and* $\mathcal{P}(\mathsf{A})$*. On the other hand, for any set of attackers for* B *where no argument is undefeated, any subset of attackers (even the empty set) would yield* B *as undefeated.*

By definition, a dialectical bonsai is not an arbitrary pruning of its associated (active) dialectical tree, but one that shares the mark of every argument; in particular, the mark of the root. A more relaxed version of this definition could require sharing only the mark of the root argument. However, for a first approach it is simpler not to allow obscurity over the marking of the rest of the arguments, since it would add unnecessary complexity. The most important property dialectical bonsai should satisfy is to warrant exactly the same arguments than non-pruned trees do.

Lemma 1 (Soundness & Completeness). *Given a dialectical bonsai $\mathcal{B}(\mathsf{A})$ of a dialectical tree $\mathcal{T}(\mathsf{A})$, A is warranted from $\mathcal{T}(\mathsf{A})$ iff A is warranted from $\mathcal{B}(\mathsf{A})$.*

Proof. *Trivial, from Definition 8, the marking of the root nodes in $\mathcal{B}(\mathsf{A})$ equals the marking of the root in $\mathcal{T}(\mathsf{A})$.*

A different version of the definition for a dialectical bonsai might imply a much more complicated procedure to determine the status of the root argument. However, it would be desirable for any alternative definition to not interfere with the satisfaction of the meta-properties of soundness and completeness.

Fast-Prune Bonsai

When a dialectical tree is built, the order in which children (*i.e.*, attackers) are generated is relevant, as it could lead to very different results if a pruning technique like 1U were applied (as shown in Example 2). In our approach, each argument has an associated strength value, therefore once all the children of a given inner node were gathered, they are sorted from the strongest to the weakest. In this way, we always seek for the strongest attackers first in order to find an undefeated argument as fast as possible, to then be able to cut the remaining siblings off. This strategy is called *fast prune*.

Definition 9 (Fast-Prune Bonsai). *Given a DAF τ and a dialectical bonsai $\mathcal{B}(\mathsf{A})$ for an active dialectical tree $\mathcal{T}(\mathsf{A})$, let B be an inner node in $\mathcal{B}(\mathsf{A})$ with a set of attackers Γ in $\mathcal{T}(\mathsf{A})$ such that the subset of attackers in $\mathcal{B}(\mathsf{A})$ is $\Gamma' \subseteq \Gamma$. Given the argument strength function $\mu(\cdot)$, $\mathcal{B}(\mathsf{A})$ is a* ***fast-prune bonsai*** *iff $\exists \mathsf{B}_k \in \Gamma', \mathfrak{m}(\mathsf{B}_k, \lambda_k, \mathcal{B}(\mathsf{A})) = \mathbf{U}$ implies both:*

1. $\forall \mathsf{B}_{j \neq k} \in \Gamma', \mu(\mathsf{B}_k) \leq \mu(\mathsf{B}_j), \mathfrak{m}(\mathsf{B}_j, \lambda_j, \mathcal{B}(\mathsf{A})) = \mathbf{D}$
2. $\forall \mathsf{B}_x \in \Gamma \setminus \Gamma', \mu(\mathsf{B}_x) \leq \mu(\mathsf{B}_k)$.

Another way of reading this definition is to think that the strongest attackers B_j for B have to be considered within $\mathcal{B}(\mathsf{A})$, until an undefeated attacker is found; then, those attackers B_x that were left out of the bonsai are weaker than (or as strong as) the attackers for B in $\mathcal{B}(\mathsf{A})$.As in Definition 7, Γ' containing all defeated arguments receives no special treatment, since any subset would preserve the marking for B. Procedurally, however, the algorithm would have to check all the arguments, from the strongest to the weakest, to finally find out that none of them is undefeated. Hence, in practice, when all arguments in Γ are defeated, it holds that $\Gamma' = \Gamma$. Another variant would be to stop checking arguments statuses whenever a certain strength threshold is met, *e.g.*, not to check arguments with a strength value below 0.2. In that case, we would have that $\Gamma \subset \Gamma$ and could prune even further, at the risk of losing soundness.

Proposition 2. *A fast-prune bonsai for a dialectical tree $\mathcal{T}$ is a 1U pruning for $\mathcal{T}$.*

Proof. *A fast-prune bonsai requires, for any set of attackers with at least one undefeated argument: (1) to have only one undefeated argument B_k (2) for B_k to be weaker than the remaining (defeated) arguments. Condition 1 is equivalent to the requirement for a 1U pruning.*

Example 5. Consider the potential tree for A depicted below, in (a). Assuming that D and H are not active right now, we have the active tree shown in (b). Applying fast prune over this active tree (which keeps the strength values from the potential) we have that the strongest attacker of A is C. This argument has two children, where G is the strongest one. Then, we consider G, whose only attacker is J, therefore G ends up defeated and its siblings should be evaluated. Again, F is defeated by leaf I, and C has no more attackers, thus it ends up undefeated. This means that C's siblings can be pruned off, yielding the bonsai in (c). Note that every node in the bonsai preserves its marking; in particular the root.

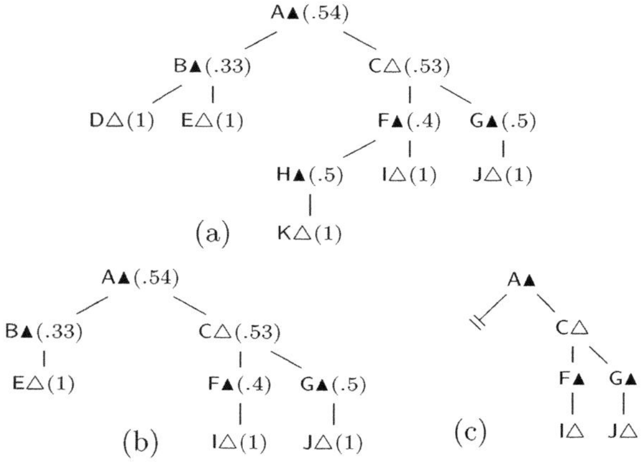

Example 6. Consider the potential tree depicted below, in (a). Assuming a world in which C_3 and C_4 are inactive, we have the active tree in (b), where A is now defeated. The strongest attackers for A are B_1 and B_3. Due to lexicographic order, B_1 is expanded, whose one attacker C_1 is undefeated, hence B_1 is defeated. Then, we seek for the second stronger attacker for A, which is B_3, and it is a leaf, *i.e.*, it is undefeated. Therefore, B_2 is cut off and the root is marked as defeated. The resulting bonsai is depicted in (c). Note that the bonsai could get even smaller if it were composed just by A and B_3.

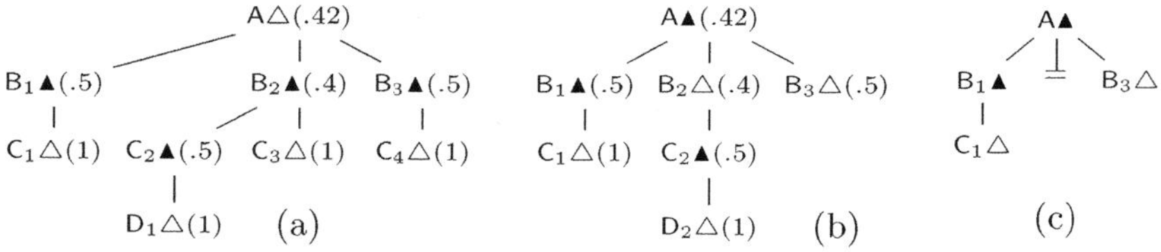

5 Empirical Results

In this section, the performance of the fast-prune bonsai (FP) is measured from a simulation. The analysis involves comparing the FP bonsai against a *blind-prune bonsai* (BP), *i.e.*, a pruning technique imitating the behaviour of a typical dialectical tree construction procedure, using no guide, like DeLP. The comparison between the two bonsai will help us analyse the improvement achieved by FP bonsai. The simulation consists of the following steps:

1. *Generation of a random DAF with U arguments:* by creating a graph of U nodes and over U edges;
2. *Strength computation:* performed following the formula in Definition 5 using dynamic programming techniques by reutilising subtrees, which implies big time savings. Strength computation allows for storing trees (*i.e.*, arguments and attacks), as a precompilation [8]. Since generally in a real-world scenario the construction of arguments takes time, we simulated this by introducing a very small time penalty of 0.001 seconds each time an argument is built;
3. *Loop 500 times:*
   (a) *Deactivation of arguments:* a certain percentage of randomly chosen arguments is deactivated.
   (b) *Selection of a query:* a random active argument is selected to act as a query.
   (c) *Computation of warrant:*
      – *Fast-prune bonsai:* an algorithm following Definition 9 computes warrant by using precompiled arguments and attacks; as for strength calculation, the time penalty per argument is introduced.
      – *Blind-prune bonsai:* an algorithm following the usual procedure in Definition 7 computes warrant by building arguments and attacks on-the-fly, which includes the time penalty.

Hence, each simulation generates a DAF, deactivates some arguments, and performs 500 queries. Since results are DAF-dependent, this simulation was ran 1000 times and the deactivation ratio was set to 10%. The results for different DAF sizes can be seen in Table 1, where "U" indicates the size (*i.e.*, number of arguments) of the DAF, "str(t)" the average amount of time taken to compute strength values, "FP(t)" (resp., "BP(t)") the average amount of seconds taken to compute the 500 queries using fast-prune (resp., blind-prune) bonsai, "B/F" indicates the *online speedup* obtained considering queries only (without time needed for *offline* strength computation), "str(#)" (resp., "BP(#)") reflects the average amount of arguments generated to compute strength (resp., BP bonsai.)

Table 1. Results

U	str(t)	FP(t)	BP(t)	B/F	str(#)	BP(#)
200	0.86	0.029	2.02	70	807	1822
300	1.41	0.031	2.06	66	1329	1852
500	2.35	0.039	2.08	53	2219	1861

A question that might arise after this description is why tolerating a time penalty while building the BP bonsai if the argument base can be precompiled. The answer to this is split in two: (1) we are trying to simulate the usual procedure to compute warrant (such as the one used in DeLP) which does not perform any precompilation; (2) precompilation of arguments and attacks involves computing potential trees and from there computation of strength would involve a very small overhead: traversing the tree from the leaves to the root. The original question however remains: how does FP perform against BP when both work over precompiled argument bases? The answer can be found in Table 2. Here, we consider the time taken to precompile knowledge ("c(t)"), which is equivalent to compute strength. Speedup ("Bc/Fc") now considers the total time to answer queries plus precompilation time. Note that time to compute BP bonsai ("BP(t)") has no penalty, thus modifying the per-query speedup ("B/F").

Table 2. Precompiled argument base

U	c(t)	FP(t)	BP(t)	Bc/Fc
200	0.008	0.029	0.076	2.27
300	0.013	0.031	0.08	2.11
500	0.016	0.039	0.091	1.95

Regarding the amount of space needed to store potential trees, we only need to store the potential graph, as each potential tree is the spanning of this graph from a given argument. For instance, a DAF of 500 arguments and 500 attacks with a maximum argument size of N kilobytes would require $500 \times N + 500 \times M$, where M is just a few bytes representing an attack: both arguments IDs plus attack direction. For this problem size, considering arguments of a maximum size of 5 kilobytes (which is approximately two pages of plain text), the amount of storage necessary to keep the precompiled information would remain below 5 megabytes.

6 Discussion

We have presented an approach to accelerate the computation of dialectical trees (and thus warrant) within a dynamic abstract argumentative setting. Such dynamic frameworks allow for the representation of active and inactive arguments, being the former the only ones to be considered to compute warrant. This characteristic leads to the consideration of potential dialectical trees (containing every argument in the universal set), as well as active trees (containing only currently active arguments.) Therefore, every time the situation changes, warrants would have to be recalculated. There is a high degree of uncertainty regarding how the fluctuation of active arguments affects previously computed warrants. Nonetheless, the information codified by potential trees can be successfully used to speed up the construction of active trees. To this end, we calculate a measure of strength for each argument using potential trees. This is an advantage, as strengths are calculated just once, and not upon every change in the world, as it would be done if strength were calculated over active trees. Strength computation, even when performed once, is indeed expensive. However, empirical testing has shown that the time taken to compute argument strength is amortised within a reasonable time window.

The fast-prune heuristic technique is guided by argument strength and attempts to maximise pruning by looking to find undefeated arguments as soon as possible, which implies that their siblings can be cut off. Computing argument strength allows us to store precompiled arguments and attacks, which would be used later when answering queries. Hence, significant speedup (shown as "B/F" in Table 1) is not only a result of traversing smaller trees, but also due to the non-calculation of arguments and attacks. We have also compared the amount of arguments created by both approaches. In the case of FP, arguments are built at the stage of strength computation; as for BP, every query triggers the construction of a tree and every argument in it. Table 1 shows both quantities in columns "str(#)"and "BP(#)", where the latter is the accumulated amount after 500 queries. Finally, when considering precompiled arguments and attacks for both approaches, speedup remains considerable, as illustrated by column "Bc/Fc" in Table 2, where is shown that FP responds twice as fast than BP.

Experimental results show that each fast-prune bonsai is a significantly smaller version of its associated non-pruned active tree, providing a meaningful speedup. This is an asset in massive argumentation domains, like the WWW [14], where repeatedly looking for counter-arguments is expensive. Another scenario where our approach would perform well can be found in the context of multi-agent systems. An argumentation-based agent with goals expressed as a set of warranted arguments [1] would be computing warrant several times per cycle. In such a setting, precompiling and storing potential trees would yield a clear advantage: a twice-as-fast response.

7 Related Work

In [13] the subject of efficient computation of dialectical trees was already addressed. However, the approach is mostly declarative and it does not include any concrete techniques to perform the pruning, let alone empirical testing.

In [6] an approach for speeding up the construction of argumentation trees is presented. The authors also use information from the argument knowledge base beforehand, to speed up the argumentation process. This argument compilation produces a hyper-graph of the inconsistent subsets, which, together with a special algorithm, can be used for efficiently building argument trees, saving time for conflict detection. In our work, we try to avoid the evaluation of attackers whenever possible. Both approaches follow parallel paths and could be combined.

Potential trees storage to avoid their reconstruction was considered similarly to [8]. The main drawback is the amount of storage space needed, keeping one tagged potential tree per argument in the universal set.

In order to deal with the warrant recalculation, some work has been recently done on dynamics in argumentation. In [7], the authors propose a series of principles to determine under what conditions an extension does not change when faced to a change in the framework. This study on the impact of change over extensions is based on arguments graphs and complementary to ours.

References

1. Amgoud, L., Devred, C., Lagasquie, M.C.: A constrained argumentation system for practical reasoning. In: AAMAS 2008: Proceedings of the 7th International Joint Conference on Autonomous Agents and Multiagent Systems (May 12-16 (2008)
2. Baroni, P., Dunne, P.E., Giacomin, M.: Computational properties of resolution-based grounded semantics. In: IJCAI, pp. 683–689 (2009)
3. Baroni, P., Giacomin, M.: On Principle-Based Evaluation of Extension-Based Argumentation Semantics. Artif. Intell. 171, 675–700 (2007)
4. Bench-Capon, T.J.M., Dunne, P.E.: Argumentation in artificial intelligence. Artif. Intell. 171, 619–641 (2007)
5. Besnard, P., Hunter, A.: A logic-based theory of deductive arguments. Artif. Intell. 128(1-2), 203–235 (2001)
6. Besnard, P., Hunter, A.: Knowledgebase compilation for efficient logical argumentation. In: KR, pp. 123–133 (2006)
7. Boella, G., Kaci, S., van der Torre, L.: Dynamics in argumentation with single extensions: Abstraction principles and the grounded extension. In: Sossai, C., Chemello, G. (eds.) ECSQARU 2009. LNCS, vol. 5590, pp. 107–118. Springer, Heidelberg (2009)
8. Capobianco, M., Chesñevar, C.I., Simari, G.R.: Argumentation and the dynamics of warranted beliefs in changing environments. JAAMAS 11, 127–151 (2005)
9. Chesñevar, C.I., Simari, G.R., García, A.J.: Pruning search space in defeasible argumentation. In: ATAI, pp. 46–55 (2000)
10. Dung, P.M.: On the acceptability of arguments and its fundamental role in non-monotonic reasoning, logic programming and n-person games. Artif. Intell. 77(2), 321–358 (1995)
11. García, A.J., Simari, G.R.: Defeasible logic programming: An argumentative approach. TPLP 4(1-2), 95–138 (2004)
12. Matt, P.A., Toni, F.: A game-theoretic measure of argument strength for abstract argumentation. In: JELIA, pp. 285–297 (2008)
13. Chesñevar, C.I., Simari, G.R., Godo, L.: Computing dialectical trees efficiently in possibilistic defeasible logic programming. In: Baral, C., Greco, G., Leone, N., Terracina, G. (eds.) LPNMR 2005. LNCS (LNAI), vol. 3662, pp. 158–171. Springer, Heidelberg (2005)
14. Rahwan, I.: Mass argumentation and the semantic web. Journal of Web Semantics 6(1), 29–37 (2008)
15. Rotstein, N., Moguillansky, M., García, A., Simari, G.: A Dynamic Argumentation Framework. In: COMMA, pp. 427–438 (2010)
16. Rotstein, N.D., Moguillansky, M.O., Simari, G.R.: Dialectical abstract argumentation: A characterization of the marking criterion. In: IJCAI, pp. 898–903 (2009)

On Warranted Inference in Argument Trees Based Framework

Safa Yahi

LSIS-CNRS, UMR 6168
IUT d'Aix-en-Provence
Université de la Méditerranée
safa.yahi@univmed.fr

Abstract. In this paper, we focus on logical argumentation introduced by Besnard and Hunter. First, we consider the so-called warranted inference which is based on the dialectical principle that is widely used in the literature of argumentatation. More precisely, we compare warranted inference with respect to the most frequently used coherence based approaches from flat belief bases in terms of productivity. It turns out that warranted inference is incomparable, w.r.t. productivity, with almost the coherence based approaches considered in this paper. Also, although too productive in some situations, warranted inference does not entail some very desirable conclusions which correspond to those which can be entailed from each consistent formula. Then, we introduce a new inference relation where the key idea is that the support of a counter-argument must not entail the conclusion of the objected argument which is quite intuitive. We show then that this inference relation ensures the inference of the previous desirable conclusions. Besides, we suggest to distinguish two levels of attacks: strong attacks and weak attacks. We propose then to weight our new inference relation based on the structure of the argument tree and also by taking into account the level strength of attacks.

1 Introduction

Argumentation is a key approach that can be used in several situations like conflicts handling [2], negotiation [3], decision making [15], etc. There are a number of proposals for logic-based formalisations of argumentation. In this paper, we focus on the argument trees based framework proposed by Besnard and Hunter [8,5]. Especially, we consider the so-called warranted inference [6,8] which is based on the dialectical principle that is widely adopted in the literature on argumentatation. Warranted inference has been recently studied from a complexity point of view in [14]. Also, it has been applied, for instance, in the case of inconsistency management policies in relational databases [17]. However, to the best of our knowledge, there is no study that compares warranted inference with coherence based approaches unlike other logical argumentation frameworks [1,9,5].

S. Benferhat and J. Grant (Eds.): SUM 2011, LNAI 6929, pp. 191–204, 2011.
© Springer-Verlag Berlin Heidelberg 2011

In this paper, we compare warranted inference with respect to the most frequently used coherence based approaches from flat belief bases in terms of productivity. Surprisingly, this inference relation does not always entail very desirable conclusions which are derived using a too cautious inference relation while it is too productive in some situations. So, we introduce a new inference relation that we call rational warranted inference where the key idea is that the support of a counter-argument must not entail the conclusion of the objected argument that is is quite natural. We show then that this inference relation garantees to deduce the previous desirable conclusions. Besides, we suggest to distinguish two levels of attacks: strong attacks and weak attacks. We propose then to weight our new inference relation based on the structure of the argument tree and also by taking into account the level strength of attacks. The rest of the paper is structured as follows. First, we give a brief background on logical argumentation and the most popular coherence based approaches from flat belief bases where no preference relation is considered over beliefs. Then, we compare warranted inference with these coherence based-approaches. After that, we introduce a new class of counter-arguments, namely rational defeaters and rational undercuts and we give a number of their properties. Based on them, we introduce the notion of rational warranted inference and compare it with the other inference relations. Finally, we present weighted form of R-warranted inference before concluding the paper and giving some of our perspectives.

2 A Brief Background on Logical Argumentation

We consider a finite set of propositional variables denoted by lower case Roman letters $a, b, c, \ldots$. Formulae are denoted by lower case Greek letters $\alpha, \beta, \gamma, \ldots$ while finite sets of formulae are denoted by upper case Greek letters Φ, Ψ, Δ, ... The symbols $\top$ and $\bot$ denote tautology and contradiction respectively. As to the symbol $\models$, it denotes classical propositional inference.

In [5], Besnard and Hunter assume a belief base (a finite set of formulae) Σ that they use throughout their definitions. Moreover, they suppose that every subset of Σ is given an enumeration $< \phi_1, \ldots, \phi_n >$ of its elements, which they call its canonical enumeration. Please, note that this constraint is satisfied when any arbitrary total ordering is imposed over Σ. Now, we are ready to recall what is logical argumentation starting by the notion of an argument.

Definition 1. *An* ***argument*** *is a pair* $< \Phi, \alpha >$ *such that:* Φ *is a consistent subset of* Σ, $\Phi \models \alpha$ *and* Φ *is a minimal subset of* Σ *satisfying the previous properties.* $< \Phi, \alpha >$ *is said to be an argument for* α *where* Φ *is called the* ***support*** *of the argument and* α *its consequent or its* ***claim****.*

Some arguments encompass others. This is captured with the notion of more conservative arguments.

Definition 2. *An argument* $< \Phi, \alpha >$ *is* ***more conservative than*** *an argument* $< \Psi, \beta >$ *iff* $\Phi \subseteq \Psi$ *and* $\beta \models \alpha$*.*

Example 1. $< \{a\}, a \vee b >$ is more conservative than $< \{a, a \Rightarrow b\}, b >$.

Then, a more conservative argument can be seen as more general in the sense that it is less demanding on the support and less specific with respect to the claim. Now, let us consider counter-arguments which include defeaters, undercuts and rebuttals. Defeaters are arguments whose claim refutes the support of another argument [19,21,22] while undercuts are arguments which directly oppose the support of others. As to rebuttals, they capture the most direct form of a conflict between arguments which occurs when two arguments have opposite consequences.

Definition 3. *Let $< \Phi, \alpha >$ be an argument.*

- *A* ***defeater*** *for $< \Phi, \alpha >$ is an argument $< \Psi, \beta >$ s.t. $\beta \models \neg(\phi_1 \wedge \ldots \wedge \phi_n)$ where $\{\phi_1, \ldots, \phi_n\} \subseteq \Phi$.*
- *An* ***undercut*** *for $< \Phi, \alpha >$ is an argument $< \Psi, \neg(\phi_1 \wedge \ldots \wedge \phi_n) >$ where $\{\phi_1, \ldots, \phi_n\} \subseteq \Phi$.*
- *An argument $< \Psi, \beta >$ is a* ***rebuttal*** *for $< \Phi, \alpha >$ iff $\beta \equiv \neg\alpha$.*

Example 2. Let $\Sigma = \{a, a \rightarrow \neg b, c, c \rightarrow \neg a, \neg a \rightarrow b\}$ be a belief base. Let us consider the argument $< \{a, a \rightarrow \neg b\}, \neg b >$. Then,

- The argument $< \{c, c \rightarrow \neg a\}, \neg a >$ is an undercut .
- The argument $< \{c, c \rightarrow \neg a, \neg a \rightarrow b\}, b >$ is a rebuttal.

One can easily see that both undercuts and rebuttals are defeaters. Moreover, an undercut for an argument need not be a rebuttal for that argument and vice versa. Now, according to [5], we have the following properties:

Property 1. *If $< \Psi, \beta >$ is a defeater for $< \Phi, \alpha >$ then there exists an undercut for $< \Phi, \alpha >$ which is more conservative than $< \Psi, \beta >$.*

Property 2. *If $< \Psi, \beta >$ is a maximally conservative defeater of $< \Phi, \alpha >$ then $< \Psi, \beta' >$ is an undercut of $< \Phi, \alpha >$ for some β' which is logically equivalent with β.*

The two previous properties are of a special interest. In fact, they point to undercuts of an arguments as candidates to be representative of all the defeaters of that argument. In particular, maximally conservative undercuts are even better candidates. An argument $< \Psi, \beta >$ is a maximally conservative undercut of $< \Phi, \alpha >$ iff $< \Psi, \beta >$ is an undercut of $< \Phi, \alpha >$ such that there is no an undercut for $< \Phi, \alpha >$ which is strictly more conservative than $< \Psi, \beta >$. Note that the claim of a maximally conservative undercut of an argument is nothing than the negation of the full support of the considered argument.

Property 3. *Let $< \Psi, \neg(\phi_1 \wedge \ldots \wedge \phi_n) >$ be an undercut for $< \Phi, \alpha >$. Then $< \Psi, \neg(\phi_1 \wedge \ldots \wedge \phi_n) >$ is a maximally conservative undercut for $< \Phi, \alpha >$ iff $\Phi = \{\phi_1, \ldots, \phi_n\}$.*

In order to avoid redundancy, the notion of canonical undercuts has been introduced.

Definition 4. *An argument* $< \Psi, \neg(\phi_1 \wedge \ldots \wedge \phi_n) >$ *is a* ***canonical undercut*** *for* $< \Phi, \alpha >$ *iff* $< \phi_1, \ldots, \phi_n >$ *is the canonical enumeration of* Φ*.*

Example 3. Let $\Sigma = \{b, c, a, c \rightarrow \neg a \wedge \neg b\}$ be a belief base.

Let us suppose that the canonical enumeration of Σ is as follows: $< c, a, c \rightarrow \neg a \wedge \neg b, b >$ and let us consider the argument $< \{a, b\}, a \wedge b >$.

- The argument $< \{c, c \rightarrow \neg a \wedge \neg b\}, \neg a >$ is not a canonical undercut.
- The argument $< \{c, c \rightarrow \neg a \wedge \neg b\}, \neg(b \wedge a) >$ is not a canonical undercut.
- The argument $< \{c, c \rightarrow \neg a \wedge \neg b\}, \neg(a \wedge b) >$ is a canonical undercut.

The following property is of a relevant interest: It shows that the set of all the canonical undercuts of an argument represent all the defeaters of that argument.

Property 4. *If* $< \Psi, \beta >$ *is a defeater for* $< \Phi, \alpha >$ *then there exists a canonical undercut for* $< \Phi, \alpha >$ *which is more conservative than* $< \Psi, \beta >$*.*

Based on the notion of canonical undercuts, an argument tree shows the various ways an argument of interest can be objected, as well as how its defeaters can themselves be objected, and so on.

Definition 5. *An* ***argument tree*** *for* α *is a tree where the nodes are arguments such that:*

1. *The root is an argument for* α*.*
2. *For no node* $< \Psi, \beta >$ *with ancestor nodes* $< \Psi_1, \beta_1 >, \ldots, < \Psi_n, \beta_n >$, Ψ *is a subset of* $\Psi_1 \cup \ldots \cup \Psi_n$*. This avoids circularity.*
3. *The children nodes of a node* N *consist of all its canonical undercuts which verify (2).*

Example 4. Let $\Sigma = \{a, a \rightarrow b, c \wedge d, c \rightarrow \neg a, \neg d \vee \neg a\}$ be a belief base. Then, the argument tree associated with the argument $< \{a, a \rightarrow b\}, b >$ is given by figure 1.

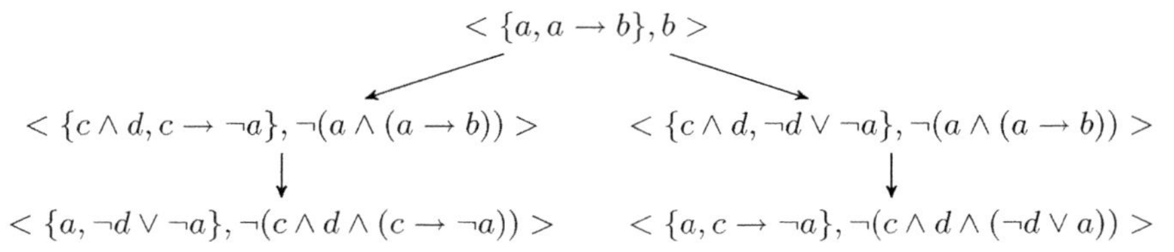

Fig. 1. Argument tree

In [6], the authors introduce the notion of judge function for determining whether an argument tree is warranted.

Definition 6. *An argument tree is said to be* ***warranted*** *iff the argument at its root is marked as undefeated where each node is marked as undefeated iff all its children are defeated. Otherwise, it is marked as defeated.*

Note that this definition has been adopted from dialectical tree marking for defeasible logic programming [12].

Example 5. The argument tree given by figure 1 is warranted.

3 A Refresher on Coherence-Based Approaches from Flat Belief Bases

Coherence-based approaches [20] can be considered as a two step process consisting first in generating some consistent subbases and then using classical inference from some of them according to a given entailment principle [18]. The most frequently used entailment principles are the universal (or skeptical) inference where each consequent is a classical consequent of all the considered consistent subbases, the existential (or credulous) inference where each consequent is a classical consequent of at least one considered consistent subbase and the argumentative inference where each conclusion is credulously inferred but its negation is not. In the context of flat belief bases, the most popular sets of consistent subbases are $MaxCons(\Sigma)$ and $CardCons(\Sigma)$ which correspond respectively to the set of all the maximal (with respect to set inclusion) consistent subbases of Σ and the set of all its maximal, with respect to set cardinality, consistent subbases.

We can also consider the set of all the consistent subbases of Σ. Note that we discard the empty set (which is a consistent subbase) in the case where Σ contains at least one consistent formula. Let $Cons(\Sigma)$ denotes the previous set.

Based on $Cons(\Sigma)$, $MaxCons(\Sigma)$, $CardCons(\Sigma)$ and the three previous entailment principles, we obtain nine inference relations defined as follows.

Definition 7.
- *A formula φ is a* ***Uni-Cons*** *conclusion (resp.* ***Uni-MaxCons, Uni-Card****) of Σ iff $\forall B \in Cons(\Sigma)$ (resp. $MaxCons(\Sigma), CardCons(\Sigma)$), $B \models \varphi$.*
- *A formula φ is an* ***Exi-Cons*** *conclusion (resp.* ***Exi-MaxCons, Exi-Card****) of Σ iff $\exists B \in Cons(\Sigma)$ (resp. $MaxCons(\Sigma), CardCons(\Sigma)$), $B \models \varphi$.*
- *A formula φ is an* ***Arg-Cons*** *conclusion (resp.* ***Arg-MaxCons, Arg-Card****) of Σ iff $\exists B \in Cons(\Sigma)$ (resp. $MaxCons(\Sigma), CardCons(\Sigma)$) , $B \models \varphi$ and $\nexists B' \in Cons(\Sigma)$ (resp. $MaxCons(\Sigma), CardCons(\Sigma)$) such that $B' \models \neg\phi$.*

These inference relations have been compared in terms of productivity in [10]. Roughly speaking, an inference relation I_1 is said to be at least as productive as I_2 iff each conclusion of I_2 is derived using I_1. Equivalently, we can say that

I_2 is at least as cautious as I_1. In this paper, we complete the study achieved in [16] by the following results based essentially on the fact that logical inference is monotonic.

Proposition 1. *Exi-Cons inference is equivalent to Exi-MaxCons inference. In addition, Arg-Cons inference and Arg-MaxCons inference are equivalent.*

Combining the results of [16] and the previous proposition, we obtain figure 2 where an arrow $I \rightarrow J$ means that inference relation I is less productive (or more cautious) than the inference relation J.

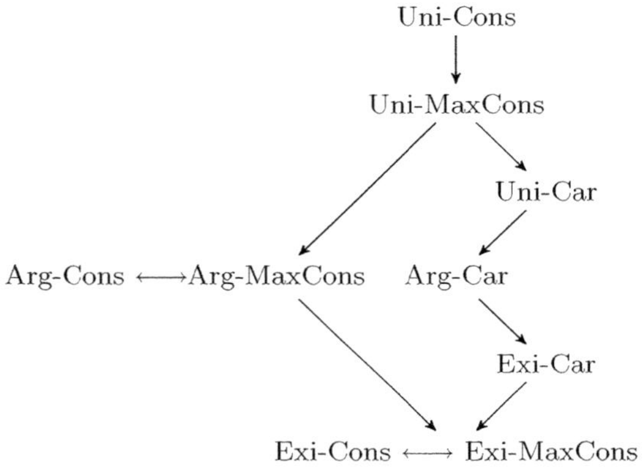

Fig. 2. Productivity of coherence based approaches

Finally, let us mention that the so-called argumentative inference introduced in [4] has been shown to be equivalent Arg-MaxCons inference in the same paper.

4 Warranted Inference Properties

In this section, we study some properties of warranted argument tree based inference which we will simply call in the following warranted inference.

Definition 8. *A formula φ is a* ***warranted*** *conclusion of a belief base Σ, denoted by $\Sigma \models_W \varphi$, iff there exists a warranted argument tree for φ.*

In particular, we compare it in terms of productivity with respect to the coherence-based approaches that we have recalled in the previous section. Let CBA denotes the set of all these inference relations.

First of all, we show that some desirable conclusions that correspond to those entailed using Uni-Cons inference which is strictly less productive than all the coherence-based approaches considered in this paper, are not warranted conclusion.

Proposition 2. *Warranted inference is not at least as productive as Uni-Cons inference.*

This proposition is shown via the following example.

Example 6. Let $\Sigma = \{a \wedge b, \neg a \wedge b\}$ be a belief base. Clearly, the formula b is a Uni-Cons conclusion of Σ. Indeed, Σ has only two non empty consistent subbases namely $\{a \wedge b\}$ and $\{\neg a \wedge b\}$ which both classically infer b. Now, let us consider the arguments of b. We have two arguments: $A_1 :< \{a \wedge b\}, b >$ and $A_2 :< \{\neg a \wedge b\}, b >$. Neither the argument tree of A_1 nor the argument tree of A_2 are warranted. Thus, b is not a warranted conclusion.

Notice that from Proposition 2 and the fact that Uni-Cons inference is strictly more cautious than all the other coherence based approaches considered in this paper, we can directly derive that warranted inference is not at least as productive as any inference relation from CBA.

Surprisingly, warranted inference verifies the following property.

Proposition 3. *Given a belief base, both a formula and its negation can be entailed using warranted inference.*

To check this proposition, it suffices to consider the following example.

Example 7. Let $\Sigma = \{a \wedge b, \neg a \wedge c, \neg b \wedge \neg c\}$ be a belief base. Let us consider the argument tree given by figure 3.

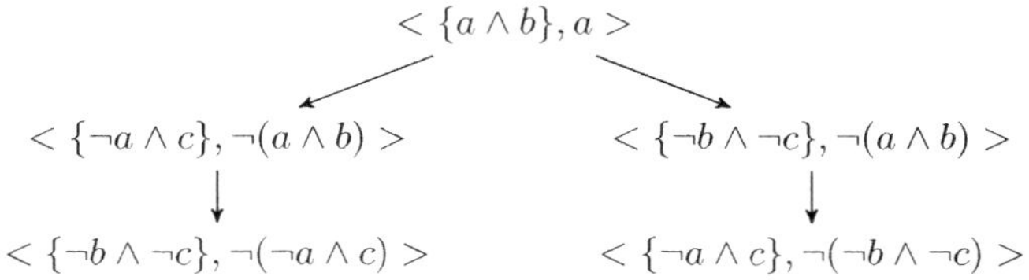

Fig. 3. An argument Tree for a

This argument tree is warranted, so a is a warranted conclusion of Σ. Similarly, the argument tree having $< \{\neg a \wedge c\}, \neg a >$ as a root is warranted which implies that $\neg a$ is also a warranted conclusion of Σ.

Now, coherence-based approaches using universal or argumentative principles clearly do not infer a conclusion and its negation unlike warranted inference. Then, we obtain the following proposition.

Proposition 4. *Each inference relation using universal or argumentative principle from CBA is not more productive than warranted inference.*

The same thing holds with Exi-Car inference.

Proposition 5. *Exi-Car inference is not more productive than warranted inference.*

Example 8. Let $\Sigma = \{a, \neg a \wedge b, \neg a \wedge c, \neg a \wedge d, \neg b \wedge \neg c \wedge \neg d\}$. $CardCons(\Sigma) = \{\{\neg a \wedge b, \neg a \wedge c, \neg a \wedge d\}\}$, so a is not an Exi-Card conclusion of Σ. However, a is a warranted conclusion of Σ.

Besides, a formula φ is a warranted conclusion implies that there exists an argument $< \Phi, \varphi >$ for φ where Φ is a consistent subbase of Σ. Then, φ is an Exi-Cons conclusion of Σ and also an Exi-MaxCons conclusion of Σ.

All the results presented in this section are summarized by Figure 4

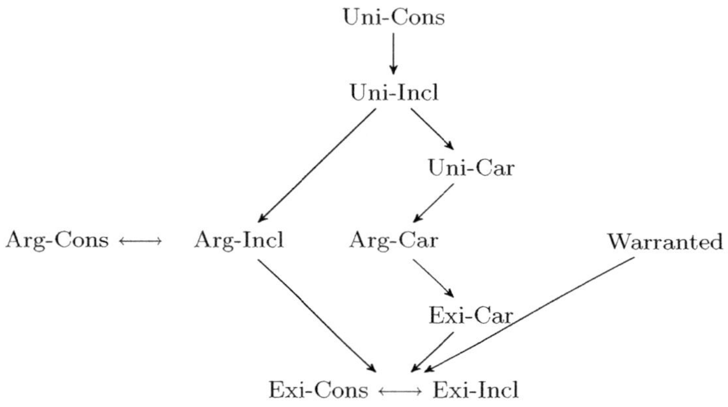

Fig. 4. Productivity results

So, to conclude this section, given a belief base where each consistent piece of information (formula) infers a conclusion φ, warranted inference easily fails to capture such a natural conclusion despite deriving a conclusion and its negation in some situations. In the following, we will propose a new inference relation which captures these desirable conclusions. For this aim, we first introduce a new class of intuitive counter-argument and show some of them properties.

5 A New Class of Counter-Arguments

In this section, we introduce a new class of counter-arguments (defeaters and undercuts) where the additional condition, which we find quite natural and intuitive, tells that the support of a counter-argument must not entail the claim of the argument it challenges. This notion is captured through what we roughly call rational defeaters and rational undercuts.

Definition 9. *Let* $< \Phi, \alpha >$ *be an argument.*

- *An* ***R-defeater*** *(R for Rational) for* $< \Phi, \alpha >$ *is an argument* $< \Psi, \beta >$ *s.t.* $\beta \models \neg(\phi_1 \wedge \ldots \wedge \phi_n)$ *where* $\{\phi_1, \ldots, \phi_n\} \subseteq \Phi$ *and* $\Psi \not\models \alpha$.
- *An* ***R-undercut*** *of an argument* $< \Phi, \alpha >$ *is an argument* $< \Psi, \neg(\phi_1 \wedge \ldots \wedge \phi_n) >$ *where* $\{\phi_1, \ldots, \phi_n\} \subseteq \Phi$ *and* $\Psi \not\models \alpha$.

Notice that, by definition, the support of a rebuttal does not entail the claim of the argument it objects (since it entails its negation). Thus, we do not need to introduce a new class of R-rebuttals. In addition, the relations between defeaters, undercuts and rebuttals are maintained with respect to R-defeaters, R-undercuts and rebuttals. Now, let us consider the following property of R-defeaters which is directly derived from their definition.

Proposition 6. *If* $< \Gamma, \delta >$ *is an R-defeater for* $< \Psi, \beta >$ *which is in turn an R-defeater for* $< \Phi, \alpha >$ *then* $\Gamma \not\models \neg\Phi$.

This property is especially interesting in the sense that no argument can be, at the same time, an R-defeater and a direct R-defender (R-defeater for an R-defeater) for the same argument contrary to what happens with respect to classical defeaters (but not with respect to classical undercuts) as mentioned in [7]. Better yet, a subbase of a belief base can not be used as both the support of a direct R-defender and the support of an R-defeater for a given argument unlike what occurs in the case of classical defeaters and also classical undercuts. Especially, a direct R-defender of an argument can not refute the conclusion of the defended argument counter to classical defeaters and classical undercuts.

In addition, Property 1 and Property 2 can be extended to R-defeaters and R-undercuts as shown by the following two propositions.

Proposition 7. *If* $< \Psi, \beta >$ *is an R-defeater for* $< \Phi, \alpha >$ *then there exists an R-undercut for* $< \Phi, \alpha >$ *which is more conservative than* $< \Psi, \beta >$.

In fact, by definition, $< \Psi, \beta >$ is an R-defeater for $< \Phi, \alpha >$ implies that $< \Psi, \beta >$ is a defeater for $< \Phi, \alpha >$. Thus, according to Property 1, there exists an undercut $< \Gamma, \gamma >$ for $< \Phi, \alpha >$ which is more conservative than $< \Psi, \beta >$. Since $\Psi \not\models \alpha$ and $\Gamma \subseteq \Psi$, we deduce that $\Gamma \not\models \alpha$. So, $< \Gamma, \gamma >$ is an R-undercut for $< \Psi, \alpha >$.

Proposition 8. *If* $< \Psi, \beta >$ *is a maximally conservative R-defeater of* $< \Phi, \alpha >$ *then* $< \Psi, \beta' >$ *is an R-undercut of* $< \Phi, \alpha >$ *for some* β' *which is logically equivalent with* β.

This proof can be easily adapted from the proof of Theorem 4.2 in [5].

Thus, following Proposition 7 and Proposition 8, the set of all the R-defeaters of an argument can be represented by the set of all the R-undercuts of this argument. This set can be refined again as we will see in the following using maximally conservative R-undercuts.

Proposition 9. *Let* $< \Psi, \neg(\phi_1 \wedge \ldots \wedge \phi_n) >$ *be an R-undercut for* $< \Phi, \alpha >$. *Then* $< \Psi, \neg(\phi_1 \wedge \ldots \wedge \phi_n) >$ *is a maximally conservative R-undercut for* $< \Phi, \alpha >$ *iff* $\Phi = \{\phi_1, \ldots, \phi_n\}$.

The proofs of Theorem 5.2 and Theorem 5.4 in [5] can be easily adapted to prove the previous proposition.

Definition 10. *An argument* $< \Psi, \neg(\phi_1 \wedge \ldots \wedge \phi_n) >$ *is a* ***canonical R-undercut*** *for* $< \Phi, \alpha >$ *iff it is an R-undercut for* $< \Phi, \alpha >$ *and* $< \phi_1, \ldots, \phi_n >$ *is the canonical enumeration of* Φ.

Moreover, we can easily show the following proposition which is of a particular attention in the sense that it tells us that the set of all canonical R-undercuts of an argument represent all its R-defeaters.

Property 5. *If* $< \Psi, \beta >$ *is an R-defeater for* $< \Phi, \alpha >$ *then there exists a canonical R-undercut for* $< \Phi, \alpha >$ *which is more conservative than* $< \Psi, \beta >$.

Using the notion of canonical R-undercuts that we have presented, we propose a new class of argument trees.

Definition 11. *An* ***R-argument tree*** *for* α *is an argument tree for* α *where the children of each node are its canonical R-undercuts.*

6 Rational Warranted Inference Properties

In this section, we apply on rational warranted inference the study we achieved on warranted inference. Notice that what we call rational warranted inference is defined as follows.

Definition 12. *A formula* φ *is an* ***R-warranted*** *conclusion of a belief base* Σ, *denoted by* $\Sigma \models_{RW} \varphi$, *iff there exists a warranted R-argument tree for the formula* φ.

Firstly, the good news is that the expected Uni-Cons conclusions are R-warranted conclusions too.

Proposition 10. *R-warranted inference is at least as productive as Uni-Cons inference.*

Indeed, a formula φ is a Uni-Cons conclusion of Σ means that φ is a logical conclusion of each consistent formula in Σ. So, any formula of Σ can be the support of an argument for φ. Moreover, such an argument does not admit any R-defeater since any subset of Σ logically entails φ. Besides, unlike Uni-Cons conclusions, there exist Uni-MaxCons conclusions which are not R-warranted ones.

Proposition 11. *R-warranted inference is not at least as productive as Uni-MaxCons inference.*

The following example proves our proposition.

Example 9. Let $\Sigma = \{a\wedge b, a\wedge c, \neg a\wedge b, \neg a\wedge d\}$. Since, $MaxCons(\Sigma) = \{\{a\wedge b, a\wedge c\}, \{\neg a \wedge b, \neg a \wedge d\}\}$, the formula b is a Uni-MaxCons conclusion of Σ. However, b admits only two arguments in Σ, namely $< \{a \wedge b\}, b >$ and $< \{\neg a \wedge b\}, b >$ whose the corresponding R-argument trees are as follows (where the symbol $\Box$ denotes the negation of the support of the objected argument).

1. $< \{a \wedge b\}, b > \leftarrow < \{\neg a \wedge d\}, \Box > \leftarrow < \{a \wedge c\}, \Box > \leftarrow < \{\neg a \wedge b\}, \Box >$
2. $< \{\neg a \wedge b\}, b > \leftarrow < \{a \wedge c\}, \Box > \leftarrow < \{\neg a \wedge b\}, \Box > \leftarrow < \{a \wedge b\}, \Box >$

None of these R-argument trees is warranted which implies that b is not an R-warranted conclusion.

However, R-warranted inference still may infer a formula and its negation.

Proposition 12. *Given a belief base, both a formula and its negation can be entailed using R-warranted inference.*

Example 10. Let $\Sigma = \{a \wedge b, \neg b \wedge c, \neg c \wedge d, \neg d \wedge e, \neg e \wedge \neg a\}$ be a belief base and let us consider the arguments $< \{a \wedge b\}, a >$ and $< \{\neg e \wedge \neg a\}, \neg a >$. One can easily check that the corresponding R-argument trees are warranted which means that both a and $\neg a$ are R-warranted conclusions.

Notice that R-warranted inference applied on the belief base of example 7 does not entail neither a nor $\neg a$. So, we can derive the following proposition.

Proposition 13. *R-warranted inference and warranted inference are incomparable in terms of productivity.*

So, we can easily show that R-warranted inference checks the following properties as we have done regarding to warranted inference.

Proposition 14.
- *R-Warranted inference is not at least as productive as any inference relation from CBA/{Uni-Cons}.*
- *Each inference relation using universal or argumentative principle from CBA is not more productive than R-warranted inference.*
- *Exi-Car inference is not more productive than R-warranted inference.*
- *R-warranted inference is strictly less productive than Exi-Cons inference and Exi-MaxCons inference.*

Now, we are ready to extend the taxonomy given by figure 2 by including warranted and R-warranted inferences as shown by figure 5.

7 Weighted R-warranted Inference

As we have previously seen, R-warranted inference only tells us whether a conclusion is acceptable or not. However, we argue that is is important, in many situations like decision making, negotiation, persuasion, etc, to know to what

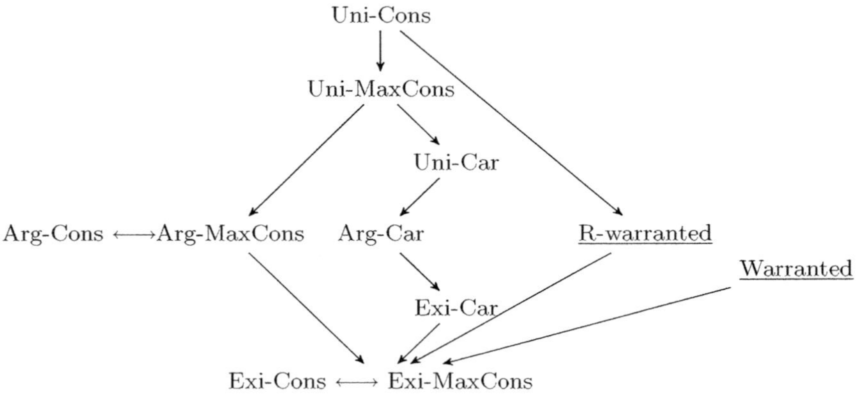

Fig. 5. Productivity results

extent a given conclusion is acceptable. That is why we propose to weight R-warranted inference. One possible way to achieve such a purpose is to consider the so-called categorisers proposed by Besnard and Hunter which are mappings from argument trees to numbers [5]. The number given by a categoriser tries to capture the strength of an argument based on its argument tree. The authors propose a very intuitive categorizer, namely the h-categorizer such that an argument tree of root R is affected a number $h(R)$ defined recursively as follows: $h(A) = \frac{1}{1+\sum_{i=1}^{n} h(A_i)}$ where A_i's are the canonical undercuts of A and $h(A_j) = 1$ for each leave node A_j. The intuition behind the h-categoriser is that the more undercuts an argument has, the less its strength is and the more undercuts there are to the undercuts of an argument, the more its strength is.

In this paper, we propose to extend the h-categoriser by taking into account two levels of strength attacks. So, let us first introduce the motivation of these two levels via the following example:

Example 11. Let $\Sigma = \{a \wedge c, a \rightarrow b, \neg a, \neg c\}$ be a belief base. Let us consider the argument $A_1 :< \{a \wedge c, a \rightarrow b\}, b >$. This argument has two R-undercuts $A_2 :< \{\neg a\}, \neg(a \wedge c \wedge a \rightarrow b) >$ and $A_3 :< \{\neg c\}, \neg(a \wedge c \wedge a \rightarrow b) >$. Intuitively, the attack of A_2 is stronger than the attack of A_3. In fact, the sub-formula c is not indispensable to infer b using $\{a \wedge c, a \rightarrow b\}$. In other words, we can forget c from the support of the argument and still infering b. This is not the case with respect to the sub-formula a which is vital to infer b.

Then, roughly speaking, we say that the attack of an (R-)undercut $< \Psi, \beta >$ for $< \Phi, \alpha >$ is **strong** iff $< \Psi, \beta >$ attacks a part from Φ which is vital to infer α. Otherwise, we say that is a **weak** attack.

Now, supposing two numbers S and W such that $0 < W < S \leq 1$, we extend the h-categorizer by h2-categorizer (2 for two levels) as follows: $h2(A) = \frac{1}{1+\sum_{i=1}^{n} C(A_i) * h(A_i)}$ where A_i's are the canonical (R)-undercuts of A and

$h2(A_j) = 1$ for each leave node A_j. In addition, $C(A_i) = S$ if the attack of A_i against $< \Phi, \alpha >$ is strong and $C(A_i) = W$ otherwise.

8 Discussion and Perspectives

In this paper, we have compared the inference relation based on the so-called warranted argument trees with respect to the most frequently used coherence based approaches from flat belief bases in terms of cautiousness. It turned out that this inference relation does not entail some very natural conclusions although it is, in some situations, too permissive by infering both a formula and its negation. Then, we have introduced a new inference relation that we have called rational warranted inference and we have shown that it guarantees to entail all conclusions which are classically derived from each consistent formula in the belief base. In fact, the key idea is that an argument whose the support entails the claim of another argument can not be its counter-argument with respect to this conclusion which is quite intuitive for us.

In [13], Gorogiannis and Hunter propose a number of postulates to attack relations in logical argumentation. Given five arguments A, B, A', B' and C, these postulates are as follows:

- (D_0) if $A \equiv A'$ and $B \equiv B'$ then A attacks B iff A' attacks B'
- (D_1) if A attacks B then $Claim(A)$ is incoherent with $Support(B)$
- (D_2) if A attacks B and $Claim(C) \equiv Claim(A)$ then C attacks B
- (D_3) if A attacks B and $Support(B) = Support(C)$ then A attacks C
- (D_4) if we consider only the arguments that can be generated from a belief base and find that no two arguments attack each other, then the knowledge base must be consistent.

Among these postulates, neither D_2 nor D_3 are satisfied. Indeed, in the case of (D_2), let $A :< \{a \wedge \neg b\}, \neg(a \wedge b) >$, $B :< \{a \wedge b\}, b >$ and $C :< \{\neg a \wedge b\}, \neg(a \wedge b) >$. According to our new attack relation, A attacks B. However, C does not attack B which seems intuitive since B is an argument for b while C entails also b. As to the postulate (D_3), let $A :< \{\neg a \wedge b\}, \neg(a \wedge b) >$, $B :< \{a \wedge b\}, b >$ and $C :< \{a \wedge b\}, b >$. Clearly, A attacks B w.r.t. our attack relation. Nevertheless, A does not attack C which w.r.t. the same attack relation. Once again, this seems natural for us since C is an argment for b and A entails b.

Now, this work calls for several perspectives. First of all, we are interested in knowing how one can adapt rational warranted inference to prevent entailing a conclusion and its negation. A possible way is to modify the R-argument tree construction by imposing that each argument must be, in addition, coherent with all the arguments it indirectly defends in the argument tree. Also, we plan to extend logical argumentation to the case of pre-ordered belief bases. Another perspective is to refine our two levels of attacks strength by adapting the notion of "degree of undercut" [6] which is defined only with respect to the supports. In addition, we plain to instantiate abstract argumentation [11] using the new attack relation we have introduced in this paper.

References

1. Amgoud, L., Besnard, P.: A formal analysis of logic-based argumentation systems. In: Deshpande, A., Hunter, A. (eds.) SUM 2010. LNCS, vol. 6379, pp. 42–55. Springer, Heidelberg (2010)
2. Amgoud, L., Cayrol, C.: Inferring from inconsistency in preference-based argumentation frameworks. J. Autom. Reasoning 29(2), 125–169 (2002)
3. Amgoud, L., Dimopoulos, Y., Moraitis, P.: A unified and general framework for argumentation-based negotiation. In: AAMAS, p. 158 (2007)
4. Benferhat, S., Dubois, D., Prade, H.: Argumentative inference in uncertain and inconsistent knowledge bases. In: UAI, pp. 411–419 (1993)
5. Besnard, P., Hunter, A.: A logic-based theory of deductive arguments. Artificial Intelligence. 128(1-2), 203–235 (2001)
6. Besnard, P., Hunter, A.: Comparing and Rationalizing Arguments. In: Elements of Argumentation. MIT Press, Cambridge (2008)
7. Besnard, P., Hunter, A.: Logical Argumentation. In: Elements of Argumentation. The MIT Press, Cambridge (2008)
8. Besnard, P., Hunter, A.: Argumentation based on classical logic. In: Rahwan, I., Simari, G. (eds.) Argumentation in Artificial Intelligence (2009)
9. Cayrol, C.: On the relation between argumentation and non-monotonic coherence-based entailment. In: IJCAI, pp. 1443–1448 (1995)
10. Cayrol, C., Lagasquie-Schiex, M.-C.: Non-monotonic syntax-based entailment: A classification of consequence relations. In: Froidevaux, C., Kohlas, J. (eds.) ECSQARU 1995. LNCS, vol. 946, pp. 107–114. Springer, Heidelberg (1995)
11. Dung, P.M.: On the acceptability of arguments and its fundamental role in non-monotonic reasoning, logic programming and n-person games. Artif. Intell. 77(2), 321–358 (1995)
12. García, A.J., Simari, G.R.: Defeasible logic programming: an argumentative approach. Theory Pract. Log. Program. 4, 95–138 (2004)
13. Gorogiannis, N., Hunter, A.: Instantiating abstract argumentation with classical logic arguments: Postulates and properties. Artif. Intell. 175(9-10), 1479–1497 (2011)
14. Hirsch, R., Gorogiannis, N.: The complexity of the warranted formula problem in propositional argumentation. J. Log. Comput. 20(2), 481–499 (2010)
15. Kakas, A.C., Moraitis, P.: Argumentation based decision making for autonomous agents. In: AAMAS, pp. 883–890 (2003)
16. Lagasquie-Schiex, M.-C.: Contribution à l'étude des relations d'inférence non-monotone combinant inférence classique et préférences. Thèse de doctorat, Université Paul Sabatier, Toulouse, France (Décembre 1995)
17. Martinez, M.V., Hunter, A.: Incorporating classical logic argumentation into policy-based inconsistency management in relational databases. In: The Uses of Computational Argumentation Symposium, AAAI 2009 Fall Symposium Series (2009)
18. Pinkas, G., Loui, R.P.: Reasoning from inconsistency: A taxonomy of principles for resolving conflict. In: KR, pp. 709–719 (1992)
19. Pollock, J.L.: How to reason defeasibly. Artif. Intell. 57(1), 1–42 (1992)
20. Resher, N., Manor, R.: On inference from inconsistent premises. Theory and Decision 1, 179–219 (1970)
21. Verheij, B.: Automated argument assistance for lawyers. In: ICAIL, pp. 43–52 (1999)
22. Vreeswijk, G.: Abstract argumentation systems. Artif. Intell. 90(1-2), 225–279 (1997)

Uncertainty Handling in Quantitative BDD-Based Fault-Tree Analysis by Interval Computation

Christelle Jacob[1], Didier Dubois[2], and Janette Cardoso[1]

[1] Institut Supérieur de l'Aéronautique et de l'Espace (ISAE), DMIA department, Campus Supaéro, 10 avenue Édouard Belin - Toulouse

[2] Institut de Recherche en Informatique de Toulouse (IRIT), ADRIA department, 118 Route de Narbonne 31062 Toulouse Cedex 9, France

{jacob@isae.fr,dubois@irit.fr,cardoso@isae.fr}

Abstract. In fault-tree analysis, probabilities of failure of components are often assumed to be precise. However this assumption is seldom verified in practice. There is a large literature on the computation of the probability of the top (dreadful) event of the fault-tree, based on the representation of logical formulas in the form of binary decision diagrams (BDD). When probabilities of atomic propositions are ill-known and modelled by intervals, BDD-based algorithms no longer apply to the computation of the top probability interval. This paper investigates this question for general Boolean expressions, and proposes an approach based on interval methods, relying on the analysis of the structure of the Boolean formula. The considered application deals with the fault-tree-based analysis of the reliability of aircraft operations.

Keywords: Binary Decision Diagrams, interval analysis, fault-tree.

1 Introduction

In aviation business, maintenance costs play an important role because they are comparable to the cost of the plane itself. Another important parameter is reliability. A reliable aircraft will have less down-time for maintenance and this can save a lot of money for a company. No company can afford neglecting the reliability and safety of the aircraft, which necessitates a good risk management. Airbus has started a project called @MOST to cater for above mentioned issues. It is focused on the reduction of maintenance costs and hence improving the quality of its products.

This paper deals the probability evaluation of fault-trees through Binary Decision Diagrams (BDDs). This is the most usual approach to risk management in large-scale systems, for which operational software is available. At present, dependability studies are carried out by means of fault-tree analysis from the models of the system under study. This method requires that all probabilities of elementary component failures or relevant states be known in order to compute the probability of some dreadful events. But in real life scenarios, these

S. Benferhat and J. Grant (Eds.): SUM 2011, LNAI 6929, pp. 205–218, 2011.
© Springer-Verlag Berlin Heidelberg 2011

probabilities are never known with infinite precision. Therefore in this paper we investigate an approach to solve cases where such probabilities are imprecise. More specifically, we are concerned with the more general problem of evaluating the probability of a Boolean expression in terms of the probabilities of its literals, imprecisely known via intervals.

2 Dependability Studies and Fault-Tree Analysis

One of the objectives of safety analysis is to evaluate the probabilities of dreadful events. There are two ways to find the probability of an event: by an experimental approach or by an analytical approach.

The experimental approach consists in approximating the probability by computing a relative frequency from tests: if we realize N experiments (where N is a very large number) in the same conditions, and we observe n times the event e, the quotient n/N provides an approximation of the probability $\mathrm{P}(e)$. Indeed, this probability can be defined by: $P(e) = \lim_{N \to +\infty} n/N$. In practice, it is very difficult to observe events such as complex scenarios involving multiple, more elementary, events several times.

The analytical approach can be used when the dreadful event is described as a Boolean function F of atomic events. Such dreadful event probability is thus computed from the knowledge of probabilities of atomic events, that are often given by some experts. In a Boolean model of a system, a variable represents the state of an elementary component, and formulas describe the failures of the system as function of those variables. It is interesting to know the minimal sets of failures of elementary components that imply a failure of the studied system, in order to determine if this system is safe enough or not.

If only probability of these minimal sets are of interest, the corresponding Boolean formulas are monotonic and only involve positive literals representing the failures. However, in this paper we consider more general Boolean formulas where both positive and negative literals appear. In fact such kind of non-monotonic formulas may actually appear in practical reliability analysis as shown later.

2.1 The Probability of a Boolean Formula

In the following, we denote Boolean variables by A_i and literals by L_i ($L_i = A_i$ or its negation $\neg A_i$). There are two known methods to calculate the probability of a Boolean formula, according to the way the formula is written.

Boolean Formula as a Sum of Minterms. A Boolean formula can be written as a disjunction of minterms (maximal consistent conjunctions of literals). By definition, minterms are mutually exclusive, so the probability of a dreadful event is equivalent to:

$$P(F) = \sum_{\pi \in Minterms(F)} P(\pi) \tag{1}$$

Boolean methods of risk analysis often make the hypothesis that *all atomic events are stochastically independent.* So, the previous equation becomes:

$$P(F) = \sum_{\pi \in Minterms(F)} \prod_{L_i \in \pi} P(L_i) \tag{2}$$

But this method requires that all minterms of F be enumerated, which is very costly in computation time. In practice, an approximation of this probability is obtained by making some simplifications, namely:
– using the monotonic envelope of the formula instead of using the formula itself.
– using minimal cut sets with high probabilities only.
– using the k first terms of the Sylvester-Poincaré development, also known as inclusion-exclusion principle (letting each X_i stand for product of literals):

$$\begin{gathered} P(X_1 \vee ... \vee X_n) = \sum_{i=1}^{n} P(X_i) - \sum_{i=1}^{n-1} \sum_{j=i+1}^{n} P(X_i \wedge X_j) \\ + \sum_{i=1}^{n-2} \sum_{j=i+1}^{n-1} \sum_{k=j+1}^{n} P(X_i \wedge X_j \wedge X_k) - ... + (-1)^{n+1} P(X_1 \wedge ... \wedge X_n) \end{gathered} \tag{3}$$

Shannon Decomposition of a Boolean Formula. Shannon decomposition consists of building a tree whose leaves correspond to mutually disjoint conjunctions of literals.

Definition 1. *Let us consider a Boolean function F on a set of variables $\mathcal{X}$, and A a variable in $\mathcal{X}$. The* Shannon decomposition *of F related to A is obtained by the following pattern:*

$$F = (A \wedge F_{A=1}) \vee (\neg A \wedge F_{A=0}) \tag{4}$$

$F_{A=0}$ and $F_{A=1}$ are mutually exclusive, so the probability of the formula F is:

$$P(F) = (1 - P(A)) \cdot P(F_{A=0}) + P(A) \cdot P(F_{A=1}) \tag{5}$$

For a chosen ordering of variables of a set $\mathcal{X}$, the recursive application of the Shannon decomposition for each variable appearing in a function F built on $\mathcal{X}$ gives a binary tree, called *Shannon tree.* Each internal node of this tree can be read as an *if-then-else (ite) operator*: it contains a variable A and has two edges. One edge points towards the node encoded by the positive cofactor $F_{A=1}$ and the other one towards the negative cofactor $F_{A=0}$. The leaves of the tree are the truth values 0 or 1 of the formula. The expression obtained by the product of variables going down from the root to a leaf 1 is a minterm, and the sum of all those products gives an expression of the Boolean function.

Example 1. The Shannon tree for the formula $F = A \vee (S \wedge C)$ with order $A > S > C$ is represented on Fig. 1.a (the dotted lines represent the *else* edges). Notice that by convention, edges in Shannon trees are not arrows, but these graphs are directed and acyclic from the top to the bottom.

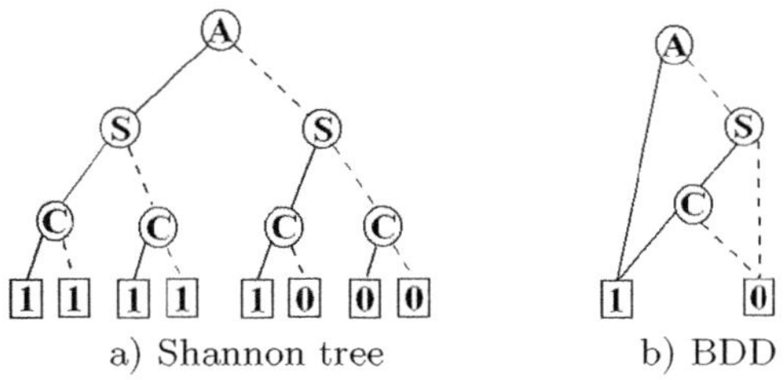

Fig. 1. Formula $A \vee (S \wedge C)$ with order $A > S > C$

But such a representation of the data is exponential in memory space; this is the reason why *Binary decision diagrams* [3] are introduced: they reduce the size of the tree by means of reduction rules.

Binary Decision Diagrams. A Binary Decision Diagram (BDD) is a representation of a Boolean formula in the form of a directed acyclic graph whose nodes are Boolean variables (see Fig. 1.b). A BDD has a root and two terminal nodes, one labeled by 1 and the other by 0, representing the truth values of the function. Each *path* from the root to terminal node 1 can be seen as a product of literals, and the disjunction of all those conjunctions of literals gives a representation of the function. For a given ranking of the variables, it is unique up to an isomorphism. The size of the BDD is directly affected by the ranking of variables, and some heuristics can be used to optimize its size.

A variable can be either *present* (in a positive or negative polarity) in a path, or *absent*:

- a variable is present in a positive polarity in the corresponding product if the path contains the *then-edge* of a node labeled by this variable;
- a variable is present in the negative polarity in the corresponding product if the path contains the *else-edge* of a node labeled by this variable;
- a variable is absent if the path does not contain a node labeled by this variable.

Let $\mathscr{P}$ be the set of paths in the BDD that reach terminal leaf 1, and $\mathscr{A}_p$ the set of literals contained in a path p of $\mathscr{P}$. If a literal L is present positively (resp. negatively $L = \neg A$) in the path $p \in \mathscr{P}$, we will write $L \in \mathscr{A}_p^+$ (resp. $L \in \mathscr{A}_p^-$). According to those notations, the probability of the formula F described by a BDD is given by the equation:

$$P(F) = \sum_{p \in \mathscr{P}} \left[\prod_{L_i \in \mathscr{A}_p^+} P(A_i) \prod_{L_j \in \mathscr{A}_p^-} (1 - P(A_j)) \right] \tag{6}$$

Now assume probabilities are partially known. The most elementary approach for dealing with imprecise BDDs is to apply interval analysis. It would enable robust conclusions to be reached, even if precise values of the input probabilities are not known, but their ranges are known.

3 Interval Arithmetic and Interval Analysis

Interval analysis is a method developed by mathematicians since the 1950s and 1960s [7] as an approach to computing bounds on rounding or measurement errors in mathematical computation. It can also be used to represent some lack of information. The main objective of interval analysis is to find the upper and lower bounds, $\overline{f}$ and $\underline{f}$, of a function f of n variables $\{x_1, ..., x_n\}$, knowing the intervals containing the variables: $x_1 \in [\underline{x_1}, \overline{x_1}]$, ...,$x_n \in [\underline{x_n}, \overline{x_n}]$. For a continuous function f, with $x_i \in [\underline{x_i}, \overline{x_i}]$, the image of a set of intervals $f([\underline{x_1}, \overline{x_1}], ..., [\underline{x_n}, \overline{x_n}])$ is an interval.

The basic operations of *interval arithmetic* that are generally used for interval analysis are, for two intervals $[a, b]$ and $[c, d]$ with $a, b, c, d \in \mathbb{R}$ and $b \geq a, d \geq c$:

$$Addition : [a, b] + [c, d] = [a + c, b + d] \tag{7}$$

$$Subtraction : [a, b] - [c, d] = [a - d, b - c] \tag{8}$$

$$Multiplication : [a, b] \cdot [c, d] = [min(ac, ad, bc, bd), max(ac, ad, bc, bd)] \tag{9}$$

$$Division : \frac{[a, b]}{[c, d]} = [min(\frac{a}{c}, \frac{a}{d}, \frac{b}{c}, \frac{b}{d}), max(\frac{a}{c}, \frac{a}{d}, \frac{b}{c}, \frac{b}{d})] \tag{10}$$

The equation of multiplication can be simplified for the intervals included in the positive reals: $PositiveMultiplication : [a, b] \cdot [c, d] = [ac, bd]$.

3.1 Naive Interval Computations and Logical Dependency

The major limitation in the application of naive computation using interval arithmetics to more complex functions is the *dependency problem*. The dependency problem comes from the repetition of the same variable in the expression of a function. It causes some difficulties to compute the exact range of the function. It must be pointed out that the dependency here is a *functional* dependency notion, contrary to *stochastic* dependence assumed in section 2.1 and leading to formula 2.

Let us take a simple example in order to illustrate the dependency problem: we consider the function $f(x) = x + (1 - x)$, with $x \in [0, 1]$. Applying addition (7) and subtraction (8) rules of interval arithmetic gives that $f(x) \in [0, 2]$, while this function is always equals to 1. Some artificial uncertainty is created by considering x and $(1 - x)$ as two different, functionally independent, variables.

a) $F = A \Leftrightarrow B$ b) $G = A \vee B$

Fig. 2. BDD representation

Consider another example to illustrate the dependency problem on a BDD. We consider the formula of the equivalence $F = A \Leftrightarrow B = (\neg A \vee B) \wedge (\neg B \vee A)$ whose BDD is represented by $(A \wedge B) \vee (\neg A \wedge \neg B)$ and pictured on Fig. 2.a.

The probability of F knowing that $P(A) = a$, $P(B) = b$ applying eq. (3) is:

$$P(F) = a \cdot b + (1 - a)(1 - b) \tag{11}$$

The values of a and b are not known, the only information available is that $a \in [\underline{a}, \overline{a}]$ and $b \in [\underline{b}, \overline{b}]$, with $[\underline{a}, \overline{a}]$ and $[\underline{b}, \overline{b}]$ included in $[0, 1]$.

The goal is to compute the interval associated to $P(F) \in [\underline{P(F)}, \overline{P(F)}]$. If we directly apply the equation of addition (7) and multiplication (9) of two intervals to the expression of $P(F)$, we obtain that:

$$[\underline{P(F)}, \overline{P(F)}] = [\underline{ab}, \overline{ab}] + (1 - [\underline{a}, \overline{a}])(1 - [\underline{b}, \overline{b}]) = [\underline{ab}, \overline{ab}] + [1 - \overline{a}, 1 - \underline{a}] \cdot [1 - \overline{b}, 1 - \underline{b}])$$

$$[\underline{P(F)}, \overline{P(F)}] = [\underline{ab} + (1 - \overline{a})(1 - \overline{b}), \overline{ab} + (1 - \underline{a})(1 - \underline{b})] \tag{12}$$

This result is obviously wrong because we use two different values of the same variable ($\underline{a}$ and $\overline{a}$ for a, $\underline{b}$ and $\overline{b}$ for b) when calculating $\underline{P(F)} = \underline{ab} + (1 - \overline{a})(1 - \overline{b})$ as well $\overline{P(F)} = \overline{ab} + (1 - \underline{a})(1 - \underline{b})$.

Finally, consider $G = A \vee B$, with $a = P(A)$ and $b = P(B)$; its BDD representation (Fig. 2.b) is $G_{BDD} = A \vee (\neg A \wedge B)$ with $P(G_{BDD}) = a + (1 - a)b$, while the Sylvester-Poincaré decomposition gives $P(G) = a + b - ab$. Both $P(G)$ and $P(G_{BDD})$ display a dependency problem because variable a appears twice with different signs in these expressions (in $P(G)$ it happens also with b).

3.2 Interval Analysis Applied to BDDs

One way to solve the logical dependency problem is to factorize $f(x_1, ..., x_n)$ in such a way that variables appear only once in the factorized equation and guess the monotonicity of the resulting function. For the previous example $P(G)$ can be factorized as $1 - (1 - a)(1 - b)$ where both a and b appear only once. The function is clearly increasing with a and b, hence $P(G) \in [\underline{a} + \underline{b} - \underline{ab}, \overline{a} + \overline{b} - \overline{ab}]$, substituting the same value in each place. Unfortunately, it is not always possible to do so, like in the above example of the equivalence connective, hence we must resort to some alternative ways, by analyzing the function.

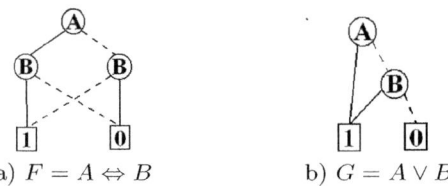

Configurations

Given n intervals $[\underline{x_i}, \overline{x_i}], i = 1, \ldots, n$, a n-tuple of values z in the set $\mathcal{X} = \times_i \{\underline{x_i}, \overline{x_i}\}$ is called a *configuration* [6], [5]. An extremal configuration z^j, is obtained by selecting one interval end for each component of the n-tuple: z^j has the form $(x_1^{c_1}, ..x_n^{c_n}), c_n \in \{0, 1\}$ with $x_i^0 = \underline{x_i}$ and $x_i^1 = \overline{x_i}$. The set of extremal configurations is denoted by $\mathcal{H} = \times_i \{\underline{x_i}, \overline{x_i}\}$, and $| \mathcal{H} |= 2^n$.

A locally *monotonic function* f [5] is such that the function, obtained by fixing all variables x_i but one, is monotonic with respect to the remaining variable $x_j, j \neq i$. Extrema of such f are attained for extremal configurations. Then the range $f([\underline{x_1}, \overline{x_1}], ..., [\underline{x_n}, \overline{x_n}])$ can be obtained by testing the 2^n extremal configurations. If the function is monotonically increasing (resp. decreasing) with respect to x_j, the lower bound of this range is attained for $x_j = \underline{x_j}$ (resp. $x_j = \overline{x_j}$) and the upper bound is attained for $x_j = \overline{x_j}$ (resp. $x_j = \underline{x_j}$). The monotonicity study of a function is thus instrumental for interval analysis.

The Probability of a Boolean Formula

As far as Boolean functions are concerned, some results can be useful to highlight. The goal is to compute the smallest interval $[\underline{P(F)}, \overline{P(F)}]$ for the probability $P(F)$ of a formula, knowing the probability intervals of its variables. First of all, if a variable A appears only once in a Boolean function F, the monotonicity of $P(F)$ with respect to $P(A)$ is known: it will increase if A appears positively, and decrease if it appears negatively ($\neg A$). More generally, if a variable A appears several times, but only as a positive (resp. negative) literal, then $P(F)$ is increasing (resp. decreasing) with respect to $P(A)$. The dependency problem present when both terms of the form $1 - P(A)$ and $P(A)$ appear in the expression of the function, which is an important issue for the application of interval analysis to probabilistic BDDs.

Monotonicity of the Probability of a BDD Representation

One important thing to notice about the formula (6) of probability computation from a BDD, is that it is a multilinear polynomial, hence it is locally monotonic.

Let us consider the Shannon decomposition of a formula F for a variable A_i, $i \in [\![1, ..., n]\!]$. Formula (5) with $A = A_i$ can be written as: $P(F) = (1 - P(A_i)) \cdot P(F_{A_i=0}) + P(A_i) \cdot P(F_{A_i=1})$, where $P(A_i)$ appears twice. It can be written as:

$$P(F) = P(F_{A_i=0}) + P(A_i) \cdot [P(F_{A_i=1}) - P(F_{A_i=0})] \tag{13}$$

In order to study the local variations of the function $P(F)$ following $P(A_i)$, we fix all others A_j,$j \in [\![1, ..., n]\!]$, $j \neq i$; the partial derivative of equation (13) with respect to $P(A_i)$ is:

$$\frac{\partial P(F)}{\partial P(A_i)} = P(F_{A_i=1}) - P(F_{A_i=0})$$

$[P(F_{A_i=1}) - P(F_{A_i=0})]$ is a function of some $P(A_j)$, $j \neq i$, so it does not depend upon $P(A_i)$, for any $i \in [\![1, ..., n]\!]$. The monotonicity of function $P(F)$ with respect to $P(A_i)$ depends on the sign of $[P(F_{A_i=1}) - P(F_{A_i=0})]$; if constant, the function is monotonic with respect to $P(A_i)$. We can deduce that if $P(A_i) \in [\underline{a}, \overline{a}]$, the tightest interval for $P(F)$ is:

- $[P(F_{A_i=0})+\underline{a}\cdot(P(F_{A_i=1})-P(F_{A_i=0})), P(F_{A_i=0})+\overline{a}\cdot(P(F_{A_i=1})-P(F_{A_i=0}))]$ if $P(F_{A_i=1}) \geq P(F_{A_i=0})$
- $[P(F_{A_i=0})+\overline{a}\cdot(P(F_{A_i=1})-P(F_{A_i=0})), P(F_{A_i=0})+\underline{a}\cdot(P(F_{A_i=1})-P(F_{A_i=0}))]$ if $P(F_{A_i=1}) \leq P(F_{A_i=0})$

Knowing the monotonicity of a function makes the determination of its range straightforward. For some functions, the monotonicity can be more easily seen in other formats than BDD. But finding the sign of $P(F_{A_i=1}) - P(F_{A_i=0})$ is not so simple, as it depends on some other A_j. If we are able to find this sign for each variable of the function, we can find its exact (tight) range right away.

Consider again the equivalence connective $F = A \Leftrightarrow B$ of section 3.1, where events A and B are associated with the following intervals, respectively $[\underline{a}, \overline{a}] = [0.3, 0.8]$ and $[\underline{b}, \overline{b}] = [0.4, 0.6]$. We know that the function is locally monotonic, so the range of $P(F)$ in eq. (11) will be obtained at vertices of the domain of the variables a and b. To find the exact bounds of $P(F)$, since a and b appear both with positive and negative signs in eq. (11), we will have to explore the 2^2 configurations:

$$\begin{aligned}
z_1 &= (\underline{a}, \underline{b}), P_{z_1}(F) = \underline{a}\cdot\underline{b} + (1-\underline{a})(1-\underline{b}) = 0.3\cdot 0.4 + 0.7\cdot 0.6 = 0.54\\
z_2 &= (\underline{a}, \overline{b}), P_{z_2}(F) = \underline{a}\cdot\overline{b} + (1-\underline{a})(1-\overline{b}) = 0.3\cdot 0.6 + 0.7\cdot 0.4 = 0.46\\
z_3 &= (\overline{a}, \underline{b}), P_{z_3}(F) = \overline{a}\cdot\underline{b} + (1-\overline{a})(1-\underline{b}) = 0.8\cdot 0.4 + 0.2\cdot 0.6 = 0.44\\
z_4 &= (\overline{a}, \overline{b}), P_{z_4}(F) = \overline{a}\cdot\overline{b} + (1-\overline{a})(1-\overline{b}) = 0.8\cdot 0.6 + 0.2\cdot 0.4 = 0.56
\end{aligned}$$

$\underline{P(F)} = \min_{i=1,\dots,4} P_{z_i}(F) = 0.44$ and $\overline{P(F)} = \max_{i=1,\dots,4} P_{z_i}(F) = 0.56$. The exact result is $P(F) \in [0.44, 0.56]$, while using expression (12), we find $[\underline{P(F)}, \overline{P(F)}] = [0.2, 0.9]$. If we neglect dependencies, we introduce artificial uncertainty in the result. The exact result is much tighter than the one based on naive interval computation. This is why it is so important to find ways to optimize the bounds of the uncertainty range in interval analysis.

We can apply full-fledged interval analysis to binary Boolean connectives and compare the results with the ones obtained by applying naive interval computation (Table 1). For those tests, we took the same input probabilities as for the example presented in section 3.1, Fig. 2.a: $P(A) \in [0.3, 0.8]$, $P(B) \in [0.4, 0.6]$. It is obvious that the two results are the same only when each variable appears once in the probability $P(F)$, e.g for $F = A \wedge B$, $P(F) = ab$. For all other cases, we get a tighter interval by testing all extreme bounds of the input intervals. The more redundancy of variables there will be in a formula, the more naive interval computation will give ineffective results, moreover irrelevant in some cases; e.g $F = A \vee B$ or $F = A \Rightarrow B$ where we get intervals with values higher than 1. On the contrary, the approach based on local monotony gives the exact range of the Boolean function F.

Table 1. Comparison between naive interval computation and full-fledged interval analysis

Connective	Formula	Function	Naive	Exact
OR	$A \vee B$	$a + b - ab$	[0.22, 1.28]	[0.58,0.92]
OR	$A \vee (B \wedge \neg A)$	$a + b(1 - a)$	[0.38,1.22]	[0.58,0.92]
OR	$A \vee B$	$1 - (1 - a)(1 - b)$	[0.58, 0.92]	[0.58,0.92]
AND	$A \wedge B$	ab	[0.12,0.48]	[0.12,0.48]
IMPLIES	$\neg A \vee (A \wedge B)$	$1 - a + ab$	[0.32,1.18]	[0.52,0.88]
EQUIVALENCE	$(A \wedge B) \vee (\neg A \wedge \neg B)$	$ab + (1 - a)(1 - b)$	[0.2,0.9]	[0.44,0.56]
ExOR	$(A \wedge \neg B) \vee (\neg A \wedge B)$	$a(1 - b) + b(1 - a)$	[0.2,0.9]	[0.44,0.56]

3.3 Beyond Two Variables

However, the study of monotonicity can be also very complicated. For instance, the function *2 out of 3* is given by the formula:

$$F = (A \wedge \neg B \wedge \neg C) \vee (\neg A \wedge B \wedge \neg C) \vee (\neg A \wedge \neg B \wedge C)$$

The partial derivative with respect to a of the probability of this formula is :

$$\frac{\partial}{\partial a} P(F) = (1 - b)(1 - c) - b(1 - c) - c(1 - b) = 1 - 2b - 2c + 3bc$$

Partial derivatives with respect to b and c are similar. It is more difficult to find the sign of such a derivative, because we have a 2-place function that is not monotonic and has a saddle point (see Fig. 3.a). On fig. 3.b we can see the level cuts of the curve, $\frac{\partial}{\partial a} P(F) = \alpha$.

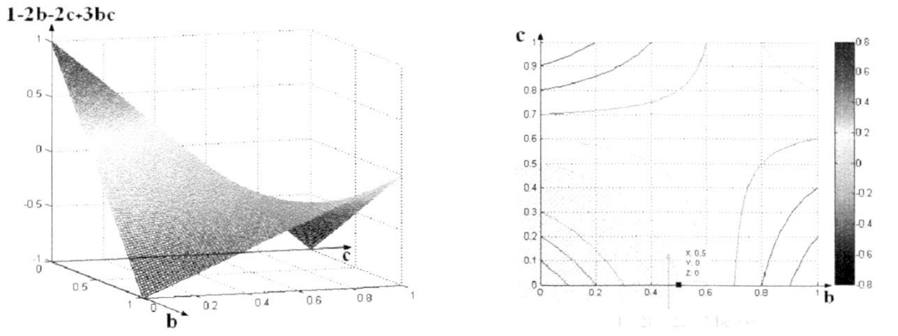

Fig. 3. a) Partial derivative $1 - 2b - 2c + 3bc$ b) Level cuts of $1 - 2b - 2c + 3bc$

The derivative is null for $1 - 2b - 2c + 3bc = 0 \Leftrightarrow b = \frac{1-2c}{1-3c}$, that is the equation of an hyperbola. The positive region of $\frac{\partial}{\partial a} P(F)$ is delimited by this hyperbola, starting from the $(0, 0, 1)$ point.

The study of specific Boolean expressions like the *2 out of 3* can be used as a heuristic for speeding up the computation of more complex formulas.

4 Algorithm for Interval Analysis Applied to BDDs

In this section, we will present an algorithm corresponding to the method of imprecise probability computations for BDDs described in the section 3.2. This algorithm is able to compute the exact bounds of the probability of a Boolean function, given the interval ranges of its atomic probabilities.

The input of the algorithm is a Boolean function F where a probability interval $[v.lb, v.up]$ is associated to each variable v of F (the lower/upper bound of its probability interval). The variable v of the BDD representation of F is characterized by the following additional attributes: *path.value* (= 1 if the variable appears as a positive literal in this path, = 0 otherwise) and *type* $\in \{0,1,2\}$. The output is a text file with the probability interval of F, $[P_F.lb, P_F.up]$.

The BDD is represented by a set Pa of paths with the following format:

$$< v_1^{path_1} : value, \ldots, v_{n_1}^{path_1} : value > \cdots < v_1^{path_m} : value, \ldots, v_{n_m}^{path_m} : value > \quad (14)$$

where m is the number of paths and $n_i, i = 1, \ldots, m$ is the number of literals (positive or negative) in a path i. For example, $Pa = < A : 1 >< S : 0, C : 1 >$ for the BDD represented in fig. 1.b.

The algorithm consists of three main steps: in the first step, an Aralia file is parsed for the variables and Boolean formula F of the dreadful event, and a corresponding BDD is generated. In the next step, the parsed variables are split into three categories:

- **Type 0:** Variables that only appear negatively in the Boolean formula
- **Type 1:** Variables that only appear positively in the Boolean formula
- **Type 2:** Variables present in the Boolean formula along with their negation.

We need to find the *configuration* that determines the minima and maxima for the probability of F. We know the exact corresponding bounds of the input probabilities for the 2 first categories from section 3.2, so the optimal configuration is known for these variables:

- if $v.type = 0$, $v.ub$ is used for calculating $P_F.lb$ and $v.lb$ is used for $P_F.ub$,
- if $v.type = 1$, $v.lb$ is used for calculating $P_F.lb$ and $v.ub$ is used for $P_F.ub$,
- if $v.type = 2$, $P_F.lb$ (as well as $P_F.ub$) can be reached for $v.lb$ or $v.up$, and all possible extreme values for v must be explored. The total number of these tuples of bounds (configurations) is 2^k, where k is the number of variables classified as Type 2; hence the problem is at most NP-hard.

The last step consists of BDD-based calculations. These calculations are carried out by considering all m paths leading to leaf 1 from the top of the BDD.

Let $\boldsymbol{z}^j = (v_{1j}^{c_1}, \ldots, v_{nj}^{c_n}), j \in 1, \ldots, 2^n, c_i \in \{0,1\}$ be a configuration (section 3.2). For each configuration $\boldsymbol{z}^j$ we calculate $P_{\boldsymbol{z}^j}(F)$ $=P(F[v_{1j}^{c_1}, \ldots, v_{nj}^{c_n}])$. The extremal values of $P(F$ are obtained by exploring all extremal configurations: $\underline{P(F)} = \min_{i=1,\ldots,2^n} P_{z_i}(F)$ and $\overline{P(F)} = \max_{i=1,\ldots,2^n} P_{z_i}(F)$.

The notation introduced in eq. (6) is extended to take the category of a variable into account. Let $\mathcal{V}_l$ be the set of variables belonging to the path l; $\mathcal{V}_l^{2+}$ (resp. $\mathcal{V}_l^{2-}$) the set of positive (resp. negative) literals of Type 2 in path l; $\mathcal{V}_l^{+}$ (resp. $\mathcal{V}_l^{-}$) the set of variables of Type 1 (resp. Type 0) in path l.

Let us consider now a configuration $(v_1^{c_1}, \dots, v_k^{c_k})$ for variables of Type 2, with $v.c = v.lb$ if $v^c = 0$ and $v.c = v.ub$ if $v^c = 1$:

$$C_l(v_1^{c_1}, \dots, v_k^{c_k}) = \prod_{v \in V_l^{2+}} v.c \cdot \prod_{v \in V_l^{2-}} (1 - v.c).$$

We have then:

$$\underline{P(F)} = \sum_{l=1,\dots,m} \Big(\prod_{v \in \mathcal{V}_l^+} v.lb \cdot \prod_{v \in \mathcal{V}_l^-} (1 - v.ub) \cdot \min_{i=1,\dots,2^k} C_l(v_1^{c_1}, \dots, v_k^{c_k}) \Big) \text{ and}$$

$$\overline{P(F)} = \sum_{l=1,\dots,m} \Big(\prod_{v \in \mathcal{V}_l^+} v.ub \cdot \prod_{v \in \mathcal{V}_l^-} (1 - v.lb) \cdot \max_{i=1,\dots,2^k} C_l(v_1^{c_1}, \dots, v_k^{c_k}) \Big)$$

that extends eq. (6) when the probability of the atomic events is given by an interval.

An algorithm that performs the sorting of the variables and computes optimal intervals has been encoded in C++ language, and based on BDD packages named CUDD [10] and BuDDy [12]. BuDDy explores the order of variables in order to optimize the size of the BDD.

5 Application to Fault-Tree Analysis

In Fault-Tree Analysis theory, modelling conventions are generally such that all variables appear only positively in the Boolean formula of the top event: no variable appears with a negation, so the probabilistic formula is monotonic and increasing. But in practice, there are several cases where some variables can appear negatively, and sometimes even both negatively and positively, so that the top formula can be non-monotonic:

- Some negations are introduced due to compilation: this is clear in BDDs and also in fault-trees obtained from so-called Mode Automata [8]. In this case, the expression is still monotonic as long as the Boolean formula could also be expressed without negative literals (e.g. the connective OR).
- State modeling: in some systems, it is necessary to use variables that model some special states, or modes, which no longer represent a failure, and the global formula may depend on such variables and their negation. It is not necessary increasing with respect to these variables.
- Exclusive Failures: sometimes, failures cannot physically occur simultaneously, they are then represented by mutually exclusive events or failure modes. Mutual exclusion implies non-monotonicity.

In practice, those kind of variables are very few compared to "usual" failures; hence, the algorithm will only have NP-hard complexity for them, and be linear for all other variables.

5.1 Case Study

The A320 electrical and hydraulic generation systems were used for the first experiments in model-based safety analysis using the Altarica language [1] where each node (represented by an icon in fig. 4) is a mode automaton [8]. The mode automata are compiled into Boolean formulae using the algorithm presented in [1] allowing for an automatic generation of fault trees from the Altarica model. The Rudder Control System controls the rudder in order to rotate the aircraft around the yaw axis. It is composed of:

- three primary calculators (P1, P2, P3), a secondary calculator (S1) and a emergency autonomous equipment, constituted by a Back-up Control Module (BCM) and 2 Back-up Power Supply (BPS_B, BPS_Y),
- three servo-commands: Green (G), Blue (B), Yellow (Y),
- two electric sources (Bus 2PP, Bus 4PP) and three hydraulic sources (Hyd Green, Hyd Blue, Hyd Yellow).

The AltaRica model of the Rudder Control System [2] using the workshop *Cécilia OCAS* from Dassault is presented in fig. 4.

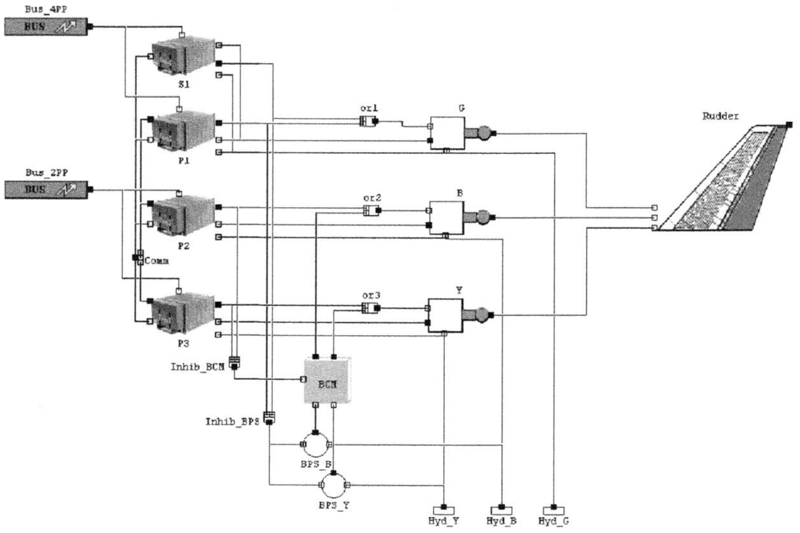

Fig. 4. Rudder's OCAS model

This model is used for safety analysis and it is has also been used as part of the operational reliability analysis [11]. It is genuinely non-monotonic because of the explicit use of states that do not refer to failures. The failures (elementary events) taken into account are: loss of a component (e.g. B.loss means loss of the Y servo-command), a hidden failure and an active failure (that can occur in

S1 and BCM). An excerpt from the detailed generated fault tree in the Aralia format, for the loss of the first primary calculator P1 is given on Fig. 5. Sub-trees are automatically generated and given names (DTN3578, etc.), one per line in the figure.

```
P1.Status.hs := ((-B.loss & DTN3578) | (B.loss & DTN3617));
DTN3578 := ((-BCM.active_failure & DTN3503) | (BCM.active_failure & DTN3577));
DTN3503 := ((-BPS_B.active_failure & DTN3500) | (BPS_B.active_failure & DTN3502));
DTN3500 := ((-BPS_Y.active_failure & DTN3483) | (BPS_Y.active_failure & DTN3499));
DTN3483 := ((-Bus_2PP.loss & DTN3482) | (Bus_2PP.loss & DTN3480));
DTN3482 := ((-Hyd_B.loss & DTN3478) | (Hyd_B.loss & DTN3481));
DTN3478 := ((-Hyd_Y.loss & DTN3475) | (Hyd_Y.loss & DTN3477));
DTN3475 := (P1.loss & DTN3474);
DTN3474 := ((-P2.loss & -S1.active_failure) | (P2.loss & DTN3473));
DTN3473 := ((-P3.loss & DTN3471) | (P3.loss & DTN3472));
```

Fig. 5. Excerpt from the fault tree of failure of P1: & is the conjunction, | the disjunction and – the negation

The probabilities of some variables are known: P(Hyd_i.loss)= e^{-4}, P(Bus_iPP)= e^{-3}. But for some others, only an interval containing the probability values is known: i.[lb,ub] $=[0.15, 0.25], i \in \{Y, B, G\}$, P_i[lb,ub]$=[0, e^{-2}], i = 1, \ldots, 3$, S1.Active_failure[lb,ub]=[0.1, 0.4] and BCM.Active_failure[lb,up]= [0.15, 0.345].

Using the algorithm described in section 4 on the whole fault-tree (that is non-monotonic) we for instance obtain the interval $I_1 = [0.01689, 0.01691]$ for the event *Loss of the Rudder* represented by F, whereas a wider interval $I_2 =$ [0.00972; 0.0269] is obtained when logical dependencies are not taken into account (applying directly equations (7) and(9) as in equation 12, section 3.1). It can be noticed that the interval I_1 is much tighter than I_2.

6 Conclusion

If naive computations using interval arithmetics are applied directly to the BDD-based expression of the probability of Boolean formula, variables and their negation will often appear, and the resulting interval is too imprecise and sometime totally useless. We presented in this paper an algorithm that allows to calculate the exact range of the probability of any formula. We pointed out that there are two cases when a variable and its negation appear in a Boolean expression: the case when there exists an equivalent expression where each variable appear with the same sign, and its probability is then monotonic in terms of the probability of atomic events; the case where such an equivalent expression does not exist. Then this probability will not be monotonic in terms of some variables, and the interval computation has NP-hard theoretical complexity.

Even if in practical fault-trees the latter situation does not prevail, the potentially exponential computation time can make it inapplicable to very big systems, so some heuristics are under study in order to tackle this issue, for example, methods to check monotonicity of the obtained numerical functions prior to running the interval calculation, or devising approximate calculation schemes when probabilities of faults are very small.

In future works, our aim is to generalize the approach to fuzzy intervals, using α-cuts. Besides, we must exploit reliability laws with imprecise failure rates, especially exponential laws, so as to generate imprecise probabilities of atomic events. Later on, we can also extend this work by dropping the independence assumption between elementary faults. One idea can be to use *Frechet bounds* to compute a bracketing of the probability of a Boolean formula when no dependence assumption can be made.

Acknowledgments. This work is supported by @MOST Prototype, a joint project of Airbus, LAAS, ONERA and ISAE. The authors would like to thank Christel Seguin, Chris Papadopoulos for their support in the project and Antoine Rauzy for his advice on mode automata.

References

1. Arnold, A., Griffault, A., Point, G., Rauzy, A.: The Altarica language and its semantics. Fundamenta Informaticae 34, 109–124 (2000)
2. Bernard, R., Aubert, J.-J., Bieber, P., Merlini, C., Metge, S.: Experiments on model-based safety analysis: flight controls. In: IFAC Workshop on Dependable Control of Discrete Systems, Cachan, France (2007)
3. Bryant, R.E.: Graph-Based Algorithms for Boolean Function Manipulation. IEEE Transactions on Computers C-35(8), 677–691 (1986)
4. Dutuit, Y., Rauzy, A.: Exact and Truncated Computations of Prime Implicants of Coherent and non-Coherent Fault Trees within Aralia. Reliability Engineering and System Safety 58, 127–144 (1997)
5. Fortin, J., Dubois, D., Fargier, H.: Gradual Numbers and Their Application to Fuzzy Interval Analysis. IEEE trans. Fuzzy Systems 16, 388–402 (2008)
6. Moore, R.: Methods and applications of interval analysis. SIAM Studies in Applied Mathematics, Philadelphia (1979)
7. Moore, R.E., Kearfott, R.B., Cloud, M.J.: Introduction to Interval Analysis. Society for Industrial & Applied Mathematics, U.S. (2009)
8. Rauzy, A.: Mode automata and their compilation into into fault trees. Reliability Engineering and System Safety 78, 1–12 (2002)
9. Siegle, M.: BDD extensions for stochastic transition systems. In: Kouvatsos, D. (ed.) Proc. of 13th UK Performance Evaluation Workshop, Ilkley/West Yorkshire, pp. 9/1-9/7 (July 1997)
10. Somenzi, F.: University of Colorado, `http://vlsi.colorado.edu/~fabio/CUDD/`
11. Tiaoussou, K., Kanoun, K., Kaaniche, M., Seguin, C., Papadopoulos, C.: Operational Reliability of an Aircraft with Adaptive Missions. In: 13th European Workshop on Dependable Computing (2011)
12. `http://drdobbs.com/cpp/184401847`

A Branching Time Logic with Two Types of Probability Operators

Zoran Ognjanović[1], Dragan Doder[2], and Zoran Marković[1]

[1] Matematički institut SANU, Kneza Mihaila 36, 11000 Beograd, Serbia
{zorano,zoranm}@mi.sanu.ac.rs
[2] Mašinski fakultet, Kraljice Marije 16, 11120 Beograd, Serbia
ddoder@mas.bg.ac.rs

Abstract. We introduce a propositional logic whose formulas are built using the language of CTL^*, enriched by two types of probability operators: one speaking about probabilities on branches, and one speaking about probabilities of sets of branches with the same initial state. An infinitary axiomatization for the logic, which is shown to be sound and strongly complete with respect to the corresponding class of models, is proposed.

1 Introduction

Interest in temporal reasoning came from theoretical and practical points of view. Logicians [5,6,30] investigated consequences of different assumptions about the structure of time, while temporal formalisms can be used in computer science to reason about properties of programs [11,29]. In both cases discrete linear and branching time logics have been extensively studied. Linear temporal logics are suitable for specification and verification of universal properties of all executions of programs. On the other hand, the branching time approach is appropriate to analyze nondeterministic computations described in the form of execution trees. In the later framework a state (a node) may have many successors. Then, it is natural to attach probabilities to the corresponding transitions and to analyze the corresponding discrete time Markov chains as the underlying structures. All this led to probabilistic branching temporal logic [2,3,17,18,21,35]. The mentioned papers mainly investigate semantical properties of the logics and do not offer any axiomatic system. The only exception is [35], where the logic with a very restricted language is presented. A more detailed overview on the topic is presented in Section 5, when we will be able to precisely formulate relevant notions and connections between them using the formalism from Section 2.

In this paper we consider a propositional discrete probabilistic branching temporal logic (denoted pBTL). We use a logical language which allows us to formulate statements that combine temporal and qualitative probabilistic features. Thus, the statements as "in at least half of paths α holds in at least a third of states" and "if α holds in the next moment, then the probability of α is positive" are expressible in our logic. To the best of our knowledge, the former

S. Benferhat and J. Grant (Eds.): SUM 2011, LNAI 6929, pp. 219–232, 2011.
© Springer-Verlag Berlin Heidelberg 2011

sentence is not expressible in any of existing logics. The language for pBTL is obtained by adding temporal operators $\bigcirc$ ("next"), A (universal path operator) and U ("until"), as well as the two types of probability operators, $P^p_{\geqslant r}$ and $P^s_{\geqslant r}$ ($r \in \mathbb{Q} \cap [0,1]$), to the classical propositional language. The temporal operators are well known from other formalizations of branching time logics, while the intended meaning of $P^s_{\geqslant r}\alpha$ ($P^p_{\geqslant r}\alpha$) is "the probability that α is true on a randomly chosen branch is at least r" ("the probability that α holds on a particular branch is at least r"). The superscript s in $P^s_{\geqslant r}$ (p in $P^p_{\geqslant r}$) indicates that the probability depends only on a time instant - state (on a chosen branch - path).

We present a class of suitable models for the pBTL-language and an infinitary axiomatization, for which we prove strong completeness theorem ("every consistent set of formulas is satisfiable", in contrast to weak completeness: "every consistent formula is satisfiable"). Up to our knowledge it is the first such result reported in literature. The corresponding proof uses ideas (the Henkin construction) presented in [7,8,23,24,25,26,27,31].

The rest of the paper is organized as follows. In Section 2 we define syntax and semantics for pBTL. Section 3 introduces an infinitary axiomatization for the logic, which is proved to be strongly complete in Section 4. Comparison with the related work is discussed in Section 5. Section 6 contains concluding remarks and directions for further work.

2 Syntax and Semantics

Let $\mathcal{P}$ be at most countable set of propositional letters. The set of formulas For of the logic pBTL is the smallest set which satisfies the following conditions:

- $\mathcal{P} \subseteq For$,
- if $\alpha, \beta \in For$, then $\alpha \wedge \beta, \neg\alpha \in For$,
- if $\alpha, \beta \in For$, then $\bigcirc\alpha, \alpha U \beta, A\alpha \in For$,
- if $\alpha \in For$ and $r \in \mathbb{Q} \cap [0,1]$, then $P^p_{\geqslant r}\alpha, P^s_{\geqslant r}\alpha \in For$.

Intuitively, the operators mean:

- $\bigcirc\alpha$: α holds in the next time instant on a particular branch,
- $\alpha U \beta$: α holds in every time instant (on a particular branch) until β becomes true,
- $A\alpha$: α holds on every branch which passes through the current state,
- $P^p_{\geqslant r}\alpha$: the probability that α holds at a randomly chosen time instant on a particular branch is at least r, and
- $P^s_{\geqslant r}\alpha$: the probability of branches (with a particular initial time instant) on which α holds is at least r".

A formula is a *state formula* if it is a boolean combination of propositional letters, formulas of the form $P^s_{\geqslant r}\alpha$ and formulas of the form $A\alpha$. We denote the set of all state formulas by St. For $n \in \omega$, we define $\bigcirc^{n+1}\alpha$ as $\bigcirc(\bigcirc^n\alpha)$. If T is a set of formulas, then $\bigcirc T$ denotes $\{\bigcirc\alpha | \alpha \in T\}$, and AT denotes $\{A\alpha | \alpha \in T\}$. The temporal operators F (sometime), G (always) and E (existential path quantifier) are defined as follows:

- $F\alpha$ is $\top U\alpha$,
- $G\alpha$ is $\neg F\neg\alpha$,
- $E\alpha$ is $\neg A\neg\alpha$.

Also, in order to simplify notation, we introduce the following convention:

- $P^p_{<r}\alpha$ is $\neg P^p_{\geqslant r}\alpha$, $P^p_{\leqslant r}\alpha$ is $P^p_{\geqslant 1-r}\neg\alpha$, $P^p_{>r}\alpha$ is $\neg P^p_{\leqslant r}\alpha$ and $P^p_{=r}\alpha$ is $P^p_{\geqslant r}\alpha \wedge P^p_{\leqslant r}\alpha$,
- $P^s_{<r}\alpha$, $P^s_{\leqslant r}\alpha$, $P^s_{>r}\alpha$ and $P^s_{=r}\alpha$ are defined in a similar way.

An example of a formula is

$$EG\alpha \rightarrow P^s_{\geqslant\frac{1}{2}} P^p_{\geqslant\frac{1}{3}}\alpha,$$

which can be read as: "if there exists a path on which the formula α always holds, then on at least a half of paths α holds in at least a third of time instants".

Definition 1. *A* model $\mathcal{M}$ *is any tuple* $\langle S, v, R, \Sigma, Prob^{state}, Prob^{path}\rangle$ *such that:*

- *S is a non-empty set of states (time instants),*
- *$v : S \times \mathcal{P} \longrightarrow \{0,1\}$ assigns a truth labelling to every state.*
- *R is a binary relation on S, which is total (for every $s \in S$ there is $t \in S$ such that sRt),*
- *Σ is a set of ω-sequences $\sigma = s_0, s_1, s_2, \ldots$ of states from S, such that $s_i R s_{i+1}$, for all $i \in \omega$. A* path *is an element of Σ. We assume that Σ is suffix-closed, i.e., if $\sigma = s_0, s_1, s_2, \ldots$ is a path and $i \in \omega$, the sequence $s_i, s_{i+1}, s_{i+2}, \ldots$ is also a path.*
- *$Prob^{state}$ associates to every $s \in S$, a probability space $Prob_s = \langle H_s, \mu_s\rangle$ such that:*
  - *H_s is an algebra of subsets of $\Sigma_s = \{\sigma \in \Sigma \mid \sigma_0 = s\}$, i.e., it contains Σ_s and it is closed under complements and finite union,*
  - *$\mu_s : H_s \longrightarrow [0,1]$ is a finitely additive probability measure, i.e.,*
    - $\mu_s(H_s) = 1$, *and*
    - $\mu_s(X \cup Y) = \mu_s(X) + \mu_s(Y)$, *whenever X and Y are disjoint.*
- *$Prob^{path}$ associates to every $\sigma \in \Sigma$, $\sigma = s_0, s_1, \ldots, s_i, s_{i+1}, s_{i+2}, \ldots$, a probability space $Prob_\sigma = \langle A_\sigma, \mu_\sigma\rangle$ such that:*
  - *A_σ is an algebra of subsets of $S_\sigma = \{\pi \in \Sigma \mid \pi = s_i, s_{i+1}, s_{i+2}, \ldots, for\ i \in \omega\}$,*
  - *$\mu_\sigma : A_\sigma \longrightarrow [0,1]$ is a finitely additive probability measure.*

Let $\sigma = s_0, s_1, s_2, \ldots$ In the rest of the paper, we will use the following abbreviations:

- $\sigma_{\geq i}$ is the path $s_i, s_{i+1}, s_{i+2}, \ldots$
- σ_i is the state s_i.

Definition 2. *Let $\mathcal{M} = \langle S, v, R, \Sigma, Prob^{state}, Prob^{path}\rangle$ be any model. The* satisfiability relation $\models$ *(we denote the fact that a formula α is satisfied at a path σ in a model $\mathcal{M}$ by $\mathcal{M}, \sigma \models \alpha$) is defined recursively as follows:*

- *if $p \in \mathcal{P}$, then $\mathcal{M}, \sigma \models p$ iff $v(s_0, p) = 1$,*
- *$\mathcal{M}, \sigma \models \neg\alpha$ iff $\mathcal{M}, \sigma \not\models \alpha$,*
- *$\mathcal{M}, \sigma \models \alpha \wedge \beta$ iff $\mathcal{M}, \sigma \models \alpha$ and $\mathcal{M}, \sigma \models \beta$,*
- *$\mathcal{M}, \sigma \models \bigcirc\alpha$ iff $\mathcal{M}, \sigma_{\geq 1} \models \alpha$,*
- *$\mathcal{M}, \sigma \models A\alpha$ iff for every path π, if $\sigma_0 = \pi_0$ then $\mathcal{M}, \pi \models \alpha$.*
- *$\mathcal{M}, \sigma \models \alpha U \beta$ iff there is some $i \in \omega$ such that $\mathcal{M}, \sigma_{\geq i} \models \beta$ and for each $j \in \omega$, if $0 \leq j < i$ then $\mathcal{M}, \sigma_{\geq j} \models \alpha$,*
- *$\mathcal{M}, \sigma \models P^{s}_{\geqslant r}\alpha$ iff $\mu_{\sigma_0}\{\pi \in \Sigma_{\sigma_0} \mid \mathcal{M}, \pi \models \alpha\} \geqslant r$,*
- *$\mathcal{M}, \sigma \models P^{p}_{\geqslant r}\alpha$ iff $\mu_{\sigma}\{\pi \in S_{\sigma} \mid \mathcal{M}, \pi \models \alpha\} \geqslant r$.*

Note that the satisfiability of any state formula (for example $P^{s}_{\geqslant r}\alpha$) depends only on the initial state of the path, while the other formulas are path-dependent.

If $\mathcal{M} = \langle S, v, R, \Sigma, Prob^{state}, Prob^{path}\rangle$ is a model and $\sigma \in \Sigma$, we will denote:

- $[\alpha]^{path}_{\mathcal{M},\sigma} = \{\pi \in S_\sigma \mid \mathcal{M}, \pi \models \alpha\}$, and
- $[\alpha]^{state}_{\mathcal{M},s} = \{\pi \in \Sigma_s \mid \mathcal{M}, \pi \models \alpha\}$.

The possible problems in Definition 2 are that for an α the sets $[\alpha]^{path}_{\mathcal{M},\sigma}$ and $[\alpha]^{state}_{\mathcal{M},s}$ might not be in A_σ and in H_s, respectively. To overcome this, in the rest of the paper we will consider only so-called measurable models.

Definition 3. *A model $\mathcal{M} = \langle S, v, R, \Sigma, Prob^{state}, Prob^{path}\rangle$ is* measurable *if the following conditions are satisfied:*

- *$[\alpha]^{path}_{\mathcal{M},\sigma} \in A_\sigma$, for every $\alpha \in For$,*
- *$[\alpha]^{state}_{\mathcal{M},s} \in H_s$, for every $\alpha \in For$.*

We will denote the probabilistic branching-time temporal logic characterized by the class of all measurable models by pBTL_{Meas}.

The expression $\mathcal{M}, \sigma \models T$ denotes the fact that $\mathcal{M}, \sigma \models \alpha$, for every $\alpha \in T$. A formula α is satisfiable if there is a path σ in a model $\mathcal{M}$ such that $\mathcal{M}, \sigma \models \alpha$. A formula is valid if $\mathcal{M}, \sigma \models \alpha$ for every model $\mathcal{M}$ and every path σ of $\mathcal{M}$. We write $T \models \alpha$ ("α is a semantical consequence of T"), if for every model $\mathcal{M}$ and every σ in $\mathcal{M}$, if $\mathcal{M}, \sigma \models T$, then $\mathcal{M}, \sigma \models \alpha$.

3 Axiomatization

Propositional axioms

A1. all the tautologies of the classical propositional logic

Temporal axioms

A2. $\bigcirc(\alpha \rightarrow \beta) \rightarrow (\bigcirc\alpha \rightarrow \bigcirc\beta)$
A3. $\neg \bigcirc \alpha \leftrightarrow \bigcirc\neg\alpha$
A4. $\alpha U\beta \leftrightarrow \beta \vee (\alpha \wedge \bigcirc(\alpha U\beta))$
A5. $p \rightarrow Ap,\ \ p \in \mathcal{P}$
A6. $Ep \rightarrow p,\ \ p \in \mathcal{P}$
A7. $A\alpha \rightarrow \alpha$
A8. $A(\alpha \rightarrow \beta) \rightarrow (A\alpha \rightarrow A\beta)$
A9. $A\alpha \rightarrow AA\alpha$
A10. $E\alpha \rightarrow AE\alpha$

Probabilistic axioms ($x \in \{p, s\}$)

A11. $P^x_{\geq 0}\alpha$
A12. $P^x_{\leq s}\alpha \rightarrow P^x_{<t}\alpha,\ t > s$
A13. $P^x_{<s}\alpha \rightarrow P^x_{\leq s}\alpha$
A14. $(P^x_{\geq s}\alpha \wedge P^x_{\geq r}\beta \wedge P^x_{\geq 1}(\neg\alpha \vee \neg\beta)) \rightarrow P^x_{\geq \min(1,s+r)}(\alpha \vee \beta)$
A15. $(P^x_{\leq s}\alpha \wedge P^x_{<r}\beta) \rightarrow P^x_{<s+r}(\alpha \vee \beta),\ s + r \leq 1$

Axioms about probability and temporality

A16. $G\alpha \rightarrow P^p_{\geq 1}\alpha$
A17. $A\alpha \rightarrow P^s_{\geq 1}\alpha$
A18. $P^s_{\geq r}\alpha \rightarrow AP^s_{\geq r}\alpha$
A19. $EP^s_{\geq r}\alpha \rightarrow P^s_{\geq r}\alpha$

Inference rules

R1. from $\{\alpha, \alpha \rightarrow \beta\}$ infer β
R2. from α infer $\bigcirc\alpha$
R3. from α infer $A\alpha$
R4. from the set of premises

$$\{\gamma \rightarrow \neg((\wedge^i_{k=0} \bigcirc^k \alpha) \wedge \bigcirc^{i+1}\beta) \mid i \in \omega\}$$

infer $\gamma \rightarrow \neg(\alpha U\beta)$
R5. from the set of premises

$$\{\beta \rightarrow \bigcirc^m P^x_{\geq r-\frac{1}{k}}\alpha \mid k \in \omega, k \geq \frac{1}{r}\}$$

infer $\beta \rightarrow \bigcirc^m P^x_{\geq r}\alpha$ (for any $m \in \omega$ and $x \in \{p, s\}$)

Let us briefly discuss some of the above axioms and rules. By the axiom A1 and the inference rule R1 (Modus ponens), pBTL extends the classical propositional logic. The axioms A2–A4 are standard axioms of discrete linear-time temporal logic, while the axioms A5–A10 concern the non-linear aspect of the temporal logic [34]. Probabilistic axioms captures the basic properties of probability: non-negativity and finite additivity. The last group of axioms concerns mixing of probabilistic and temporal reasoning.

The inference rules R2 and R3 are the variants of modal Necessitation. They can be applied only to theorems. The rules R4 and R5 are infinitary inference rules. The former one characterizes the until operator, while the later one intuitively says that if the probability is arbitrarily close to r, then it is at least r.

We say that a formula α is deducible from a set T of formulas, and write $T \vdash \alpha$, if there is an at most countable sequence of formulas $\alpha_0, \alpha_1, \ldots, \alpha$, such that every α_i is an axiom or a formula from T, or it is derived from the preceding formulas by an inference rule (with the exception that R2 and R3 can be applied to theorems only). That sequence is called the proof of α from T. The formula α is a theorem, denoted by $\vdash \alpha$, if it is deducible from the empty set. A set T of formulas is consistent if there is at least one formula which is not deducible from T; otherwise it is inconsistent. A consistent set T of sentences is said to be maximally consistent if for every $\alpha \in For$, either $\alpha \in T$ or $\neg\alpha \in T$.

It is easy to prove soundness of the proposed axiomatic system (with respect to the considered class of models), using a straightforward induction on the length of the inference.

4 Completeness

In this section, some straightforward parts of the proof are omitted because of limited space.

Theorem 1 (Deduction theorem). *If T is a set of formulas, φ is a formula, and $T, \varphi \vdash \psi$, then $T \vdash \varphi \to \psi$.*

Proof. The proof is on the the transfinite induction on the length of the inference. We will only consider the case when we apply the inference rule R4.

If $T, \varphi \vdash \gamma \to \neg(\alpha U \beta)$ is obtained by the inference rule R4, then $T, \varphi \vdash \gamma \to \neg((\wedge_{k=0}^{i} \bigcirc^k \alpha) \wedge \bigcirc^{i+1}\beta)$, for all $i \in \omega$. By the induction hypothesis, we have $T \vdash \varphi \to (\gamma \to \neg((\wedge_{k=0}^{i} \bigcirc^k \alpha) \wedge \bigcirc^{i+1}\beta))$ (for all $i \in \omega$). From A1 we obtain $T \vdash (\varphi \wedge \gamma) \to (\neg((\wedge_{k=0}^{i} \bigcirc^k \alpha) \wedge \bigcirc^{i+1}\beta)))$, for all $i \in \omega$. Applying the inference rule R4 we conclude $T \vdash (\varphi \wedge \gamma) \to (\neg(\alpha U \beta))$. Finally, by A1 we obtain $T \vdash \varphi \to (\gamma \to \neg(\alpha U \beta))$.

The cases when ψ is a theorem and when we apply Modus ponens are standard, while the cases when we apply the inference rules R2 and R3 are trivial, since they can be applied to theorems only. In the case when we apply R5, the proof is similar to the considered case (R4). □

Lemma 1. *Let α, β, γ be formulas.*

1. *the following inference rule is derivable: from the set of formulas*

$$\{\gamma \to \bigcirc^i \beta \mid i \in \omega\}$$

   infer $\gamma \to G\beta$,
2. *if $\vdash \alpha$, then $\vdash G\alpha$,*
3. $\vdash G \bigcirc \alpha \leftrightarrow \bigcirc G\alpha$,

4. $\vdash (\bigcirc\alpha \rightarrow \bigcirc\beta) \rightarrow \bigcirc(\alpha \rightarrow \beta)$,
5. $\vdash \bigcirc(\alpha \wedge \beta) \leftrightarrow (\bigcirc\alpha \wedge \bigcirc\beta)$,
6. $\vdash \bigcirc(\alpha \vee \beta) \leftrightarrow (\bigcirc\alpha \vee \bigcirc\beta)$,
7. $G\alpha \vdash \bigcirc^i\alpha$ *for every* $i \geq 0$,
8. *if* $T \vdash \alpha$, *where* T *is a set of formulae, then* $\bigcirc T \vdash \bigcirc\alpha$.
9. *for* $j \geq 0$, $\bigcirc^j\beta, \bigcirc^0\alpha, \ldots, \bigcirc^{j-1}\alpha \vdash \alpha U\beta$,
10. *if* T *is a set of formulas and* $T \vdash \alpha$, *then* $AT \vdash A\alpha$.
11. $\vdash G\alpha \leftrightarrow \alpha \wedge \bigcirc G\alpha$,
12. $\vdash G(\alpha \rightarrow \beta) \rightarrow (G\alpha \rightarrow G\beta)$,
13. $\vdash G(\alpha \rightarrow \bigcirc\alpha) \rightarrow (\alpha \rightarrow G\alpha)$,
14. $\vdash (G(\alpha \rightarrow \alpha_1) \wedge (\alpha U\beta)) \rightarrow (\alpha_1 U\beta)$,
15. $\vdash (G(\beta \rightarrow \beta_1) \wedge (\alpha U\beta)) \rightarrow (\alpha U\beta_1)$,
16. $\vdash F\alpha \leftrightarrow F\neg\neg\alpha$
17. $\vdash \alpha U\beta \rightarrow F\beta$.

Proof. (1) is an immediate consequence of R4, obtained by replacing α and β with $\top$ and $\neg\beta$, respectively. (2) follows from (1) and R2.

For the proof of (3), (8) and (9) we refer the reader to [26], while the proof of (10) can be found in [7].
(14) Note that by (9) we have:

- $G(\alpha \rightarrow \alpha_1) \vdash \neg(\alpha_1 U\beta) \rightarrow \neg((\bigwedge_{k=0}^{i-1} \bigcirc^k \alpha_1) \wedge \bigcirc^i\beta)$, for every $i \geq 0$
- $G(\alpha \rightarrow \alpha_1) \vdash \neg(\alpha_1 U\beta) \rightarrow ((\bigwedge_{k=0}^{i-1} \bigcirc^k \alpha_1) \rightarrow \neg \bigcirc^i \beta)$, for every $i \geq 0$
- $G(\alpha \rightarrow \alpha_1) \vdash \neg(\alpha_1 U\beta) \rightarrow ((\bigwedge_{k=0}^{i-1} \bigcirc^k \alpha) \rightarrow \neg \bigcirc^i \beta)$, for every $i \geq 0$
- $G(\alpha \rightarrow \alpha_1) \vdash \neg(\alpha_1 U\beta) \rightarrow \neg((\bigwedge_{k=0}^{i-1} \bigcirc^k \alpha) \wedge \bigcirc^i\beta)$, for every $i \geq 0$
- $G(\alpha \rightarrow \alpha_1) \vdash \neg(\alpha_1 U\beta) \rightarrow \neg((\alpha U\beta))$, by R4

Thus, the statement holds. The statement (15) can be proved in a similar way, while (16) follows from the definition of $F\alpha = \top U\alpha$ and the previous steps. (17) follows directly from (14), taking $\alpha_1 = \top$. The remaining statements are easy consequences of the temporal part of the above axiomatization. □

Note that Lemma 1 states that some of the formulas and inference rules, proposed as the part of some (weakly) complete axiomatic systems [4,32,34] for temporal reasoning, hold in our logic. Thus, the temporal part of our axiomatization is sufficient to capture the semantical properties of the operators $\bigcirc$, A and U.

Theorem 2. *Every consistent set* T *of formulas can be extended to a maximal consistent set* T^*.

Proof. Let us assume that $For = \{\alpha_i \mid i \in \omega\}$. The maximally consistent set T^* is defined recursively, as follows:

1. $T_0 = T$.
2. If α_i is consistent with T_i, then $T_{i+1} = T_i \cup \{\alpha_i\}$.
3. If α_i is not consistent with T_i, then:

(a) Otherwise, if α_i has the form $\gamma \rightarrow \neg(\alpha U \beta)$, then

$$T_{i+1} = T_i \cup \{\gamma \rightarrow ((\wedge_{k=0}^{n_0} \bigcirc^k \alpha) \wedge \bigcirc^{n_0+1}\beta)\},$$

where n_0 is a positive integer such that T_{i+1} is consistent.

(b) Otherwise, if α_i is of the form $\gamma \rightarrow \bigcirc^m P^x_{\geq r}\beta$, for $x \in \{p, s\}$, then

$$T_{i+1} = T_i \cup \{\gamma \rightarrow \neg \bigcirc^m P^x_{\geq r - \frac{1}{n_1}}\beta\}$$

where n_1 is a positive integer such that T_{i+1} is consistent.

(c) Otherwise, $T_{i+1} = T_i$.

4. $T^* = \bigcup_{n \in \omega} T_n$.

Let us prove the existence of the number n_0 in 3(a). If we suppose that $\gamma \rightarrow ((\wedge_{k=0}^{n} \bigcirc^k \alpha) \wedge \bigcirc^{n+1}\beta)$ is not consistent with T_i, for every $n \in \omega$, then, by Theorem 1, $T_i \vdash \neg(\gamma \rightarrow ((\wedge_{k=0}^{n} \bigcirc^k \alpha) \wedge \bigcirc^{n+1}\beta))$, for every $n \in \omega$. By A1 we obtain $T_i \vdash \gamma \rightarrow \neg((\wedge_{k=0}^{n} \bigcirc^k \alpha) \wedge \bigcirc^{n+1}\beta)$, for every $n \in \omega$. By R4 we have $T_i \vdash \gamma \rightarrow \neg(\alpha U \beta)$, which contradicts the assumption. The proof of the existence of the number n_1 in 3(b) is similar.

It is easy to show that T_i is consistent for every i, and that for each $\alpha \in For$, either $\alpha \in T^*$ or $\neg\alpha \in T^*$.

Note that deductive closeness of T^* would imply its consistency: $T^* \vdash \bot$ would imply $\bot \in T^*$, thus there would exist i such that $\bot \in T_i$, which is impossible. In order to prove that T^* is deductively closed, it is sufficient to prove that it is closed under the inference rules, since all instances of axioms are obviously in T^*. We will only prove closeness under the inference rule R4, since the case when we consider R5 is similar, while the other cases are trivial.

Suppose that $\gamma \rightarrow \neg(\alpha U \beta) \notin T^*$, while $\gamma \rightarrow \neg((\wedge_{k=0}^{i} \bigcirc^k \alpha) \wedge \bigcirc^{i+1}\beta) \in T^*$ for every $i \in \omega$. By maximality of T^*, $\neg(\gamma \rightarrow \neg(\alpha U \beta)) \in T^*$, or, equivalently, $\gamma \wedge (\alpha U \beta) \in T^*$. Consequently, $\gamma \in T^*$ and $\alpha U \beta \in T^*$, so there are $m, n \in \omega$ such that $\gamma \in T_m$ and $\alpha U \beta \in T_n$. If $\gamma \rightarrow \neg(\alpha U \beta) = \alpha_l$, then, by the construction of T^*, there is n_0 such that $\gamma \rightarrow ((\wedge_{k=0}^{n_0} \bigcirc^k \alpha) \wedge \bigcirc^{n_0+1}\beta) \in T_l$. By Lemma 1(9), $T_l \vdash \alpha U \beta$. Consequently, $T_{\max\{l,m,n\}}$, which is in contradiction with consistency of $T_{\max\{l,m,n\}}$. □

We define the equivalence relation $\sim$ on the set of maximally consistent sets of formulas as follows:

$$T_1^* \sim T_2^* \text{ iff } T_1^* \cap St = T_2^* \cap St.$$

The equivalence class of T^* is $[T^*] = \{T_1^* \mid T_1^* \sim T^*\}$.

A canonical model $\mathcal{M}^* = \langle S, v, R, \Sigma, Prob^{state}, Prob^{path}\rangle$ is defined in the following way:

- $S = \{[T^*] \mid T^*$ is maximally consistent set of formulas$\}$,
- $v([T^*], p) = 1$ iff $T^* \vdash p$, $p \in \mathcal{P}$,
- $[T_1^*]R[T_2^*]$ if there exist $T_3^* \sim T_1^*$, $T_4^* \sim T_2^*$ such that $T_4^* = \{\alpha \mid \bigcirc \alpha \in T_3^*\}$,

– Σ is the set of paths $[T_0^*], [T_1^*], [T_2^*], \ldots$ such that $T_{i+1}^* = \{\alpha | \bigcirc \alpha \in T_i^*\}$, for all $i \in \omega$. If the sequence $\{T_i^*\}_{i\in\omega}$ determines a path σ, we will write $\sigma(i)$ for T_i^*,
– $Prob^{path}$ is defined as follows: for every $\sigma = [T_0^*], [T_1^*], [T_2^*], \ldots$, $Prob_\sigma = \langle A_\sigma, \mu_\sigma \rangle$ is a probability space such that:
  • $A_\sigma = \{[\alpha]_\sigma \mid \alpha \in For\}$, where $[\alpha]_\sigma = \{\sigma_{\geqslant} i \mid T_i^* \vdash \alpha, i \in \omega\}$,
  • $\mu_\sigma([\alpha]_\sigma) = \sup\{r \in \mathbb{Q} \cap [0,1] \mid T_0^* \vdash P^p_{\geqslant r}\alpha\}$,
– $Prob^{state}$ is defined as follows: for every $\sigma = [T_0^*], [T_1^*], [T_2^*], \ldots$, the probability space $Prob_\sigma = \langle A_\sigma, \mu_\sigma \rangle$ is determined by the following conditions: that:
  • $H_s = \{[\alpha]_s \mid \alpha \in For\}$, where $[\alpha]_s = \{\pi \mid \pi(0) \sim T_0^*,\ \pi(0) \vdash \alpha\}$,
  • $\mu_s([\alpha]_s) = \sup\{r \in \mathbb{Q} \cap [0,1] \mid T_0^* \vdash P^s_{\geqslant r}\alpha\}$.

Theorem 3. *$\mathcal{M}^*$ is a* pBTL*-model.*

Proof. Note that definitions of v and μ_s depend on the chosen element of equivalence class. We will show that the definition of $\mathcal{M}^*$ is correct:

– v is well defined, since $\mathcal{P} \subseteq St$, so $T_1^* \vdash p$ iff $T_2^* \vdash p$, whenever $T_1^* \sim T_2^*$, $p \in \mathcal{P}$.
– The definition of R is correct. Namely, using Temporal axioms, one can show that the properties of consistency and maximality transfer from T^* to $\{\alpha | \bigcirc \alpha \in T^*\}$. Moreover, R is obviously a total relation.
– A_σ is an algebra of sets. It is easy to show that $S_\sigma = [\top]_\sigma$, $[\alpha]_\sigma^c = [\neg\alpha]_\sigma$ and $[\alpha]_\sigma \cup [\beta]_\sigma = [\alpha \vee \beta]_\sigma$. Similarly, H_s is an algebra of sets.
– The function μ_s is well defined, since any formula of the form $P^s_{\geqslant r}\alpha$ is a state formula, so it belongs to a maximally consistent set T_1^* if and only if it belongs to any other maximally consistent set $T_2^* \in [T_1^*]$. Consequently, $\sup\{r \in \mathbb{Q} \cap [0,1] \mid T_1^* \vdash P^s_{\geqslant r}\alpha\} = \sup\{r \in \mathbb{Q} \cap [0,1] \mid T_2^* \vdash P^s_{\geqslant r}\alpha\}$.

By the axiom A11, $\mu_s(\alpha) \geqslant 0$, for every $\alpha \in For$. By R3, $\vdash A\top$, so, by A17, $T^* \vdash P^s_{\geqslant 1}\top$, for every maximally consistent set T^*. Since $H_s = [\top]_s$, we obtain $\mu_s(H_s) = 1$. Similarly, $\mu_\sigma(A_\sigma) = 1$ (by Lemma 1(2) and A16).

For the proof of finite additivity of μ_s and μ_σ, we refer the reader to [26], where a similar result is proved. □

Note that, since each $[T^*]$ may contain many maximally consistent sets, it is possible that one state belongs to several paths.

Theorem 4 (Strong completeness theorem). *Every consistent set of formulas is satisfiable.*

Proof. Let T be a consistent set of formulas, and let $\mathcal{M}^*$ be the model constructed above. We will prove that for every $\alpha \in For$, $\mathcal{M}^*, \sigma \models \alpha$ iff $\alpha \in \sigma(0)$.

If α is a propositional letter, this is immediate consequence of the definition of v. The proof in the cases when α is a negation or a conjunction is standard. For the proof in the cases when α is of the form $\bigcirc\beta$ or $\beta U \gamma$, we refer the reader to [26], where the similar proofs are presented.

Let $\alpha = A\beta$. If $\mathcal{M}^*, \sigma \not\models A\beta$, then there exists $\pi \in \Sigma_{\sigma_0}$ such that $\mathcal{M}^*, \pi \models \neg\beta$. By the induction hypothesis we obtain $\neg\beta \in \pi(0)$, so $\beta \notin \pi(0)$. By Axiom A7, $A\beta \notin \pi(0)$. From $\pi(0) \sim \sigma(0)$ and $A\beta \in St$, we conclude $A\beta \notin \sigma(0)$. For the other direction, suppose that $\mathcal{M}^*, \sigma \models A\beta$. Then for all $\pi \in \Sigma_{\sigma_0}$, $\mathcal{M}^*, \sigma \models \beta$. Consequently, by the induction hypothesis, for all $\pi \in \Sigma_{\sigma_0}$, $\beta \in \pi(0)$. If $A\beta \notin \sigma(0)$, using Temporal axioms one can show that there exists $\rho \in \Sigma_{\sigma_0}$ such that $\beta \notin \rho(0)$, which contradicts the assumption.

Let $\alpha = P^s_{\geq r}\beta$ (in the case when $\alpha = P^p_{\geq r}\beta$ the proof is similar). Suppose that $\mathcal{M}^*, \sigma \models P^s_{\geq r}\beta$. If $\sup\{t \in \mathbb{Q} \cap [0,1] \mid P^s_{\geq t}\beta \in \sigma(0)\} = r$, then $P^s_{\geq r}\beta \in \sigma(0)$, by the maximality of $\sigma(0)$ and the rule R3. If $\sup\{t \in \mathbb{Q} \cap [0,1] \mid P^s_{\geq t}\beta \in \sigma(0)\} > r$, then there exists $q \in \mathbb{Q} \cap (r, \sup\{t \in \mathbb{Q} \cap [0,1] \mid P^s_{\geq t}\beta \in \sigma(0)\}]$ such that $P^s_{\geq q}\beta \notin \sigma(0)$. By deductively closeness of $\sigma(0)$, $P^s_{\geq r}\beta \in \sigma(0)$. On the other hand, if $P^s_{\geq r}\beta \in \sigma(0)$, then $\mu_s(\{\pi \mid \pi(0) \sim \sigma(0),\ \pi(0) \vdash \beta\}) = \sup\{t \in \mathbb{Q} \cap [0,1] \mid P^s_{\geq t}\beta \in \sigma(0)\} \geq r$. By the induction hypothesis, $\{\pi \mid \pi(0) \sim \sigma(0),\ \pi(0) \vdash \beta\} = \{\pi \mid \pi(0) \sim \sigma(0),\ \mathcal{M}^*, \pi \models \beta\}$, so $\mathcal{M}^*, \sigma \models P^s_{\geq r}\beta$.

Let T^* be a maximally consistent set such that $T \subseteq T^*$. If $\sigma = [T^*], [\{\alpha | \bigcirc\alpha \in T^*\}], [\{\alpha | \bigcirc^2 \alpha \in T^*\}] \ldots$, then $\mathcal{M}^*, \sigma \models T$. □

Note that, by the proof of the previous theorem, $[\alpha]_\sigma = \{\sigma_{\geqslant} i \mid T^*_i \vdash \alpha, i \in \omega\} = \{\pi \in S_\sigma \mid \mathcal{M}^*, \pi \models \alpha\} = [\alpha]^{path}_{\mathcal{M},\sigma}$. Similarly, $[\alpha]_s = [\alpha]^{state}_{\mathcal{M}^*,\sigma}$, so $\mathcal{M}^*$ is a measurable model.

Corollary 1. *If α is a formula and T is a set of formulas, then $T \models \alpha$ implies $T \vdash \alpha$.*

Proof. Let $T \models \alpha$. Then $T \cup \{\neg\alpha\}$ is not satisfiable. By Theorem 4, $T \cup \{\neg\alpha\} \vdash \bot$, and, by Theorem 1, $T \vdash \alpha$.

5 Related Work

The branching-time logic PCTL for reasoning about time and probability is described in [17]. The underlying temporal logic is Computational Tree Logic CTL (Emerson, Clark, Sistla [10]). The statements of the form: "after a request for service there is at least a 98% probability that the service will be carried out within 2 seconds" are expressible in the language of PCTL. Formulas are interpreted over discrete time Markov chains and algorithms for checking satisfiability of formulas by a given Markov chain are described. No axiomatization is presented. The logic follows the division of CTL into state formulas and path formulas. The classical propositional language is enriched in the following way:

– $\alpha U^{\leq t}\beta$ and $\alpha \mathcal{U}^{\leq t}\beta$ are path formulas, if α and β are state formulas, and $t \in \omega \cup \{\infty\}$. The intuitive meaning of $\alpha U^{\leq t}\beta$ is similar to the meaning of $\alpha U \beta$, with the exception that β has to become true within t time instances (for $t = \infty$, $U^{\leq t}$ and U coincide). The relation of $\alpha \mathcal{U}^{\leq t}\beta$ to $\alpha \mathcal{U}\beta \equiv \alpha U \beta \vee G\alpha$ is analogous.

- $\alpha U_{>r}^{\leq t}\beta$ and $\alpha \mathcal{U}_{>r}^{\leq t}\beta$ are state formulas, if α and β are path formulas, and $t \in \omega \cup \{\infty\}$. The meaning of those formulas is given by the satisfiability relation (formulation is adopted according to our terminology):
  $\mathcal{M}, \sigma \models \alpha U_{>r}^{\leq t}\beta$ iff $\mu_{\sigma_0}(\{\pi \mid \sigma_0 = \pi_0,\ \mathcal{M}, \pi \models \alpha U^{\leq t}\beta\}) > r$,

The formulas of PCTL are expressible in our language. For example:

- $\alpha U^{\leq n}\beta$ may be written as $\beta \vee \bigvee_{i=1}^{n}((\wedge_{k=0}^{i-1} \bigcirc^k \alpha) \wedge \bigcirc^i\beta)$,
- $\alpha U_{>r}^{\leq n}\beta$ may be written as $P_{>r}^{s}(\beta \vee \bigvee_{i=1}^{n}((\wedge_{k=0}^{i-1} \bigcirc^k \alpha) \wedge \bigcirc^i\beta))$.

On the other hand, our operator $P_{\geqslant r}^{p}$ is not expressible in PCTL. Also, boolean combinations of state and path formulas are not PCTL-formulas.

A more expressive branching-time logic denoted PCTL* is described in [2]. The underlying temporal logic is CTL* with path quantifiers replaced by probabilities ($P_{=1}$, $P_{>0}$). Thus, the propositional language is extended with:

- state formulas: $P_{\geq r}\alpha$ (α is a path formula),
- path formulas: $\bigcirc\alpha$, $\alpha U\beta$ (α and β are state formulas).

According to definition of satisfiability, their probability operator $P_{\geqslant r}$ corresponds to our operator $P_{\geqslant r}^{s}$, while our operator $P_{\geqslant r}^{p}$ is not expressible in PCTL*. Similarly as in PCTL, the conjunction of a state formula and a path formula is not a formula. No axiomatization for PCTL* is given.

The paper [3] presents model-checking algorithms for extensions of PCTL and PCTL* that involve non-determinism.

A probabilistic modal logic PPL is introduced in [35]. It allows applying probabilities to sequences of formulas (giving so called path expressiveness). A Gentzen-style axiom system is presented and proved to be sound and complete. Probabilities are expressed using terms (similarly as in [12]). The language allows linear combinations of terms of the form $P(\alpha_1, \ldots, \alpha_n)$ which means "the probability of the sequence of formulas." Iteration of probabilities in a term is allowed. The formula $P(\alpha_1, \ldots, \alpha_n) \geq r$ is expressible in our logic as $P_{\geqslant r}^{s}(\bigwedge_{i=1}^{n} \bigcirc^i\alpha_i)$. On the other hand, formulas of PPL can not express probability within a path ($P_{\geqslant r}^{p}$). Also, the temporal operators are not definable in PPL. Although our language does not allow linear combinations of probabilities, combining the techniques from [8,9,28], where arithmetical operations are built into the syntax of probabilistic logic, with the ideas presented in this paper, would lead to a logic in which formulas of PPL are expressible.

In [18] and [21] propositional logics that use the languages of CTL and CTL* are presented. The probabilities are not expressible in syntax, but the formulas are interpreted over Markov systems which can simulate the execution of probabilistic programs.

The papers [15,19,22] introduce real-time interval logics that can be used in design of an embedded real-time systems. The infinite intervals are considered in [16].

The language of the logic presented in [13] is based on the propositional dynamic logic, and the main objects are programs. Probabilistic operators can be applied on a limited class of formulas, and the completeness problem is not solved. A fragment of [13]) is considered in [20]. A dynamic generalization of the logic of qualitative probabilities from [33] is presented in [14]. Completeness is proved using an infinitary rule, similarly as in our approach.

6 Conclusion and Future Work

We have introduced the propositional probabilistic branching time logic pBTL that enables us to formulate (and combine) both purely temporal statements and the expressions such as: "in at least half of paths α holds in at least a third of states". The formulas are interpreted over models that involve a class of probability measures assigned to states, and a class of probability measures assigned to paths. We have proved that the infinitary axiomatic system for pBTL is sound and strongly complete.

One of the main axiomatization issues for temporal logics with the operators $\bigcirc$ and G, and for real valued probability logics is the non-compactness phenomena. The set of formulas $\{P^s_{>0}\alpha\} \cup \{P^s_{\leqslant \frac{1}{n}}\alpha \mid n \in \omega\}$ and $\{G\alpha\} \cup \{\bigcirc^n \neg\alpha \mid n \in \omega\}$ are finitely satisfiable but they are not satisfiable. It is well known that, in the absence of compactness, any finitary axiomatization would be incomplete. Thus, infinitary axiomatic systems are the only way to establish strong completeness.

The temporal fragment of pBTL uses the language of CTL^*. The restricted class of models (without probabilities) corresponds to the class of models of so-called $\forall LT$ logic from [34] (compare Lemma 1 and the axiomatic system from [34]). The paper [32] solved the problem of (weak) completeness of Full Computation Tree Logic (with the class of models satisfying the desirable properties FC (Fusion closed) and LC (Limit closed)), extending the axiomatization of $\forall LT$. Thus, the question of extending the temporal part of our axiomatization, with the aim to obtain completeness of probabilistic Full Computation Tree Logic, naturally arise.

Also, we believe that there are several other promising ways to extend the results presented here, along the lines of our previous research:

- Combining the techniques from this paper and [7] may lead to the first-order extension of pBTL. That logic would be not only of theoretical interest, since the set of all valid formulas is not recursively enumerable [1], and no complete finitary axiomatization is possible in that undecidable framework. In this situation, a complete (even if infinitary) axiomatization would be of great practical significance.
- A branching time logic in which linear combinations of probabilities are expressible could be developed combining the ideas presented here with the ideas from [8,9,28]. The formulas of the logic presented in [35] would be expressible in the resulting language (see Section 5).
- It is well known that CTL and CTL^* are decidable [11]. We expect that, similarly as it is done in [26] for probabilistic linear time logic, it is possible to adapt the corresponding procedures to prove decidability of the logic presented here.

Acknowledgements. The work presented here was partially supported by the Serbian Ministry of Education and Science (projects ON174026 and III44006). The authors would like to thank Kosta Došen for several useful suggestions.

References

1. Abadi, M.: The power of temporal proofs. Theoretical Computer Science 65, 35–83 (1989)
2. Aziz, A., Singhal, V., Balarin, F., Brayton, R.K., Sangiovanni-Vincentelli, A.L.: It usually works: The temporal logic of stochastic systems. In: Wolper, P. (ed.) CAV 1995. LNCS, vol. 939, Springer, Heidelberg (1995)
3. Bianco, A., de Alfaro, L.: Model checking of probabilistic and nondeterministic systems. In: Thiagarajan, P.S. (ed.) FSTTCS 1995. LNCS, vol. 1026, pp. 499–512. Springer, Heidelberg (1995)
4. Burgess, J.: Axioms for tense logic. I. "Since" and "until". Notre Dame Journal of Formal Logic 23(4), 367–374 (1982)
5. Burgess, J.: Logic and time. The Journal of Symbolic Logic 44(4), 566–582 (1979)
6. Burgess, J.: Basic tense logic. In: Gabbay, D., Guenthner, F. (eds.) Handbook of Philosophical Logic, vol. II, pp. 89–133. D. Reidel Publishing Compaany, Dordrecht (1984); Kopetz, H., Kakuda, Y. (eds.) Dependable Computing and Fault-Tolerant Systems. Responsive Computer Systems, Vol. 7 pp. 30–52. Springer, Heidelberg (1993)
7. Doder, D., Ognjanović, Z., Marković, Z.: An Axiomatization of a First-order Branching Time Temporal Logic. Journal of Universal Computer Science 16(11), 1439–1451 (2010)
8. Doder, D., Marković, Z., Ognjanović, Z., Perović, A., Rašković, M.: A Probabilistic Temporal Logic That Can Model Reasoning about Evidence. In: Link, S., Prade, H. (eds.) FoIKS 2010. LNCS, vol. 5956, pp. 9–24. Springer, Heidelberg (2010)
9. Doder, D., Marinković, B., Maksimović, P., Perović, A.: A Logic with Conditional Probability Operators. Publications de l'Institut Mathématique, Nouvelle Série, Beograd 87(101), 85–96 (2010)
10. Emerson, E., Clarke, E.: Using branching time logic to synthesize synchronization skeletons. Sci. Comput. Program. 2, 241–266 (1982)
11. Emerson, E.: Temporal and Modal Logic. In: van Leeuwen, J. (ed.) Handbook of Theoretical Computer Science, Volume B: Formal Models and Sematics, pp. 995–1072. North-Holland Pub. Co./MIT Press (1990)
12. Fagin, R., Halpern, J., Megiddo, N.: A logic for reasoning about probabilities. Information and Computation 87(1–2), 78–128 (1990)
13. Feldman, Y.: A decidable propositional dynamic logic with explicit probabilities. Information and Control 63, 11–38 (1984)
14. Guelev, D.P.: A propositional dynamic logic with qualitative probabilities. Journal of Philosophical Logic 28(6), 575–605 (1999)
15. Guelev, D.P.: Probabilistic neighbourhood logic. In: Joseph, M. (ed.) FTRTFT 2000. LNCS, vol. 1926, pp. 264–275. Springer, Heidelberg (2000)
16. Guelev, D.P.: Probabilistic Interval Temporal Logic and Duration Calculus with Infinite Intervals: Complete Proof Systems. Logical Methods in Computer Science 3(3), 1–43 (2007)
17. Hansson, H., Jonsson, B.: A logic for reasoning about time and reliability. Formal Aspect of Computing 6(5), 512–535 (1994)

18. Hart, S., Sharir, M.: Probabilistic temporal logics for finite and bounded models. In: 16th ACM Symposium on Theory of Computing, pp. 1–13. ACM, New York (1984): Extended version In: Information and Control 70 (2/3), 97-155 (1986)
19. Hung, D.V., Chaochen, Z.: Probabilistic Duration Calculus for Continuous Time. Formal Aspects of Computing 11(1), 21–44 (1999)
20. Kozen, D.: A probabilistic PDL. Journal of Computer and System Sciences 30, 162–178 (1985)
21. Lehmann, D., Shelah, S.: Reasoning with time and chance. Information and Control 53, 165–198 (1982)
22. Liu, Z., Ravn, A.P., Sorensen, E.V., Chaochen, Z.: A Probabilistic Duration Calculus. In: Kopetz, H., Kakuda, Y. (eds.) Dependable Computing and Fault-Tolerant Systems. Responsive Computer Systems, vol. 7, pp. 30–52. Springer, Heidelberg (1993)
23. Marković, Z., Ognjanović, Z., Rašković, M.: A Probabilistic Extension of Intuitionistic Logic. Mathematical Logic Quarterly 49, 415–424 (2003)
24. Ognjanović, Z., Rašković, M.: Some probability logics with new types of probability operators. Journal of Logic and Computation 9(2), 181–195 (1999)
25. Ognjanović, Z., Rašković, M.: Some first-order probability logics. Theoretical Computer Science 247(1-2), 191–212 (2000)
26. Ognjanović, Z.: Discrete Linear-time Probabilistic Logics: Completeness, Decidability and Complexity. Journal of Logic Computation 16(2), 257–285 (2006)
27. Ognjanović, Z., Perović, A., Rašković, M.: Logics with the Qualitative Probability Operator. Logic Journal of IGPL 16(2), 105–120 (2008)
28. Perović, A., Ognjanović, Z., Rašković, M., Marković, Z.: A Probabilistic Logic with Polynomial Weight Formulas. In: Hartmann, S., Kern-Isberner, G. (eds.) FoIKS 2008. LNCS, vol. 4932, pp. 239–252. Springer, Heidelberg (2008)
29. Pnueli, A.: The Temporal Logic of Programs. In: Proceedings of the 18th IEEE Symposium Foundations of Computer Science (FOCS 1977), pp. 46–57 (1977)
30. Prior, A.: Time and Modality. Oxford University Press, Oxford (1957)
31. Rašković, M., Marković, Z., Ognjanović, Z.: A Logic with Approximate Conditional Probabilities that can Model Default Reasoning. International Journal of Approximate Reasoning 49(1), 52–66 (2008)
32. Reynolds, M.: An axiomatization of full computation tree logic. The Journal of Symbolic Logic 66(3), 1011–1057 (2001)
33. Segerberg, K.: Qualitative probability in a modal setting. In: Fenstad, J.E. (ed.) Proceedings of the Second Scandinavian Logic Symposium. North-Holland, Amsterdam (1971)
34. Stirling, C.: Modal and temporal logic. Handbook of Logic in Computer Science 2, 477–563 (1992)
35. Tzanis, E., Hirsch, R.: Probabilistic Logic over Paths. Electronic Notes in Theoretical Computer Science 220(3), 79–96 (2008)

A Partition-Based First-Order Probabilistic Logic to Represent Interactive Beliefs

Alessandro Panella and Piotr Gmytrasiewicz

University of Illinois at Chicago
Department of Computer Science
Chicago, IL 60607
{apanella,piotr}@cs.uic.edu

Abstract. Being able to compactly represent large state spaces is crucial in solving a vast majority of practical stochastic planning problems. This requirement is even more stringent in the context of multi-agent systems, in which the world to be modeled also includes the mental state of other agents. This leads to a hierarchy of beliefs that results in a continuous, unbounded set of possible interactive states, as in the case of Interactive POMDPs. In this paper, we describe a novel representation for interactive belief hierarchies that combines first-order logic and probability. The semantics of this new formalism is based on recursively partitioning the belief space at each level of the hierarchy; in particular, the partitions of the belief simplex at one level constitute the vertices of the simplex at the next higher level. Since in general a set of probabilistic statements only partially specifies a probability distribution over the space of interest, we adopt the maximum entropy principle in order to convert it to a full specification.

1 Introduction

One of the main problems to be faced in the field of stochastic planning is the curse of dimensionality. Traditional methods based on the enumeration of the state, action, and observation spaces have shown to be unpractical for all but the simplest settings. Factorizing the description of the domain into "features", like in a Bayesian network, has been a prominent direction of research that has led to outstanding results [3,19]. Yet, in many real-world problems this approach does not suffice, because of the large number of such features. Hence the need to lift the representation from the propositional level to a more abstract level, by exploiting the synergy between first-order logic (FOL) and probability theory, allowing to compactly summarize the regularities of the domain and the interactions between objects. Several applications of this paradigm to MDPs can be found in literature (e.g. [4,22,25]). On the other hand, only little work has surfaced that focuses on lifted first-order inference for representing and solving POMDPs [23,26].

A compact representation of the domain is even more necessary in the context of decision making in partially observable, multi-agent environments, in which an agent needs to model the mental states (beliefs, preferences, and intentions)

S. Benferhat and J. Grant (Eds.): SUM 2011, LNAI 6929, pp. 233–246, 2011.
© Springer-Verlag Berlin Heidelberg 2011

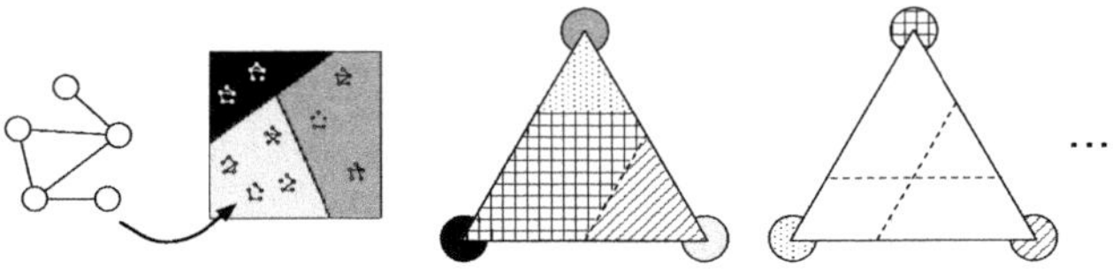

Fig. 1. Qualitative representation of the belief hierarchy. (a) Real state of the world (objects and relations.) (b) Set of possible world states. (c) Agent i's belief simplex over partitions of world states. (d) Agent j's belief simplex over partitions of i's simplex. Note that the partitions at one level are the vertices of the simplex at the next level; this association is represented by matching colors (b-c) and patterns (c-d.)

of other agents, in addition to the physical world. In particular, we focus on the Interactive POMDP (I-POMDP) framework [8], in which the agent maintains a belief about the other agents' *types*, intended as the set of private information involved in their decision making. Each type includes the agent's own *belief*, which is a probability distribution over the state of the world and, recursively, other agents' types. In this paper, we focus on the representation of such interactive beliefs, limiting the discussion to a setting with two agents, i and j. The generalization to scenarios with more than two agents is straightforward. Because of the impossibility of representing infinitely nested beliefs in a finite space, Gmytrasiewicz and Doshi [8] define finitely nested I-POMDPs as a specialization of the infinitely nested ones.

Let us denote as $\Delta(\cdot)$ the regular simplex over the set given as argument. The *interactive state space* at nesting level n for agent i, denoted $IS_{i,n}$, is inductively defined as:

$$\begin{aligned} IS_{i,0} &= S \\ IS_{i,1} &= S \times \Delta(IS_{j,0}) \\ &\vdots \\ IS_{i,n} &= S \times \Delta(IS_{j,n-1}) \end{aligned} \tag{1}$$

One problem of I-POMDPs is that the set of possible beliefs of the other agent is uncountable and unbounded as soon as nesting level 2 is reached [6]. This makes it impossible to even represent interactive beliefs, in that they are not computable functions. For this reason, being able to abstract over the regularities of the interactive state space in order to provide a finite representation is of utmost benefit.

In this paper, we describe for the first time a First-Order Probabilistic Language to express interactive beliefs. The approach is conceived in the context of I-POMDPs, but grows out to be a general representation of nested probability distributions that can be applied in a variety of contexts in multi-agent systems. The main idea is to *recursively partition the belief space into regions, building the belief simplex at the next level over the partitions of the belief simplex at the lower level*, as intuitively depicted in Fig. 1. In this way, we can provide a compact, finite representation of interactive beliefs.

A belief is represented as a set of (probabilistic) sentences, each associated with a probability. These sentences are not required to be non-overlapping, in that the agent should be free to express his belief about any arbitrary statement. As a result of this augmented freedom, a belief base constitutes in general only a set of constraints rather than a full specification of a probability distribution. For this reason, we adopt the well-known *maximum entropy principle* in order to provide a unique distribution associated with the belief base.

The paper is organized as follows. In Sect. 2 we provide a brief survey of the existing related work. Section 3 presents our contribution, describing the probabilistic-logical framework for interactive beliefs in a bottom-up fashion and providing examples to clarify the concepts. Section 4 concludes the paper and hints at directions for future research.

2 Related Work

The integration of first-order logic and probability theory has been an important area of research since the mid and late 80's. Nilsson [18] proposes a probabilistic logic in which a probability value is attached to logic sentences, either propositional or first-order, belonging to a probabilistic knowledge base. He devises a linear problem that, given any query sentence, computes its probability intervals that are consistent with knowledge base. Some years later, Bacchus [2] and Halpern [9] describe how first-order probabibilistic logic comes in two flavors: to express probabilities on possible worlds, such as in the sentence "Tweety flies with probability 0.9," and to express statistical knowledge, as in "90% of birds fly." Subsequent work in probabilistic languages has mostly adopted the first type of semantics. This early approaches provide theoretical basis for the field, but lack practical inference algorithms.

The work on probabilistic logic has evolved in what has been recently named Statistical Relational AI (Star AI), that includes a number of different approaches, of which we report a few examples. Koller and Pfeffer [15] define Probabilistic Relational Models (PRMs), borrowing the semantics from relational databases. Like databases, PRMs model complex domains in terms of entities and properties. Moreover, by incorporating a directed causal relations like in Bayesian networks, PRMs allow to express uncertainty over the properties of entities and the relations among them (relational uncertainty.) Markov Logic Networks [20] are collections of first-order logic formulae that are assigned a weight. The atomic formulae appearing in such set constitute the vertices of a Markov Network, whose edges correspond to the logical connectives. The weights determine the potential function assigned to each groups of vertices that compare in the same original formula. In [17] the authors introduce Bayesian Logic (BLOG), a generalization of Bayesian network similar to RPMs that assume an open universe, i.e. the number of objects is unknown and potentially infinite. BLOG models are described by means of a generative semantics that allows to deal with domains of unknown size. Another line of work studies probabilistic logic programs [16] and relational probabilistic conditionals [14]. These approaches adopt the maximum entropy principle to provide semantics to probabilistic knowledge bases.

The study interactive belief systems has been subject of substantial research in the field of game theory and multi-agent systems in general, especially since the introduction of games of incomplete information [12]. Several works [11] study the use of modal logic and its derivations in order to describe players' knowledge about other players' knowledge. Probabilistic extension to modal logic have been proposed [7,10,24], and are based on commonly known prior probability distribution on possible worlds and accessibility relation. Aumann [1] describes an approach that embeds knowledge and probabilistic beliefs in the context of interactive epistemology. The work on Interactive POMDPs [8,5] rejects the common knowledge assumption and proposes a hierarchy of probabilistic beliefs that we take as our starting point, as already described in the introductory section.

3 Probabilistic First-Order Logic to Represent Interactive Beliefs

We begin this section by describing the logic setup. The semantics of first-order logic used here assumes closed universe and unique names. Let $Q = \{q_1, q_2, \ldots, q_{|Q|}\}$ be a *set of predicates*, and let $\rho : Q \rightarrow \mathbb{N}$ be a function that associates each predicate to its *arity*. Given a *domain* (set of constants) $D = \{d_1, d_2, \ldots d_{|D|}\}$, define the *set of ground predicates* G (the Herbrand base), corresponding to each possible instantiation of predicates in the domain, i.e. $G = \{q(d_1, \ldots, d_{\rho(q)}) : q \in Q, d_i \in D \;\; \forall i = 1, \ldots, \rho(q)\}$. An interpretation of Q in domain D is a function $\sigma : G \rightarrow \{T, F\}$ that assigns a truth value to every ground predicate. The set of possible *states of the world* S corresponds to the set of all possible interpretations.

Given a first-order logic sentence ϕ, we denote as $S(\phi)$ the subset of S for which ϕ is true, i.e. the set of models of ϕ (under the usual definition of FOL entailment.) Given a set of FOL sentences Φ, we denote as $S(\Phi)$ the collection of sets of models of the formulae in Φ, i.e. $S(\Phi) = \{S(\phi) : \phi \in \Phi\}$.

3.1 Level 0 Beliefs

In this section, we describe how to represent 0-th level beliefs, i.e. the belief an agent holds about the state of the world. The approach is similar to the one described in [18], that we take as our starting point.

Definition 1 (Level-0 Belief Base). *A Level-0 Belief Base (L0-BB) $\mathcal{B}^{i,0}$ for agent i is a set of pairs of the form $\langle \phi_k, \alpha_k \rangle$, for $k = 1, 2, \ldots, m$, where ϕ_k is a sentence in first-order logic, and α_k a real number between 0 and 1.*

For each pair, α_k intuitively represents i's degree of belief about sentence ϕ_k. In addition to simple pairs, we allow universally quantified expressions of the type $\forall \mathbf{x} \langle \phi(\mathbf{x}), \alpha \rangle$ to appear in the belief base, where $\mathbf{x} = \langle x_1, \ldots, x_l \rangle$ is a tuple of logical variables that are free in FOL formula ϕ. Semantically, this expression is equivalent to the set of pairs resulting from the propositionalization of ϕ, that is:

$$\forall \mathbf{x} \ \langle \phi(\mathbf{x}), \alpha \rangle \equiv \left\{ \langle \textsc{Subst}(\mathbf{x}/\mathbf{d}, \phi), \alpha \rangle : \mathbf{d} \in D^{|\mathbf{x}|} \right\} , \tag{2}$$

where $\textsc{Subst}(\mathbf{x}/\mathbf{d}, \phi))$ represent the FOL sentence resulting from ϕ by substituting the tuple of logical variables $\mathbf{x}$ with domain elements $\mathbf{d}$, adopting the notation used in [21]. In the following, we will consider L0-BB's in which all universally quantified pairs have been proposizionalized.

Let $\Phi_{\mathcal{B}} = \{\phi_1, \phi_2, \ldots, \phi_m\}$ be the set of FOL sentences involved in the pairs of a L0-BB $\mathcal{B}^{i,0}$. As mentioned earlier, we do not require the elements of $\Phi_{\mathcal{B}}$ to identify non-overlapping regions of S. Instead, we compute the partitions that are induced by such regions, i.e. every possible overlap, and form a probability distribution on such partitioning. We hence define the set of logical partitions as

$$\Psi_{\mathcal{B}} = \left\{ \bigwedge_{\phi \in \Phi_I} \phi \setminus \bigvee_{\phi \in I \Phi_I^C} \phi : \Phi_I \subseteq \Phi_{\mathcal{B}} \cup \{\top\} \right\} , \tag{3}$$

where $\Phi_I^C = (\Phi_{\mathcal{B}} \cup \top) \setminus \Phi_I$. We denote as $\Psi_{\mathcal{B}}(\phi)$ the set of partitions whose union is ϕ:

$$\Psi_{\mathcal{B}}(\phi) = \{\psi \in \Psi_{\mathcal{B}} : S(\psi) \subseteq S(\phi)\} \tag{4}$$

The concept of satisfiability of a belief base is formalized in the following definitions.

Definition 2 (Satisfiability). *Given a L0-BB $\mathcal{B}^{i,0}$, a probability distribution $p_{i,0}$ over the set of logical partitions $\Psi_{\mathcal{B}}$ is said to* satisfy $\mathcal{B}^{i,0}$ *if, for all $\phi_k \in \Phi_{\mathcal{B}}$, it is true that*

$$\sum_{\psi \in \Psi_{\mathcal{B}}(\phi)} p_i(S(\psi)) = \alpha_k \tag{5}$$

Definition 3 (Consistency). *A L0-BB $\mathcal{B}^{i,0}$ is said to be* consistent *(or* satisfiable*) if there exists a probability distribution $p_{i,0}$ over $\Psi_{\mathcal{B}}$ that satisfies it.*

In general, there exist multiple distributions $p_{i,0}$ satisfying a L0-BB $\mathcal{B}^{i,0}$.[1] This is due to the fact that a L0-BB constitutes in general only a partial specification of a probability distribution over S, as noted by Nilsson [18]. There are different ways to cope with this indeterminacy. A "skeptical" approach is to compute the upper and lower bounds of the probability of each state, and consider such intervals. A more "credulous" solution is to pick one probability distribution among the ones that are consistent with the L0-BB. We follow the latter direction by choosing the maximum entropy (max-ent) distribution [13].

Given a L0-BB $\mathcal{B}^{i,0}$, the max-ent probability distribution $p_{i,0}$ over $\Psi_{\mathcal{B}}$ that satisfies $\mathcal{B}^{i,0}$ is given by the solution to the following optimization problem:

$$\max_{p_{i,0}} \left(- \sum_{\psi \in \Psi_{\mathcal{B}}} p_{i,0}(S(\psi)) \log p_{i,0}(S(\psi)) \right) \tag{6}$$

[1] In fact, it can be shown that a L0-BB either has a unique model or admits an uncountably infinite set of models.

subject to

$$\begin{aligned} \textstyle\sum_{\psi\in\Psi_{\mathcal{B}}(\phi)} p_{i,0}(S(\psi)) &= \alpha_k \qquad && \forall\, \phi \in \Phi_{\mathcal{B}} \\ \textstyle\sum_{\psi\in\Psi_{\mathcal{B}}} p_{i,0}(S(\psi)) &= 1 && \\ p_{i,0}(S(\psi)) &\geq 0 && \forall \psi \in \Psi_{\mathcal{B}} \end{aligned} \tag{7}$$

Hence, a L0-BB $\mathcal{B}^{i,0}$ represents the probability space $(S(\Psi_{\mathcal{B}}), 2^{S(\Psi_{\mathcal{B}})}, p_{i,0})$, were $p_{i,0}$ is the max-ent probability distribution just defined. Slightly abusing notation, we will sometimes write $p_{i,0}(\psi)$ rather than $p_{i,0}(S(\psi))$ when this is not source of ambiguity. We clarify the concepts presented in this section by introducing a simple running example.

Example (Grid world). In a world consisting of an $n \times n$ grid, an agent i wants to tag a moving target j. The agent knows his own position, but is uncertain about the target's. The predicate $jPos(x,y)$ indicates the target position, where x and y are integers representing the coordinates of a location on the grid, i.e. $0 \leq x, y < n$. Obviously, the target occupies one and only one position in the grid; this can be expressed by the following FOL sentence:

$$\exists x, y\Big(jPos(x,y) \wedge \neg\exists w, z\big(jPos(x,y) \wedge jPos(w,z) \wedge (x \neq y \vee w \neq z)\big)\Big) \tag{8}$$

Instead of including this fact in the belief base, we assume the agent implicitly knows it, as a limitation on possible worlds. We introduce the auxiliary deterministic predicates $geq(x,k) \equiv x \geq k$ and $leq(x,k) \equiv x \leq k$. Say the agent is interested, in a particular moment in time, to the horizontal location of the target in the grid with respect to the center, i.e. whether the target is in either the left or right half of the map. A plausible L0-BB representing i's belief is:

$$\mathcal{B}^{i,0} = \begin{array}{|c|}\hline \langle \exists x, y(jPos(x,y) \wedge leq(x, \lfloor n/2 \rfloor)),\ 0.8\rangle \\ \hline \langle \exists x, y(jPos(x,y) \wedge geq(x, \lfloor n/2 \rfloor)), 0.5\rangle \\ \hline \end{array} \tag{9}$$

We denote the two FOL sentences in the L0-BB as ϕ_0 and ϕ_1, respectively. The partitioning $\Psi_{\mathcal{B}} = \{\psi_0, \psi_1, \psi_2\}$ induced on the state space is shown in Fig. 2-a. In this case, there is only one probability distribution over $\Psi_{\mathcal{B}}$ that is consistent with the belief base, namely $p_{i,0} = (0.5, 0.3, 0.2)$.

3.2 Level 1 Beliefs

After having described how to represent the beliefs of agent i about the state of the world using first-order logic and probability, we now introduce how to specify agent i's beliefs about agent j's beliefs about S. In order to do so, agent i needs a language that allows him to abstract over the space of agent j's 0-th level beliefs, in the same way first-order logic provides abstraction over the set of states of the world in a L0-BB. We call this language *Level-0 First-Order Probabilistic Logic* (L0 FOPL), in that it provides a way to describe j's level 0 beliefs.

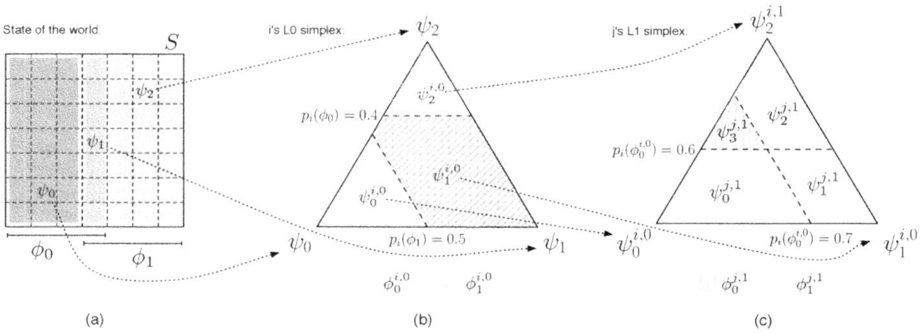

Fig. 2. Example of nested First-Order Belief Base

Definition 4 (L0 FOPL). *The language $\mathscr{L}^{j,0}$ of j's level 0 probabilistic statements is recursively defined as:*

1. *$P_j(\phi) \vartriangle \beta$ is a formula, where ϕ is a formula in first-order logic, $\beta \in [0,1]$, and $\vartriangle \in \{<, \leq, =, \geq, >\}$;*
2. *If $\phi^{j,0}$ is a formula, then $\neg\phi^{j,0}$ is a formula;*
3. *If $\phi_1^{j,0}$ and $\phi_2^{j,0}$ are formulae, then $\phi_1^{j,0} \wedge \phi_2^{j,0}$, $\phi_1^{j,0} \vee \phi_2^{j,0}$, $\phi_1^{j,0} \Rightarrow \phi_2^{j,0}$ are formulae;*
4. *If $\phi^{j,0}$ is a formula, then $\exists\mathbf{x}(\phi^{j,0})$ and $\forall\mathbf{x}(\phi^{j,0})$ are formulae, where $\mathbf{x}$ is a subset of the free logical variables of $\phi^{j,0}$;*
5. *A sentence is a formula with no free variables.*

Agent i assigns degrees of belief to some sentences of $\mathscr{L}^{j,0}$. This is represented as a Level 1 Belief Base, defined in the following.

Definition 5 (Level 1 Belief Base). *A* Level 1 Belief Base *(L1-BB) $\mathcal{B}^{i,1}$ for agent i is a collection of pairs $\langle\phi_k^{j,0}, \alpha_k\rangle$, for $k = 1, 2, \ldots, m$, where $\phi_k^{j,0}$ is a sentence of $\mathscr{L}^{j,0}$ and $\alpha_k \in [0,1]$.*

For the sake of presenting the semantics of a L1-BB, we will assume that the quantified statements about j's probabilities (of the type of rule 4 in Definition 4) are expanded into propositional form over the elements of the domain. The basic idea behind the semantics of a L1-BB is that *the partitions over the state of the world correspond to the vertices of the simplex of j's belief about the state of the world.* In turn, agent i maintains a distribution over the partitions of such simplex. This mechanism in intuitively depicted in Fig. 3.

We now formalize this process. From the belief base, let us define the set $\Phi_{\mathcal{B}}^{j,0}$ of j's level 0 probabilistic statements about which i holds a degree of belief in $\mathcal{B}^{i,0}$ (Fig 3-a). Formally, we have:

$$\Phi_{\mathcal{B}}^{j,0} = \{\phi^{j,0} : \langle\phi^{j,0}, \alpha\rangle \in \mathcal{B}^{i,1} \text{ for some } \alpha \in [0,1]\} \tag{10}$$

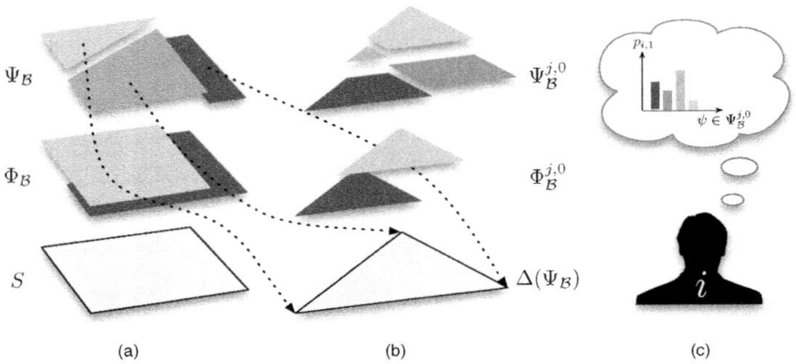

Fig. 3. Intuitive representation of the partition-based belief hierarchy and corresponding notation. (a) States of the world. (b) Agent j's 0-th level beliefs. (c) Agent i maintains a distribution over $\Psi^{j,0}_{\mathcal{B}}$.

In turn, we also define the set of FOL sentences $\Phi_{\mathcal{B}}$ that appear in some of j's level-0 probabilistic statements. This will allow us to describe the space of j's 0-th level beliefs that agent i considers (Fig 3(b)). Formally, we have:

$$\Phi_{\mathcal{B}} = \{\phi : \phi \circ \phi^{j,0} \text{ for some } \phi^{j,0} \in \Phi^{j,0}_{\mathcal{B}}\} \ , \tag{11}$$

where the circle symbol is read "occurs in" and is recursively defined as:

1. If $\phi^{j,0}$ is of the form $P_j(\phi) \vartriangle \alpha$, then $\phi \circ \phi^{j,0}$;
2. If $\phi \circ \phi^{j,0}_u$, then $\phi \circ (\neg\phi^{j,0}_u)$;
3. If $\phi \circ \phi^{j,0}_u$, then $\phi \circ (\phi^{j,0}_u \wedge \phi^{j,0}_v)$ and $\phi \circ (\phi^{j,0}_u \vee \phi^{j,0}_v)$, for any $\phi^{j,0}_v$.

We denote as $\bigcirc(\phi^{j,0})$ the set of of $\phi \in \Phi_{\mathcal{B}}$ such that $\phi \circ \phi^{j,0}$. The set $\Psi_{\mathcal{B}}$ of partitions of S induced by $\Phi_{\mathcal{B}}$ is defined as in (3), and the set $\Psi_{\mathcal{B}}(\phi)$ as in (4). Let us denote as $\Delta(\Psi_{\mathcal{B}})$ the regular simplex whose vertices are the elements of $\Psi_{\mathcal{B}}$. This simplex is the set of all possible probability distributions that j may hold about the sentences in $\Psi_{\mathcal{B}}$. The semantics of $\mathscr{L}^{j,0}$ is defined on such simplex, as described in the following definition.

Definition 6 (L1-BB Probabilistic entailment). *Given a L1-BB $\mathcal{B}^{i,1}$, the probabilistic entailment ($\models$) of a sentence $\phi^{j,0}$ (such that $\bigcirc(\phi_{j,0}) \in \Phi_{\mathcal{B}}$) by a point $p_{j,0} \in \Delta_{\mathcal{B}}(\Psi_{\mathcal{B}})$ is recursively defined as:*

1. $p_{j,0} \models (P_j(\phi) \vartriangle \beta)$ if and only if $\sum_{\psi\in\Psi_{\mathcal{B}}(\phi)} p_{j,0}(S(\psi)) \vartriangle \beta$;
2. $p_{j,0} \models \neg\phi^{j,0}$ if and only if $p_{j,0} \not\models \phi^{j,0}$;
3. $p_{j,0} \models (\phi^{j,0}_u \wedge \phi^{j,0}_v)$ if and only if $p_{j,0} \models \phi^{j,0}_u$ and $p_{j,0} \models \phi^{j,0}_v$;
4. $p_{j,0} \models (\phi^{j,0}_u \vee \phi^{j,0}_v)$ if and only if $p_{j,0} \models (\neg\phi^{j,0}_u \wedge \neg\phi^{j,0}_v)$;

If $p_{j,0} \models \phi^{j,0}$, $p_{j,0}$ is said to be a *model* of sentence $\phi^{j,0}$. It is easy to see that the models of a probabilistic sentence $\phi^{j,0}$ are the points in the continuous region of the simplex that is identified by $\phi^{j,0}$. We refer to the set of models of a sentence

$\phi^{j,0}$ as $[\Delta(\Psi_{\mathcal{B}})](\phi^{j,0})$, to remark that it corresponds to a subset of the simplex $\Delta(\Psi_{\mathcal{B}})$. To make the notation lighter, we will usually refer to the set of models as $\Delta_{\mathcal{B}}(\phi^{j,0})$. We add the extra symbol $\top^{j,0}$ to the language $\mathscr{L}^{j,0}$ as a shorthand for $P_j(\top) = 1$. Clearly, $\Delta_{\mathcal{B}}(\top^{j,0}) = \Delta(\Psi_{\mathcal{B}})$ (the whole simplex.)

Similarly to (3), we now define the logical probabilistic partitioning induced by $\Phi_{\mathcal{B}}^{j,0}$ as:

$$\Psi_{\mathcal{B}}^{j,0} = \left\{ \bigwedge_{\phi^{j,0} \in \Phi_I^{j,0}} \phi^{j,0} \setminus \bigvee_{\phi^{j,0} \in \Phi_I^{j,0,C}} \phi^{j,0} : \Phi_I^{j,0} \subseteq (\Phi_{\mathcal{B}}^{j,0} \cup \top^{j,0}) \right\}, \tag{12}$$

where $\Phi_I^{j,0,C} = (\Phi_{\mathcal{B}}^{j,0} \cup \top^{j,0}) \setminus \Phi_I^{j,0}$.

The sets $\Delta_{\mathcal{B}}(\psi^{j,0})$ for $\psi^{j,0} \in \Psi_{\mathcal{B}}^{j,0}$ are non-overlapping regions that together cover the simplex $\Delta_{\mathcal{B}}(\Psi_{\mathcal{B}})$ entirely, hence they form a partitioning of such space.

We denote as $\Psi_{\mathcal{B}}^{j,0}(\phi^{j,0})$ the set of elements of $\Psi_{\mathcal{B}}^{j,0}$ that correspond to subsets of $\Delta_{\mathcal{B}}(\phi^{j,0})$, where $\phi^{j,0} \in \Phi_{\mathcal{B}}^{j,0}$. Formally,

$$\Psi_{\mathcal{B}}^{j,0}(\phi^{j,0}) = \{\psi^{j,0} \in \mathcal{B}^{j,0} : \Delta_{\mathcal{B}}(\psi^{j,0}) \subseteq \Delta_{\mathcal{B}}(\phi^{j,0})\} \tag{13}$$

We now describe the max-ent distribution over the belief simplex induced by a L1-BB. Note that we are defining a distribution over the set of j's distribution about the state of the world, which is a continuous space. Nevertheless, instead of defining a probability directly over the simplex $\Delta(\Psi_{\mathcal{B}})$, we will use the space of partitions of this simplex, namely $\Delta_{\mathcal{B}}(\Psi_{\mathcal{B}}^{j,0})$. This allows the agent to represent his first level interactive belief as a discrete probability distribution, rather than a continuous one. Formally, we define the probability space

$$\left(\Delta_{\mathcal{B}}(\Psi_{\mathcal{B}}^{j,0}), 2^{\Delta_{\mathcal{B}}(\Psi_{\mathcal{B}}^{j,0})}, p_{i,1}\right) \ , \tag{14}$$

where $p_{i,1}$ is the result of the following optimization problem:

$$\max \left(- \sum_{\psi^{j,0} \in \Psi_{\mathcal{B}}^{j,0}} p_{i,1}(\Delta_{\mathcal{B}}(\psi^{j,0})) \log p_{i,1}(\Delta_{\mathcal{B}}(\psi^{j,0})) \right) \tag{15}$$

subject to:

$$\begin{array}{rll} \sum_{\psi^{j,0} \in \Psi_{\mathcal{B}}^{j,0}(\phi_k^{j,0})} p_{i,1}(\Delta_{\mathcal{B}}(\psi^{j,0})) = \alpha_k & & \forall\, \phi_k^{j,0} \in \Phi_{\mathcal{B}}^{j,0} \\ p_{i,1}(\Delta_{\mathcal{B}}(\psi^{j,0})) \geq 0 & & \forall\, \psi^{j,0} \in \Psi_{\mathcal{B}}^{j,0} \\ \sum_{\psi^{j,0} \in \Psi_{\mathcal{B}}^{j,0}} p_{i,1}(\Delta_{\mathcal{B}}(\psi^{j,0})) = 1 & & \end{array} \tag{16}$$

As before, we will sometimes write $p_{i,1}(\psi^{j,0})$ when this slight abuse of notation does not generate ambiguities.

Example (Grid world, cont'd). In the grid world example introduced before, suppose the moving target j is an agent on its own, maintaining a probability distribution over S and over i's belief about S. We assume j knows that i is concerned about his horizontal position w.r.t. the center. A possible L1-BB is:

$$B^{j,1} = \begin{array}{|c|}\hline \langle P_i(\phi_0) \geq 0.4, 0.4\rangle \\ \hline \langle P_i(\phi_1) > 0.5, 0.7\rangle \\ \hline \end{array} \quad (17)$$

where ϕ_0 and ϕ_1 are the same as in (9). We identify the set

$$\Phi_{\mathcal{B}}^{i,0} = \{P_i(\phi_0) \geq 0.5, P_i(\phi_1) < 0.4\} = \{\phi_0^{i,0}, \phi_1^{i,0}\} \quad (18)$$

The probabilistic sentences $\phi_0^{i,0}$ and $\phi_1^{i,0}$ induce three partitions on i's L0 belief simplex, as shown in Fig. 2-b. Again, there is only one consistent probability distribution over $\Phi_{\mathcal{B}}^{i,0}$, namely $p_{j,1} = (0.3, 0.1, 0.6)$.

3.3 Level *n* Beliefs

In this section, we follow the same steps as for the L1-BB and generalize the approach to any level of nesting. Intuitively, we need to represent agent i's degree of belief about agent j's beliefs about i's beliefs about... and so on, down to level 0 beliefs about the state of the world. In order to do so, we build a logical-probabilistic framework that allows the definition of a probability distribution over the other agent's $(n-1)$-th level beliefs in a recursive fashion. The result will be that, by partitioning the belief simplices at each level of the hierarchy, we can provide a finite representation of the interactive beliefs at any level of nesting. The intuition behind this process is that *the partitions of the belief simplex at any level $n-1$ corresponds to the vertices of the simplex at level n*. Since we need to specify the beliefs over some simplex at level $n-1$, we begin the description by formally defining the language of *Level $n-1$ First-Order Probabilistic Logic* (L($n-1$) FOPL) for agent j.

Definition 7 (L($n-1$) FOPL). *Given a set of predicate symbols Q and a domain D, the language $\mathscr{L}^{j,n-1}$ of j's level $n-1$ probabilistic statements is recursively defined as:*

1. *$P_j(\phi^{i,n-2}) \vartriangle \beta$ is a formula, where $\phi^{i,n-2}$ is a formula of language $\mathscr{L}^{i,n-2}$, $\beta \in [0,1]$, and $\vartriangle \in \{<, \leq, =, \geq, >\}$;*
2. *If $\phi^{j,n-1}$ is a formula, then $\neg\phi^{j,n-1}$ is a formula;*
3. *If $\phi_1^{j,n-1}$ and $\phi_2^{j,n-1}$ are formulae, then $\phi_1^{j,n-1} \wedge \phi_2^{j,n-1}$, $\phi_1^{j,n-1} \vee \phi_2^{j,n-1}$, $\phi_1^{j,n-1} \Rightarrow \phi_2^{j,n-1}$ are formulae;*
4. *If $\phi^{j,n-1}$ is a formula, then $\exists\mathbf{x}(\phi^{j,n-1})$ and $\forall\mathbf{x}(\phi^{j,n-1})$ are formulae, where $\mathbf{x}$ is a subset of the free logical variables of $\phi^{j,n-1}$;*
5. *A sentence is a formula with no free variables.*

For convenience, we use the symbol $\top^{j,n-1}$ as a shorthand for $P_j(\top^{i,n-2}) = 1$. We represent i's beliefs about some sentences of the language $\mathscr{L}^{j,n-1}$ as a Level n Belief Base.

Definition 8 (Level n Belief Base). *A* Level n Belief Base (Ln-BB) $\mathcal{B}_i^n$ *for agent i is a collection of pairs $\langle \phi_k^{j,n-1}, \alpha_k \rangle$, for $k = 1, 2, \ldots, m$, where $\phi^{j,n-1}$ is a sentence of $\mathscr{L}^{j,n-1}$ and $\alpha_k \in [0, 1]$.*

In order to be able to define the probabilistic semantics of a Ln-BB, we propositionalize each quantified probabilistic statement down the nesting hierarchy.

A Ln-BB represents agent i's degree of belief about a number of probabilistic statements regarding j's beliefs. These are the level $n-1$ probabilistic statements that appear in the tuples of the Ln-BB. Formally, we define the set of such statements as:

$$\Phi_{\mathcal{B}}^{j,n-1} = \{\phi^{j,n-1} : \langle \phi^{j,n-1}, \alpha \rangle \in \mathcal{B}^{i,n} \text{ for some } \alpha \in [0, 1]\} \tag{19}$$

Note that this is analogous to the set $\Phi_{\mathcal{B}}^{j,0}$ defined in (10) for a L1-BB. As in the previous subsection, we now need to provide a notion of probabilistic entailment for the language $\mathscr{L}^{j,n-1}$. We consider the regular simplex whose vertices are i's $(n-2)$-th level probabilistic statements that appear in the belief base. Formally, we define the set $\Phi_{\mathcal{B}}^{i,n-2}$ of sentences of $\mathscr{L}^{i,n-2}$ that occur in some element of $\Phi_{\mathcal{B}}^{j,n-1}$. Mathematically,

$$\Phi_{\mathcal{B}}^{i,n-2} = \{\phi^{i,n-2} : \phi^{i,n-2} \circ \phi^{j,n-1} \text{ for some } \phi^{j,n-1}\} \ , \tag{20}$$

where again we use the $\circ$ operator introduced in the previous section, generalized to $\mathscr{L}^{j,n-1}$ (we omit the full definition for conciseness.) We now consider the set of logical partitions induced by $\Phi_{\mathcal{B}}^{i,n-2}$:

$$\Psi_{\mathcal{B}}^{i,n-2} = \left\{ \bigwedge_{\phi^{i,n-2} \in \Phi_I^{i,n-2}} \phi^{i,n-2} \setminus \bigvee_{\phi^{i,n-2} \in \Phi_I^{i,n-2,C}} \phi_h^{i,n-2} \right. \\ \left. : \Phi_I^{i,n-2} \subseteq (\Phi_{\mathcal{B}}^{i,n-2} \cup \top^{i,n-2}) \right\} \ , \tag{21}$$

where $\Phi_I^{i,n-2,C} = (\Phi_{\mathcal{B}}^{i,n-2} \cup \top^{i,n-2}) \setminus \Phi_I^{i,n-2}$.

The regular simplex that has the elements of $\Psi_{\mathcal{B}}^{i,n-2}$ as vertices, denoted $\Delta(\Psi_{\mathcal{B}}^{i,n-2})$, is the space of j's $n-1$ level probability distributions. Hence, it represents the set over which the Ln-BB of agent i induces a max-ent distribution (remember that we are defining probability distributions over probability distributions over...) Instead of considering a distribution over this continuous space, we consider a distribution over partitions of such space. To do so, we first need the notion of probabilistic entailment for this case, that is a straightforward generalization of the level 1 entailment defined in the previous section, and is not reported here for brevity.

A distribution $p_{j,n-1}$ that entails a sentence $\phi^{j,n-1}$ is said to be a *model* of $\phi^{j,n-1}$. The set of models of $\phi^{j,n-1}$ is denoted as $[\Delta(\Psi_{\mathcal{B}}^{i,n-2})](\phi^{j,n-1})$, and is usually abbreviated as $\Delta_{\mathcal{B}}(\phi^{j,n-1})$.

Each element of $\Phi_{\mathcal{B}}^{j,n-1}$ corresponds therefore to a region of the simplex $\Delta(\Psi_{\mathcal{B}}^{i,n-2})$. In order to obtain the max-ent probability distribution encoded in the Ln-BB we need to define the probability space given by the partitions induced by the sentences $\phi^{j,n-1} \in \Phi_{\mathcal{B}}^{j,n-1}$. The set of logical partitions induced by $\Phi_{\mathcal{B}}^{j,n-1}$ is defined as in (21), by substituting $(n-2)$ with $(n-1)$, and i with j. We do not report the complete definition for brevity.

As before, we also define the set of logical partitions $\Psi_{\mathcal{B}}^{j,n-1}(\phi^{j,n-1})$ whose union is the set $\Delta_{\mathcal{B}}(\phi^{j,n-1})$. At this point, we are ready to introduce the max-ent probability distribution over j's belief partitions given by the Ln-BB. To this sake, we define the probability space

$$\left(\Delta_{\mathcal{B}}(\Psi_{\mathcal{B}}^{j,n-1}), 2^{\Delta_{\mathcal{B}}(\Psi_{\mathcal{B}}^{j,n-1})}, p_{i,n}\right) , \tag{22}$$

where $p_{i,n}$ is the solution to the following optimization problem:

$$\max_{p_{i,n}} \left(- \sum_{\psi^{j,n-1} \in \Psi_{\mathcal{B}}^{j,n-1}} p_{i,n}(\psi^{j,n-1}) \log p_{i,n}(\psi^{j,n-1}) \right) \tag{23}$$

subject to the constraints:

$$\begin{aligned} \sum_{\substack{\psi^{j,n-1} \in \\ \Psi_{\mathcal{B}}^{j,n-1}(\phi_k^{j,n-1})}} p_{i,n}(\psi^{j,n-1}) &= \alpha_k \qquad \forall\, \phi_k^{j,n-1} \in \Phi_{\mathcal{B}}^{j,n-1} \\ \sum_{\psi^{j,n-1} \in \Psi_{\mathcal{B}}} p_{i,n}(\psi^{j,n-1}) &= 1 \\ p_{i,n}(\psi^{j,n-1}) &\geq 0 \qquad \forall\, \psi^{j,n-1} \in \Psi_{\mathcal{B}}^{j,n-1} \end{aligned} \tag{24}$$

Above, we use the abbreviated notation $p_{i,n}(\psi^{j,n-1})$ instead of $p_{i,n}(\Delta_{\mathcal{B}}(\psi^{j,n-1}))$.

Example (Grid world, cont'd). We borrow the structure of the grid world example seen for the L1-BB. Assume now that agent i models the target j as a rational agent, who is in turn maintaining a belief over i's beliefs. The L2-BB for agent i we consider is:

$$B^{i,2} = \boxed{\begin{array}{l} \langle P_j(P_i(\phi_0) \geq 0.4) < 0.4, 0.2\rangle \\ \langle P_j(P_i(\phi_1) > 0.5) < 0.7, 0.6\rangle \end{array}} \tag{25}$$

We consider the set:

$$\Phi_{\mathcal{B}}^{j,1} = \{P_j(\phi_0^{i,0}) < 0.4, P_j(\phi_1^{i,0}) < 0.7\} = \{\phi_0^{j,1}, \phi_1^{j,1}\} , \tag{26}$$

where $\phi_0^{i,0}$ and $\phi_1^{i,0}$ are defined in (18). This set induces four non-empty partitions on j's level 1 belief simplex, as shown in Fig. 2-c. We now compute the max-ent distribution $p_{i,2}$ over the four partitions:

$$\max_{p_{i,2}} \left(- \sum_{k=0}^{3} p_{i,2}(\psi_k^{j,1}) \log p_{i,2}(\psi_k^{j,1}) \right), \quad \text{s.t.} \tag{27}$$

$$\begin{bmatrix} 0 & 0 & 1 & 1 \\ 1 & 0 & 0 & 1 \\ 1 & 1 & 1 & 1 \end{bmatrix} \begin{bmatrix} p_{i,2}(\psi_0^{j,1}) \\ p_{i,2}(\psi_1^{j,1}) \\ p_{i,2}(\psi_2^{j,1}) \\ p_{i,2}(\psi_3^{j,1}) \end{bmatrix} = \begin{bmatrix} 0.2 \\ 0.6 \\ 1 \end{bmatrix} \tag{28}$$

The resulting probability distribution is: $p_{i,2} = (0.48, 0.32, 0.08, 0.12)$.

4 Conclusion and Future Work

In this paper, we have contribute a novel theoretical framework to compactly represent a hierarchy of interactive beliefs exploiting first-order logic and probability theory. The main idea is to partition the belief simplex at each level of the hierarchy, and let the simplex at level $n-1$ constitute the vertices of the simplex at level n. We have shown that, by recursively partitioning the belief simplices, the representation of the interactive space is finite, thus overcoming the unboundedness of the space of distributions that is typical of standard, enumeration-based representations.

There are several directions for future research. First, we will develop a feasible implementation of our proposed theoretical system and will evaluate the computational costs, both of exact and approximate inference techniques. In particular, we intend to study the ties between our approach and existing first-order probabilistic systems, such as Relational Probabilistic Models and Markov Logic Networks, and possibly extend them towards the interactive beliefs semantics presented in this paper.

Second, we want to embed this novel representation of interactive beliefs in decision making algorithms. One possible application is to extend the work of Sanner and Kersting [23] on First-Order POMDPs to interactive settings. In particular, we believe that our partition-based interactive belief system is suitable to be embedded in decision making frameworks such as Interactive POMDPs. In fact, the optimal value function for (I-)POMDPs divides the belief simplex in partitions corresponding to the optimal policy for each such region. Hence, we want to explore the use of interactive first-order belief bases to recursively represent the relevant belief partitions of the other agents.

References

1. Aumann, R.J.: Interactive epistemology II: Probability. International Journal of Game Theory 28(3), 301–314 (1999)
2. Bacchus, F.: Representing and reasoning with probabilistic knowledge: a logical approach to probabilities. MIT Press, Cambridge (1990)
3. Boutilier, C., Poole, D.: Computing optimal policies for partially observable decision processes using compact representations. In: Proceedings of the National Conference on Artificial Intelligence, pp.1168–1175 (1996)
4. Boutilier, C., Reiter, R., Price, B.: Symbolic Dynamic Programming for First-Order MDPs. In: IJCAI 2001, pp. 690–700. Morgan Kaufmann, San Francisco (2001)

5. Doshi, P., Gmytrasiewicz, P.J.: Monte Carlo sampling methods for approximating interactive POMDPs. J. Artif. Int. Res. 34, 297–337 (2009)
6. Doshi, P., Perez, D.: Generalized point based value iteration for interactive POMDPs. In: Proceedings of the 23rd National Conference on Artificial Intelligence, vol. 1, pp. 63–68. AAAI Press, Menlo Park (2008)
7. Fagin, R., Halpern, J.Y.: Reasoning about knowledge and probability. J. ACM 41, 340–367 (1994)
8. Gmytrasiewicz, P.J., Doshi, P.: A framework for sequential planning in multi-agent settings. Journal of Artificial Intelligence Research 24, 24–49 (2005)
9. Halpern, J.Y.: An analysis of first-order logics of probability. In: Proceedings of the 11th International Joint Conference on Artificial Intelligence, vol. 2, pp. 1375–1381. Morgan Kaufmann Publishers Inc., San Francisco (1989)
10. Halpern, J.Y.: Reasoning about Uncertainty. MIT Press, Cambridge (2003)
11. Halpern, J.Y., Moses, Y.: Knowledge and common knowledge in a distributed environment. J. ACM 37, 549–587 (1990)
12. Harsanyi, J.C.: Games with Incomplete Information Played by "Bayesian" Players, I-III. Part I. The Basic Model. Management Science 14(3), 159–182 (1967)
13. Jaynes, E.T.: Information Theory and Statistical Mechanics. Physical Review Online Archive (Prola) 106(4), 620–630 (1957)
14. Kern-Isberner, G., Thimm, M.: Novel semantical approaches to relational probabilistic conditionals. In: Proceedings of the 12th International Conference on the Principles of Knowledge Representation and Reasoning (2010)
15. Koller, D., Pfeffer, A.: Probabilistic frame-based systems. In: Proceedings of the 15th National Conference on Artificial Intelligence (AAAI), pp. 580–587 (1998)
16. Lukasiewicz, T., Kern-Isberner, G.: Probabilistic logic programming under maximum entropy. In: Hunter, A., Parsons, S. (eds.) ECSQARU 1999. LNCS (LNAI), vol. 1638, pp. 279–292. Springer, Heidelberg (1999)
17. Milch, B., Marthi, B., Russell, S.J., Sontag, D., Ong, D.L., Kolobov, A.: Blog: Probabilistic models with unknown objects. In: Probabilistic, Logical and Relational Learning (2005)
18. Nilsson, N.J.: Probabilistic logic. Artif. Intell. 28, 71–88 (1986)
19. Poupart, P.: Exploiting structure to efficiently solve large scale Partially Observable Markov Decision Processes. Ph.D. thesis, University of Toronto, Toronto, Ont., Canada (2005), aAINR02727
20. Richardson, M., Domingos, P.: Markov logic networks. Mach. Learn. 62, 107–136 (2006)
21. Russell, S., Norvig, P.: Artificial Intelligence: A Modern Approach, 3rd edn. Pearson Education, London (2010)
22. Sanner, S., Boutilier, C.: Practical solution techniques for first-order MDPs. Artif. Intell. 173, 748–788 (2009)
23. Sanner, S., Kersting, K.: Symbolic dynamic programming for first-order POMDPs. In: AAAI 2010, pp. 1140–1146 (2010)
24. Shirazi, A., Amir, E.: Factored models for probabilistic modal logic. In: Proceedings of the 23rd National Conference on Artificial Intelligence, vol. 1, pp. 541–547. AAAI Press, Menlo Park (2008)
25. Wang, C., Joshi, S., Khardon, R.: First order decision diagrams for relational MDPs. J. Artif. Int. Res. 31, 431–472 (2008)
26. Wang, C., Khardon, R.: Relational partially observable MDPs. In: AAAI 2010, pp. 1153–1157 (2010)

Patterns Discovery for Efficient Structured Probabilistic Inference

Lionel Torti, Christophe Gonzales, and Pierre-Henri Wuillemin

Université Pierre et Marie Curie,
Laboratoire d'Informatique de Paris 6,
75252 Paris Cedex 05, France
firstname.lastname@lip6.fr

Abstract. In many domains where experts are the main source of knowledge, e.g., in reliability and risk management, a framework well suited for modeling, maintenance and exploitation of complex probabilistic systems is essential. In these domains, models usually define closed-world systems and result from the aggregation of multiple patterns repeated many times. Object Oriented-based Frameworks (OOF) such as Probabilistic Relational Models thus offer an effective way to represent such systems. OOFs define patterns as classes and substitute large Bayesian networks (BN) by graphs of instances of these classes. In this framework, Structured Inference avoids many computation redundancies by exploiting class knowledge, hence reducing BN inference times by orders of magnitude. However, to keep modeling and maintenance costs low, OOF classes often encode only generic situations. More complex situations, even those repeated many times, are only represented by combinations of instances. In this paper, we propose to determine such combination patterns and exploit them as classes to speed-up Structured Inference. We prove that determining an optimal set of patterns is NP-hard. We also provide an efficient algorithm to approximate this set and show numerical experiments that highlight its practical efficiency.

1 Introduction

Bayesian networks (BN) [19] are a valued framework for reasoning under uncertainty and their popularity stimulated the need for handling problems of ever increasing size. However BNs turn out to be inadequate for large scale real-world applications due to high design and maintenance costs [18,20]. Indeed, defining a BN requires to specify explicitly probabilistic dependencies and conditional probabilities over the whole set of its random variables. This may lead to unrealistic modeling costs when dealing with complex systems. Furthermore, BN's design is static: any change in the topology of their graphical structure induces significant update costs.

Solving these problems has been the main concern of several BN extensions using the object-oriented paradigm [15,18]. Besides, first-order logic extensions were proposed to offer more expressive power than the propositional framework

S. Benferhat and J. Grant (Eds.): SUM 2011, LNAI 6929, pp. 247–260, 2011.
© Springer-Verlag Berlin Heidelberg 2011

offered by BNs [13,14]. Learning being a critical problem when exploiting BNs over large knowledge bases, entity-relationship extensions were also proposed for relational learning [8,10]. These extensions are all allegedly considered as First-Order Probabilistic Models (FOPM) or as Knowledge Based Construction Models.

During the last decade, the Probabilistic Graphical Model (PGM) community has worked actively on FOPMs and object-oriented models have been somewhat neglected: since the introduction of Object-Oriented Bayesian Networks [3,15], the amount of contributions on object-oriented PGMs has actually been relatively small [1,2,8]. However, in many industrial areas, efficient frameworks for the construction of large-scale complex systems are strongly needed and, in domains like risk management or monitoring of complex industrial processes, this usually boils down to experts modeling large-scale BNs by aggregating hierarchically small network fragments repeated many times. In addition, all the relations between these fragments are usually fully specified, thus resulting in modeling "closed worlds". For these domains, object-oriented frameworks seem more suitable than first-order logic extensions. In particular, the "closed world" assumption strongly degrades the behavior of lifted inference in FOPM.

Object-oriented frameworks assume that many parts of a large BN are similar and can thus be described as instances of a generic class defined only once. This scheme induces low construction costs. In addition, maintenance costs are kept as low as possible since a modification in a class definition updates many areas of the BN at once. Furthermore, repetitions of structures in the BN (multiple instances of the same class) can speed-up inference by performing computations within classes, caching them and using the cache for all their instances. This process allows algorithms like *Structured Variable Elimination* (SVE) to outperform classical BN inference engines by orders of magnitude [21].

In this paper, we propose an enhancement of structured inference for Probabilistic Relational Models [7,25]. In real world applications, instances are often combined and form patterns repeated many times throughout the network. By using a frequent subgraph pattern mining algorithm, it is possible to discover such combinations and exploit them to speed-up structured inference. However, mining optimally such patterns is time expensive. In this paper, we both provide a structured inference algorithm for PRMs exploiting patterns and a mining heuristic fast enough for efficient inference.

The paper is organized as follows: Section 2 recalls the basics of object-oriented frameworks using PRMs. Section 3 generalizes structured inference. Section 4 shows the complexity of mining patterns and provides an approximate algorithm for such mining. Experiments reported in Section 5 show the practical efficiency of our approach. Finally, concluding remarks are given in Section 6.

2 Description of PRMs

Using PRMs as an object-oriented framework can be surprising as they were first proposed for relational learning [5]. However, it is important to remember

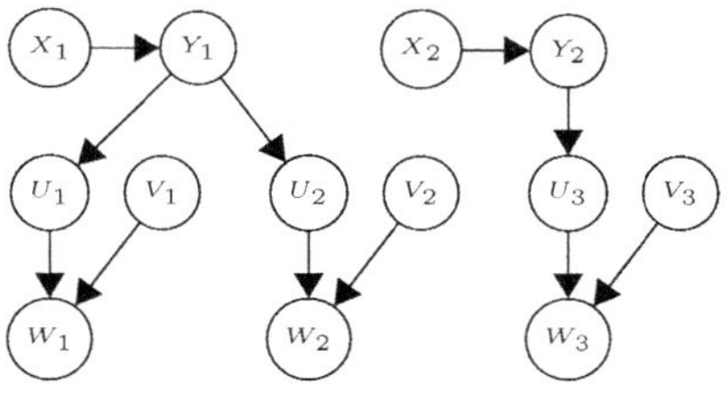

(a) A Bayesian network. The gray areas do not belong to the BN specification

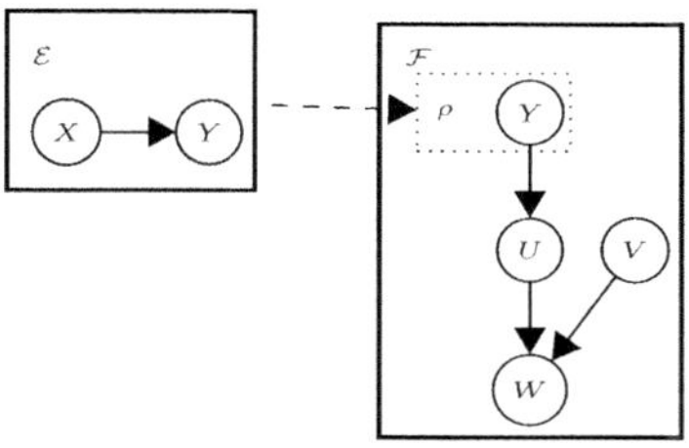

(b) Two connected classes $\mathcal{E}$ and $\mathcal{F}$

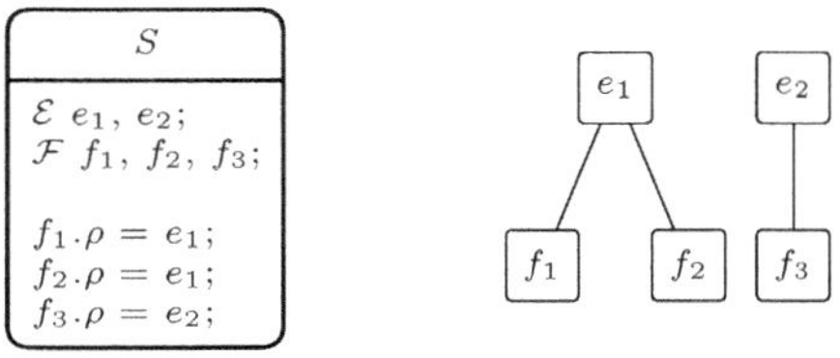

(c) System declaration and relational skeleton for the BN of Figure 1(a)

Fig. 1. Representation of a BN as a PRM: analysis of the BN reveals the use of two recurrent patterns (a), which are confined in two classes (b). Hence, a system equivalent to the BN may be built (c).

that PRMs are an extension of Object Oriented Bayesian Networks [21,25] and, thus, they offer a sound object-oriented framework.

Due to a lack of space, we only present briefly and incompletely the PRM framework [7,25]. Fig. 1.(a) shows a BN encoding relations between two different kinds of patterns (variables $\{X_i, Y_i\}$ and $\{U_j, V_j, W_j\}$). Variables whose names begin with the same letter share identical conditional probability tables (CPT). Object-oriented representations aim to abstract each pattern as a generic entity (a *class*) that encapsulates all the relations between the variables of the pattern. So, in Fig. 1.(b), class $\mathcal{E}$ encapsulates precisely variables X_i and Y_i as well as their probabilistic relations (arc (X_i, Y_i)) and their CPTs. The pattern of variables U_j, V_j, W_j cannot be directly encapsulated in a class since the CPTs of variables U_j are conditional to some variables Y_k (e.g., the CPT of U_3 is $P(U_3|Y_2)$ according to Fig. 1.(a)). Hence classes must have a mechanism allowing to reference variables outside themselves. In PRMs, this mechanism is called a *reference slot*. A reference slot ρ of a class $\mathcal{C}$ is a local name for another class $\mathcal{D}$ allowing $\mathcal{C}$ to access its variables. As shown in Fig. 1.(c), the original BN can then be built up from the PRM: it is sufficient to create two instances, say e_1 and e_2, of class $\mathcal{E}$ as well as three instances f_1, f_2, f_3 of $\mathcal{F}$ and connect them using one edge per reference slot.

Definition 1 (Class, Attribute, Reference slot). *In order to define classes, one needs the following cross-definitions:*

- *A* class *is a quadruple* $\langle \mathbf{A}(\mathcal{C}), \mathbf{R}(\mathcal{C}), G(\mathcal{C}), \mathbf{P}(\mathcal{C})\rangle$ *where:*
- $\mathbf{A}(\mathcal{C})$ *is a set of attributes. An attribute* $\mathcal{C}.X \in \mathbf{A}(\mathcal{C})$ *is a random variable.* $\mathcal{C}$ *is called the resident class of* X.
- $\mathbf{R}(\mathcal{C})$ *is a set of reference slots.* $\mathcal{C}.\rho \in \mathbf{R}(\mathcal{C})$ *is a surrogate for another class, say* $\mathcal{D}$*, giving access to all its attributes and reference slots from within* $\mathcal{C}$. $Range(\rho)$ *denotes class* $\mathcal{D}$*. The set* $\overline{\mathbf{A}(\mathcal{C})}$ *of all the attributes reachable by way of reference slots or within* $\mathcal{C}$ *is called the closure of* $\mathcal{C}$.
- $G(\mathcal{C}) = (\overline{\mathbf{A}(\mathcal{C})}, E)$ *is a Directed Acyclic Graph (DAG) where* $E \subseteq \overline{\mathbf{A}(\mathcal{C})} \times \mathbf{A}(\mathcal{C})$ *: only the attributes of* $\mathcal{C}$ *have parents in this DAG.*
- $\mathbf{P}(\mathcal{C}) = \{P(X|\Pi_X), X \in \mathbf{A}(\mathcal{C})\}$ *is the set of CPTs of attributes* $X \in \mathbf{A}(\mathcal{C})$ *conditionally to their parents in* $G(\mathcal{C})$.

Classes are not meant to be used as is, but through instances. For example, a class may represent various failure odds of a cooling system in a nuclear power plant and, when modeling a given power plant, such class is instantiated for each occurrence of the cooling system in the whole plant.

Definition 2 (Instance, System). *Let* $\mathcal{B}$ *be a BN,*

- *An* instance c *of a class* $\mathcal{C}$ *is a subset of the random variables of* $\mathcal{B}$ *whose relations are described by* $\mathcal{C}$. $c.X$ *(resp.* $c.\rho$*) refers to the instantiation of* $\mathcal{C}.X \in \mathbf{A}(\mathcal{C})$ *(resp.* $\mathcal{C}.\rho \in \mathbf{R}(\mathcal{C})$*) in* c*. By abuse of notation, we denote the sets of such instantiations as* $\mathbf{A}(c)$ *and* $\mathbf{R}(c)$ *respectively.*
- *A system* S *is the representation of* $\mathcal{B}$ *in the PRM framework: it is a finite set of instances such that* $\forall i \in S, \forall \rho \in \mathbf{R}(i), \exists j \in S$ *such that* $Range(i.\rho) = j$ *and such that there is a one-to-one mapping between random variables of* $\mathcal{B}$ *and the set of all attributes declared in the instances of* S.

The graph representing instances by nodes and connections between range and resident instances by edges is called the relational skeleton of S.

Definition 2 enforces the "closed world" feature of systems, i.e., they are finite sets of instances with all reference slots properly defined. As mentioned in the introduction, this constraint is reasonable for complex systems of many domains. For instance, to reason on industrial milk fermenters, pipe connections need to be fully specified.

3 Structured Inference

Determining the probabilities of random variables given evidence is the most common query performed in probabilistic graphical models. There exists a wide range of inference algorithms to compute these distributions. They often rely in some way to a Variable Elimination scheme [4,17]. The basic idea consists of marginalizing out random variables one by one from the joint distribution until there only remains the variables of interest. Dechter's Variable Elimination (VE) is representative of this class of algorithms. It first fills a pool of functions called *potentials* with the CPTs representing the decomposition of the joint distribution. Then, eliminating some variable X_j from the joint probability just amounts to extract from the pool all the potentials involving X_j, multiply them and sum-up the result over all the values of X_j, and insert back the resulting potential into the pool. Conditional probabilities $P(\mathbf{X}|\mathbf{e})$ are computed similarly by first adding to the pool some potentials representing the additional knowledge brought by evidence $\mathbf{e}$.

The above scheme is efficient and can be used in PRMs by applying it on their grounded BN. However, by processing random variables separately, VE is unable to exploit the structural repetitions in the graphical model to avoid computation redundancies. The aim of Structured Inference is to fill this gap [5,21] and Object-Oriented frameworks provide a simple and effective way to achieve this goal. Indeed, consider an attribute A of a class $\mathcal{C}$ such that all of its children also belong to $\mathcal{C}$ and let $c_1, \ldots, c_k$ be some instances of $\mathcal{C}$ in which no attribute received any evidence. Then it is easy to see that eliminating attributes $A \in \mathbf{A}(c_i)$ in the grounded BN produces precisely the same computations for all the instances c_i, $i = 1 \ldots, k$. In this case, eliminating attribute A within class $\mathcal{C}$, i.e., *at class level*, and updating accordingly all the relevant instances before constructing the grounded BN avoids the redundancies involved by eliminating A in each c_i, i.e., *at instance level*. This process is called *Structured Inference* and the gain brought by this approach usually reduces computation times by orders of magnitude.

More Formally, an attribute $A \in \mathbf{A}(\mathcal{C})$ is called an *inner* or *internal* attribute if all of its children also belong to $\mathbf{A}(\mathcal{C})$, otherwise A is called an *outer* attribute. In addition, the attributes referenced in $\mathbf{R}(\mathcal{C})$ are called *non-resident*. For instance, in Fig. 1.b, attributes X, U, V, W are internal, Y is an outer attribute of class $\mathcal{E}$ and $\rho.Y$ is a non-resident attribute of $\mathcal{F}$. Class-level elimination corresponds to the elimination of all the inner attributes (using any inference algorithm). As such, it amounts to substitute the pool of potentials $\mathbf{P}(\mathcal{C})$ of class $\mathcal{C}$ defined over all of its inner, outer and non-resident attributes by a new set of potentials $\mathbf{P}'(\mathcal{C})$ defined only over the outer and non-resident attributes. The pool of potentials corresponding to any instance c of $\mathcal{C}$ is thus substituted by $\mathbf{P}'(c)$ if no inner attribute in c received any evidence, else it is kept to $\mathbf{P}(c) \cup \{\text{potentials(evidence)}\}$ (because evidence may induce different distributions from one instance to another).

4 PRM's Patterns Discovery

4.1 Problem and Complexity

Marginalizing-out internal nodes at *class level* is the key to Structured Inference efficiency as it reduces significantly redundant computations. However, not all redundancies can be identified by this scheme: let $\mathcal{C}, \mathcal{D}$ be two classes and let $X \in \mathbf{A}(\mathcal{C})$, $Y \in \mathbf{A}(\mathcal{D})$ be two attributes such that the only non-resident child of $\mathcal{C}.X$ is $\mathcal{D}.Y$. Then X cannot be eliminated at class level because it is not internal. However, if we consider a "new" class $\mathcal{F}$ defined by compound $(\mathcal{C}, \mathcal{D})$, attribute $\mathcal{F}.X$ is no longer an outer attribute since $Y \in \mathbf{A}(\mathcal{F})$. Hence, pairs of instances (c, d) of $\mathcal{C}$ and $\mathcal{D}$ that fit the definition of $\mathcal{F}$ can be considered as instances of $\mathcal{F}$ in which X is internal, thus eligible for class-level elimination. Note however that not all pairs (c, d) are necessarily eligible: in Fig. 1, pairs (e_1, f_1) and (e_1, f_2) cannot be both considered as instances of compound $(\mathcal{E}, \mathcal{F})$ as e_1 would be counted twice in the grounded BN. Pair (e_1, f_3) is neither eligible because there is no edge between e_1 and f_3 in the system. The key idea is that finding effective compounds/instance-class reassignments should speed-up Structured Inference since it increases the opportunities for class-level eliminations. It is most convenient to search them in the following graph:

Definition 3 (Boundary graph). *A boundary graph is an undirected graph* $\mathcal{BG} = (\mathcal{I}, \mathcal{E})$, *where*

- $\mathcal{I}$ *is a set of vertices representing instances;*
- $\mathcal{E} \subseteq \mathcal{I} \times \mathcal{I}$ *is a set of edges such that* $\exists (c, d) \in \mathcal{E}$ *iff*
  $$\mathbf{L}_{cd} = \bigcup_{X \in \mathbf{A}(C)} (\Pi_{c.X} \cap \mathbf{A}(d)) \;\cup\; \bigcup_{X \in \mathbf{A}(d)} (\Pi_{d.X} \cap \mathbf{A}(c)) \neq \emptyset.$$
  Edge (c, d) *is labeled by* $\mathbf{L}_{cd}$.

An edge (c, d) of the boundary graph and its label define precisely the attributes that should be eliminated at class level if (c, d) was considered as an instance of a compound. So two pairs of instances (c_1, d_1) and (c_2, d_2) of classes $\mathcal{C}$ and $\mathcal{D}$ should not be considered as instances of the same compound if $\mathbf{L}_{c_1 d_1} \neq \mathbf{L}_{c_2 d_2}$. Fig. 2 illustrates two boundary graphs for which different compound classes will be mined. In this case, we can see that Y is an outer attribute for compound $\{c_1, d_1\}$ while being an inner attribute for compound $\{c_2, d_2\}$. This suggests the following definition:

Definition 4 (Dynamic class). *Let* $\mathcal{BG}$ *be a boundary graph. A dynamic class* $\widehat{\mathcal{F}}$ *in* $\mathcal{BG}$ *is a pair* $(\mathcal{F}, \mathbf{B})$, *where:* $\mathcal{F}$ *is a compound class;* $\mathbf{B} \subseteq \mathbf{A}(\mathcal{F})$ *is the set of all the outer attributes of* $\mathcal{F}$. *Set* $\mathbf{B}$ *is called* $\widehat{\mathcal{F}}$*'s boundary.*

Hence, given a dynamic class $\widehat{\mathcal{F}}$, all the nodes in $\mathbf{A}(\widehat{\mathcal{F}}) \setminus \mathbf{B}$ are internal and can be eliminated at class level whereas nodes in $\mathbf{B}$ are referenced by other instances and can only be eliminated at instance level. So, to improve structured inference, we shall search the boundary graph for frequent subgraphs, i.e., subgraphs repeated many times, create their corresponding dynamic class, substitute each subgraph by one instance of its dynamic class and, finally, apply an inference algorithm

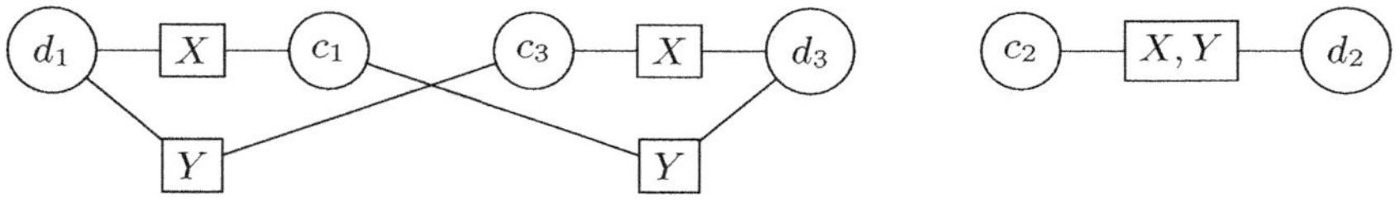

Fig. 2. Different possible connections between instances, resulting in different labels (square nodes). We can see that compound $\{c_1, d_1\}$ is different from compound $\{c_2, d_2\}$.

like SVE. However substitutions must be performed carefully: it may actually happen that the occurrences of frequent subgraphs share some nodes. In this case, only one of these occurrences can be substituted else some instances of the "original" system would be counted several times (see (e_1, f_1) and (e_1, f_2) in Fig. 1). Hence the following rule:

Rule 1. *In the boundary graph, substituted subgraphs cannot share any node (i.e. any instance of the original PRM).*

Optimizing structured inference thus amounts to searching for the "best" set of dynamic classes and subgraph substitutions satisfying Rule 1. Unfortunately, as shown in the following proposition, this problem is NP-hard[1]:

Proposition 1. *The following problem is NP-hard:*
Instance: *a PRM, a boundary graph, an integer* $K \geq 0$.
Question: *is there a set of dynamic classes and boundary subgraph substitutions of these classes such that the number of operations (multiplications and summations) performed by structured inference is smaller than* K*?*

In a sense, this proposition is not very surprising since determining the minimal number of operations in variable elimination algorithms such as SVE or VE is equivalent to determining an optimal elimination sequence, which is known to be NP-hard [22]. In addition, determining all the occurrences of a given subgraph in a graph is NP-hard as well [6]. Finally, given a set of dynamic classes and their subgraph occurrences in the boundary graph, determining which ones should be substituted amounts to solve an *Independent Set* problem in which each vertex represents a boundary subgraph and edges link vertices corresponding to overlapping boundary subgraphs. Again, this problem is NP-hard [6]. However, the proof of Proposition 1 shows that finding the best dynamic classes/substitutions remains NP-hard even in cases where inference in the grounded BN is polynomial (singly-connected BNs). We shall however present in the next subsection an efficient approximate algorithm for determining an effective set of dynamic classes.

4.2 An Approximate Algorithm

The problem of finding frequent patterns in labeled graphs has received many contributions in the literature, although the aim is somewhat different in that it consists of finding subgraphs that appear in many graphs of a database of labeled graphs [12,16,26]. However, the connection with our problem is sufficiently high that techniques from this domain can be borrowed to solve our problem. In this paper, we suggest to use a variant of gSpan [26].

The idea consists of creating a search tree $\mathbb{T}$ as follows: each node $N(\widehat{\mathcal{D}})$ of the tree represents a pair $(\widehat{\mathcal{D}}, \mathbb{O}(\widehat{\mathcal{D}}))$ where $\widehat{\mathcal{D}}$ is a dynamic class and $\mathbb{O}(\widehat{\mathcal{D}})$ is the set of its instances in the boundary graph $\mathcal{BG}$. In other words, $\mathbb{O}(\widehat{\mathcal{D}})$ is the set of subgraphs of the $\mathcal{BG}$ that fit $\widehat{\mathcal{D}}$. Tree $\mathbb{T}$ is initialized with all the dynamic

[1] Proof can be found at http://agrum.lip6.fr/doku.php?id=sum2011

classes corresponding to 1-edge subgraphs of $\mathcal{BG}$. In $\mathbb{T}$, nodes at level $k+1$ are derived from those at level k by extending their associated subgraph in $\mathcal{BG}$ with one of their adjacent node in $\mathcal{BG}$. As a consequence, each node of $\mathbb{T}$ represents a dynamic class whose boundary subgraph is connected and whose set of instances is nonempty. The whole tree thus reveals precisely all the possible substitutions that can be applied to the PRM. More precisely, $\mathbf{V} = \cup_{N(\widehat{\mathcal{D}})\in\mathbb{T}}\mathbb{O}(\widehat{\mathcal{D}})$ represents the set of substitutions. There just remains to select among $\mathbf{V}$ the "*best*" substitutions possible. To do so, we must enforce Rule 1. It is easily done by observing that each node of $\mathcal{BG}$ can only belong to one dynamic class and, more precisely, to one instance of this dynamic class. Hence, if we create a graph $G = (V, E)$ in which each node of V represents a given element of $\mathbf{V}$, i.e., a subgraph of $\mathcal{BG}$, and each edge $(v_1, v_2) \in E$ represents the fact v_1 and v_2 have a nonempty intersection in $\mathcal{BG}$, then any subset $W \subseteq V$ such that no pair of nodes of W are adjacent in G corresponds to a set of substitutions satisfying Rule 1. In other words, there is a one-to-one mapping between the *Independent Sets* of G and the sets of substitutions satisfying Rule 1. Of course, some substitutions are better than others because they induce higher speed-ups in Structured Inference (see the $\beta_{\widehat{\mathcal{D}}}/\gamma_{\widehat{\mathcal{D}}}$ ratio below). So by weighting nodes of V according to the speed-up improvements they induce, the "*best*" substitutions we look for correspond to solutions of a *Max Weighted Independent Set* problem [9].

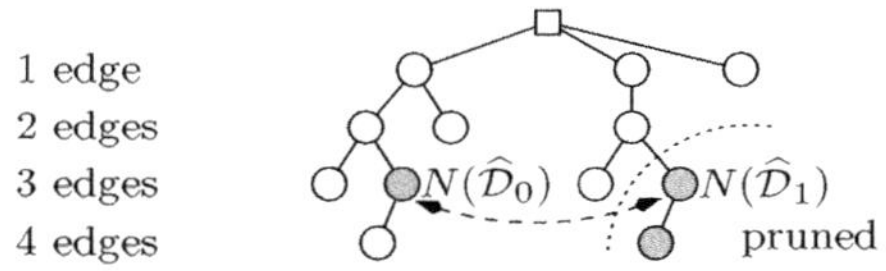

Fig. 3. Dynamic class search tree $\mathbb{T}$

Of course, the size of $\mathbb{T}$ is exponential and, thus, some pruning is necessary. Pruning rules will be described in the next subsection. But, to guaranty their efficiency, we shall construct $\mathbb{T}$ in such a way that the "best" dynamic classes are constructed first. For this purpose, gSpan defines a linear order that ensures that the more promising the node the smaller its index in the order and suggests to sort all the nodes of each level of $\mathbb{T}$ according to this order [26]. Thus, parsing $\mathbb{T}$ in a depth-first search (DFS) manner guarantees that the "most promising" dynamic classes are constructed first. This leads to the following algorithm:

4.3 Pruning Rules

For the first pruning rule, note that the descendants of a node in $\mathbb{T}$ define the possible extensions of its corresponding dynamic class. Hence, if another node of the search tree corresponds to the same dynamic class (say, e.g., that $\widehat{\mathcal{D}}_1$ is the same class as $\widehat{\mathcal{D}}_0$), then both nodes and their descendants represent identical dynamic classes. So, $N(\widehat{\mathcal{D}}_1)$ and its descendants can be safely pruned from the search. Determining whether two nodes represent the same dynamic class is

Input: A PRM and its boundary graph $\boldsymbol{\mathcal{BG}}$
Output: A set of dynamic classes/substitutions
$\mathbb{T} \leftarrow$ all dynamic classes of 1-edge subgraphs of $\boldsymbol{\mathcal{BG}}$
sort the nodes in $\mathbb{T}$ according to the gSpan linear order
parse $\mathbb{T}$ in a DFS manner
foreach *node* $N(\widehat{\mathcal{D}})$ *visited* **do**
    create the children of $N(\widehat{\mathcal{D}})$, sort them w.r.t. gSpan's linear order and add them to $\mathbb{T}$
    prune the "unpromising" children
end
solve a Max Weighted Independent Set
return *the set of "best" dynamic classes/substitutions*

Algorithm 1. Computing dynamic classes/substitutions

simply achieved through gSpan's canonical labeling of subgraphs (see [26] for a detailed description).

The second rule is related to the gain achievable in Structured Inference using dynamic classes: nodes $N(\widehat{\mathcal{D}})$ that define classes whose subgraph substitutions do not speed-up Structured Inference can be pruned. To estimate the gain in speed, recall that, by Rule 1, only a subset of the subgraphs of $\mathbb{O}(\widehat{\mathcal{D}})$ can be substituted in $\boldsymbol{\mathcal{BG}}$ by instances of $\widehat{\mathcal{D}}$. Let $s_{\widehat{\mathcal{D}}}$ denote the cardinal of this subset. The number of operations (multiplications, additions) performed by Structured Inference on these substitutions is equal to $w_{\widehat{\mathcal{D}}} + s_{\widehat{\mathcal{D}}} \times \overline{w}_{\widehat{\mathcal{D}}}$, where $w_{\widehat{\mathcal{D}}}$ and $\overline{w}_{\widehat{\mathcal{D}}}$ denote the number of operations necessary to eliminate $\widehat{\mathcal{D}}$'s inner nodes at class level and $\widehat{\mathcal{D}}$'s outer nodes at instance level respectively. Now remember that, in tree $\mathbb{T}$, $\widehat{\mathcal{D}}$ corresponds to a 1-edge extension of its parent $\pi(\widehat{\mathcal{D}})$. So, the subgraphs of $\mathbb{O}(\widehat{\mathcal{D}})$ that were not substituted are 1-edge extensions of subgraphs of $\mathbb{O}(\pi(\widehat{\mathcal{D}}))$. Assuming that they were all substituted as instances of $\pi(\widehat{\mathcal{D}})$, their eliminations by Structured Inference would have cost $w_{\pi(\widehat{\mathcal{D}})} + (|\mathbb{O}(\widehat{\mathcal{D}})| - s_{\widehat{\mathcal{D}}}) \times \overline{\overline{w}}_{\widehat{\mathcal{D}}}$ where $\overline{\overline{w}}_{\widehat{\mathcal{D}}} = \overline{w}_{\pi(\widehat{\mathcal{D}})} + k_{\widehat{\mathcal{D}}}$ and $k_{\widehat{\mathcal{D}}}$ corresponds to the elimination of the edge added to $\pi(\widehat{\mathcal{D}})$. So the total cost incurred by the exploitation of $N(\widehat{\mathcal{D}})$ is $\beta_{\widehat{\mathcal{D}}} = w_{\widehat{\mathcal{D}}} + w_{\pi(\widehat{\mathcal{D}})} + s_{\widehat{\mathcal{D}}} \times \overline{w}_{\widehat{\mathcal{D}}} + (|\mathbb{O}(\widehat{\mathcal{D}})| - s_{\widehat{\mathcal{D}}}) \times \overline{\overline{w}}_{\widehat{\mathcal{D}}}$ whereas, by just exploiting $\pi(\widehat{\mathcal{D}})$, it would have been $\gamma_{\widehat{\mathcal{D}}} = w_{\pi(\widehat{\mathcal{D}})} + |\mathbb{O}(\widehat{\mathcal{D}})| \times \overline{\overline{w}}_{\widehat{\mathcal{D}}}$. So, class $\widehat{\mathcal{D}}$ is unattractive for inference and $N(\widehat{\mathcal{D}})$ may be pruned whenever $\alpha_{\widehat{\mathcal{D}}} = \beta_{\widehat{\mathcal{D}}} - \gamma_{\widehat{\mathcal{D}}} = w_{\widehat{\mathcal{D}}} + s_{\widehat{\mathcal{D}}} \times (\overline{w}_{\widehat{\mathcal{D}}} - \overline{\overline{w}}_{\widehat{\mathcal{D}}}) > 0$. Finally, note that $s_{\widehat{\mathcal{D}}}, w_{\widehat{\mathcal{D}}}, \overline{\overline{w}}_{\widehat{\mathcal{D}}}, k_{\widehat{\mathcal{D}}}$ can be estimated quickly: as shown in the preceding subsection, $s_{\widehat{\mathcal{D}}}$ can be estimated by solving a *Max Independent Set* problem induced by $\mathbb{O}(\widehat{\mathcal{D}})$. To estimate $w_{\widehat{\mathcal{D}}}$, it is sufficient to compute a junction tree of $\widehat{\mathcal{D}}$'s DAG [23], eliminating only inner nodes, and to sum-up the sizes of its cliques. Eliminating the remaining variables provides an estimation of $\overline{w}_{\widehat{\mathcal{D}}}$. $k_{\widehat{\mathcal{D}}}$ can be estimated similarly.

Note however that $\mathbb{T}$ is not α-decreasing, i.e., it may happen that $\alpha_{\widehat{\mathcal{D}}} > 0$ for a given node $N(\widehat{\mathcal{D}})$, but not for some of its descendants. This property results from the fact that, in these descendants the number of inner nodes may be far higher than that in $\widehat{\mathcal{D}}$, hence decreasing $w_{\widehat{\mathcal{D}}}$ (dropping constraints on the junction tree's elimination order) as well as $\overline{w}_{\widehat{\mathcal{D}}}$ (the inner nodes do not belong to the boundary). The α-non-decreasing property does not allow for a clear pruning rule. In the paper, we used the following rule: whenever a node in $\mathbb{T}$ had an $\alpha_{\hat{\mathcal{D}}} > 0$, we pruned the node and its descendants.

5 Experimental Results

We now describe different set of experiments that highlight the gain in inference speed resulting from the combination of structured inference and pattern mining. In each experiment, we compared our new algorithm (subsequently denoted as PD for Pattern Discovery) with Structured Variable Elimination (SVE), the standard inference algorithm for structured inference [24], and also with Variable Elimination (VE), a classic and standard probabilistic inference algorithm for Bayesian Networks [4]. Response times reported for PD take into account both pattern mining and inference. For experiments using VE, results include both grounding and inference time. It is important to note that our experiments included no evidence. This choice was motivated by the fact that the structure of the network varies drastically given evidence. Our goal here was to show how pattern mining can improve inference when there exist repetitions in the network. Moreover, evidence is not a good indicator of repetitions as it can either be identically applied in each pattern, thus preserving repetition, or applied randomly, thus breaking the structure. Experiences 1 and 2 show the results of our new approach on networks with and without repetitions, hence providing a good insight of PD's performance. All our experiments were performed on an Intel Xeon at 2.7 Ghz. The source code of our PRMs implementation, the inference algorithm and the generation algorithms can be found in the aGrUM project[2].

The key to understand these experimentations lies in the generation of the benchmarked PRMs. High level frameworks such as PRMs offer a wide variety of generation methods. Here, our primary concern was the generation of PRMs in which we could control the amount of structure repetition in order to prove that, when confronted to a large amount of pattern repetitions, i) a substantial speed gain can be achieved and ii) our approach does not suffer from a prohibitive pattern mining cost. Our generator takes the following parameters as inputs: *domain* is the domain size of each attribute; min_{attr} is the number of attributes common to all classes; max_{attr} is the number of attributes in each class; c is the minimal number of classes; max_{ref} is the maximal number of reference slots allowed per class; n is the number of instances in the system.

The PRM's generation process is performed as follows: first, we generate an interface[3] with min_{attr} attributes which will be implemented by all classes and will be the slot type of each reference slot in each class. Next, for all $k \in [0, \ldots, \max_{ref}]$, a class with precisely k reference slots is created. Then, if $max_{ref} < c$, we generate new classes until exactly c classes have been created.

[2] http://agrum.lip6.fr

[3] If a class implements a given interface, then it guarantees the existence of the attributes and reference slots defined in that interface [25].

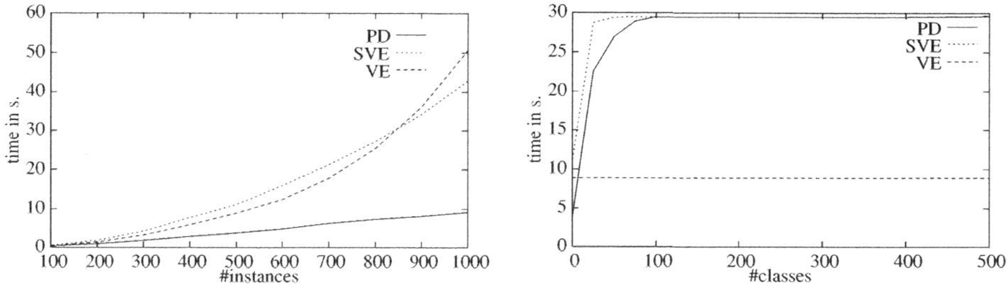

(a) First set of experiments (b) Second set of experiments

Fig. 4. Structural repetition is an important factor for PD's performance. Unsurprisingly, performance decrease dramatically for systems with no structural repetition.

For those new classes, the number of reference slots is chosen randomly between 0 and max_{ref}. Finally, we generate a DAG S representing the relational skeleton of our generated system: each node represents an instance and an arc $i \rightarrow j$ represents the fact that there exists $\rho \in \mathbf{R}(j)$ such that $i = j.\rho$. For a given node i with π_i parents in S, we instantiate a class randomly chosen among all the classes with precisely π_i reference slots. A given class C is generated as follows: we first create a DAG G_C with max_{attr} nodes, we then add to C k reference slots and max_{attr} attributes. Dependencies between attributes are defined using G_C. For each reference slot ρ, we create a slot chain $\rho.A$, where $A \in \mathbf{A}(\mathcal{I})$ is chosen randomly among all the attributes in $\mathbf{A}(\mathcal{I})$. The slot chain is then added as a parent of an attribute of C chosen randomly. DAGs are generated using the algorithm provided in [11].

In our first set of experiments, we generated systems with an increasing number of instances. Each class contains 15 attributes ($max_{attr} = 15$), each attribute's domain size is equal to 4 ($domain = 4$) and each class has at most 4 incoming arcs ($max_{ref} = 4$). Finally, the minimal amount of classes required was set to $c = 5$, which implies that there are precisely $max_{ref} + 1 = 5$ classes in each system. These experiments highlight the behavior of PD when many repetitions can be found in the system. Fig. 4(a) shows the response times of PD, SVE and VE when no evidence is observed and with a number of instances varying from 100 to 1000. Clearly, in this case, PD significantly outperforms both VE and SVE.

An important factor is the ratio of PD's inference time over that of SVE. The gain of PD against SVE and that of SVE against VE are due to the presence of structural repetition in the generated networks. It can be seen that SVE's complexity is less impacted by the size of the system than VE's complexity. But for small systems with small classes, SVE does not guarantee a considerable speed gain. By exploiting pattern mining, PD significantly increases the gain obtained by repetition. Thus, where SVE does not perform well compared to VE, PD infers larger patterns that can drastically increase performance. In our first experiments, there is enough structure to see the possible gain provided by

Table 1. Third experiment: patterns mining efficiency for PD. Values are averages. Inst. stands for instances, attr. for attributes and pat. for pattern.

#inst.	#pat.	pat. repetition	max pat. repetition	#inst. per pat.	max inst. per pat.	% of attr. in a pat.
200	11.88	2.92	6.26	2.15	4.08	37.29%
400	24.68	3.40	10.46	2.25	4.71	47.20%
600	36.35	3.91	15.92	2.36	5.25	55.90%
800	46.51	4.50	20.25	2.45	5.62	64.09%
1000	54.19	5.25	30.07	2.62	6.12	75.54%

our new approach. Yet, we must also consider cases where there are few or even no structural repetitions. The amount of pattern repetitions can be influenced by the number of classes, so if we increase that number we should observe a less favorable ratio between PD's and SVE's inference time against VE. This is the purpose of our second set of experiments.

In our second set of experiments we generated systems with an increasing number of classes ($c \in [0, 500]$) and 500 instances. The remaining parameters are equal to those of the first experiment. The goal here is twofold: we want to show that, when no structure is exploitable, there is no overhead in proceeding with the pattern mining and that pattern repetitions is critical for PD's performance. Fig. 4(b) shows that when the number of classes increases dramatically, the speed gain induced by PD and SVE are considerably less significant. If we compare those results with those obtained by VE, we see that PD and SVE are considerably counter-performing. To anyone familiar with structural inference, this is an unsurprising result and these results can be explained by the fact that the elimination order used by PD and SVE (inner attributes before outer attributes) is in most cases suboptimal. If PD and SVE show better results than VE in Fig. 4(a) it is only because the gain resulting from the reduction of redundant computations compensate the suboptimal elimination order. Fortunately, detecting repetition is trivial in an object-oriented framework as the amount of instantiations of each class is a good indicator of structural repetition. The presence of evidence is also a good indicator, as different evidence will break down the structure and thus reduce the amount of repetition in the network. We can easily switch to classic inference if needed by detecting situations which would lead to counter-performing results: few instantiations of each classes, heavy evidence, seemingly random evidence. Finally, we observe no over-cost due to pattern mining. This is also an unsurprising result as our pruning rules take into account frequencies and cut the mining process when such value is too low (here the minimal frequency allowed was set to 2).

In our third experiment, we analyze the amount of patterns found by PD with the parameters from experiment 1 ($max_{attr} = 15, domain = 4, max_{ref} = 4, c = 5, max_{ref} + 1 = 5$). The results of this experiment are summarized in Tab. 1. A noticeable point is the low number of instances in each pattern. This is a consequence of our pruning rule which was designed to be strict. It favors smaller patterns because larger ones are in most cases less cost effective (they often induce a larger clique than an optimal elimination order would) and because they are less frequent. In general, discovered patterns consisted of few small patterns largely repeated and many different patterns less repeated. The latter were used to fill-in the gaps in the structure once the main patterns were applied. If we consider the last column of Tab. 1 we can see that the larger a system, the more the attributes covered. The fact that the coverage increases with the system size explains why the inference time of PD increases linearly with the system size: the large number of usable patterns compensates the complexity induced by the number of instances.

To conclude our experiments, we applied PD to a classic BN: the Pigs network. This network is remarkable in that it only contains two distinct CPTs, which are represented in our framework by two classes. The network in itself is too small to point out any significant gain in inference time, however it is still interesting to analyze the patterns found by PD. Our approach mined 14 different patterns. On average, they are repeated 11 times and the maximal amount of repetitions equals 45. Only patterns with 2 instances are found. Discovered patterns cover up to 69% of the 441 attributes present in the Pigs network. As for our previous results, our pruning rules favor smaller patterns since larger ones tend to be less cost effective and less frequent. While the size of the Pigs network does not enable to point out the efficiency of our approach in terms of inference time, the existence of such structures and the results we obtained over random networks can help conclude to the efficiency of our approach. We can also point out its usefulness w.r.t. modeling: by pointing out frequent patterns in a system we can infer new classes which can then be used by experts for modeling purposes.

6 Conclusion

In this paper, we showed that mining patterns can significantly alleviate inference costs. Although finding the optimal set of patterns is NP-hard, we provided an efficient approximate mining algorithm. Our experimental results confirm that this approach can lead to a significant improvement of inference tasks in PRM. But there is still room for improving inference in PRMs. For instance, our approach, especially its pruning, can still be improved. In addition, many refinements of the PRM framework like class inheritance, structural uncertainty or multiple references, should be used to speed-up inference.

References

1. Bangsø, O.: Object Oriented Bayesian Networks. Ph.D. thesis. Aalborg University (March 2004)
2. Bangsø, O., Sønderberg-Madsen, N., Jensen, F.: A bn framework for the construction of virtual agents with human-like behaviour. In: Proc. of PGM 2006 (2006)
3. Bangsø, O., Wuillemin, P.H.: Top-down construction and repetitive structures representation in Bayesian networks. In: Proc. of FLAIRS 2000, pp. 282–286 (2000)
4. Dechter, R.: Bucket elimination: A unifying framework for reasoning. Artificial Intelligence 113, 41–85 (1999)

5. Friedman, N., Getoor, L., Koller, D., Pfeffer, A.: Learning probabilistic relational models. In: Proc. of IJCAI 1999 (1999)
6. Garey, M., Johnson, D.: Computers and Intractability: A Guide to the Theory of NP-Completeness. W.H. Freeman, New York (1979)
7. Getoor, L., Friedman, N., Koller, D., Pfeffer, A., Taskar, B.: Probabilistic relational models. In: Getoor, L., Taskar, B. (eds.) An Introduction to Statistical Relational Learning. ch. 5. MIT Press, Cambridge (2007)
8. Getoor, L., Koller, D., Taskar, B., Friedman, N.: Learning probabilistic relational models with structural uncertainty. In: ICML 2000 Workshop on Attribute-Value and Relational Learning: Crossing the Boundaries (2000)
9. Halldórsson, M.: Approximations of weighted independent set and hereditary subset problems. Journal of Graph Algorithms and Applications (2000)
10. Heckerman, D., Meek, C., Koller, D.: Probabilistic models for relational data. Tech. rep., WA: Microsoft Corporation, Redmond (2004)
11. Ide, J.S., Cozman, F.G., Ramos, F.T.: Generating random Bayesian networks with constraints on induced width. In: Proc. of ECAI 2004, pp. 323–327 (2004)
12. Inokuchi, A., Washio, T., Motoda, H.: A general framework for mining frequent subgraphs from labeled graphs. Fundamenta Informaticae (2005)
13. Jaeger, M.: Relational Bayesian networks. In: Proc. of UAI 1997 (1997)
14. Kersting, K., Raedt, L.D.: Bayesian logic programs. Technical report no. 151, Institute for Computer Science, University of Freiburg, Germany (April 2001)
15. Koller, D., Pfeffer, A.: Object-oriented Bayesian networks. In: Proc. of AAAI 1997, pp. 302–313 (1997)
16. Kuramochi, M., Karypis, G.: Frequent subgraph discovery. In: Proc. of ICDM 2001 (2001)
17. Madsen, A., Jensen, F.: LAZY propagation: A junction tree inference algorithm based on lazy inference. Artificial Intelligence 113(1–2), 203–245 (1999)
18. Mahoney, S., Laskey, K.: Network engineering for complex belief networks. In: Proc. of UAI 1996 (1996)
19. Pearl, J.: Probabilistic Reasoning in Intelligent Systems: Networks of Plausible Inference. Morgan Kaufmann, San Francisco (1988)
20. Pfeffer, A., Koller, D., Milch, B., Takusagawa, K.: SPOOK: A system for probabilistic object-oriented knowledge representation. In: Proc. of UAI 1999 (1999)
21. Pfeffer, A.: Probabilistic Reasoning for Complex Systems. Ph.D. thesis. Stanford University (2000)
22. Rose, D., Lueker, G., Tarjan, R.: Algorithmic aspects of vertex elimination on graphs. SIAM J. on Computing 5, 266–283 (1976)
23. Rose, D.: Triangulated graphs and the elimination process. J. Math. Analysis and Applications (1970)
24. Torti, L., Wuillemin, P.H.: Structured value elimination with d-separation analysis. In: Proc. of FLAIRS 2010, pp. 122–127 (2010)
25. Torti, L., Wuillemin, P.H., Gonzales, C.: Reinforcing the object-oriented aspect of probabilistic relational models. In: Proc. of PGM 2010 (2010)
26. Yan, X., Han, J.: gSpan: Graph-based substructure pattern mining. In: Proc. of ICDM 2002 (2002)

Indirect Elicitation of NIN-AND Trees in Causal Model Acquisition

Yang Xiang, Minh Truong, Jingyu Zhu, David Stanley, and Blair Nonnecke

University of Guelph, Canada
yxiang@socs.uoguelph.ca

Abstract. To specify a Bayes net, a conditional probability table, often of an effect conditioned on its n causes, needs to be assessed for each node. Its complexity is generally exponential in n and hence how to scale up is important to knowledge engineering. The non-impeding noisy-AND (NIN-AND) tree causal model reduces the complexity to linear while explicitly expressing both reinforcing and undermining interactions among causes. The key challenge to acquisition of such a model from an expert is the elicitation of the NIN-AND tree topology. In this work, we propose and empirically evaluate two methods that indirectly acquire the tree topology through a small subset of elicited multi-causal probabilities. We demonstrate the effectiveness of the methods in both human-based experiments and simulation-based studies.

1 Introduction

To specify a Bayes net (BN), a conditional probability table (CPT), needs to be assessed for each non-root node. A BN is often constructed in the causal direction, where a CPT is about an effect conditioned on its n causes. In general, specifying a CPT has the complexity exponential in n. Noisy-OR [Pearl(1988)] and a number of extensions, e.g., [Heckerman and Breese(1996), Galan and Diez(2000), Lemmer and Gossink(2004)] reduce the complexity to linear, but are limited to the reinforcing causal interaction.

The NIN-AND tree [Xiang and Jia(2007)] causal model, as well as its special case [Maaskant and Druzdzel(2008)], extends noisy-OR and explicitly encodes reinforcing and undermining causal interactions, as well as their mixture. Its specification consists of a linear (in n) number of probability parameters and a linear sized tree topology. Its default independence assumptions may be flexibly relaxed to trade efficiency for expressiveness. That is, by relaxing the assumptions incrementally and specifying more parameters, any CPT can be encoded.

The key challenge to specifying a NIN-AND tree causal model is the acquisition of the tree topology, which encodes types of causal interactions among causes. Elicitation of the tree topology requires nontrivial training of a domain expert on the syntax and semantics of NIN-AND tree causal models, and demands nontrivial mental exercise by the expert to articulate the partial order of causal interactions among causes. Usability of NIN-AND tree causal modeling will be enhanced if such training and mental exercise can be avoided during model acquisition.

We accomplish this by proposing two model acquisition methods that bypass direct elicitation of the NIN-AND tree topology. Instead, a small subset of causal probabilities

S. Benferhat and J. Grant (Eds.): SUM 2011, LNAI 6929, pp. 261–274, 2011.
© Springer-Verlag Berlin Heidelberg 2011

in the order of $O(n^2)$ or $O(n^3)$ are elicited, from which a NIN-AND tree topology is generated. From these probabilities and the tree topology, a NIN-AND tree causal model is defined and the corresponding CPT can be constructed. We show that the acquired CPT is a good approximation of the underlying true CPT.

The remainder of the paper is organized as follows: Background on NIN-AND tree causal models is covered in Sect. 2. The task of NIN-AND tree acquisition and the assumption underlying this work are presented in Sect. 3. In Sect. 4 and 5, we propose two novel techniques for the task. Setup of human-based experiments for evaluation is described in Sect. 6 and results are presented in Sect. 7. They are followed in Sect. 8 by simulation-based studies. Sect. 9 draws the conclusion.

2 NIN-AND Tree Causal Models

An uncertain cause is a cause that can produce an effect but does not always do so. We denote a binary effect variable by $e \in \{e^+, e^-\}$, where e^+ denotes $e = true$, and a set of binary cause variables of e by $X = \{c_1, ..., c_n\}$, where $c_i \in \{c_i^+, c_i^-\}$ $(i = 1, ..., n)$.

A single-causal success is an event where c_i caused e to occur successfully when all other causes are absent. We denote the event by $e^+ \leftarrow c_i^+$ and its probability by $P(e^+ \leftarrow c_i^+)$. For instance, smoking causing lung cancer is denoted by $lc^+ \leftarrow smk^+$. A single-causal failure, where e is false when c_i is true and all other causes of e are false, is denoted by $e^+ \not\leftarrow c_i^+$. A multi-causal success is an event where a set $X = \{c_1, ..., c_n\}$ $(n > 1)$ of causes caused e, and is denoted by $e^+ \leftarrow c_1^+, ..., c_n^+$ or $e^+ \leftarrow \underline{x}^+$. Denote the set of all causes of e by C.

CPT $P(e|C)$ relates to probabilities of causal events as follows: If $C = \{c_1, c_2, c_3\}$, then $P(e^+|c_1^+, c_2^-, c_3^+) = P(e^+ \leftarrow c_1^+, c_3^+)$. C is assumed to include a leaky variable (if any) to capture causes not represented explicitly, and hence $P(e^+|c_1^-, c_2^-, c_3^-) = 0$.

Causes reinforce each other if collectively they are at least as effective as when some are active. For example, radiotherapy and chemotherapy are reinforcing causes for curing cancer. If collectively causes are less effective, they undermine each other. Living with mother and living with wife are undermining causes for the happiness of a man, as often observed. If $C = \{c_1, c_2\}$, and c_1 and c_2 undermine each other, the following hold: $P(e^+|c_1^-, c_2^-) = 0$, $P(e^+|c_1^+, c_2^-) > 0$, $P(e^+|c_1^-, c_2^+) > 0$,

$$P(e^+|c_1^+, c_2^+) < min(P(e^+|c_1^+, c_2^-), P(e^+|c_1^-, c_2^+)).$$

The following Def.1 defines the two types of causal interactions generally.

Definition 1. *Let $R = \{W_1, W_2, ...\}$ be a partition of a set X of causes, $R' \subset R$ be any proper subset of R, and $Y = \cup_{W_i \in R'} W_i$. Sets of causes in R* **reinforce** *each other, iff*

$$\forall R' \; P(e^+ \leftarrow \underline{y}^+) \leq P(e^+ \leftarrow \underline{x}^+).$$

Sets of causes in R **undermine** *each other, iff* $\forall R' \; P(e^+ \leftarrow \underline{y}^+) > P(e^+ \leftarrow \underline{x}^+)$.

Reinforcement and undermining occur between individual causes as well as sets of them. When the interaction is between individual causes, each W_i is a singleton. Otherwise, each W_i can be a generic set. For instance, consider $X = \{c_1, c_2, c_3, c_4\}$, $W_1 =$

$\{c_1, c_2\}$, $W_2 = \{c_3, c_4\}$, $R = \{W_1, W_2\}$, where c_1 and c_2 reinforce each other, and so do c_3 and c_4. But sets W_1 and W_2 can undermine each other.

Disjoint sets of causes $W_1, ..., W_m$ satisfy failure conjunction iff

$$(e^+ \not\leftarrow \underline{w}_1^+, ..., \underline{w}_m^+) = (e^+ \not\leftarrow \underline{w}_1^+) \wedge ... \wedge (e^+ \not\leftarrow \underline{w}_m^+).$$

That is, when causes collectively fail to produce the effect, each must have failed to do so. They also satisfy failure independence iff

$$P((e^+ \not\leftarrow \underline{w}_1^+) \wedge ... \wedge (e^+ \not\leftarrow \underline{w}_m^+)) = P(e^+ \not\leftarrow \underline{w}_1^+) \, ... \, P(e^+ \not\leftarrow \underline{w}_m^+). \quad (1)$$

Disjoint sets of causes $W_1, ..., W_m$ satisfy success conjunction iff

$$(e^+ \leftarrow \underline{w}_1^+, ..., \underline{w}_m^+) = (e^+ \leftarrow \underline{w}_1^+) \wedge ... \wedge (e^+ \leftarrow \underline{w}_m^+).$$

That is, collective success requires individual effectiveness. They also satisfy success independence iff

$$P((e^+ \leftarrow \underline{w}_1^+) \wedge ... \wedge (e^+ \leftarrow \underline{w}_m^+)) = P(e^+ \leftarrow \underline{w}_1^+) \, ... \, P(e^+ \leftarrow \underline{w}_m^+). \quad (2)$$

It has been shown [Xiang and Jia(2007)] that causes are undermining when they satisfy success conjunction and independence. Hence, undermining can be modeled by a direct NIN-AND gate (Fig. 1, left). Its root nodes (top) are single-causal successes, and its leaf node (bottom) is the multi-causal success in question. Success conjunction is expressed by AND gate, and success independence by disconnection of root nodes other than through the gate. The probability of the leaf event can be computed by Eqn. (2). Similarly, causes are reinforcing when they satisfy failure conjunction and independence. Hence, reinforcement can be modeled by a dual NIN-AND gate (Fig. 1, middle). The leaf event probability is obtained by Eqn. (1).

By organizing multiple direct and dual NIN-AND gates in a tree, both reinforcement and undermining, as well as their mixture at multiple levels can be expressed in a NIN-AND tree model. A simple example is given below and more can be found in [Xiang and Jia(2007)]. Consider $C = \{c_1, c_2, c_3\}$, where c_1 and c_3 undermine each other, but collectively they reinforce c_2. Assuming event conjunction and independence, their causal interaction (a two-level mixture of reinforcement and undermining) relative to the event $e^+ \leftarrow c_1^+, c_2^+, c_3^+$ can be expressed by the NIN-AND tree in Fig. 1 (right). The top gate is direct and the bottom gate (the leaf gate) is dual. The link downward from node $e^+ \leftarrow c_1^+, c_3^+$ has a white oval end (a negation link) and

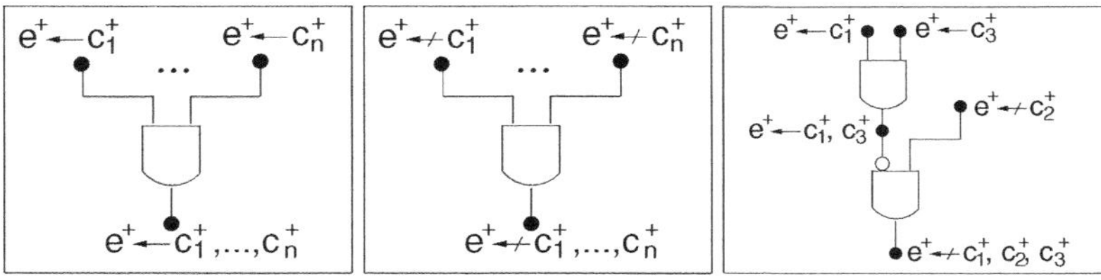

Fig. 1. Direct (left), dual (middle) NIN-AND gates, and a NIN-AND tree (right)

negates the event. All other links are forward links. Probability of the leaf event can be computed by Eqn. (1) and (2). For instance, from single-causal probabilities for root events, $P(e^+ \leftarrow c_1^+) = 0.85$, $P(e^+ \leftarrow c_2^+) = 0.8$, $P(e^+ \leftarrow c_3^+) = 0.7$, probability $P(e^+ \not\leftarrow c_1^+, c_2^+, c_3^+)$ is derived:

$$\begin{aligned} P(e^+ \leftarrow c_1^+, c_3^+) &= P(e^+ \leftarrow c_1^+)P(e^+ \leftarrow c_3^+) = 0.595 \\ P(e^+ \not\leftarrow c_1^+, c_2^+, c_3^+) &= P(e^+ \not\leftarrow c_1^+, c_3^+)P(e^+ \not\leftarrow c_2^+) \\ &= (1 - P(e^+ \leftarrow c_1^+, c_3^+))(1 - P(e^+ \leftarrow c_2^+)) = 0.081 \end{aligned}$$

Furthermore, using a more sophisticated algorithm [Xiang(2010a)], the CPT in Table 1 can be obtained from the NIN-AND tree and these parameters.

Table 1. CPT of the example NIN-AND tree model

$P(e^+ \mid c_1^-, c_2^-, c_3^-)$	0	$P(e^+ \mid c_1^-; c_2^+, c_3^-)$	0.8	$P(e^+ \mid c_1^+, c_2^-, c_3^+)$	0.595	$P(e^+ \mid c_1^+, c_2^+, c_3^-)$	0.97
$P(e^+ \mid c_1^+, c_2^-, c_3^-)$	0.85	$P(e^+ \mid c_1^-; c_2^-, c_3^+)$	0.7	$P(e^+ \mid c_1^-, c_2^+, c_3^+)$	0.94	$P(e^+ \mid c_1^+, c_2^+, c_3^+)$	0.919

Variables in a NIN-AND tree model can generally be multi-valued [Xiang(2010b)]. Assumptions on event conjunction and independence can also be relaxed, in which case some root events will be multi-causal. In this work, we focus on binary effect and causes, and on models whose root events are single-causal.

3 Acquisition of NIN-AND Tree Models

As illustrated above, a NIN-AND tree model over e and C consists of its tree topology as well as a single-causal probability for each $c_i \in C$. In general, a NIN-AND tree causal model M is a tuple $M = (e, C, T, PS)$, where e is the effect, C is the set of all causes of e, T is a NIN-AND tree, and PS is the set of single causal probabilities one for each cause in C. From M, a CPT $P(e|C)$ can be uniquely constructed. M and $P(e|C)$ are said to be *consistent*.

Furthermore, NIN-AND tree causal models $M = (e, C, T, PS)$ and $M' = (e, C, T', PS')$ are said to be *structurally consistent* if T and T' are isomorphic. M and M' are said to be *consistent* if they are consistent with the same CPT.

To acquire M, its tree topology T may be elicited directly from the expert. To complete such a task, the expert must have a thorough understanding of the syntax and semantics of NIN-AND tree models, in order to assess and articulate the partial order of causal interactions among causes and cause groups. This demands an nontrivial amount of training of the domain expert before elicitation and nontrivial mental exercise of the expert during elicitation.

To ease these burdens for model acquisition, we investigate the idea to bypass direct tree elicitation. Instead, we elicit a small number of multi-causal probabilities (in addition to the single-causal probabilities PS), and generate T from elicited probabilities. Our work is based on the following assumption:

Assumption 1. *Let $P_t(e|C)$ be the (true) CPT that characterizes the probabilistic relation over an effect e and its causes C, such that the following hold:*

1. *There exists a NIN-AND tree causal model $M_t = (e,C,T,PS)$ that is consistent with $P_t(e|C)$.*
2. *A domain expert is able to approximately assess all single-causal probabilities and some multi-causal probabilities relative to $P_t(e|C)$.*

The first condition is justified by the observation that reinforcement and undermining capture intuitive patterns of causal interaction, and reinforcement based causal models, such as Noisy-OR, have been widely applied. The second condition is justified by knowledge engineering practice in building BNs. Note that the condition does not require the expert to assess all multi-causal probabilities, nor to assess them accurately.

In the following, we investigate two alternative techniques to generate tree topology based on structure elimination (SE) and pairwise causal interaction (PCI).

4 Generate NIN-AND Tree by Structure Elimination

The SE technique builds on minimal NIN-AND tree models [Xiang et al(2009a)] and their enumeration [Xiang et al(2009b)]. Models $M = (e,C,T,PS)$ and $M' = (e,C,T',PS)$ may be consistent even though they are not structurally consistent. By limiting T and T' within the space of minimal NIN-AND trees, model consistency implies structure consistency in general. This means that a unique minimal tree exists for each pattern of causal interactions among a set of causes.

Definition 2. *Let T be a NIN-AND tree. If T contains a gate t that outputs to a gate g of the same type (direct or dual), delete t and connect its inputs to g. Apply such deletion until no longer possible. The resultant NIN-AND tree is minimal.*

The uniqueness of minimal NIN-AND trees allows them to be enumerated explicitly, e.g., using the two-step enumeration algorithm in [Xiang et al(2009b)]. For binary effect and causes, if $|C| = 4$, there are 52 minimal NIN-AND trees. For $|C| = 5,6,7$, the number is 472, 5504, 78416, respectively.

We **propose** the SE technique as follows. Denote $n = |C|$. First, a set PS_e of n single-causal probabilities, e.g., $P_e(e^+|c_i^+)$, are elicited from the expert, where subscript e denotes ‘elicited’. Then the set TM of minimal NIN-AND trees over C are enumerated. Combining each $T \in TM$ with PS_e, a set NM_e of NIN-AND tree models is obtained. In general, a unique CPT over e and C can be constructed from each model in NM_e. A set CPT_e of CPTs is thus defined. Note that there is a one-to-one mapping between TM and NM_e, and generally also between NM_e and CPT_e.

Subsequently, the expert is asked to assess some multi-causal probabilities. Let $P_e(e^+|c_i^+,c_j^+,c_k^+)$ be elicited from an expert, and $P'(e^+|c_i^+,c_j^+,c_k^+)$ be from a CPT $P'(e|C) \in CPT_e$. If $P'(e^+|c_i^+,c_j^+,c_k^+)$ differs significantly from $P_e(e^+|c_i^+,c_j^+,c_k^+)$, $P'(e|C)$ is deemed to be *inconsistent* with the true CPT, and the NIN-AND tree model corresponding to $P'(e|C)$ is eliminated from the candidate set NM_e. Based on such comparison of CPTs in CPT_e and elicited multi-causal probabilities, all models in NM_e except one, $M_e = (e,C,T_e,PS_e)$, will be eliminated. M_e is returned as the indirectly elicited model and T_e is the indirectly elicited NIN-AND tree. Below, we investigate several variations for elicitation and elimination procedures:

[Threshold based sequential elimination] Since elicitation from an expert is sequential, it is natural to interleave model elimination with elicitation. Elicitation and elimination proceed in rounds. Each round starts with elicitation of a multi-causal probability, followed by elimination of one or more inconsistent NIN-AND tree models. The process continues until a single model in NM_e remains in the last round.

The elimination operation requires a threshold s. Only when difference $\delta = |P_e(e^+|c_i^+,c_j^+,c_k^+) - P'(e^+|c_i^+,c_j^+,c_k^+)| > s$, $P'(e|C)$ is deemed inconsistent with the true CPT. However, choosing the adequate threshold value is difficult in practice for the reason below.

By Assumption 1, the expert assessment of single-causal probabilities PS_e is approximate. Hence, none of the models in NM_e is consistent with the true model M_t. Furthermore, by assumption, an elicited multi-causal probability may also differ from the corresponding true probability. Hence, δ above contain elicitation errors. If s is set too low, even if a model $M \in NM_e$ is structurally consistent with the true model M_t, it may still be eliminated because δ exceeds s. On the other hand, if s is set too high, multiple models structurally inconsistent with the true model M_t may pass each round, and no single model can be selected in the last round.

[Bounded sequential elimination] Elicitation and elimination proceed in K rounds, where K is the number of multi-causal probabilities to be elicited, is predetermined, and can be varied based on expert availability. In each round, after elicitation of a multi-causal probability, its difference δ from each CPT in CPT_e is calculated, a given number of models in NM_e with the minimum δ values are retained, and the other models are eliminated. The number of models retained in each round decreases over consecutive rounds, and it is one for the Kth round.

The threshold is no longer needed, and its drawback is avoided. Instead, a set of K bounds is used, one for the number of retained models in each round. For example, if $K = 4$, numbers of models retained in succeeding rounds can be 16, 8, 4, and 1.

One limitation is that the model returned may depend on the order in which the K multi-causal probabilities are elicited. The NIN-AND tree model $M \in NM_e$ that is structurally consistent with M_t (such M is unique whenever single-causal probabilities by $P_t(e|C)$ are distinct) may be eliminated in an earlier round. This occurs when the probability elicited in the current round is not distinguishing, and too many models in NM_e have similar, small δ values: If the bound for the current round is m, the model M may be eliminated because its δ value is slightly larger than that of the model ranked m. Whereas if multi-causal probabilities were elicited in another order, M may be retained in each round and returned in the end.

[Simultaneous elimination] Only one round of elicitation and elimination is conducted. A set PM_e of K multi-causal probabilities are first elicited. Its root-mean-square (rms) distance from the corresponding set PM' of multi-causal probabilities determined by each CPT in CPT_e is calculated:

$$d(PM_e,PM') = \sqrt{\frac{1}{K}\sum_{i=1}^{K}(P_e(e^+|\underline{x}_i^+) - P'(e^+|\underline{x}_i^+))^2} \quad (3)$$

The model in NM_e with the minimum distance will be returned.

The method overcomes the limitation on threshold or elicitation order by the two alternative procedures. It is thus used in the further investigation of the SE technique. Although any multi-causal probabilities may be used with the SE technique, in the remainder of the paper, we assume that they are triple-causal.

5 Generate NIN-AND Tree by Pairwise Causal Interaction

The PCI technique builds on the pairwise causal interaction function defined by a NIN-AND tree [Xiang et al(2009a)].

Proposition 1. *Let T be a minimal NIN-AND tree for effect e and its causes C. Then T defines a function pci from pairs of distinct causes $\{c_i, c_j\} \subset C$, where $i \neq j$, to the set $\{rif, udm\}$, where rif stands for reinforcing and udm for undermining.*

The pci function signifies explicitly the causal interaction between each pair of causes. For instance, the NIN-AND tree in Fig. 1 (right) defines the function: $pci(c_1, c_2) = rif, pci(c_1, c_3) = udm, pci(c_2, c_3) = rif$.

Let TM be the set of all minimal NIN-AND trees over n causes. Then each NIN-AND tree $T \in TM$ has a distinct pci function (exhaustively confirmed for $n = 3, ..., 10$). Hence, a NIN-AND tree can be identified from a given pci function.

Based on this idea, we **propose** the PCI technique for generating a NIN-AND tree as follows: First, elicit a set PS_e of single-causal probabilities from the expert, and enumerate the set TM, as done in the SE technique. From TM, a set $PCIF$ of pci functions, one for each NIN-AND tree $T \in TM$ is defined. Then, a set PD_e of all double-causal probabilities (a total of $n\ (n-1)/2$ values) are elicited from the expert.

From PS_e and PD_e, a pci function $pci_e()$ can be determined according to Def. 1. For example, suppose the CPT in Table 1 is the true CPT, elicited single-causal probabilities include $P_e(e^+ \leftarrow c_2^+) = 0.82$, $P_e(e^+ \leftarrow c_3^+) = 0.67$, and elicited double-causal probabilities include $P_e(e^+ \leftarrow c_2^+, c_3^+) = 0.91$. From $P_e(e^+ \leftarrow c_2^+, c_3^+) > P_e(e^+ \leftarrow c_2^+)$ and $P_e(e^+ \leftarrow c_2^+, c_3^+) > P_e(e^+ \leftarrow c_3^+)$, the function value $pci(c_2, c_3) = rif$ can be determined.

Subsequently, the derived $pci_e()$ is compared against functions in $PCIF$. If $pci_e()$ matches $pci'() \in PCIF$, then the NIN-AND tree $T' \in TM$ that produces $pci'()$ will be returned.

The key operation of the PCI technique is the derivation of $pci_e()$ function from PS_e and PD_e. Below, we consider how to carry out the operation in practice. For any pair of causes c_i and c_j, $pci(c_i, c_j) \in \{rif, udm\}$. By Def. 1, $pci(c_i, c_j) = rif$ iff

$$P(e^+ \leftarrow c_i^+, c_j^+) \geq \max(P(e^+ \leftarrow c_i^+), P(e^+ \leftarrow c_j^+)), \tag{4}$$

and $pci(c_i, c_j) = udm$ iff

$$P(e^+ \leftarrow c_i^+, c_j^+) < \min(P(e^+ \leftarrow c_i^+), P(e^+ \leftarrow c_j^+)). \tag{5}$$

Therefore, in theory, it suffices to compare $P(e^+ \leftarrow c_i^+, c_j^+)$ and $P(e^+ \leftarrow c_i^+)$, and use the outcome to determine the value for $pci(c_i, c_j)$.

In practice, however, due to elicitation errors, it is possible that

$$P_e(e^+ \leftarrow c_i^+) < P_e(e^+ \leftarrow c_i^+, c_j^+) < P_e(e^+ \leftarrow c_j^+).$$

For example, if $P_t(e^+ \leftarrow c_i^+) = 0.6$, $P_t(e^+ \leftarrow c_j^+) = 0.9$, and c_i undermines c_j, we have $P_t(e^+ \leftarrow c_i^+, c_j^+) = 0.54$. Elicited values, however, may be

$$P_e(e^+ \leftarrow c_i^+) = 0.56 < P_e(e^+ \leftarrow c_i^+, c_j^+) = 0.59 < P_e(e^+ \leftarrow c_j^+) = 0.93$$

due to elicitation errors. Similarly, when c_i reinforces c_j, we have $P_t(e^+ \leftarrow c_i^+, c_j^+) = 1 - (0.4 * 0.1) = 0.96$, while elicited values may be

$$P_e(e^+ \leftarrow c_i^+) = 0.56 < P_e(e^+ \leftarrow c_i^+, c_j^+) = 0.91 < P_e(e^+ \leftarrow c_j^+) = 0.93.$$

When these happen, comparing $P_e(e^+ \leftarrow c_i^+, c_j^+)$ against one of $P_e(e^+ \leftarrow c_i^+)$ and $P_e(e^+ \leftarrow c_j^+)$ has a 0.5 chance to assign pci function value incorrectly. Comparing against both is not even feasible, because Eqn. (4) and (5) will both fail. To address this issue, we develop the following algorithm:

1. If Eqn. (4) holds for elicited probabilities, assign $pci(c_i, c_j) = rif$.
2. Else if Eqn. (5) holds for elicited probabilities, assign $pci(c_i, c_j) = udm$.
3. Else if
$$|P(e^+ \leftarrow c_i^+, c_j^+) - \min(P(e^+ \leftarrow c_i^+), P(e^+ \leftarrow c_j^+))|$$
$$< |P(e^+ \leftarrow c_i^+, c_j^+) - \max(P(e^+ \leftarrow c_i^+), P(e^+ \leftarrow c_j^+))|,$$
assign $pci(c_i, c_j) = udm$.
4. Else assign $pci(c_i, c_j) = rif$.

The algorithm handles normal cases (1 and 2) according to Eqn. (4) and (5). When elicitation errors fail these equations (cases 3 and 4), the pci function value is determined by assuming small errors. For the first example above, $pci(c_i, c_j) = udm$ will be assigned correctly due to case 3. For the second example, $pci(c_i, c_j) = rif$ will be assigned due to case 4.

It is possible that a derived function $pci_e() \notin PCIF$. That is, there exists no NIN-AND tree model that would produce the function $pci_e()$. The $pci_e()$ is said to be *invalid*. When this occurs, we apply a method in [Xiang(2010a)]: A valid pci function $pci_e^*()$ in *PCIF* which differs from $pci_e()$ the least will be selected, and its corresponding NIN-AND tree model will be returned as the indirectly elicited model.

6 Experimental Setup

To evaluate the effectiveness of SE and PCI techniques, human-based experiments are conducted, using an approach that extends that in [Zagorecki and Druzdzel(2004)]. A true causal model is simulated, from which a human is trained into an expert. A subset of causal probabilities are then elicited from the expert, from which a NIN-AND tree model is generated using the SE or PCI technique. The rms distance between the discovered model and the true model (similar to Eqn. (3)) is then measured to evaluate the effectiveness of these techniques. The experiment is organized into three stages elaborated below.

The first is *expert training*, during which each human participant is trained into an expert. A simulated NIN-AND tree model $M_t = (e,C,T,PS)$ is used as the true model, from which the true CPT $P_t(e|C)$ is constructed. Given the presence of a subset $X \subseteq C$ of active causes, an example $(e,\underline{x}^+)$, where $e \in \{e^+,e^-\}$, is generated by stochastic simulation from causal probability $P_t(e^+ \leftarrow \underline{x}^+)$. After seeing a sufficient number of examples for a sufficient number of distinctive $\underline{x}^+$ (detailed below), the participant is deemed to be an expert on model M_t.

To ensure that a participant's knowledge on M_t is obtained entirely from the training, and is not biased by outside experience, we presented M_t to be about phenomena from an imaginary planet. A software Environment Simulator (ES) is implemented accordingly to allow a participant to specify active causes $\underline{x}^+$ and observe simulated effects e. Note that this setup ensures condition 1 of Assumption 1.

The second stage is *elicitation*, during which a subset of causal probabilities $P_e(e^+ \leftarrow \underline{x}^+)$ are elicited from the expert. As stated in Assumption 1, generally, $P_e(e^+ \leftarrow \underline{x}^+) \neq P_t(e^+ \leftarrow \underline{x}^+)$. Their difference has so far been referred to as *elicitation error*, but in fact is the combination of two sources of errors.

1. Sampling error: Assuming $P_e(e^+ \leftarrow \underline{x}^+)$ is based on observed relative frequency $F(e^+ \leftarrow \underline{x}^+) = N(e^+ \leftarrow \underline{x}^+)/N(\underline{x}^+)$, where $N(e^+ \leftarrow \underline{x}^+)$ is the number of observations of example $(e^+,\underline{x}^+)$ and $N(\underline{x}^+)$ is the number of observations of $\underline{x}^+$, we have $F(e^+ \leftarrow \underline{x}^+) \neq P_t(e^+ \leftarrow \underline{x}^+)$ because $N(\underline{x}^+)$ is finite.
2. Retention-Articulation (RA) error: The participant may not be able to retain and articulate either $N(e^+ \leftarrow \underline{x}^+)$ and $N(\underline{x}^+)$, or $F(e^+ \leftarrow \underline{x}^+)$ accurately [Kahneman et al(1982)].

To ensure condition 2 of Assumption 1, both the sampling error and RA error need to be controlled. To control sampling error, we setup ES to enforce the requirement $N(\underline{x}^+) \geq 100$ for each $P_e(e^+ \leftarrow \underline{x}^+)$ to be elicited. That is, the participant must have a sufficient number of observations of $\underline{x}^+$ during training.

To control RA error, for each distinct $\underline{x}^+$, the frequency pair $F(e^+ \leftarrow \underline{x}^+)$ and $F(e^- \leftarrow \underline{x}^+)$ observed during the training stage is shown in a stacked bar graph (Fig. 2). The bar graph helps to reduce the RA error by providing a visual hint for the observed $F(e^+ \leftarrow \underline{x}^+)$. Yet, it does not eliminate RA error as it is visual, while $P_e(e^+ \leftarrow \underline{x}^+)$ is elicited numerically.

The final stage is *discovery*, during which the set of $P_e(e^+|\underline{x}^+)$ elicited is used to generate a NIN-AND tree model M_e.

Participants are recruited from university students (second year or above). Each participant is trained with a distinct true model $M_t = (e,C,T,PS)$. All models used have $|C| = 4$, but they differ in both T and PS.

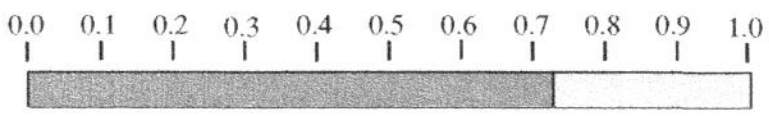

Fig. 2. A stacked bar graph where $F(e^+ \leftarrow \underline{x}^+) = 0.72$

Our objective is to evaluate the effectiveness of SE and PCI techniques. To facilitate the evaluation, we compare them against direct elicitation of each causal probability (all 15 parameters in $P_t(e|C)$). We refer to it as the *direct numerical* (DN) technique. For SE, we elicit 8 parameters (4 single-causal and 4 triple-causal). For PCI, we elicit 10 parameters (4 single-causal and 6 double-causal).

7 Experimental Results

Each data set consists of a number of causal probabilities elicited from one participant. A data set for evaluation of DN, SE, or PCI technique contains 15, 8 or 10 elicited probabilities, respectively, and the number of data sets collected are 23, 29, 29, respectively.

From the true CPT used to simulate training examples for a participant and probabilities elicited from the participant, the elicitation error (Section 6) of the participant is measured by the rms distance between the true CPT and elicited probabilities. The mean and standard deviation of elicitation errors over all participants are shown in Table 2 (column 4). Th elicitation error consists of sampling and RA errors (Section 6). From ES log of examples generated for training a participant and the true CPT used in example generation, the sampling error of training examples is measured by the rms distance between example frequencies and the true CPT. From the log of examples generated for training a participant and elicited probabilities, RA error of the participant is measured by rms distance between example frequencies and elicited probabilities. The means and standard deviations of sampling and RA errors over all participants are also shown in the table (columns 2 and 3). It can been seen that our elicitation aid by stacked bar graphs has effective control of the RA error. Hence, the elicitation error is composed mainly of the sampling error.

The DN technique directly elicits a CPT from the expert, which we refer to as the *CPT elicited with the DN technique*. On the other hand, for each data set collected for SE evaluation, the SE technique is applied to generate a NIN-AND tree model, from which a CPT is constructed. We refer to it as the *CPT elicited with the SE technique*. The *CPT elicited with the PCI technique* is similarly defined.

For each data set, the CPT elicited by the corresponding technique is compared against the true CPT used to drive expert training, and the rms distance between the two CPTs is calculated. For each of DN, SE, and PCI technique, the mean and standard deviation over the corresponding data sets are summarized in Table 3.

Results from all three techniques are comparable. Note that PCI technique depends on single and double-causal probabilities (10), SE technique depends on single and triple-causal probabilities (8), while DN technique depends on all causal probabilities (15). Hence, the results demonstrate that both SE and PCI techniques improve efficiency in CPT acquisition while maintaining comparable accuracy.

Table 2. Mean (μ) and standard deviation (σ) of errors over all participants

	Sampling Errors	RA Errors	Elicitation Errors
μ	0.0293	0.0076	0.0301
σ	0.0096	0.0038	0.0099

Table 3. Mean (μ) and standard deviation (σ) of model distance by DN, SE and PCI techniques

	DN	SE	PCI
μ	0.0301	0.0356	0.0281
σ	0.0099	0.0343	0.0146

8 Simulation Study

Due to resource involved in human-based experiments, large numbers of participants and multiple setups are not feasible. To compensate this limitation, we enhanced human experiments with simulation-based studies.

For the DN technique, we simulated a true model $M_t = (e, C, T_t, PS_t)$ and constructed the true CPT $P_t(e|C)$ from M_t. For each subset $X \subseteq C$ of active causes, K examples $(e, \underline{x}^+)$ are stochastically generated from $P_t(e^+|\underline{x}^+)$. The elicited probability $P_e(e^+|\underline{x}^+)$ is simulated as the ratio between the number of examples $(e^+, \underline{x}^+)$ and K. This is justified by two observations. First, the elicitation errors in human experiments are made up mainly by sampling errors (Table 2). Second, as we decrease K, the elicitation error $|P_e(e^+|\underline{x}^+) - P_t(e^+|\underline{x}^+)|$ will increase. Hence, simulated elicitation errors can be well controlled through K.

After the elicited CPT $P_e(e|C)$ is thus simulated, we calculate the rms distance between $P_e(e|C)$ and $P_t(e|C)$. We repeat the above for W true models, and the effectiveness of the DN technique is evaluated by the mean distance from the W trials.

For the PCI technique, the true model $M_t = (e, C, T_t, PS_t)$ and true CPT $P_t(e|C)$ are simulated as above. A set $PS_e = \{P_e(e|c_i^+)\}$ of single-causal elicited probabilities and a set $PD_e = \{P_e(e|c_i^+, c_j^+)\}$ of double-causal elicited probabilities are simulated from $P_t(e|C)$. Applying the PCI technique to PS_e and PD_e, an indirectly elicited model $M_e = (e, C, T_e, PS_e)$ is generated.

From M_e, the elicited CPT $P_e(e|C)$ is constructed and the rms distance between $P_e(e|C)$ and $P_t(e|C)$ calculated. The effectiveness of the PCI technique is evaluated by repeating the above for W true models, and obtaining the mean distance.

For the SE technique, a set PS_e of single-causal elicited probabilities and a set $PT_e = \{P_e(e|c_i^+, c_j^+, c_k^+)\}$ of triple-causal elicited probabilities are simulated from $P_t(e|C)$. The set of all NIN-AND tree models $NM_e = \{(e, C, T, PS_e)\}$ are obtained by enumeration. Note that each model $M \in NM_e$ has a distinct NIN-AND tree topology T, but has the same PS_e. An indirectly elicited NIN-AND tree model M_e is then selected from NM_e if its corresponding CPT has the minimum distance from PT_e.

From M_e, CPT $P_e(e|C)$ is constructed and the rms distance between $P_e(e|C)$ and $P_t(e|C)$ is calculated. The SE technique is evaluated by the mean distance from simulation over W true models.

In simulation studies for the three techniques, we used $K = 100$ and $W = 1000$. $K = 100$ is chosen so that magnitudes of simulated elicitation errors are similar to those observed in the human-based study. $W = 1000$ is used as higher W values do not show significant difference in outcomes. For each technique, simulations are run for each of $n = |C| = 4, 5, 6, 7$. Table 4 shows the number of causal probabilities simulated for each technique and each n value.

Table 4. Number of simulated causal probabilities used by DN, SE and PCI studies

n	# CPT probs	# probs for DN	# probs for SE	# probs for PCI
4	16	15	8	10
5	32	31	15	15
6	64	63	26	21
7	128	127	42	28

The second column shows the number of independent probability parameters in $P(e|C)$, which is 2^n. The third column shows the number of elicited probabilities simulated by DN evaluation, which is $2^n - 1$, because NIN-AND tree models satisfy $P(e^+|\underline{c}^-)$ $= 0$. The fourth column shows the count for SE evaluation, which is $n + C(n,3)$. The last column shows the count for PCI evaluation, which is $n + C(n,2)$.

Results from simulation-based studies are summarized in Table 8. Means and standard deviations of model distances for the three techniques are shown in columns 2, 3, 4, 5, 7, 8. Columns 6 and 9 show percentages of models indirectly elicited by SE and PCI that recover true tree topology T_t. The last column shows percentages of indirectly elicited *pci* functions that are invalid.

Table 5. Model distance by DN, SE and PCI techniques from simulation study

n	DN (μ)	DN (σ)	SE (μ)	SE (σ)	Rcv (%)	PCI (μ)	PCI (σ)	Rcv (%)	Ivad (%)
4	0.0363	0.0099	0.0470	0.0485	79.6	0.0352	0.0340	98.5	0.9
5	0.0368	0.0086	0.0352	0.0268	86.5	0.0369	0.0397	98.1	0.5
6	0.0364	0.0076	0.0317	0.0215	88.2	0.0338	0.0237	95.7	2.2
7	0.0356	0.0076	0.0311	0.0183	85.8	0.0344	0.0284	94.2	3.6

The mean distances for DN indicate the magnitudes of simulated elicitation errors in the studies of all three techniques, since the same $K = 100$ value is used. Note that the magnitudes are slightly higher than that observed in human-based experiments (Table 2).

Comparing columns 6 and 9, PCI technique performs better than SE in recovering true NIN-AND tree topology. On the other hand, although SE technique is less accurate in tree recovery, the mean model distance and standard deviation for $n = 5,6,7$ are slightly smaller than PCI. This observation shows that given the existence of elicitation errors, multiple NIN-AND tree models may generate similar CPTs, and the SE technique is robust under such condition. We attribute the reverse performance difference when $n = 4$, i.e., $SE(\mu) > PCI(\mu)$, to the number of elicited probabilities used (8 for SE and 10 for PCI).

Overall, SE and PCI techniques achieved the comparable model distance in comparison with DN technique, while requiring a much less number of elicited probabilities. In general, the number of probabilities to be elicited by the DN technique is $O(2^n)$. The number is $O(n^3)$ for SE and $O(n^2)$ for PCI. The performance of PCI technique makes it particularly attractive: It achieves about the same elicitation accuracy while requiring the smallest number of elicitations. For instance, when $n = 7$, DN requires 127, SE requires 42, while PCI requires only 28.

Finally, column 10 shows that although elicitation errors sometimes cause failure in constructing the pci function, our fault-tolerance method recovers from the failure well. Not only a valid NIN-AND tree model is returned under the failure condition, but the model is sufficiently close to the true model (shown by columns 7 and 8).

9 Conclusion

NIN-AND tree causal models provide an efficient tool for CPT acquisition in construction of Bayes nets. Direct elicitation of such a model involves elicitation of a number (linear in n) of single-causal probabilities, and a NIN-AND tree (of a size linear in n). The tree elicitation step requires nontrivial training of an expert on the syntax and semantics of these models, as well nontrivial mental exercise by the expert to identify correctly the partial order of interactions among causes.

In this work, we investigate the novel idea to substitute direct elicitation of a NIN-AND tree with elicitation of some multi-causal probabilities. The NIN-AND tree is then automatically generated based on elicited probabilities. We propose two alternative techniques that implement this idea with low-order multi-causal probabilities. Our human-based and simulation-based studies demonstrated the feasibility of the idea. These techniques eliminate above-mentioned expert training and demanding mental exercise, while remaining efficient. Numbers of probabilities to be elicited are $O(n^3)$ and $O(n^2)$ for (triple-causal based) SE and PCI, respectively.

The main assumption these techniques depend on is the expert's ability to approximately assess required causal probabilities. Elicitation error can be decomposed into sampling error and RA error. The RA error may be reduced through training and/or technical aids, although detailed investigation is beyond the scope of this work. Sampling error may be controlled by the number of examples observed for each causal combination (i.e., $\underline{x}^+$). Our experiments have shown that 100 examples per causal combination is sufficient for our techniques to work well.

Acknowledgements. Financial support from NSERC, Canada to the first author is acknowledged.

References

[Galan and Diez(2000)] Galan, S., Diez, F.: Modeling dynamic causal interaction with Bayesian networks: temporal noisy gates. In: Proc. 2nd Inter. Workshop on Causal Networks, pp. 1–5 (2000)

[Heckerman and Breese(1996)] Heckerman, D., Breese, J.: Causal independence for probabilistic assessment and inference using Bayesian networks. IEEE Trans. on System, Man and Cybernetics 26(6), 826–831 (1996)

[Kahneman et al(1982)] Kahneman, D., Slovic, P., Tversky, A. (eds.): Judgment under uncertainty: heuristics and biases. Cambridge University Press, Cambridge (1982)

[Lemmer and Gossink(2004)] Lemmer, J., Gossink, D.: Recursive noisy OR - a rule for estimating complex probabilistic interactions. IEEE Trans. on System, Man and Cybernetics, Part B 34(6), 2252–2261 (2004)

[Maaskant and Druzdzel(2008)] Maaskant, P., Druzdzel, M.: An independence of causal interactions model for opposing influences. In: Jaeger, M., Nielsen, T. (eds.) Proc. 4th European Workshop on Probabilistic Graphical Models, Hirtshals, Denmark, pp. 185–192 (2008)

[Pearl(1988)] Pearl, J.: Probabilistic Reasoning in Intelligent Systems: Networks of Plausible Inference. Morgan Kaufmann, San Francisco (1988)

[Xiang(2010a)] Xiang, Y.: Acquisition and computation issues with NIN-AND tree models. In: Myllymaki, P., Roos, T., Jaakkola, T. (eds.) Proc. 5th European Workshop on Probabilistic Graphical Models, Finland, pp. 281–289 (2010a)

[Xiang(2010b)] Xiang, Y.: Generalized non-impeding noisy-AND trees. In: Proc. 23th Inter. Florida Artificial Intelligence Research Society Conf., pp. 555–560 (2010b)

[Xiang and Jia(2007)] Xiang, Y., Jia, N.: Modeling causal reinforcement and undermining for efficient cpt elicitation. IEEE Trans. Knowledge and Data Engineering 19(12), 1708–1718 (2007)

[Xiang et al(2009a)] Xiang, Y., Li, Y., Zhu, J.: Towards effective elicitation of NIN-AND tree causal models. In: Godo, L., Pugliese, A. (eds.) SUM 2009. LNCS (LNAI), vol. 5785, pp. 282–296. Springer, Heidelberg (2009a)

[Xiang et al(2009b)] Xiang, Y., Zhu, J., Li, Y.: Enumerating unlabeled and root labeled trees for causal model acquisition. In: Gao, Y., Japkowicz, N. (eds.) AI 2009. LNCS (LNAI), vol. 5549, pp. 158–170. Springer, Heidelberg (2009b)

[Zagorecki and Druzdzel(2004)] Zagorecki, A., Druzdzel, M.: An empirical study of probability elicitation under Noisy-OR assumption. In: Proc. 17th Inter. Florida Artificial Intelligence Research Society Conf., pp. 880–885 (2004)

Change in Argumentation Systems: Exploring the Interest of Removing an Argument

Pierre Bisquert, Claudette Cayrol,
Florence Dupin de Saint-Cyr, and Marie-Christine Lagasquie-Schiex

IRIT, Université Paul Sabatier, 31062 Toulouse Cedex 9, France
{bisquert,ccayrol,bannay,lagasq}@irit.fr

Abstract. This article studies a specific kind of change in an argumentation system: the removal of an argument and its interactions. We illustrate this operation in a legal context and we establish the conditions to obtain some desirable properties when removing an argument.

Keywords: change in argumentation, reasoning with uncertain and inconsistent information.

1 Introduction

Argumentation is a very active research area, in particular for its applications concerning reasoning ([10,2]) or negotiation between agents ([4]). It allows to model the exchange of arguments between several agents (dialog), but also allows a single agent to manage incomplete and potentially contradictory information. Hence, argumentation is a way to handle uncertainty about the outcome of a dialog or the conclusion of a reasoning process. The arguments thus emitted are in interaction, generally by means of an attack relation representing the conflicts between arguments (for example, when the conclusion of an argument contradicts an assumption of another one).

Argumentation theory proposes several methods for drawing a conclusion about a set of interacting arguments. One of these methods is the study of *"extensions"*, sets of arguments that are said *acceptable* (*i.e.* a set able to defend itself collectively while avoiding internal conflicts). Another method is the study of the individual status of each argument determined by its membership to one or all extensions. Formal frameworks were proposed for representing argumentation systems, in particular [10] which allows to handle the arguments like purely abstract entities connected by binary relations.

Although dynamics of argumentation systems has been recently explored by several works ([6,7,5,11]), the removal of an argument has scarcely been mentioned. However, there exist practical applications. First of all, a speaker can need *to occult* some argument, in particular when he does not want, or is not able, to present this argument in front of a given audience[1]; it is then necessary to know what would be the output of the speaker's argumentation system without this argument: that can be achieved by a removal in his initial system. In addition, this same audience can force the speaker *to remove* an argument, in particular when this last is regarded as illegal in the context. Note

[1] Social norms, or the will to avoid providing information to an adversary, etc.

S. Benferhat and J. Grant (Eds.): SUM 2011, LNAI 6929, pp. 275–288, 2011.
© Springer-Verlag Berlin Heidelberg 2011

that, when the argument is uttered, it is not discarded by default because the audience may not know that this argument is illegal; however, it could be removed later, when proved to be illicit. Moreover, the removal turns out to be useful in order to evaluate *a posteriori* the impact of a precise argument on the output of the system. In particular for evaluating the quality of a dialog, it is important to be able to differentiate the unnecessarily uttered arguments from the decisive ones (see [3]: an argument is decisive if its removal makes it possible to change the conclusion of the dialog). Lastly, it can be interesting to know how to guarantee that one or more arguments are accepted by removing a minimal set of arguments. Note that the removal of an argument X cannot always be reduced to the addition of a new argument Y attacking X, in particular because it may happen that an attacked argument remains acceptable. Furthermore, it is more economic to remove an argument rather than to add one, which might progressively overload the system. We thus propose to study theoretically the impact that the removal of one argument may produce on the initial set of extensions of an argumentation system.

The article is organized as follows. An example illustrating the interests of removing an argument is presented in Section 2. Section 3 gives a brief state of the art about argumentation. Section 4 exposes some properties of the extensions and of the status of some arguments when a particular argument is removed. Lastly, Section 5 establishes the links with related works and concludes this article.

2 Illustrative Example

We describe a *four players game* example inspired from the one given by Brewka in [8]. This game involves two speakers (the prosecutor and the lawyer) and two listeners (the judge and the jury). Although the discussion concerns only the two speakers, we choose to also model the audience in order to be able to study the dialog from the point of view of a neutral external observer. The presence of a judge makes it possible to illustrate a case of permanent removal: the objection. Let us note that this example can be expanded easily with more than two speakers.

2.1 Presentation of the Game

This game takes place during an oral hearing, gathering four entities which play quite distinct roles and which interact in order to determine whether an argument is acceptable or not.

- the *prosecutor* (P) wants to make accept a particular argument Q which is the subject of the hearing by the court. He has his own argumentation system in which he can *occult* arguments threatening Q, *i.e.*, withdraw temporarily some arguments which could prevent Q from being accepted. He can also occult some other arguments not threatening Q, but considered, for example, irrelevant or dangerous in front of a particular jury (according to his strategy of argumentation).

- the *defense lawyer* (D) possesses also its own argumentation system[2], he behaves like the prosecutor, but he tries to make the argument Q defeated.
- the *judge* ensures that the argumentation process takes place under good conditions. He gets involved when one participant makes an objection; he can then accept this objection (thus force the corresponding argument to be deleted), or reject it.
- the *jury*[3] has the last word. Its role is to listen to the prosecutor's and lawyer's arguments and to draw from them a conclusion concerning the acceptability of the argument Q. The jury begins the game with an argumentation system containing only the argument Q and supplement it with the arguments presented successively by the prosecutor and the lawyer (if these arguments are not cancelled by objections). When the hearing is closed (*i.e.* when neither the prosecutor nor the lawyer can give new arguments), the jury can determine whether Q is acceptable or not.

In our example, the subject of the argumentation is the guilt of the defendant concerning the murder of his wife. Table 1 summarizes the set of existing arguments concerning this example and their distribution between the prosecutor and the defense lawyer.

Table 1. Arguments concerning the case of Mr. X

	Argument	Known by
1	*Mr. X is guilty of premeditated murder of Mrs. X, his wife.*	P & D
2	*The defendant has an alibi, his business associate having solemnly sworn that he had seen him at the time of the murder.*	D
3	*The close working business relationships between Mr X. and his associate induce suspicions about his testimony.*	P
4	*Mr. X loves his wife so extremely that he asked her to marry him twice. Now, a man who loves his wife could not be her murderer.*	P & D
5	*Mr. X has a reputation for being promiscuous.*	P
6	*The defendant would not have had any interest to kill his wife, since he was not the beneficiary of the enormous life insurance she had contracted.*	P
7	*The defendant is a man known to be venal and his "love" for a very rich woman could be only lure of profit.*	D

2.2 Arguments of the Prosecutor

Let us examine the arguments of the prosecutor. He knows only two arguments attacking his thesis (Argument 1): Arguments 6 and 4. The prosecutor is not over worried about

[2] In order to have a genuine confrontation, it is necessary that the prosecutor and the lawyer share some arguments. However, these arguments are dealt differently by the two speakers: often considered as positive for one and negative for the other one.

[3] Although being a group of persons, the jury is considered to be only one decision-making entity.

4 because 5 enables him to defend his thesis against it. The prosecutor knows, on the other hand, no argument which can defeat 6. Not being able to find what could beat this argument, and hoping that the lawyer is not informed about it, the prosecutor decides to occult it in order to ensure the acceptability of his thesis in his argumentation system.

2.3 Arguments of the Defense

Now let us examine the arguments of the defense lawyer who aims at preventing the acceptability of Argument 1. The lawyer has two arguments attacking directly 1, namely, 4 and 2. While 2 is not attacked (as far as he knows), it is not the same for 4 which is attacked by 7; unable to find something to oppose to 7, the lawyer thus prefers to occult 7 (in order to be sure that 1 will be rejected), hoping that the prosecutor will not utter it.

2.4 The Oral Hearing in Front of the Court

Now that we know the arguments of the two speakers, we can consider the exchanges between them during the hearing (see Table 2).

Table 2. Successive turns of the hearing

Turn	Player	Action
0	Prosecutor	1
1	Defender	2
2	Prosecutor	3
3	Defender	4
4	Prosecutor	5
5	Defender	Objection
6	Judge	Sustained
7	Prosecutor	Close
8	Defender	Close
9	Jury	Deliberation

Turn 0 establishes the subject of the dialog; it is a mandatory stage fixing the argument that the prosecutor and the lawyer will try to make, respectively, accept or reject.

Turns 1 to 4 are "normal" exchanges of arguments between the speakers, these arguments are used by the jury to build its argumentation system.

Turn 5 introduces the objection process, *i.e.*, opposition by the adverse party to an argument considered as illegal[4]. Here, the defense lawyer utters an objection against Argument 5 because it is based on *hearsay*.

The validity of the objection is examined at **Turn 6**: the judge has to decide if the argument presented in Turn 4 is illegal (by referring to the protocol in force in this context). The judge has chosen to sustain the objection requested by the lawyer, which introduces the mechanism of removal. Indeed, an objection indicates that the targeted

[4] Arguments illegality criteria are defined by the protocol governing the hearing and may evolve according to the context; nevertheless arguments that are fallacious, irrelevant or obtained by hearsay, are assumed to be illegal.

argument should not be taken into account anymore nor registered in the official report. Note that adding a new argument is not equivalent to the removal of an argument since addition increases the number of arguments, hence the complexity of the system. Moreover, adding an argument that attacks the illegal argument does not guarantee the rejection of the latter, especially through mechanisms, such as defense, that can lead the illegal argument to remain accepted. Removing the illegal argument thus ensures impossibility of taking this argument into account anymore. Still during Turn 6, the jury proceeds to the removal of the objected argument from his argumentation system, and both speakers are occulting this argument definitely.

Turns 7 to 9 are closing the hearing: none of the two speakers has any new argument to present, which they successively indicate by the action *"Close"*. The deliberation of the jury follows, in order to determine if the subject of the hearing (Argument 1) is accepted or not.

The result of this deliberation will be given in the following section after some reminders about argumentation theory and a formalization of the example.

3 Formal Framework

The work presented in this article uses the formal framework proposed by [10].

Definition 1 (Argumentation System). *An* argumentation system *is a pair* $\langle \mathbf{A}, \mathbf{R} \rangle$, *where* $\mathbf{A}$ *is a finite nonempty set of arguments and* $\mathbf{R}$ *is a binary relation on* $\mathbf{A}$, *called* attack relation. *Let* $A, B \in \mathbf{A}$, $A\mathbf{R}B$ *means that* A *attacks* B. $\langle \mathbf{A}, \mathbf{R} \rangle$ *will be represented by a graph whose vertices are the arguments and whose arcs correspond to* $\mathbf{R}$.

The acceptable set of arguments ("extensions") are determined according to a given semantics whose definition is usually based on the following concepts:

Definition 2 (Conflict-free set, defense and admissibility). *Let* $A \in \mathbf{A}$ *and* $\mathcal{S} \subseteq \mathbf{A}$,

- $\mathcal{S}$ *is* conflict-free *iff there does not exist* $A, B \in \mathcal{S}$ *such that* $A\mathbf{R}B$.
- $\mathcal{S}$ defends an argument A *iff each attacker of* A *is attacked by an argument of* $\mathcal{S}$. *The set of the arguments defended by* $\mathcal{S}$ *is denoted* $\mathcal{F}(\mathcal{S})$; $\mathcal{F}$ *is called the* characteristic function *of* $\langle \mathbf{A}, \mathbf{R} \rangle$.
- $\mathcal{S}$ *is an* admissible *set iff it is conflict-free and it defends all its elements.*

In this article, we restrict our study to the most traditional semantics proposed by [10]:

Definition 3 (Acceptability semantics). *Let* $\mathcal{E} \subseteq \mathbf{A}$,

- $\mathcal{E}$ *is a* preferred extension *iff* $\mathcal{E}$ *is a maximal admissible set (with respect to set inclusion* $\subseteq$*).*
- $\mathcal{E}$ *is* the only grounded extension *iff* $\mathcal{E}$ *is the least fixed point (with respect to* $\subseteq$*) of the characteristic function* $\mathcal{F}$.
- $\mathcal{E}$ *is a* stable extension *iff* $\mathcal{E}$ *is conflict-free and attacks any argument not belonging to* $\mathcal{E}$.

The status of an argument is determined by its presence in the extensions of the selected semantics. For example, an argument can be "accepted sceptically" (resp. "credulously") if it belongs to all the extensions (resp. at least to one extension) and be "rejected" if it does not belong to any extension.

We now recall the definition given by [9] for the removal of an argument and its interactions:

Definition 4 (Removing an argument). *Let* $\langle \mathbf{A}, \mathbf{R} \rangle$ *be an argumentation system.* Removing an argument $Z \in \mathbf{A}$ *interacting with other arguments is a change operation, denoted* $\ominus_I^a$*, providing a new argumentation system such that:*

$$\langle \mathbf{A}, \mathbf{R} \rangle \ominus_I^a Z = \langle \mathbf{A} \setminus \{Z\}, \mathbf{R} \setminus \mathcal{I}_z \rangle$$

where $\mathcal{I}_z = \{(Z, X) \mid (Z, X) \in \mathbf{R}\} \cup \{(X, Z) \mid (X, Z) \in \mathbf{R}\}$ *is the set of interactions concerning* Z[5].

The set of extensions of $\langle \mathbf{A}, \mathbf{R} \rangle$ is denoted by $\mathbf{E}$ (with $\mathcal{E}_1, \ldots, \mathcal{E}_n$ standing for the extensions). A change creates a new argumentation system $\langle \mathbf{A}', \mathbf{R}' \rangle$ represented by a graph $\mathcal{G}'$, whose set of extensions is denoted by $\mathbf{E}'$ (with $\mathcal{E}'_1, \ldots, \mathcal{E}'_m$ standing for the extensions). It is assumed that the change does not concern semantics, *i.e.*, the semantics remains the same after the change.

Note that if an argumentation system $\langle \mathbf{A}', \mathbf{R}' \rangle$ is obtained by removing an argument Z in the argumentation system $\langle \mathbf{A}, \mathbf{R} \rangle$, then $\langle \mathbf{A}, \mathbf{R} \rangle$ can be obtained by adding Z to $\langle \mathbf{A}', \mathbf{R}' \rangle$. The study of the duality of addition with respect to removal is left for future work, along with another kind of duality, evoked below, concerning the change properties.

A change operation has an impact on the structure of the set of extensions and thus on the status of particular arguments. The reader may refer to [9] for a presentation of these properties and for their detailed analysis in the case of the addition of an argument. Among all these properties, one may find for example the expansive change that occurs when the number of extensions remains the same, whereas each extension of $\mathcal{G}'$ includes strictly an extension of $\mathcal{G}$, and any extension of $\mathcal{G}$ is strictly included in an extension of $\mathcal{G}'$[6].

Definition 5 (Expansive change). *The change from* $\mathcal{G}$ *to* $\mathcal{G}'$ *is expansive*[7] *iff*

$$\begin{cases} (1)\ \mathbf{E} \neq \varnothing, |\mathbf{E}| = |\mathbf{E}'|, \\ (2)\ \forall \mathcal{E}'_j \in \mathbf{E}', \exists \mathcal{E}_i \in \mathbf{E}, \mathcal{E}_i \subset \mathcal{E}'_j \text{ and} \\ (3)\ \forall \mathcal{E}_i \in \mathbf{E}, \exists \mathcal{E}'_j \in \mathbf{E}', \mathcal{E}_i \subset \mathcal{E}'_j \end{cases}$$

Example 1

Under preferred semantics, the change $\ominus_I^a Z$ with $\mathcal{I}_z = \{(Z, C), (D, Z)\}$ is expansive because $\mathbf{E} = \{\{A\}\}$ and $\mathbf{E}' = \{\{A, C\}\}$.

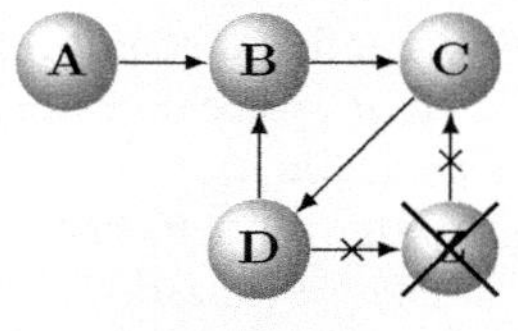

5 In the symbol $\ominus_I^a$, the a stands for "argument" and I for "interactions", meaning that the removal concerns an argument and its interactions.

6 The notation $\subset$ stands for strict set-inclusion.

7 We give here a more restrictive definition than the one given by [9] (the third condition has been added). Strict inclusion is used in order to avoid overlap with other properties.

In the following, we introduce a property which could be considered as dual of the previous one. Indeed, a narrowing change occurs when the number of extensions remains the same while any extension of $\mathcal{G}'$ is strictly included in an extension of $\mathcal{G}$ and any extension of $\mathcal{G}$ includes strictly an extension of $\mathcal{G}'$.

Definition 6 (Narrowing change). *The change from $\mathcal{G}$ to $\mathcal{G}'$ is narrowing iff*

$$\begin{cases} (1)\ \mathbf{E} \neq \varnothing, |\mathbf{E}| = |\mathbf{E}'|, \\ (2)\ \forall \mathcal{E}'_j \in \mathbf{E}', \exists \mathcal{E}_i \in \mathbf{E}, \mathcal{E}'_j \subset \mathcal{E}_i \textit{ and} \\ (3)\ \forall \mathcal{E}_i \in \mathbf{E}, \exists \mathcal{E}'_j \in \mathbf{E}', \mathcal{E}'_j \subset \mathcal{E}_i \end{cases}$$

Example 2

The change $\ominus_I^a Z$ with $\mathcal{I}_z = \{(B, Z)\}$ is narrowing under the preferred and stable semantics because $\mathbf{E} = \{\{A, C, Z\}, \{A, D, Z\}\}$ and $\mathbf{E}' = \{\{A, C\}, \{A, D\}\}$, and also under the grounded semantics because $\mathbf{E} = \{\{A, Z\}\}$ and $\mathbf{E}' = \{\{A\}\}$.

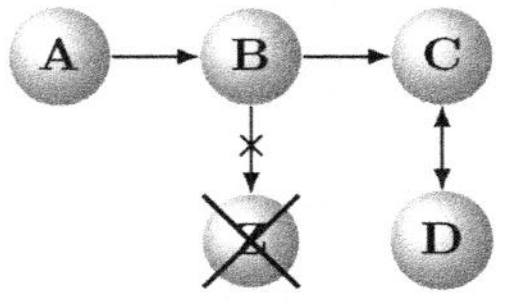

Back to the Example

The previous definitions enable us to deal with our example. Indeed, the argumentation systems of the prosecutor and lawyer at the beginning of the hearing are represented by Turn 0 of Table 3 and, for each semantics recalled in Definition 3, the prosecutor's (resp. lawyer's) system admits only one extension $\mathcal{E} = \{1, 3, 5\}$ (resp. $\mathcal{E} = \{2, 4\}$). Note that it could be the case that each agent uses her own semantics since her reasoning (and thus her argumentation system and semantics) is personal. Nevertheless, in the current example, we consider that all agents use the same semantics because it seems natural to assume that the prosecutor and the lawyer know which semantics the jury is using and so they use the same one.

Let us note that at the beginning of the hearing, some existing attacks between arguments may not belong to any of the two speakers' argumentation systems; here, for example, the attack from 3 to 2, observable at Turn 1 of Table 3, was not present for the prosecutor nor for the lawyer at Turn 0. Nevertheless, at each turn, the latter update their argumentation systems when they encounter a new argument that they did not know (the jury is supposed to ignore every argument at the beginning but after that, it proceeds in the same way). Table 3 shows the evolution of the various argumentation systems throughout the hearing.

At the deliberation time, the jury must decide upon the case of the hearing, namely Argument 1. For that, the jury should determine its status (accepted or rejected) by computing the extension(s) of its argumentation system with respect to a selected semantics. Whatever the semantics adopted by the jury among those recalled in Definition 3, its argumentation system has only one single extension $\mathcal{E} = \{3, 4\}$. Thus the jury can found the defendant "not guilty" since 1 do not belong to this extension.

Let us note that the removal of 5, to which the objection is related, has an influence on the acceptability of 1. Indeed, if the objection had been rejected, 5 could have defended 1 and ensured its presence in the extension, making it possible to convict the defendant.

Table 3. Argumentation systems throughout the hearing (a diamond is surrounding the current turn argument)

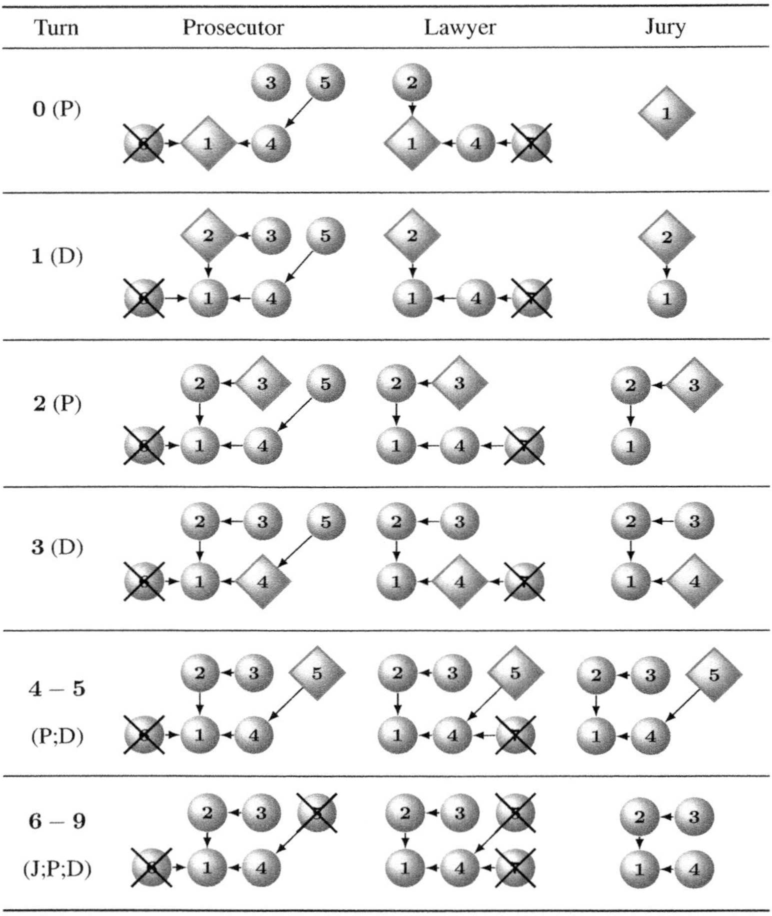

Moreover, let us notice that the lawyer was quite right to occult 7 because by doing so he has saved his client.

4 First Steps towards a Decision of Removal

In this section, we study some properties characterizing the operation of removal. This kind of work may help a user to decide with full knowledge, in which situation and how to make a removal according to its strategic objectives. Note that the properties presented here are the first results of our study of the removal operation and will have to be deepened.

4.1 Some Properties Concerning "Monotony"

In the framework of change in argumentation, "monotony" is related to the conservation of the extensions. More precisely, [9] defines it as follows: a change from $\mathcal{G}$ to $\mathcal{G}'$ satisfies monotony iff any extension of $\mathcal{G}$ is included into at least one extension of $\mathcal{G}'$. The following property gives the conditions under which a set of arguments that was jointly accepted remains so after the change.

Proposition 1 (Sufficient conditions for monotony/non-monotony). *When removing an argument Z (according to Definition 4) under the preferred, stable or grounded semantics,*

- *if* $\exists \mathcal{E}_i \in \mathbf{E}$ *such that* $Z \in \mathcal{E}_i$ *then* $\exists \mathcal{E}_j \in \mathbf{E}$ *such that* $\forall \mathcal{E}' \in \mathbf{E}'\ \mathcal{E}_j \not\subseteq \mathcal{E}'$;
- *if* $\nexists \mathcal{E}_i \in \mathbf{E}$ *such that* $Z \in \mathcal{E}_i$ *then* $\forall \mathcal{E}_j \in \mathbf{E}\ \exists \mathcal{E}' \in \mathbf{E}'$ *such that* $\mathcal{E}_j \subseteq \mathcal{E}'$.

Since before the change, the removed argument may belong to an extension, it can be interesting to also consider the conditions for a "weak monotony" (*i.e.*, conservation of an extension without taking the removed argument into account), as defined by [6]:

Proposition 2. *When removing an argument Z, if Z does not attack any argument then*

- $\forall \mathcal{E}$ *preferred extension of* $\mathcal{G}$, $\begin{cases} \mathcal{E} \setminus \{Z\} \text{ is admissible in } \mathcal{G}' \text{ and thus} \\ \exists \mathcal{E}' \text{ preferred extension of } \mathcal{G}' \text{ s.t. } \mathcal{E} \setminus \{Z\} \subseteq \mathcal{E}' \end{cases}$
- $\forall \mathcal{E}$ *stable extension of* $\mathcal{G}$, $\mathcal{E} \setminus \{Z\}$ *is a stable extension of* $\mathcal{G}'$.

Proposition 3. *When removing an argument Z under the preferred, stable or grounded semantics, if Z does not attack any argument of* $\mathcal{G}$*, then* $\forall \mathcal{E}$ *extension of* $\mathcal{G}$ *such that* $Z \notin \mathcal{E}$, $\mathcal{E}$ *is an extension of* $\mathcal{G}'$.

4.2 Some Properties of the Expansive Change

An expansive change increases the size of the extensions and thus allows to obtain a larger number of arguments in the extensions. The following properties give the conditions under which a change cannot be expansive.

Proposition 4. *It is impossible to have an expansive change* $\ominus_I^a$ *under stable semantics.*

Proposition 5. *When removing Z under the preferred or grounded semantics,*

if this change is expansive then $\begin{cases} Z \text{ does not belong to any extension of } \mathcal{G}, \\ \text{and } Z \text{ attacks at least one element of } \mathcal{G}. \end{cases}$

4.3 Some Properties of the Narrowing Change

The narrowing change may be considered as dual of the expansive change since it decreases the size of the extensions. This can be desirable when one wishes to reduce the possibilities of argumentation of the adverse party. The following property merges three properties (one for each semantics) that provide a necessary condition for obtaining a narrowing change.

Proposition 6. *When removing Z under the preferred, stable or grounded semantics, if the change is narrowing then there exists one extension* $\mathcal{E}$ *of* $\mathcal{G}$ *s.t.* $Z \in \mathcal{E}$.

5 Discussion and Conclusion

In this article, we studied a particular kind of change in argumentation: the removal of an argument and its interactions. First, we have presented an example coming from the legal world. This example illustrates the need to remove arguments in an argumentation system. In this application, at least two reasons are invoked in order to remove an argument: namely *objection* and *occultation*. After having pointed out the theoretical bases of argumentation, we have been able to model our example while showing the impact of these removals. Then we have studied some properties of the operation of removal.

Although the removal of an argument has been given little attention in the literature, at least three papers have focused on it. Namely, [6] has studied the removal of arguments and attacks (called "abstraction") in a quite particular case since the authors were interested in the situations where there exists only one single extension which one wants to preserve at identical after removal[8]. The results given by [6] make it possible to characterize the set of attacks to remove in order to "conserve the extension" under a given semantics (generally the grounded semantics). This paper also characterizes the set of interactions relating an argument to the argumentation system when the removal of this argument should respect the property of conservation. This may give birth to promising lines of research when developing further the study of the properties defined by [9] applied to removal.

[5] deals with the question (called *"enforcement"*) of how to modify an argumentation system so as to guarantee that a given set of arguments would belong to an extension. The modifications they considered are additions of arguments (with their associated interactions) and semantics switches. Results of impossibility and results concerning monotony are proposed. The authors stress that the removal of argument presents little interest for the problem considered, since it would offer the trivial solution of removing all the arguments that differs from those which one wants to guarantee. However, with a minimal change criterion, it could be interesting to compute the minimal set of arguments to remove so as to guarantee a given set of arguments. It is one prospect of our work.

Let us note finally that [9] gives also examples of removal illustrating various properties of change. Besides, this same article shows that doing a parallel between addition in an argumentation system and revision in the sense of [1] (*AGM*) is not convenient (the formalisms are different and the concept of consistency which is central in the work of *AGM* does not have any equivalent in argumentation). For the same reasons, a parallel between removal of argument and the *AGM* contraction is not really meaningful (even if some concepts of *AGM* have inspired our work).

Several issues are to be specified and improved, we next describe some future orientations of our research.

- Many properties about change in argumentation are to be discovered or deepened, particularly for the removal change.
- Intuitively, it seems that an objected argument, and thus removed one, makes nevertheless its effect on the audience; the jury cannot instantaneously delete this

[8] This "conservation of the extension" preserves all the arguments except for the removed argument.

argument from its mind and is likely to be influenced about it. It would be interesting to study the impact that such an argument can have on the preferences, or on the moral values, of the jury.
- The narrowing change seems to be a dual property of the expansive change. This concept of *duality* between operations and changes should be studied more deeply.
- In the illustrative example, we have seen that it may be beneficial not to reveal some arguments. One of our prospects is to characterize situations, in the way of [12], where it is crucial to select which arguments to reveal or to hide. This will allow us to develop strategies to maximize the chances that the audience accepts a specific argument. Furthermore, it would be interesting to focus particularly on the cases where the participants do not share the same semantics, and on the strategic choices which might arise consequently.

Acknowledgements. We would like to thank the reviewers for their help and valuable suggestions.

References

1. Alchourrón, C., Gärdenfors, P., Makinson, D.: On the logic of theory change: partial meet contraction and revision functions. Journal of Symbolic Logic 50, 510–530 (1985)
2. Amgoud, L., Cayrol, C.: Inferring from inconsistency in preference-based argumentation frameworks. International Journal of Automated Reasoning 29(2), 125–169 (2002)
3. Amgoud, L., Dupin de Saint-Cyr, F.: Extracting the core of a persuasion dialog to evaluate its quality. In: Sossai, C., Chemello, G. (eds.) ECSQARU 2009. LNCS (LNAI), vol. 5590, pp. 59–70. Springer, Heidelberg (2009)
4. Amgoud, L., Maudet, N., Parsons, S.: Modelling dialogues using argumentation. In: Proc. of ICMAS, pp. 31–38 (2000)
5. Baumann, R., Brewka, G.: Expanding argumentation frameworks: Enforcing and monotonicity results. In: Proc. of COMMA, pp. 75–86. IOS Press, Amsterdam (2010)
6. Boella, G., Kaci, S., van der Torre, L.: Dynamics in argumentation with single extensions: Abstraction principles and the grounded extension. In: Sossai, C., Chemello, G. (eds.) ECSQARU 2009. LNCS (LNAI), vol. 5590, pp. 107–118. Springer, Heidelberg (2009)
7. Boella, G., Kaci, S., van der Torre, L.: Dynamics in argumentation with single extensions: Attack refinement and the grounded extension. In: Proc. of AAMAS, pp. 1213–1214 (2009)
8. Brewka, G.: Dynamic argument systems: A formal model of argumentation processes based on situation calculus. Journal of Logic and Computation 11(2), 257–282 (2001)
9. Cayrol, C., Dupin de Saint Cyr, F., Lagasquie-Schiex, M.-C.: Change in Abstract Argumentation Frameworks: Adding an Argument. Journal of Artificial Intelligence Research 38, 49–84 (2010)
10. Dung, P.M.: On the acceptability of arguments and its fundamental role in nonmonotonic reasoning, logic programming and n-person games. Artificial Intelligence 77(2), 321–358 (1995)
11. Moguillansky, M.O., Rotstein, N.D., Falappa, M.A., García, A.J., Simari, G.R.: Argument theory change through defeater activation. In: Proc. of COMMA 2010, pp. 359–366. IOS Press, Amsterdam (2010)
12. Rahwan, I., Larson, K., Tohmé, F.: A characterisation of strategy-proofness for grounded argumentation semantics. In: Proc. of IJCAI 2009, pp. 251–256 (2009)

Annex: Proofs

Proof of Proposition 1. For the first item of this proposition, under any semantics, if there exists an extension $\mathcal{E} \in \mathbf{E}$ such that $Z \in \mathcal{E}$ then $\forall \mathcal{E}' \in \mathbf{E}'$, $\mathcal{E} \not\subset \mathcal{E}'$, since, the change being a suppression, Z does not belong to any extension of $\mathcal{G}'$. For the second item of this proposition, we consider each semantics separately:

Preferred Semantics: let us suppose that Z does not belong to any extension of $\mathcal{G}$. We show that any extension $\mathcal{E}$ of $\mathcal{G}$ is admissible in $\mathcal{G}'$. Let $\mathcal{E} \in \mathbf{E}$:

- $\mathcal{E}$ is conflict-free in $\mathcal{G}$ and thus still conflict-free in $\mathcal{G}'$.
- Let us show that $\mathcal{E}$ defends its elements in $\mathcal{G}'$. If $X \in \mathcal{E}$ such that X is attacked by Y in $\mathcal{G}'$, then X is also attacked by Y in $\mathcal{G}$, but $X \in \mathcal{E}$, therefore it is defended by an argument T which attacks Y in $\mathcal{G}$. Since, we had assume that $Z \notin \mathcal{E}$, we know that $T \neq Z$, therefore $T \in A'$ and thus T attacks also Y in $\mathcal{G}'$. Thus, $\mathcal{E}$ defends X in $\mathcal{G}'$.

$\mathcal{E}$ is thus admissible. In conclusion, since $\mathcal{E}$ is admissible in $\mathcal{G}'$, it is included in one of the preferred extensions of $\mathcal{G}'$.

Stable Semantics: let us suppose that Z does not belong to any extension of $\mathcal{G}$. We show that any stable extension $\mathcal{E}$ of $\mathcal{G}$ is a stable extension in $\mathcal{G}'$. Let $\mathcal{E} \in \mathbf{E}$:

- $\mathcal{E}$ is conflict-free in $\mathcal{G}$ and thus still conflict-free in $\mathcal{G}'$.
- If $Y \in A'$ and $Y \notin \mathcal{E}$ then $Y \in A$ and $Y \notin \mathcal{E}$. Since the extension $\mathcal{E}$ is stable in $\mathcal{G}$, $\mathcal{E}$ attacks Y in $\mathcal{G}$. Therefore, there exists $T \in \mathcal{E}$ such that T attacks Y. As we had assumed that $Z \notin \mathcal{E}$, we know that $Z \neq T$, so T attacks Y in $\mathcal{G}'$.

$\mathcal{E}$ is thus stable in $\mathcal{G}'$, hence $\mathcal{E} \in \mathbf{E}'$ so $\mathcal{E}$ is included in a stable extension of $\mathcal{G}'$.

Grounded Semantics: *Case where* $\mathbf{E} = \{\{\}\}$: Let us proceed similarly to preferred semantics: it is known that $\mathbf{E}'$ is nonempty (since we are under the grounded semantics). Thus there exists $\mathcal{E}' \in \mathbf{E}'$. Since $\mathcal{E} = \varnothing \subseteq \mathcal{E}'$, hence the proposition is true.
Case where $\mathbf{E} \neq \{\{\}\}$: Let us suppose that Z does not belong to the grounded extension of $\mathcal{G}$. It is enough to show that the extension $\mathcal{E}$ of $\mathcal{G}$ is included in the grounded extension $\mathcal{E}'$ of $\mathcal{G}'$. We know, thanks to Definition 1, that the binary relation $\mathbf{R}$ is finite. However, according to [10], if $\mathbf{R}$ is finite then $\mathcal{E} = \bigcup_{I \geq 1} \mathcal{F}^i(\varnothing)$ and $\mathcal{E}' = \bigcup_{I \geq 1} \mathcal{F}'^i(\varnothing)$. Let us prove by induction on $i \geq 1$ that $\mathcal{F}^i(\varnothing) \subseteq \mathcal{F}'^i(\varnothing)$.

- $i = 1$: for any argument Y, if $Y \in \mathcal{F}(\varnothing)$ then Y is not attacked in $\mathcal{G}$. Removing Z does not change anything about that, Y is thus not attacked in $\mathcal{G}'$, and thus $Y \in \mathcal{F}'(\varnothing)$.
- Induction assumption (for $1 \leq I \leq p, \mathcal{F}^i(\varnothing) \subseteq \mathcal{F}'^i(\varnothing)$): Let $\mathcal{S} = \mathcal{F}^p(\varnothing)$ and $\mathcal{S}' = \mathcal{F}'^p(\varnothing)$. First of all, let us prove that $\mathcal{F}(\mathcal{S}) \subseteq \mathcal{F}'(\mathcal{S})$. Let $Y \in \mathcal{F}(\mathcal{S})$. By definition, $\mathcal{F}(\mathcal{S}) \subseteq \mathcal{E}$, therefore $Y \in \mathcal{E}$. If Y is attacked by X in $\mathcal{G}'$ then Y is attacked by X in $\mathcal{G}$. But since $Y \in \mathcal{F}(\mathcal{S})$, $\mathcal{S}$ defends Y, therefore $\exists T \in \mathcal{S}$ such that T attacks X in $\mathcal{G}$. By assumption, $Z \notin \mathcal{E}$, therefore $Z \notin \mathcal{S}$, therefore $T \neq Z$ and thus $T \in \mathbf{A}'$. Thus, $\mathcal{S}$ defends Y in $\mathcal{G}'$. Thus $Y \in \mathcal{F}'(\mathcal{S})$.

  We have just shown that $\mathcal{F}(\mathcal{S}) \subseteq \mathcal{F}'(\mathcal{S})$ and we also have, using the induction assumption, $\mathcal{S} \subseteq \mathcal{S}'$. Knowing that $\mathcal{F}'$ is monotonous (by definition), we have $\mathcal{F}(\mathcal{S}) = \mathcal{F}^{p+1}(\varnothing) \subseteq \mathcal{F}'(\mathcal{S}) \subseteq \mathcal{F}'(\mathcal{S}') = \mathcal{F}'^{p+1}(\varnothing)$. Therefore, $\mathcal{E} \subseteq \mathcal{E}'$. ∎

Proof of Proposition 2. **Preferred semantics:** Let us suppose that $\mathcal{E} \setminus \{Z\}$ is not admissible in $\mathcal{G}'$. $\mathcal{E}$ being an extension of $\mathcal{G}$, there is no conflict in $\mathcal{E} \setminus \{Z\}$, therefore it exists an argument $Y \in \mathcal{E} \setminus \{Z\}$ such that Y is not defended by $\mathcal{E} \setminus \{Z\}$ in $\mathcal{G}'$. Thus there exists an argument $T \in \mathcal{G}'$ such that T attacks Y in $\mathcal{G}'$. Since we are removing Z, Z can be neither Y, nor T, therefore T also attacks Y in $\mathcal{G}$. Moreover, Y cannot be defended by Z in $\mathcal{G}$ since Z does not attack any argument, therefore Y is not defended by $\mathcal{E} \setminus \{Z\}$ in $\mathcal{G}$, and thus $\mathcal{E} \setminus \{Z\}$ is not admissible in $\mathcal{G}$, which contradicts our starting assumption. Thus, $\mathcal{E} \setminus \{Z\}$ is admissible in $\mathcal{G}'$ and is thus contained in a preferred extension of $\mathcal{G}'$.

Stable Semantics: let Y be an argument such that $Y \notin \mathcal{E} \setminus \{Z\}$, and $Y \in \mathcal{G}'$. Then, $Y \neq Z$ and thus $Y \notin \mathcal{E}$. However, $\mathcal{E}$ is a stable extension of $\mathcal{G}$, therefore $\mathcal{E}$ attacks Y in $\mathcal{G}$. As Z does not attack any argument, $\mathcal{E} \setminus \{Z\}$ attacks also Y in $\mathcal{G}$. Besides since the change is a removal, $\mathcal{E} \setminus \{Z\}$ attacks also Y in $\mathcal{G}'$ and thus $\mathcal{E} \setminus \{Z\}$ is stable in $\mathcal{G}'$. ■

Lemma 1. *When removing an argument Z under the preferred semantics, if Z does not attack any argument, any extension of $\mathcal{G}'$ is admissible in $\mathcal{G}$.*

Proof of Lemma 1. Let $\mathcal{E}'$ be a preferred extension of $\mathcal{G}'$. $\mathcal{E}'$ is conflict-free in $\mathcal{G}'$ and thus in $\mathcal{G}$ also. If an argument $Y \in \mathcal{E}'$ is attacked in $\mathcal{G}$ by another argument X then $X \neq Z$ and $X \in \mathcal{G}'$, therefore Y is also attacked by X in $\mathcal{G}'$. $\mathcal{E}'$ is a preferred extension of $\mathcal{G}'$ which contains Y, hence $\mathcal{E}'$ attacks X in $\mathcal{G}'$, so in $\mathcal{G}$. ■

Proof of Proposition 3. **Preferred semantics:** let $\mathcal{E}$ be a preferred extension of $\mathcal{G}$. According to Proposition 2, there exists an extension $\mathcal{E}'$ of $\mathcal{G}'$ such that $\mathcal{E} \setminus \{Z\} \subseteq \mathcal{E}'$. However, $Z \notin \mathcal{E}$, therefore $\mathcal{E} = \mathcal{E} \setminus \{Z\} \subseteq \mathcal{E}'$. In addition, according to Lemma 1, since Z does not attack any argument, any extension of $\mathcal{G}'$ is admissible in $\mathcal{G}$, therefore there exists an extension $\mathcal{E}_i$ of $\mathcal{G}$ such that $\mathcal{E}' \subseteq \mathcal{E}_i$. Thus $\mathcal{E} \subseteq \mathcal{E}' \subseteq \mathcal{E}_i$. However, $\mathcal{E}$ is a maximal admissible set for set-inclusion in $\mathcal{G}$. Thus $\mathcal{E} = \mathcal{E}' = \mathcal{E}_i$. Thus, $\mathcal{E}$ is an extension of $\mathcal{G}'$.

Stable semantics: it is directly due to Proposition 2 and to the fact that $\mathcal{E} \setminus \{Z\} = \mathcal{E}$.

Grounded semantics: let $\mathcal{E}$ be the single grounded extension of $\mathcal{G}$ and let $Z \in \mathcal{G}$ be an argument such that $Z \notin \mathcal{E}$. According to Proposition 1, we get $\mathcal{E} \subseteq \mathcal{E}'$, where $\mathcal{E}'$ is the grounded extension of $\mathcal{G}'$.

It remains to establish that $\mathcal{E}' \subseteq \mathcal{E}$. For this purpose, we show that $\mathcal{E}$ is a fixpoint of $\mathcal{F}'$ and since $\mathcal{E}'$ is the least one, we have $\mathcal{E}' \subseteq \mathcal{E}$. Thus let us show that $\mathcal{F}'(\mathcal{E}) = \mathcal{E}$.

- First, let us prove that $\mathcal{F}'(\mathcal{E}) \subseteq \mathcal{E}$. Let $Y \in \mathcal{F}'(\mathcal{E})$ then $Y \neq Z$ and $\mathcal{E}$ defends Y in $\mathcal{G}'$. Since Z does not attack any argument, the only attackers of Y in $\mathcal{G}$ are those of $\mathcal{G}'$, therefore $\mathcal{E}$ defends Y in $\mathcal{G}$ and $Y \in \mathcal{F}(\mathcal{E}) = \mathcal{E}$.
- Conversely, let us show now that $\mathcal{E} \subseteq \mathcal{F}'(\mathcal{E})$. Let $Y \in \mathcal{E} = \mathcal{F}(\mathcal{E})$ then $\mathcal{E}$ defends Y in $\mathcal{G}$. We know that $Z \notin \mathcal{E}$ thus $Y \neq Z$. Since Z attacks no argument, $\mathcal{E}$ thus defends Y in $\mathcal{G}'$ and $Y \in \mathcal{F}'(\mathcal{E})$.

Thus $\mathcal{E} = \mathcal{F}'(\mathcal{E})$ and we have $\mathcal{E}' \subseteq \mathcal{E}$. In conclusion, $\mathcal{E}$ is an extension of $\mathcal{G}'$. ■

Proof of Proposition 4. Let us suppose that there exists an expansive suppression. It is thus assumed that $\mathbf{E} \neq \varnothing$, $|\mathbf{E}| = |\mathbf{E}'|$, and for any extension $\mathcal{E}'$ of $\mathcal{G}'$, there exists an extension $\mathcal{E}$ of $\mathcal{G}$ such that $\mathcal{E} \subset \mathcal{E}'$. Let us consider any extension $\mathcal{E}'_j$ of $\mathcal{G}'$ then there exists an extension $\mathcal{E}_i$ of $\mathcal{G}$ such that $\mathcal{E}_i \subset \mathcal{E}'_j$. Thus there exists an argument $Y \in \mathcal{E}'_j$ such that $Y \notin \mathcal{E}_i$. Let us note that $Y \in \mathcal{G}$ since we are in the case of the suppression of an argument. $\mathcal{E}_i$ being stable in $\mathcal{G}$, there exists an argument $T \in \mathcal{E}_i$ such that T attacks Y in $\mathcal{G}$. However, $\mathcal{E}_i \subset \mathcal{E}'_j$, therefore $T \in \mathcal{E}'_j$, and, by assumption, $Y \in \mathcal{E}'_j$. Thus T attacks Y in $\mathcal{G}'$ and thus $\mathcal{E}'_j$ is not conflict-free, which contradicts our starting assumption. ■

Proof of Proposition 5. The first item of this proposition comes directly from the definition of the expansive change and Proposition 1, both for preferred semantics or grounded semantics.

For the second item, **Preferred semantics:** let us suppose that there exists an expansive change and that Z does not attack any argument from $\mathcal{G}$. It is thus supposed that $\mathbf{E} \neq \varnothing$, $|\mathbf{E}| = |\mathbf{E}'|$ and for any extension $\mathcal{E}'$ of $\mathcal{G}'$, there exists an extension $\mathcal{E}$ of $\mathcal{G}$ such that $\mathcal{E} \subset \mathcal{E}'$. Due to the first item of Proposition 5, we know that Z does not belong to any extension of $\mathcal{G}$. If Z does not attack any argument of $\mathcal{G}$ then $\forall \mathcal{E} \in \mathbf{E}$, $\mathcal{E} \setminus \{Z\} = \mathcal{E}$ and, according to Proposition 2, $\mathcal{E}$ is an admissible set in $\mathcal{G}'$. Therefore, $\mathcal{E} \subseteq \mathcal{E}'$, where $\mathcal{E}'$ is a maximal admissible set of $\mathcal{G}'$. However, according to Lemma 1, since Z does not attack anything, $\mathcal{E}'$ is also an admissible set in $\mathcal{G}$, therefore $\mathcal{E}' \subseteq \mathcal{E}$ and thus $\mathcal{E} = \mathcal{E}'$, which contradicts our starting assumption.

Grounded semantics: let us suppose that there exists an expansive change and that Z attacks no argument of $\mathcal{G}$. According to item 1 of Proposition 5, Z does not belong to the grounded extension of $\mathcal{G}$ and, due to Proposition 3, it holds that $\mathcal{E} = \mathcal{E}'$, where $\mathcal{E}'$ is the grounded extension of $\mathcal{G}'$, which contradicts the expansive change. ■

Proof of Proposition 6. **Grounded semantics:** let us suppose that Z does not belong to any extension of $\mathcal{G}$. According to Proposition 1, we have $\mathcal{E} \subseteq \mathcal{E}'$, where $\mathcal{E}$ (resp. $\mathcal{E}'$) is the single grounded extension of $\mathcal{G}$ (resp. $\mathcal{G}'$), which is contradictory with the definition of the narrowing change. **Preferred and stable semantics:** let us suppose that Z does not belong to any extension of $\mathcal{G}$. According to Proposition 1, $\forall \mathcal{E} \in \mathbf{E}$, $\exists \mathcal{E}' \in \mathbf{E}'$, $\mathcal{E} \subseteq \mathcal{E}'$. However, the change being narrowing, $\mathbf{E} \neq \varnothing$ and $\mathbf{E}' \neq \varnothing$. Let $\mathcal{E}_i \in \mathbf{E}$ be an extension, thus it exists an extension $\mathcal{E}'_j \in \mathbf{E}'$ such that $\mathcal{E}_i \subseteq \mathcal{E}'_j$. In addition, still due to the definition of the narrowing change, there exists an extension $\mathcal{E}_k \in \mathbf{E}$ such that $\mathcal{E}'_j \subset \mathcal{E}_k$. We get $\mathcal{E}_i \subset \mathcal{E}_k$. In the case of the **Preferred semantics**, $\mathcal{E}_i$ is not a maximal admissible set and consequently, is not an extension of $\mathcal{G}$, which contradicts our assumption. In the case of the **Stable semantics**, each stable extension being also preferred, this is also impossible under the stable semantics. ■

Incorporating Domain Knowledge and User Expertise in Probabilistic Tuple Merging

Fabian Panse and Norbert Ritter

Universität Hamburg, Vogt-Kölln Straße 33, 22527 Hamburg, Germany
{panse,ritter}@informatik.uni-hamburg.de
http://vsis-www.informatik.uni-hamburg.de/

Abstract. Today, probabilistic databases (PDB) become helpful in several application areas. In the context of cleaning a single PDB or integrating multiple PDBs, duplicate tuples need to be merged. A basic approach for merging probabilistic tuples is simply to build the union of their sets of possible instances. In a merging process, however, often additional domain knowledge or user expertise is available. For that reason, in this paper we extend the basic approach with aggregation functions, knowledge rules, and instance weights for incorporating external knowledge in the merging process.

Keywords: probabilistic data, tuple merging, external knowledge.

1 Introduction

In recent time, the need for probabilistic databases grows in many real-world applications [17,18,8,15]. In general, for certain databases as well as for probabilistic databases duplicates are pervasive problems of data quality [7]. To solve this problem duplicates have to be identified and merged. Strategies for resolving data conflicts in a merge of certain tuples is extensively discussed in the literature [3,11]. However, there is only a low attention on the merge of probabilistic tuples, so far. Nevertheless, if probabilistic source data are given, the degree of uncertainty which has to be resolved during the merging process is higher than in the merge of certain tuples. On the other hand, probabilistic data models provide new capabilities for handling conflicts in the merging process. Thus, tuple merging becomes also more powerful [6,16]. In [12] we introduce a basic approach for merging the instance data of probabilistic tuples which is conceptually based on the set union operator. In real duplicate elimination scenarios, however, often a lot of domain knowledge or user expertise is available. This knowledge cannot be included in our simple merging approach. For that reason, we extend this approach by enabling the user to define aggregation functions for single attributes, and instance weights as well as knowledge rules for whole instances. The incorporation of external knowledge is an important property, because in several scenarios a simple union of all possible instances does not correspond with the semantics of some attributes (see motivating example below), or the set of possible instances can be evidently reduced.

S. Benferhat and J. Grant (Eds.): SUM 2011, LNAI 6929, pp. 289–302, 2011.
© Springer-Verlag Berlin Heidelberg 2011

(i) Source Data: t_1 and t_2

	name	producer	stock
I_1	Twix	Maas Inc.	15
I_2	Dwix	Nestle	20

	name	producer	stock
I_3	Twix	Mars Inc.	6
I_4	Raider	Mars Inc.	8

(ii) Basic Approach:

name	producer	stock
Twix	Maas Inc.	15
Dwix	Nestle	20
Twix	Mars Inc.	6
Raider	Mars Inc.	8

(iii) Extended Approach:

name	producer	stock
Twix	Mars Inc.	21
Twix	Mars Inc.	23
Raider	Mars Inc.	23
~~Dwix~~	~~Mars Inc.~~	~~26~~
~~Twix~~	~~Mars Inc.~~	~~26~~
~~Dwix~~	~~Mars Inc.~~	~~28~~
~~Raider~~	~~Mars Inc.~~	~~28~~

r : (producer='Mars Inc.') $\rightarrow$ (stock < 25)

Fig. 1. The possible instances of t_1 (I_1 and I_2) and t_2 (I_3 and I_4) (i), the instances resulting from merging $\{t_1, t_2\}$ with the basic approach (ii), and the instances resulting from merging $\{t_1, t_2\}$ whilst taking external knowledge into account (iii)

As a motivating example, we consider a merge of the two base-tuples t_1 and t_2 as shown in Figure 1. Both tuples have two possible instances ($\{I_1, I_2\}$ for t_1 and $\{I_3, I_4\}$ for t_2) and are defined on a schema *inventory* with the three attributes *name*, *producer* and *stock*. Both tuples represent the same product (and hence are duplicates), but the stock information of each tuple belongs to different orders. Therefore, in this scenario neither 15, 20, 6 nor 8 items of this product, but rather 21, 23, 26 or 28 items are available. As a consequence, the true stock value of this product results from the sum of the stocks of both base-tuples instead of being the stock of one of them. Moreover, the responsible user knows that the producer name of the second tuple ('Mars Inc.') is the correct one. Thus, this value is chosen for all possible instances of the merged tuple. Finally, it is known that the company never bought more than 25 items of an article produced by 'Mars Inc.'. Hence some of the resultant instances can be excluded for sure (see knowledge rule r). In conclusion, the result of the extended approach is much more accurate than the result of the basic approach. Moreover, by using the extended approach the final values of all attributes are nearly known for sure ('Mars Inc.' with certainty 1, 'Twix' and '23' with certainty 2/3).

The main contributions of this paper are:

- a discussion about different kinds of external knowledge and in which way these can be incorporated into the merging process.
- a detailed description of a merging approach extended by aggregation functions, knowledge rules, and instance weights. Moreover, we show that this approach is a generalization of existing methods for merging certain tuples and is a generalization of our basic approach based on the set union operator.
- a discussion about the characteristics of the extended merging approach.

The outline of the paper is as follows: First we present some basics on probabilistic data and duplicate elimination including our basic approach for probabilistic tuple merging (Section 2). Then we present the types of external knowledge we handle in this work (Section 3). In Section 4, we discuss aggregation functions, knowledge rules, and instance weights in more detail, before introducing the extended version of our probabilistic tuple merging approach in Section 5. Finally, we present related work in Section 6 and conclude in Section 7.

2 Basics

In this section, we introduce a definition of probabilistic tuples, before we present the concept of duplicate elimination including our basic approach for merging probabilistic tuples [12].

2.1 Probabilistic Tuples

In this paper, we primarily focus on the merge of tuples in relational tuple-independent probabilistic databases defined on an arbitrary probability measure P. This class of databases includes BID-tables [4], x-relations [2] without lineage (e.g. base x-relations without external lineage), and U-relations [9] where tuples with different TIDs do not share same variables.

Due to duplicates are most often detected in base relations (data cleaning) or in combining multiple independent source (data integration) duplicate elimination in tuple-independent PDBs already covers a wide space of real-world scenarios. To be independent from the used representation system (BID, MayBMS, etc.), we consider a probabilistic tuple within the *possible world semantics* [4]. Thus, similar to the ULDB model [2], we define a probabilistic tuple as a set of possible mutually exclusive instances, also denoted as tuple alternatives [2] and define a probabilistic relation (referred to as $\mathcal{R}^p$) as a set of probabilistic tuples.

Definition 1 (Probabilistic Tuple): *Let $sch(\mathcal{R})$ be a relation schema with the domain $dom(\mathcal{R})$. A probabilistic tuple t defined on $sch(\mathcal{R})$ is a set of possible instances $pI(t) = \{I_1, \dots, I_k\}$ where each instance is an ordinary tuple $I_i \in dom(\mathcal{R})$. Moreover, each instance $I \in pI(t)$ is assigned with a probability $P(t[I]) > 0$ where $t[I]$ is the event that I is the true instance of t. Trivially, all possible instances of t are mutually exclusive: $(\forall I_1, I_2 \in pI(t)) : P(t[I_1] \mid t[I_2]) = 0$. The probability that t exist is: $P(t) = \sum_{I \in pI(t)} P(t[I]) \leq 1$.*

Since all tuples are independent of each other, the true instantiation of one tuple does not depend on the true instantiation of another tuple:

$$(\forall t_1, t_2 \in \mathcal{R}^p, t_1 \neq t_2) : (\forall I_1 \in pI(t_1), I_2 \in pI(t_2)) : P(t_1[I_1] \mid t_2[I_2]) = P(t_1[I_1])$$

To make some of the considerations of this paper easier, we introduce the null-instance $I_\perp$ which represents the case a probabilistic tuple does not exist ($P(t[I_\perp]) = 1 - P(t)$). The null-instance is schemaless, i.e. $\pi_A(I_\perp) = I_\perp$ for every valid set of attributes A and $I_\perp \times S = I_\perp$ for every relation S. For simplification, we also define the set $pI_\perp(t)$:

$$pI_\perp(t) = \begin{cases} pI(t) \cup \{I_\perp\}, & \text{iff } P(t) < 1 \\ pI(t), & \text{else} \end{cases} \tag{1}$$

Note that the null instance $I_\perp$ and hence the set $pI_\perp(t)$ are only virtual and not stored in the database.

In the rest of the paper, we represent the set of possible instances of a probabilistic tuple by an own table (one row per instance). Figure 2.1 shows an example of a tuple modeling the movie 'Crash' with two possible instances I_1 and I_2.

	title	year	studio	total	P
I_1	Crash	2004	WB	12k	0.5
I_2	Cash	2005	US	15k	0.2
$I_\perp$		———			0.3

Fig. 2. A probabilistic tuple t with $pI_\perp(t) = \{I_1, I_2, I_\perp\}$, $I_1 = $("Crash",2004,"WB",12k), $I_2 = $ ("Cash",2005,"US",15k), $P(t[I_1]) = 0.5$, $P(t[I_2]) = 0.2$, and $P(t[I_\perp]) = 0.3$

2.2 Duplicate Elimination

In the first duplicate elimination step multiple representations of same real-world entities are detected [7]. The result of this step is a partitioning of the set of input tuples into duplicate cluster (one cluster for each real-world entity).

In the second step, all tuples of one duplicate cluster are merged to a single one. This step is usually denoted as tuple merging [19] or data fusion [3]. In [12] we gave a first discussion on tuple merging in probabilistic data. We split probabilistic tuple merging into two steps: (a) a merging of instance data and (b) a merging of tuple membership. In this paper, we consider only merging of tuples defined in same contexts and hence we simply include membership merging in instance merging by using the null-instance $I_\perp$. In the following, we always consider a single cluster and hence denote the merged tuple as t_μ.

Basic Merging Approach. The basic approach for probabilistic tuple merging we present in [12] is based on the set union operator. This means that an instance is possible for the merged tuple, if it was possible for at least one base-tuple. Since the merging is not associative, if the resultant probability is simply computed by the probabilities of the base-tuples, we assign a weight $w(t)$ to each base-tuple t and define that the weight of a merged tuple $t_\mu = \mu(\{t_1, \ldots, t_k\})$ results from the sum of the weights of its base-tuples ($w(t_\mu) = w(t_1) + \ldots + w(t_k)$). If tuple merging is considered within the context of data integration, the reliabilities of the corresponding sources can be used as tuple weights. Probabilities are computed by a weighted average. Let t_μ be the tuple merged from the base-tuples of cluster C, our basic approach of probabilistic tuple merging can be formalized as:

$$pI_\perp(t_\mu) = \bigcup_{t \in C} pI_\perp(t) \quad (2) \qquad \forall I \in pI_\perp(t_\mu),\ P(t_\mu[I]) = \sum_{t \in C} \frac{w(t)}{w(t_\mu)} P(t[I]) \quad (3)$$

3 Domain Knowledge and User Expertise

In this paper, we consider two different kinds of external knowledge: (i) domain knowledge which is generally applicable for a specific domain, as for example the information that stock values have to be summed up (see motivating example) and (ii) user expertise which is only applicable for individual items or groups of items (sources, tuples, values, etc.), as for example the information that the studio 'WB' does not produce movies for adults. Domain knowledge usually concerns metadata like the correct semantics of relational schemas or the correct scope of attribute domains. Therefore, domain knowledge does not change over time very frequently and hence acquired once, it can be used for multiple merging processes. In contrast, user expertise often concerns only the given instance data (see example above). Thus, a user can be very competent for the data of one merging process and incompetent for the data of another, even if both processes work on equivalent schemas. For example, a user can be an expert for horror movies and a non-expert for romantic movies.

With respect to its effects on the merge result, external knowledge can be furthermore classified into the following four types:

1. Knowledge about specific semantics of individual attributes or sets of attributes. For example, the knowledge that the numbers of sold tickets stored in duplicate entries in a box office list belong to different points of time and hence the maximum value has to be chosen. Knowledge of this type is usually domain knowledge.
2. Knowledge about the true instance of one attribute value or a set of attribute values. For example, the user knows that one of the given values is the true one, or the user knows that the true value is missing and introduces it himself. Knowledge of this type usually results from user expertise.
3. Knowledge required for excluding some combinations of attribute values for sure. For example, such knowledge can be based on physical rules (e.g. a studio cannot have produced movies before it was founded), economical rules (e.g. a salary cap), or private guidelines (e.g. a specific company never buys more than 100 items from a single article per month). Knowledge of this type can be domain knowledge, user expertise or a combination of both.
4. Knowledge about new evidence or further evidence, or a user's own degree of belief. For example, at merging time a person is known to life rather in Italy than in France. Knowledge of this type is most often user expertise.

A modeling of knowledge about tuple correlations (e.g. a specific attribute has to be unique) implies a definition of new tuple dependencies. Since we restrict to tuple-independent PDBs, this is out of the scope of this paper.

4 Methods for Incorporating External Knowledge

For incorporating domain knowledge and user expertise, we resort to three classical concepts which have already been partially used in the merge of certain data: user-defined aggregation functions, user-defined knowledge rules, and user-defined weights of possible instances.

Aggregation functions can be used to assign specific semantics to concrete attributes (knowledge of Type 1) or to define the true value for a concrete attribute by hand (knowledge of Type 2). In contrast, knowledge rules are excellently suited for excluding instances which violate a given pattern of regulations (knowledge of Type 3). Instance weights can be used to accommodate new evidence (knowledge of Type 4).

Aggregation functions are already used for resolving conflicts in the fusion of certain data by Bleiholder et al. [3]. Note that we use the concept of aggregation for another purpose. In the merge of certain data for each attribute only a single value can be stored. Thus, conflicting values need to be resolved and the usage of aggregation functions is often mandatory, even if the user does not know how to aggregate these values at best. In contrast, such a conflict resolution is not required by using a probabilistic data model as the target model because all possible values can be stored simultaneously. We use these functions only for incorporating available context information. Therefore, in our approach these functions should be only used, if this information is given for sure (or at least very likely). This in turn implies that a lot of aggregation functions listed in [3] are not suitable for our purpose (e.g. First, Last, Random, etc.), because they do not express a certain kind of knowledge.

Knowledge rules are already used by Whang et al. [19] for preventing invalid merging results and hence for detecting the best merging easier. In general, we use these rules for same purposes, because we use them to avoid invalid instances.

In the rest of this section, we take a closer look at these three concepts and how they can be used to express a certain kind of knowledge.

4.1 Aggregation Functions

Aggregation functions are a simple and adequate method for incorporating external knowledge into a tuple merging process. For aggregation we consider functions as defined for conflict resolution in certain data [3]. This set of functions can be classified into deciding functions which choose one of the given values and mediating functions which create a new value.

Moreover, aggregation functions can be of a simple or complex nature. A simple aggregation function only takes the values of the considered attribute into account or is a constant function which does not need any input at all. In contrast, complex aggregation functions also consider the values of other attributes (from input as well as output) as own input. Thus, they aggregate a set of given input values depending from the result of aggregating other attributes or from the initial values of other attributes. As an example, consider a deciding function with takes the value for attribute a_i that occurs as most often with the value already chosen for attribute a_j.

Simple Aggregation Functions. A simple aggregation function aggregates only the values of the attribute it is defined for or returns a constant. Let f_i be a simple aggregation function which aggregates the input values $A_i^I = \{v_1, \ldots, v_m\}$ of an attribute a_i to the single output value $v_{f_i} \in dom(a_i)$, $f_i(A_i^I)$ is defined as:

$$f_i : dom(a_i)^m \rightarrow dom(a_i) \qquad f_i : \{v_1, \ldots, v_m\} \mapsto v_{f_i}$$

If for aggregating a set of instances M only simple functions are used (each for another attribute), each function can be applied independently and the output instance I_O results by the cross product of all the functions' output values. Let $A = \{a_1, \ldots, a_n\}$ be a set of attributes an aggregation function is defined for (f_i for a_i), the result from aggregating the input set $\pi_A(M)$ with $f_1 - f_n$ is:

$$I_O = f_1(\pi_{a_1}(M)) \times f_2(\pi_{a_2}(M)) \times \ldots \times f_n(\pi_{a_n}(M)) = (v_{f_1}, v_{f_2}, \ldots, v_{f_n})$$

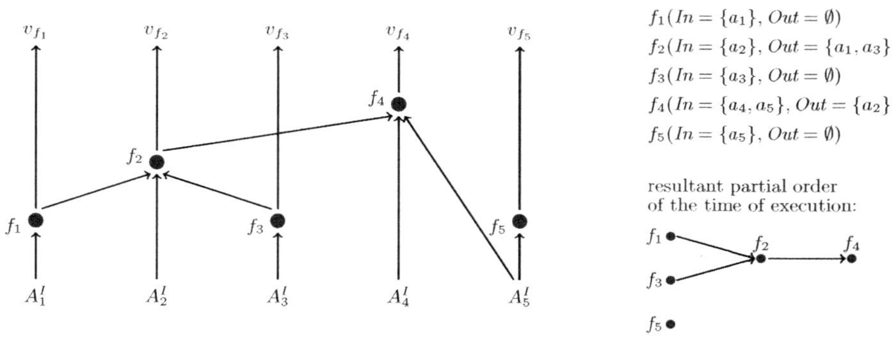

Fig. 3. Set of five sample aggregation functions $\{f_1, \ldots, f_5\}$ along with the dependencies between their input $A_i^I = \pi_{a_i}(M)$ and output $v_{f_i} = f_i(In, Out)$ (left side) and the resultant partial order of their time of execution (right side, below)

Complex Aggregation Functions. A complex aggregation function also consider the input values and/or output values of other functions for producing its own output. Thus, we need a more general definition of aggregation functions. In the following, a function f_i aggregating the values of attribute a_i is described by a set of attributes In which values are used from the input data[1] and by a set of attributes Out which values are used from the output data. Note, for each $a_j \in In$ a set of input values is used, i.e. A_j^I, but for each $a_j \in Out$ only a single output value is processed, i.e. v_{f_j}. Thus, $f_i(In, Out)$ is defined as:

$$f_i : \{A_j^I \mid a_j \in In\} \times \{v_{f_j} \mid a_j \in Out\} \mapsto v_{f_i}$$

Note, by using this description, a function f_i is simple, if the set In contains at most the attribute a_i and the set Out is empty: $f_i(In = \{a_i\}, Out = \emptyset)$.

The combined execution of aggregation functions becomes much more complicated, if complex functions are involved. Certainly, an output value of one function has to be produced before it can be serve as the input for another function. In general, a given set of complex aggregation functions has to be executed according to the partial order of the dependencies between their input values and output values. This fact is illustrated in Figure 3 where the dependencies between the input values and output values of five aggregation functions (left side) as well as their resultant partial order (right side) are depict.

The combined execution of a set of complex aggregation functions F for aggregating the set of instances M to the output instance I_O using the set of attributes A as input is in the following denoted as $I_O = \mathfrak{F}(F, \pi_A(M))$.

4.2 Knowledge Rules

Compared to aggregation functions, knowledge rules enhance the capability to incorporate knowledge further on in two ways: First, whereas aggregation

[1] The input of a function can also contain metadata like the age of a value [3].

functions only influence the set of attributes $\{a_l, \ldots, a_n\}$ for which such a function is defined, by the usage of knowledge rules conditions for whole instances can be specified. Second, it is in the nature of aggregation functions that they choose a single value and exclude all other ones. Nevertheless, such a restrictive knowledge is often not available (knowledge of Type 3), but instead we can only exclude a single value (or few values) to be the true one. Such restrictions of the set of possible instances, however, cannot be realized by aggregation functions.

Knowledge rules are logical rules of inference (*premises* $\rightarrow$ *conclusions*) which take premises and return conclusions. A rule is violated by an instance, if for this instance all premises are valid, but the conclusions are not. In this way, impossible instances can be excluded from the merging result.

As an example we consider a combination of the general domain knowledge that studios cannot have produced movies before they were founded and the specific user expertise that the studio 'WB' (Warner Bros. Entertainment) was founded in 1923. Thus, we can conclude that each instance having the value 'WB' as studio name and having a value year lower than 1923 cannot be true:

$$\text{rule } r_1 : \text{studio='WB'} \rightarrow \text{year} \geq 1923 \tag{4}$$

A knowledge rule can use values from the output (the instances of the merged tuple) as well as values from the input (the instances of the base-tuples). One meaningful example is the condition that a combination of values for an attribute set A is only valid for the merged tuple, if it was valid for at least one base-tuple:

$$\text{rule } r_2 : I \in \pi_A(pI(t_\mu)) \rightarrow (\exists t \in C) : I \in \pi_A(pI(t)) \tag{5}$$

Knowledge rules are applied to each possible instance individually. Instances violating one or more of the defined rules can be excluded to be the true one and hence are removed from the merging result.

4.3 Instance Weights

In our basic approach (Section 2.2), we use tuple weights for (a) making the merging process associative and (b) allowing an assignment of different degrees of trust to individual sources. To make the merging process more adaptable to further evidence known at merging time, we also allow a definition of weights on instance level: $w(t, I)$ is the weight of instance I for tuple t. Thus, the user can prefer a base-instance I_1 to another base-instance I_2 ($w(t, I_1) > w(t, I_2)$) or can exclude a base-instance I from the merging process for sure ($w(t, I) = 0$) without manipulating the original probabilities. Typically, weights are assigned for each instance individually and hence represent user expertise. Nevertheless, weights can be also assigned by a given pattern (e.g. a weight w is assigned to instances satisfying a specific condition derived from the semantics of the considered universe of discourse) and hence also can be used to express domain knowledge.

We define the weight of a tuple t as the expected weight of its instances: $w(t) = \sum_{I \in pI(t)} w(t, I)P(t[I])$. For ensuring associativity, the weight of a merged tuple is still the sum of the tuple it is merged from. Moreover, all instances of the merged tuple are weighted equally, i.e. the new evidence is already incorporated in the resultant probabilities.

5 Extended Approach for Probabilistic Tuple Merging

For incorporating external knowledge according to the possible world semantics, the aggregation functions have to be applied to each possible combination (so called *merging lists*) of the base-tuples' instances (one instance per base-tuple) individually. Each merging list $M = \{I_1, \dots, I_k\}$ contains as many instances as base-tuples to be merged (in this case k). We consider the instances in a merging list to be sorted by their corresponding base-tuples, meaning that I_i originates from tuple t_i. Performing aggregation on a single merging list is denoted as fusion. Knowledge rules are applied to the merging lists' fused instances.

Let $C = \{t_1, \dots, t_k\}$ be a set of base-tuples to be merged. Let $w(t, I)$ be the weight defined for instance $I \in pI(t), t \in C$ and let $w(t_\mu) = \sum_{t \in C} w(t)$ be the total weight of all base-tuples. Moreover, let $N = \{r_1, \dots, r_q\}$ be a set of knowledge rules and let $A = \{a_1, \dots, a_{l-1}, a_l, \dots, a_n\}$ be the attributes of the considered schema, where for each of the attributes $A_2 = \{a_l, \dots, a_n\}$ an aggregation function is defined for (f_i for a_i). Our tuple merging approach extended with aggregation functions, knowledge rules, and instance weights is performed by the following steps (the first two steps are illustrated in Figure 4):

1. **Divide the Input.** First, all merging lists are built:

$$\mathcal{M}(\{t_1, t_2, \dots, t_k\}) = \{\{I_1, I_2 \dots, I_k\} \mid I_i \in pI_\perp(t_i), t_i \in C\} \tag{6}$$

2. **Apply Aggregation.** Then, each merging list $M \in \mathcal{M}(C)$ is fused by applying the set of aggregation functions $F = \{f_l, \dots, f_n\}$ (defined for attributes $\{a_l, \dots, a_n\}$) having the attribute set $A_I \subseteq A$ as input. For the attributes without any aggregation function (attributes $A_1 = \{a_1, \dots, a_{l-1}\}$) all possible values are taken into account (recall $I_\perp \times S = I_\perp$ and $\pi_A(I_\perp) = I_\perp$):

$$\forall M \in \mathcal{M}(C), \quad \mu(M) = \pi_{A_1}(M) \times \mathfrak{F}(F, \pi_{A_I}(M)) \tag{7}$$

During this step the probabilities of the resultant instances can be directly computed. Let $P(M) = \prod_{I \in M} P(I)$ be the probability[2] of the merging list M. The probability of an instance I_μ dependent on M results in:

$$P(I_\mu \mid M) = \frac{1}{w(t_\mu)} \times \sum\nolimits_{I_i \in M, \pi_{A_1}(I_i) = \pi_{A_1}(I_\mu)} w(t_i, I_i) \tag{8}$$

Note, the probabilities of duplicate instances eliminated by the relational projection operator in Formula 7 are added up. Duplicate instances resulting from the fusion of different merging lists are handled in Step 4.

3. **Apply Rules.** Third, the set of knowledge rules N is checked for each instance resulting from fusing the merging list M. If an instance is invalid for at least one rule, this instance is removed from the fusion result $\mu(M)$.

[2] Due to all tuples are independent of each other, the probability of M is equal to the product of the probabilities of all its instances $I \in M$.

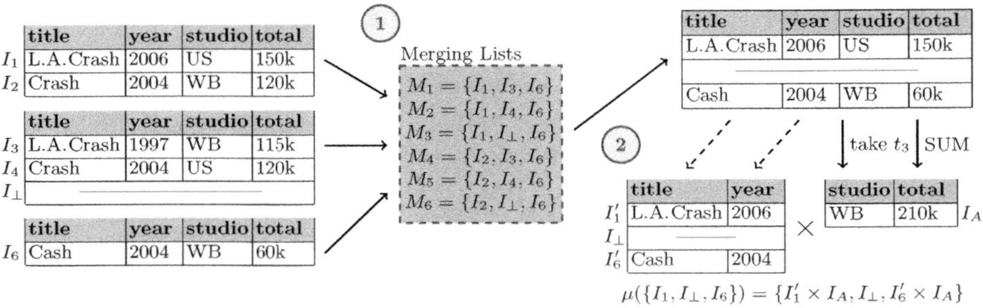

Fig. 4. Building of all merging lists (Step 1) and fusing the merging list $M_3 = \{I_1, I_\perp, I_6\}$ (Step 2). The total values are added up (mediating function). For the studio name the value from the instance of tuple t_3 is chosen (deciding function). For the movie title and the production all values are taken into account.

4. **Combine Results.** Finally, the fusion results of all merging lists are combined to the final set of possible instances $pI_\perp(t_\mu)$ and all probabilities $P(t_\mu[I]), \forall I \in pI_\perp(t_\mu)$ are computed. If instances are eliminated by knowledge rules, a final normalization need to be applied.

$$pI_\perp(t_\mu) = \bigcup_{M \in \mathcal{M}(C)} \mu(M) \qquad (9) \qquad\qquad P(t_\mu[I]) = \sum_{M \in \mathcal{M}(C)} P(I \mid M)\, P(M) \qquad (10)$$

Note, if for all attributes an aggregation function is defined ($l = 1$), for each merging list a single possible instance results (we denote this setting a full-aggregation). Otherwise, for each merging list at most as many instances as base-tuples can result (one for each of the merged instances).

If the instance data of each base-tuple is certain (each tuple has exact one possible instance), only one merging list is built. If in addition a full-aggregation is applied, from tuple merging a single possible instance results. Thus, this approach is a generalization of conflict resolution used for tuple merging in certain databases. In contrast, if no aggregation function is defined, the result contains each instance possible for at least one base-tuple. Therefore, this approach is also a generalization of our basic approach for probabilistic tuple merging.

In the example of Figure 4, three tuples of a relation *box office* are merged. Two aggregation functions are defined. The total amount of sold tickets results in the sum of all values of attribute *total* (mediating function). Moreover, the user knows that the *studio* name of the third tuple is correct. For that reason, this value is chosen for all instances (deciding function). For the attributes *title* and *year* no functions are specified. Thus, all possible values are taken into account.

5.1 Scalability

Given a duplicate cluster of size k. Assume that each tuple has averagely l possible instances (including $I_\perp$). From the basic approach at most $k \times l$ instances result. In contrast, in the extended approach l^k merging lists are built. Thus, if aggregation functions are used, at most $k \times l^k$ instances can be result. This is an increase to the basic approach of l^{k-1} times. Experience has shown that cluster rarely have more than 5 tuples and uncertainty often can be adequate modeled by around 10 possible instances. Certainly, this increase is still tremendous ($5 \cdot 10^5$ instead of 50), but can be flexibly reduced to a desired amount of data by only taking the most likely (resultant or base) instances into account. An important reflection of future work is to compute the most probable resultant instances more efficiently by pruning of irrelevant merging lists first.

5.2 Characteristics of Merging Approaches

In [12], we introduce a set of characteristics of merging approaches which are useful in many merging scenarios. Some of these characteristics are:

- **Independence of the Merging Order:** The tuple resulting from merging multiple base-tuples should be independent from the merging order. This requirement is important, if tuple merging is considered as a part of a data integration process and the integration is performed in a pairwise fashion instead of integrating all sources at one time. This independence is given, if tuple merging is associative:

$$\mu(T) = \mu(\{\mu(T \setminus T_i), \mu(T_i)\}) \text{ for all } T_i \subset T \tag{11}$$

- **Stability:** We denote a merging to be stable, if from deduplicating a duplicate free relation the relation itself results. This property is given, if tuple merging is idempotent ($\mu(\{t\}) = t$).

- **SP-Query Consistency:** Query consistency means that the result of querying a merged tuple should be equal to merging the tuples resulting from querying the individual base-tuples. Since we only consider queries on single tuples, joins and set-based operators are not taken into account. Moreover, aggregation functions change the tuples' schema. Thus, a consistency w.r.t. queries with aggregations is generally not possible and we restrict to the probabilistic equivalences of the algebraic operations selection and projection (we use the definition of the world-set algebra [9]). In the following, we define this class of queries as SP-Queries. Let $\mathcal{R}^p$ be a probabilistic relation and let Q be the set of all possible SP-Queries which can be formulated on $\mathcal{R}^p$, the requirement of SP-Query consistency is formalized as:

$$(\forall T \subseteq \mathcal{R}^p, \forall q \in Q) : \mu(\{q(t) \mid t \in T\}) = q(\mu(T)) \tag{12}$$

  SP-Query consistency is very useful to reduce the dataflow between source databases and target database, because irrelevant data (tuples in case of selection, attributes in case of projection) can be already excluded at the source databases' sides. This is especially important, if data are paid by amount. However, for doing that there must be a certain attribute serving as real-world identifier. Otherwise, the duplicate detection result can be influenced by an early performing of SP-Queries.

Now, we discuss the influences on these characteristics by using aggregation functions, knowledge rules, and instance weights:

Basic Approach: The basic approach is independent of the merging order, SP-Query consistent and stable.

Ext. Approach + Instance Weights: Instance weights only affect stability, but non-stability is even the goal of taking further evidence into account.

Ext. Approach + Aggregation Functions: The approach extended with aggregation functions (simple or complex) is associative, if all these functions are associative and stable, if all these functions are idempotent. Both properties are satisfied by aggregation functions usually used for modeling attribute semantics (e.g. sum, max). The extended approach is not consistent with selections, because a removing of tuples and instances definitely influences the set of values to be aggregated. This approach, however, is consistent with projections, if either only simple aggregation functions are used or none of the excluded attributes is required as input for one of the aggregation functions defined for the remaining attributes.

Ext. Approach + Knowledge Rules: The approach extended with knowledge rules is stable, if none of the base-instances violates a rule, i.e. the source data is consistent with the set of given rules. Independence of merging order is not satisfied, if an intermediate result violates a rule which would not be violated by the final result. This can only happen, if new values arise and hence only, if knowledge rules are used in combination with aggregation functions. Nevertheless, the most useful aggregation functions are monotonically increasing (sum, max) or monotonically decreasing (min). Thus, rules restricting an attribute domain to a range with an lower and an upper bound (e.g. the stock value is between 10 and 100) do not pose a problem. Finally, this approach is consistent with selection, but not consistent with projection, in cases one of the excluded attributes is used in a rule.

In summary, the merging approach extended with aggregation functions, knowledge rules, and instance weights guarantees independence of the merging order in the most useful scenarios (associative aggregation functions and meaningful rules). Stability can be ensured, if wanted by using only tuple-uniformly weighted instances and cleaning data first. Consistency with selection cannot be ensured in general, if aggregation functions are used. In contrast, if attributes are projected carefully, consistency with projection can be achieved easily.

6 Related Work

Tuple merging in certain data is considered in different works [5,3,11,19]. Since in certain data only single values can be stored, conflicts always have to be resolved by applying aggregation functions. In contrast, because we process probabilistic data, in our approach such functions are not mandatory, but a helpful capability to incorporate external knowledge into the merging process.

Robertson et al. [14] consider tuple merging within a transposition of certain data. Merging of two tuples with contrary instance data is not provided (in such cases both tuple are denoted to be *non mergeable*).

DeMichiel [6] and Tseng et al. [16] use *partial values* (resp. *probabilistic values*) to resolve conflicts between certain values by taking multiple possible instances into account. Consequently, these approaches already produce uncertain data as result data. This is similar to our basic approach for instance merging if each base-tuple is considered to be certain. Nevertheless, both approaches consider conflict resolution on an attribute by attribute basis. Dependencies between possible attribute values are not considered.

Andritsos et al. [1] define queries on multiple conflicting duplicates. Thus instead of merging the tuples of each cluster into a single one, query results are derived from sets of mutual exclusive base-tuples. Since to each cluster's tuple a probability can be assigned, this approach is mostly identical to our basic approach applied to certain base-tuples. However, concepts for incorporating external knowledge are not provided.

Van Keulen et al. [17] store conflicting duplicates in a probabilistic database and use user feedback to resolve these conflicts at query time. Thus, they do not directly incorporate the user expertise into the merging process.

A merging of tuples representing uncertain information is proposed by Lim et al. [10], but instead of probability theory this approach is based on the *Dempster-Shafer theory of evidence* and hence is not applicable for probabilistic data.

None of these studies, however, allows probabilistic data as source data.

7 Conclusion

Many applications naturally produce probabilistic data. For integrating probabilistic data from multiple sources in a consistent way or to clean a single database duplicate tuples need to be identified and merged. We consider duplicate detection in probabilistic data in [13] and introduce a basic approach for merging the instance data of probabilistic tuples in [12]. In this paper, we focus on the incorporation of external knowledge (domain knowledge or user expertise) in the merging process making the merging result more accurate. For that purpose, we generalize our basic approach by incorporating aggregation functions, knowledge rules, and instance weights. Finally, we analyze in which way these extensions influence several important characteristics of merging processes.

In future research we aim to make the merging process more adaptable for individual needs. The merging approach based on the set union operator produces data correct as possible. Moreover, this approach is associative and SP-Query consistent. Nevertheless, the more base-tuples are merged, the more possible instances result. Thus, the merged tuple becomes more and more uncertain. For that reason, we need new methods enabling the user to make a trade-off between correctness and certainty being best for him. Finally, we are working on techniques making the extended merging approach more efficient.

References

1. Andritsos, P., Fuxman, A., Miller, R.J.: Clean Answers over Dirty Databases: A Probabilistic Approach. In: ICDE, p. 30–41 (2006)
2. Benjelloun, O., Sarma, A.D., Halevy, A., Widom, J.: ULDBs: Databases with Uncertainty and Lineage. In: VLDB, pp. 953–964 (2006)
3. Bleiholder, J., Naumann, F.: Data fusion. ACM Comput. Surv. 41(1) (2008)
4. Dalvi, N.N., Suciu, D.: Efficient query evaluation on probabilistic databases. VLDB J. 16(4), 523–544 (2007)
5. Dayal, U.: Processing Queries Over Generalization Hierarchies in a Multidatabase System. In: VLDB, pp. 342–353 (1983)
6. DeMichiel, L.G.: Resolving Database Incompatibility: An Approach to Performing Relational Operations over Mismatched Domains. IEEE Trans. Knowl. Data Eng. 1(4), 485–493 (1989)
7. Elmagarmid, A.K., Ipeirotis, P.G., Verykios, V.S.: Duplicate Record Detection: A Survey. IEEE Trans. Knowl. Data Eng. 19(1), 1–16 (2007)
8. Khoussainova, N., Balazinska, M., Suciu, D.: Probabilistic event extraction from rfid data. In: ICDE, pp. 1480–1482 (2008)
9. Koch, C.: MayBMS: A System for Managing Large Uncertain and Probabilistic Databases. In: Managing and Mining Uncertain Data. Springer, Heidelberg (2009)
10. Lim, E.-P., Srivastava, J., Shekhar, S.: An Evidential Reasoning Approach to Attribute Value Conflict Resolution in Database Integration. IEEE Trans. Knowl. Data Eng. 8(5), 707–723 (1996)
11. Motro, A., Anokhin, P.: Fusionplex: Resolution of Data Inconsistencies in the Integration of Heterogeneous Information Sources. Information Fusion 7(2), 176–196 (2006)
12. Panse, F., Ritter, N.: Tuple Merging in Probabilistic Databases. In: MUD, pp. 113–127 (2010)
13. Panse, F., van Keulen, M., de Keijzer, A., Ritter, N.: Duplicate Detection in Probabilistic Data. In: NTII, pp. 179–182 (2010)
14. Robertson, E., Wyss, C.M.: Optimal Tuple Merge is NP-Complete. Technical Report TR599, IUCS (2004)
15. Suciu, D., Connolly, A., Howe, B.: Embracing Uncertainty in Large-Scale Computational Astrophysics. In: MUD, pp. 63–77 (2009)
16. Tseng, F.S.-C., Chen, A.L.P., Yang, W.-P.: Answering Heterogeneous Database Queries with Degrees of Uncertainty. Distributed and Parallel Databases 1(3), 281–302 (1993)
17. van Keulen, M., de Keijzer, A.: Qualitative effects of knowledge rules and user feedback in probabilistic data integration. VLDB J. 18(5), 1191–1217 (2009)
18. Wang, D.Z., Michelakis, E., Franklin, M.J., Garofalakis, M., Hellerstein, J.M.: Probabilistic declarative information extraction. In: ICDE, pp. 173–176 (2010)
19. Whang, S.E., Benjelloun, O., Garcia-Molina, H.: Generic entity resolution with negative rules. VLDB J. 18(6), 1261–1277 (2009)

Qualitative Reasoning about Incomplete Categorization Rules Based on Interpolation and Extrapolation in Conceptual Spaces

Steven Schockaert[1] and Henri Prade[2]

[1] Department of Applied Mathematics and Computer Science, Ghent University, Belgium
steven.schockaert@ugent.be
[2] Institut de Recherche en Informatique de Toulouse (IRIT), Université Paul Sabatier, Toulouse, France
prade@irit.fr

Abstract. Various forms of commonsense reasoning may be used to cope with situations where insufficient knowledge is available for a given purpose. In this paper, we rely on such a strategy to complete sets of symbolic categorization rules, starting from background information about the semantic relationship of different properties and concepts. Our solution is based on Gärdenfors conceptual spaces, which allow us to express semantic relationships with a geometric flavor. In particular, we take the inherently qualitative notion of betweenness as primitive, and show how it naturally leads to patterns of interpolative reasoning. Both a semantic and a syntactic characterization of this process is presented, and the computational complexity is analyzed. Finally, some patterns of extrapolative reasoning are sketched, based on the notions of betweenness and parallelism.

1 Introduction

Applications of AI paradigms are often hampered by the fact that the knowledge bases on which they rely are incomplete. The causes for incompleteness are many. When a knowledge base is built by an expert, for instance, the expert may have forgotten to add some relevant information, or he may have deliberately chosen to omit certain pieces of information (e.g. because he is ignorant about them, or because a complete description of the domain at hand would be infeasible). Along similar lines, when knowledge is extracted (semi-)automatically from some corpus, incompleteness may result from limitations of the extraction methods that are used, or from the absence of relevant documents in the corpus.

Example 1. Consider the following rule base about housing options:

$$\textit{bungalow} \rightarrow \textit{medium} \qquad \textit{bungalow} \rightarrow \textit{detached} \tag{1}$$

$$\textit{mansion} \rightarrow \textit{very-large} \qquad \textit{mansion} \rightarrow \textit{detached} \tag{2}$$

$$\textit{large} \wedge \textit{detached} \rightarrow \textit{comf} \vee \textit{lux} \qquad \textit{large} \wedge \textit{row-house} \rightarrow \textit{comf} \tag{3}$$

$$\textit{small} \wedge \textit{detached} \rightarrow \textit{bas} \vee \textit{comf} \qquad \textit{mansion} \rightarrow \textit{excl} \tag{4}$$

S. Benferhat and J. Grant (Eds.): SUM 2011, LNAI 6929, pp. 303–316, 2011.
© Springer-Verlag Berlin Heidelberg 2011

where *excl* (exclusive), *lux* (luxurious), *comf* (comfortable) and *bas* (basic) refer to different comfort levels.

The rule base (1)–(4) is incomplete, as e.g. nothing is stated about the size of villas and cottages, nor is anything asserted about the comfort level of medium-sized or semi-detached houses. In the face of incomplete knowledge, human reasoning heavily relies on information about *similar* situations [2]. For instance, as medium-sized houses are somewhat similar to both large and small houses, their comfort level should also be somewhat similar to that of large houses and that of small houses. Formalizing this kind of commonsense inferences seems a promising way of alleviating the incompleteness of knowledge bases.

Existing formalisms such as case-based reasoning, or fuzzy rules, are based on the premise that if $\alpha \rightarrow \beta$ is a valid rule and α is sufficiently similar to α', then whenever α' holds, some β' which is similar to β should be the case [15,3,18]. However, such approaches put high demands on the availability of quantitative background information. Indeed, often there is no principled way of determining, given a particular α', how similar β' and β should be before we can consider β' to be among the possible conclusions of α' (at a given confidence level). For instance, knowing only that cottages are similar to bungalows[1], what, if anything, can we actually derive about the size of cottages from (1)–(4)? Moreover, it is hard to come up with a similarity measure that always yields plausible results, and even if there were a unique way of measuring similarity among literals, it is usually not clear how similarity among conjunctions of literals should be evaluated.

The aim of this paper is to introduce a qualitative model for commonsense reasoning which does not (directly) rely on measuring similarity, but rather relies on notions such as *betweenness* and *direction of change*. For instance, as a villa can be seen as being conceptually *between* a bungalow and a mansion, we may conclude that the size of a villa should be between that of a mansion and that of a bungalow, i.e. *medium-sized* or *large* or *very-large*. In addition to such interpolative inference patterns, we may also consider extrapolative patterns: as the change from *mansion* to *bungalow* goes in the same direction as the change from *bungalow* to *cottage*, and as bungalows are known to be smaller than mansions, we could conclude that cottages are smaller than bungalows, i.e. that the size of a cottage is *very-small* or *small* or *medium-sized*.

There is a need for a principled and general approach to interpolative and extrapolative reasoning that goes beyond ad-hoc techniques, similar in spirit to what was done for exception-tolerant reasoning within the realm of nonmonotonic reasoning [9]. To this end, we propose a formalization based on the theory of conceptual spaces [6], which identifies (natural) properties with (convex) regions in an appropriate geometric space. Within such conceptual spaces, the

[1] The terms *cottage* and *bungalow* may be used with somewhat different meanings. Throughout the examples, we use *cottage* for a small detached house, and *bungalow* for a somewhat larger detached house, in accordance with `http://en.wikipedia.org/wiki/Single-family_detached_home` (accessed on April 25, 2011).

aforementioned notions of betweenness and direction of change have a clear geometric interpretation. As a result, using conceptual spaces, we can make explicit under which assumptions the considered inference patterns are valid. Although conceptual spaces are crucial to justify our approach, in practical applications, we do not actually require that the conceptual spaces representation of properties are available. In particular, the inference mechanism itself will only require qualitative knowledge about how these representations are spatially related.

The paper is structured as follows. In the next section, we explain how a propositional rule base can be seen as the approximation of a mapping between conceptual spaces. By making the assumption that this mapping preserves geometrical relations such as betweenness, different commonsense inference relations can be obtained. In Section 3, we focus on betweenness and show how it leads to a form of interpolative reasoning. We first present a semantic approach and then provide its syntactic counterpart and an analysis of the computational complexity. Section 4 illustrates how extrapolative inference relations can be obtained in a similar way. Finally, an overview of related work is provided.

2 A Functional View on Propositional Knowledge

2.1 Mappings between Attribute Spaces

Let $A_1, ..., A_n$ be finite sets of atomic properties, where each A_i corresponds to a certain type of properties (e.g. colors), and the elements of A_i correspond to labels describing particular properties of the corresponding type (e.g. red, green, orange). The labels in A_i are assumed to be jointly exhaustive and pairwise disjoint (JEPD). Note that each element $(a_1, ..., a_n)$ from the Cartesian product $\mathcal{A} = A_1 \times ... \times A_n$ then corresponds to a maximally descriptive specification of the properties that some object may satisfy. We will refer to the sets A_i as attribute spaces.

Example 2. Consider the following attribute spaces:

$$\begin{aligned}
A_1 &= \{\textit{cottage}, \textit{bungalow}, \textit{villa}, \textit{mansion}, \textit{bedsit}, \textit{studio}, \textit{one-bed-ap}, \\
&\quad \textit{two-bed-ap}, \textit{three-bed-ap}, \textit{loft}, \textit{penthouse}\} \\
A_2 &= \{\textit{detached}, \textit{semi-detached}, \textit{row-house}, \textit{apartment}\} \\
A_3 &= \{\textit{very-small}, \textit{small}, \textit{medium}, \textit{large}, \textit{very-large}\} \\
A_4 &= \{\textit{basic}, \textit{comfortable}, \textit{luxurious}, \textit{exclusive}\}
\end{aligned}$$

We will sometimes use abbreviations of these labels, as e.g. in Example 1.

We consider propositional rules of the form $\beta \to \gamma$, where β and γ are propositional formulas, built in the usual way from the set of atoms $A_1 \cup ... \cup A_n$ and the connectives $\wedge$ and $\vee$. We say that an element $(a_1, ..., a_n) \in \mathcal{A}$ is a model of a rule $\beta \to \gamma$, written $(a_1, ..., a_n) \models_{\mathcal{A}} \beta \to \gamma$ iff the corresponding propositional interpretation $\{a_1, ..., a_n\}$ is a model of $\beta \to \gamma$, where we see interpretations as sets containing all atoms that are interpreted as true. For $\beta_1 \to \gamma_1$ and $\beta_2 \to \gamma_2$

rules, we say that $\beta_1 \rightarrow \gamma_1$ entails $\beta_2 \rightarrow \gamma_2$, written $\beta_1 \rightarrow \gamma_1 \models_{\mathcal{A}} \beta_2 \rightarrow \gamma_2$ if for every $\omega \in \mathcal{A}$, $\omega \models_{\mathcal{A}} \beta_1 \rightarrow \gamma_1$ implies $\omega \models_{\mathcal{A}} \beta_2 \rightarrow \gamma_2$, and analogously for entailment between sets of formulas. Note that the notion of entailment we consider is classical entailment, modulo the assumption that the propositions in each set A_i are JEPD.

In the following, we consider a *rule base* R over $\mathcal{A}$ which encodes knowledge about how different attribute spaces are related to each other. In particular, we consider subconcept relations ("every bungalow is a house") and other categorization rules ("a small, rural, detached house is a cottage"). Different types of rules, such as association rules, causal rules, or deontic rules may require mechanisms that are different from the ones we present in this paper. Our goal is to generate new categorization rules from a given set of categorization rules R, using generic meta-principles. In this respect, our aim is similar to that of System P [9], where (other) generic meta-principles are used to support exception-tolerant reasoning.

Let $B_1, ..., B_s$ be the attribute spaces whose labels occur in the antecedent of rules in R, and $C_1, ..., C_k$ the attribute spaces whose labels occur in the consequent of rules in R. Each knowledge base R can equivalently be expressed as a function f_R from subsets of $\mathcal{B} = B_1 \times ... \times B_s$ to subsets of $\mathcal{C} = C_1 \times ... \times C_k$, defined for $X \subseteq \mathcal{B}$ as

$$f_R(X) = \bigcap \{Y \in 2^{\mathcal{C}} \mid R \models_{\mathcal{A}} (\bigvee_{(x_1,...,x_s) \in X} \bigwedge_{i=1}^{s} x_i) \rightarrow (\bigvee_{(y_1,...,y_k) \in Y} \bigwedge_{i=1}^{k} y_i)\}$$

In other words, knowing that the state of the attribute spaces in $\mathcal{B}$ is among those in X, the possible states of the attribute space in $\mathcal{C}$, given R, are exactly those in $f_R(X)$.

Example 3. Consider the set of rules R from Example 1 and the attribute spaces from Example 2, where we have $\mathcal{B} = A_1 \times A_2 \times A_3$ and $\mathcal{C} = A_2 \times A_3 \times A_4$. Then we find that a large detached villa is either comfortable or luxurious:

$$f_R(\{(\mathit{villa}, \mathit{det}, \mathit{large})\}) = \{(\mathit{det}, \mathit{large}, \mathit{comf}), (\mathit{det}, \mathit{large}, \mathit{lux})\}$$

Similarly, we find that a bungalow is detached and medium-sized, while we find no restrictions on the possible comfort levels:

$$f_R(\{(\mathit{bun}, x, y) \mid x \in A_2, y \in A_3\}) = \{(\mathit{det}, \mathit{medium}, z) \mid z \in A_4\}$$

We may see f_R as an approximate (i.e. incomplete) model of the world, which may be refined as soon as new information becomes available. In particular, for two $2^{\mathcal{B}} \rightarrow 2^{\mathcal{C}}$ functions f and f' which are monotone w.r.t. set inclusion, we say that f is a refinement of f', written $f \leq f'$, iff

$$\forall X \subseteq \mathcal{B} \,.\, f(X) \subseteq f'(X) \tag{5}$$

Hence at the semantic level, completing the rule base R amounts to refining the corresponding function f_R.

2.2 Mappings between Conceptual Spaces

Our approach is motivated by the idea that knowledge about the cognitive meaning of the labels in the attribute spaces can be used to refine the rule base R in an appropriate way. In particular, we assume that only qualitative knowledge about the cognitive relationships of the labels is available. This qualitative knowledge can be given a precise, geometric meaning using the theory of conceptual spaces.

The theory of conceptual spaces [6] is centered around the assumption that a natural property is represented as a convex region in some, typically high-dimensional space. Formally, a conceptual space is the Cartesian product $Q_1 \times ... \times Q_m$ of a number of quality dimensions, each of which corresponds to a certain quality (i.e. an atomic, cognitively meaningful attribute). A typical example is the conceptual space of colors, which can be described using the qualities hue, saturation and intensity. Labels to describe colors, in some natural language, are then posited to correspond to convex regions in this conceptual space. Note that while e.g. *red* may be an atomic property at the symbolic level, at the cognitive level it is defined in terms of more primitive notions. As is common [7], we will identify conceptual spaces with Euclidean spaces. Cognitive similarity can then be determined by evaluating the Euclidean distance in the corresponding conceptual space. Note that the context-dependent nature of concept similarity can be addressed by appropriately rescaling the relevant quality dimensions [6].

Each attribute space A_i may be seen as a tessellation of some conceptual space in convex regions. More generally, the elements of $\mathcal{B}$ and $\mathcal{C}$ will also correspond to convex regions in conceptual spaces. We will denote these conceptual spaces respectively by $\mathfrak{B}$ and $\mathfrak{C}$. The mapping f_R, induced by the knowledge base R, can then be seen as a mapping from subsets of $\mathfrak{B}$ to subsets of $\mathfrak{C}$. As explained above, f_R represents an approximate model of the real world. More precisely, we take the view that f_R is the approximation of an unknown mapping m from points of $\mathfrak{B}$ to points of $\mathfrak{C}$, i.e. for $X \subseteq \mathcal{B}$, we then have $reg(f_R(X)) \supseteq \{m(p) \mid p \in reg(X)\}$, where we write $reg(X)$ for the geometric representation of X as a region in $\mathfrak{B}$ or $\mathfrak{C}$. The actual conceptual spaces $\mathfrak{B}$ and $\mathfrak{C}$, and a fortiori the mapping m, are inaccessible in most applications. For instance, we cannot assume that a precise definition of a loft is available, or even an exhaustive enumeration of the qualities on which such a definition would depend. On the one hand, we thus assume the existence of a precise, but unknown mapping m between $\mathfrak{B}$ and $\mathfrak{C}$. On the other hand, given our finite vocabulary, we can only describe approximate models. Let us write $\widehat{f}$ for the most informative approximation that can be described using the available labels, i.e. $\widehat{f}$ is the $2^{\mathcal{B}} \rightarrow 2^{\mathcal{C}}$ mapping defined for $X \subseteq \mathcal{B}$ by

$$\widehat{f}(X) = \{y \in \mathcal{C} \mid x \in X, m^*(reg(x)) \cap reg(y) \neq \emptyset\}$$

where the $2^{\mathfrak{B}} \rightarrow 2^{\mathfrak{C}}$ mapping m^* is defined as the pointwise extension of m.

All we know about $\widehat{f}$ is that $\widehat{f} \leq f_R$ holds. By making further assumptions about the nature of the mapping m, we will be able to derive further restrictions on $\widehat{f}$, which will, in turn, allow us to determine some refinement $\widehat{f_R}$, satisfying $\widehat{f} \leq \widehat{f_R} \leq f_R$. In particular, the fact that we restrict R to contain only *categorization rules* means that the relevant quality dimensions of $\mathfrak{C}$ are a subset of those of $\mathfrak{B}$. In other words, categorization rules, as we use the term, express relationships that follow from the cognitive representation of concepts, rather than e.g. from observations about the world. Consider, for instance, the rule that bungalows are of medium size. The conceptual space in which bungalows are defined includes a quality dimension which refers to the size, whereas the conceptual space in which medium size is defined clearly is a unidimensional space with size as the only quality dimension. On the other hand, the conceptual space in which *detached house* is defined may consist of the same qualities as the conceptual space in which *bungalow* is defined, but the relative importance of size may be higher in the latter case. This view suggests that m can be described as a combination of operations such as projection and scaling, and more in particular, that m satisfies most of the properties of an affine transformation.

We now consider betweenness and parallelism in $\mathfrak{B}$ and $\mathfrak{C}$. In contrast to distance or similarity, betweenness and parallelism remain invariant under affine transformations. Using betweenness as primitive thus leads to an approach which is more robust to context changes. Let us write $bet(p,q,r)$ to denote that q lies between p and r (on the same line), and $par(p,q,r,s)$ to denote that the vectors $\overrightarrow{pq}$ and $\overrightarrow{rs}$ point in the same direction. The fact that point q is between points p and r means that for every point x it holds that $d(q,x) \leq \max(d(p,x), d(r,x))$, and in particular, that whenever p and r are close to a prototype of some concept, then q is close to it as well. In this sense, we may see $bet(p,q,r)$ as a way to express that whatever relevant properties p and r have in common, p and q have them in common as well (identifying points in a conceptual space with instances). On the other hand, $par(p,q,r,s)$ intuitively means that to arrive at s, r needs to be changed in the same direction as p needs to be changed to arrive at q. We also consider a notion of comparative distance, writing $d(p,q) < d(r,s)$ to denote that the distance between p and q is smaller than the distance between r and s. The validity of interpolative and extrapolative inference will be tied to the following postulates ($p,q,r,s \in \mathfrak{B}$):

(bet1) $bet(p,q,r) \Rightarrow bet(m(p), m(q), m(r))$
(bet2) $bet(p,q,r) \wedge d(p,q) < d(r,q) \Rightarrow d(m(p), m(q)) < d(m(r), m(q))$
(par1) $par(p,q,r,s) \Rightarrow par(m(p), m(q), m(r), m(s))$
(par2) $par(p,q,r,s) \wedge d(p,q) < d(r,s) \Rightarrow d(m(p), m(q)) < d(m(r), m(s))$

Each of these four postulates are valid whenever m is an affine mapping. In this paper, we mainly consider Postulate (**bet1**), as is discussed in detail in the next section. The remaining postulates are briefly discussed in Section 4.

3 Basic Interpolative Reasoning

3.1 Betweenness

We assume that extra-logical information is available about the betweenness of properties from the same attribute space. For example, we may intuitively think of a studio to be between a bedsit and a one-bedroom apartment. Note that this notion of betweenness acts on labels, assumed to belong to an attribute space A_i, and thus on convex regions, whereas (**bet1**) deals with betweenness of points. As is well known, notions such as collinearity and betweenness can be extended to regions in different ways [1]. We will consider the following two notions for regions A, B, and C:

$$\overline{bet}(A, B, C) \quad \text{iff} \quad \exists q \in B \,.\, \exists p \in A \,.\, \exists r \in C \,.\, bet(p, q, r)$$
$$\underline{bet}(A, B, C) \quad \text{iff} \quad \forall q \in B \,.\, \exists p \in A \,.\, \exists r \in C \,.\, bet(p, q, r)$$

In other words, $\overline{bet}(A, B, C)$ holds if B overlaps with the convex hull of $A \cup C$, whereas $\underline{bet}(A, B, C)$ holds if B is included in this convex hull. Note in particular that both relations are reflexive w.r.t. the first two arguments, in the sense that $\overline{bet}(A, A, C)$, as well as symmetric, in the sense that $\overline{bet}(A, B, C) \equiv \overline{bet}(C, B, A)$ (and similar for $\underline{bet}$). However, in contrast to points, transitivity does not necessarily hold for regions, e.g. from $\underline{bet}(A, B, C)$ and $\underline{bet}(B, C, D)$ we cannot infer that $\underline{bet}(A, B, D)$, which indeed agrees with the geometric interpretation in terms of convex hulls. In the terminology of rough set theory [12], $\overline{bet}$ and $\underline{bet}$ correspond to upper and lower approximations of betweenness. In the following, we will often identify labels with the corresponding regions, writing e.g. $\underline{bet}(a, b, c)$ for $\underline{bet}(reg(a), reg(b), reg(c))$.

Example 4. We may consider that $\overline{bet}$(*three-bed-ap*, *loft*, *penthouse*) holds but not $\underline{bet}$(*three-bed-ap*, *loft*, *penthouse*), which corresponds to the view that some, but not all lofts are conceptually between a three-bedroom apartment and a penthouse. On the other hand, we may consider that all studios are between bedsits and one-bedroom apartments, and thus that both $\overline{bet}$(*bedsit*, *studio*, *one-bed-ap*) and $\underline{bet}$(*bedsit*, *studio*, *one-bed-ap*) hold.

Regarding the applicability of our approach, an important question is where the specification of $\underline{bet}$ and $\overline{bet}$ comes from. Depending on the application, such relations may be specified by an expert, or derived indirectly from available metadata [17]. For instance, [7] suggests to start from pairwise similarity judgements between instances, and use multi-dimensional scaling to obtain coordinates for them in a Euclidean space. Representations of concepts can then be obtained by determining the corresponding Voronoi tessellation, after which the relations $\overline{bet}$ and $\underline{bet}$ can be evaluated by straightforward geometric calculations. Another method might be to consider that instance q is between instance p and instance r if for every known instance x, it holds that $d(q, x) \leq \max(d(p, x), d(r, x))$, and then lift this betweenness for instances to concepts, by considering that $\overline{bet}(a, b, c)$ (resp. $\underline{bet}(a, b, c)$) holds for concepts a, b and c iff some (resp. every) known instance of b is between some known instance of a and some known instance of c. Such data-driven approaches rely on the assumption that a lot of information is available about the relationship between labels from the same attribute space, while little information is available about the relationship between labels from different attribute spaces. It is also interesting to note that such approaches convert similarity scores, which are sensitive to changes in context, to betweenness information, which is in principle robust against context changes.

3.2 Semantic Characterization

To perform interpolative reasoning based on rules with conjunctions in the antecedent, we need to lift the relations $\overline{bet}$ and $\underline{bet}$ to betweenness relations on Cartesian products of attribute spaces. Here the intuition of $\overline{bet}$ and $\underline{bet}$ as upper and lower approximations of betweenness becomes important. Indeed, in general, betweenness for a vector of labels cannot be reduced to betweenness for the labels in the respective components. In particular, notice that when $\underline{bet}(a_1, b_1, c_1)$ and $\underline{bet}(a_2, b_2, c_2)$ hold, we do not necessarily have that (b_1, b_2) is between (a_1, a_2) and (c_1, c_2). Indeed, even for points in a Euclidean space of dimension two or more, betweenness in each dimension does not entail collinearity. First consider a Cartesian product $D_1 \times ... \times D_l$ of attribute spaces such that no two attribute spaces D_i and D_j rely on the same quality dimensions, i.e. such that every element of $D_1 \times ... \times D_l$ corresponds to a non-empty region in some conceptual space. We call such attribute spaces orthogonal (expressing logical independence). The relation $\underline{bet}$ is then defined for elements of $D_1 \times ... \times D_l$ as

$$\underline{bet}(\mathbf{a}, \mathbf{b}, \mathbf{c}) \quad \text{iff} \quad \mathbf{a} = \mathbf{b} \vee \mathbf{b} = \mathbf{c} \vee \big(\exists j \,.\, (\forall i \neq j \,.\, a_i = b_i = c_i) \wedge \underline{bet}(a_j, b_j, c_j)\big)$$

where we write e.g. a_i for the ith component of $\mathbf{a}$. Note that this is a conservative approach, and that indeed $\underline{bet}$ can still be seen as a lower approximation of betweenness, under the assumption of orthogonality. On the other hand, when some attribute spaces rely on the same quality dimensions, nothing can be derived about the betweenness of elements from $D_1 \times ... \times D_l$, except for the trivial cases where $\mathbf{b} = \mathbf{a}$ or $\mathbf{b} = \mathbf{c}$.

Example 5. The quality dimensions underlying attribute spaces A_1 and A_2 from Example 2 clearly overlap. For example, it is not possible for a *bungalow* to also be an *apartment*, or for a *loft* to be a *row-house*. On the other hand, we may consider that attribute spaces A_2 and A_3 are orthogonal. Note that this orthogonality holds irrespective of whether there actually exist apartments that are *very-large*. What is important is that nothing in the definition of an *apartment* prevents it from possibly being *very-large*. As a result, we can derive e.g. that

$$\underline{bet}\big((\textit{apartment}, \textit{small}), (\textit{apartment}, \textit{large}), (\textit{apartment}, \textit{very-large})\big)$$

holds but not e.g.

$$\underline{bet}\big((\textit{bungalow}, \textit{detached}), (\textit{bungalow}, \textit{semi-detached}), (\textit{bungalow}, \textit{row-house})\big)$$

For any Cartesian product $D_1 \times ... \times D_l$ of attribute spaces, $\overline{bet}$ is defined as

$$\overline{bet}(\mathbf{a}, \mathbf{b}, \mathbf{c}) \quad \text{iff} \quad \forall j \,.\, \overline{bet}(a_j, b_j, c_j)$$

which is in accordance with the idea of $\overline{bet}$ as an upper approximation of betweenness. We can now consider the sets $\overline{bet}(X_1, X_2)$ and $\underline{bet}(X_1, X_2)$ of all elements between X_1 and X_2, with X_1 and X_2 subsets of $D_1 \times ... \times D_l$:

$$\overline{bet}(X_1, X_2) = \{\mathbf{b} \,|\, \mathbf{a} \in X_1, \mathbf{c} \in X_2, \overline{bet}(\mathbf{a}, \mathbf{b}, \mathbf{c})\}$$
$$\underline{bet}(X_1, X_2) = \{\mathbf{b} \,|\, \mathbf{a} \in X_1, \mathbf{c} \in X_2, \underline{bet}(\mathbf{a}, \mathbf{b}, \mathbf{c})\}$$
$$\cup \{(b_1, ..., b_{i-1}, x, b_{i+1}, ..., b_l) \,|\, \mathbf{b} \in \underline{bet}(X_1^{\downarrow i}, X_2^{\downarrow i}), 1 \leq i \leq l, x \in D_i\}$$

where

$$(a_1, ..., a_{i-1}, a_{i+1}, ..., a_l) \in X_j^{\downarrow i} \quad \text{iff} \quad \forall x \in D_i \,.\, (a_1, ..., a_{i-1}, x, a_{i+1}, ..., a_l) \in X_j$$

Intuitively, $\mathbf{b}$ is in $\overline{bet}(X_1, X_2)$ if it is between (w.r.t. $\overline{bet}$) some element from X_1 and some element from X_2. The definition of $\underline{bet}(X_1, X_2)$ is slightly more complex to correctly address the case where some of the attribute spaces $D_1, ..., D_l$ are not orthogonal. In such a case, $\underline{bet}(\mathbf{a}, \mathbf{b}, \mathbf{c})$ will only hold for $\mathbf{b} = \mathbf{a}$ or $\mathbf{b} = \mathbf{c}$. However, when $(a_1, ..., a_{i-1}, a_{i+1}, ..., a_l) \in X_1^{\downarrow i}$ and $(c_1, ..., c_{i-1}, c_{i+1}, ..., c_l) \in X_2^{\downarrow i}$, we know that $(b_1, ..., b_{i-1}, x, b_{i+1}, ..., b_l)$ will geometrically be between X_1 and X_2 for every $x \in D_i$ if $(b_1, ..., b_{i-1}, b_{i+1}, ..., b_l)$ is between $(a_1, ..., a_{i-1}, a_{i+1}, ..., a_l)$ and $(c_1, ..., c_{i-1}, c_{i+1}, ..., c_l)$. Along similar lines, it should be noted that e.g. $\mathbf{b}$ may geometrically be between $\mathbf{a}_1 \cup ... \cup \mathbf{a}_p$ and $\mathbf{c}_1 \cup ... \cup \mathbf{c}_q$ even if $\mathbf{b}$ is not between $\mathbf{a}_i$ and $\mathbf{c}_j$ for any i and j. In this sense, the definition of $\underline{bet}(X_1, X_2)$ is again conservative, and may be further refined if information is available about the betweenness of disjunctions of labels.

Example 6. Let $X_1 = \{(x, det, small) \,|\, x \in A_1\}$ and $X_2 = \{(x, det, large) \,|\, x \in A_1\}$, and assume that A_2 and A_3 are orthogonal (considering that the size of a house is irrelevant in deciding whether it is e.g. detached or not), while A_1 and A_2 are not. First note that

$$\underline{bet}(\{(det, small)\}, \{(det, large)\}) = \{(det, small), (det, medium), (det, large)\}$$

from which we find $\underline{bet}(X_1, X_2) = \{(x, det, y) \,|\, x \in A_1, y \in \{small, medium, large\}\}$
From Postulate (**bet1**), we know that for all $A_1, A_2 \in 2^{\mathfrak{B}}$

$$m^*(bet(A_1, A_2)) \subseteq bet(m^*(A_1), m^*(A_2))$$

where $bet(A_1, A_2)$ is the true set of points between A_1 and A_2 and m^* is the pointwise extension of m as before. As a result, we also find that for $X_1, X_2 \in 2^{\mathcal{B}}$

$$\widehat{f}(\underline{bet}(X_1, X_2)) \subseteq \overline{bet}(\widehat{f}(X_1), \widehat{f}(X_2)) \tag{6}$$

This observation allows us to improve our approximation of $\widehat{f}$ to the most conservative refinement $\widehat{f_R}$ which satisfies the constraint (6), i.e. we define $\widehat{f_R}$ to be the largest fixpoint, w.r.t. the ordering $\leq$ defined in (5), of

$$\widehat{f_R}(\{\mathbf{x}\}) = f_R(\{\mathbf{x}\}) \cap \bigcap \{\overline{bet}(\widehat{f_R}(Y), \widehat{f_R}(Z)) | \mathbf{x} \in \underline{bet}(Y, Z)\}$$

and $\widehat{f_R}(X) = \bigcup_{\mathbf{x} \in X} \widehat{f_R}(\{\mathbf{x}\})$. The existence of this unique largest fixpoint follows from the well-known Knaster-Tarski theorem.

Example 7. Let us determine the comfort level of a medium-sized detached villa:

$$\begin{aligned}
&\widehat{f_R}(\{(villa, det, med)\}) \\
&\subseteq \widehat{f_R}(\underline{bet}(X_1, X_2)) \subseteq \overline{bet}(\widehat{f_R}(X_1), \widehat{f_R}(X_2)) \subseteq \overline{bet}(f_R(X_1), f_R(X_2)) \\
&= \overline{bet}(\{(det, small, bas), (det, small, comf)\}, \{(det, large, comf), (det, large, lux)\}) \\
&= \{(det, x, y) \,|\, x \in \{small, med, large\}, y \in \{bas, comf, lux\}\}
\end{aligned}$$

where X_1 and X_2 are as defined in Example 6. Furthermore, we also have

$$\widehat{f_R}(\{(\textit{villa}, \textit{det}, \textit{med})\}) \subseteq f_R(\{(\textit{villa}, \textit{det}, \textit{med})\}) = \{(\textit{det}, \textit{med}, x) \mid x \in A_4\}$$

Together we thus find

$$\widehat{f_R}(\{(\textit{villa}, \textit{det}, \textit{med})\}) \subseteq \{(\textit{det}, \textit{med}, \textit{bas}), (\textit{det}, \textit{med}, \textit{comf}), (\textit{det}, \textit{med}, \textit{lux})\}$$

3.3 Syntactic Characterization

In this section, we develop a syntactic counterpart of the mapping $\widehat{f_R}$. First, we introduce the connectives $\overline{\oplus}$ and $\underline{\oplus}$ which capture the upper and lower approximations of betweenness, and which act on formulas in disjunctive-normal form (DNF). First we consider conjunctions of atoms $\beta = b_1 \wedge ... \wedge b_l$ and $\gamma = c_1 \wedge ... \wedge c_m$, where each b_i and c_j belongs to some attribute space. Recall that such a conjunction of atoms is consistent iff it contains at most one atom from each attribute space. We define

$$\beta \underline{\oplus} \gamma \equiv \beta \vee \gamma \vee (\delta_1 \wedge \delta_2 \wedge \bigvee \{b \in A_i \mid \underline{bet}(a, b, c)\})$$

if (i) the attribute spaces underlying $b_1, ..., b_l$ and $c_1, ..., c_l$ are orthogonal, (ii) β and γ are consistent and (iii) a, c, δ_1 and δ_2 exist such that $\beta \equiv \delta_1 \wedge a$, $\gamma \equiv \delta_2 \wedge c$, where a and c belong to the same attribute space and $\delta_1 \wedge \delta_2$ is consistent. Otherwise, we define $\beta \underline{\oplus} \gamma \equiv \beta \vee \gamma$. If β and γ are consistent, connective $\overline{\oplus}$ is defined as

$$\beta \overline{\oplus} \gamma \equiv \bigvee \{y_1 \wedge ... \wedge y_s \mid y_i \in A_i \wedge \overline{bet}(x_i, y_i, z_i)\}$$

where $x_i, z_i \in A_i$, $\beta \equiv \delta_1 \wedge x_1 \wedge ... \wedge x_s$, $\gamma \equiv \delta_2 \wedge z_1 \wedge ... \wedge z_s$, and none of the atoms from δ_1 is in the same attribute space as an atom from δ_2. If β or γ is inconsistent, we define $\beta \underline{\oplus} \gamma \equiv \beta \vee \gamma$. Finally, we define

$$(\beta_1 \vee ... \vee \beta_n) \overline{\oplus} (\gamma_1 \vee ... \vee \gamma_m) \equiv \bigvee_{i,j} (\beta_i \overline{\oplus} \gamma_j)$$

$$(\beta_1 \vee ... \vee \beta_n) \underline{\oplus} (\gamma_1 \vee ... \vee \gamma_m) \equiv \bigvee_{i,j} (\beta_i \underline{\oplus} \gamma_j)$$

We now define an interpolative inference relation $\vdash^i$ as:

1. If $R \models_{\mathcal{A}} \beta \rightarrow \gamma$ then $R \vdash^i \beta \rightarrow \gamma$.
2. If $R \vdash^i \beta_1 \rightarrow \gamma_1$ and $R \vdash^i \beta_2 \rightarrow \gamma_2$ then $R \vdash^i (\beta_1 \underline{\oplus} \beta_2) \rightarrow (\gamma_1 \overline{\oplus} \gamma_2)$
3. If $R \vdash^i \beta_1 \rightarrow \gamma_1$, $R \vdash^i \beta_2 \rightarrow \gamma_2$ and $\{\beta_1 \rightarrow \gamma_1, \beta_2 \rightarrow \gamma_2\} \models_{\mathcal{A}} \beta_3 \rightarrow \gamma_3$ then $R \vdash^i \beta_3 \rightarrow \gamma_3$

The first rule ensures that $\vdash^i$ refines the classical consequence relation, while the second rule allows for a form of interpolation: for every β between β_1 and β_2, we should be able to conclude that something between γ_1 and γ_2 is the case. Finally, the third rule ensures that the set of conclusions of R is closed under classical deduction (modulo $\mathcal{A}$).

Example 8. Consider again the comfort level of a medium-sized detached villa. From the rules in Example 1, we find

$$R \vdash^i small \wedge det \rightarrow bas \vee comf \qquad R \vdash^i large \wedge det \rightarrow comf \vee lux$$

which together leads to

$$R \vdash^i (small \wedge det)\underline{\otimes}(large \wedge det) \rightarrow (bas \vee comf)\overline{\otimes}(comf \vee lux)$$

which is equivalent to $R \vdash^i (small \vee med \vee large) \wedge det \rightarrow bas \vee comf \vee lux$ entailing $R \vdash^i villa \wedge det \wedge med \rightarrow bas \vee comf \vee lux$, which indeed corresponds to what we have found in Example 7.

In general, the following soundness and completeness result can be shown.

Proposition 1. *Let $X \subseteq \mathcal{B}$ and $Y \subseteq \mathcal{C}$. It holds that $\widehat{f_R}(X) \subseteq Y$ iff*

$$R \vdash^i \Big(\bigvee_{(x_1,\ldots,x_s)\in X} \bigwedge_i x_i \Big) \rightarrow \Big(\bigvee_{(y_1,\ldots,y_k)\in Y} \bigwedge_i y_i \Big)$$

Somewhat surprisingly, the computational complexity of interpolative inference is quite high in general.

Proposition 2. *Let R be as before and let $b_1, \ldots, b_r$ and c be atoms from different attribute spaces. The problem of deciding whether $R \vdash^i b_1 \wedge \ldots \wedge b_r \rightarrow c$ is PSPACE-hard.*

Having a few large attribute spaces is computationally more desirable than having a large number of small attribute spaces. In particular, if the number of attribute spaces is bounded by a constant, then $\mathcal{A}$ contains a polynomial number of vectors, from which we can prove the following result.

Proposition 3. *Let R be as before, and let ϕ and ψ be propositional formulas over $\mathcal{A}$. If the number of attribute spaces is bounded by a constant, the problem of deciding whether $R \vdash^i \phi \rightarrow \psi$ is in P.*

4 Refinements and Extrapolative Reasoning

The approach that was presented in the previous section suggests a general scheme for studying interpolative and extrapolative inference: starting from particular constraints on the mapping m and a particular type of ontological background information, we may derive constraints on $\widehat{f}$, which translate to constraints on f_R and thus lead to a particular inference relation. In this section, we further illustrate this general idea.

Considering the rules of Example 1, then nothing can be concluded about the size of a cottage using the inference relation $\vdash^i$. However, if we consider bungalows to be between cottages and mansions, then the fact that bungalows are known to be smaller than mansions should intuitively enable to conclude that cottages are smaller than bungalows. The reason that $\vdash^i$ falls short in this respect is that such conclusions depend on extrapolation rather than interpolation. Such extrapolative inferences also rely on Postulate (**bet1**), but they require background knowledge of a slightly different nature. Specifically, let us define $\underline{bet'}$ as

$$\underline{bet'}(A, B, C) \quad \text{iff} \quad \forall p \in A \,.\, \exists q \in B \,.\, \exists r \in C \,.\, bet(p, q, r)$$

Note that $\underline{bet'}(A, B, C)$ expresses some form of betweenness which focuses on area A instead of B. Accordingly, the role played by the convex hull in Section 3 is replaced by a notion of conical extension. For $b, c \in \mathcal{B}$ and $y, z \in \mathcal{C}$, we define

$$\overline{con}(y, z) = \{x \mid \overline{bet}(x, y, z)\} \qquad \underline{con}(b, c) = \{a \mid \underline{bet'}(a, b, c)\}$$

where geometrically e.g. $a \in \underline{con}(b, c)$ if region a is included in a conical extension of region b, in the directions determined by region c. Postulate (**bet1**) entails that for all $b, c \in \mathcal{B}$

$$\widehat{f}(\underline{con}(b, c)) \subseteq \overline{con}(\widehat{f}(b), \widehat{f}(c))$$

which leads to an extrapolative inference relation $\vdash^e$ such that for the rules of Example 1, we find

$$R \vdash^e \textit{cottage} \rightarrow \textit{small} \vee \textit{medium}$$

due to the rules *bungalow* $\rightarrow$ *medium* and *mansion* $\rightarrow$ *very-large*. The idea of conical extension can be generalized to reasoning based on information about parallel directions (i.e. analogies), assuming Postulate (**par1**). The relation *par* can be extended to regions in four different ways, depending on which of the four arguments we focus on. In each case, we get $\overline{par}(A, B, C, D)$ iff

$$\exists p \in A, q \in B, r \in C, s \in D \,.\, par(p, q, r, s)$$

If we focus on the first argument, for instance, we get $\underline{par}(A, B, C, D)$ iff

$$\forall p \in A \,.\, \exists q \in B, r \in C, s \in D \,.\, par(p, q, r, s)$$

The corresponding notion of conical extension is given for $x, y, z \in \mathcal{C}$ and $b, c, d \in \mathcal{B}$ by

$$\overline{con}(x, y, z) = \{u \mid \overline{par}(u, x, y, z)\} \qquad \underline{con}(b, c, d) = \{a \mid \underline{par}(a, b, c, d)\}$$

From (**par1**), we can derive the following constraint:

$$\widehat{f}(\underline{con}(b, c, d)) \subseteq \overline{con}(\widehat{f}(b), \widehat{f}(c), \widehat{f}(d)) \tag{7}$$

Example 9. Consider the following rules:

$$R = \{\textit{mansion} \rightarrow \textit{excl}, \textit{bungalow} \rightarrow \textit{comf}, \textit{three-bed-ap} \rightarrow \textit{lux}\}$$

Assuming furthermore that $\underline{par}$(*penthouse*, *three-bed-ap*, *mansion*, *bungalow*), we obtain from (7) an analogical-like inference relation $\vdash^a$ which is such that $R \vdash^a$ *penthouse* $\rightarrow$ *lux* $\vee$ *excl*. Note in particular how the first two rules together determine the direction of the change in comfort level when going from *mansion* to *bungalow*, which was asserted to be the same as when going from *penthouse* to *three-bed-ap*.

Finally, note that using Postulates (**bet2**) and (**par2**), refined inference relations can be obtained, which take comparative distance information into account.

5 Related Work

The problem of interpolation from rules, and more generally similarity-based reasoning, has mainly been studied in a quantitative way, based on fuzzy set theory [15,3] or neural networks [18] among others. In the propositional setting, the idea of interpolation and extrapolation has been studied in [4], but from a rather different angle. In particular, the paper discusses how the belief that certain propositions hold at certain moments in time can be extended to beliefs about other moments in time, using persistence assumptions as a starting point. The general idea of augmenting the incomplete specification of a function based on some rationality postulates also seems related to some of the work on multi-criteria decision making, such as the well-known family of ELECTRE methods [14], or for instance the approach presented in [8]. In contrast to our paper, the aforementioned works do not rely on any ontological information (other than some meta-knowledge in the form of rationality postulates). On the other hand, the importance of extra-logical information is well-recognized in settings such as belief revision (e.g. in the form of epistemic entrenchment orderings). Recently, the use of ontological background information in the context of merging conflicting multi-source information has also been advocated [16]. Apart from the work on conceptual spaces, the idea of assuming a spatial representation to reason about concepts also underlies [11], where an approach to integrate heterogeneous databases is proposed based on spatial relations between concepts. Finally, there is some resemblance between our inference procedure and the early work on qualitative reasoning about physical systems [5,10], which deal with monotonicity constraints like "if the value of x increases, then (all things being equal) the value of y decreases". Our inference procedure differs from these approaches as the domains we reason about do not need to be linearly ordered. Moreover, in the special case of linearly ordered domains, we assume no prior information about which partial mappings are increasing and which are decreasing.

6 Conclusions

We have studied the problem of interpolating and extrapolating propositional knowledge about concepts, as a vehicle to formalize commonsense approaches for dealing with incompleteness. We have introduced a general semantics, which relates such inferences to the existence of an affine mapping between conceptual spaces. A knowledge base is then seen as an approximation of this mapping, which can be further refined by exploiting different types of ontological background information with a geometric flavor. We have zoomed in on one particular case, namely interpolative reasoning based on knowledge about betweenness of properties. We have syntactically characterized the resulting inference relation, and analyzed the computational complexity. Finally, we have outlined how information about betweenness or parallelism can be exploited to support extrapolative inference. In that respect, a deeper investigation of links between the proposed approach and analogical reasoning [13] would be worth of interest.

Acknowledgments. Steven Schockaert was funded as a postdoctoral fellow by the Research Foundation – Flanders (FWO).

References

1. Billen, R., Clementini, E.: Semantics of collinearity among regions. In: Chung, S., Herrero, P. (eds.) OTM-WS 2005. LNCS, vol. 3762, pp. 1066–1076. Springer, Heidelberg (2005)
2. Collins, A., Michalski, R.: The logic of plausible reasoning: A core theory. Cognitive Science 13(1), 1–49 (1989)
3. Dubois, D., Prade, H., Esteva, F., Garcia, P., Godo, L.: A logical approach to interpolation based on similarity relations. International Journal of Approximate Reasoning 17(1), 1–36 (1997)
4. Dupin de Saint-Cyr, F., Lang, J.: Belief extrapolation (or how to reason about observations and unpredicted change). Artif. Intell. 175(2), 760–790 (2011)
5. Forbus, K.D.: Qualitative process theory. Artif. Intell. 24(1-3), 85–168 (1984)
6. Gärdenfors, P.: Conceptual Spaces: The Geometry of Thought. MIT Press, Cambridge (2000)
7. Gardenfors, P., Williams, M.: Reasoning about categories in conceptual spaces. In: Int. Joint Conf. on Artificial Intelligence, pp. 385–392 (2001)
8. Gérard, R., Kaci, S., Prade, H.: Ranking alternatives on the basis of generic constraints and examples: a possibilistic approach. In: International Joint Conference on Artifical Intelligence, pp. 393–398 (2007)
9. Kraus, S., Lehmann, D., Magidor, M.: Nonmonotonic reasoning, preferential models and cumulative logics. Artificial Intelligence 44(1-2), 167–207 (1990)
10. Kuipers, B.: Qualitative simulation. Artificial Intelligence 29(3), 289–338 (1986)
11. Lehmann, F., Cohn, A.G.: The EGG/YOLK reliability hierarchy: semantic data integration using sorts with prototypes. In: Int. Conf. on Information and Knowledge Management, pp. 272–279 (1994)
12. Pawlak, Z.: Rough Sets: Theoretical Aspects of Reasoning about Data. Kluwer Academic Publishers, Dordrecht (1992)
13. Prade, H., Schockaert, S.: Completing rule bases in symbolic domains by analogy making. In: 7th Conf. of the European Society for Fuzzy Logic and Technology (2011)
14. Roy, B.: The outranking approach and the foundations of ELECTRE methods. Theory and Decision 31, 49–73 (1991)
15. Ruspini, E.: On the semantics of fuzzy logic. International Journal of Approximate Reasoning 5, 45–88 (1991)
16. Schockaert, S., Prade, H.: An inconsistency-tolerant approach to information merging based on proposition relaxation. In: AAAI Conf. on Artificial Intelligence, pp. 363–368 (2010)
17. Schockaert, S., Prade, H.: Interpolation and extrapolation in conceptual spaces: A case study in the music domain. In: Proceedings of the 5th International Conference on Web Reasoning and Rule Systems (2011)
18. Sun, R.: Robust reasoning: integrating rule-based and similarity-based reasoning. Artificial Intelligence 75(2), 241–295 (1995)

A Change Model for Credibility Partial Order

Luciano H. Tamargo, Marcelo A. Falappa,
Alejandro J. García, and Guillermo R. Simari

National Council of Scientific and Technical Research (CONICET)
Artificial Intelligence Research & Development Laboratory (LIDIA)
Universidad Nacional del Sur (UNS), Bahía Blanca, Argentina
{lt,mfalappa,ajg,grs}@cs.uns.edu.ar

Abstract. In a multi-agent system (MAS), an agent may often receive information through a potentially large number of informants. We will consider the case where the informants are independent agents who have their own interests and, therefore, are not necessarily completely reliable; in this setup, it will be natural for some agent to believe an informant more than other. The use of the notion of credibility will allow agents to organize their peers in a partial order that will reflect the relative credibility of their informants. It is also natural that the assigned credibility will change dynamically, leading to changes in the associated partial order. We will investigate the problem of updating the credibility order to reflect the change in the perceived agent's credibility, seeking to define a complete change theory over the agents' trust and reputation. The focus will be on the characterization and development of change operators (expansion, contraction, and revision) for modeling the dynamics of this partial order of agents. These operators, characterized through postulates and representation theorems, can be used to dynamically modify the credibility of informants to reflect a new perception of informant's plausibility, or admit the arrival of a new agent to the system.

1 Introduction and Motivation

In a multi-agent system (MAS), an agent may often receive information through informants. Although the number of these informants is dependent on the particular application, many real life situations present themselves as web related, thus involving a potentially large number of informant agents. For instance, the Amazon Mechanical Turk is an example of *crowdsourcing* [5] that coordinates the use of human intelligence to perform tasks that computers are unable to do yet, like finding specific objects in pictures. Clearly, this complex entity can be perceived as a massive multi-agent system where there is a need for establishing a partial order among the capabilities of the agents with the goal of maximizing the quality and precision of the result. The partial order among the participants could be affected by the result of training tasks or just by the performance of the agents. This in turn motivates the work on the formalization of the handling of the set of information providers, considering the changing situation regarding the perceived quality of the results obtained. The crowdsourcing model for

S. Benferhat and J. Grant (Eds.): SUM 2011, LNAI 6929, pp. 317–330, 2011.
© Springer-Verlag Berlin Heidelberg 2011

solving tasks can be applied to many problems where the agents in a massive multi-agent system can be seen as competing towards finding the best solution.

Multi-Source Belief Revision (MSBR) can be defined as belief revision performed by a single agent that can obtain new beliefs from multiple sources of information. In [18] we have introduced an epistemic model for MSBR that considers both beliefs and meta-information representing the credibility of the belief's source. We have investigated how the agent's belief base can be rationally modified when the agent receives information from other agents that can have different degrees of credibility. Thus, the main contribution was the definition based on the AGM model [1] of different belief change operators that use the credibility of informant agents in order to make decisions regarding what information prevails. These operators were defined through constructions and representation theorems.

It is important to note that the revision operator that we proposed in [18] is similar to the revision operator proposed in [3]. However, these operators are built in a different way. In [3], the epistemic state is represented by a possibility distribution which is a mapping from the set of classical interpretations or worlds to the [0,1] interval. This distribution represents the degree of compatibility of the interpretations with the available information and the revision is done over the possibility distribution. This revision modifies the ranking of interpretations so as to give priority to the input information.

In [18] we have required a total order among agents, but we have shown that this assumption can be generalized by considering a partial order over the set of agents. Also, we have assumed that this order is fixed; however, we have observed that the order relation can be changed without affecting the definition of the operators.

In this paper, we will extend the applications of the framework proposed in [18] considering environments in which the credibility of agents varies. That is, we will investigate the problem of updating the partial order among agents to reflect the change in credibility, seeking to define a complete change theory over the agents' trust and reputation. The use of a credibility order will allow agents to rank their peers according to the credibility they assign to each other. In [18], this order was used to define a rational way to compare beliefs for decision making during the knowledge revision process.

Our goal is therefore to formalize change operators over the credibility order. These operators will provide the capability of dynamically modifying the credibility of informants to reflect a new perception of the informant's plausibility, or extend the set of informants by admitting the arrival of a new agent to the system. While the concepts of trust and reputation are complex, we will take the position here that they can be seen as a kind of credibility value that the agents assign to each other. The approach taken here is to combine belief revision formalisms with trust and reputation maintenance techniques for agents in a distributed environment. Although there exist relevant works in both areas (*e.g.*, Multi-Agent Belief Revision [14,13,10,6] and Trust and

Reputation [15,16,9,2]) using these techniques independently, their combination in one formalism is novel.

The need for such a formalism can be exemplified by a real-world scenario. Consider the case of an agent that needs weather information. It is highly possible that there will be several weather-forecasting informants available to the agent, *e.g.*, there might be a number of web pages providing such services. However, given a particular time and date, different informants could predict different conditions. In such a case the agent would act based on the forecast of the most reliable informant and then use historic information on the reliability of the forecasts to update these relations.

Since its publication, the AGM paradigm has been widely accepted as a standard framework for belief revision. With the aim of modeling the dynamics of trust and reputation of agents in a system, and following AGM's style, we will develop change operators (expansion, contraction, and revision). This will give the agents the capability of updating the order relation representing the reputation of their peers. AGM's impact was due in part to the representation theorems (also called axiomatic characterizations) of contraction and revision, characterizing operations in terms of algorithms and a set of postulates. In the same spirit, the operators proposed in this research will be characterized by postulates and representation theorems. Work relevant to this effort began in [17].

An interesting critique to AGM's formalism was put forward in [11]. The authors observed that there is no reason to believe that the relative strength of the agent's beliefs will be maintained after a change. This implies that iterated belief change in the AGM framework as originally presented in [1] cannot be correctly defined. There are some models of iterated belief change [4,7,8] that generate a new order from the previous order, but their capabilities are limited. One of the main advantages of the model proposed in [18] (defined on belief bases) is that, after a belief change is effected, new changes can be applied because the plausibility of beliefs is preserved after a change.

The model presented here is more general because the change in the belief's order can be produced for different reasons:

- the addition of a new informant,
- the elimination of an informant, or
- a change in the credibility order among the informants.

That is, a belief change is not the only way in which the credibility order of informants can be altered and, consequently not the only way the strength of beliefs can be affected.

The rest of this paper is structured as follows. Next, in Section 2 we introduce a model for representing plausibility relations among informants in the context of a multi-agent system. In Section 3 we define the change operators on the credibility order: expansion, contraction, and revision. Finally, in Section 4 we offer our conclusions and the future work that lies ahead.

2 Representation of Informant Credibility Relations

Let us begin assuming that we have an universal set of informants, $\mathbb{A}$, and that, of these informants, some are to be considered more reliable than others. This means that in any case when two distinct informants provide an agent with contradictory information the information provided by the more credible will be preferred. Thus, the agent must have a mechanism to order the set $\mathbb{A}$. We begin its formalization with the following structure.

Definition 1. *Given a set of informants $\mathbb{A}$, a* generator set over $\mathbb{A}$ *is a binary relation G on $\mathbb{A}$, ($G \subseteq \mathbb{A}^2$). An informant A_i is less credible than an informant A_j according to G if $(A_i, A_j) \in G^*$, where G^* represents the reflexive transitive closure of G.*

Graphically, we will represent a generator set as a directed graph, where the informants in $\mathbb{A}$ label the nodes. The tuples in G are represented as directed arcs: for each tuple $(A_i, A_j) \in G$ we add an arc from node A_i to node A_j. For example, in Figure 1 we can see the graphic representation of the generator set $G = \{(A_1, A_3), (A_1, A_2), (A_2, A_6), (A_3, A_4), (A_3, A_5), (A_4, A_7), (A_5, A_7), (A_6, A_8), (A_7, A_8)\}$.

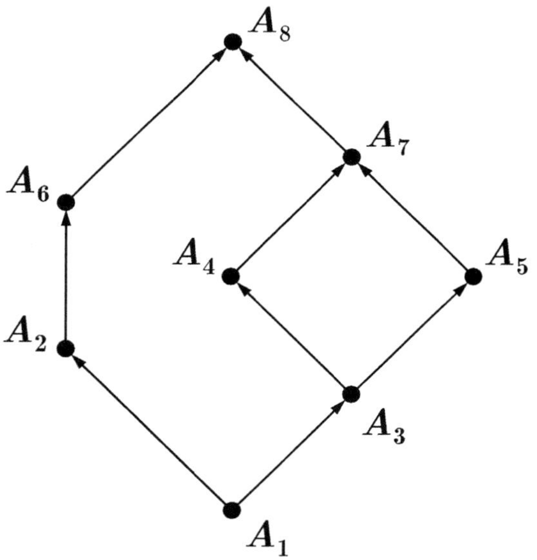

Fig. 1. A graph representation of a generator set

The relation G^*, as the reflexive transitive closure of G, is the smallest preorder containing G. Although it is desirable that G^* be a partial order over $\mathbb{A}$; the preceding definition does not fulfill the requirement. We address this matter in the following definition.

Definition 2. *A generator set $G \subseteq \mathbb{A}^2$ is said to be* sound *if G^* is a partial order over $\mathbb{A}$.*

Example 1. For example, the generator set $G_1 = \{(A_1, A_2), (A_2, A_3), (A_1, A_4)\}$ is sound. However, $G_2 = G_1 \cup \{(A_3, A_1)\}$ is *not* sound because $(A_1, A_3) \in G_2^*$ and $(A_3, A_1) \in G_2^*$, violating the antisymmetry condition for partial orders.

For a relation to be a partial order it must obey reflexivity, antisymmetry and transitivity. Given a generator set G its reflexive transitive closure, G^*, will obey reflexivity and transitivity. However if antisymmetry is not respected then there is at least one pair of distinct informants, A_i and A_j such that $(A_i, A_j), (A_j, A_i) \in G^*$. This would mean that both A_i is less trustworthy than A_j and that A_j is less trustworthy than A_i. Since these beliefs are contradictory, believing them simultaneously would lead the agent to an inconsistent belief status. For this reason we require for the generator set to be sound.

Throughout the discussions in the remainder of this paper we will sometimes speak of a tuple as being *entailed* by a generator set. This is a shorthand for saying that the tuple belongs to the reflexive transitive closure of the generator set, formally:

Definition 3. *We will say that a tuple (A_i, A_j) is* entailed *by a generator set G if $(A_i, A_j) \in G^*$.*

Example 2. The tuple (A_1, A_4) is entailed by the generator set $G = \{(A_1, A_2), (A_2, A_3), (A_3, A_4)\}$. When we represent a generator set graphically we will show entailed tuples of interest by using a dashed line as can be seen in *Figure 2 (a).*

A generator set may contain tuples that, if removed, would still be entailed by the remaining tuples. In this case we say that the tuple is *redundant* with respect

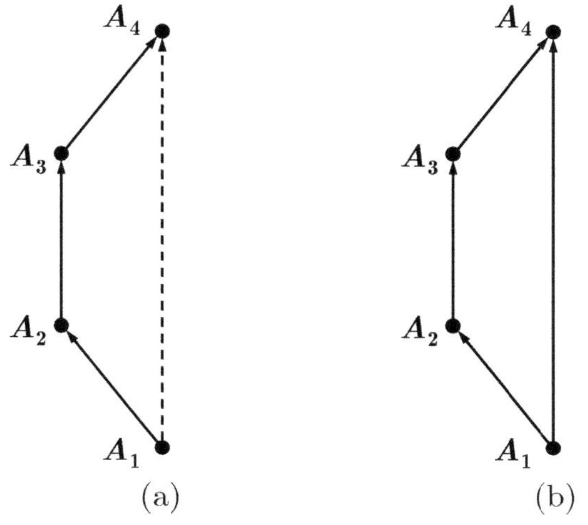

Fig. 2. (a) An entailed tuple, (b) A redundant tuple

to the generator set. We may also say that the generator set itself is redundant because it contains a redundant tuple. Formally:

Definition 4. *Given a tuple* (A_i, A_j) *in a generator set* G*, it is said that* (A_i, A_j) *is* redundant *in* G *if* $(A_i, A_j) \in (G \setminus \{(A_i, A_j)\})^*$*. A generator set is said to be* redundant *if it contains a redundant tuple. Otherwise, the generator set is said to be* non-redundant.

Example 3. The generator set $G = \{(A_1, A_2), (A_2, A_3), (A_3, A_4), (A_1, A_4)\}$ is redundant because it contains the redundant tuple (A_1, A_4) *(See Figure 2 (b))*.

3 Change Operators for Credibility Partial Order

In this section we will consider the definition of change operators on the credibility partial order. We will define operators for expansion, contraction, and revision, providing the appropriated postulates and constructions for each of them.

3.1 Expansion Operator for Credibility

Let us assume an agent learns that, of a pair of informants, one is more reliable than the other. This would warrant the modification of its knowledge accordingly. For this purpose, we define the operator $\oplus : \mathcal{P}(\mathbb{A}^2) \times \mathbb{A}^2 \longrightarrow \mathcal{P}(\mathbb{A}^2)$. This operator adds new tuples to a generator set in order to establish relations between informants. Given a pair of informants and a generator set, this function returns a new generator set in which said agents are now related. According to this new generator set we may say that the first informant is "less reliable" than the second.

Postulates for the Expansion Operator

E1-Success: $(A_i, A_j) \in (G \oplus (A_i, A_j))^*$
Determining new relations among informants is a costly process for the agent. Consequently, a desirable property of expansions is that the new relation learnt will be a member of the agents beliefs.

E2-Inclusion: $G^* \subseteq (G \oplus (A_i, A_j))^*$
This postulate reflects the fact that the agent will not loose beliefs during the expansion.

E3-Vacuity: if $(A_i, A_j) \in G^*$ then $(G \oplus (A_i, A_j))^* = G^*$
What this postulate states is that the equality between the original set and the product of the expansion will occur when the expansion is by a relation already entailed by the generator set, *i.e.*, there is no information to be lost or gained by the addition of redundant data.

E4-Commutativity: $((G \oplus (A_k, A_l)) \oplus (A_i, A_j))^* = ((G \oplus (A_i, A_j)) \oplus (A_k, A_l))^*$
The order in which tuples are added to the generator set does not affect the final, closed relation. This is important because sometimes we will use $G \oplus H$ as

a shorthand for the expansion of G by every tuple in H. Such is the case of the following postulate.

E5-Extensionality: if $H^*{=}I^*$ then $(G \oplus H)^* = (G \oplus I)^*$
The expansion of a generator set by two sets whose reflexive transitive closure is equal yields generator sets whose closure is also equal.

E6-Conditional Soundness Preservation: if G is a sound generator set and $(A_j, A_i) \notin G^*$ then $G \oplus (A_i, A_j)$ is a sound generator set.

Construction
The construction of expansions on credibility relations is formally defined as follows.

Definition 5. *Given a pair of informants $A_i, A_j \in \mathbb{A}$ and generator set $G \subseteq \mathbb{A}^2$, we define the expansion of G by (A_i, A_j) as*

$$G \oplus (A_i, A_j) = G \cup \{(A_i, A_j)\}$$

Expansion does not preserve soundness *per se*, but is conditioned as stated in the postulate. This property is a consequence of the definition of sound generator sets and the definition of expansion that we have provided.

3.2 Contraction Operator for Credibility

At the beginning of the previous subsection, we said that an agent may need to assert the fact that one informant is less reliable than another. In a similar fashion the opposite may also become true, *i.e.*, we may wish to reflect the fact that an informant is no longer more reliable than another. For this purpose we define a contraction operator $\ominus : \mathcal{P}(\mathbb{A}^2) \times \mathbb{A}^2 \longrightarrow \mathcal{P}(\mathbb{A}^2)$.

Assume we have a pair of informants A_i and A_j and a generator set G such that $(A_i, A_j) \in G^*$. The basic task of the $\ominus$ function is to construct a new generator set in which this is no longer the case while losing as little information as possible. However we cannot simply remove the pair (A_i, A_j) from G. In fact, (A_i, A_j) may not even be in G. Care must be taken to also remove pairs that, through transitivity, would entail the pair (A_i, A_j) in G^*. As long as there is a path in the generator set from A_i to A_j, (A_i, A_j) will be found in its transitive closure. It is therefore necessary to eliminate a set of pairs so that no path is left from A_i to A_j in G. This set will be required to be minimal.

Postulates for the Contraction Operator

C1-Success: if $A_i \neq A_j$ then $(A_i, A_j) \notin (G \ominus (A_i, A_j))^*$
A tuple cannot be entailed by the generator set resulting from its contraction. In the case of $A_i = A_j$, the tuple will trivially be in the reflexive transitive closure of any generator set due to reflexivity.

C2-Inclusion: $(G \ominus (A_i, A_j))^* \subseteq G^*$

If a tuple is entailed by a generator set, then its contraction by said tuple removes at least one element from the set: the tuple itself. The sets are equal in the case in which $(A_i, A_j) \notin G^*$.

C3-Uniformity: if for all $G' \subseteq G$, $(A_i, A_j) \in (G')^*$ if and only if $(A_p, A_t) \in (G')^*$ then $G \ominus (A_i, A_j) = G \ominus (A_p, A_t)$.
This property establishes that if two tuples (A_i, A_j) and (A_p, A_t) are entailed by exactly the same subsets of G, then the contraction of G by (A_i, A_j) should be equal to the contraction of G by (A_p, A_t).

C4-Core retainment: If $(A_p, A_t) \in G$ and $(A_p, A_t) \notin G \ominus (A_i, A_j)$ then there exist a set G' such that $G' \subseteq G$, $(A_i, A_j) \notin (G')^*$ but $(A_i, A_j) \in (G' \cup \{(A_p, A_t)\})^*$.
The tuples that we give up in order to contract G by (A_i, A_j) should all be such that they contributed to the fact that G, but not $G - (A_i, A_j)$, entails (A_i, A_j).

C5-Soundness Preservation: if G is a sound generator set then $G \ominus (A_i, A_j)$ is a sound generator set.

Construction
In this subsection we will introduce a construction for contractions on credibility relations. However, before we do so, we will need to present a few concepts.

First let us briefly review the concept of path. We say that a set of tuples P is a *path* from A_i to A_j if $(A_i, A_j) \in P$, or $(A_i, A_k) \in P$ and there is a path from A_k to A_j in P. We say that P is a *nonredundant path* from A_i to A_j if it is a path from A_i to A_j and there is no path P' from A_i to A_j, such that $P' \subset P$.

Definition 6. *Given a pair of informants $A_i, A_j \in \mathbb{A}$ and generator set $G \subseteq \mathbb{A}^2$, we define the* path set *from A_i to A_j in G, and we will note it G_{ij}, as $G_{ij} = \{C \subseteq G : C$ is a nonredundant path from A_i to A_j in $G\}$.*

Notice that according to this definition the path set from A_i to A_j in a generator set G is a set of sets. Each set represents a path from A_i to A_j. In the contraction of G by (A_i, A_j), in order to avoid the occurrence of this tuple, none of these paths may remain complete. Therefore, we need a selection mechanism to decide which tuples will be erased from each path in G_{ij}.

Definition 7. *Given a path set G_{ij}, we say that γ is a* cut function *for G_{ij} if and only if:*

1. *$\gamma(G_{ij}) \subseteq \bigcup(G_{ij})$.*
2. *For each $C \in G_{ij}$, $C \neq \emptyset$, $C \cap \gamma(G_{ij}) \neq \emptyset$.*

Now we may present our definition of contraction.

Definition 8. *Given a pair of informants $A_i, A_j \in \mathbb{A}$ and generator set $G \subseteq \mathbb{A}^2$, we define the contraction for credibility of G by (A_i, A_j) as*

$$G \ominus (A_i, A_j) = G \setminus \gamma(G_{ij})$$

It is important to note that we have defined a family of contraction operators following other established formalisms of belief revision [1,12]. The specification of the cut function will allow the introduction of different possibilities.

Example 4. Consider the generator set $G = \{(A_1, A_2), (A_1, A_3), (A_1, A_8), (A_8, A_2), (A_2, A_3)\}$ of the agent A_8. Then, suppose A_8 wants to contract G by (A_1, A_2) using "$\ominus$". Note that, there exist two paths from A_1 to A_2 in G. That is, $G_{12} = \{\{(A_1, A_2)\}, \{(A_1, A_8), (A_8, A_2)\}\}$. Then, in order to avoid the occurrence of (A_1, A_2), the *cut function* selects the tuples that will be erased from each path in G_{12}. Note that in the second path the *cut function* can select the two tuples or either one of these, depending on its specification. Suppose that (A_1, A_8) is selected by the *cut function* from the second path. Then, $G \ominus (A_1, A_2) = \{(A_1, A_3), (A_8, A_2), (A_2, A_3)\}$ *(See Figure 3).*

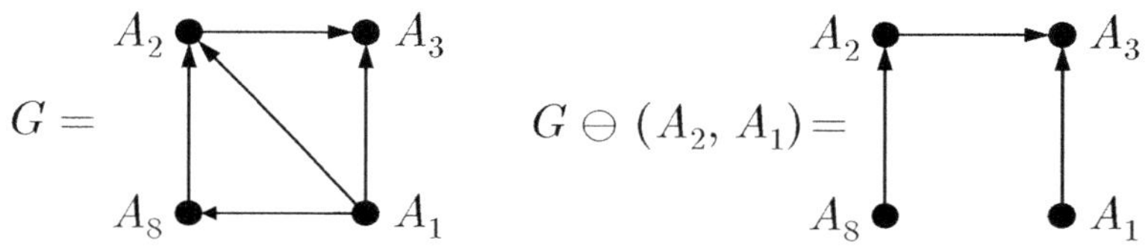

Fig. 3. Contraction operator

Notice that here, in contrast to the case of expansion, the soundness preservation property of contraction is not conditioned. This is due to the way we define contraction. Since contraction is basically a process of elimination, it is impossible for this operation to introduce cycles if there were none with which to begin.

Next, we give a proposition used in the representation theorem of contraction operator (Theorem 1). This Theorem gives a summary of the properties of the contraction operator.

Proposition 1. *$G_{ij} = G_{pt}$ if and only if for all subsets G' of G: $(A_i, A_j) \in (G')^*$ if and only if $(A_p, A_t) \in (G')^*$.*

Proof
We will use *reductio by absurdum.*
($\Rightarrow$) Suppose that there is some subset B of G such that $(A_i, A_j) \in B^*$ and $(A_p, A_t) \notin B^*$. Then, there is some path P of G_{ij} such that $P \subseteq B$. Since $P \subseteq B$ and $(A_p, A_t) \notin B^*$, we have $(A_p, A_t) \notin P^*$, so that $P \notin G_{pt}$. Then $P \in G_{ij}$ and $P \notin G_{pt}$ contrary to $G_{ij} = G_{pt}$.
($\Leftarrow$) Suppose that $G_{ij} \neq G_{pt}$. We may assume that there is some $P \in G_{ij}$ such that $P \notin G_{pt}$. There are two cases:

- $(A_p, A_t) \notin P^*$: then we have $(A_i, A_j) \in P^*$ and $(A_p, A_t) \notin P^*$, showing that the conditions of the proposition are not satisfied.

- $(A_p, A_t) \in P^*$: then it follows from $P \notin G_{pt}$ that there is some P' such that $P' \subset P$ and $(A_p, A_t) \in (P')^*$. We than have $(A_p, A_t) \in (P')^*$ and $(A_i, A_j) \notin (P')^*$, showing that the conditions of the proposition are not satisfied.

Theorem 1. *Let G be a generator set and let "$\ominus$" be a contraction operator. "$\ominus$" is a* contraction for credibility *for G iff it satisfies* **C1, ..., C4**, *i.e., it satisfies* success, inclusion, uniformity *and* core retainment.

Proof
• *Postulates to Construction.* We need to show that if an operator $(-)$ satisfies the enumerated postulates, then it is possible to build an operator in the way specified in the theorem $(\ominus)$. Let "γ" be a function such that, for every generator set G ($G \subseteq \mathbb{A}^2$) and for every tuple (A_i, A_j), it holds that:
[Hypothesis] $\gamma(G_{ij}) = G \setminus G - (A_i, A_j)$.
We must show:
– Part A.

1. "γ" is a well defined function.
2. $\gamma(G_{ij}) \subseteq \bigcup(G_{ij})$.
3. For each $C \in G_{ij}$, $C \neq \emptyset$, $C \cap \gamma(G_{ij}) \neq \emptyset$.

– Part B. "$\ominus$" is equal to "$-$", that is, $G \ominus (A_i, A_j) = G - (A_i, A_j)$.

Part A.

1. "γ" is a well defined function.

Let (A_i, A_j) and (A_p, A_t) be such that $G_{ij} = G_{pt}$. We need to show $\gamma(G_{ij}) = \gamma(G_{pt})$. It follows from $G_{ij} = G_{pt}$, by Prop. 1, for all subset G' of G, $(A_i, A_j) \in (G')^*$ iff $(A_p, A_t) \in (G')^*$. Thus, by **uniformity**, $G - (A_i, A_j) = G - (A_p, A_t)$. Then, by the definition of γ adopted in the hypothesis, $\gamma(G_{ij}) = \gamma(G_{pt})$.

2. $\gamma(G_{ij}) \subseteq \bigcup(G_{ij})$.
Let $(A_p, A_t) \in \gamma(G_{ij})$. By the definition of γ adopted in the hypothesis $(A_p, A_t) \in G \setminus G - (A_i, A_j)$. Thus, $(A_p, A_t) \in G$ and $(A_p, A_t) \notin G - (A_i, A_j)$. It follows by **core retainment** that there is some $G' \subseteq G$ such that $(A_i, A_j) \notin (G')^*$ but $(A_i, A_j) \in (G' \cup \{(A_p, A_t)\})^*$. Then, there is some finite subset G'' of G' such that $(A_i, A_j) \in (G'' \cup \{(A_p, A_t)\})^*$. Since $(A_i, A_j) \notin (G')^*$ we have $(A_i, A_j) \notin (G'')^*$. It follows from $(A_i, A_j) \notin (G'')^*$ and $(A_i, A_j) \in (G'' \cup \{(A_p, A_t)\})^*$ that there is some path that contains (A_p, A_t). Hence, $(A_p, A_t) \in \bigcup(G_{ij})$.

3. For each $C \in G_{ij}$, $C \neq \emptyset$, $C \cap \gamma(G_{ij}) \neq \emptyset$.
Let $\emptyset \neq C \in G_{ij}$, we need to show that $C \cap \gamma(G_{ij}) \neq \emptyset$. We should prove that, there exists $(A_p, A_t) \in C$ such that $(A_p, A_t) \in \gamma(G_{ij})$. By **success**, $(A_i, A_j) \notin (G - (A_i, A_j))^*$. Since $C \neq \emptyset$ then $(A_i, A_j) \in C^*$ and $C \nsubseteq G - (A_i, A_j)$; *i.e.*, there is some (A_p, A_t) such that $(A_p, A_t) \in C$ and $(A_p, A_t) \notin G - (A_i, A_j)$. Since $C \subseteq G$ it follows that $(A_p, A_t) \in (G \setminus G - (A_i, A_j))$; *i.e.*, by the definition of γ adopted in the hypothesis $(A_p, A_t) \in \gamma(G_{ij})$. Therefore, $C \cap \gamma(G_{ij}) \neq \emptyset$.

Part B. "$\ominus$" is equal to "$-$", that is, $G \ominus (A_i, A_j) = G - (A_i, A_j)$.

Let "$\ominus$" a contraction operator defined as $G \ominus (A_i, A_j) = G \setminus \gamma(G_{ij})$ and γ defined as in the hypothesis.

($\supseteq$) Let $(A_p, A_t) \in G - (A_i, A_j)$. Then, $(A_p, A_t) \in (G - (A_i, A_j))^*$. It follows by **inclusion** that $(G - (A_i, A_j))^* \subseteq G^*$ and $(A_p, A_t) \in G^*$. It follows from $(A_p, A_t) \in G - (A_i, A_j)$ and $(A_p, A_t) \in G^*$ that $(A_p, A_t) \notin (G^* \setminus G - (A_i, A_j))$. Since $G \subseteq G^*$, then $(A_p, A_t) \notin (G \setminus G - (A_i, A_j))$. Thus, by the definition of γ adopted in the hypothesis, $(A_p, A_t) \notin \gamma(G_{ij})$. Hence, $(A_p, A_t) \in G \ominus (A_i, A_j)$.

($\subseteq$) Let $(A_p, A_t) \in G \ominus (A_i, A_j)$. By definition $(A_p, A_t) \in G \setminus \gamma(G_{ij})$. Then, $(A_p, A_t) \in G$ and $(A_p, A_t) \notin \gamma(G_{ij})$. Thus, by the definition of γ adopted in the hypothesis, $(A_p, A_t) \notin G \setminus G - (A_i, A_j)$. Hence, $(A_p, A_t) \in G - (A_i, A_j)$.

• *Construction to Postulates.* Let $\ominus$ be a contraction for credibility for G. We need to show that it satisfies the four conditions of the theorem.

(C1) *Success:* if $A_i \neq A_j$, then $(A_i, A_j) \notin (G \ominus (A_i, A_j))$.

Proof. Suppose to the contrary that $A_i \neq A_j$ and $(A_i, A_j) \in (G \ominus (A_i, A_j))$. There is then a path $P \in G_{ij}$ such that $P \subseteq G \ominus (A_i, A_j)$. It follows from $A_i \neq A_j$ that $P \neq \emptyset$. By clause (2) of Definition 7, there is some $(A_p, A_t) \in P$ such that $(A_p, A_t) \in \gamma(G_{ij})$. By Definition 8, $(A_p, A_t) \notin (G \ominus (A_i, A_j))$, contrary to $(A_p, A_t) \in P$ with $P \subseteq G \ominus (A_i, A_j)$.

(C2) *Inclusion:* $(G \ominus (A_i, A_j))^* \subseteq G^*$.

Proof. Straightforward by definition.

(C3) *Uniformity:* If for all $G' \subseteq G$, $(A_i, A_j) \in (G')^*$ if and only if $(A_p, A_t) \in (G')^*$ then $G \ominus (A_i, A_j) = G \ominus (A_p, A_t)$.

Proof. Suppose that for all subset G' of G, $(A_i, A_j) \in (G')^*$ if and only if $(A_p, A_t) \in (G')^*$. By Proposition 1, $G_{ij} = G_{pt}$. Since "γ" is a well defined function then $\gamma(G_{ij}) = \gamma(G_{pt})$. Therefore, by Definition 8 $G \ominus (A_i, A_j) = G \ominus (A_p, A_t)$.

(C4) *Core retainment*: If $(A_p, A_t) \in G$ and $(A_p, A_t) \notin G \ominus (A_i, A_j)$ then there exist a set G' such that $G' \subseteq G$, $(A_i, A_j) \notin (G')^*$ but $(A_i, A_j) \in (G' \cup \{(A_p, A_t)\})^*$.

Proof. Suppose $(A_p, A_t) \in G$ and $(A_p, A_t) \notin G \ominus (A_i, A_j)$. Then, by Definition 8, $(A_p, A_t) \in \gamma(G_{ij})$. By Definition 7 of cut function, $\gamma(G_{ij}) \subseteq \bigcup(G_{ij})$, so that there is some path P such that $(A_p, A_t) \in P \in G_{ij}$. Let $X = P \setminus \{(A_p, A_t)\}$. Then, since P is minimal, $(A_i, A_j) \notin (P)^*$ but $(A_i, A_j) \in (P \cup \{(A_p, A_t)\})^*$.

3.3 Revision Operator for Credibility

Suppose that an agent learns that an informant is less reliable than another. The agent's current generator set should be modified to reflect this new information. However, it would be convenient if the generator set were also modified, when necessary, so that the opposite can no longer hold. That is to say, if up to now the agent believed that the second informant was less reliable then this should be retracted.

For this purpose we define the revision operator $\otimes : \mathcal{P}(\mathbb{A}^2) \times \mathbb{A}^2 \longrightarrow \mathcal{P}(\mathbb{A}^2)$. Assume we have a pair of informants A_i and A_j and a generator set G, and the agent now has reason to believe that A_i is less reliable than A_j. The basic task of the $\otimes$ operator is to construct a new generator set in which (A_i, A_j) is entailed but (A_j, A_i) is not.

Postulates for the Revision Operator

R1-Success: $(A_i, A_j) \in (G \otimes (A_i, A_j))^*$.
This is basically a consequence of the definition given for revision and the success postulate for expansion.

R2-Inclusion: $(G \otimes (A_i, A_j))^* \subseteq (G \oplus (A_i, A_j))^*$.
This is due to the fact that expansion simply inserts the new tuple into the generator set while revision may need to remove tuples before adding the new one. The border case of equality presents itself when $(A_j, A_i) \notin G^*$.

R3-Soundness Preservation: if G is a sound generator set then $G \otimes (A_i, A_j)$ is a sound generator set.

The main aim of the revision operator is to hold soundness in the generator set revised.

R4-Uniformity: If for all $G' \subseteq G$, $\{(A_i, A_j)\} \cup G'$ is not sound if and only if $\{(A_p, A_t)\} \cup G'$ is not sound then $G \cap (G \otimes (A_i, A_j)) = G \cap (G \otimes (A_p, A_t))$.

This postulate determines that if two tuples (A_i, A_j) and (A_p, A_t) are inconsistent with the same subsets of G then G revised by those tuples should preserve the same tuples from G.

R5-Core retainment: If $(A_p, A_t) \in G$ and $(A_p, A_t) \notin G \otimes (A_i, A_j)$ then there exist a set G' such that $G' \subseteq G$, $(A_i, A_j) \notin (G')^*$ but $(A_i, A_j) \in (G' \cup \{(A_p, A_t)\})^*$.

The intuition behind this postulate is similar to that of the *core-retainment* postulate (C4) for contractions introduced above.

Construction

In this subsection, we will introduce a construction of revisions on credibility relations.

Definition 9. *Given a pair of informants $A_i, A_j \in \mathbb{A}$ and generator set $G \subseteq \mathbb{A}^2$, we define the revision for credibility of G by (A_i, A_j) as*

$$G \otimes (A_i, A_j) = (G \ominus (A_j, A_i)) \oplus (A_i, A_j)$$

Example 5. Consider the generator set $G = \{(A_1, A_2), (A_1, A_3), (A_8, A_2), (A_2, A_3)\}$ of the agent A_8. Then, suppose A_8 wants to revise G by (A_2, A_1) using "$\otimes$". Since $(A_1, A_2) \in G^*$ then it is necessary to contract G by (A_1, A_2) and then expand G by (A_2, A_1). Thus, $G \otimes (A_2, A_1) = \{(A_1, A_3), (A_8, A_2), (A_2, A_3), (A_2, A_1)\}$ *(See Figure 4).*

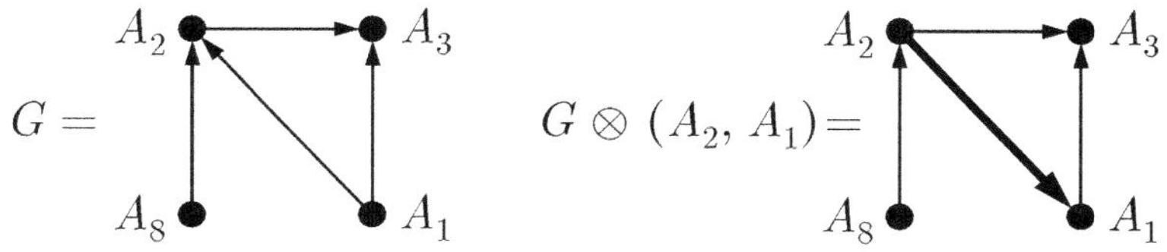

Fig. 4. Revision operator

The following result enunciate an interesting property of the revision operator.

Theorem 2. *Let G be a generator set and let "$\otimes$" be a revision operator. "$\otimes$" is a* revision for credibility *for G if and only if it satisfies* **R1, ..., R5**, *i.e., it satisfies* success, inclusion, soundness preservation, uniformity *and* core retainment.

Proof: This proof is analogous to the proof of Theorem 1.

Again here, as in the case of contraction, soundness preservation is not conditioned. In the case that the new tuple to be inserted, (A_i, A_j) were to complete a cycle, the previous contraction of (A_j, A_i) would insure that there is no link between A_j and A_i. Hence, it is impossible for revision to introduce cycles.

4 Conclusions and Future Work

In this work, we have introduced a model for representing credibility relations among informants in the context of a potentially massive multi-agent system. This was done through the use of *generator sets* which, when sound, establish a partial order over the set of an agent's informants. Thus, when faced with contradicting information, the agent can solve the inconsistency by believing the more reliable informant as determined by the generator set.

Given the dynamic nature of multi-agent systems, we also have proposed operators for the modification of the plausibility relation. These change operators (expansion, contraction and revision) were defined over the generator sets. They were formally characterized through postulates, analogous in some cases to those of the AGM model, and unique in others. Furthermore, we have proved representation theorems for the more important changes (contractions and revisions). Finally, a construction was provided for each of these change operators, thus completing their definition.

As future work, we will extend the framework adding a reliability measure to the relation tuples. With this, the cut function can be specified to determine a rational way for cutting the tuples during the revision process. Furthermore, based on the reliability, a non-prioritized revision operator for credibility partial orders can be defined.

References

1. Alchourrón, C., Gärdenfors, P., Makinson, D.: On the logic of theory change: Partial meet contraction and revision functions. J. of Symbolic Logic 50(2), 510–530 (1985)
2. Barber, K.S., Kim, J.: Belief revision process based on trust: Simulation experiments. In: Proceedings of Autonomous Agents 2001 Workshop on Deception, Fraud, and Trust in Agent Societies, pp. 1–12 (2001)
3. Benferhat, S., Dubois, D., Prade, H., Williams, M.A.: A practical approach to revising prioritized knowledge bases. Studia Logica 70(1), 105–130 (2002)
4. Boutilier, C.: Iterated revision and minimal change of conditional beliefs. Journal of Philosophical Logic 25(3), 262–305 (1996)
5. Brabham, D.C.: Crowdsourcing as a model for problem solving: An introduction and cases. Convergence: The International Journal of Research into New Media Technologies 14(1), 75–90 (2008)
6. Cantwell, J.: Resolving conflicting information. Journal of Logic, Language and Information 7(2), 191–220 (1998)
7. Darwiche, A., Pearl, J.: On the logic of iterated belief revision. Artificial Intelligence 89, 1–29 (1997)
8. Delgrande, J.P., Dubois, D., Lang, J.: Iterated revision as prioritized merging. In: 10th Int. Conf. on Principles of Knowledge Representation and Reasoning, UK, pp. 210–220 (2006)
9. Dellarocas, C.: The digitalization of word-of-mouth: Promise and challenges of online reputation mexhanisms. In: Management Science (2003)
10. Dragoni, A., Giorgini, P., Puliti, P.: Distributed belief revision versus distributed truth maintenance. In: Proceedings of the Sixth IEEE International Conference on Tools with Artificial Intelligence (TAI 1994), pp. 499–505. IEEE Computer Society Press, Los Alamitos (1994)
11. Friedman, N., Halpern, J.: Belief Revision: A Critique. Journal of Logic, Language and Information 8(4), 401–420 (1999)
12. Hansson, S.O.: Kernel contraction. Journal of Symbolic Logic 59(3), 845–859 (1994)
13. Kfir-Dahav, N.E., Tennenholz, M.: Multi-agent belief revision. In: Theoretical Aspects of Rationality and Knowledge: Proceeding of the Sixth Conference (TARK 1996), pp. 175–196. Morgan Kaufmann Publishers Inc., San Francisco (1996)
14. Liu, W., Williams, M.: A framework for multi-agent belief revision, part i: The role of ontology. In: Foo, N.Y. (ed.) AI 1999. LNCS, vol. 1747, pp. 168–179. Springer, Heidelberg (1999)
15. Sabater, J., Sierra, C.: Review on computational trust and reputation models. Artificial Intelligence Review 24(1), 33–60 (2005)
16. Sabater, J., Sierra, C.: Regret: A reputation model for gregarious societies. In: Proceedings of the Fourth Workshop on Deception, Fraud and Trust in Agent Societies, pp. 61–69 (2001)
17. Simari, P.D., Falappa, M.A.: Revision of informant plausibility in multi-agent systems. Journal of Computer Science and Technology 2(5) (2001)
18. Tamargo, L.H., García, A.J., Falappa, M.A., Simari, G.R.: Modeling knowledge dynamics in multi-agent systems based on informants. In: The Knowledge Engineering Review, KER (2010) (in print)

Conflict-Aware Historical Data Fusion

Vladimir Zadorozhny and Ying-Feng Hsu

Graduate Program of Information Science and Technology
University of Pittsburgh
Pittsburgh, PA
Tel.: 1 412 624 9411
{vladimir,yfhsu}@sis.pitt.edu

Abstract. Historical data reports on numerous events for overlapping time intervals, locations, and names. As a result, it may include severe data conflicts caused by database redundancy that prevent researchers from obtaining the correct answers to queries on an integrated historical database. In this paper, we propose a novel conflict-aware data fusion strategy for historical data sources. We evaluated our approach on a large-scale data warehouse that integrates historical data from approximately 50,000 reports on US epidemiological data for more than 100 years. We demonstrate that our approach significantly reduces data aggregation error in the integrated historical database.

1 Introduction

Efficient interdisciplinary research requires consolidation of large amounts of *historical data* from disparate data sources in different subject areas. For example, epidemiological data analysis often relies upon knowledge of population dynamics, climate changes, migration of biological species, drug development, etc. As another example, consider the task of exploring long-term and short-term social changes that require consolidating a comprehensive set of data on social-scientific, health, and environmental dynamics.

The historical data reports on events of interest occurring within various time intervals. As a result, it may include severe *data conflicts* that prevent researchers from obtaining the correct answers to queries on an integrated historical database. It is common to have multiple concurrent reports about the same event within *overlapping time intervals*. For example, we may have hundreds of reports from different authorities about cases of measles in LA for the year 1900. We may also have multiple reports on historical statistics for *overlapping locations*. A cumulative report on the total number of measles cases for the state of California may differ considerably from the available reports on the total number of measles cases in California cities. Another challenge is *overlapping names*: evolving concepts may be reported under different names and categories co-existing at different time intervals. For example, many 19th century reports on yellow fever were actually referring to cases of hepatitis. In 1947, viral hepatitis was classified as hepatitis A and hepatitis B; that distinction was not immediately reflected in the epidemiological records. Determining the correct number of cases from all of those reports is problematic.

S. Benferhat and J. Grant (Eds.): SUM 2011, LNAI 6929, pp. 331–345, 2011.
© Springer-Verlag Berlin Heidelberg 2011

Consideration of only non-overlapping reports may result in significant underestimation; at the same time, by ignoring the overlaps, we risk over-estimating that number.

Resolving historical data conflicts requires efficient data fusion strategies. Research on the data fusion for the data integration systems is relatively recent. Our work is one of the first attempts to systematically investigate the challenge of large-scale historical data fusion. Towards this end, the paper has three primary contributions: (a) development of conflict-aware data fusion for efficient aggregation of historical data, (b) simulation-based study of the tradeoffs between the data fusion solutions and data accuracy, (c) evaluation of the solutions in a large-scale integrated framework that includes historical data from heterogeneous sources in different subject areas. The last contribution utilizes our ongoing effort on the design and development of Tycho, an integrated epidemiological data warehouse. This work is undertaken in collaboration with the University of Pittsburgh School of Public Health. Currently, Tycho consolidates information from approximately 50,000 reports on United States epidemiological data for more than 100 years. This work enabled us to address realistic historical data fusion challenges that go far beyond purely academic interests.

2 Background and Related Work

In this paper we investigate issues of efficient data fusion from heterogeneous historical data sources. Our proposed approach is of general applicability to large-scale Data Integration Systems that address two major challenges: (1) *heterogeneous data* and (2) *conflicting data*. Disparate data sources can describe the same application domain using different schemas, which causes *schema heterogeneity*. Various techniques for resolving schema heterogeneity include schema matching [24], data exchange [19], and model management [4]. Another type of heterogeneity occurs at the *instance level*. Data sources can represent the same entity in different ways due to the lack of standard formats. In this case, the integrated database may include duplicate records which do not share a common key. The duplicates may also occur as the result of transcription errors, or incomplete information. Resolving instance heterogeneity typically requires using duplicate detection algorithms based on various record similarity metrics. [16] provides a comprehensive review of this area.

Resolving data heterogeneities has been the focus of active research and development for more than two decades [9,20]. There are numerous tools on the market for efficient mapping of data sources in a homogenous schema with proper data cleaning (eliminating typos, misspellings, and formatting errors), standardization of names, conversion of data types, duplicate elimination, etc. A separate body of research deals with Web data integration, which includes our work on accessing heterogeneous Web data sources [33,34,35].

Conflicting data may occur in an integrated data set even after the data heterogeneities have been resolved. While each data source may provide a consistent data set, the integrated data can violate application integrity constraints -- resulting in numerous data conflicts. Consider Fig. 1, which shows a fragment of an integrated employment database. Tuples *t1* and *t4* were extracted from independent data sources *s1* and *s2*, correspondingly. The integrated table violates an application requirement that each employee receives only one salary. This constraint is expressed as a functional dependency *emp_name* → *salary*. Violation of this constraint makes it problematic to obtain consistent answers to aggregate queries. For example, we cannot find a correct value of total salary for all employees. We say that tuples *t1* and *t4* are conflicting with respect to the functional dependency constraint.

	emp_name	salary
t1	Smith	1000
t2	Brown	2000
t3	Jones	3000
t4	Smith	4000

Integrity Constraint: emp_name → salary
(employee can receive only one salary)

Tuples t1 and t4 are conflicting.
DB is inconsistent.

Query:
select sum(salary)
from Emp

Consistent answer?

Fig. 1. Example of an inconsistent integrated database

Assuming that we consider only one employer and that there is only one Smith among the employees, this inconsistency may occur if *s1* and *s2* refer to different time intervals where Smith indeed received different salaries. If *s1* reports Smith's salary from 1/2004 to 12/2004 and *s2* reports Smith's new salary as of 1/2005 – 12/2005, then we can easily obtain the correct total salary value within a specific time interval. Consider now a different scenario in which *s1* reports Smith's salary from 1/2004 to 12/2004 while *s2* reports it from 5/2004 to 12/2005, i.e. the reporting time intervals overlap. Obtaining the correct total salary in this case is not possible since we do not know Smith's actual salary during the intersection of the reported time intervals (i.e., from 5/2004 to 12/2004). Most likely, this situation indicates that there was an accounting error either in *s1* or in *s2*, which can be repaired by consulting related employer documents.

In summary, data sources *s1* and *s2* report on historical data about salaries. Overlapping reporting periods for someone's salary are unexpected and indicate an error. Such an error can be fixed using readily-available additional information. The class of historical data sources considered in this paper also report on events of interest occurring within various time intervals. However, overlapping reporting periods are expected. For example, it is common to have multiple concurrent reports about the same event within the same time interval. We may have hundreds of reports from different authorities about measles cases in LA for the year 1900.

Thus, as in the employment database, historical tuples reporting the same event for overlapping time intervals conflict and may prevent us from obtaining the correct result to a query on an integrated historical database. At the same time, the nature of conflicts in historical data differs from the nature of the conflict considered in Fig. 1. Historical data conflicts do not necessary imply an inconsistency in the integrated database. If the overlapping historical reports are accurate, the conflicts reflect *data redundancy* that prevents us from obtaining *accurate* aggregate query results. The inconsistency may be caused by inaccurate reports. For example, two historical tuples may report different values for the same time interval. Historical data conflicts cannot be easily resolved using readily-available external information, as was the case with Smith's salary. Moreover, the number of conflicting tuples in the integrated historical data can be quite large.

The amount of research in the area of data conflict resolution and querying inconsistent data is considerable. See [7,15] and [5,6] for a comprehensive review of the current state of the art. Generally speaking, data conflict handling strategies are classified in three groups: conflict ignoring, conflict avoidance, and conflict resolution. Conflict ignoring is a straightforward approach that may result in inconsistent query results. For example, in Fig. 1 the conflict ignoring strategy would return a total salary of $10,000. Conflict avoidance utilizes some data preference logic that can be applied uniformly to all conflicting data (i.e., ignore all conflicting data from the data source *s1*). In Fig. 1, tuple *t1* would be ignored as coming from source *s1*, which would produce a total salary of $9000. Finally, conflict resolution performs a fusion of each individual conflict. For example, in Fig. 1 we could consider a weighted average of salaries from conflicting tuples, or choose a more recent salary. . Alternatively, we could introduce some uncertainty and consider *ranges* of possible aggregate values, e.g., for the total salary it could be between $6000 and $9000. This conflict can also be resolved using metadata about data source accuracy [32] and freshness [30]. Another approach exploits dependencies between data sources, where information from one source can be re-used in another source [31]. Source dependencies are common in historical data, where publications reflect recorded history from older reports. Conflict resolution is the most advanced approach: it is the focus of our work as well.

Early research on handling inconsistencies was mostly theoretical and did not relate this problem directly to data integration [21]. Data inconsistency as a key integrity constraint violation was considered in [2]. Consistent query answering that ignores inconsistent data that violates integrity constraints was introduced in [11]. This approach is related to more recent research on query transformation for consistent query answering [28]. An alternative approach is based on inconsistent database repair, producing a minimally different – yet consistent -- database that satisfies integrity constraints [8,29]. A notable body of related research considered different classes of queries and constraints for both query transformation and database repair strategies [1,12]. An interesting research direction was to investigate query transformation and database repair using logic programming techniques [3]. Another promising research direction is related to consistent aggregate query answering for data analysis. This involves research on aggregation constraints on numerical data [17, 18]. A separate body of research on temporal databases explored mechanisms for managing time-varying information [14,22,26]. We are not aware of any works in this area that systematically consider fusing temporal data.

In summary, there is a considerable ongoing work around Data Integration Systems focused on resolving data heterogeneities. There is also a less substantial body of relatively recent research exploring the concept of data fusion from the perspective of data consistency, data completeness, and computational complexity. To the best of our knowledge, the problem of large-scale fusion of redundant historical data has not been systematically addressed in related research.

3 Problems in Historical Data Fusion

Historical data fusion explores *conflicting tuples* and their impact on the accuracy of results of aggregate queries over the historical database. Next, we discuss three major types of conflicts that may occur between the historical tuples: (1) *temporal conflicts*, (2) *spatial conflicts*, and (3) *naming conflicts*.

Temporal Conflicts. Fig. 2 shows an example of a historical database including data references for total number of cases of measles in NYC (tuples *t1*, *t4*). Note that we use this and following examples for illustration purposes only and they do not represent any actual disease occurrences. We cannot simply add the values of *t1* and *t4* to find the total number of cases of measles, as *t1* and *t4* have overlapping time intervals. We say that there is a temporal conflict between *t1* and *t4*.

Spatial Conflicts. Fig. 2 also shows an example of two data references for total number of cases of smallpox in the state of New York (tuple *t2*), and the corresponding total cases of smallpox in New York City (tuple *t3*). Although time intervals of *t2* and *t3* do not overlap, we cannot simply add up their corresponding data values to obtain the total number of smallpox cases in the state of New York. Tuple *t2* refers to the total number of smallpox cases reported for the state of New York. Meanwhile, it is unknown if this includes all New York City cases reported in *t3*. There is a spatial conflict between *t2* and *t3*.

Naming Conflicts. Historical data sources often report on conditions that evolve over time and can be noted under different names and categories within different time intervals. Moreover, different names for the same condition can co-exist within the same time interval. As we already mentioned in the introduction, many 19th century reports on yellow fever were actually referring to cases of hepatitis.

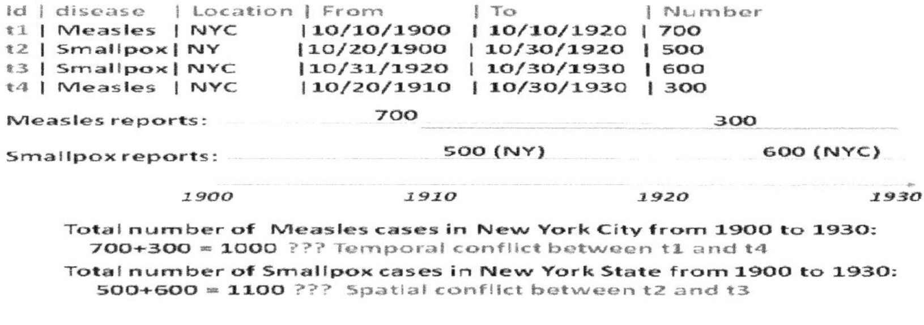

Fig. 2. Example of temporal and spatial conflicts

A historical database is *redundant* if it includes two or more conflicting tuples. As we mentioned in section 2, redundancy does not necessary imply inconsistency, which may be caused by presence of inaccurate reports. In this paper, we focus on handling temporal conflicts, although our approach can also be extended to deal with other conflict types. More specifically, we consider a novel approach for *conflict-aware historical data fusion*.

4 Conflict-Aware Fusion of Historical Data

A redundant historical database includes conflicting tuples. As a result, there is a risk of double-counting and over-estimating the value of the aggregate data. At the same time, non-redundant databases ignore conflicting tuples and take into account only non-overlapping reports, which may result in notable underestimations of the aggregate value. Consider again the conflicting measles reports from Fig. 2. Fig. 3 refines the impact on the estimation accuracy of ignoring the conflict (keeping both reports) and enforced non-redundancy (excluding one of the reports from consideration). Counting both of them will overestimate the actual number of measles cases for the period of 10/10/1900-10/30/1930. Ignoring report *R1* will leave the cases between 10/10/1900 and 10/20/1910 unreported (uncovered) and thus underestimate the actual number. Ignoring report *R2* will underestimate the actual number by excluding from consideration the cases between 10/10/1920 and 10/30/1930.

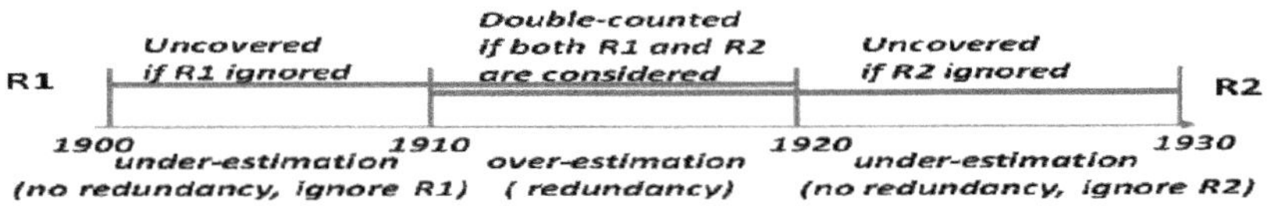

Fig. 3. Effect of redundancy and enforced non-redundancy

We propose a concept of *conflict degree (CD)* to assess the risk of misestimation of values reported in conflicting historical tuples.

4.1 Measuring Conflict Degree

Without losing generality, we will represent conflicting tuples as triples *(From, To, Value)*, where *Value* is a historical statistic (number of events) reported within the [*From, To*] time interval. Fig. 4 shows two scenarios for redundant databases DB1 and DB2. Here, we show only the time intervals of conflicting tuples annotated with corresponding reported values. Consider Scenario (a) first. In both cases of redundant databases, the time overlap is equal, but the relative contributions of conflicting tuples in the aggregated total value estimate are different. Both DB1 and DB2 can be split into two non-redundant snapshots including only one of the conflicting tuples. The total value estimated over non-redundant snapshots of DB1 is 100 in both cases, while corresponding values estimated over non-redundant snapshots of DB2 are 10 and 100. For DB2, the difference between the non-redundant estimates is greater, thus we conclude that degree of conflict between tuples in DB1 is higher than degree of conflict between DB2 tuples. Consider now Scenario (b), where the time overlaps of conflicting tuples are different while reported values are equal. The difference between total values estimated for non-redundant database snapshots is the same in both cases. Meanwhile, it is safer to assume that the risk of double counting is higher in the case of conflicting tuples with a larger time overlap (DB1). Thus, we expect to see that the conflict degree is higher for DB1.

To sum up, we would like to define a conflict degree measure between two historical tuples with the following characteristics:

- Conflict degree ranges within [0,1] interval;
- Tuples with non-overlapping time intervals have a conflict degree of 0;
- Tuples with overlapping time intervals have a higher conflict degree if their reported values are closer to each other (i.e., relative contributions of the reports are comparable)[1].
- Tuples with comparable reporting values have a higher conflict degree if their time overlap is higher.

  We also alternative definitions of the conflict degree measure. For example, a relative contribution of two conflicting tuples can be defined in such a way that closer values reduce the conflict. The expected fusion strategy in this case would favor reports with similar event distributions.

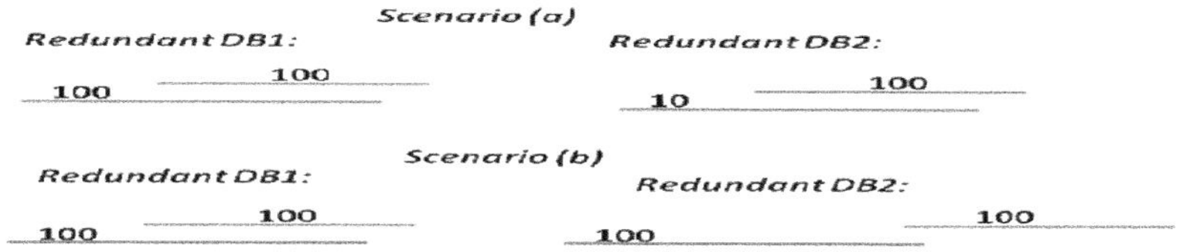

Fig. 4. To explanation of conflict degrees

Below, we define the conflict degree measure reflecting those characteristics. Informally, the conflict degree should account for both relative contributions and time overlaps of the conflicting tuples. We define a *relative contribution* of two conflicting tuples $r1 = (F1,T1,V1)$ and $r2 = (F2,T2,V2)$ as

$$RC(r1,r2) = 1 - |V1-V2|/(V1+V2).$$

Here, we assume that at least one of the two, $V1$ and $V2$, is non-zero. For example, consider Scenario (a) in Fig. 4. The relative contribution of conflicting tuples in DB1 is $1 - |100-100|/100+100 = 1$, while the relative contribution of conflicting tuples in DB2 is $1 - |10-100|/100+100 = 0.45$.

In order to define a relative time overlap of two conflicting tuples, we introduce several operations similar to some of the interval operations from [14]. *Length* of time interval $t = [F,T]$ (denoted $|t|$) is the difference T-F plus one, i.e. $|t| = T-F+1$. *Unit time interval* ***1*** is a time interval whose length is equal to one: $|\mathbf{1}| = 1$ (i.e, unit time interval has equal *From* and *To* components). Thus, $|t|$ is a number of unit time intervals covered by t. Time interval $t = [F,T]$ is *consistent* iff $F \leq T$. Below we consider only consistent time intervals if not stated otherwise. Time intervals $t1 = [F1,T1]$ and $t2 = [F2, T2]$ overlap if $F1 \leq F2 \leq T1$. For two time intervals

[1] Under alternative definitions of the conflict degree a relative contribution of two conflicting tuples could be defined in such a way that closer values reduce the conflict. The expected fusion strategy in this case would favor reports with similar event distributions. We plan to explore this approach in future work.

t1 = *[F1,T1]* and *t2* = *[F2, T2]*, we define their *sum t1+t2* and *intersection t1 o t2* as follows

$$t1 + t2 = [min(F1,F2), max(T1,T2)]; \quad t1 \text{ o } t2 = [max(F1,F2), min(T1,T2)].$$

Note that the result of *t1 o t2* may be an inconsistent time interval with a zero or negative intersection length. This would indicate that there is a gap between the time intervals *t1* and *t2*. We will use this feature to define *relative overlap RO* of two consistent time intervals *t1* = *[F1,T1]* and *t2* = *[F2, T2]* as follows:

$$RO(t1, t2) = max(|t1 \text{ o } t2|/|t1+t2|, 0),$$

i.e., *RO(t1, t2)* = *0* means that *t1* and *t2* do not overlap. Fig. 5 illustrates the introduced time interval operations.

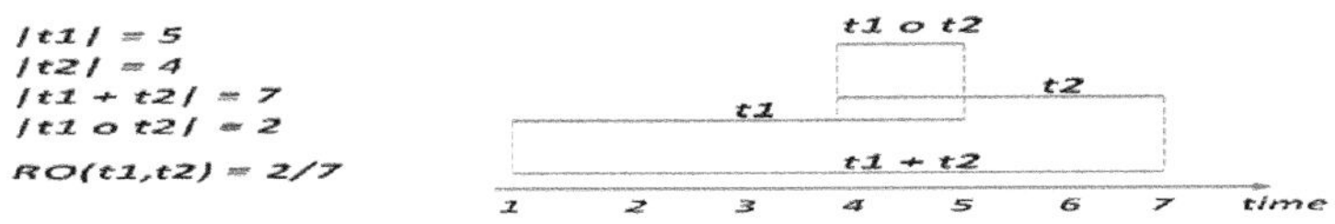

Fig. 5. Time interval operations

We define the relative time overlap *RO(r1,r2)* of two historical tuples *r1* and *r2* to be equal to the relative overlap of their corresponding time intervals. Next, we define the conflict degree *CD(r1,r2)* of two historical tuples *r1* and *r2* as the following exponential function:

$$CD(r1, r2) = RO(r1, r2) \times e^{k(1-RC(r1,r2))(RO(r1,r2)-1)}$$

Assuming that r1 and r2 are known from their context, we can specify the *CD* in a more compact format:

$$CD = RO \times e^{k(1-RC)(RO-1)}$$

This definition captures the desired characteristics of a *CD* measure. Fig. 6 illustrates the behavior of the *CD* function as we change *RO*, *RC* and *k*. In general, a higher value of *RO* implies a higher *CD*. Meanwhile, the lower value of *RC* slows down the rate at which the *CD* grows as the *RO* increases. A higher value of *k* amplifies the impact of *RC* on the *CD* value.

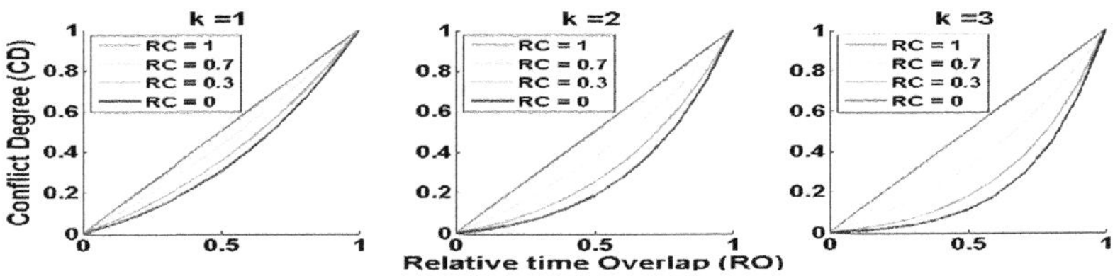

Fig. 6. Behavior of CD measure

4.2 Conflict-Aware Data Fusion

Using the *CD* measure, we can assess a redundant database with respect to conflict degrees between its tuples. Moreover, the *CD* measure allows us to better aggregate tuples performing *conflict-aware data fusion*. The idea is to set up a target database that can tolerate a certain value of conflict degrees (*CD threshold*) between its tuples. The expectation is that we can define an optimal *CD* threshold value that will minimize the misestimation error caused by underestimations due to uncounted events as well as the overestimation due to double counting. Next we will elaborate on characteristics of conflict-aware historical data fusion. The questions that we are going to address are (1) how does conflict-aware data fusion perform under different application constraints; (2) is there a practical way to find an optimal conflict threshold that minimizes the data fusion misestimation error?

Characteristics of Conflict-Aware Data Fusion. In order to explore characteristics of the conflict-aware data fusion techniques in the context of the aforementioned questions, we performed a simulation-based study using Matlab Version 7.11. Fig. 7 summarizes the simulation set-up. First, we randomly generated different numbers of events of interest per recording interval within a reasonably large span of time. We specify the time duration in units of minimal recording intervals. The size of the minimal recording interval (e.g, day, week, month, etc.) does not impact the results of our study. Then, we consider a maximal time duration of 1000 minimal recording intervals. The number of events can be reflected in multiple reports. Each report collects the number of events within a certain time interval (report duration or length).

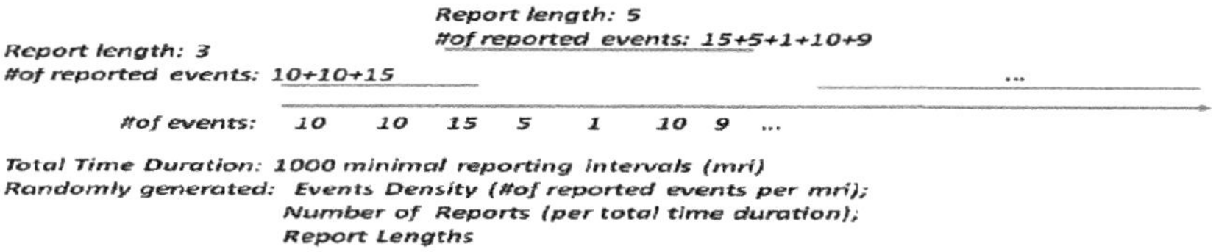

Scenario	Expected Events Density	Expected Number of Reports	Expected Report Lengths
Few short reports on sparse events	20	20	20
Few long reports on sparse events	20	20	100
Many short reports on sparse events	20	100	20
Many long reports on sparse events	20	100	100
Few short reports on dense events	100	20	20
Few long reports on dense events	100	20	100
Many short reports on dense events	100	100	20
Many long reports on dense events	100	100	100

Fig. 7. Simulation setup and scenarios

We assume that each event corresponds to the same data reference, i.e., overlapping reports result in conflicting tuples. We used normal distributions to configure (1) number of events of interest per minimal recording interval; (2) number of reports per total time duration, and (3) report durations (lengths). We selected several significantly different configurations to explore various real-life scenarios. They are summarized in the lower section of Fig. 7 with corresponding expected values (*mu*) of the simulated parameters. We expect that many lengthy reports on densely populated events would result in numerous high-degree conflicts. Meanwhile, a few short reports on sparsely populated events would hardly produce any conflict at all. The expectation is that a *proper data fusion strategy would combine information from multiple reports, minimizing misestimation of the actual number of events within the time duration.*

We performed simulations changing the acceptable CD threshold within the [0, 1] range with a step of 0.01. We aggregated only reports with a conflict degree below the CD threshold value and calculated the misestimation error in each case. For every combination of simulation scenario and CD threshold, we performed multiple simulation runs. Fig. 8 shows frequency histograms of relative misestimation errors for each scenario in Fig. 7 as we change the CD threshold (names of the axes are shown on the bottom left subplot). We observe that the number and scale of the misestimation errors differ notably for different event and report densities. Note that ***the report density has a higher impact on the error dynamics than the event density***. Fig. 9 also shows heat maps under the histogram plots that help us to observe the dynamics of the underestimations (positive relative errors) and overestimations (negative relative errors). The overestimation error due to double-counting from overlapping reports grows considerably as the report density increases. The underestimation error results from uncovered events and manifests itself more clearly in the scenarios with lower report densities. In all cases, the CD threshold has a notable impact on the misestimation error dynamics.

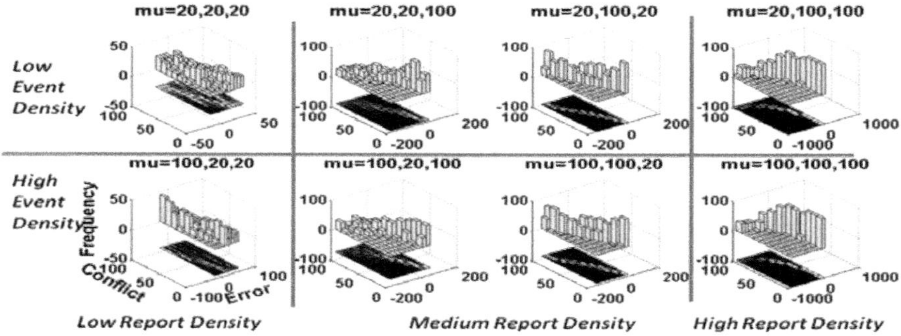

Fig. 8. Frequency histograms for error dynamics

Fig. 9 plots the CD threshold versus absolute value of the relative error in order to clarify how exactly they are related. For all scenarios, we observe similar error dynamics: the error decreases before some *critical value* of the CD threshold; after that, the error increases. This is an expected behavior: the initial error decrease is due to increased event coverage from a larger number of aggregated reports and, as a consequence, reduced underestimation errors. Meanwhile, after the critical CD threshold value is reached, the overlapping reports start accumulating overestimation errors due to double-counting. Here, we make another important observation: ***each scenario is associated with an optimal CD threshold that minimizes the misestimation error***. This also provides inspiration for a feasible way to estimate an optimal CD threshold, as we explain next.

Estimating Optimal CD Threshold. Let us repeat two important observations noted above: (1) the report density has a higher impact on the error dynamics than the event density, (2) each scenario is associated with an optimal CD threshold that minimizes the misestimation error. In real life, we do not have information about actual numbers of events per reporting interval (actual event density). The information available in historical databases includes only reported event numbers from a set of potentially conflicting reports. The challenge is to define the optimal CD threshold for each group of conflicting reports that would minimize the misestimation error without knowledge of actual event numbers.

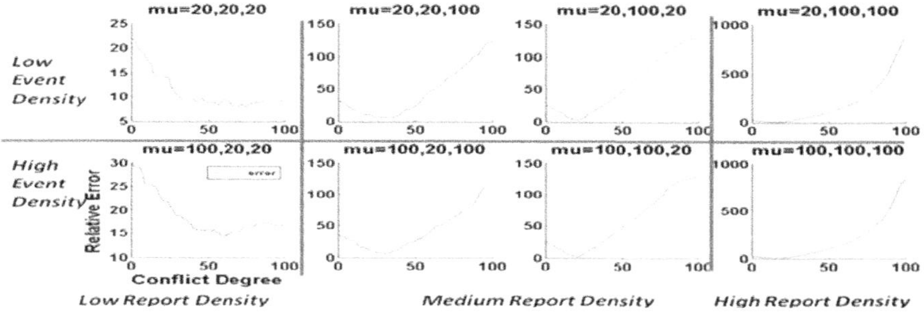

Fig. 9. Impact of CD threshold on error dynamics

We propose to use a Monte-Carlo simulation-based method to estimate such optimal CD thresholds for groups of conflicting reports corresponding to data references in the historical database. For a group of conflicting reports, our method will perform repeated random sampling of events. The sampling probability distribution can be adjusted using event density estimated from the conflicting reports in the group. For example, we can estimate minimal and maximal reported event densities to achieve uniform event sampling. Alternatively, we can estimate mean and standard deviation of reported event densities to perform normal event sampling. We expect the CD threshold to converge to an optimal value as the number of sampling-based simulation runs increases.

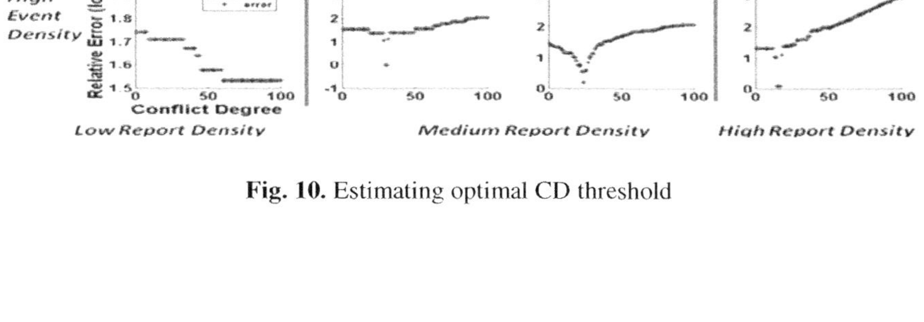

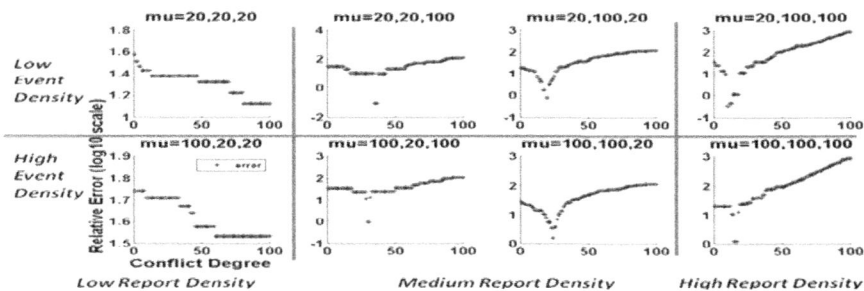

Fig. 10. Estimating optimal CD threshold

We tested this approach using our simulation testbed. Instead of randomly generating reports for a random event distribution, we were fixing different (random) report configurations. As before, each fixed report configuration was generated using a predefined event distribution. However, this time we assumed that the actual numbers of events are unknown; therefore, the impact of the CD threshold on the misestimation error was evaluated using randomly sampled events. Fig. 10 shows results of this evaluation for a uniform event sampling and 100 simulation runs. In order to make the dynamic more obvious, we used a log10 error scale. We observe that, in each of the considered scenarios, we converge on an optimal CD threshold value. In order to estimate the quality of this convergence, please refer to the optimal CD threshold plots for the actual number of events (Fig. 9). As we can see, the convergence process is consistent with the actual dynamics.

5 Conflict-Aware Data Warehousing

In the previous section, we suggested a simulation-based approach to estimate an optimal CD threshold value for groups of overlapping reports without the knowledge of actual event distributions. Such estimation can be performed periodically as a part of the data warehouse loading procedure for each data reference in the integrated historical database. The value of the estimated optimal CD threshold can be used to maximize accuracy of the query results. Fig. 11 illustrates this approach. Here, we considered two groups of conflicting reports on cases of measles and smallpox in NYC. For each of those groups, we evaluated an optimal Conflict Degree that minimizes a missestimation error of data aggregation. The value of the estimated optimal CD threshold was used to maximize accuracy of the query results.

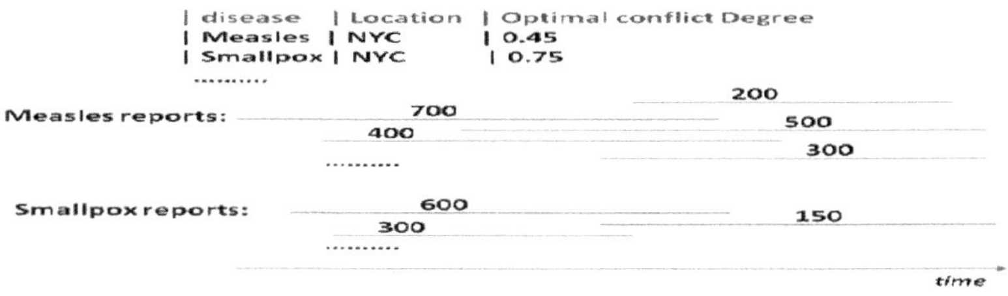

Fig. 11. Estimating optimal Conflict Degree in an integrated data warehouse

We implemented and tested our approach in Tycho, - a large integrated epidemiological data warehouse that includes historical data from numerous heterogeneous sources. As a part of our ongoing efforts, we have completed the first stage of Tycho data integration including heterogeneity resolution and data validation. We have integrated information from approximately 50,000 reports on United States epidemiological data for more than 100 years. The original reports are digitized and represented as semi-structured Excel spreadsheets with heterogeneous data formats and multiple transcription errors. After performing the heterogeneity resolution, we loaded more than 26 million records into the Tycho data warehouse. Then we

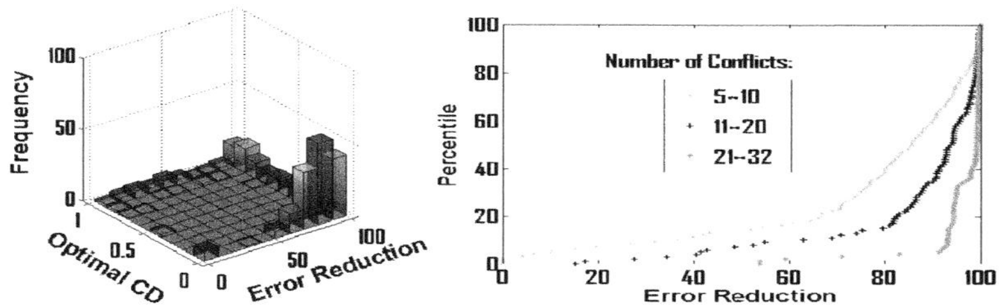

Fig. 12. Conflict-aware error reduction in Tycho

performed an aggregate assessment of conflicting data in the data warehouse. For this assessment we considered only temporal conflicts. We evaluated the numbers of conflicting reports per time span. We observed up to 32 conflicts occurring over the time spans of days, weeks, months, and years.

Fig. 12 shows results of conflict-aware data fusion for a representative set of Tycho reports. We selected three hundred Tycho data references including up to 32 conflicts, determined corresponding optimal CD threshold, and estimated error reduction due to conflict-awareness. We observe up to sixty percent in error reduction, - a notable improvement in the estimation accuracy.

6 Conclusion

We have considered a novel conflict-aware approach to historical data fusion. Our experimental results demonstrate high efficiency of the proposed approach. In future works we will further explore the performance and accuracy tradeoffs associated with the various conflict-aware query evaluation strategies. We will design and develop a query optimizer that efficiently utilizes those tradeoffs. We will further investigate conflict-aware data fusion for *spatial* and *naming conflicts*. We will investigate alternative historical data fusion strategies based on different notions of maximal likelihood estimation to estimate the event distribution from a set of reported numbers of events.

Acknowledgements. We are thankful to Div Sharma, John Grefenstette, Willem Aysbert Van Panhuis, Shawn Brown, and Dave Koenig for their help and collaboration on this project. We are thankful to John Grant, Felix Naumann, Patrick Manning, Louiqa Raschid and anonymous reviewers for their valuable comments on a draft of this paper.

References

1. Afrati, F., Kolaitis, P.: Repair Checking in Inconsistent Databases: Algorithms and Complexity. In: Proc. of ICDT (2009)
2. Agarwal, S., Keller, A., Wiederhold, G., Saraswat, K.: Flexible Relation: An Approach for Integrating Data from Multiple, Possibly Inconsistent Databases. In: Proc. of ICDE (1995)
3. Arenas, M., Bertossi, L., Chomicki, J.: Specifying and Querying Database Repairs using Logic Programs with Exceptions. In: Proc. of FQAS (2000)

4. Bernstein, P., Melnik, S.: Model Management 2.0: Manipulating Richer Mappings. In: Proc. of ACM SIGMOD (2007)
5. Bertossi, L.: Consistent Query Answering in Databases. ACM SIGMOD Record 35(2) (2006)
6. Bertossi, L., Chomicki, J.: Query Answering in Inconsistent Databases. In: Logics for Emerging Applications of Databases. Springer, Heidelberg (2003)
7. Bleiholder, J., Naumann, F.: Data Fusion. ACM Computing Surveys 41(1) (2008)
8. Bohannon, P., Flaster, M., Fan, W., Rastorgi, R.: A Cost-based Model and Effective Heuristic for Repairing Constraints by Value Modification. In: Proc. of ACM SIGMOD (2005)
9. Brodie, M.: Data Integration at Scale: From Relational Data Integration to Information Ecosystems. In: Proc. of AINA (2010)
10. Brodie, M.: Data Management Challenges in Very Large Enterprises. In: Proc. of VLDB (2002)
11. Bry, F.: Query Answering in Information Systems with Integrity Constraints. In: Proc. of IICIS (1997)
12. Caroprese, L., Greco, S.: Active Integrity Constraints for Database Consistency Maintenance. IEEE TKDE 21(7) (2009)
13. Chomicki, J., Staworko, S., Marcinkowski, J.: Computing Consistent Query Answers Using Conflict Hypergraph. In: Proc. of CIKM (2004)
14. Date, J., Darwen, H., Lorentzos: Temporal Data and the Relational Model. Morgan Kaufmann, San Francisco (2003)
15. Dong, X., Naumann, F.: Data Fusion - Resolving Data Conflicts for Integration. In: PVLDB, vol. 2(2) (2009)
16. Elmagarmid, A., Ipeirotis, P., Verykios, V.: Duplicate Record Detection: A Survey. IEEE TKDE 19(1) (2007)
17. Flesca, S., Furfaro, F., Parisi, F.: Querying and Repairing Inconsistent Numerical Databases. ACM TODS 35(2) (2010)
18. Flesca, S., Furfaro, F., Parisi, F.: Consistent Query Answers on Numerical Databases Under Aggregate Constraints. In: Bierman, G., Koch, C. (eds.) DBPL 2005. LNCS, vol. 3774, pp. 279–294. Springer, Heidelberg (2005)
19. Fagin, R., Kolaitis, P., Popa, L.: Data Exchange: Getting to the Core. ACM TODS 30(1) (2005)
20. Haas, L.: Beauty and the Beast: The Theory and Practice of Information Integration. In: Schwentick, T., Suciu, D. (eds.) ICDT 2007. LNCS, vol. 4353, pp. 28–43. Springer, Heidelberg (2006)
21. Imelinski, T., Lipski, W.: Incomplete Information in Relational Databases. Journal of ACM 31(4) (1984)
22. Jensen, C., Snograss, R.: Temporal Data Management. IEEE TKDE 11(1) (1999)
23. Kay, S.: Fundamentals of Statistical Signal Processing: Estimation Theory. Prentice-Hall, Englewood Cliffs (1993)
24. Rahm, E., Bernstein, P.: A Survey of Approaches to Automatic Schema Matching. The VLDB Journal 10(4) (2001)
25. Senn, S.: Overstating the Evidence - Double Counting in Meta-analysis and Related Problems. BMC Medical Research Methodology 9(10) (2009)
26. Snodgrass, R.: Developing Time-oriented Database Applications in SQL. Morgan Kaufmann, San Francisco (2000)
27. Staworko, S., Chomicki, J.: Consistent Query Answers in the Presence of Universal Constraints. Inf. Syst. 35(1) (2010)

28. Wijsen, J.: Consistent Query Answering under Primary Keys: A Characterization of Tractable Queries. In: Proc. of ICDT (2009)
29. Wijsen, J.: Database repairing using updates. ACM TODS 30(3) (2005)
30. Dong, X.L., Berti-Equille, L., Srivastava, D.: Truth Discovery and Copying Detection in a Dynamic World. In: PVLDB, vol. 2(1) (2009)
31. Dong, X.L., Berti-Equille, L., Srivastava, D.: Integrating Conflicting Data: The Role of Source Dependence. In: PVLDB, vol. 2(1) (2009)
32. Yin, X., Han, J., Yu, P.: Truth Discovery with Multiple Conflicting Information Provided on the Web. In: Proc. of SIGKDD (2007)
33. Zadorozhny, V., Raschid, L., Gal, A.: Scalable Catalog Infrastructure for Managing Access Costs and Source Selection in Wide Area Networks. International Journal of Cooperative Information Systems 17(1) (2008)
34. Zadorozhny, V., Gal, A., Raschid, L., Ye, Q.: AReNA: Adaptive Distributed Catalog Infrastructure Based On Relevance Networks. In: Proc. of VLDB (2005)
35. Zadorozhny, V., Bright, L., Vidal, M.E., Raschid, L., Urhan, T.: Efficient Evaluation of Queries in a Mediator for WebSources. In: Proc. of ACM SIGMOD (2002)

Extended Galois Derivation Operators for Information Retrieval Based on Fuzzy Formal Concept Lattice

Yassine Djouadi[1,2]

[1] University of Tizi-Ouzou,
BP 17, RP, 15000 Tizi-Ouzou, Algeria
[2] IRIT, Université Paul Sabatier,
118 Route de Narbonne, 31062 Toulouse Cedex 09, France
ydjouadi@mail.ummto.dz

Abstract. The obvious analogy between an *Objects*×*Properties* binary relationship (called a formal context) and a binary *Documents*×*Terms* incidence matrix has led to a growing interest for the use of formal concept analysis (FCA) in information retrieval (IR). The main advantage of using FCA for IR is the possibility of creating a conceptual representation of a given document collection in the form of a lattice. Also, potentials of FCA for IR have been highlighted by a number of research studies since its inception. It turns out that almost all existing FCA-based IR approaches rely on: i) a Boolean *Documents*×*Terms* incidence matrix and, ii) the use of the classical Galois connection initially proposed by Wille. In such a case, there is no way for expressing weighted queries as well as there is no way for ranking query results that is usually done by query refinement or empirical navigation trough the concept lattice. In this paper we first enlarge the use of FCA for IR to the fuzzy setting which allows for a fuzzy incidence matrix and weighted queries. For instance, an incidence matrix may now allow for (normalized) numerical entries that may be achieved using the well known *term frequency* measure. Furthermore, it is worth noticing that in the existing approaches, user queries are restricted to the conjunctive form. Thus, another contribution consists in considering, in an original way, the use of other Galois derivation operators (namely the possibility, necessity and dual sufficiency) in order to express disjunction and negation in queries.

1 Introduction

The main aim of Formal Concept Analysis (FCA for short) is to extract interesting clusters of knowledge, called formal concepts, from a particular representation of data, called formal context. The original idea of FCA has been introduced by Wille [19] and is becoming increasingly popular among various methods of conceptual data analysis and knowledge processing. In the classical setting [19, 12], a formal context consists of a (crisp) binary relationship between a set of objects and a set of properties. Whereas a formal concept consists of a

S. Benferhat and J. Grant (Eds.): SUM 2011, LNAI 6929, pp. 346–358, 2011.
© Springer-Verlag Berlin Heidelberg 2011

pair $\langle objects, properties \rangle$, where the set of *objects* is referred to as the extent, and the set of *properties* as the intent. They uniquely determine each other. The family of all formal concepts is a complete lattice. The classical model [19, 12] relies on two assumptions: i) properties are *Boolean*: a property is true or false for an object (and thus it is supposed to always *apply*); ii) information is *complete*: it is always known if a property holds or not for an object. As an effective method for data analysis, formal concept analysis has been widely applied to many fields like psychology, sociology, anthropology, medicine, biology, linguistics, etc.

The obvious analogy between an $Objects \times Properties$ binary relationship (i.e. a formal context) closely related to FCA theory and a binary $Documents \times Terms$ incidence matrix closely related to information retrieval has led to a growing interest for the use of the theory of FCA and the underlying notion of concept lattice in information retrieval. In this case, the documents correspond to formal objects and indexing terms (descriptors, thesaurus elements, etc..) correspond to the properties [17].

Almost all existing information retrieval approaches based on FCA rely on the use of the lattice structure of the formal concepts [17]. It is also worth noticing that the use of general lattice structures (i.e. not especially concept lattices) in information retrieval has been early addressed. Indeed, in the 1960's other retrieval approaches were considered besides the usual vector space model among them lattice representations [18]. The theoretical consolidation of FCA on a mathematical (algebraic) model has given rise to a practical and efficient use of formal concept lattices in information retrieval [17], [3], [16], [5], [15], [11]. Nowadays, with the recent advances on FCA theory, retrieval models based on concept lattices outperform significantly classical information retrieval approaches [3].

One of the most natural applications of concept lattices in information retrieval is *query refinement* which relies on the two fundamental observations stated hereafter [15]:

1. A formal concept c of a concept lattice may be seen as a pair $\langle answer, query \rangle$ where the $query$ corresponds to the intent of c whereas the $answer$ corresponds to the extent of c.
2. Following edges departing upward (resp. downward) from a query (i.e. a formal concept) produces all minimal refinements (resp. enlargements) of the query w.r.t. the collection of documents from which the concept lattice has been built.

It may be remarked that almost all encountered information retrieval approaches based on FCA assume:

1. A Boolean incidence matrix (Boolean formal context).
2. Boolean queries (i.e. the terms of the queries can take only the weights 0 or 1).
3. Queries limited to conjunctive expressions.
4. No ranked results (the different results are achieved through an ad hoc navigation among the concept lattice structure without supplying any order between them).

Concerning the first and second points, it is well known that Boolean retrieval systems [4] have several limitations. One of the most important limitation is that only documents that satisfy a query exactly are retrieved. Concerning the third point, the "and" operator may appear too restrictive because it fails even in the case when all its arguments except one are satisfied which is counter intuitive with the principle of information retrieval.

This paper gives proposals for the above mentioned drawbacks. We first propose to enlarge classical FCA to the fuzzy setting which allows for fuzzy incidence matrices and for weighted queries. Secondly, we give a fuzzy extension of possibilistic Galois derivation operator in order to consider the disjunction as well as the negation in the expression of the queries.

The paper is organized as follows. The next section gives a background on formal concept analysis. The possibility-theoretic view of FCA is also presented in this section. In section 3, we gives the extension of possibilistic derivation operators to the fuzzy setting. In the next section, we give our proposals about the use of possibilistic derivation operators in order to allow disjunction and negation whereas the section 5 gives an illustrative example.

2 Formal Concept Analysis: A Survey

2.1 Classical Settings

Let $\mathcal{D}$ be a finite set of documents ($\mathcal{D} = \{d_1, d_2, \ldots, d_n\}$) and $\mathcal{T}$ be a finite set of index terms (descriptors) ($\mathcal{T} = \{t_1, t_2, \ldots, t_m\}$). A (crisp) binary formal context is a triple $\mathcal{K} := (\mathcal{D}, \mathcal{T}, \mathcal{R})$ where $\mathcal{R}$ corresponds to a crisp binary relationship functionally given as $\mathcal{R}: \mathcal{D} \times \mathcal{T} \longrightarrow \{\bot, \top\}$, where $\mathcal{R}(d,t) = \top$ (resp. $\mathcal{R}(d,t) = \bot$) means that document d is related to (resp. is not related) the term t. The relation $\mathcal{R}$ is usually represented as a table with rows corresponding to documents, columns corresponding to properties (or conversely). The entries in the table indicate whether a document satisfies or does not satisfy the corresponding term (generally by using cross mark or respectively blank mark). We shall also use in the rest of this paper an equivalent notation $d\mathcal{R}t$ which means that document d satisfies term t. Let $R(d) = \{t \in \mathcal{T} \mid d\mathcal{R}t\}$ be the set of properties satisfied by document d. In a polymorphic notation, let $R(t) = \{d \in \mathcal{D} \mid d\mathcal{R}t\}$ be the set of documents that satisfy term t.

By extending singleton operators $R(.)$ to powerset operators (also called Galois derivation operators) between $2^{\mathcal{D}}$ and $2^{\mathcal{T}}$, we can establish relationships between subsets of documents and subsets of terms. The derivation operator which is at the basis of formal concept analysis is here called *sufficiency* operator as in [10, 9] and denoted $(.)^{\triangle}$. It is given as follows. For a set of terms T, we define the set $T^{\triangle}$ of documents that satisfy all terms in T as:

$$\begin{aligned} T^{\triangle} &= \{d \in \mathcal{D} \mid \forall t \in \mathcal{T} (t \in T \Rightarrow d\mathcal{R}t)\} \\ &= \{d \in \mathcal{D} \mid T \subseteq R(d)\} \\ &= \textstyle\bigcap_{t \in T} R(t) \end{aligned} \tag{1}$$

The set $D^{\triangle}$ of terms that are satisfied by all documents in D is dually defined. Thus, a formal concept is a pair $\langle D, T \rangle$ s.t. $D^{\triangle} = T$ and $T^{\triangle} = D$.

An important feature of formal concept analysis is that the pair $((.)^{\triangle}, (.)^{\triangle})$ of derivation operators forms a Galois connection [19, 12]. A further interesting feature is that a "*closure*" property of the composition $(.)^{\triangle\triangle}$ of the derivation operators is implicitly conveyed.

The main objective in formal concept analysis consists of inducing all formal concepts. Given a formal context $\mathcal{K}$, the set of all formal concepts is naturally equipped with a partial order (denoted $\preceq$) defined as: $\langle D_1, T_1\rangle \preceq \langle D_2, T_2\rangle$ iff $D_1 \subseteq D_2$ (or, equivalently, $T_2 \subseteq T_1$). In [12] authors have proved that the set of all formal concepts ordered with $\preceq$ forms a complete lattice, called the *concept lattice* of $\mathcal{K}$ and denoted $\mathcal{L}(\mathcal{K})$. Its structure is given by the following theorem.

Theorem 1. [12]. *The concept lattice $\mathcal{L}(\mathcal{K})$ is a complete lattice in which infimum and supremum are given by:*

$$\bigwedge_{j\in J}\langle D_j, T_j\rangle = \langle \bigcap_{j\in J} D_j, (\bigcup_{j\in J} T_j)^{\triangle\triangle}\rangle \quad , \quad \bigvee_{j\in J}(D_j, T_j) = \langle (\bigcup_{j\in J} D_j)^{\triangle\triangle}, \bigcap_{j\in J} T_j\rangle$$

2.2 Possibility-Theoretic View of Formal Concept Analysis

Dubois et al [9] and more recently Dubois and Prade [8] have given a possibility-theoretic [20] reading of FCA which allows for the use of three other operators, namely possibility $(.)^{\Pi}$, necessity $(.)^{\mathrm{N}}$ and dual sufficiency $(.)^{\nabla}$ operators beside the sufficiency one $(.)^{\triangle}$. In this spirit, we have already highlighted the interest of the necessity operator for formal context decomposition [6] and have also given the appropriate (minimal) requirements for building sound compositions of the four operators w.r.t. to the closure property in the fuzzy settings [7]. Note that the four powerset operators have been also considered in qualitative data analysis by [10, 13]. We recall these three operators:

- T^{Π} is the set of documents that satisfy at least one term in T:

$$\begin{aligned} T^{\Pi} &= \{d \in \mathcal{D} \mid T \cap R(d) \neq \emptyset\} \\ &= \{d \in \mathcal{D} \mid \exists t \in T : \; d\mathcal{R}t\} \end{aligned} \tag{2}$$

- T^{N} is the set of documents such that any term satisfied by *one* of them is necessarily in T:

$$\begin{aligned} T^{\mathrm{N}} &= \{d \in \mathcal{D} \mid R(d) \subseteq T\} \\ &= \{d \in \mathcal{D} \mid \forall t \in T \;\; (d\mathcal{R}t \Rightarrow t \in T)\} \end{aligned} \tag{3}$$

- T^{∇} is the set of documents that do not satisfy at least one term in $\overline{T}$:

$$\begin{aligned} T^{\nabla} &= \{d \in \mathcal{D} \mid T \cup R(d) \neq \mathcal{D}\} \\ &= \{d \in \mathcal{D} \mid \exists t \in \overline{T}, \, d\overline{\mathcal{R}}t\} \end{aligned} \tag{4}$$

Operators D^{Π}, D^{N}, D^{∇} are dually defined.

3 Extending Possibilistic Derivation Operators to the Fuzzy Setting

Let us recall that almost all existing FCA-based information retrieval approaches are restricted to Boolean conjunctive queries that address Boolean incidence matrix. A fuzzy FCA-based approach which allows conjunctive weighted queries is also proposed in [14]. However, this approach assumes the use of a specific fuzzy implication, namely the Rescher-Gaïnes implication ($p \rightarrow q = 1$ if $p \leqslant q$ 0 otherwise). We are not aware about the use of the disjunction and negation. Thus this section generalizes a Boolean querying process to the fuzzy setting which allows for the use of a fuzzy incidence matrix as well as weighted queries, whereas our proposal for considering the disjunction and negation is illustrated in the next section.

A fuzzy formal context is a tuple $\mathcal{K} = (L, \mathcal{D}, \mathcal{T}, \mathcal{R})$ where the fuzzy incidence matrix $\mathcal{R} \in L^{\mathcal{D}\times\mathcal{T}}$ is defined as a mapping : $\mathcal{D} \times \mathcal{T} \longrightarrow L$ (generally $L = [0, 1]$). The generalization of the four derivation operators (i.e. sufficiency, possibility, necessity and dual sufficiency) to the fuzzy setting arises quite naturally from their crisp counterpart. Since Expressions 1 and 3 of sufficiency and necessity operators are based on inclusions, their generalization are naturally based on fuzzy implications. Expressions 2 and 4 correspond to conditions of non emptiness for the intersection of subsets. Their generalization are naturally given by the largest value over t (or d) of a fuzzy conjunction of the membership degrees. Their definitions are given as:

Definition 1. *The generalization of sufficiency, possibility, necessity and dual sufficiency powerset operators to the fuzzy setting is defined for a fuzzy set $\widetilde{T} \in L^{\mathcal{T}}$ as (and similarly defined for a fuzzy set $\widetilde{D} \in L^{\mathcal{D}}$):*

$$\widetilde{T}^{\triangle}(d) = \bigwedge_{t\in\mathcal{T}} \left(\widetilde{T}(t) \rightarrow \mathcal{R}(d,t)\right) \tag{5}$$

$$\widetilde{T}^{\Pi}(d) = \bigvee_{t\in\mathcal{T}} \left(\widetilde{T}(t) * \mathcal{R}(d,t)\right) \tag{6}$$

$$\widetilde{T}^{\mathrm{N}}(d) = \bigwedge_{t\in\mathcal{T}} \left(\mathcal{R}(d,t) \rightarrow \widetilde{T}(t)\right) \tag{7}$$

$$\widetilde{T}^{\nabla}(d) = \bigvee_{t\in\mathcal{T}} \left(\neg\widetilde{T}(t) * \neg\mathcal{R}(d,t)\right) \tag{8}$$

In the above definition, the $\rightarrow$ operator denotes a fuzzy implication operator, which is decreasing (in the broad sense) in its first component and increasing (in the broad sense) in its second component and verifies boundaries conditions ($0 \rightarrow 0 = 0 \rightarrow 1 = 1 \rightarrow 1 = 1$ and $1 \rightarrow 0 = 0$). Whereas, the connective denoted $*$ with a fuzzy conjunction semantics, is a binary increasing operator (in the broad sense) which verifies identity condition $(p * 1) = p$ and boundaries conditions ($0 * 0 = 0 * 1 = 1 * 0 = 0$). Note that $*$ is not necessarily a t-norm.

Like for the crisp setting, a fuzzy formal concept consists of a pair $\langle\widetilde{D}, \widetilde{T}\rangle$ s.t. $\widetilde{D}^{\triangle} = \widetilde{T}$ and $\widetilde{T}^{\triangle} = \widetilde{D}$. It is also important to point out that the generalization of formal concept analysis theory to the fuzzy setting implies some algebraic requirements on the fuzzy implication $\rightarrow$ used in expression 5 (see [7]) Indeed, depending on the choice of "$\rightarrow$", the fuzzy operator $(.)^{\triangle\triangle}$ could not fulfill the closure property (recalled below) and consequently may not form a Galois connection. Under the appropriate requirements, the set of all fuzzy formal concepts is a complete lattice called fuzzy concept lattice [1].

Definition 2. *Given a universe $\mathcal{U}$, a mapping $\Phi : L^{\mathcal{U}} \longrightarrow L^{\mathcal{U}}$ is a fuzzy closure operator iff $\forall\ U, V \in L^{\mathcal{U}}$ it satisfies :*
(1): $U \subseteq V \Longrightarrow \Phi(U) \subseteq \Phi(V)$
(2): $U \subseteq \Phi(U)$
(3): $\Phi\big(\Phi(U)\big) = \Phi(U)$

4 Flexible Querying Based of Fuzzy Formal Concepts

This section enlarges FCA-based information retrieval approaches to the fuzzy setting and illustrates also that such approaches may take advantage of the other possibilistic operators rather than the sufficiency operator.

4.1 Conjunctive Queries

Let $\mathcal{K} := (\mathcal{D}, \mathcal{T}, \mathcal{R})$ be a formal context s.t. $\mathcal{R}$ is a Boolean incidence matrix, $D \subseteq \mathcal{D}$, and $T \subseteq \mathcal{T}$. Let $\hat{Q}(T) \equiv t_1 \wedge t_2 \wedge \ldots \wedge t_k$ denotes a conjunctive query built on the set T. According to Salton's Boolean retrieval model, we define the satisfaction of $\hat{Q}(T)$ by a set of documents D as: $D \models \hat{Q}(T) \Leftrightarrow D = \bigcap_{t \in T}\{t\}^{\triangle} = T^{\triangle}$.

Let $\widetilde{D} \in L^{\mathcal{D}}$ and $\widetilde{T} \in L^{\mathcal{T}}$. A fuzzy (weighted) conjunctive query $\hat{Q}(\widetilde{T})$ is of the form $t_1^{w_1} \wedge t_2^{w_2} \wedge \ldots \wedge t_k^{w_k}$ where $\widetilde{T}(t_i) = w_i$. The satisfaction of $\hat{Q}(\widetilde{T})$ by a fuzzy set $\widetilde{D}$ of documents arises quite naturally from its crisp counterpart. That is: $\widetilde{D} \models \hat{Q}(\widetilde{T}) \Leftrightarrow \widetilde{D} = \widetilde{T}^{\triangle}$.

Relatively to a fuzzy concept lattice $\mathcal{L}(\mathcal{K})$, let us consider the two functions $Ext(c)$ and $Int(c)$ that correspond to the intent and the extent of the fuzzy formal concept c. Thus, a fuzzy conjunctive query is satisfied by c iff $Ext(c) = \widetilde{T}^{\triangle}$ or $Int(c) = \widetilde{T}^{\triangle\triangle}$. Since $(.)^{\triangle\triangle}$ is a fuzzy closure operator (i.e. $(\widetilde{T}^{\triangle\triangle})^{\triangle\triangle} = \widetilde{T}^{\triangle\triangle}$ and $(\widetilde{T}^{\triangle})^{\triangle\triangle} = \widetilde{T}^{\triangle}$), it is easy to conclude that for any fuzzy conjunctive query $\hat{Q}(\widetilde{T})$ there exists one and just one fuzzy formal concept satisfying $\hat{Q}(\widetilde{T})$.

According to the expression of $\widetilde{T}^{\triangle}$ namely $\widetilde{T}^{\triangle}(d) = \bigwedge_{t \in \mathcal{T}} \big(\widetilde{T}(t) \rightarrow \mathcal{R}(d,t)\big)$, it is obvious that the result of a given query depends of the choice of a fuzzy implication. However some eligible implications (e.g. Goguen implication) make the fuzzy concept lattice infinite which is counter intuitive with the idea of information retrieval based on lattice representation and *navigation*. For this purpose, the following proposition advocates the use of the Gödel implication that makes the concept lattice finite (a deep investigation about the conveyed semantics of fuzzy implications is left for further research).

Proposition 1. *Let $\mathcal{K} = (L, \mathcal{D}, \mathcal{T}, \mathcal{R})$ be a fuzzy formal context where L is an arbitrary scale (not necessarily finite). Then, the fuzzy concept lattice is a finite set under Gödel implication.*

Proof. It is a well known result that the set of all intents and the set of all extents are isomorphic complete lattices [2]. Thus, it is sufficient to prove that one of them is a finite set, namely the set of all extents. Thus, we have:

$$\widetilde{D}^{\triangle\triangle}(d) = \bigwedge_{t \in \mathcal{T}} (\widetilde{D}^{\triangle}(t) \to \mathcal{R}(d,t))$$

$$= \bigwedge_{t \in \mathcal{T}} \left(\begin{cases} 1 & \text{if } \widetilde{D}^{\triangle}(t) \leqslant \mathcal{R}(d,t) \\ \mathcal{R}(d,t) & \text{otherwise} \end{cases} \right)$$

Thus, necessarily $\widetilde{D}^{\triangle\triangle}(d) \in \{\mathcal{R}(d_i, t_j)\} \cup \{1\}$ s.t. $i \leqslant n$ and $j \leqslant m$ where n and m are finite since $n = |\mathcal{D}|$ and $m = |\mathcal{T}|$. This implies that the set of all extents is finite. □

4.2 Disjunctive Queries

Let $\check{Q}(T) \equiv t_1 \vee t_2 \vee \ldots \vee t_k$ denotes a disjunctive query in the crisp setting. The satisfaction of $\check{Q}(T)$ by a set of documents D is easily expressed by means of the possibility operator as: $D \models \check{Q}(T) \Leftrightarrow D \subseteq \bigcup_{t \in T} \{t\}^{\Pi} = T^{\Pi}$. Note that $\{t\}^{\Pi} = \{t\}^{\triangle}$ since $\{t\}$ is a singleton.

A fuzzy disjunctive query $\check{Q}(\widetilde{T})$ is of the form $t_1^{w_1} \vee t_2^{w_2} \vee \ldots \vee t_k^{w_k}$ where $\widetilde{T}(t_i) = w_i$. As a generalization of the crisp setting, the satisfaction of $\check{Q}(\widetilde{T})$ by by a fuzzy set $\widetilde{D}$ is defined as: $\widetilde{D} \models \check{Q}(\widetilde{T}) \Leftrightarrow \widetilde{D} \subseteq \widetilde{T}^{\Pi}$ where $\subseteq$ denotes the standard fuzzy set inclusion (defined by the pointwise inequality of the membership functions i.e. $\widetilde{D} \subseteq \widetilde{T}^{\Pi} \Leftrightarrow \forall d \in \mathcal{D} : \widetilde{D}(d) \leqslant \widetilde{T}^{\Pi}(d)$.

Relatively to a fuzzy concept lattice $\mathcal{L}(\mathcal{K})$, a fuzzy disjunctive query $\check{Q}(\widetilde{T})$ is satisfied by a fuzzy formal concept $c \in \mathcal{L}(\mathcal{K})$ iff $Ext(c) \subseteq \widetilde{T}^{\Pi}$. Let $\mathcal{B}(\check{Q}(\widetilde{T}))$ denotes the set of all formal concepts c s.t. c satisfies $\check{Q}(\widetilde{T})$. The following proposition establishes a significant property for information retrieval purpose.

Proposition 2. *Let $\widetilde{T} \in L^{\mathcal{T}}$ and $\check{Q}(\widetilde{T})$ corresponds to a disjunctive query. The set $\mathcal{B}(\check{Q}(\widetilde{T})) \bigcup \{\langle \widetilde{T}^{\Pi}, (\widetilde{T}^{\Pi})^{\triangle} \rangle\}$ is a complete lattice.*

Proof. We prove first that the composition $((.)^{\Pi})^{\triangle}$ is not a closure operator. We do this by giving the counter example illustrated in Table 1. Applying the possibility operator to the set $\{t_2, t_3\}$ we get $\{t_2, t_3\}^{\Pi} = \{d_1, d_2\}$. Applying then the sufficiency operator we get $\{d_1, d_2\}^{\triangle} = \{t_1\}$. Consequently, it appears that $\{t_2, t_3\} \nsubseteq (\{t_2, t_3\}^{\Pi})^{\triangle}$. That is, $((.)^{\Pi})^{\triangle}$ is not a closure operator.

On the other hand, since: i) $\mathcal{B}(\check{Q}(\widetilde{T}))$ is a bounded sublattice of $\mathcal{L}(\mathcal{K})$ and, ii) $\{\langle \widetilde{T}^{\Pi}, (\widetilde{T}^{\Pi})^{\triangle} \rangle\}$ is an upper bound of $\mathcal{B}(\check{Q}(\widetilde{T}))$, it comes that $\mathcal{B}(\check{Q}(\widetilde{T})) \bigcup \{\langle \widetilde{T}^{\Pi}, (\widetilde{T}^{\Pi})^{\triangle} \rangle\}$ is a complete lattice. □

Table 1. Counter example for the composition $((.)^{\Pi})^{\triangle}$

$\mathcal{R}_1$	t_1	t_2	t_3	t_4
d_1	×	×		
d_2	×		×	
d_3				×

4.3 Negation in Queries

In the following, let us use the notations $(.)^{\triangle}_{\overline{\mathcal{R}}}$, $(.)^{\Pi}_{\overline{\mathcal{R}}}$, $(.)^{N}_{\overline{\mathcal{R}}}$, $(.)^{\nabla}_{\overline{\mathcal{R}}}$ when the considered derivation operators apply to the complementary fuzzy formal context $\overline{\mathcal{K}} = (L, \mathcal{D}, \mathcal{T}, \overline{\mathcal{R}})$ where $\overline{\mathcal{R}}(d,t) = \neg\mathcal{R}(d,t)$ and $\neg$ is an involutive negation (i.e. $\neg\neg p = p$). The following proposition establishes useful correspondences between the four fuzzy possibilistic derivation operators given in Section 3.

Proposition 3. *Let the fuzzy lattice* $\boldsymbol{L} = (L, \rightarrow, *, \neg)$. *The following properties hold* $\forall\ \widetilde{T} \in L^{\mathcal{T}}$ *(similarly* $\forall\ \widetilde{D} \in L^{\mathcal{D}}$*) if the condition (R0):* $p \rightarrow q = \neg(p * \neg q)$ *is satisfied:*

$$(P1): \quad \widetilde{T}^{\triangle} = \overline{\overline{\widetilde{T}}^{\nabla}} \qquad\qquad (P1'): \quad \widetilde{T}^{\nabla} = \overline{\overline{\widetilde{T}}^{\triangle}}$$

$$(P2): \quad \widetilde{T}^{N} = \overline{\overline{\widetilde{T}}^{\Pi}} \qquad\qquad (P2'): \quad \widetilde{T}^{\Pi} = \overline{\overline{\widetilde{T}}^{N}}$$

$$(P3): \quad \widetilde{T}^{\triangle} = \overline{\widetilde{T}^{\Pi}_{\overline{\mathcal{R}}}} \qquad\qquad (P3'): \quad \widetilde{T}^{\triangle} = \overline{\widetilde{T}}^{N}_{\overline{\mathcal{R}}}$$

Proof. We just prove (P1), the proofs for the remaining properties are similar.

$$\overline{\widetilde{T}}^{\nabla}(d) = \bigvee_{t \in \mathcal{T}} \left(\neg\neg\widetilde{T}(t) * \neg\mathcal{R}(d,t)\right)$$

$$\Longrightarrow \overline{\widetilde{T}}^{\nabla}(d) = \bigvee_{t \in \mathcal{T}} \neg\left(\widetilde{T}(t) \rightarrow \mathcal{R}(d,t)\right), \text{ by using (R0)}$$

$$\Longrightarrow \overline{\widetilde{T}}^{\nabla}(d) = \neg \bigwedge_{t \in \mathcal{T}} \left(\widetilde{T}(t) \rightarrow \mathcal{R}(d,t)\right)$$

$$\Longrightarrow \overline{\overline{\widetilde{T}}^{\nabla}}(d) = \bigwedge_{t \in \mathcal{T}} \left(\widetilde{T}(t) \rightarrow \mathcal{R}(d,t)\right), \text{ since } \neg \text{ is involutive}$$

$$\Longrightarrow \overline{\overline{\widetilde{T}}^{\nabla}}(d) = \widetilde{T}^{\triangle}(d)$$

□

In order to express negation of both conjunctive and disjunctive queries, one may take advantage of Proposition 3. Indeed, from property $(P3)$, it comes that $\overline{\widetilde{T}^{\triangle}} = \widetilde{T}^{\Pi}_{\overline{\mathcal{R}}}$. Consequently, fuzzy sets of documents $\widetilde{D}$ satisfying the negation of a conjunctive query $\overline{\hat{Q}(\widetilde{T})}$ are the ones that satisfy the disjunctive query $\check{Q}(\widetilde{T})$

which is built upon the formal context $\overline{\mathcal{K}} = (L, \mathcal{D}, \mathcal{T}, \overline{\mathcal{R}})$ with $\overline{\mathcal{R}}(d,t) = \neg\mathcal{R}(d,t)$. Since the satisfaction of a disjunctive query has been already treated in Section 4.2, it remains to build once the concept lattice of $\overline{\mathcal{K}}$ and then apply the results of Proposition 3 to obtain the set (i.e. a complete lattice w.r.t Proposition 3) of formal concepts satisfying a negative disjunctive query.

From property $(P3)$, it comes also that $\widetilde{T}^{\Pi} = \widetilde{T}^{\triangle}_{\overline{\mathcal{R}}}$. Consequently, the fuzzy sets of documents satisfying a negative disjunctive query $\overline{\check{Q}(\widetilde{T})}$ are the ones that satisfy the conjunctive query $\hat{Q}(\widetilde{T})$ which is built upon the formal context $\overline{\mathcal{K}} = (L, \mathcal{D}, \mathcal{T}, \overline{\mathcal{R}})$. Once the concept lattice of $\overline{\mathcal{K}}$ has been built, from the results presented in Section 4.1, it comes that there is one and just one formal concept $c \in \mathcal{L}(\overline{\mathcal{K}})$ satisfying a negative disjunctive query.

5 Illustrative Example

Example 1. We give an example which illustrates the interest of Proposition 2. Let us consider the incidence matrix $\mathcal{R}_2$ given in Table 2 and a disjunctive query $\check{Q}(T) \equiv t_1 \vee t_2 \vee t_3$. All formal concepts c s.t $Ext(c) \subseteq \{t_1, t_2, t_3\}^{\Pi} = \{d_2, d_3, d_4, d_5\}$ satisfy this query (for instance, $\langle\{d_3, d_4\}, \{t_2\}\rangle$). One may also remark that the upper bound $\left(\{t_1, t_2, t_3\}^{\Pi}, (\{t_1, t_2, t_3\}^{\Pi})^{\triangle}\right)$ is not a formal concept. Figure 1 depicts the whole concept lattice related to the incidence matrix $\widetilde{\mathcal{R}}_2$. Whereas, according to Proposition 2, Figure 2 illustrates in dashed lines the (complete) sublattice of formal concepts that satisfy the above disjunctive query.

Table 2. Boolean incidence matrix

$\mathcal{R}_2$	t_1	t_2	t_3	t_4	t_5
d_1					×
d_2	×				
d_3		×			
d_4		×	×		
d_5			×	×	×

Example 2. Let us now consider a fuzzy incidence matrix $\mathcal{R}_3$ illustrated in the left part of Table 3. For instance, let us take $\neg p = 1 - p$. Thus, the right part of Table 3 illustrates $\overline{\mathcal{R}_3}$ ($\overline{\mathcal{R}_3}(d.t) = 1 - \mathcal{R}_3(d,t)$). We want to retrieve formal concepts satisfying the negation of the weighted conjunctive query $t_1^{0.0} \wedge t_2^{0.3} \wedge t_3^{0.1}$. For this purpose we have to build the concept lattice of $\overline{\mathcal{R}_3}$ which corresponds to the whole lattice illustrated in Figure 3 (except the node in red color). The formal concepts satisfying the considered query are the ones that satisfy the disjunctive query $t_1^{0.0} \vee t_2^{0.3} \vee t_3^{0.1}$ and are s.t. $Ext(c) \subseteq (t_1^{0.0} t_2^{0.3} t_3^{0.1})^{\Pi} = (d_1^{0.2} d_2^{1} d_3^{0.3} d_3^{0.7})$. This corresponds to the (complete) sublattice represent in dashed lines. Note that we have considered the Gödel implication and the corresponding min t-norm $*$.

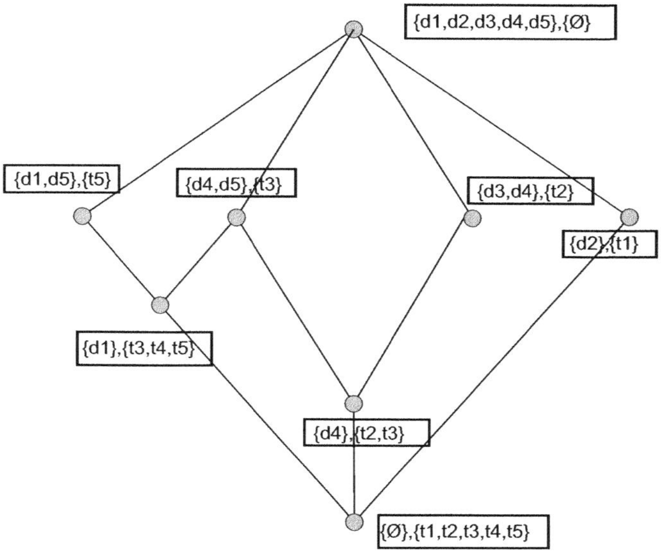

Fig. 1. Concept lattice corresponding to the incidence matrix $\mathcal{R}_2$

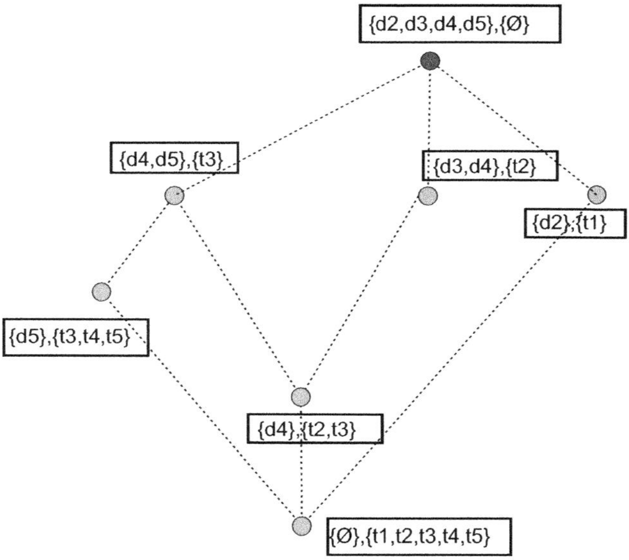

Fig. 2. Lattice of formal concepts satisfying the disjunctive query $t_1 \vee t_2 \vee t_3$

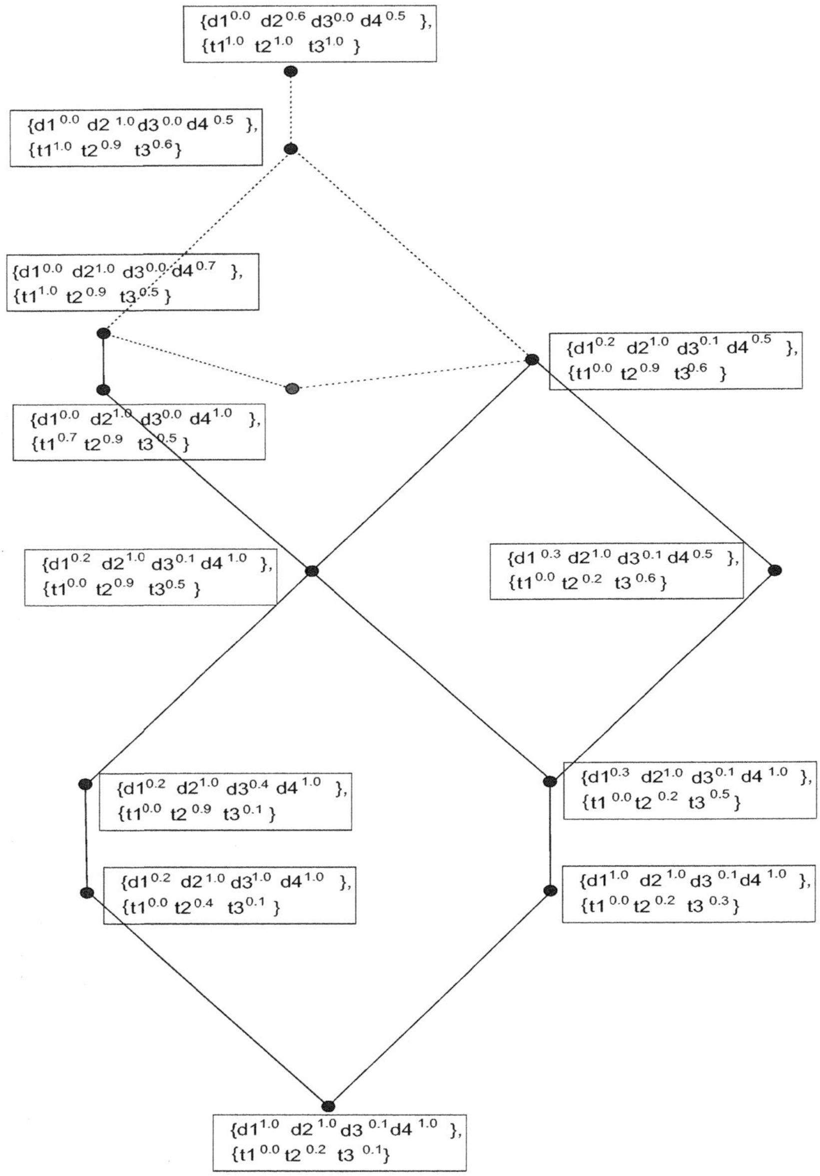

Fig. 3. Formal concepts satisfying a negative conjunctive query

Table 3. Fuzzy incidence matrix $\mathcal{R}_3$ and $\overline{\mathcal{R}_3}$

$\mathcal{R}_3$	t_1	t_2	t_3	$\overline{\mathcal{R}_3}$	t_1	t_2	t_3
d_1	1	0.8	0.7	d_1	0.0	0.2	0.3
d_2	0.0	0.1	0.4	d_2	1	0.9	0.6
d_3	1	0.6	0.9	d_3	0.0	0.4	0.1
d_4	0.3	0.1	0.5	d_4	0.7	0.9	0.5

6 Conclusion

Contributions of this paper are manifold. First, it extends the framework of FCA-based information retrieval approaches to the fuzzy setting which allows for fuzzy incidence matrices as well as weighted queries in a sound truth structure. Fuzzy settings may be useful for information retrieval purpose since it allows to obtain relevant incidence matrices using (normalized) numerical values such the well known term frequency measure.

Another original contribution consists of considering possibilistic derivation operators in order to express disjunction and negation in queries. We have showed that the results of a disjunctive query is also a complete lattice whereas the negation is easily expressed by means of a proposition that establishes correspondences between possibilistic derivation operators.

As a perspective, we intend to address the problem of ranking the query results in a deterministic (lattice-based) way. We intend also to consider the case where the entries of the incidence matrix are partially or even completely unknown.

References

[1] Bělohlávek, R.: Fuzzy Galois connections. Math. Logic Quart. 45, 497–504 (1999)

[2] Bělohlávek, R., Vychodil, V.: What is a fuzzy concept lattice. In: Proceedings CLA 2005 Concept Lattices and their Applications, Olomounc, Czech Republic, pp. 34–45 (2005)

[3] Carpineto, C., Romano, G.: Exploiting the potential of concept lattices for information retrieval with credo. Journal of Universal Computer Science 10(8), 985–1013 (2004)

[4] Carpineto, C., Romano, G.: Effective reformulation of boolean queries with concept lattices. In: FQAS, pp. 83–94 (1998)

[5] Dau, F., Ducrou, J., Eklund, P.W.: Concept similarity and related categories in searchsleuth. In: Eklund, P., Haemmerlé, O. (eds.) ICCS 2008. LNCS (LNAI), vol. 5113, pp. 255–268. Springer, Heidelberg (2008)

[6] Djouadi, Y., Dubois, D., Prade, H.: Possibility theory and formal concept analysis: Context decomposition and uncertainty handling. In: Hüllermeier, E., Kruse, R., Hoffmann, F. (eds.) IPMU 2010. LNCS (LNAI), vol. 6178, pp. 260–269. Springer, Heidelberg (2010)

[7] Djouadi, Y., Prade, H.: Possibility-theoretic extension of derivation operators in formal concept analysis. Fuzzy Optimization and Decision Making (submitted, 2011)

[8] Dubois, D., Prade, H.: Possibility theory and formal concept analysis in information systems. In: Proc. IFSA 2009, International Fuzzy Systems Association World Congress, Lisbon, Portugal, pp. 1021–1026 (2009)
[9] Dubois, D., de Saint Cyr, F.D., Prade, H.: A possibilty-theoretic view of formal concept analysis. Fundamenta Informaticae 75(1-4), 195–213 (2007)
[10] Düntsch, I., Gediga, G.: Approximation operators in qualitative data analysis. In: de Swart, H., Orłowska, E., Schmidt, G., Roubens, M. (eds.) Theory and Applications of Relational Structures as Knowledge Instruments. LNCS, vol. 2929, pp. 214–230. Springer, Heidelberg (2003)
[11] ElQadi, A., Aboutadjine, D., Ennouary, Y.: Formal concept analysis for information retrieval. International Journal of Computer Science and Information Security 7(2), 119–125 (2010)
[12] Ganter, B., Wille, R.: Formal Concept Analysis. Springer, Heidelberg (1999)
[13] Gediga, G., Düntsch, I.: Modal-style operators in qualitative data analysis. In: Proc. ICDM 2002 IEEE Int. Conf. on Data Mining, pp. 9–12 (2002)
[14] Latiri, C.C., Elloumi, S., Chevallet, J.P., Jaouay, A.: Extension of fuzzy Galois connection for information retrieval using a fuzzy quantifier. In: Proc. AICCSA 2003 Int. Conf. on Computer Systems and Applications, Tunis, Tunisia, pp. 81–91 (2003)
[15] Messai, N., Devignes, M.D., Napoli, A., Smaïl-Tabbone, M.: Many-valued concept lattices for conceptual clustering and information retrieval. In: ECAI 2008 - 18th European Conference on Artificial Intelligence, Patras, Greece, July 21-25, pp. 127–131 (2008)
[16] Nauer, E., Toussaint, Y.: Classification dynamique par treillis de concepts pour la recherche d'information sur le web. In: Proc. Conférence en Recherche d'Infomations et Applications - CORIA 2008, Trégastel, France, March 12-14, pp. 71–86 (2008)
[17] Priss, U.: Lattice-based information retrieval. Knowledge Organization 27(3), 132–142 (2000)
[18] Salton, G.: Automatic information organization and retrieval. McGraw-Hill, New York (1968)
[19] Wille, R.: Restructuring Lattice Theory: an Approach Based on Hierarchies of Concepts. In: Rival, I. (ed.) Ordered Sets. Reidel, Dordrecht (1982)
[20] Zadeh, L.: Fuzzy sets as a basis for a possibility theory. Fuzzy Sets and Systems 1(1), 3–28 (1978)

Processing Fuzzy Queries in a Peer Data Management System Using Distributed Fuzzy Summaries

Olivier Pivert, Grégory Smits, and Allel Hadjali

Irisa – Enssat, University of Rennes 1
Technopole Anticipa 22305 Lannion Cedex France
pivert@enssat.fr, gregory.smits@univ-rennes1.fr,
hadjali@enssat.fr

Abstract. In this paper, we consider the situation where a fuzzy query is submitted to distributed data sources. In order to save bandwith and processing cost, we propose an approach whose aim is to forward the query to the most relevant sources only. An efficient fuzzy-cardinality-based technique for summarizing each data source is described. The approach we propose consists in estimating the relevance of a source with respect to a user query, based on its associated summary. Some experiments illustrate the efficiency of the approach.

1 Introduction

In recent years, many research works have acknowledged the need for flexible ways to access information. A typical example is the "top-k query" approach where a user can incorporate vague terms in his/her query and the system aims at providing the user with the best k answers. A more general approach is that based on fuzzy set theory, which allows for a large variety of flexible terms and connectives. Another important recent phenomenon in the world of information systems is the emergence of decentralized approaches to data sharing, illustrated in particular by peer-to-peer systems (P2P). A PDMS (Peer Data Management System) consists of a set of peers, each of which acts as an information integration component [1]. Queries submitted to one peer are answered by local data as well as by data that is reachable along paths of mappings through the network of peers. In such a context, due to the bandwith costs related to query propagation, it becomes crucial to assess data sources in order to optimize query processing. Having available summaries of the different source contents enables to propagate the query only to the (more or less) relevant sources. In this paper, we tackle the problem of fuzzy query processing in a context of large-scale distributed relational databases. The main objective is to determine as efficiently as possible the set of databases which are likely to provide "good answers" to a fuzzy query submitted by a user. We study a specific type of "fuzzy summaries" — based on the concept of fuzzy cardinality — that proves useful for this purpose. In the following, we assume that the sources share the same schema. Otherwise, one would need to have available some mappings between schema sources, but this data mediation issue is beyond the scope of the paper.

The remainder of the paper is organized as follows. Section 2 is devoted to a reminder about fuzzy queries whereas Section 3 describes a database summarization technique

S. Benferhat and J. Grant (Eds.): SUM 2011, LNAI 6929, pp. 359–372, 2011.
© Springer-Verlag Berlin Heidelberg 2011

based on the use of fuzzy partitions of the domains and the computation of fuzzy cardinalities. In Section 4, we describe the principle of the computation of an estimated relevance degree for a source wrt to a given fuzzy query. Section 5 presents an application scenario of this technique in the context of a PDMS. Section 6 presents some experimental results which illustrate the efficiency of the approach. Related work is discussed in Section 7, whereas Section 8 recalls the main contributions of the paper and outlines some perspectives for future work.

2 Fuzzy Queries

Regular sets allow for the definition of Boolean predicates. In an analogous way, gradual predicates (or conditions) can be associated with fuzzy sets [2] aimed at describing classes of objects with vague boundaries. A fuzzy predicate P can be modeled as a function μ_P (usually of triangular or trapezoidal shape) from one (or several) domain(s) X to the unit interval. The degree $\mu_P(x)$ represents the extent to which element x satisfies the vague predicate P (or equivalently the extent to which x belongs to the fuzzy set of objects which match the fuzzy concept P). A fuzzy predicate can also compare two attributes using a gradual comparison operator such as "more or less equal".

Conjunction (resp. disjunction) is interpreted by means of a triangular norm $\top$ (resp. co-norm $\bot$), for instance the minimum or the product (resp. the maximum or the probabilistic sum). As to negation, it is interpreted as: $\forall x,\ \mu_{\neg P}(x) = 1 - \mu_P(x)$. Weighted conjunction and disjunction as well as weighted averaging operators can be used to assign a different importance to each of the predicates (see [3] for more details).

The operations from relational algebra can be extended to fuzzy relations by considering fuzzy relations as fuzzy sets on the one hand and by introducing gradual predicates in the appropriate operations on the other hand. The definitions of these extended operators can be found in [4]. As an illustration, we give the definition of the fuzzy selection hereafter, where r denotes a (fuzzy or crisp) relation, t is a tuple from r, and *cond* is a fuzzy predicate.

$$\mu_{sel(r,\, cond)}(t) = \top(\mu_r(t),\ \mu_{cond}(t)).$$

The language called SQLf described in [5] extends SQL so as to support fuzzy queries. Here, we just describe the base block in SQLf since this is all we need for our purpose. The principal differences w.r.t. SQL affect mainly two aspects :

- the calibration of the result, which can be achieved through a number of desired answers (k), a minimal level of satisfaction (δ), or both, and
- the nature of the authorized conditions as mentioned previously.

Therefore, the base block is expressed as:

select [**distinct**] [$k \mid \delta \mid k,\, \delta$] *attributes* **from** *relations* **where** *fuzzy-condition*.

3 Summarizing a Relation

We first recall the principle of the fuzzy-cardinality-based approach to database summarization introduced in [6]. Let r be a (non fuzzy) relation involving attributes A,

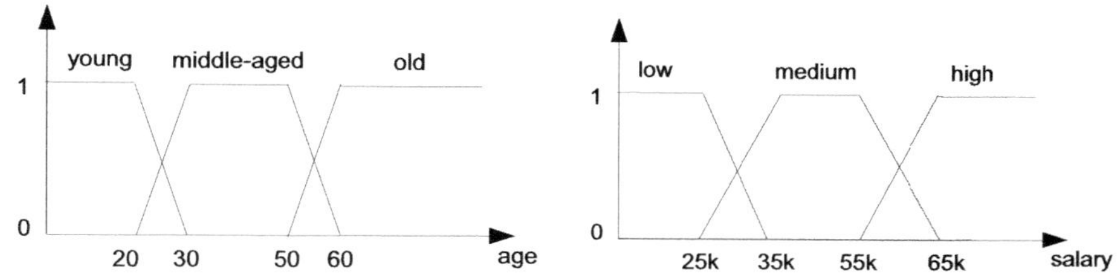

Fig. 1. Fuzzy partitions of the domains of attribute *age* (left) and *salary* (right)

B, and C (for notational simplicity, we use only three attributes, which is sufficiently general for discussing the main issues, but the relation may involve any number of attributes). Let (a_i, b_j, c_k) denote a tuple of r. Let D_A, D_B, D_C be the attribute domains. We assume that each domain is equipped with a fuzzy partition $(A_1, A_2, \ldots, A_{na})$, $(B_1, B_2, \ldots, B_{nb})$, $(C_1, C_2, \ldots, C_{nc})$ respectively (cf. Fig. 1). Each fuzzy set in a partition is assumed to be normalized and of trapezoidal form. Each partition is ordered, and a fuzzy set, say A_i, can only overlap with its predecessor A_{i-1} or/and its successor A_{i+1} (when they exist).

We further assume that a finite scale (with $m + 1$ levels) is used for assessing the membership degrees, namely $1 = \sigma_1 > \ldots > \sigma_m > 0$. Each level corresponds to a different possible understanding of A_r as the crisp level-cut $(A_r)_{\sigma_i}$.

From relation r, we build a new relation r_{su} (for "r summarized") by a procedure involving two main steps which are now described.

3.1 The Labelling Step

We replace each tuple $\langle a_i, b_j, c_k\rangle$ by one or several tuples of fuzzy sets $\langle A_r, B_s, C_t\rangle$ subject to the constraint: $A_r(a_i) > 0$ [1], $B_s(b_j) > 0$, $C_t(c_k) > 0$. Thus $\langle a_i, b_j, c_k\rangle$ may be replaced by one tuple $\langle A_r, B_s, C_t\rangle$ if all the three degrees of membership are equal to 1, or by several (up to $2^3 = 8$) in case one or several of the element(s) in the tuple belong to two fuzzy sets. For instance, if $A_r(a_i) = 1$, $B_s(b_j) = 0.8$, $B_{s+1}(b_j) = 0.2$, $C_{t-1}(c_k) = 0.6$, $C_t(c_k) = 0.4$, four tuples are produced:

$$\langle A_r,\ 0.8/B_s,\ 0.6/C_{t-1}\rangle,\ \langle A_r,\ 0.8/B_s,\ 0.4/C_t\rangle,$$
$$\langle A_r,\ 0.2/B_{s+1},\ 0.6/C_{t-1}\rangle,\ \langle A_r,\ 0.2/B_{s+1},\ 0.4/C_t\rangle$$

where we keep track of the membership degrees (A_r stands for $1/A_r$). This corresponds to all the possible "readings" of the tuple $\langle a_i, b_j, c_k\rangle$ in terms of the vocabulary provided by the fuzzy partitions. In the context we consider, it is not necessary to store the summarized relation r_{su}. The only additional data that has to be stored is the fuzzy cardinalities, whose computation is described in the following subsection.

3.2 Fusion Step and Computation of Fuzzy Cardinalities

We want to know how many tuples from r are A_r, are B_s, are C_t, are A_r and B_s, ..., are A_r and B_s and C_t, and this, for all the fuzzy labels. It is then necessary to compute

[1] $A_r(a_i)$ denotes the membership degree of a_i in A_r.

the different cardinalities related to each linguistic label and to the diverse conjunctive combinations of these labels.

All the tuples of the form $\langle x/A_r,\ y/B_s,\ z/C_t\rangle$ which are identical with respect to the three labels are fused into one tuple $\langle A_r,\ B_s,\ C_t\rangle$ of r_{su}. At the same time, we compute the cardinalities F_{A_r}, F_{B_s}, F_{C_t}, $F_{A_rB_s}$, $F_{A_rC_t}$, $F_{B_sC_t}$, $F_{A_rB_sC_t}$ where F_{A_r} (resp. F_{B_s}, ..., $F_{A_rB_sC_t}$) is a fuzzy set defined on the integers $\{0, 1, ...\}$ which represents the fuzzy number of tuples which are somewhat A_r (resp. B_s, ..., A_r and B_s and C_t) and which are fused into the considered tuple. In the following, a fuzzy cardinality is represented as $F_{A_r} = 1/c_1 + \sigma_2/c_2 + ... + \sigma_f/c_f$, where c_i is the number of tuples in the concerned relation that are A_r with a degree at least equal to σ_i. Each cardinality is computed incrementally in the following way. At the beginning $F_{A_r} = 1/0$. Let:

$$F_{A_r} = 1/c_1 + \sigma_2/c_2 + \ldots + \sigma_k/c_k + \ldots + \sigma_f/c_f$$

be the current value of F_{A_r}. Let us consider a new tuple whose A-value rewrites A_r. Let x' be the degree attached to A_r in this tuple. F_{A_r} must then be modified. If $x' = 1$, F_{A_r} becomes:

$$1/(c_1 + 1) + \sigma_2/(c_2 + 1) + \ldots + \sigma_k/(c_k + 1) + \ldots + \sigma_f/(c_f + 1)$$

If $x' < 1$, there are two cases. If $\exists i,\ x' = \sigma_i$ then F_{A_r} is modified into:

$$1/c_1 + \sigma_2/c_2 + \ldots + \sigma_{i-1}/(c_{i-1}) + \sigma_i/(c_i + 1) + \ldots + \sigma_f/(c_f + 1).$$

Otherwise, $\exists j,\ \sigma_j > x' > \sigma_{j+1}$ and F_{A_r} becomes:

$$1/c_1 + \ldots + \sigma_j/c_j + x'/1 + \sigma_{j+1}/(c_{j+1} + 1) + \ldots + \sigma_f/(c_f + 1).$$

If, for the computation of F_{A_r} (resp. F_{B_s} and F_{C_t}), one takes into account the value x' (resp. y' the degree related to B_s, and z' the degree related to C_t), the computation of $F_{A_rB_s}$ (resp. $F_{A_rC_t}$, $F_{B_sC_t}$ and $F_{A_rB_sC_t}$), takes into account the value $min(x',\ y')$ (resp. $min(x',\ z')$, $min(y',\ z')$, $min(x',\ y',\ z')$), thus reflecting the fact that the tuple to fuse is both A_r and B_s (resp. A_r and C_t, B_s and C_t, A_r and B_s and C_t).

Table 1. Tuples from *emp* (left) and their degrees (right)

#e	name	age	sal(k$)
17	Smith	51	65
76	Martin	40	45
26	Jones	24	19
12	Green	39	32
19	Duncan	28	24
8	Brown	54	57
31	Harris	29	18
9	Davis	61	15
44	Howard	22	45
23	Lewis	62	59

μ_{yg}	μ_{ma}	μ_{old}	μ_{low}	μ_{med}	μ_{high}
0	0.9	0.1	0	0	1
0	1	0	0	1	0
0.6	0.4	0	1	0	0
0	1	0	0.3	0.7	0
0.2	0.8	0	1	0	0
0	0.6	0.4	0	0.8	0.2
0.1	0.9	0	1	0	0
0	0	1	1	0	0
0.8	0.2	0	0	1	0
0	0	1	0	0.6	0.4

Example 1. Let us consider a relation *emp* of schema (#*e*, *name*, *age*, *salary*) describing employees of a company, cf. Table 1 (left). Let us consider the fuzzy partition on the domain of attribute *age* (resp. *salary*) given in Figure 1. The degrees associated with the tuple values from *emp* are given in Table 1 (right). The fuzzy cardinalities which summarize relation *emp* are given in Table 2. ◇

Table 2. Fuzzy summary associated with relation *emp*

young	0.8/1 + 0.6/2 + 0.2/3 + 0.1/4
middle-aged	1/2 + 0.9/3 + 0.9/4 + 0.8/5 + 0.6/6 + 0.4/7 + 0.2/8
old	1/2 + 0.4/3 + 0.1/4
low	1/4 + 0.3/5
medium	1/2 + 0.8/3 + 0.7/4 + 0.6/5
high	1/1 + 0.4/2 + 0.2/3
{*young*, *low*}	0.6/1 + 0.2/2 + 0.1/3
{*young*, *medium*}	0.8/1
{*young*, *high*}	1/0
{*middle-aged*, *low*}	0.9/1 + 0.8/2 + 0.4/3 + 0.3/4
{*middle-aged*, *medium*}	1/1 + 0.7/2 + 0.6/3 + 0.2/4
{*middle-aged*, *high*}	0.9/1 + 0.2/2
{*old*, *low*}	1/1
{*old*, *medium*}	0.6/1 + 0.4/2
{*old*, *high*}	0.4/1 + 0.2/2 + 0.1/3

3.3 Constructing the Summary of a Data Source

The summarization process of a data source relies on a first step of tuple interpretation in terms of the fuzzy vocabulary. This task has a complexity which is linear in the size of the data source. The second step concerns the update of the fuzzy cardinalities for all possible conjunctions of predicates taken from the set of interpretations for the concerned tuple. For each tuple this step has a maximum complexity of 2^A where A is the number of attributes. Section 6, dedicated to the evaluation of this approach on a concrete example, shows that the time needed to compute such summaries is acceptable, all the more so as these summaries can be updated incrementally. For example, it takes less than 7 minutes to compute the cardinalities associated with a database containing 46,069 tuples and 10 attributes.

A strategy to reduce the size of the summaries and the time needed to compute them, is to focus on attributes or predicates that have been frequently combined by users in their previously submitted queries. Such a workload can be used to initialize a first version of the summaries that can then be easily refined when new queries are submitted. Using such a progressive construction of summaries, a new query that is not related to one of the stored fuzzy cardinalities is forwarded to all possible data sources. The results returned are then used to (incrementally) update the summaries.

4 Estimating the Relevance of a Data Source

As explained before, one wants to estimate the relevance of a data source S with respect to a given query Q, i.e., to assess the extent to which S may provide "good answers" to

Q. This covers two aspects: a qualitative one (how satisfactory are the answers returned by S) and a quantitative one (how many answers are returned by S). Both these aspects are captured by the fuzzy cardinality attached to the result of Q in S. Its estimation — whose principle was introduced in [7] — is presented hereafter.

4.1 Principle of the Approach

Let P_A be a fuzzy predicate from the user query (expressed in a language such as SQLf [5]). It is assumed that P_A concerns attribute A. Let $G_i(A)$ be the fuzzy partition defined on the domain of attribute A in relation r_i present in the data source S_i. The objective is to estimate the fuzzy cardinality relative of the fuzzy set of answers to P_A in S_i, using the fuzzy cardinalities attached to the labels from $G_i(A)$. Along with his/her query, the user may specify either a threshold $\delta \in \{\sigma_1, \ldots, \sigma_m\}$ (then, a data source will be selected only if it contains tuples whose satisfaction degree with respect to the user query is at least δ) or a number k (then, only the k "best" data sources will be selected, a source being all the better as the average satisfaction degree of its tuples with respect to the query is high). Let $L_i(P_A)$ be the set gathering the predicates of $G_i(A)$ which have a nonempty intersection with P_A ($L_i(P_A)$ is nonempty since $G_i(A)$ is a partition.) We assume that each domain is bounded. Then, every α-cut is a closed interval, even if the membership function is a right- or a left-shoulder. One can easily derive the number of tuples which have a certain degree μ for a given label, from the fuzzy cardinality attached to this label. For instance, if one has the fuzzy cardinality $F_{P_j} = 1/7 + 0.8/16 + 0.7/21$, one deduces that 7 elements in r_i get degree 1 for P_j, 9 $(16 - 7)$ get degree 0.8, and 5 $(21 - 16)$ get degree 0.7.

4.2 Single Predicate Case

Let us first consider the case where query Q involves a single fuzzy predicate P.

Qualitative Thresholding. If δ is specified, it is then necessary to evaluate only the α-cuts such that $\alpha \geq \delta$. For a given source S_i, a given α from $\{\sigma_1, \ldots, \sigma_m\}$ and a given fuzzy predicate P_j from $L_i(P)$, we denote by $inter_j^i(\alpha)$ the interval which corresponds to the intersection between the α-cut of P and that of P_j, and $card_j^i(\alpha)$ its associated estimated cardinality. We denote by $card^i(\alpha)$ the overall estimated cardinality of the α-cut of P in S_i. From the different $card^i(\alpha)$, one can build the estimated fuzzy cardinality $F^i_{P,\,\delta}$ associated with P in S_i. The algorithm is as follows.

for each source S_i **do**
  for $\alpha := \sigma_m$ **down to** δ **do**
    $card^i(\alpha) = 0$;
    for each P_j in $L_i(P)$ **do**
      compute $inter_j^i(\alpha)$;
      estimate $card_j^i(\alpha)$;
    endfor;
    compute $card^i(\alpha)$ from the $card_j^i(\alpha)$'s;
  endfor;

build the fuzzy cardinality $F^i_{P,\,\delta}$;
endfor;
forward Q only to the sources such that $card^i(\delta) > 0$;

Notice that this algorithm does not induce any access to the data from S_i, but only simple computations involving membership functions and fuzzy cardinalities, which means that its processing time is negligible with respect to the cost of actually evaluating Q against S_i. Notice also that a more demanding view would be to select only the sources which contain *at least n tuples* whose degree is at least δ. Then, the condition $card^i(\delta) > 0$ is replaced by $card^i(\delta) \geq n$, where n is a user-specified parameter. The following example illustrates the way the approach works. We assume that the scale used for assessing the membership degrees is $\{0, 0.2, 0.4, 0.6, 0.8, 1\}$.

Example 2. Let us consider the user-defined predicate P_A of support $[310,\ 410]$ and core $[330,\ 350]$ (cf. Figure 2) and the user-specified threshold $\delta = 0.4$. It is assumed that the domain of A is a subset of the integers. We have:

$$\text{1-cut}(P_A) = [330, 350],\quad \text{0.8-cut}(P_A) = [326, 362],$$
$$\text{0.6-cut}(P_A) = [322, 374],\quad \text{0.4-cut}(P_A) = [318, 386].$$

We assume that for the source S_i, one has $L_i(P) = \{P_1,\ P_2\}$ (cf. Figure 2) with

- P_1 a predicate of support [250, 390] and of core [300, 340], such that: $F_{P_1} = 1/70 + 0.8/161 + 0.6/317 + 0.4/555$, which gives:

  1-cut(P_1) = [300, 340]; $card = 70$; 0.8-cut(P_1) = [290, 350]; $card = 161$;
  0.6-cut(P_1) = [280, 360]; $card = 317$; 0.4-cut(P_1) = [270, 370]; $card = 555$.

  From F_{P_1} and the shape of μ_{P_1}, one can also deduce that:
  - 70 elements are in $[300,\ 340]$,
  - 91 $(= 161 - 70)$ elements are in $[290, 300[\cup]340,\ 350]$,
  - 156 $(= 317 - 161)$ elements are in $[280,\ 290[\cup]350,\ 360]$,
  - 238 $(= 555 - 317)$ elements are in $[270,\ 280[\cup]360,\ 370]$.
- P_2 a predicate of support [340, 470] and of core [390, 420], such that: $F_{P_2} = 1/0 + 0.8/82 + 0.4/363$, which gives:

  1-cut(P_2) = $[390,\ 420]$; $card = 0$; 0.8-cut(P_2) = [380, 430]; $card = 82$;
  0.6-cut(P_2) = [370, 440]; $card = 82$; 0.4-cut(P_2) =[360, 450]; $card = 363$.

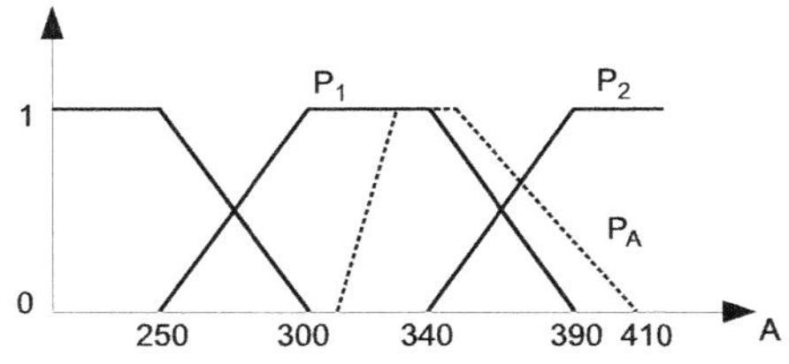

Fig. 2. A fuzzy partition of the domain of A and a user predicate P_A

From F_{P_2} and the shape of μ_{P_2}, one can also deduce that:

- 0 elements are in $[390,\ 420]$,
- 82 elements are in $[380,\ 390[\cup]420,\ 430]$,
- 0 $(= 82 - 0)$ elements are in $[370,\ 380[\cup]430,\ 440]$,
- 281 $(= 363 - 82)$ elements are in $[360,\ 370[\cup]440,\ 450]$.

The estimation of the fuzzy cardinality relative to P_A based on those relative to P_1 and P_2 is performed as follows.

- $\alpha = 1$:
  $inter_1^i(1) = [330, 340]$;
  $card_1^i(1) = \lfloor 70 \times \frac{340-330+1}{340-300+1} \rfloor = 19$ (here, one assumes that the elements are uniformly distributed in the core of P_1);
  $inter_2^i(1) = \emptyset$; $card_2^i(1) = 0$; $card^i(1) = card_1^i(1) + card_2^i(1) = 19$;
- $\alpha = 0.8$:
  $inter_1^i(0.8) = [326, 350]$;
  $card_1^i(0.8) = \lfloor \frac{340-326+1}{340-300+1} + \frac{161-70}{2} \rfloor = 71$ (here, one assumes that the elements of degree 0.8 are evenly distributed on the left and the right shoulders of the function, i.e., in $[290, 300[$ and $]340, 350]$);
  $inter_2^i(0.8) = \emptyset$; $card_2^i(0.8) = 0$; $card^i(0.8) = card_1^i(0.8) + card_2^i(0.8) = 71$;
- $\alpha = 0.6$:
  $inter_1^i(0.6) = [322, 360]$;
  $card_1^i(0.6) = \lfloor \frac{340-322+1}{340-300+1} + \frac{161-70}{2} + \frac{317-161}{2} \rfloor = 156$;
  $inter_2^i(0.6) = [370, 374]$; $card_2^i(0.6) = 0$;
  $card^i(0.6) = card_1^i(0.6) + card_2^i(0.6) = 156$;
- $\alpha = 0.4$:
  $inter_1^i(0.4) = [318, 370]$;
  $card_1^i(0.4) = \lfloor \frac{340-318+1}{340-300+1} + \frac{161-70}{2} + \frac{317-161}{2} + \frac{555-317}{2} \rfloor = 282$;
  $inter_2^i(0.4) = [360, 386]$; $card_2^i(0.4) = \lfloor \frac{386-380+1}{389-380+1}) \times \frac{82-0}{2} + \frac{363-82}{2} \rfloor = 169$
  Since $inter_1^i(0.4) \cap inter_2^i(0.4) = [360,\ 370] \neq \emptyset$, one must be careful not to count the same elements twice. Using F_{P_1}, the estimated number of elements in $[360,\ 370]$ is: $\lfloor \frac{555-317}{2} \rfloor = 119$. Using F_{P_2}, it is: $\lfloor \frac{363-82}{2} \rfloor = 141$. A solution is to use the average of these two estimations, i.e. 130, and we get: $card^i(0.4) = card_1^i(0.4) + card_2^i(0.4) - 130 = 321$.

Finally, the estimation of $F^i_{P_A, 0.4}$ is: $1/19 + 0.8/71 + 0.6/156 + 0.4/321$.$\diamond$

Quantitative Thresholding. Let k be the number of answers (the best possible ones) that the user intends to obtain. The idea is to rank-order the sources according to their relevance wrt the query, and to select the best ones, to which the query will be forwarded. Two methods for ranking the sources may be thought of:

- lexicographic ordering of the sources so that the satisfaction degree is prioritary over the associated cardinality. For instance, if we have $F_P^1 = 1/3+0.8/5+0.6/72$, $F_P^2 = 1/7 + 0.8/7 + 0.6/7$, and $F_P^3 = 0.8/51 + 0.6/172$, we get $S_2 \succ S_1 \succ S_3$.

- ordering based on the relative scalar cardinality S^i_P which "synthesizes" the fuzzy cardinality F^i_P attached to a source S_i. For instance, the fuzzy cardinality computed in Example 2 may be synthesized into:

$$S^i_{P_A} = \frac{19 + 0.8 \times (71 - 19) + 0.6 \times (156 - 71) + 0.4 \times (321 - 156)}{321} \approx 0.55.$$

We get: $S^1_P = 0.62$, $S^2_P = 1$, and $S^3_P = 0.66$, which yields $S_2 \succ S_3 \succ S_1$.

The first strategy favors the sources which provide good quality answers (even if not many) whereas the second technique mixes the qualitative and the quantitative aspects in its assessment of a source. The generic algorithm is as follows.

for each source S_i **do**
  for $\alpha := \sigma_m$ **down to** σ_1 **do**
    $card^i(\alpha) = 0$;
    for each P_j in $L_i(P)$ **do**
      compute $inter^i_j(\alpha)$; estimate $card^i_j(\alpha)$;
    endfor;
    compute $card^i(\alpha)$ from the $card^i_j(\alpha)$'s;
  endfor;
  build the fuzzy cardinality F^i_P;
endfor;
rank-order the sources S_i;

Again, no disk access to the tuples from S_i is necessary. The last step is to select the sources to which the query will be forwarded, i.e. to determine the smallest number k' such the k' best sources provide together at least k answers. This can be done using the fuzzy cardinality associated with each source.

4.3 Case of a Conjunctive Fuzzy Query

When the user query involves several fuzzy predicates, we proceed as follows. Let us suppose that the query involves a fuzzy predicate P_A on attribute A and another P_B on attribute B. Let us consider the source S_i. As before, one first determines the set of predicates from $G_i(A)$ which intersect P_A — denoted by $L_i(P_A)$ — and those from $G_i(B)$ which intersect P_B — denoted by $L_i(P_B)$. Let us recall that we have available the fuzzy cardinality $F^i_{P_j P_k}$ related to S_i for every $P_j \in L_i(P_A)$ and $P_k \in L_i(P_B)$.

For every α of the scale considered, and for every P_j, P_k from $L_i(P_A) \times L_i(P_B)$, one computes the cardinality of the intersection between the α-cut of P_A (resp. P_B) and that of P_j (resp. P_k) divided by the cardinality of the α-cut of P_j (resp. P_k). These computations rest on the same principle as in the single predicate case. Let us denote by γ_A (resp. γ_B) the ratios obtained. Let us denote by $\rho = card^i_{jk}(\alpha)$ the cardinality associated with α in $F^i_{P_j P_k}$. The estimated contribution of (P_j, P_k) to the cardinality associated with α in $F^i_{P_A P_B}$ is:

$$c^i_{jk}(\alpha) = \gamma_A \times \gamma_B \times \rho$$

This corresponds to assuming that if γ_A (resp. γ_B) percent of the items which are P_j (resp. P_k) to a degree α are also P_A (resp. P_B) to a degree α, and if ρ items satisfy $P_j \wedge P_k$ to a degree α, then $\gamma_A \times \gamma_B \times \rho$ items satisfy $P_A \wedge P_B$ to a degree α. In other words, the (rather strong) assumption underlying this formula is that the proportion of elements which are P_A (resp. P_B) among those which are P_j (resp. P_k) is equal to the proportion of elements which are P_A (resp. P_B) among those which are $P_j \wedge P_k$.

Finally, the overall estimated cardinality $c^i(\alpha)$ associated with α in $F^i_{P_A P_B}$ is computed in the same way as in the single predicate case, taking care of not counting twice the elements which belong to two adjacent labels.

5 Application Scenario

Let us now outline the way a fuzzy query can be processed in the context of a PDMS organized according to a super-peer architecture.

5.1 The X-Peer Architecture

In the XPeer architecture [8], an information system plays the role of a peer. A set of peers sharing the same schema are grouped into a cluster where the mediated schema for the set is managed by a cluster-peer node. A cluster-peer is also called super-peer in the data mediation literature [9].

While a cluster-peer provides an integration of a set of peers within its cluster, this is insufficient for providing a large scale information system. The authors of [8] take the example of an information system for healthcare, where one needs to group together information from physicians, clinics, medicine industries as well as government resources. These different resources do not share the same schema and even the same resource may have different schemas and data access. To manage these heterogeneities, Roantree *et al.* [8] introduce two further layers in the peer hierarchy: domain-peers (a single domain-peer manages a set of cluster-peers), and a global-peer which manages the set of domain-peers. The authors define a *domain* as a set of clusters sharing the same category of information (physicians for instance). They also define a special node called a *domain peer* for each domain, which acts as the entry point for this domain.

5.2 Processing Strategy

Let us assume that each cluster-peer stores the schema mappings associated with every of its descendant nodes, as well as the summaries attached to them. When a peer receives a fuzzy query Q (associated either with a minimal satisfaction threshold δ or a quantitative threshold k), it forwards it to its cluster-peer which assesses every of its descendant nodes (data sources) by means of the estimation technique described in Section 4. The fuzzy query is forwarded to the peers corresponding to the selected sources (either the k best ones or those which are able to provide tuples whose satisfaction degree is at least equal to δ), which processes it and forwards the result to its cluster-peer. When the cluster-peer which forwarded the query has received all the results, it merges them into an ordered list and returns it via the peer which initiated the query.

6 Experimentation

We consider 46,089 ads provided by `www.eurotaxglass.fr` about second hand cars distributed over 6 databases. Each ad is described with the following schema (*idAds*, *model*, *description*, *year*, *mileage*, *price*, *make*, *length*, *height*, *nbseats*, *consumption*, *acceleration*, *co2emission*) and a common sense vocabulary has been defined on the 10 last attributes with 73 fuzzy predicates. A super node, acting as an access node, stores the summaries of these 6 data sources.

Summarization. To study the behavior of this fuzzy-cardinality-based summarization, we have conducted some experiments. Figures 3 shows how the computation time of the fuzzy cardinalities (left) as well as the space needed to store them in a database (right) increase according to a varying number of tuples between 5,000 and 46,089. It confirms a predictable phenomenon, i.e. the convergence of the size of the summary, which can be explained by the fact that whatever the number of tuples, the number of possible combinations of properties to describe them is finite and relatively small. So, one may expect that even for very large databases, the summaries will easily fit in the memory of the super node. The time complexity of the summarization process is linear with respect to the cardinality of the database and exponential with respect to the size of the vocabulary. Thus, according to the behavior of the computation time observed on a dataset varying between 5,000 and 45,000, one can estimate that it would take approximately 170 minutes to compute the summary of a database containing one million tuples with a shared vocabulary of 73 predicates. Such a summary has to be computed only once and can then be updated incrementally. Fuzzy cardinalities have been stored in a PostgreSQL DBMS hosted in a 2.53 GHz Intel Core2 Duo computer with 4GB of memory, and one can perform up to 6 modifications per second on the summary.

As to the exponential increase of the processing time and the storage space of the fuzzy cardinalities according to the number of predicates involved in the shared vocabulary, this is not a problem in practice as, even if databases are getting larger every day, the number of attributes and subsequently, the number of predicates that can be specified through query interfaces is stable (generally around 10). Moreover, as fuzzy

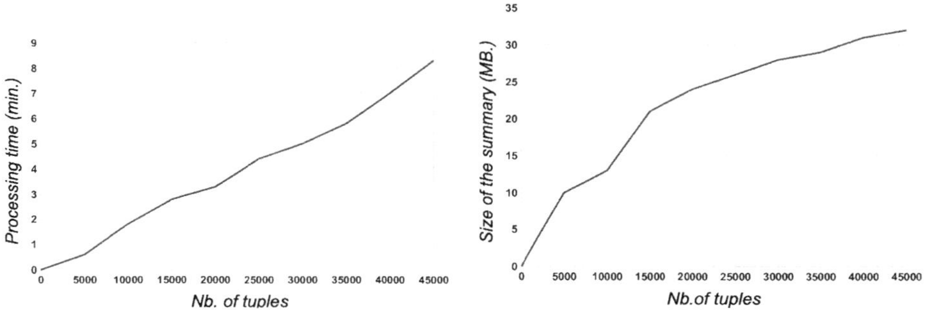

Fig. 3. Evolution of the processing time and space with respect to the number of tuples

cardinalities can be updated incrementally according to modifications performed on the databases, time complexity is not a crucial issue as this summarization step can be done off-line and dynamically maintained.

Cardinality estimation and query routing. The fuzzy cardinality of a user query is *estimated* from the data source summaries. We have evaluated the precision of this estimation process and especially the bias introduced by the hypothesis used during this estimation process (uniform distribution of the tuples over the intervals of the α-cuts). To do this, we have manually defined and submitted 30 atomic flexible queries. The estimated fuzzy cardinalities have then been compared to the actual cardinalities and we have observed a global error rate of 10.5%. Using two *a priori* defined demanding thresholds ($\alpha = 0.6$ and $k = 40$), we have then evaluated the extent to which the estimated fuzzy cardinalities are helpful for an efficient query routing among the distributed databases. For the submitted queries that can be satisfied, i.e. queries for which there exist at least k answers with a satisfaction degree of 0.6 across all the databases, 90% of them have been efficiently routed to an average of only 2.4 sources instead of a full propagation to the six databases. This shows that fuzzy summaries constitute a useful tool for defining efficient query routing strategies.

7 Related Work

7.1 Fuzzy Queries

In [10], Saint-Paul *et al.* propose an approach to the production of linguistic summaries structured in a hierarchy, i.e., a summarization tree where the tuples from the database are rewritten using the linguistic variables involved in fuzzy partitions of the attribute domains. The deeper the summary in the tree, the finer its granularity (basically, the father node of a set of nodes is associated with a set of disjunctions of fuzzy labels). First, the tuples from the database are rewritten using the linguistic variables involved in fuzzy partitions of the attribute domains. Then, each candidate tuple is incorporated into the summarization tree and reaches a leaf node (which can be seen as a classification of the tuple). In the hierarchical structure, a level is associated with the relative proportion (crisp ratio) of data that is described by the associated summary, as well as the maximal satisfaction degree of the tuples which rewrite into that summary. Hayek *et al.* further use this framework in [11] for minimizing query routing in a P2P database context. Obviously, this approach has a lot in common with the present work (use of fuzzy summaries for optimizing the processing of fuzzy queries in a PDMS context) but the main difference lies in the type of fuzzy summaries used. The summarization approach defined in [10] has indeed a much poorer semantics than that we propose, since it does not provide a fine description of the satisfaction degrees of the tuples which rewrite into a given fuzzy label (again, only a crisp ratio and a maximal satisfaction degree are associated with a given summary in the sense of [10], whereas a fuzzy cardinality conveys much more information about the distribution of the tuples in the different α-cuts of each fuzzy label). The *approximate query answering* technique that the authors of [11] propose is thus necessarily much less precise than the estimation based on fuzzy cardinality which constitutes the heart of our approach.

7.2 Top-k and Skyline Queries

In [12], Yu *et al.* use regular histograms, maintained at a central site, to estimate the score (which corresponds to what we call relevance degree) of distributed databases wrt a top-k query and send the query to the databases that are more or less likely to involve top results (see also [13,14,15,16] where similar approaches are advocated). Some authors have also proposed to use histogram-based summaries or bitmap indexes for optimizing skyline query processing in a decentralized database context (see in particular [17,18,19]). Even though the approaches which deal with top-k or skyline queries in P2P systems have an objective in common with the present contribution — that of enriching PDMSs with flexible querying capabilities —, they differ a lot on the techniques used, since:

- top-k queries take into account a limited form of flexibility in the selection conditions (only conditions of the type "attribute = constant" are considered, and transformed into "attribute $\approx$ constant") and the goal is not to get answers which are *good enough*, but the *best* answers (even if these answers are all mediocre),
- the skyline approach rests on a Pareto ordering of the answers and looks for the elements which are not dominated by any other; only a partial order is obtained since there is no global scoring function used (unlike the fuzzy-set-based approach which assumes commensurability between the degrees coming from different predicates).

8 Conclusion

In this paper, we have proposed an approach aimed at assessing different data sources wrt to a fuzzy query in the context of peer data management systems so as to optimize query processing. The idea is to forward the query to the most relevant sources only. The approach rests on the use of fuzzy partitions of the attribute domains, which themselves constitute the basis to the construction of fuzzy summaries of the different sources involved. A fuzzy summary gathers the fuzzy cardinalities attached to the different rewritings of the tuples in terms of the fuzzy labels from the partitions considered. The assessment of a data source is based on the estimation of the fuzzy cardinality attached to the user query in this source, which is computed from the summaries and takes into account the intersection of the α-cuts of the user predicates on the one hand and the fuzzy labels involved in the summaries on the other hand.

We have shown that the fuzzy cardinalities stored in the access node can be easily and incrementally updated according to modifications performed on the data sources. As a perspective for future work, let us mention that the comparison of the estimated fuzzy cardinalities with the results returned by the data sources can also serve for updating the fuzzy partitions on which summaries are pre-computed. Indeed, in order to minimize the error rate for predicates that are frequently used and to refine the summaries involving such predicates, one can use a more detailed scale of membership degrees to get more accurate summaries or even revise the shape of the fuzzy predicates.

References

1. Herschel, S., Heese, R.: Humboldt discoverer: A semantic P2P index for PDMS. In: Proc. of DISWeb (2005)
2. Zadeh, L.: Fuzzy sets. Information and Control 8, 338–353 (1965)
3. Fodor, J., Yager, R.: Fuzzy-set theoretic operators and quantifiers. In: Dubois, D., Prade, H. (eds.) Fundamentals of Fuzzy Sets. The Handbooks of Fuzzy Sets Series, vol. 1, pp. 125–193. Kluwer Academic Publishers, Dordrecht (2000)
4. Bosc, P., Buckles, B., Petry, F., Pivert, O.: Fuzzy databases. In: Bezdek, J., Dubois, D., Prade, H. (eds.). The Handbooks of Fuzzy Sets Series, vol. 3, pp. 403–468. Kluwer Academic Publishers, Dordrecht (1999)
5. Bosc, P., Pivert, O.: SQLf: a relational database language for fuzzy querying. IEEE Transactions on Fuzzy Systems 3(1), 1–17 (1995)
6. Bosc, P., Dubois, D., Pivert, O., Prade, H., de Calmés, M.: Fuzzy summarization of data using fuzzy cardinalities. In: Proc. of IPMU 2002, pp. 1553–1559 (2002)
7. Pivert, O., Hadjali, A., Smits, G.: Estimating the relevance of a data source using a fuzzy-cardinality-based summary. In: Proc. of the 5th IEEE International Conference on Intelligent Systems (IEEE IS 2010), London, Great-Britain, pp. 96–101 (2010)
8. Bellahsene, Z., Roantree, M.: Querying distributed data in a super-peer based architecture. In: Galindo, F., Takizawa, M., Traunmüller, R. (eds.) DEXA 2004. LNCS, vol. 3180, pp. 296–305. Springer, Heidelberg (2004)
9. Wiederhold, G., Genesereth, M.R.: The basis for mediation. In: Proc. of CoopIS 1995, pp. 140–157 (1995)
10. Saint-Paul, R., Raschia, G., Mouaddib, N.: General purpose database summarization. In: Proc. of VLDB 2005, pp. 733–744 (2005)
11. Hayek, R., Raschia, G., Valduriez, P., Mouaddib, N.: Summary management in P2P systems. In: Proc. of EDBT 2008, pp. 16–25 (2008)
12. Yu, C., Philip, G., Meng, W.: Distributed top-N query processing with possibly uncooperative local systems. In: Proc. of VLDB 2003, pp. 117–128 (2003)
13. Theobald, M., Weikum, G., Schenkel, R.: Top-k query evaluation with probabilistic guarantees. In: Proc. of VLDB 2004, pp. 648–659 (2004)
14. Nejdl, W., Siberski, W., Thaden, U., Balke, W.T.: Top-k query evaluation for schema-based peer-to-peer networks. In: McIlraith, S.A., Plexousakis, D., van Harmelen, F. (eds.) ISWC 2004. LNCS, vol. 3298, pp. 137–151. Springer, Heidelberg (2004)
15. Marian, A., Bruno, N., Gravano, L.: Evaluating top- queries over web-accessible databases. ACM Trans. Database Syst. 29(2), 319–362 (2004)
16. Akbarinia, R., Martins, V., Pacitti, E., Valduriez, P.: Top-k query processing in the APPA P2P system. In: Daydé, M., Palma, J.M.L.M., Coutinho, Á.L.G.A., Pacitti, E., Lopes, J.C. (eds.) VECPAR 2006. LNCS, vol. 4395, pp. 158–171. Springer, Heidelberg (2007)
17. Hose, K., Lemke, C., Sattler, K.U.: Processing relaxed skylines in PDMS using distributed data summaries. In: Proc. of CIKM 2006, pp. 425–434 (2006)
18. Fotiadou, K., Pitoura, E.: BITPEER: continuous subspace skyline computation with distributed bitmap indexes. In: Proc. of DaMaP 2008, pp. 35–42 (2008)
19. Rocha-Junior, J.B., Vlachou, A., Doulkeridis, C., Nørvåg, K.: Agids: A grid-based strategy for distributed skyline query processing. In: Hameurlain, A., Tjoa, A.M. (eds.) Globe 2009. LNCS, vol. 5697, pp. 12–23. Springer, Heidelberg (2009)

Embedding Forecast Operators in Databases

Francesco Parisi[1], Amy Sliva[2,3], and V.S. Subrahmanian[2]

[1] Università della Calabria, Via Bucci–87036 Rende (CS), Italy
fparisi@deis.unical.it
[2] University of Maryland College Park, College Park MD 20742, USA
vs@umiacs.umd.edu
[3] Northeastern University, Boston MA 02115, USA
asliva@ccs.neu.edu

Abstract. Though forecasting methods are used in numerous fields, we have seen no work on providing a general theoretical framework to build forecast operators into temporal databases. In this paper, we first develop a formal definition of a forecast operator as a function that satisfies a suite of forecast axioms. Based on this definition, we propose three families of forecast operators called *deterministic*, *probabilistic*, and *possible worlds* forecast operators. Additional properties of coherence, monotonicity, and fact preservation are identified that these operators may satisfy (but are not required to). We show how deterministic forecast operators can always be encoded as probabilistic forecast operators, and how both deterministic and probabilistic forecast operators can be expressed as possible worlds forecast operators. Issues related to the complexity of these operators are studied, showing the relative computational tradeoffs of these types of forecast operators. Finally, we explore the integration of forecast operators with standard relational operators in temporal databases and propose several policies for answering forecast queries.

1 Introduction

Though time series analysis methods have been studied extensively over the years in many contexts [3], there has only recently been work that merges classical forecasting with standard operations in temporal databases [1,5,6]. Given the widespread use of temporal data, there are numerous applications that require such capabilities, allowing for the consistent use and application of time series forecasts in databases. A university might want to forecast research grant income (or expenditures) in the future by examining a database of research projects. A stock market firm might want to include support for various kinds of specialized forecasting algorithms that predict the values of mutual fund portfolios or a single stock over time. A government might want to forecast the number of electricity connections or other development indicators in their country over time. Such forecasts might not just be made about the future, but also used to fill in gaps about the past. For instance, using data about the number of electricity connections in Ecuador from 1990–2000 and 2002–2007, officials may want to interpolate the number of connections there might have been in 2001.

This paper is not about how to make such forecasts. Currently, in all forecasting applications, the model building and forecasting is performed outside of the database

S. Benferhat and J. Grant (Eds.): SUM 2011, LNAI 6929, pp. 373–386, 2011.
© Springer-Verlag Berlin Heidelberg 2011

system itself, rather than as a smoothly integrated process. The implementation of these forecasting models are often ad hoc in nature, and general relationships between different forecasting tasks and domains are not exploited to their full potential. Yet, the broad demand for forecasting and predictive analyses creates a need for a robust theoretical framework that can incorporate forecasting directly into temporal databases.

The field of forecasting is extensive and widely studied, with an array of general techniques [3] as well as specialized forecast models for a variety of domains, such as finance [15], epidemiology [9], politics [2,10,11,14], and even product liability claims [13]. All these methods are very different from one another, and even within a restricted domain such as the stock market, there are hundreds of forecasting models available, each with varying strengths and weaknesses. In spite of these variations, we can identify general properties of forecasting that will facilitate integration of these methods into query languages, making them available for managing and analyzing temporal databases.

In this paper, the question *"what should count as a forecast operator"* is answered by first providing a set of axioms that a forecast operator must satisfy. We assume that forecast operators apply to temporal relational databases—the main reason for this assumption is that in today's world, most (though certainly not all) temporal data is in fact stored in such databases. Subsequently, we define three classes of forecast operators—deterministic forecast (DF) operators, probabilistic forecast (PF) operators, and possible worlds forecast (PWF) operators. We show that DF operators are a special case of PF operators which in turn are a special case of PWF operators. Certain classical forecasting methods such as linear regression, polynomial regression, and logistic regression methods are all demonstrated to be special cases of this framework. Some new operators for forecasting are also developed, along with results characterizing the complexity of applying certain forecast operators. This generalized understanding of the properties and relationships of forecast operators will allow such forecasts to be incorporated into temporal databases in a consistent way, as well as provide possible transformations for choosing the best operator for a particular application.

The remainder of this paper is organized as follows. Section 2 contains two motivating examples—one about forecasting academic grant incomes, and another about electricity connections in developing countries based on real data from the World Bank. Section 3 introduces basic notation for temporal databases. Section 4 provides an axiomatic definition of a forecast operator and then defines the classes of DF, PF and PWF forecast operators. This section also develops theorems showing relationships between DF, PF and PWF operators and the complexity of specific types of operator constructions. In Section 5, query answering mechanisms are presented that incorporate forecast operators into the standard relational algebra. Finally, related work and conclusions are given in Section 6.

2 Motivating Examples

Two motivating examples are used throughout this paper. The *grants* example specifies the total dollar amount ("Amount") of grants and number of employees ("Employees") of a Math and a CS department. Here, we are interested in predicting both of these attributes. The *electricity* example is drawn from real World Bank[1] data about the total expenditures ("Expend") on electricity and the number of electricity connections ("Connections") in some Latin American countries. Here, we wish to forecast the number of electricity connections and the amount of total expenditures (which includes operating costs and capital investment).

	Year	Dept	Amount	Employees
t_1	2000	CS	6M	70
t_2	2001	CS	6.2M	70
t_3	2002	CS	7M	75
t_4	2003	CS	6M	75
t_5	2004	CS	7.3M	74
t_6	2005	CS	9M	80
t_7	2000	Math	1M	71
t_8	2001	Math	1.1M	74
t_9	2002	Math	1M	73
t_{10}	2003	Math	0.5M	66
t_{11}	2004	Math	1.5M	79
t_{12}	2006	Math	1.2M	77

The *grants* relation

	Year	Country	Connections	Expend
e_1	2000	Brazil	48,000,000	6.8B
e_2	2001	Brazil	50,200,000	7.5B
e_3	2002	Brazil	52,200,000	6.9B
e_4	2003	Brazil	53,800,000	6.3B
e_5	2004	Brazil	56,300,000	7.7B
e_6	2005	Brazil	57,900,000	10.7B
e_7	2000	Venezuela	4,708,215	7.7B
e_8	2001	Venezuela	4,877,084	5.2B
e_9	2002	Venezuela	4,998,433	4.3B
e_{10}	2003	Venezuela	5,106,783	3.3B
e_{11}	2004	Venezuela	5,197,020	3.1B
e_{12}	2005	Venezuela	5,392,500	3B

The *electricity* relation

3 Basic Notation

The forecasting framework discussed in this paper applies only to temporal databases. Therefore, some basic temporal database (DB) notation is introduced in this section. We assume the existence of a finite set **rel** of relation names, and a finite set **att** of attribute names, disjoint from **rel**. A *temporal relation schema* will be denoted as $\mathcal{S}(A_1, \ldots, A_{n-1}, A_T)$ where $\mathcal{S} \in$ **rel** and $A_1, \ldots, A_{n-1}, A_T \in$ **att**. Each attribute $A \in$ **att** is typed and has a domain $dom(A)$. Assume the existence of a special attribute A_T denoting time whose domain $dom(A_T)$ is the set of all integers (positive and negative). Also assume that each attribute is either a *variable* or *invariant* attribute. Invariant attributes do not change with time, while variable attributes might. In *grants*, "Dept" is an invariant attribute, while "Amount" and "Employees" are variable attributes. In *electricity* , "Country" is invariant, while "Connections" and "Expend" are variable.

A *temporal tuple* over $\mathcal{S}(A_1, \ldots, A_{n-1}, A_T)$ is a member of $dom(A_1) \times \cdots \times dom(A_{n-1}) \times dom(A_T)$. A *temporal relation instance* R over the relation schema $\mathcal{S}$ is a set of tuples over $\mathcal{S}$.

Given a tuple t over $\mathcal{S}(A_1, \ldots, A_n)$, we use $t[A_i]$ (where $i \in [1...n]$) to denote the value of attribute A_i in tuple t. We use *Attr*$(\mathcal{S})$ to denote the set of all attributes in $\mathcal{S}$. Given a relation schema $\mathcal{S}$, we say that schema $\mathcal{S}_e$ is an *extension* of schema $\mathcal{S}$, denoted $\mathcal{S}_e \supseteq \mathcal{S}$ iff *Attr*$(\mathcal{S}_e) \supseteq$ *Attr*$(\mathcal{S})$.

[1] Benchmarking Data of the Electricity Distribution Sector in the Latin America and Caribbean Region 1995-2005. Available at: http://info.worldbank.org/etools/lacelectricity/home.htm

Throughout the rest of the paper, we will abuse notation and write $\mathcal{S}(A_1, \ldots, A_n)$ *instead of* $\mathcal{S}(A_1, \ldots, A_{n-1}, A_T)$*, simply assuming that the last attribute in any schema is the time attribute.*

Definition 1 (Equivalence of tuples). *Let* R *be a temporal relational instance* R *over schema* $\mathcal{S}$, $\mathcal{A} \subseteq Attr(\mathcal{S})$ *a set of attributes of* $\mathcal{S}$*, and* t_1, t_2 *tuples over* $\mathcal{S}$. $t_1 \sim_{\mathcal{A}} t_2$ *iff for each* $A_i \in \mathcal{A}$, $t_1[A_i] = t_2[A_i]$*. It is easy to see that* $\sim_{\mathcal{A}}$ *is an equivalence relation—we define a* cluster *for relation* R *w.r.t. the set of attributes* $\mathcal{A}$ *to be any equivalence class under* $\sim^*_{\mathcal{A}}$*, and* $clusters(R, \mathcal{A})$ *denotes the set of clusters of* R *w.r.t.* $\mathcal{A}$*.*

The following example shows clusters w.r.t. the *grants* and *electricity* examples.

Example 1. Consider the *grants* relation and suppose $\mathcal{A} = \{Dept\}$. Then $clusters(grants, \{Dept\})$ contains two clusters $\{t_1, \ldots, t_6\}$ and $\{t_7, \ldots, t_{12}\}$. On the other hand, if the *electricity* relation and the invariant set $\mathcal{A} = \{Country\}$ are considered, there are again two clusters ($\{e_1, \ldots, e_6\}$ and $\{e_7, \ldots, e_{12}\}$) in $clusters(electricity, \{Country\})$.

4 Forecast Operator

In this section, we formally define a generic *forecast operator* for any temporal DB and identify several families of forecast operators. Intuitively, a forecast operator must take as input some historical information and a time period for which to produce a forecast, which might include the future as well as past times where data is missing. The output of a forecast operator, however, can vary dramatically in form. For instance, forecasts can contain a single unambiguous prediction (called deterministic forecasts), or a single probabilistic forecast (called a probabilistic forecast), or a set of possible situations (called a possible worlds forecast). For each of these "types" of forecasts, the content can vary widely as well. The following definition accounts for all of these classes of forecasts, but requires that they satisfy specific desired properties.

Definition 2 (Forecast Operator). *Given a temporal relation instance* R *over the schema* $\mathcal{S}$ *and a temporal interval* $\mathcal{I}$ *defined over* $dom(A_T)$*, a* forecast operator ϕ *is a mapping from* R *and* $\mathcal{I}$ *to a set of relation instances* $\{R_1, \ldots, R_n\}$ *over a schema* $\mathcal{S}_e \supseteq \mathcal{S}$ *satisfying the following axioms:*

Axiom A1. *Every tuple in each* R_i *(*$i \in [1..n]$*) has a timestamp in* $\mathcal{I}$. This axiom says that the forecast operator only makes predictions for the time interval $\mathcal{I}$.

Axiom A2. *For each relation* R_i *(*$i \in [1..n]$*) and for each tuple* $t \in R$ *such that* $t[A_T] \in \mathcal{I}$*, there is exactly one tuple* $t_i \in R_i$ *such that* $\forall A \in Attr(\mathcal{S})$, $t[A] = t_i[A]$. This axiom says that tuples of R having a timestamp in $\mathcal{I}$ are preserved by the forecast operator (though they can be extended to include new attributes of the schema $\mathcal{S}_e$).

Axiom A3. *For each timestamp* $ts \in \mathcal{I}$ *and tuple* $t \in R$*, there is relation* R_i *with* $i \in [1..n]$ *containing the (forecasted) tuple* t' *such that* $t'[A_T] = ts$ *and* $t' \sim_{\mathcal{A}} t$ *where* $\mathcal{A} \subseteq Attr(\mathcal{S})$ *is a set of invariant attributes.* This axiom says that the forecasting is complete with respect to the timestamps in $\mathcal{I}$ and original tuples in R.

Note that axioms (A1) to (A3) above are not meant to be exhaustive. They represent a minimal set of conditions that any forecast operator should satisfy. Specific forecast operators may satisfy additional properties. In addition, we reiterate that the temporal interval $\mathcal{I}$ in the above definition can represent both the future and the past, i.e., it can include times that follow and/or precede those in relation R.

Forecast operators may satisfy the following additional properties; however, they are not mandatory for definition as an operator.

Definition 3 (Coherence). *Suppose R is a temporal relational instance over a temporal relational schema $\mathcal{S}$, $\mathcal{I}$ is a temporal interval, and $\mathcal{A}$ a set of invariant attributes. A forecast operator ϕ is* coherent *w.r.t. $\mathcal{A}$ iff for each $R_i \in \phi(R, \mathcal{I}) = \{R_1, R_2, \ldots, R_n\}$, there is a bijection $\beta_i : clusters(R, \mathcal{A}) \rightarrow clusters(R_i, \mathcal{A})$ such that for each $cl \in clusters(R, \mathcal{A})$, it is the case that $\phi(cl, \mathcal{I}) = \{\beta_1(cl), \beta_2(cl), \cdots, \beta_n(cl)\}$.*

Basically, a forecast operator ϕ is coherent w.r.t. a set of attributes $\mathcal{A}$ if the result of applying ϕ on the whole relation R is equivalent to the union of the results obtained by applying ϕ on every single cluster in $clusters(R, \mathcal{A})$. For instance, consider the *electricity* example and $\mathcal{A} = \{Country\}$. In this case, a coherent forecast operator says that the number of electricity connections and the amount of expenditures in a country only depends on that country. Likewise, in the *grants* example with $\mathcal{A} = \{Dept\}$, using a coherent forecast operator implies that the amount of grants and number of employees only depend upon the department. Forecast operators are not required to be coherent because this property may not always be valid in all applications. For instance, there may be a correlation between grant amounts in the CS and Math departments (e.g., decreases in NSF funding may affect both of them proportionately). As a consequence, if the *grants* relation had an additional tuple t_{13} with information on the 2007 grant income of Math, then this may be relevant for a forecast about CS's grant income in 2007, but the coherence assumption would not allow this dependency. As such, coherence is not considered a basic forecast axiom.

Another property that forecast operators may satisfy (but are not required to) is *monotonicity*. Given a relation R, two disjoint sets $\mathcal{A}, \mathcal{B}$ of attributes[2], and two clusters $cl_1, cl_2 \in clusters(R, \mathcal{A})$, we say that $cl_1 <_{\mathcal{B}} cl_2$ iff $\forall t_1 \in cl_1, t_2 \in cl_2, B \in \mathcal{B}$ it is the case that $t_1[B] \leq t_2[B]$ We now use this ordering to define monotonicity.

Definition 4 (Monotonicity). *Let R be a temporal relational instance over a schema $\mathcal{S}$, $\mathcal{I}$ a temporal interval, and $\mathcal{A}, \mathcal{B} \subseteq Attr(\mathcal{S}) \setminus A_T$ two disjoint sets of attributes. A forecast operator ϕ is* monotonic *w.r.t. the pair $\langle \mathcal{A}, \mathcal{B} \rangle$ iff for each $R_i \in \phi(R, \mathcal{I})$, there is a bijection $\beta_i : clusters(R, \mathcal{A}) \rightarrow clusters(R_i, \mathcal{A})$ such that:*

(i) *$\forall\, cl \in clusters(R, \mathcal{A})$, $cl \sim_{\mathcal{A}} \beta_i(cl)$ (i.e., $\forall\, t_1 \in cl, t_2 \in \beta_i(cl), A \in \mathcal{A}$ it is the case that $t_1[A] = t_2[A]$); and*
(ii) *$\forall\, cl_1, cl_2 \in clusters(R, \mathcal{A})$ such that $cl_1 <_{\mathcal{B}} cl_2$, it is the case that $\beta_i(cl_1) <_{\mathcal{B}} \beta_i(cl_2)$.*

A forecast operator is monotonic if trends of attributes in $\mathcal{B}$ in the clusters w.r.t. $\mathcal{A}$ of the original relation R are preserved by the clusters w.r.t. $\mathcal{A}$ in the predicted relations $R_1, R_2, \ldots, R_n$. In the rest of this section, we study three families of forecast operators — deterministic forecasts, probabilistic forecasts, and possible world forecasts.

[2] An ordering of $Dom(B)$ for each $B \in \mathcal{B}$ is assumed.

4.1 Deterministic Forecast Operator

A deterministic forecast operator is one that returns a single relation with exactly the same schema as the input relation.

Definition 5 (Deterministic Forecast Operator). *Given a temporal relation R over the schema $\mathcal{S}$ and a temporal interval $\mathcal{I}$, a* deterministic forecast operator *(DF operator for short) δ is a forecast operator such that $\delta(R, \mathcal{I}) = \{R'\}$ with R' defined over $\mathcal{S}$.*

DF operators can be built on top of any standard time series forecast algorithm. The following example shows how simple linear regression is an instance of the class of deterministic forecast operators.

Example 2. Suppose (w.r.t. the *electricity* example) we want to forecast the amount of connections and expenditures in 2006 and 2007 using simple linear regression[3]. The function *LINREG*$(R, \mathcal{I})$ applies linear regression to each variable attribute in relation R for time interval $\mathcal{I}$. The result of *LINREG*$(electricity, [2006, 2007])$ is given below:

Year	Country	Connections	Expend
2006	Brazil	60,006,666.67	9.6B
2007	Brazil	61,989,523.81	10.157B
2006	Venezuela	5,495,630.8	1.353B
2007	Venezuela	5,623,904.6	0.473B

LINREG$(R, \mathcal{I})$ is an example of a DF operator, as it maps $electricity$ and a time interval $\mathcal{I}$ to the single relation $electricity' =$ *LINREG*$(electricity, [2006, 2007])$. In this example, *LINREG*$(R, \mathcal{I})$ also satisfies coherence w.r.t. the set $\mathcal{A} = \{Country\}$ and monotonicity w.r.t. the pair $\langle\{Country\}, \{Connections\}\rangle$.

4.2 Probabilistic Forecast Operator

Deterministic forecasts are 100% certain in their forecasts. In contrast, probabilistic forecasts also include information about the probability that a forecast is correct.

Definition 6 (Probabilistic Forecast Operator). *Given a temporal relation instance R over the schema $\mathcal{S}$ and a temporal interval $\mathcal{I}$, a* probabilistic forecast operator *(PF operator for short) μ is a forecast operator such that $\mu(R, \mathcal{I}) = \{R'\}$ with R' defined over the schema $\mathcal{S}' = Attr(\mathcal{S}) \cup \{P\}$ where $dom(P) = [0, 1]$.*

PF operators are just like DF operators except they have an additional probability attribute P. Each tuple returned by a PF operator includes the probability of that tuple being valid at the timestamp (associated with that tuple). Basically, the result of applying a PF operator can be seen as a probabilistic database [4] with tuple-level uncertainty[4]. In addition to the general axioms (A1)–(A3), we often want PF operators to satisfy a property called fact preservation.

3 The same method shown in this example would allow us to use a variety of other traditional forecasting methods, such as logistic regression, nonlinear regression, etc.

4 Extending the framework to the case of forecast operators dealing with attribute-level uncertainty is left as future work.

Property 1 (Fact Preservation). Let R be a temporal relational instance over $\mathcal{S}$ and $\mathcal{I}$ a temporal interval. PF operator μ *preserves facts* of R if for each tuple $t \in R$ such that $t[A_T] \in \mathcal{I}$, there is a tuple $t' \in R'$ with $R' \in \mu(R, \mathcal{I})$ such that $\forall A \in Attr(\mathcal{S})$, $t[A] = t'[A]$ and $t'[P] = 1$.

Axiom (A2) ensures that tuples having a timestamp in $\mathcal{I}$ are preserved by the forecast operator, i.e., for each tuple $t \in R$ such that $t[A_T] \in \mathcal{I}$ there is a certain tuple $t' \in R'$ such that t and t' have the same values in the attributes in $Attr(\mathcal{S})$. This property strengthens axiom (A2) for PF operators since it requires the additional condition that the probability values of the tuples in the resulting relation R' corresponding to those of R (preserved tuples) must be exactly 1.

The fact preservation property should be satisfied by a PF operator when the user trusts what is in the database; in other cases when the user does not trust the content of a database, he may choose to use a PF operator that does not guarantee fact preservation.

Example 3. Consider the *grants* relation. Suppose we want to forecast the amount of grants and employees for the CS and Math departments for 2006 and 2007, along with their probabilities. We may choose to apply a polynomial regression method $P\_REG(R, \mathcal{A}, \mathcal{I})$, to variable attributes in each cluster in relation R w.r.t. $\mathcal{A}$ for a time interval $\mathcal{I}$. $P\_REG(R, \mathcal{A}, \mathcal{I})$ is an operator that computes the probability that the actual value will be within one standard deviation of the forecasted value, based on a normal distribution. Assuming independence, the probability of the entire tuple is the product of the probabilities for the individual attributes.

$P\_REG(R, \mathcal{A}, \mathcal{I})$ is an example of a PF operator. It first computes the forecasted values for each cluster:

Year	Dept	Amount	Employees
2006	CS	6.929471566	74
2007	CS	6.932925939	74
2006	Math	1.051905341	73
2007	Math	1.052429721	74

The probability of each forecasted value is computed as mentioned above:

CS: $P(Amount = 6.929471566 \pm \sigma | Year = 2006) = 0.68266$
$P(Amount = 6.932925939 \pm \sigma | Year = 2007) = 0.68264$
$P(Employees = 74 \pm \sigma | Year = 2006) = 0.68268$
$P(Employees = 74 \pm \sigma | Year = 2007) = 0.68268$

Math: $P(Amount = 1.051905341 \pm \sigma | Year = 2006) = 0.68268$
$P(Amount = 1.052429721 \pm \sigma | Year = 2007) = 0.68267$
$P(Employees = 73 \pm \sigma | Year = 2006) = 0.68141$
$P(Employees = 74 \pm \sigma | Year = 2007) = 0.6776$

The final relation, $grants'$ is shown below:

Year	Dept.	Amount	Employees	Prob
2006	CS	6.929471566	74	0.46604
2007	CS	6.932925939	74	0.46603
2006	Math	1.051905341	73	0.46519
2007	Math	1.052429721	74	0.46258

It is clear that every deterministic forecast can be expressed as a probabilistic forecast. Given a DF δ, a temporal relation instance R over schema $\mathcal{S}$, and a time period $\mathcal{I}$, we can define a simple probabilistic forecast operator $\mu^{simp,\delta}(R, \mathcal{I})$ to return $\{(t, 1) \mid t \in R'\}$ where $\delta(R, \mathcal{I}) = \{R'\}$.

Theorem 1. *Suppose δ is a DF operator. Then, the following relationships are true:*

(i) $\mu^{simp,\delta}$ *is a probabilistic forecast operator.*
(ii) *If δ is coherent w.r.t. $\mathcal{A}$ (resp. monotonic w.r.t. pair $\langle \mathcal{A}, \mathcal{B} \rangle$), then $\mu^{simp,\delta}$ is coherent w.r.t. $\mathcal{A}$ (resp. monotonic w.r.t. pair $\langle \mathcal{A}, \mathcal{B} \rangle$).*
(iii) $\mu^{simp,\delta}$ *is fact-preserving.*

4.3 Possible Worlds Forecast Operator

Probabilistic forecasts still only give one value for the attributes being forecasted per time period. However, in general, there may be many possible instances of relation R at a future (or past) time point t. Possible worlds forecasts try to return not one instance as the output of a forecast, but a set of relations, each of which is a possible instance of the relation at the time being forecast.

Definition 7 (Possible Worlds Forecast Operator). *Given a temporal relation instance R over the schema $\mathcal{S}$ and a temporal interval $\mathcal{I}$, a* possible worlds forecast operator *(PWF operator for short) ω is a forecast operator such that $\omega(R, \mathcal{I}) = \{R_1, \ldots, R_n\}$ where each R_i is defined over $\mathcal{S}$ and has probability value $\mathbf{P}(R_i)$ such that (i) $\mathbf{P}(R_i) > 0$ and (ii) $\sum_{i \in [1..n]} \mathbf{P}(R_i) = 1$.*

Basically, every resulting relation instance R_i represents a possible forecasted world. Observe that axiom (A2) entails that every world includes the tuples representing facts in the temporal interval $\mathcal{I}$ that were assumed to be true in the original relation R.

Given any DF operator δ, we can define a PWF operator ω^{δ}. One possible method called the *discretized PWF w.r.t. δ*, denoted $\omega^{disc,\delta}$, is given below. Suppose R is a temporal relation over schema $\mathcal{S}$ and $\mathcal{I}$ is a temporal interval; $\omega^{disc,\delta}$ is defined as:

1. Let R' be the relation returned by $\delta(R, \mathcal{I})$. Consider each tuple $t \in R'$. For each variable attribute $A \in Attr(\mathcal{S})$, define $\mathbf{P}(\lfloor t[A] \rfloor) = \lceil t[A] \rceil - t[A]$ and $\mathbf{P}(\lceil t[A] \rceil) = 1 - \mathbf{P}(\lfloor t[A] \rfloor)$. The set of *tuple worlds* $tw(t)$ associated with any tuple $t \in R'$ is now defined to be:
   (a) $tw(t) = \{t' \mid$ for all variable attributes $A \in Attr(\mathcal{S})$, $t'[A] = \lfloor t[A] \rfloor$ or $t'[A] = \lceil t[A] \rceil$ and for all invariant attributes $B \in Attr(\mathcal{S})$, $t[B] = t'[B]\}$.
   (b) $tw(t) = \{t\}$ if $t[A_T] \in \mathcal{I}$.
2. The probability of a tuple $t' \in tw(t)$ is defined to be the product of the probabilities of all the variable attribute elements of t', i.e., if $X \subseteq \mathcal{S}$ is the set of all variable attributes in the schema of R, then $\mathbf{P}(t') = \Pi_{A \in X} \mathbf{P}(t'[A])$. If $tw(t)$ coincides with t, then $\mathbf{P}(t) = 1$.
3. The set of *relation worlds* $rw(\delta, R, \mathcal{I})$ is now defined to be the Cartesian product of all tuple worlds, i.e., $\Pi_{t \in \delta(R,\mathcal{I})} tw(t)$. Each member of $rw(\delta, R, \mathcal{I})$ is called a *relation world.* The probability of a given relation world $w \in rw(\delta, R, \mathcal{I})$ is given by $\mathbf{P}(w) = \Pi_{t' \in w} \mathbf{P}(t')$[5].

[5] This assumes that the events represented by different tuples in $\delta(R, \mathcal{I})$ are independent of one another.

4. Return $rw(\delta, R, \mathcal{I})$ and the probability distribution $\mathbf{P}$ on $rw(\delta, R, \mathcal{I})$.

Theorem 2. *Suppose δ is any deterministic forecast operator. Then, the following relationships are true:*

(i) *$\omega^{disc,\delta}$ is a PWF operator.*
(ii) *If δ is coherent w.r.t. the set of attributes $\mathcal{A}$, then $\omega^{disc,\delta}$ is coherent w.r.t. $\mathcal{A}$.*

Example 4. Let us return to the *electricity* relation and consider using the simple linear regression $LINREG(electricity, [2006, 2006])$ for just the one year 2006. The result of this operator follows immediately from Example 2 and consists of the first and third tuple in the relation $electricity'$ of Example 2. For this relation, the construction $\omega^{disc,\delta}$ creates 16 possible relation worlds. The total number of connections in Brazil in 2006 could be 60,006,666 (33%) or 60,006,667 (67%), and the corresponding number in Venezuela could be 5,495,630 (20%) or 5,495,631 (80%). The possible expenditures in Brazil are 9B (40%) or 10B (60%), and in Venezuela are 1B (64.7 %) or 2B (35.3 %). The probability of each world is the product of the probabilities of the tuples selected. As an example, for world w given below, $P(w) = (0.33 * 0.6) * (0.8 * 0.647) = 0.102$.

Year	Country	Connections	Expend
2006	Brazil	60,006,666	10B
2006	Venezuela	5,495,631	1B

It is worth noting that, as both DF and PWF operators satisfy axiom (A2), the tuples of the original relation belonging to the predicted temporal interval are preserved by DF operator δ, and then preserved by PWF operator $\omega^{disc,\delta}$ as well.

The following example shows that $\omega^{disc,\delta}$ does not preserve monotonicity.

Example 5. Assume that for countries C_1 and C_2, electricity connections are almost the same in a given year, differing only in their decimal number, as shown below:

Year	Country	Connections
2005	C_1	50,900,800.4
2005	C_2	50,900,800.8

Relation el

Year	Country	Connections
2008	C_1	50,900,802.3
2008	C_2	50,900,802.9

Relation $\delta(el, [2008, 2008])$

Suppose the result of $\delta(el, [2008, 2008])$ is the relation given above. Clearly, δ is monotonic w.r.t. the pair $\langle\{Country\}, \{Connections\}\rangle$. In contrast, $\omega^{disc,\delta}$ is not monotonic w.r.t. $\langle\{Country\}, \{Connections\}\rangle$, since there is relation world $w = \{(2008, C_1, 50, 900, 803), (2008, C_2, 50, 900, 802)\}$ in $rw(\delta, el, [2008, 2009])$ for which the number of electricity connections of C_2 is not greater than that of C_1.

The $\omega^{disc,\delta}$ construction takes exponential time to enumerate the possible relation worlds and compute the associated probability distribution; the number of tuple worlds $tw(t)$ for a tuple t is exponential in the number of variable attributes, and the total number of relation worlds is exponential in the number of tuple worlds.

Theorem 3. *Suppose R is a temporal relation instance over schema $\mathcal{S}$, $\mathcal{I}$ is a temporal interval, and $\mathcal{A} \subset Attr(\mathcal{S})$ is a set of variable attributes. For any DF operator δ, the running time of $\omega^{disc,\delta}$ is $O(2^{|\mathcal{A}|\cdot|R'|})$, where R' is the relation returned by $\delta(R, \mathcal{I})$.*

From the possible relation worlds produced by $\omega^{disc,\delta}$, a user may only be interested in examining those relations that are sufficiently probable and contain a given tuple.

Proposition 1. *Suppose R is a temporal relation instance over schema $\mathcal{S}$, $\mathcal{I}$ is a temporal interval, and δ is a polynomial-time computable* DF *operator. Given a tuple t over the schema $\mathcal{S}$ and probability threshold k, deciding whether there is a relation world $w \in rw(\delta, R, \mathcal{I})$ such that $t \in w$ and $\mathbf{P}(w) \geq k$ (or $\mathbf{P}(w) \leq k$) is in $PTIME$.*

Proof (Sketch). Let R' be the relation returned by $\delta(R, \mathcal{I})$. First check if there is tuple $t' \in R'$ such that by rounding its value, for each variable attribute A, we obtain t. If no, the answer to our decision problem is "no." Otherwise, keep this tuple t' and find a relation world w_{max} with max probability, i.e., $\forall t'' \in R'$, $t'' \neq t'$ create a maximal tuple world by choosing $t''[A] = argmax\ \mathbf{P}(t''[A])$ for all variable attributes A. If $\mathbf{P}(w_{max}) \geq k$, then the answer is "yes."

We can also convert a PF operator μ to a PWF operator. Two possible mechanisms are provided below where R is a temporal relation and $\mathcal{I}$ is a temporal interval:

(i) $\omega^{simp,\mu}(R, \mathcal{I})$ returns just one world as follows. Suppose $\mu(R, \mathcal{I}) = \{R'\}$. Then $\omega^{simp,\mu}(R, \mathcal{I}) = \{\pi_{Attr(\mathcal{S})}(R')\}$. In other words, it eliminates the probability column in R'. This one world has probability 1 according to the PWF $\omega^{simp,\mu}$.

(ii) $\omega^{ind,\mu}(R, \mathcal{I})$ operates as follows:
1. Compute $\mu(R, \mathcal{I}) = \{R'\}$ as above.
2. Let $\mathcal{W}$ be the power set of $\pi_{Attr(\mathcal{S})}(R')$.
3. For each tuple t in a relation $R_i \in \mathcal{W}$, let $\mathbf{P}(t)$ be the probability attribute of the tuple in R' whose non-probability attributes are identical to those of t. The probability of a particular relation R_i in $\mathcal{W}$ is set to $\mathbf{P}(R_i) = \Pi_{t \in R_i} \mathbf{P}(t) \times \Pi_{t' \in \pi_{Attr(\mathcal{S})}(R') \setminus R_i}(1 - \mathbf{P}(t'))$.
4. Let $\mathcal{W}'$ be the set of relations $R_i \in \mathcal{W}$ such that $\mathbf{P}(R_i) > 0$. Return $\mathcal{W}'$ together with the above probability distribution on this set.

The following theorem shows a strong relationship between a PF operator μ and the PWF operator $\omega^{simp,\mu}$.

Theorem 4. *Suppose μ is any* PF *operator. Then the following relationships are true:*

(i) $\omega^{simp,\mu}$ *is a* PWF *operator.*

(ii) *If μ is coherent w.r.t. $\mathcal{A}$ (resp. monotonic w.r.t. $\langle \mathcal{A}, \mathcal{B} \rangle$), then $\omega^{simp,\mu}$ is also coherent w.r.t. $\mathcal{A}$ (resp. monotonic w.r.t. $\langle \mathcal{A}, \mathcal{B} \rangle$).*

The above theorem holds irrespective of whether the PF operator μ is fact preserving or not. In contrast, $\omega^{ind,\mu}$ will be a PWF operator only if constructed using a fact preserving PF operator. To see this, consider a relation R containing tuple t such that its timestamp $t[A_T]$ belongs to the temporal interval $\mathcal{I}$. If PF operator $\mu(R, \mathcal{I})$ forecasts t' whose invariant attributes are identical to those of t and its probability value is $\mathbf{P}(t') < 1$, then there is a possible world returned by $\omega^{ind,\mu}$ that does not contain any tuple having invariant attributes identical to those of t. Hence, A2 would be violated.

Theorem 5. *Suppose μ is any fact preserving PF operator. Then the following relationships are true:*

(i) *$\omega^{ind,\mu}$ is a PWF operator.*
(ii) *If μ is coherent w.r.t. $\mathcal{A}$ (resp. monotonic w.r.t. $\langle \mathcal{A}, \mathcal{B} \rangle$), then $\omega^{ind,\mu}$ is also coherent w.r.t. $\mathcal{A}$ (resp. monotonic w.r.t. $\langle \mathcal{A}, \mathcal{B} \rangle$).*

Theorem 6. *Suppose R is a temporal relation instance over schema $\mathcal{S}$ and $\mathcal{I}$ is a temporal interval. For any probabilistic operator μ, the running time complexity of $\omega^{ind,\mu}$ is $O(2^{|R'|})$, where R' is the relation returned by $\mu(R, \mathcal{I})$.*

We characterize the complexity of determining whether there is a possible world returned by $\omega^{ind,\mu}$ such that it is sufficiently probable and contains a tuple of interest t.

Proposition 2. *Suppose R is a temporal relation instance over schema $\mathcal{S}$, $\mathcal{I}$ is a temporal interval, and μ is a polynomial-time computable PF operator. Given a tuple t over the schema $\mathcal{S}$ and a probability threshold k, deciding whether there is a world w returned by $\omega^{ind,\mu}$ such that $t \in w$ and $P(w) \geq k$ (or $P(w) \leq k$) is in $PTIME$.*

Proof (Sketch). First check if $t \in \pi_S(R')$, where R' is the relation returned by $\mu(R, \mathcal{I})$. If $t \notin \pi_S(R')$, then it cannot belong to $\omega^{ind,\mu}$, thus the answer is 'no'. If $t \in \pi_S(R')$, then there is at least one possible world w that contains t. The possible world w_{max} (resp. w_{min}) that contains t is constructed using a strategy similar to that in the proof of Proposition 1. Finally, verify whether $P(w_{max}) \geq k$ (or $P(w_{min}) \leq k$).

5 Query Answering with Forecasting Operators

In this section, we study the relationship between forecast operators and standard relational algebra (RA) operators. We suggest adding new operators to the relational algebra to combine classical operators with the forecast operators presented here. Each RA operator can be augmented by forecast operators by either applying the forecast operators first and then applying the RA operator, or the other way around. Before formalizing this concept, we introduce two semantics for the evaluation of RA operators (these semantics are inspired by the notions of possible and certain answers introduced in [8]).

Definition 8 (Possibility and cautious semantics). *Given two sets of temporal relation instances S_1, S_2 whose elements are defined over the schemas $\mathcal{S}_1, \mathcal{S}_2$ respectively, and a binary relational algebra operator $op(\cdot, \cdot)$,*

(i) *the* possibility semantics *for op is the set* $op^{poss}(S_1, S_2) = \bigcup_{\substack{R_1 \in S_1 \\ R_2 \in S_2}} op(R_1, R_2)$
(ii) *the* cautious semantics *for op is the set* $op^{caut}(S_1, S_2) = \bigcap_{\substack{R_1 \in S_1, \\ R_2 \in S_2}} op(R_1, R_2)$

This definition can be straightforwardly extended to the case of unary RA operators.

Definition 9 (Forecast-first and forecast-last plans). *Given two temporal relation instances R_1, R_2 over the schemes $\mathcal{S}_1, \mathcal{S}_2$, respectively, a temporal interval $\mathcal{I}$, a forecast operator ϕ, a relational algebra operator op, and semantics $sem \in \{poss, caut\}$,*

(i) *a* forecast-first *plan is defined as*
$$\Phi_{forecast\text{–}first}(R_1, R_2, \mathcal{I}, \phi, op) = op^{sem}(\phi(R_1, \mathcal{I}), \phi(R_2, \mathcal{I}))$$
(ii) *a* forecast-last *plan is defined as*
$$\Phi_{forecast\text{–}last}(R_1, R_2, \mathcal{I}, \phi, op) = \phi(op^{sem}(\{R_1\}, \{R_2\}), \mathcal{I}) = \phi(op(R_1, R_2), \mathcal{I})$$

The latter equality in the forecast-last plan follows from the fact that $op^{sem}(\cdot, \cdot)$, with $sem \in \{poss, caut\}$, is equivalent to $op(\cdot, \cdot)$ if applied to singletons. A forecast-first plan returns a set of tuples, whereas a forecast-last plan returns a set of relations.

For some classes of forecast operators, these query policies satisfy some additional properties. The following proposition follows directly from the definition of possibility and cautious semantics for a given RA operator.

Proposition 3. *Let R_1, R_2 be temporal relation instances, $\mathcal{I}$ a temporal interval, and op an RA operator. For* DF *and* PF *operators ϕ, the forecast-first plans under possibility and cautious semantics are equivalent, that is, $op^{poss}(\phi(R_1, \mathcal{I}), \phi(R_2, \mathcal{I})) = op^{caut}(\phi(R_1, \mathcal{I}), \phi(R_2, \mathcal{I}))$.*

Depending on the particular query application, the basic forecast-first plan as given in Definition 9 can be further extended to allow for more flexibility in the forecast intervals and operators. Given temporal relation instances R_1, R_2 and RA operator *op*, then we can define the following variations of the forecast-first plan:

(i) **Multiple interval plan.** Consider two temporal intervals $\mathcal{I}_1, \mathcal{I}_2$, then a *multiple interval (MI) forecast-first* plan is defined as $\Phi_{MI}(R_1, R_2, \mathcal{I}_1, \mathcal{I}_2, \phi, op) = op(\phi(R_1, \mathcal{I}_1), \phi(R_2, \mathcal{I}_2))$. Here, two distinct forecasts are made using the intervals $\mathcal{I}_1$ and $\mathcal{I}_2$ before the RA operator *op* is applied.
(ii) **Multiple operator plan.** Given a temporal interval $\mathcal{I}$ and two forecast operators ϕ_1, ϕ_2, a *multiple operator (MO) forecast-first* plan is defined as $\Phi_{MO}(R_1, R_2, \mathcal{I}_1, \mathcal{I}_2, \phi_1, \phi_2, op) = op(\phi_1(R_1, \mathcal{I}), \phi_2(R_2, \mathcal{I}))$. In this plan, two different forecast operators are applied to the same interval, and the results are used by the RA operator.
(iii) **Hybrid plan.** Given two temporal intervals $\mathcal{I}_1, \mathcal{I}_2$ and two forecast operators ϕ_1, ϕ_2. A *hybrid forecast-first* plan is defined as $\Phi_{Hybrid}(R_1, R_2, \mathcal{I}_1, \mathcal{I}_2, \phi_1, \phi_2, op) = op(\phi_1(R_1, \mathcal{I}_1), \phi_2(R_2, \mathcal{I}_2))$. This plan combines the multiple interval and multiple operator forecast-first plans.

The remainder of this section will examine the relationships between forecast operators and some RA operators, providing results on the resulting extended relational operators that could, in principle, be used for query optimization. The result below states that, for specific kinds of selection conditions, using a forecast-first plan with possibility semantics will yield a superset of the result given by a forecast-last plan, while the cautious semantics will produce a subset.

Proposition 4. *Let R be temporal relation instance over the schema $\mathcal{S}$, $\mathcal{I}$ a temporal interval, ϕ a forecast operator coherent w.r.t. $\mathcal{A} \subseteq Attr(\mathcal{S})$, and C a selection condition filtering out whole clusters only (i.e., $\sigma_C(R) = \bigcup_{cl \in CL} cl$, where $CL \subseteq clusters(R, \mathcal{A})$). Then,*

(i) $\sigma_C^{poss}(\phi(R, \mathcal{I})) \supseteq R_i$ *where* $R_i \in \phi(\sigma_C(R), \mathcal{I})$

(ii) $\sigma_C^{caut}(\phi(R,\mathcal{I})) \subseteq R_i$ *where* $R_i \in \phi(\sigma_C(R),\mathcal{I})$

For DF and PF forecast operators, the possibility and cautious semantics coincide for forecast-first plans (Proposition 3). It then follows that, under the conditions specified above, $\sigma_C(\phi(R,\mathcal{I}))$ returns the same relation as $\phi(\sigma_C(R),\mathcal{I})$.

The interaction between forecast plans and projection RA operator is as follows.

Proposition 5. *Let* R *be temporal relation instance over the schema* $\mathcal{S}$, $\mathcal{I}$ *a temporal interval,* ϕ *a forecast operator, and* $\mathcal{A} \subseteq Attr(\mathcal{S})$ *invariant attributes of* $\mathcal{S}$. *Then,*

(i) $\pi_{\mathcal{A}}^{poss}(\phi(R,\mathcal{I})) \supseteq R_i$ *where* $R_i \in \phi(\pi_{\mathcal{A}}(R),\mathcal{I})$
(ii) $\pi_{\mathcal{A}}^{caut}(\phi(R,\mathcal{I})) \subseteq R_i$ *where* $R_i \in \phi(\pi_{\mathcal{A}}(R),\mathcal{I})$

Analogously to the selection RA operator, by Proposition 3 it follows that $\pi_{\mathcal{A}}(\phi(R,\mathcal{I}))$ coincides with the result of $\phi(\pi_{\mathcal{A}}(R),\mathcal{I})$ for DF or PF forecast operators. Also for the union RA operator, the relationship between forecast-first and forecast-last plans depends on the choice of possibility or cautious semantics.

Proposition 6. *Let* R_1, R_2 *be temporal relation instances over the schema* $\mathcal{S}$, $\mathcal{I}$ *a temporal interval, and* ϕ *a forecast operator coherent w.r.t.* $\mathcal{A} \subseteq Attr(\mathcal{S})$. *If* $\pi_{\mathcal{A}}(R_1) \cap \pi_{\mathcal{A}}(R_2) = \emptyset$, *then*

(i) $\phi(R_1,\mathcal{I}) \cup^{poss} \phi(R_2,\mathcal{I}) \supseteq R_i$ *where* $R_i \in \phi(R_1 \cup R_2,\mathcal{I})$
(ii) $\phi(R_1,\mathcal{I}) \cup^{caut} \phi(R_2,\mathcal{I}) \subseteq R_i$ *where* $R_i \in \phi(R_1 \cup R_2,\mathcal{I})$

As above, $\phi(R_1,\mathcal{I}) \cup \phi(R_2,\mathcal{I})$ is equal to $\phi(R_1 \cup R_2,\mathcal{I})$ for DF and PF operators.

6 Related Work and Conclusions

Though there are numerous works on forecasting in general [3], as well as specialized forecast models for specific domains, such as finance [15], epidemiology [9], or politics [2,10,11,14,12], all these methods vary dramatically from one another. With a large array of possible statistical models, one previous attempt to better understand the relationship between these forecasting procedures is given by [7], which integrates several forecasting methods into a common mathematical framework.

There has also been some recent work on the issue of forecasting queries in databases [1,5,6]. [5] describes the *Fa* data management system that provides support for declarative predictive queries over time series data, incorporating algorithms to effectively choose the best model type and attributes for the best query performance. Another Predictive DBMS is presented in [1] which also proposes a declarative forecasting query language, including the flexibility for both automated and user-defined predictive models. [6] investigates the I/O efficiency of forecasting queries, using a skip-list to index the time series and provide access to multiple regression models at varying levels of granularity. The model of forecasting presented in this paper differs from these prior efforts by focusing on general characteristics of forecasting rather than specific queries for a limited set of potential time series analysis methods. In fact, the framework given here can serve as a generalized, unifying theory for forecasting in databases that encompasses the semantics of these other approaches.

In this paper, we first provide axioms that any forecast operator should satisfy, together with additional desirable (but not required) properties. Our methods allow us to take classical forecasting operators and categorize forecast operators into three increasingly expressive categories and then embed them as operators in a temporal database: (i) deterministic forecast operators, (ii) probabilistic forecast operators, and (iii) possible worlds forecast operators. These classes of operators all satisfy our forecasting axioms, and in some cases, additional desirable properties. We have explored several policies for combining forecast and standard relational algebra operators to answer forecast queries and started a theoretical analysis on the interaction between these operators. Future work will focus on further investigating forecast policy w.r.t. relational algebra operators and exploiting these results as a basis for optimization of forecast queries. Though forecasting is often complex, we are able to prove that many of the techniques reported in this paper are tractable.

References

1. Akdere, M., Cetintemel, U., Riondato, M., Upfal, E., Zdonik, S.: The case for predictive database systems: Opportunities and challenges. In: Proceedings of the 5th Biennial Conference On Innovative Data Systems Research (2011)
2. Bond, J., Petroff, V., O'Brien, S., Bond, D.: Forecasting turmoil in indonesia: An application of hidden markov models. In: International Studies Association Convention, Montreal, pp. 17–21 (March 2004)
3. Bowerman, B., O'Connell, R., Koehler, A.: Forecasting, Time Series and Regression, 4th edn. Southwestern College Publishers (2004)
4. Dalvi, N.N., Suciu, D.: Management of probabilistic data: foundations and challenges. In: PODS, pp. 1–12 (2007)
5. Duan, S., Babu, S.: Processing forecasting queries. In: Proceedings of the 33rd International Conference on Very Large Databases (2007)
6. Ge, T., Zdonik, S.: A skip-list aproach for efficiently processing forecasting queries. In: Proceedings of the 34th International Conference on Very Large Databases (2008)
7. Harvey, A.C.: A unified view of statistical forecasting procedures. International Journal of Forecasting 3(3), 245–275 (1984)
8. Imielinski, T., Lipski, W.: Incomplete information in relational databases. J. ACM 31(4), 761–791 (1984)
9. Jewell, N.P.: Statistics of Epidemiology. Chapman & Hall/CRC (2003)
10. Martinez, M.V., Simari, G.I., Sliva, A., Subrahmanian, V.S.: Convex: Context vectors as a similarity-based paradigm for forecasting group behaviors. IEEE Intelligent Systems (2008)
11. Schrodt, P.: Forecasting conflict in the balkans using hidden markov models. In: Proc. American Political Science Association meetings (August 31 - September 3, 2000)
12. Sliva, A., Subrahmanian, V., Martinez, V., Simari, G.: Mathematical Methods in Counterterrorism. Springer, Heidelberg (2009)
13. Stallard, E.: Product liability forecasting for asbestos-related personal injury claims: A multidisciplinary approach. In: National Institute on Aging Conference: Demography and Epidemiology: Frontiers in Population Health and Aging, Washington, D.C (2001)
14. Subrahmanian, D., Stoll, R.: Events, patterns, and analysis. In: Programming for Peace: Computer-Aided Methods for International Conflict Resolution and Prevention. Springer, Heidelberg (2006)
15. Taylor, S.J.: Modelling Financial Time Series, 2nd edn. World Scientific Publishing Company, Singapore (2007)

Investigating Ontological Similarity Theoretically with Fuzzy Set Theory, Information Content, and Tversky Similarity and Empirically with the Gene Ontology

Valerie Cross[1] and Xinran Yu[2]

[1] Computer Science and Software Engineering Department,
Miami University, Oxford, OH 45056
crossv@muohio.edu
[2] Department of Computer Science,
University of Texas, San Antonio, TX 78249
xyu@cs.utsa.edu

Abstract. This paper theoretically and empirically investigates ontological similarity. Tversky's parameterized ratio model of similarity [3] is shown as a unifying basis of many of the well-known ontological similarity measures. A new family of ontological similarity measures is proposed that allows parameterizing the characteristic set used to represent an ontological concept. The three subontologies of the prominent GO are used in an empirical investigation of several ontological similarity measures. A new ontological similarity measure derived from the proposed family is also empirically studied. A detailed discussion of the correlation among the measures is presented as well as a comparison of the effects of two different methods of determining a concept's information content, corpus-based and ontology-based.

Keywords: Semantic similarity, ontological similarity, information content, Tversky's parameterized ratio model, Gene Ontology.

1 Introduction

In ontologies, similarity measurement is needed to determine how similar one concept is to another. An ontological similarity measure [1] is a semantic similarity measure specific to assessing similarity between concepts within an ontology. Such measures have seen a proliferation in the last several years, particularly in the biomedical and bioinformatics area [2]. Just recently more new ontological similarity measures have been proposed based on intuitive combination of information-theoretic measures with Tversky's model of similarity [3]. In [4] the contrast model is used and then modified in [5] to use the parameterized ratio model of similarity. These recently proposed intuitive models are not the first examination of integrating the information-theoretic model and the Tversky models of similarity [6][7].

As new ontological similarity measures are proposed, evaluations of them against existing ones have typically used one of three approaches: mathematical analysis, domain-specific applications of them, and comparison of them to human judgments of

S. Benferhat and J. Grant (Eds.): SUM 2011, LNAI 6929, pp. 387–400, 2011.
© Springer-Verlag Berlin Heidelberg 2011

similarity [8]. By far, the primary approach, however, has been to compare them to human similarity judgments. More recently, due to application of ontological similarity within the Gene Ontology (GO) [9] to determine gene product similarity, other physical similarity measures such as sequence similarity [10] are being used for performance comparisons. Some efforts have been made on mathematical analysis of ontological similarity measures [11] [12] [6].

This paper theoretically and empirically investigates ontological similarity. Tversky's parameterized ratio model of similarity [3] is shown as a unifying basis of many of the well-known ontological similarity measures. A new family of ontological similarity measures is proposed that allows parameterizing the characteristic set used to represent an ontological concept. The three subontologies of the prominent GO are used in an empirical investigation of several ontological similarity measures. A new ontological similarity measure derived from the proposed family is also empirically studied. A detailed discussion of the correlation among the measures is presented.

Section 2 briefly introduces similarity assessment and the role of Tversky's two models of similarity. Section 3 presents the components of the framework for ontological similarity and the theoretical investigation of how these various components can be used to create existing ontological similarity measures. In section 4 an empirical investigation uses the GO to compare two historical ontological similarity measures, the two recently proposed intuitive ontological similarity measures using Tversky's models, and a new ontological similarity systematically developed from the integration of fuzzy set compatibility measures, information content and Tversky's models. Section 5 provides a summary of this research, conclusions, and directions for future efforts.

2 Similarity Measurement

In the psychological literature two main approaches to assess similarity are content models and geometric or distance models. Distance models have been found to be contrary to human similarity judgments in psychological domains since satisfying the minimality, symmetry, and the triangle inequality axioms often conflicts with human similarity judgments and also such models require quantitative continuous dimensions where often human judgments of similarity use qualitative and discrete dimensions.

The content models use characteristics which are conceptualized "as more or less discrete and common elements" to determine the similarity between objects [13]. Many of the proposed set theoretic measures in the content model category are generalized by Tversky's parameterized ratio model of similarity [3]:

$$S_{Tverksy\text{-}ratio}(X, Y) = \frac{f(X \cap Y)}{f(X \cap Y) + \alpha f(X - Y) + \beta f(Y - X)}. \tag{1}$$

In the model, X and Y represent sets describing respective objects x and y. $(X \cap Y)$ represents the common features that describe both x and y. $(X - Y)$ represents the features describing only object x. $(Y - X)$ represents the features describing only object y. The value $f(X)$ for object x is considered a measure of the overall salience of that object. Factors adding to an object's salience include "intensity, frequency, familiarity, good form, and informational content" [3]. The function f is an additive function on disjoint sets, i.e., whenever X and Y are disjoint sets and when all three terms are defined, then $f(X \cup Y) = f(X) + f(Y)$. Set cardinality is such a function.

Parameters α and β allow priority to be given to one object over the other, i.e., one object serves as the referent object to which the other object is being matched. If x is selected as the referent object, then α should be greater than β to emphasize that features describing x but not y are more important than those describing y but not x.

This measure is normalized so that $0 \leq S_{Tverksy\text{-}ratio}(X, Y) \leq 1$. With $\alpha = \beta = 1$, $S_{Tversky}$ becomes the Jaccard index [3]

$$S_{jaccard}(X, Y) = \frac{f(X \cap Y)}{f(X \cup Y)}. \tag{2}$$

With $\alpha = \beta = 1/2$, $S_{Tverksy\text{-}ratio}$ becomes Dice's coefficient of similarity [3]

$$S_{dice}(X, Y) = \frac{2 \times f(X \cap Y)}{f(X) + f(Y)}. \tag{3}$$

With $\alpha = 1$, $\beta = 0$, $S_{Tverksy\text{-}ratio}$ becomes the degree of inclusion for X, that is, the proportion of X overlapping with Y [3]. Inclusion is not symmetric since $\alpha \neq \beta$.

$$S_{inclusion}(X, Y) = \frac{f(X \cap Y)}{f(X)}. \tag{4}$$

Tversky [3] also proposed an unnormalized similarity model, the contrast model. It integrates the same components but uses different mathematical operations:

$$S_{Tverksy\text{-}contrast}(X, Y) = \theta f(X \cap Y) - \alpha f(X - Y) - \beta f(Y - X). \tag{5}$$

Goodman [14] argues that assessing similarity between objects *x* and *y* is vague and meaningless without a "frame of reference." He states " 'is similar to' functions as little more than a blank to be filled." Asking the question "How similar are *x* and *y*?" begs the answer to a subtly different question "How are *x* and *y* similar?" [15]. Thus, crucial to similarity assessment is the method to select on what features similarity is being judged. For ontological similarity, "filling in the blank" or feature selection for two concepts has varied greatly depending on the perceived objectives of the task.

3 Synthesizing Ontological Similarity

Early developers of ontological or semantic similarity measures incorporated the ontological structure into the similarity measures. How these measures relate to other existing similarity measures such as the Tversky models [3] or fuzzy set compatibility measures [16][17] was not explored. Historical ontological similarity measures are examples of such existing measures when one examines how the "blanks" are filled in, i.e., how the features are selected to describe a concept in ontology and how these features may have a fuzzy set-theoretic interpretation.

3.1 Information Content in Ontologies

Information content (IC) of a concept in an ontology is a measure of how specific the concept is. The more specific (general) it is, then the higher (lower) is its IC. The earliest method to measure IC uses an external resource such as an associated corpus for the problem domain. The corpus-based IC measure for concept *c* [18] is given as

$$IC_{corpus}(c) = -log\ p(c) \tag{6}$$

The value *p(c)*, the probability of concept c, is determined using the frequency count of the concept, i.e. the count of the number of occurrences of the concept within the corpus. The frequency of a concept is the number of occurrences in the corpus of all words representing that concept. The frequency of the concept also includes the total frequencies of all its children concepts. The taxonomic structure of the ontology, i.e., the is-a relationships, determine the children of a concept. In some ontologies such as the GO and WordNet, however, the part-of relationship is also used in the calculation of the IC value. The probability *p(c)* is calculated by dividing this total frequency count by the total number of words in the corpus. Because the formula is the negative logarithm of the probability, as the probability increases the information content decreases; therefore, concepts higher in the ontology which have a greater probability of occurring have less information content than those lower in the ontology.

A more recent method [19] uses the ontology structure and does not require an external resource, one of its advantages. Leaf concepts are most specific, they contain the most information. Root concepts are the least specific and contain the least information. The ontology-based IC [19] is defined as

$$IC_{ont}(c) = log\frac{(num\_desc(c)+1)}{max_{ont}} / log\frac{1}{max_{ont}} = 1 - \frac{log(num\_desc(c)+1)}{\log(max_{ont})} \tag{7}$$

where num_desc(c) is the number of descendants for concept *c* and max_{ont} is the maximum number of concepts in the ontology. It is normalized in [0...1] with the maximum for the leaf concepts and deceases to the minimum at the root concepts.

An extended Information Content (eIC) measure [5] uses other relationships between concepts, not only the taxonomic structuring ones historically used. The eIC(c) is a parameterized weighting of IC(c) and its total average relationship EIC(c). EIC(c) is the summation for each kind of relationship *k*, of the average of the IC(c_i) for all concepts c_i that are "at the end of a particular relation" [5], i.e., *k*, with concept *c*. How non-taxonomic relationships and their inverses are handled, i.e., how is "at the other end of a particular relation" is not clear. If inverse relationships are not ignored or inverse relationships are not clearly identified, a circular calculation of eIC could occur. Another concern is how and which non-taxonomic relationships are used.

3.2 A Synthesis of Fuzzy Set Compatibility Measures, Information Content, and Tverksy's Similarity Models

A concept within an ontology has many contexts and can be represented by its many different features, for example, its properties, its children, its parents, etc [24]. The three standard IC ontological similarity measures Resnik [18[, Lin [12], and Jiang-Conrath[20] use at most three ontological concepts within an ontology. Much more information, however, is conveyed in the structure of the ontology. Another view is to consider that each concept *c* can be represented by a fuzzy set. Which fuzzy set is used depends on how one proceeds in "filling in the blank." A variety of sets can be transformed into a fuzzy set description of concept *c* where the membership degrees are some function of the IC associated with each element in the set. For example, one fuzzy set describing the concept c uses its ancestor set and the concept itself as

$$F_{anc+(c)}(c_j) = \{IC(c_j)/c_j \mid c_j \text{ is an ancestor of c or c itself}\} \quad (9)$$

where the + indicates to include the concept *c* itself in the set. Here the function on IC is simply the identify function. This fuzzy set specifies each element c_j and its respective membership $IC(c_j)$ in the fuzzy set. Note either IC_{ont} or IC_{corpus} could be used. Here the set used to describe the concept *c* consists of other concepts, but the set could also be a set of links for a path associated with concept *c*.

Various methods to measure compatibility between fuzzy sets have been proposed [16] [17]. One of the most famous is Zadeh's partial matching index between two fuzzy sets F_1 and F_2:

$$S_{sup\text{-}min(}(F_1, F_2) = \sup \min(F_1(u), F_2(u)) \quad (10)$$

where sup selects the supremum membership degree from the intersection of the two fuzzy sets. An early IC ontological similarity measure [18] is formulated as

$$sim_{RES}(c1,c2) = max_{c \text{ in } S(c1,c2)} [IC(c)] \quad (11)$$

where $S(c1,c2)$ is the set of concepts that subsume both *c1* and *c2* and IC is determined using a corpus. It can be seen as the partial matching index given in (10) where $F_{anc+(c1)}$ and $F_{anc+(c2)}$ represent the fuzzy ancestor sets for *c1* and *c2* respectively in the ontology and max or sup is the scalar evaluator. The minimum operation between $F_{anc+(c1)}$ and $F_{anc+(c2)}$ serves as the intersection operator between the two fuzzy ancestor sets and produces the set of all common ancestors for *c1* and *c2*. The concept with the greatest membership degree, i.e., the maximum IC, in both fuzzy sets $F_{anc+(c1)}$ and $F_{anc+(c2)}$, therefore, represents the partial matching similarity between concepts *c1* and *c2*. The minimum intersection operation just takes the minimum between identical IC values for elements in both $F_{anc+(c1)}$ and $F_{anc+(c2)}$ since each concept has only one IC value within an ontology. Typically, Zadeh's partial matching or consistency measure assumes normalized fuzzy sets, i.e., each fuzzy set has one element with membership degree of 1. To satisfy this assumption, the membership degrees, i.e., IC values for each ancestor and *c1* can be divided by the IC value of *c1* and similarly for the ancestors of *c2*. This normalization results in different membership degrees for ancestor concepts in the two fuzzy sets. Then the minimum operator would be necessary when performing the intersection between the two fuzzy sets. Another option which follows the general formula for partial matching index [16] is to directly normalize the result of (11) by dividing it by the min(sup $F_{anc+(c1)}$, sup $F_{anc+(c2)}$) = min $(IC(c1), IC(c2))$. Since the original Resnik measure is used for the experiments described in Section 4, each concept within an ontology has only one IC value and no function is applied to modify the IC value used as a membership degree in a fuzzy set and no normalization of the Resnik value is performed. Other experiments are needed to see how such modifications might affect performance of a normalized partial matching measure.

A major criticism of the Resnik measure is that only the shared information between the two concepts is used and not their separate information. Lin [12] defined a measure to address this criticism. It is a normalized similarity measure related to Jiang and Conrath measure [20] and given as

$$sim_{Lin}(c1,c2)=\frac{2\times IC(c3)}{IC(c1)+IC(c2)} \quad (12)$$

where *c3* is the common ancestor with the greatest IC. The Dice similarity measure given in equation (3) is similar in form to the Lin measure in equation (12) if one simply substitutes $f(X) = IC(c)$ and interprets $f(X \cap Y)$ as $IC(c3)$. This approach has been proposed in [7] with the goal of integrating information-theoretic and set-theoretic similarity measures. However, the interpretation of Tversky's model is somewhat misleading in simply replacing *f* with IC and assuming that the common subsumer represents the intersection of the set of features of *c1* and the set of features of *c2*. What the sets of features are and how the IC measure provides an additive function as specified by Tversky's similarity model is not clearly described. The IC of a concept is based on a function of the probability of the frequencies of all its descendent concepts and not simply the intersection of common features of *c1* and *c2*.

In [6] Dice's coefficient given in equation (3) has been shown to be the basis for both the Wu-Palmer path-based semantic similarity measure [21] given as

$$sim_{Wu\text{-}Palmer}(c1,c2)=\frac{2\times len(root,c3)}{len(c1,root)+len(c2,root)} \quad (13)$$

and the Lin information-content based semantic similarity measure. Dice's coefficient establishes the connection between the Lin and Wu-Palmer measures.

Many fuzzy set compatibility measures are generalizations of Tverksy's parameterized ratio model when fuzzy set cardinality, an additive set function, is used for *f*, X and Y are replaced by fuzzy sets F_1 and F_2, and the crisp set operators are transformed to the corresponding fuzzy set operators. The Jaccard fuzzy set compatibility measure can be used to calculate IC ontological similarity between two concepts represented by their fuzzy sets of ancestors $F_{anc+(c1)}$ and $F_{anc+(c2)}$. The set intersection uses a t-norm, typically minimum, and the set union uses a t-co-norm, typically maximum. The Jaccard ontological similarity measure with the anc+ set to fill in the blank then becomes

$$sim_{JacAnc}(c1,\, c2) = \frac{\sum_{c\in F_{anc+(c1)} \cap F_{anc+(c2)}} IC(c)}{\sum_{c\in F_{anc+(c1)} \cup F_{anc+(c2)}} IC(c)} \quad (14)$$

Again the min and max fuzzy set operators do not need to be explicitly used because the membership degrees of the ancestors in the fuzzy sets representing both concepts *c1* and *c2* are simply the ancestor's IC value. The IC value of the concept in each fuzzy set is the same since it is a function of its number of descendents. As previously suggested, however, these IC values could be normalized so that the membership degree of an ancestor in each concept's fuzzy set could differ. For this approach, the min and max operators would then be needed since the IC membership degrees then differ. Ontological similarity could also be modified by describing a concept using a different set; for example, instead of the ancestor set to describe the concept, the descendent set could be used to describe each concept.

A wide variety of fuzzy set compatibility measures, many of which are fuzzy generalizations derived from Tversky's parameterized ratio model of similarity, can be and have been used as ontological similarity measures. With this model, there still can be variations depending on how each researcher decides to approach "filling in the blank", i.e., the objectives that determine exactly what set is used to describe the concept and what method is used to assign the membership degrees in the set.

3.3 Recent Ontological Similarity Based on Intuitive Uses of Tversky's Models

New ontological similarity measures proposed use an intuitive interpretation of Tversky's contrast and parameterized ratio models. The assumption is the function *f* in Tversky's models is simply replaced with an IC measure [4][5]. This assumption was used in [7] to show that the Lin semantic similarity measure is an example of Tversky's parameterized ratio model. This interpretation of *f* differs from the fuzzy set theoretic and information-content modeling of ontological similarity in [6] and presented in 3.2 where a concept is directly described by a set of elements and a function of IC is used as the membership degree of the element in the fuzzy set.

The P&S semantic similarity measure [4] uses the contrast model to formulate

$$sim_{P\&S}(c1,c2) = 3\times(IC(c3)) - IC(c1) - IC(c2) \text{ if } c1\neq c2, = 1 \text{ if } c1 = c2 \qquad (15)$$

where *c3* is the common ancestor concept with the greatest IC. It is argued that this formula represents the information theoretic counterpart of Tversky's set theoretic contrast model. The assumption made is that $f(X \cap Y)$ is the same as $IC(c3)$ where X represents a set of features describing *c1* and Y represents a set of features describing c2. Similarly, $f(X - Y)$ and $f(Y - X)$ map to $(IC(c1) - IC(c3))$ and $(IC(c1) - IC(c3))$ respectively. The parameters are set as $\theta=\alpha=\beta=1$ to produce

$$IC(c3) - (IC(c1) - IC(c3)) - (IC(c2) - IC(c3)) \qquad (16)$$

which simplifies to the measure given in equation (15) when $c1\neq c2$. When $c1=c2$, then $IC(c1)=IC(c2)=IC(c3)$; therefore, the result of equation (16) is $IC(c3)$ so they assign the value of 1 instead if the two concepts are the same.

In [22] an investigation of the P & S measure was done mathematically and empirically and showed that the P&S measure can produce negative values which cause difficulties in understanding the similarity values. Also, the measure sum_{ps} had the worst correlation with the historical ontological similarity measures in a majority of the performed experiments.

In [5] another ontological similarity measure is proposed. The FaITH (Feature and Information Theoretic) measure still assumes *f* as simply the *IC* measure and uses the same mappings for $f(X \cap Y)$, $f(X - Y)$, and $f(Y - X)$ as fro $sim_{P\&S}$. The only change is to use Tversky's parameterized ratio model of similarity instead of the contrast model. These mappings are substituted into equation (1) with $\alpha=\beta=1$ to produce

$$sim_{FaITH}(c1,c2)=\frac{IC(c3)}{IC(c1)+IC(c2)-IC(c3)} \qquad (17)$$

which is very similar to the sim_{Lin} measure in equation (12). Subtracting $IC(c3)$ from both the numerator and denominator of equation (12) produces the FaITH measure. With this modification, the sim_{FaITH} measure must always produce a smaller than or equal to (only when both are 0) value compared to sim_{Lin}. As IC(c3) approaches 0, the ratio of sim_{FaITH} / sim_{Lin} approaches 0.5. There is a strong correlation between these two measures as the empirical investigations show in the next section.

In both $sim_{P\&S}$ and sim_{FaITH}, the key assumption is that an information theoretic view of Tversky's models of similarity is possible by the simple substitution of IC for the function *f*. The $f(X \cap Y)$ which is a measure of the amount of shared elements, i.e., features, between sets *X* and *Y* in set-theoretic model is simply replaced by *IC(c3)*, the amount of shared information for *c1* and *c2*, i.e., the IC of their most specific ancestor concept *c3*. A difficulty with this interpretation is the difference between a set of features representing the intersection of two sets such as the shared set of properties between *c1* and *c2* and the shared information between c1 and *c2* as measured by the IC of their most specific ancestor *c3*. For two sets *X* and *Y*, a function *f* of their intersection does not change if more sets exist that are included as part of that intersection, i.e., that set of shared features. As explained in section 3.1 *IC(c3)*, however, representing a measure of shared information between *c1* and *c2*, does not solely depend on *IC(c1)* and *IC(c2)* but is affected by the *IC* of all of the children of *c3*. These children also have the same shared information with *c1* and *c2*. The amount of this shared information should not decrease simply because more children are added to *c3*. This difference makes a case against a simple substitution of *IC(c3)* for $f(X \cap Y)$ in both Tversky's set models of similarity.

This section has proposed fuzzy set compatibility measures, many based on the Tversky ratio model of similarity, that can be combined with information content measures to produce a general model for a wide variety of ontological similarity measures. The key considerations are what set is used to represent a concept and what method is used to determine the degree of membership of each element in the set describing a concept. To further explore this model, the following empirical study uses the fuzzy set $F_{anc+(c)}$ for concept *c* in addition to the two historical Resnik and Lin measures and the two recently proposed measures P&S and FaITH.

4 Empirical Investigations Using the Gene Ontology

The bioinformatics domain is serving as a primary impetus for the creation of new ontological similarity measures. The Gene Ontology (GO) [9] is an important ontology used in this domain. It contains concepts used to annotate genes or gene products. The GO relationships include two major types of structuring links: "is-a" and "part-of". For determining IC, the standard practice has been to use only these two types of relationships [10].

Besides being a major bioinformatics ontology, the GO contains three mutually exclusive subontologies: biological process (BP), cellular component (CC) and molecular function (MF) each with varying sizes. The CC subontology has 2636 concepts, of which 1724 are leaf concepts (approximately 65% leaves). The MF ontology has 8668 concepts, of which 6956 are leaf concepts (about 80.2%). The BP ontology has 18059 concepts, of which 8442 are leaf concepts (about 46.7%). The MF ontology is more single parent structured averaging 1.2 parents per concept while the CC and BP average 1.9 parents per concepts.

Another objective of this investigation is to use a much larger number of concept pairs than the very small number of pairs ranging from 28 to 65 used in human similarity judgment experiments with WordNet and MeSH. For each subontology,

GO concepts are randomly selected and similarities between all pairs of selected concepts within a subontology are calculated. Around 5% of each subontology is randomly selected: 100 for CC, 500 for MF and 880 for BP, resulting in 5050, 125250, and 387640 concept pairs, respectively.

A direct comparison approach is used that does not require an arbitrary gold standard for performance comparison. Because the plethora of ontological similarity measures are mostly the result of numerous researchers developing a new measure for what is perceived as a very specific objective, these measures have typically been assessed with that objective in mind. Performance comparisons are limited to a small group of measures to an experimentally developed gold standard. The typical conclusion, more often than not, is that the new measure performs as well if not better than the others in the group. Finding one gold standard to encompass the wide variety of uses for ontological similarity is a daunting challenge. In the bioinformatics domain, measuring the degree of gene similarity between genes requires an aggregation operator in addition to a semantic similarity measure. A wide variety of aggregation operators have been used to produce the final gene pair similarity. The gene product semantic similarity is then correlated with actual gene sequence similarity, a potential gold standard in the bioinformatics domain [1]. The focus here is on the ontological similarity measures themselves within the GO subontologies so that the need for the selection of various aggregation operators can be eliminated.

IC based ontological similarity measures are the focus of this study since within the bioinformatics, they have been predominantly used [1], and the primary path-based measure Wu-Palmer has been shown equivalent to the Lin measure when each path edge is weighted by the difference in the child's and parent's IC value [6]. In the following IC_{ont} given in equation (7) is used since it is simpler to calculate and has also been reported in several studies [4][5][19][23] to perform as well if not better than IC_{corpus}. The five IC ontological similarity measures selected are the Resnik (R), the Lin (L), the FaITH (F), the P&S (P), and the proposed JacAnc (J).

Performance evaluations of semantic similarity measures to human similarity judgments typically use the Pearson correlation coefficient. It measures the degree of linear relationship between two variables. The assumption is each variable is approximately normally distributed. The Spearman coefficient assumes that the variables under consideration are measured on at least a rank order or ordinal scale. It is can be viewed as the Pearson coefficient in terms of proportion of variability accounted for, only the ranks of the observations for each variable are used to calculate the Spearman coefficient. The Spearman correlation coefficient produces a 1 when the two variables being compared are monotonically related and does not require the strict linear relationship of the Pearson correlation Both coefficients have been calculated. For this investigation, each correlation coefficient calculated had a p-value < 0.001, which indicate that the result is significant.

Figures 1 through 3 are for the Pearson coefficient for the CC, MF, and BP, respectively. Figures 4 through 6 are for the Spearman. The y-axis is the coefficient value; the x-axis is the respective measure: R-Resnik, L-Lin, F-FaITH, P-P&S, and J-JacAnc. Each line shows how a measure correlates with all the other measures.

For both coefficients, the Resnik measure correlates best with the Lin and FaITH measures across all three subontologies. The FaITH measure is strongly connected to the Lin measure. It uses a consistent reduction to both the numerator and denominator of the Lin measure by the IC of the ancestor with the greatest IC. The experiments verify this showing that these two measures have 1.0 correlation with each other across all subontologies for the Spearman coefficient and have approximately the same high Pearson correlation, 0.97, across the three subontologies. Across all three subontologies, the Spearman correlation for the Lin measures with each the other three ontological similarity measures is identical to that of the corresponding Spearman correlation for FaITH.

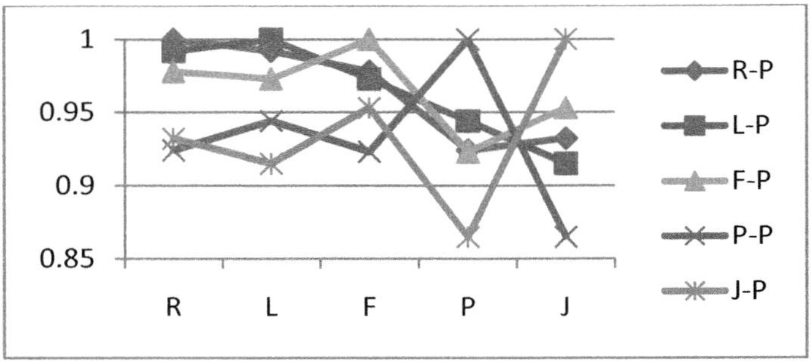

Fig. 1. Pearson Correlation between Measures for CC

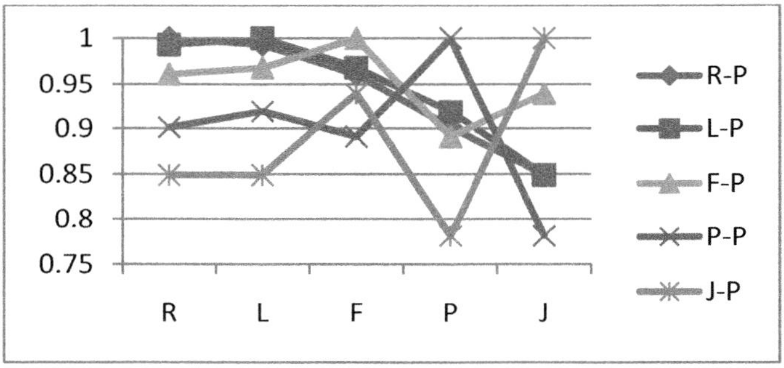

Fig. 2. Pearson Correlation between Measures for MF

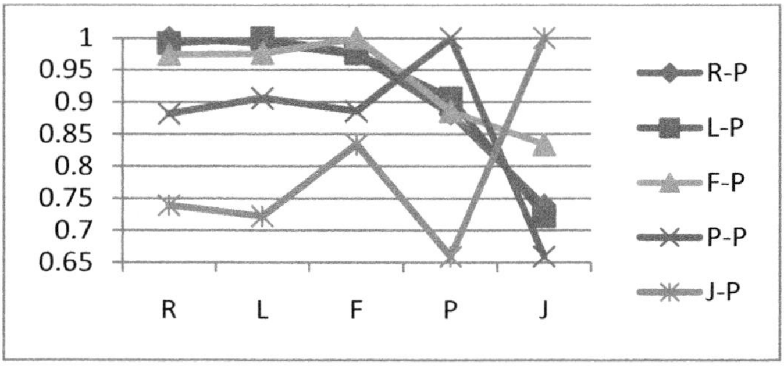

Fig. 3. Pearson Correlation between Measures for BP

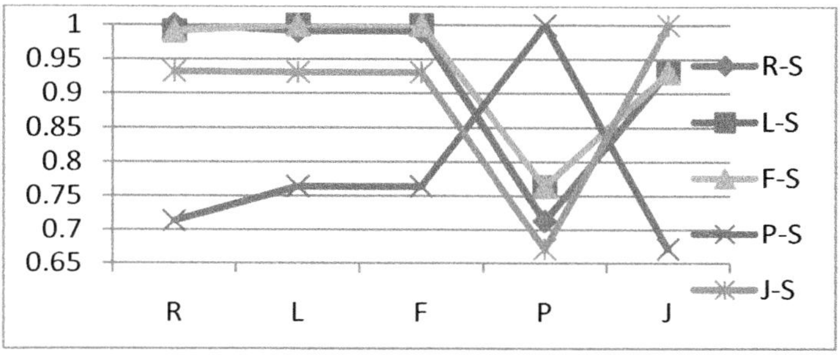

Fig. 4. Spearman Correlation between Measures for CC

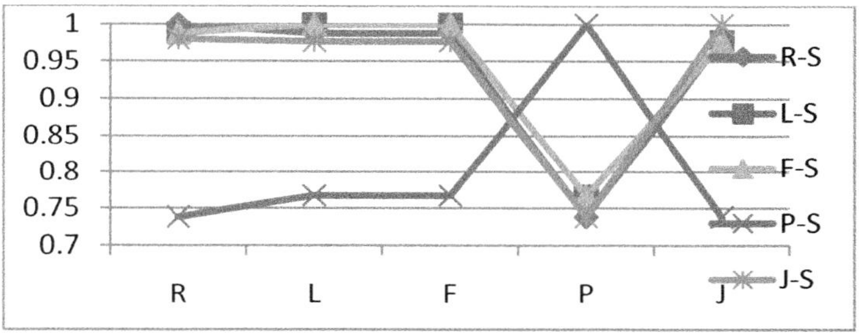

Fig. 5. Spearman Correlation between Measures for MF

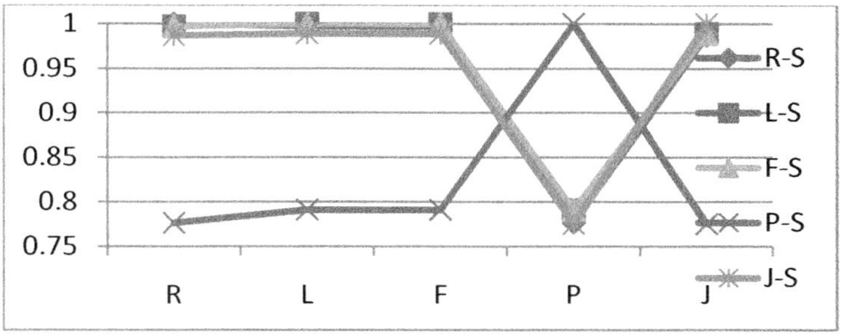

Fig. 6. Spearman Correlation between Measures for BP

Resnik correlates (Pearson) about the same with P&S and JacAnc for CC but then it correlates slightly better with the P&S for both MF and BP. The Resnik has the worst Spearman correlation with the P&S measure across all subontologies. The correlation results for Lin parallel those for the Resnik since across all three subontologies and both correlations coefficients the smallest correlation between the two is 0.988. This result occurs because Resnik and Lin measures are identical when IC_{ont} is used, i.e., $sim_{Res}(c_1, c_2) = sim_{Lin}(c_1, c_2) = 2IC(c3)/(1 + 1) = IC(c3)$.

The FaITH measure correlates (both coefficients) best with the Resnik and Lin measures for all subontologies. It has a much higher Spearman correlation with the

JacAnc measure than with the P&S measure for all subontologies. Only for Pearson and the BP does it correlate slightly better with the P&S than with the JacAnc.

The P&S measure correlates (Pearson) best with the Resnik, Lin, and FaITH measures across all subontologies, with about a 0.04 difference in the range of values. Pearson correlation with JacAnc measure is more varied with the best being for the CC and the worst being for the BP. A similar result can be seen for the Spearman coefficient with lower correlation values and a slightly bigger range in values. The JacAnc measure correlates best (Pearson) with the FaITH measures across all three subontologies. For the Spearman correlation coefficient with JacAnc, the Resnik, Lin, and FaITH have all very close coefficients for each of the subontologies.

In general Pearson correlations across all subontologies show the same patterns. P&S and JacAnc have the poorest correlation with the other three measures though there is a flip-flop between the P&S measure and the JacAnc measure for the Lin and FaITH measures for the CC and BP only. The worst correlation for P&S occurs with JacAnc. In general Spearman correlations show a similar pattern, but the P&S measure clearly correlates less with all the other measures. The distinction in the correlation values for the other four decreases going from CC to MF to BP. This result could be due to the increasing number of concept pairs from CC to MF to BP.

Generally, Resnik, Lin and FaITH are strongly correlated. The Lin and FaITH produce identical results with respect to the Spearman coefficient. From these experiments, the lowest correlation between these three measures occurs between FaITH and Res for the MF with a 0.961.

Space limitations prevent fully discussing the IC ontological similarity using IC_{corpus}. The correlation for each IC_{ont} ontological similarity with its IC_{corpus} version is summarized. The corpus used is the GOA-UniProt Version 79 database (http://www.ebi.ac.uk/GOA/). The experiments showed a very strong correlation between the IC_{ont} and IC_{corpus} versions of the ontological similarity measures. P&S has the lowest correlations across both coefficients and all subontologies (Pearson 0.74 on MF, Spearman 0.58 on MF). The JacAnc has the highest Pearson correlations for all subontologies with it lowest value being 0.976 for the BP. For the Spearman correlations, all measures but P&S had very high and similar correlation values.

With FaITH we did not use the eIC discussed in section 3.1. In [5] sim_{FaITH} + eIC always produced higher Pearson correlations with human similarity judgments than any of the others. The paper seems to indicate that the other similarity measures used IC_{ont}. Since the specific kinds of relationships used to calculate eIC are not clearly stated, IC_{ont} with only the is-a and part-of links is used in our work.

Another experiment in [5] uses a data set consisting of 36 concept pairs from the MeSH (Medical Subject Headings) ontology to evaluate the performance of sim_{FaITH} on a domain related ontology. The conclusion again was that sim_{FaITH} had the highest correlation with the human similarity judgments. It is understood that for MeSH all ontological similarity measures including sim_{FaITH} were calculated using IC_{ont}. As previously discussed, the sim_{FaITH} measure should always produce a smaller or equal value to that of sim_{Lin}. However, a problem in either the calculation of one or both of these measures is seen in [5] since in the reported similarity values for the 36 pairs, 14 of the sim_{FaITH} values are greater than the corresponding sim_{Lin} values.

4 Conclusions and Future Work

Several IC based ontological similarity measures are theoretically and empirically investigated. The motivation of the study is to establish a general fuzzy set-theoretic framework of IC ontological similarity that has as its basis Tversky's similarity models, fuzzy set compatibility measures, and information content and to examine other recently proposed measures that use an intuitive IC version of Tversky's model [4][5]. Historical ontological similarity measures are shown to be examples of fuzzy set compatibility measures when a concept is described by a fuzzy set of elements and their membership degrees are determined as a function of IC. The empirical study uses two historical measures Resnik [18] and Lin [12], two recently proposed measures, P&S [4] and FaITH [5], and JacAnc, a measure presented here and derived from the general fuzzy set-theoretic framework for IC ontological similarity. IC_{ont} is used in comparing these ontological measures. The cellular component (CC), the molecular function (MF) and the biological process (BP) of the Gene Ontology are used to empirically compare these five measures. The comparison does not use a gold standard but instead uses the correlation between their similarity values on sets of randomly selected concept pairs from each of the three GO subontologies.

The FaITH and Lin measures are shown to have a mathematical relationship validated by their very high Pearson correlation and identical Spearman correlations values. The Resnik and Lin measure are strongly correlated. This result can be partly explained by the equivalence of the two measures when the similarity between leaf concepts is being determined. The JacAnc measure shows overall the strongest correlation between its IC_{ont} and IC_{corpus} versions over all subontologies although each measure shows strong correlation between its two IC versions. It appears for JacAnc that incorporating more information to represent a concept, i.e. a concept's set of ancestors, mitigates the difference in using IC_{ont} in place of IC_{corpus}. The correlations for the two P&S versions are the smallest. Future research is to explore the use of several of these ontological similarity measures in various tasks, for example ontology alignment in order to further compare their performance.

References

1. Cross, V.: Ontological Similarity. In: Popescu, M., Xu, D. (eds.) Data Mining in Biomedicine Using Ontologies, pp. 23–43. Artech House, Norwood, MA (2009)
2. Pesquita, C., Faria, D., Falcão, A.O., Lord, P., Couto, F.M.: Semantic Similarity in Biomedical Ontologies. PLoS Comput. Biol. 5(7), e1000443, doi:10.1371/journal.p(c)bi.1000443 (2009)
3. Tversky, A.: Features of Similarity. Psychological Rev. 84, 327–352 (1977)
4. Pirrò, G., Seco, N.: Design, Implementation and Evaluation of a New Semantic Similarity Metric Combining Features and Intrinsic Information Content. In: Chung, S. (ed.) OTM 2008, Part II. LNCS, vol. 5332, pp. 1271–1288. Springer, Heidelberg (2008)
5. Pirrò, G., Euzenat, J.: A Feature and Information Theoretic Framework for Semantic Similarity and Relatedness. In: Proceedings of International Semantic Web Conference, vol. (1), pp. 615–630 (2010)

6. Cross, V.: Tversky's Parameterized Similarity Ratio Model: A Basis for Semantic Relatedness. In: Proceedings of the 2006 Conference of North American Fuzzy Information Processing Society (NAFIPS), Montreal, Canada (June 3-6, 2006)
7. Cazzanti, L., Gupta, M.R.: Information-theoretic and Set-theoretic Similarity. In: Proc. IEEE Intl. Symposium on Information Theory (2006)
8. Budanitsky, A., Hirst, G.: Semantic Distance in WordNet: An Experimental, Application-oriented Evaluation of Five Measures. In: Workshop on WordNet and Other Lexical Resources, Second meeting of the NAACL, Pittsburgh (2001)
9. The Gene Ontology Consortium, `http://www.geneontology.org/`
10. Lord, P., Stevens, R., Brass, A., Goble, C.: Investigating semantic similarity measures across the Gene Ontology: the relationship between sequence and annotation. Bioinformatics 19, 1275–1283 (2003)
11. Wei, M.: An Analysis of Word Relatedness Correlation Measures. Master's thesis, University of Western Ontario, London, Ontario (May 1993)
12. Lin, D.: An information-theoretic definition of similarity. In: Proc. of the 15th Int. Conf. on Machine Learning, pp. 296–304. Morgan Kaufmann, San Francisco (1998)
13. Attneave, F.: Dimensions of Similarity. American J. of Psychology 63, 516–556 (1950)
14. Goodman, N.: Seven strictures on similarity. In: Goodman, N. (ed.) Problems and projects, pp. 437–447. Bobbs-Merrill, New York (1972)
15. Medin, D.L., Goldstone, R.L., Gentner, D.: Respects for Similarity. Psychological Review 100(2), 254–278 (1993)
16. Cross, V.: An Analysis of Fuzzy Set Aggregators and Compatibility Measures, Ph.D. Dissertation, Computer Science and Engineering, Wright State University, Dayton, OH, 264 pages (March 1993)
17. Cross, V., Sudkamp, T.: Similarity and Compatibility in Fuzzy Set Theory: Assessment and Applications. Physica-Verlag, New York (2002) ISBN: 3-7908-1458
18. Resnik, P.: Using information content to evaluate semantic similarity in taxonomy. In: Proc. of the 14th Intl Joint Conference on Artificial Intelligence, pp. 448–453 (1995)
19. Seco, N., Veale, T., Hayes, J.: An Intrinsic Information Content Metric for Semantic Similarity in WordNet. In: ECAI, pp. 1089–1090 (2004)
20. Jiang, J., Conrath, D.: Semantic similarity based on corpus statistics and lexical taxonomy, In: Proc. of the 10th International Conference on Research (1997)
21. Wu, Z., Palmer, M.: Verb semantics and lexical selection. In: Proc. of the 32nd Annual Meeting of the Assoc. for Computational Ling, NM, Las Cruces, pp. 133–138 (1994)
22. Yu, X.: A Mathematical and Experimental Investigation of Ontological Similarity Measures and their Use in Biomedical Domains. Master's Thesis, Computer Science and Software Engineering, Miami University, Oxford OH (2010)
23. Cross, V., Sun, Y.: Semantic, Fuzzy Set and Fuzzy Measure Similarity for the Gene Ontology. In: Proceedings of the IEEE International Conference on Fuzzy Systems. Imperial College, London (2007)
24. Rodriguez, M.A., Egenhofer, M.J.: Determining Semantic Similarity among Entity Classes from Different Ontologies. IEEE Transactions on Knowledge and Data Engineering 15(2), 442–456 (2003)

Answering Threshold Queries in Probabilistic Datalog+/– Ontologies

Georg Gottlob, Thomas Lukasiewicz, and Gerardo I. Simari

Department of Computer Science, University of Oxford
Wolfson Building, Parks Road, Oxford OX1 3QD, United Kingdom
firstname.lastname@cs.ox.ac.uk

Abstract. The recently introduced Datalog+/– family of ontology languages is especially useful for representing and reasoning over lightweight ontologies, and is set to play a central role in the context of query answering and information extraction for the Semantic Web. Recently, it has become apparent that it is necessary to develop a principled way to handle uncertainty in this domain. In addition to uncertainty as an inherent aspect of the Web, one must also deal with forms of uncertainty due to inconsistency and incompleteness, uncertainty resulting from automatically processing Web data, as well as uncertainty stemming from the integration of multiple heterogeneous data sources. In this paper, we take an important step in this direction by developing the first probabilistic extension of Datalog+/–. This extension uses Markov logic networks as underlying probabilistic semantics. Here, we especially focus on scalable algorithms for answering *threshold queries*, which correspond to the question "what is the set of all atoms that are inferred from a given probabilistic ontology with a probability of at least p?". These queries are especially relevant to Web information extraction, since uncertain rules lead to uncertain facts, and only information with a certain minimum confidence is desired. We present two algorithms: a basic approach and one based on heuristics that is guaranteed to return sound results.

1 Introduction

Recently, Web search companies such as Google, Yahoo!, and Microsoft have realized that enhancing their products via the incorporation of ideas and developments from the Semantic Web (to incorporate, for instance, semantic search and complex query answering) must include principled ways in which to manage *uncertainty*. The Web is full of examples where uncertainty comes in: as an inherent aspect of Web data (such as in reviews of products or services, comments in blog posts, weather forecasts, etc.), as the result of automatically processing Web data (for instance, analyzing a document's HTML Document Object Model usually involves some degree of uncertainty), and as the result of integrating information from many different heterogeneous sources (such as in aggregator sites, which allow users to query multiple sites at once to save time). Finally, inconsistency and incompleteness are also ubiquitous as the result of over- and under-specification, respectively. In order to be applicable to web-sized data sets, any machinery developed for dealing with uncertainty in these settings must be scalable. In this paper, we take an important step in this direction by developing the first extension of Datalog$^{\pm}$ [5] by means of a probabilistic semantics based on Markov logic

S. Benferhat and J. Grant (Eds.): SUM 2011, LNAI 6929, pp. 401–414, 2011.
© Springer-Verlag Berlin Heidelberg 2011

networks [22]. The former is a recently introduced family of ontology languages that is especially useful for representing and reasoning over lightweight ontologies. This formalism is set to play a central role in the context of query answering and information extraction for the Semantic Web by means of its novel generalization of database rules and dependencies (such as tuple-generating dependencies (TGDs) and equality-generating dependencies (EGDs)) so that they can express ontological axioms. Markov logic networks, also recently developed, are a simple approach to generalizing classical logic; their relative simplicity and lack of restrictions has recently caused them to be well-received in the reasoning under uncertainty community.

The main goal of this paper, apart from introducing the novel probabilistic Datalog$^\pm$ formalism, will be to develop scalable algorithms for answering *threshold queries*, which correspond to the question "what is the set of all atoms that are inferred from a given probabilistic ontology with a probability of at least p?". These queries are especially adequate in the Web information extraction process, since uncertain rules lead to uncertain facts, and only information with a certain minimum confidence is desired. We present two algorithms: a basic approach that provides exact answers but is not scalable in the combined complexity, and an algorithm based on heuristics that is guaranteed to return sound results with a much more attractive running time.

2 Preliminaries

This section briefly recalls guarded Datalog$^\pm$and Markov logic networks.

2.1 Guarded Datalog$^\pm$

We now describe guarded Datalog$^\pm$ [5], which here includes negative constraints and (separable) equality-generating dependencies (EGDs). We first describe some preliminaries on databases and queries, and then tuple-generating dependencies (TGDs) and the concept of chase. We finally recall negative constraints and (separable) EGDs, which are other important ingredients of guarded Datalog$^\pm$ ontologies.

Databases and Queries. For the elementary ingredients, we assume data constants, nulls, and variables as follows; they serve as arguments in atomic formulas in databases, queries, and dependencies. We assume (i) an infinite universe of *data constants* Δ (which constitute the "normal" domain of a database), (ii) an infinite set of *(labeled) nulls* Δ_N (used as "fresh" Skolem terms, which are placeholders for unknown values, and can thus be seen as variables), and (iii) an infinite set of variables $\mathcal{V}$ (used in queries and dependencies). Different constants represent different values (*unique name assumption*), while different nulls may represent the same value. We assume a lexicographic order on $\Delta \cup \Delta_N$, with every symbol in Δ_N following all symbols in Δ. We denote by $\mathbf{X}$ sequences of variables $X_1, \ldots, X_k$ with $k \geqslant 0$.

We next define atomic formulas, which occur in databases, queries, and dependencies, and which are constructed from relation names and terms, as usual. We assume a *relational schema* $\mathcal{R}$, which is a finite set of *relation names* (or *predicate symbols*, or simply *predicates*). A *position* $P[i]$ identifies the i-th argument of a predicate P. A *term* t is a data constant, null, or variable. An *atomic formula* (or *atom*) $\mathbf{a}$ has the form $P(t_1, ..., t_n)$, where P is an n-ary predicate, and $t_1, ..., t_n$ are terms. We denote by $pred(\mathbf{a})$ and $dom(\mathbf{a})$ its predicate and the set of all its arguments, respectively. The latter two notations are naturally extended to sets of atoms and conjunctions of atoms. A conjunction of atoms is often identified with the set of all its atoms.

We are now ready to define the notion of a database relative to a relational schema, as well as conjunctive and Boolean conjunctive queries to databases. A *database (instance)* D for a relational schema $\mathcal{R}$ is a (possibly infinite) set of atoms with predicates from $\mathcal{R}$ and arguments from Δ. Such D is *ground* iff it contains only atoms with arguments from Δ. A *conjunctive query (CQ)* over $\mathcal{R}$ has the form $Q(\mathbf{X}) = \exists\mathbf{Y}\, \Phi(\mathbf{X}, \mathbf{Y})$, where $\Phi(\mathbf{X}, \mathbf{Y})$ is a conjunction of atoms with the variables $\mathbf{X}$ and $\mathbf{Y}$, and possibly constants, but without nulls. Note that $\Phi(\mathbf{X}, \mathbf{Y})$ may also contain equalities but no inequalities. A *Boolean CQ (BCQ)* over $\mathcal{R}$ is a CQ of the form $Q()$. We often write a BCQ as the set of all its atoms, having constants and variables as arguments, and omitting the quantifiers. Answers to CQs and BCQs are defined via *homomorphisms*, which are mappings $\mu\colon \Delta \cup \Delta_N \cup \mathcal{V} \rightarrow \Delta \cup \Delta_N \cup \mathcal{V}$ such that (i) $c \in \Delta$ implies $\mu(c) = c$, (ii) $c \in \Delta_N$ implies $\mu(c) \in \Delta \cup \Delta_N$, and (iii) μ is naturally extended to atoms, sets of atoms, and conjunctions of atoms. The set of all *answers* to a CQ $Q(\mathbf{X}) = \exists\mathbf{Y}\, \Phi(\mathbf{X}, \mathbf{Y})$ over a database D, denoted $Q(D)$, is the set of all tuples $\mathbf{t}$ over Δ for which there exists a homomorphism $\mu\colon \mathbf{X} \cup \mathbf{Y} \rightarrow \Delta \cup \Delta_N$ such that $\mu(\Phi(\mathbf{X}, \mathbf{Y})) \subseteq D$ and $\mu(\mathbf{X}) = \mathbf{t}$. The *answer* to a BCQ $Q()$ over a database D is *Yes*, denoted $D \models Q$, iff $Q(D) \neq \emptyset$.

Tuple-Generating Dependencies. Tuple-generating dependencies (TGDs) describe constraints on databases in the form of generalized Datalog rules with existentially quantified conjunctions of atoms in rule heads; their syntax and semantics are as follows. Given a relational schema $\mathcal{R}$, a *tuple-generating dependency (TGD)* σ is a first-order formula of the form $\forall\mathbf{X}\forall\mathbf{Y}\, \Phi(\mathbf{X}, \mathbf{Y}) \rightarrow \exists\mathbf{Z}\, \Psi(\mathbf{X}, \mathbf{Z})$, where $\Phi(\mathbf{X}, \mathbf{Y})$ and $\Psi(\mathbf{X}, \mathbf{Z})$ are conjunctions of atoms over $\mathcal{R}$ called the *body* and the *head* of σ, denoted $body(\sigma)$ and $head(\sigma)$, respectively. A TGD is *guarded* iff it contains an atom in its body that involves all variables appearing in the body. We usually omit the universal quantifiers in TGDs. Such σ is satisfied in a database D for $\mathcal{R}$ iff, whenever there exists a homomorphism h that maps the atoms of $\Phi(\mathbf{X}, \mathbf{Y})$ to atoms of D, there exists an extension h' of h that maps the atoms of $\Psi(\mathbf{X}, \mathbf{Z})$ to atoms of D. All sets of TGDs are finite here.

Query answering under TGDs, i.e., the evaluation of CQs and BCQs on databases under a set of TGDs is defined as follows. For a database D for $\mathcal{R}$, and a set of TGDs Σ on $\mathcal{R}$, the set of *models* of D and Σ, denoted $mods(D, \Sigma)$, is the set of all (possibly infinite) databases B such that (i) $D \subseteq B$ (ii) every $\sigma \in \Sigma$ is satisfied in B. The set of *answers* for a CQ Q to D and Σ, denoted $ans(Q, D, \Sigma)$, is the set of all tuples $\mathbf{a}$ such that $\mathbf{a} \in Q(B)$ for all $B \in mods(D, \Sigma)$. The *answer* for a BCQ Q to D and Σ is *Yes*, denoted $D \cup \Sigma \models Q$, iff $ans(Q, D, \Sigma) \neq \emptyset$. Note that query answering under general TGDs is undecidable [2], even when the schema and TGDs are fixed [4].

The two problems of CQ and BCQ evaluation under TGDs are LOGSPACE-equivalent [7,14,10,8]. Moreover, the query output tuple (QOT) problem (as a decision version of CQ evaluation) and BCQ evaluation are AC_0-reducible to each other. Henceforth, we thus focus only on the BCQ evaluation problem, and any complexity results carry over to the other problems. We also recall that query answering under TGDs is equivalent to query answering under TGDs with only single atoms in their heads. In the sequel, we thus assume w.l.o.g. that every TGD has a single atom in its head.

The Chase. The *chase* was introduced to enable checking implication of dependencies [19], and later also for checking query containment [14]. It is a procedure for repairing a database relative to a set of dependencies, so that the result of the chase satisfies the dependencies. By "chase", we refer both to the chase procedure and to its output. The TGD chase works on a database through so-called TGD *chase rules* (an extended chase with also equality-generating dependencies is discussed below). The TGD chase rule comes in two flavors: *restricted* and *oblivious*, where the restricted one applies TGDs only when they are not satisfied (to repair them), while the oblivious one always applies TGDs (if they produce a new result). We focus on the oblivious one here; the *(oblivious) TGD chase rule* defined below is the building block of the chase.

TGD CHASE RULE. Consider a database D for a relational schema $\mathcal{R}$, and a TGD σ on $\mathcal{R}$ of the form $\Phi(\mathbf{X}, \mathbf{Y}) \rightarrow \exists \mathbf{Z}\, \Psi(\mathbf{X}, \mathbf{Z})$. Then, σ is *applicable* to D if there exists a homomorphism h that maps the atoms of $\Phi(\mathbf{X}, \mathbf{Y})$ to atoms of D. Let σ be applicable to D, and h_1 be a homomorphism that extends h as follows: for each $X_i \in \mathbf{X}$, $h_1(X_i) = h(X_i)$; for each $Z_j \in \mathbf{Z}$, $h_1(Z_j) = z_j$, where z_j is a "fresh" null, i.e., $z_j \in \Delta_N$, z_j does not occur in D, and z_j lexicographically follows all other nulls already introduced. The *application of* σ on D adds to D the atom $h_1(\Psi(\mathbf{X}, \mathbf{Z}))$ if not already in D. ■

The chase algorithm for a database D and a set of TGDs Σ consists of an exhaustive application of the TGD chase rule in a breadth-first (level-saturating) fashion, which leads as result to a (possibly infinite) chase for D and Σ. Formally, the *chase of level up to* 0 of D relative to Σ, denoted $\mathit{chase}^0(D, \Sigma)$, is defined as D, assigning to every atom in D the *(derivation) level* 0. For every $k \geqslant 1$, the *chase of level up to* k of D relative to Σ, denoted $\mathit{chase}^k(D, \Sigma)$, is constructed as follows: let $I_1, \ldots, I_n$ be all possible images of bodies of TGDs in Σ relative to some homomorphism such that (i) $I_1, \ldots, I_n \subseteq \mathit{chase}^{k-1}(D, \Sigma)$ and (ii) the highest level of an atom in every I_i is $k - 1$; then, perform every corresponding TGD application on $\mathit{chase}^{k-1}(D, \Sigma)$, choosing the applied TGDs and homomorphisms in a (fixed) linear and lexicographic order, respectively, and assigning to every new atom the *(derivation) level* k. The *chase* of D relative to Σ, denoted $\mathit{chase}(D, \Sigma)$, is then defined as the limit of $\mathit{chase}^k(D, \Sigma)$ for $k \rightarrow \infty$.

The (possibly infinite) chase relative to TGDs is a *universal model*, i.e., there exists a homomorphism from $\mathit{chase}(D, \Sigma)$ onto every $B \in \mathit{mods}(D, \Sigma)$ [8,4]. This result implies that BCQs Q over D and Σ can be evaluated on the chase for D and Σ, i.e., $D \cup \Sigma \models Q$ is equivalent to $\mathit{chase}(D, \Sigma) \models Q$. In the case of guarded TGDs Σ, such BCQs Q can be evaluated on an initial fragment of $\mathit{chase}(D, \Sigma) \models Q$ of constant depth $k \cdot |Q|$, and thus be done in polynomial time in the data complexity.

Negative Constraints. Another crucial ingredient of Datalog$^\pm$ for ontological modeling are *negative constraints* (or simply *constraints*), which are first-order formulas of the form $\forall \mathbf{X} \Phi(\mathbf{X}) \rightarrow \bot$, where $\Phi(\mathbf{X})$ is a conjunction of atoms (not necessarily guarded). We usually omit the universal quantifiers, and we implicitly assume that all sets of constraints are finite here. Adding negative constraints to answering BCQs Q over databases and guarded TGDs is computationally easy, as for each constraint $\forall \mathbf{X} \Phi(\mathbf{X}) \rightarrow \bot$, we only have to check that the BCQ $\Phi(\mathbf{X})$ evaluates to false; if one of these checks fails, then the answer to the original BCQ Q is positive, otherwise the negative constraints can be simply ignored when answering the original BCQ Q.

Equality-Generating Dependencies. A further important ingredient of Datalog$^{\pm}$ for modeling ontologies are *equality-generating dependencies* (or *EGDs*) σ, which are first-order formulas of the form $\forall \mathbf{X}\, \Phi(\mathbf{X}) \rightarrow X_i = X_j$, where $\Phi(\mathbf{X})$, called the *body* of σ, denoted $body(\sigma)$, is a (not necessarily guarded) conjunction of atoms, and X_i and X_j are variables from $\mathbf{X}$. We call $X_i = X_j$ the *head* of σ, denoted $head(\sigma)$. Such σ is satisfied in a database D for $\mathcal{R}$ iff, whenever there exists a homomorphism h such that $h(\Phi(\mathbf{X}, \mathbf{Y})) \subseteq D$, it holds that $h(X_i) = h(X_j)$. We usually omit the universal quantifiers in EGDs, and all sets of EGDs are finite here.

An EGD σ on $\mathcal{R}$ of the form $\Phi(\mathbf{X}) \rightarrow X_i = X_j$ is *applicable* to a database D for $\mathcal{R}$ iff there exists a homomorphism $\eta\colon \Phi(\mathbf{X}) \rightarrow D$ such that $\eta(X_i)$ and $\eta(X_j)$ are different and not both constants. If $\eta(X_i)$ and $\eta(X_j)$ are different constants in Δ, then there is a *hard violation* of σ (and, as we will see below, the *chase* fails). Otherwise, the result of the application of σ to D is the database $h(D)$ obtained from D by replacing every occurrence of a non-constant element $e \in \{\eta(X_i), \eta(X_j)\}$ in D by the other element e' (if e and e' are both nulls, then e precedes e' in the lexicographic order). The *chase* of a database D, in the presence of two sets Σ_T and Σ_E of TGDs and EGDs, respectively, denoted $chase(D, \Sigma_T \cup \Sigma_E)$, is computed by iteratively applying (1) a single TGD once, according to the standard order and (2) the EGDs, as long as they are applicable (i.e., until a fixpoint is reached). To assure that adding EGDs to answering BCQs Q over databases and guarded TGDs along with negative constraints does not increase the complexity of query answering, all EGDs are assumed to be *separable* [5]. Intuitively, separability holds whenever: *(i)* if there is a hard violation of an EGD in the chase, then there is also one on the database w.r.t. the set of EGDs alone (i.e., without considering the TGDs); and *(ii)* if there is no chase failure, then the answers to a BCQ w.r.t. the entire set of dependencies equals those w.r.t. the TGDs alone (i.e., without the EGDs).

Guarded Datalog+/– Ontologies. A *(guarded) Datalog$^{\pm}$ ontology* consists of a (finite) database D, a finite set of (guarded) TGDs Σ_T, a finite set of negative constraints Σ_C, and a finite set of EGDs Σ_E that are separable from Σ_T.

Example 1. Consider the following set of TGDs and EGDs describing a simple ontology regarding a real estate information extraction system for the Web:

– $F_1 : ann(X, label), ann(X, price), visible(X) \rightarrow priceElem(X)$.
If X is annotated as a label, as a price, and is visible, then it is a price element.

– $F_2 : ann(X, label), ann(X, priceRange), visible(X) \rightarrow priceElem(X)$.
If X is annotated as a label, as a price range, and is visible, then it is a price element.

– $F_3 : priceElem(E), group(E, X) \rightarrow forSale(X)$.
If E is a price element and is grouped with X, then X is for sale.

– $F_4 : forSale(X) \rightarrow \exists P\, price(X, P)$.
If X is for sale, then there exists a price for X.

– $F_5 : hasCode(X, C), codeLoc(C, L) \rightarrow loc(X, L)$.
If X has postal code C, and C's location is L, then X's location is L.

– F_6 : $hasCode(X, C) \rightarrow \exists L\ codeLoc(C, L), loc(X, L)$.
If X has postal code C, then there exists L such that C has location L and so does X.

– F_7 : $loc(X, L1), loc(X, L2) \rightarrow L1 = L2$.
If X has the locations $L1$ and $L2$, then $L1$ and $L2$ are the same.

Formulas F_1 to F_6 are TGDs, while F_7 is an EGD. Clearly, all TGDs except for F_5 are guarded. In order to illustrate the chase, assume that we have the following atoms in the ontology: *codeLoc*(*ox1*, *central*), *codeLoc*(*ox1*, *south*), *codeLoc*(*ox2*, *summertown*), *hasCode*(*prop1*, *ox2*), *ann*(*e1*, *price*), *ann*(*e1*, *label*), *visible*(*e1*), and *group*(*e1*, *prop1*). Consider the chase relative to these atoms and the above formulas F_1 to F_7 excluding F_6; some of the atoms introduced are:

– *priceElem*(*e1*), by application of F_1;

– *forSale*(*prop1*), by application of F_3;

– *price*(*prop1*, z_1), by application of F_4, with $z_1 \in \Delta_N$.

Consider next the ontology obtained from the one above by adding the two atoms *loc*(*prop1*, *ox1*) and *loc*(*prop1*, *ox2*). Here, the EGD F_7 now leads to a failure in the chase, since there are two different locations associated with *prop1*.

2.2 Markov Logic Networks

Markov logic networks (MLNs) [22] combine first-order logic with Markov networks (MNs; or Markov random fields) [21]. We now provide a brief introduction to both.

Markov Networks. A Markov network (MN) is a probabilistic model that represents a joint probability distribution over a (finite) set of random variables $X = \{X_1, ..., X_n\}$. Each random variable X_i may take on *values* from a finite *domain* $Dom(X_i)$. A *value* for $X = \{X_1, \ldots, X_n\}$ is a mapping $x\colon X \rightarrow \bigcup_{i=1}^{n} Dom(X_i)$ such that $x(X_i) \in Dom(X_i)$; the *domain* of X, denoted $Dom(X)$, is the set of all values for X. An MN is similar to a Bayesian network (BN) in that it includes a graph $G = (V, E)$ in which each node corresponds to a variable, but, differently from a BN, the graph is undirected; in an MN, two variables are connected by an edge in G iff they are conditionally dependent. Furthermore, the model contains a *potential function* ϕ_i for each (maximal) clique in the graph; potential functions are non-negative real-valued functions of the values of the variables in each clique (called the *state* of the clique). In this work, we will assume the *log-linear* representation of MNs, which involves defining a set of *features* of such states; a feature is a real-valued function of the state of a clique (we will only consider binary features in this work). Given a value $x \in Dom(X)$ and a feature f_j for clique j, the probability distribution represented by an MN is given by $P(X = x) = \frac{1}{Z} \exp(\sum_j w_j \cdot f_j(x))$, where j ranges over the set of cliques in the graph G, and $w_j = \log \phi_j(x_{\{j\}})$ (here, $x_{\{j\}}$ is the state of the j-th clique). The term Z is a normalization constant to ensure that the values given by the equation above are in $[0, 1]$; it is given by $Z = \sum_{x \in Dom(X)} \exp(\sum_j w_j \cdot f_j(x))$. Probabilistic inference in MNs is intractable; however, approximate inference mechanisms, such as Markov Chain Monte Carlo, have been developed and successfully applied.

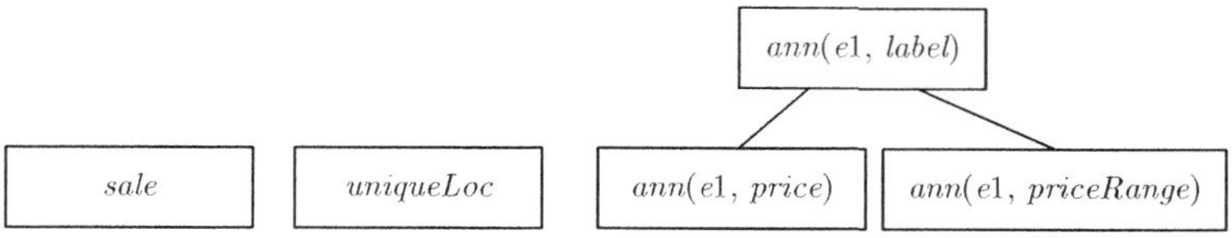

Fig. 1. The graph representation of the MLN from Example 2

Markov Logic Networks. The main idea behind Markov logic networks (MLNs) is to provide a way to soften the constraints imposed by a set of classical logic formulas. Instead of considering worlds that violate some formulas to be impossible, we wish to make them less probable. A Markov logic network is a finite set L of pairs (F_i, w_i), where F_i is a formula in first-order logic, and w_i is a real number. Such a set L, along with a finite set of constants $C = \{c_1, ..., c_m\}$, defines a Markov network $M_{L,C}$ that contains: (i) one binary node corresponding to each element of the Herbrand base of the formulas in L (i.e., all possible ground instances of the atoms), where the node's value is 1 iff the atom is true; and (ii) one feature for every possible ground instance of a formula in L. The value of the feature is 1 iff the ground formula is true, and the weight of the feature is the weight corresponding to the formula in L. From this characterization and the description above of the graph corresponding to an MN, it follows that $M_{L,C}$ has an edge between any two nodes corresponding to ground atoms that appear together in at least one formula in L. Furthermore, the probability of $x \in Dom(X)$ given this ground MLN is $P(X = x) = \frac{1}{Z}\exp(\sum_j w_j \cdot n_j(x))$, where $n_i(x)$ is the number of ground instances of F_i made true by x, and Z is defined analogously as above. This formula can be used in a generalized manner to compute the probability of any setting of a subset of random variables $X' \subseteq X$, as we will see below.

Example 2. The following is a simple example of a Markov logic network. Later on, we will use this MLN in combination with the ontology presented in Example 1.

- ψ_1: $(ann(X, label) \wedge ann(X, price), 0.3)$;
- ψ_2: $(ann(X, label) \wedge ann(X, priceRange), 0.4)$;
- ψ_3: $(sale, 0.8)$;
- ψ_4: $(uniqueLoc, 1.1)$.

The graphical representation of this MLN w.r.t. the ground atoms $ann(e1, label)$, $ann(e1, price)$, $ann(e1, priceRange)$, $sale$, and $uniqueLoc$ (obtained by grounding the formulas w.r.t. the set of constants $\{e1\}$) is shown in Fig. 1. This MLN represents a probability distribution over the possible Boolean values for each node. Given that there are five ground atoms, there are $2^5 = 32$ possible settings of the variables in the MLN. The normalizing factor Z is the sum of the probabilities of all possible worlds, which is computed as shown above by summing the exponentiated sum of weights times the number of ground formulas satisfied, yielding $Z \approx 127.28$. Similarly, the probability that a formula, such as $ann(e1, label)$, holds is the sum of the probabilities that all the satisfying worlds hold, which in this case is $\frac{87.82}{127.28} \approx 0.6903$.

3 Syntax and Semantics of Probabilistic Guarded Datalog$^\pm$

Considering the basic setup from Sections 2.1 and 2.2, we now present the language of probabilistic guarded Datalog$^\pm$.

3.1 Syntax

As in Section 2.1, we assume an infinite universe of data constants Δ, an infinite set of labeled nulls Δ_N, and an infinite set of variables $\mathcal{V}$. Furthermore, we assume a finite set of random variables X, as in Section 2.2. Informally, a probabilistic guarded Datalog$^\pm$ ontology consists of a finite set of probabilistic atoms, guarded TGDs, negative constraints, and separable EGDs, along with a Markov logic network.

Definition 1. A *(probabilistic) scenario* λ is a (finite) set of pairs $\langle X_i, x_i\rangle$, where $X_i \in X$, $x_i \in \mathit{Dom}(X_i)$, and the X_i's are pairwise distinct. If $|\lambda| = |X|$, then λ is a *full* probabilistic scenario. If every random variable X_i has a Boolean domain, then we also abbreviate λ by the set of all X_i such that $\langle X_i, \mathit{true}\rangle \in \lambda$.

Intuitively, a probabilistic scenario will be used to describe an event in which the random variables in an MLN are compatible with the settings of the random variables described by λ, i.e., each X_i has the value x_i.

Definition 2. If a is an atom, σ_T is a TGD, σ_C is a negative constraint, σ_E is an EGD, and λ is a probabilistic scenario, then: (i) $a : \lambda$ is a *probabilistic atom*; (ii) $\sigma_T : \lambda$ is a *probabilistic TGD (pTGD)*; (iii) $\sigma_C : \lambda$ is a *probabilistic (negative) constraint*; and (iv) $\sigma_E : \lambda$ is a *probabilistic EGD (pEGD)*. We also refer to probabilistic atoms, TGDs, (negative) constraints, and EGDs as *annotated formulas*.

Intuitively, annotated formulas hold whenever the events associated with their probabilistic annotations occur. A *probabilistic Datalog*$^\pm$ *ontology* is of the form $\Phi = (O, M)$, where O is a set of probabilistic atoms, TGDs, (negative) constraints, and EGDs, and M is a Markov logic network. In the sequel, we implicitly assume that every such $\Phi = (O, M)$ is *separable*, which means that Σ^ν_E is separable from Σ^ν_T, for every $\nu \in \mathit{Dom}(X)$, where Σ^ν_T (resp., Σ^ν_E) is the set of all TGDs (resp., EGDs) σ such that (i) $\sigma : \lambda \in O$ and (ii) λ is contained in the set of all $\langle X_i, \nu(X_i)\rangle$ with $X_i \in X$.

Example 3. Consider the guarded Datalog$^\pm$ ontology from Example 1, and the Markov logic network M from Example 2. Both share the atoms with predicate symbol "ann"; in the following, we build a probabilistic guarded Datalog$^\pm$ ontology $\Phi = (O, M)$ by having these atoms as a part of the MLN only, as shown below:

- $F'_1 : \mathit{visible}(X) \rightarrow \mathit{priceElem}(X)$: $\{\mathit{ann}(X, \mathit{label}), \mathit{ann}(X, \mathit{price})\}$;
- $F'_2 : \mathit{visible}(X) \rightarrow \mathit{priceElem}(X)$: $\{\mathit{ann}(X, \mathit{label}), \mathit{ann}(X, \mathit{priceRange})\}$;
- $F'_3 : \mathit{priceElem}(E), \mathit{group}(E, X) \rightarrow \mathit{forSale}(X)$: $\{\mathit{sale}\}$;
- $F'_7 : \mathit{loc}(X, L1), \mathit{loc}(X, L2) \rightarrow L1 = L2$: $\{\mathit{uniqueLoc}\}$.

Furthermore, F'_4 and F'_6 are the same as in Example 1, but with the annotation "$\emptyset$": these formulas hold irrespective of the setting of the random variables of the MLN.

3.2 Semantics

The semantics of probabilistic Datalog$^{\pm}$ ontologies is given w.r.t. probabilistic distributions over *interpretations* of the form $\mathcal{I} = \langle D, \nu \rangle$, where D is a database over $\Delta \cup \Delta_N$, and $\nu : \mathcal{X} \rightarrow D(\mathcal{X})$ is a function that maps random variables to values in their domain. In the following, we abbreviate "$true : \lambda$" with "λ".

Definition 3. An interpretation $\mathcal{I} = \langle D, \nu \rangle$ *satisfies* an annotated formula $F : \lambda$, denoted $\mathcal{I} \models F : \lambda$, iff whenever $\nu(X) = x$, for all $\langle X, x \rangle \in \lambda$, then $D \models F$.

A *probabilistic interpretation* is then a probability distribution Pr over the set of all possible interpretations such that only a finite number of interpretations are mapped to a non-zero value. The probability of an annotated formula $F : \lambda$, denoted $Pr(F : \lambda)$, is the sum of all $Pr(\mathcal{I})$ such that $\mathcal{I}$ satisfies $F : \lambda$.

Definition 4. Let Pr be a probabilistic interpretation, and $F : \lambda$ be an annotated formula. We say that Pr *satisfies* (or is a *model* of) $F : \lambda$ iff $Pr(F : \lambda) = 1$. Furthermore, Pr is a model of a probabilistic Datalog$^{\pm}$ ontology $\Phi = (O, M)$ iff: (i) Pr satisfies all annotated formulas in O, and (ii) $1 - Pr(\mathit{false} : \lambda) = Pr_M(\lambda)$ for all full probabilistic scenarios λ, where $Pr_M(\lambda)$ is the probability of $\bigwedge_{\langle X_i, x_i \rangle \in \lambda}(X_i = x_i)$ in the MLN M (and computed in the same way as $P(X = x)$ in Section 2.2).

In the rest of this paper, we are especially interested in computing the probabilities associated with atoms in a probabilistic Datalog$^{\pm}$ ontology, as defined next.

Definition 5. Let $\Phi = (O, M)$ be a probabilistic Datalog$^{\pm}$ ontology, and a be a ground atom that is constructed from predicates and data constants in Φ. The *probability* of a in Φ, denoted $Pr^{\Phi}(a)$, is the infimum of $Pr(a : \{\})$ subject to all probabilistic interpretations Pr such that $Pr \models \Phi$.

Intuitively, an atom will have the probability that results from summing the probabilities of all full scenarios under which the resulting universal model contains the atom.

Threshold Queries. The focus of this paper is on queries that request the set of all atoms that have probability at least p, where p is specified as an input.

Definition 6. Let $\Phi = (O, M)$ be a probabilistic Datalog$^{\pm}$ ontology, and $p \in [0, 1]$. A *threshold query* is of the form $Q = \langle \Phi, p \rangle$, and the *set of answers* to Q contains all ground atoms a such that $Pr^{\Phi}(a) \geq p$.

The following is an example of such a query over the running example.

Example 4. Consider the probabilistic Datalog$^{\pm}$ ontology $\Phi = (O, M)$ from Example 3, and the threshold query $Q = \langle \Phi, 0.45 \rangle$. All the atoms listed in Example 1 have probability 1, and so belong to the answers to Q. To compute the probabilities of the other atoms in the chase, we sum the probabilities of the full scenarios that make the atoms true, yielding, e.g., $Pr(\mathit{priceElem}(\mathit{elem1})) = 0.492$ and $Pr(\mathit{forSale}(\mathit{prop1})) = 0.339$; clearly, the former belongs to the output, while the latter does not.

Before exploring algorithms for threshold queries, we conclude this section with a result regarding the complexity of finding the set of answers.

Algorithm 1. basicThresh($\Phi = (O, M), p$)

Initialize π_{out} as a mapping from atoms to values in $[0, 1]$ (default value is zero);
for $i = 1$ to $2^{|X|}$ do begin // i ranges over all possible probabilistic scenarios
    $\lambda := \mathit{computeFullScenario}(M, i)$;
    $O_\lambda := \mathit{getRelevantFormulas}(O, \lambda)$; // O_λ contains formulas "activated" by λ
    $ch := \mathit{computeChase}(O_\lambda)$;
    for each atom $a \in ch$ do
        $\pi_{out}(a) := \pi_{out}(a) + Pr_M(\lambda)$;
end;
remove from π_{out} all atoms a such that $\pi_{out}(a) < p$;
return π_{out}.

Fig. 2. Computes the answer π_{out} (including probabilities) to a threshold query $Q = \langle \Phi, p \rangle$

Theorem 1. *Let $\Phi = (O, M)$ be a probabilistic Datalog$^\pm$ ontology. Answering a threshold query $\langle \Phi, p \rangle$ is in PTIME in the data complexity.*

Proof (sketch). This result is an immediate consequence of the following observations: (i) M is fixed when considering the data complexity (and therefore inference is done in constant time); (ii) each (of the polynomially many) threshold queries for a ground atom can be answered by means of a constant number (in the data complexity) of BCQs to guarded Datalog$^\pm$ ontologies; and (iii) answering BCQs in guarded Datalog$^\pm$ is in PTIME in the data complexity [5].

4 Algorithms for Answering Threshold Queries

First, we explore a basic algorithm for answering threshold queries exactly, and then go on to propose a heuristic algorithm based on an annotated chase graph.

4.1 A Basic Algorithm

Fig. 2 shows the pseudocode for the basicThresh algorithm. The basic approach taken by this algorithm is to cycle through all possible settings of the random variables in X; for each one, the algorithm obtains the formulas whose probabilistic annotations are satisfied by the current scenario, and computes the chase w.r.t. this set of formulas. The probability of each atom in the chase is then updated; note that, since each probabilistic scenario is disjoint from the others being considered, the probabilities can be computed by summing the individual results (cf. Example 4). The following theorem shows the correctness and running time of this algorithm.

Theorem 2. *Let $\Phi = (O, M)$ be a probabilistic Datalog$^\pm$ ontology, $k \geq 1$, and $p \in [0, 1]$: (i) Every atom $a : \pi(a)$ in the output of Algorithm* basicThresh *belongs to the output of threshold query $\langle \Phi, p \rangle$; (ii) if an atom $a : prob(a)$ belongs to the output of threshold query $\langle \Phi, p \rangle$, then it also belongs to the output of Algorithm* basicThresh*; (iii)* heurThresh *runs in time $O\left(2^{|X|} * (m + |O| + c)\right)$, where m is the cost of computing the probability of a given scenario in the MLN M, and c is the cost of computing the chase w.r.t. O.*

Algorithm 2. heurThresh($\Phi = (O, M), p, k$)
1. Initialize π_{out} as a mapping from atoms to values in $[0, 1]$ (default value is zero);
2. for $i = 1$ to k do begin // i ranges over the k most probable scenarios
3. $\quad \lambda_i := computeMAP(M, i)$; // λ_i is the i-th most probable scenario
4. $\quad O_{\lambda_i} := getRelevantFormulas(O, \lambda_i)$; // O_{λ_i} contains formulas relevant in λ_i
5. end;
6. $O^* := \bigcup_{i=1}^{k} O_{\lambda_i}$;
7. $ch := computeChase(O^*)$;
8. $G_{ch} := chaseGraph(ch)$;
9. Annotate each node n in G_{ch} with a Boolean vector v_n of length k, initialized to all *true*;
10. for $\ell = 1$ to $numLevels(G_{ch})$ do begin
11. $\quad$ for each node a_j in level ℓ of G_{ch} do
12. $\quad\quad$ for $i = 1$ to k do
13. $\quad\quad\quad$ if $\exists$ node $p \in parents(a_j)$ such that either $(v_p[i] = false)$ or
14. $\quad\quad\quad\quad$ (a_j is obtained from $parents(a_j)$ by applying a TGD $F \notin O_{\lambda_i}$) then
15. $\quad\quad\quad\quad$ $v_p[i] = false$;
16. end;
17. for each node a_j in G_{ch} do begin
18. $\quad$ for $i = 1$ to k do
19. $\quad\quad \pi_{out}(a_j) :=$ sum of all $Pr_M(\lambda_i)$ such that $v_{a_j}[i] = true$;
20. $\quad$ if $\pi_{out}(a_j) < p$ then remove a_j from π_{out};
21. end;
22. return π_{out}.

Fig. 3. Returns π_{out}, a mapping that assigns a lower bound probability to all atoms that appear in the chase relative to at least one of the k most probable scenarios

We next present a heuristic algorithm, which is sound, but not complete.

4.2 An Annotated Chase Graph-Based Algorithm

In Fig. 3, we present a heuristic algorithm based on annotating the chase graph with information about the probabilistic scenarios. The main idea in this algorithm is to avoid going through all $2^{|X|}$ fully specified probabilistic scenarios; instead, it chooses the k most probable ones and computes the probabilities of atoms w.r.t. this subset of scenarios. Since the algorithm is working with a subset of all possible scenarios, the resulting probabilities will be lower bounds instead of exact; though this is clearly enough to decide to include an atom in the result, as we will see below (Theorem 3), it means that the algorithm is in general not complete.

The first steps are similar to those in basicThresh, but instead of going on to compute the chase of every subset associated with a probabilistic scenario, heurThresh computes the set O^* (line 6) of formulas that appear given at least one of the top-k probabilistic scenarios, and computes the chase w.r.t. this set. Every node in the chase graph (i.e., every atom) then receives an annotation that consists of a Boolean vector of size k; intuitively, this vector will store information regarding whether or not the atom belongs to the chase w.r.t. each scenario. The for-loop in lines 10 to 15 contains the work necessary to update these annotations correctly given two cases: an annotation for an atom in a scenario becomes false if (i) one of the parents has a *false* annotation, or (ii) if the edge is labeled with a formula that is not relevant in the scenario. Finally, the atoms receive a probability consisting of the sum of all the probabilities w.r.t. Markov logic network M for each scenario that has a label of *true*, and those atoms with probability less than p are removed from the output.

Example 5. Let Q be the query from Example 4, and let $k = 3$. Running the algorithm heurThresh on this input, we obtain that the probability of both $priceElem(elem1)$ and $forSale(prop1)$ has a lower bound of 0.253; therefore, $k = 3$ is not enough to determine that the former should be part of the output, and it is left out.

Even though, as seen in Example 5, some atoms may not be returned when they should have been, note that all the atoms that have probability 1 will always be returned, since the "$\emptyset$" annotations are satisfied by any scenario. The following theorem shows that this algorithm is sound, and also provides its running time.

Theorem 3. *Let $\Phi = (O, M)$ be a probabilistic Datalog$^\pm$ ontology, $k \geq 1$, and $p \in [0, 1]$: (i) Every atom a in the output of Algorithm* heurThresh *belongs to the output of threshold query $\langle \Phi, p \rangle$; (ii) If a has probability $prob(a)$ in Φ, we have that $\pi_{out}(a) \leq prob(a)$; (iii)* heurThresh *runs in time $O(k * (m + |O| + c))$, where m is the cost of computing the i-th most probable scenario in the MLN M, and c is the cost of computing the chase w.r.t. O.*

5 Related Work

Ontology languages, rule-based systems, and their integrations are central for the Semantic Web [3]. Although many approaches exist to tight, loose, or hybrid integrations of ontology languages and rule-based systems, and to generalizations of ontology languages by the ability to express rules, to our knowledge, Datalog$^\pm$ [5] is the first work on how to generalize database rules and dependencies so that they can express ontological axioms. The development of Datalog$^\pm$ was thus quite timely given that there are recently strong interests in the Semantic Web community on highly scalable formalisms for the *Web of Data*, which would benefit greatly from applying technologies and results from databases. As a consequence of this lack of development in this direction, to our knowledge, there is also no formalism that combines: (i) ontology languages, (ii) database technologies, and (iii) the management of probabilistic uncertainty.

Probabilistic ontology languages (combining (i) and (iii)) in the literature (see in particular [18] for a recent survey) can especially be classified according to the underlying ontology language, the supported forms of probabilistic knowledge, and the underlying probabilistic semantics. Some early approaches [12] generalize the description logic $\mathcal{ALC}$ and are based on propositional probabilistic logics, while others [16] generalize the tractable description logics CLASSIC and $\mathcal{FL}$, and are based on Bayesian networks as underlying probabilistic semantics. The recent approach in [17], generalizing the expressive description logics $\mathcal{SHIF}(\mathbf{D})$ and $\mathcal{SHOIN}(\mathbf{D})$ behind the sublanguages *OWL Lite* and *OWL DL*, respectively, of the Web ontology language *OWL* [20], is based on probabilistic default logics, and allows for rich probabilistic terminological and assertional knowledge. Other recent approaches [23] generalize OWL by probabilistic uncertainty as in multi-entity and standard Bayesian networks.

The combination of (i) ontology languages (including description logics (DLs) [1]) with (ii) rule systems from databases (such as *Datalog* [6]) recently plays a central role in the development of the Semantic Web [3]. Significant research efforts focus on hybrid integrations of rules and ontologies, called *description logic programs*, which are of the form $KB = (L, P)$, where L is a description logic knowledge base, and P is a finite set of rules involving either queries to L in a loose integration, or concepts and roles from L as unary and binary predicates, respectively, in a tight integration (see [9] for a recent survey). Many of these tight integrations of rule systems and ontology languages are generalizations of ontology languages by the ability to express rules.

Probabilistic databases (combining (ii) and (iii)) are a new and rapidly evolving research area motivated by the presence of uncertainty in data management scenarios, such as data integration, sensor readings, or information extraction from unstructured sources. Key challenges in probabilistic data management are (1) to design probabilistic database formalisms that can compactly represent large sets of possible interpretations of uncertain data together with their probability distributions, (2) to develop uncertainty-aware data manipulation languages akin to relational algebra for classical relational databases, and (3) to efficiently evaluate queries on very large probabilistic data [15]. Promising advances are currently pursued in the MayBMS project at EPFL and Oxford [13] on the first two challenges, and in the SPROUT project at Oxford on scalable query processing in probabilistic databases [11].

6 Summary and Outlook

In this work, we have extended the Datalog$^\pm$ language with probabilistic uncertainty, based on Markov logic networks. This is an important first step in continuing the recent generalization of database rules and dependencies that began with the Datalog$^\pm$ family of languages so that ontological axioms can be expressed. As we have discussed, managing uncertainty in a principled way is fundamental to both query answering and information extraction in the Semantic Web.

There remains much work to be done in this line of research. First of all, we plan to implement and evaluate our formalism and algorithms both on synthetic and real-world data. Further research efforts will involve, among others, identifying subsets of our language towards developing algorithms with increased scalability, and developing novel ranking techniques based on ontology-based preferences.

Acknowledgments. This work was supported by the European Research Council under the EU's 7th Framework Programme (FP7/2007-2013)/ERC grant 246858 – DIADEM, by a Yahoo! Research Fellowship, and by a Google Research Award. G. Gottlob is a James Martin Senior Fellow, and also gratefully acknowledges a Royal Society Wolfson Research Merit Award. The work was carried out in the context of the James Martin Institute for the Future of Computing. Many thanks also to the reviewers of this paper for their useful and constructive comments, which have helped to improve this work.

References

1. Baader, F., Calvanese, D., McGuinness, D.L., Nardi, D., Patel-Schneider, P.F. (eds.): The Description Logic Handbook: Theory, Implementation, and Applications. Cambridge University Press, Cambridge (2003)
2. Beeri, C., Vardi, M.Y.: The implication problem for data dependencies. In: Even, S., Kariv, O. (eds.) ICALP 1981. LNCS, vol. 115, pp. 73–85. Springer, Heidelberg (1981)
3. Berners-Lee, T., Hendler, J., Lassila, O.: The Semantic Web. Sci. Amer. 284, 34–43 (2002)
4. Calì, A., Gottlob, G., Kifer, M.: Taming the infinite chase: Query answering under expressive relational constraints. In: Proc. KR 2008, pp. 70–80. AAAI Press, Menlo Park (2008)
5. Calì, A., Gottlob, G., Lukasiewicz, T., Marnette, B., Pieris, A.: Datalog+/-: A family of logical knowledge representation and query languages for new applications. In: Proc. LICS 2010, pp. 228–242. IEEE Computer Society, Los Alamitos (2010)
6. Ceri, S., Gottlob, G., Tanca, L.: What you always wanted to know about Datalog (and never dared to ask). IEEE Trans. Knowl. Data Eng. 1, 146–166 (1989)
7. Chandra, A.K., Merlin, P.M.: Optimal implementation of conjunctive queries in relational data bases. In: Proc. STOC 1977, pp. 77–90. ACM Press, New York (1977)
8. Deutsch, A., Nash, A., Remmel, J.B.: The chase revisited. In: Proc. PODS 2008, pp. 149–158. ACM Press, New York (2008)
9. Drabent, W., Eiter, T., Ianni, G., Krennwallner, T., Lukasiewicz, T., Małuszyński, J.: Hybrid reasoning with rules and ontologies. In: Bry, F., Małuszyński, J. (eds.) Semantic Techniques for the Web. LNCS, vol. 5500, pp. 1–49. Springer, Heidelberg (2009)
10. Fagin, R., Kolaitis, P.G., Miller, R.J., Popa, L.: Data exchange: Semantics and query answering. Theor. Comput. Sci. 336(1), 89–124 (2005)
11. Fink, R., Olteanu, D., Rath, S.: Providing support for full relational algebra in probabilistic databases. In: Proc. ICDE 2011. IEEE Computer Society, Los Alamitos (2011)
12. Heinsohn, J.: Probabilistic description logics. In: Proc. UAI 1994, pp. 311–318. Morgan Kaufmann, San Francisco (1994)
13. Huang, J., Antova, L., Koch, C., Olteanu, D.: MayBMS: A probabilistic database management system. In: Proc. SIGMOD 2009, pp. 1071–1074. ACM Press, New York (2009)
14. Johnson, D.S., Klug, A.C.: Testing containment of conjunctive queries under functional and inclusion dependencies. J. Comput. Syst. Sci. 28(1), 167–189 (1984)
15. Koch, C., Olteanu, D., Re, C., Suciu, D.: Probabilistic Databases. Morgan-Claypool, San Francisco (2011)
16. Koller, D., Levy, A., Pfeffer, A.: P-CLASSIC: A tractable probabilistic description logic. In: Proc. AAAI 1997, pp. 390–397. AAAI Press/The MIT Press (1997)
17. Lukasiewicz, T.: Expressive probabilistic description logics. Artif. Intell. 172, 852–883 (2008)
18. Lukasiewicz, T., Straccia, U.: Managing uncertainty and vagueness in description logics for the Semantic Web. J. Web Sem. 6, 291–308 (2008)
19. Maier, D., Mendelzon, A.O., Sagiv, Y.: Testing implications of data dependencies. ACM Trans. Database Syst. 4(4), 455–469 (1979)
20. Patel-Schneider, P.F., Hayes, P., Horrocks, I.: OWL Web Ontology Language. W3C Recommendation (February 10 2004), `http://www.w3.org/TR/owl-semantics/`
21. Pearl, J.: Probabilistic Reasoning in Intelligent Systems: Networks of Plausible Inference. Morgan Kaufmann, San Francisco (1988)
22. Richardson, M., Domingos, P.: Markov logic networks. Mach. Learn. 62, 107–136 (2006)
23. Yang, Y., Calmet, J.: OntoBayes: An ontology-driven uncertainty model. In: Proc. CIMCA/IAWTIC 2005, pp. 457–463. IEEE Computer Society, Los Alamitos (2005)

Coherent Top-k Ontology Alignment for OWL EL

Jan Noessner, Mathias Niepert, and Heiner Stuckenschmidt

KR & KM Research Group,
Universität Mannheim
Mannheim, Germany
{jan,mathias,heiner}@informatik.uni-mannheim.de

Abstract. The integration of distributed information sources is a key challenge in data and knowledge management applications. Instances of this problem range from mapping schemas of heterogeneous databases to object reconciliation in linked open data repositories. In this paper, we approach the problem of aligning description logic ontologies. We focus particularly on the problem of computing coherent alignments, that is, alignments that do not lead to unsatisfiable classes in the resulting merged ontologies. We believe that considering coherence during the alignment process is important as it is this logical concept that distinguishes ontology alignment from other data integration problems. Depending on the heterogeneity of the ontologies it is often more reasonable to generate alignments with at most k correspondences because not every entity has a matchable counterpart. We describe both greedy and optimal algorithms for computing coherent top-k alignments between OWL EL ontologies and assess their performance relative to state-of-the-art matching systems.

1 Introduction

The growing number of heterogeneous knowledge bases on the web has made data integration systems a key technology for sharing and accumulating distributed information sources. In this paper, we focus on the problem of aligning description logic ontologies. Due to the explicit semantics of ontologies, alignment systems can take advantage of the logical concepts of coherence and consistency. Ensuring complete coherency and consistency is especially important in the area of ontology merging, where two ontologies are merged to one single ontology using the generated reference alignment.

Ontology debugging, for instance, is the process of efficiently finding and eliminating incoherencies. Several approaches to this problem were presented in [21] where the debugging process was based on the computation of minimal conflict sets. Similar concepts and algorithms have been used to debug pre-computed ontology alignments [13]. Their algorithm scales well for few conflict sets in the alignment, but if the number of conflict sets increase, the performance decreases significantly. In [15] they build a set of hard and soft markov logic rules to reduce the incoherency of the alignment. Although the performance is still high for

S. Benferhat and J. Grant (Eds.): SUM 2011, LNAI 6929, pp. 415–427, 2011.
© Springer-Verlag Berlin Heidelberg 2011

many conflict sets and most of the incoherencies are filtered out, the delivered alignments are not guaranteed to be coherent. In both, [13] and [15] a threshold is used to pre-select correspondences and a reasoner is needed to pre-calculate certain axioms.

Currently, most state-of-the-art matching systems such as Falcon [10], Aroma [5], and AgreementMaker [4] generate incoherent alignments [7]. To the best of our knowledge, only two of the matching systems that participated in the ontology alignment evaluation initiative (OAEI) of 2010 reduce the degree of alignment incoherence. While the semantic verification algorithm [11] of ASMOV reduces incoherence in a post-processing step CODI [17] employs incoherence reducing rules during the alignment process. Both matching systems, however, do not guarantee the final alignments to be coherent [7]. Another matching system not participating in the OAEI but focusing on coherent alignments is PROMPT [18]. It provides the user with different interactive views on the ontologies and aids the merging process by pointing out logical conflicts.

Depending on the heterogeneity of the ontologies it is often more sensible to generate alignments with at most k correspondences because not every entity in one ontology has a matchable counterpart in the other. Top-k algorithms are common in the area of information retrieval and ranking and have recently been applied in more structured data management systems. In the context of database schema matching, for instance, [8] presented an approach to computing the best k schema mappings.

With this paper, however, we present an *optimal coherent* top-k ontology matching algorithm, that is, an algorithm that generates optimal coherent alignments of size at most k. Compared to [13] our approach will still perform well for large number of conflict sets. The strength of the approach lies in its ability to incorporate arbitrary confidence values which could have been for example computed by other matching applications. Hence, the top-k algorithms are not intended to compete with existing matching systems but rather to complement their strength in deriving high-quality confidence values.

We present both a greedy and an optimal algorithm for computing coherent top-k alignments. The optimal algorithm utilizes the existence of a set of materialization rules for the description logic $\mathcal{EL}^{++}$ [1,12] without nominals and concrete domains, and formulates the alignment tasks as linear optimization problems. To reduce the complexity of these problems, the algorithm combines a cutting plane inference and a delayed column generation algorithm originally developed in the context of Markov logic [19].

We conduct extensive experiments to evaluate the accuracy and efficiency of both the greedy and optimal top-k algorithms. We also compare the coherence, recall, precision, and F_1 scores of the computed alignments with those generated by various state-of-the-art matching systems.

2 Description Logics

Description logics (DLs) are a family of knowledge representation languages [3]. They provide the logical formalism for ontologies and the Semantic Web. We

Table 1. The description logic $\mathcal{EL}^{++}$ without nominals and concrete domains

Name	Syntax	Semantics
top	$\top$	$\Delta^{\mathcal{I}}$
bottom	$\bot$	$\emptyset$
conjunction	$C \sqcap D$	$C^{\mathcal{I}} \cap D^{\mathcal{I}}$
existential restriction	$\exists r.C$	$\{x \in \Delta^{\mathcal{I}} \mid \exists y \in \Delta^{\mathcal{I}} : (x, y) \in r^{\mathcal{I}} \wedge y \in C^{\mathcal{I}}\}$
GCI	$C \sqsubseteq D$	$C^{\mathcal{I}} \subseteq D^{\mathcal{I}}$
RI	$r_1 \circ \ldots \circ r_k \sqsubseteq r$	$r_1^{\mathcal{I}} \circ \ldots \circ r_k^{\mathcal{I}} \subseteq r^{\mathcal{I}}$

focus on the DL $\mathcal{EL}^{++}$ which captures the expressivity of numerous real-world ontologies. $\mathcal{EL}^{++}$ is the description logic on which the web ontology language profile OWL 2 EL is based [1]. Reasoning tasks such as consistency and instance checking can be performed in polynomial time. Therefore, $\mathcal{EL}^{++}$ is practical for applications employing ontologies with large numbers of properties and classes. It is possible to express disjointness of complex concept descriptions as well as range and domain restrictions [2] and role inclusion axioms (RIs) allow the expression of role hierarchies $r \sqsubseteq s$ and transitive roles $r \circ r \sqsubseteq r$.

$\mathcal{EL}^{++}$ concept descriptions are defined recursively by a set of constructors, starting with a set $\mathsf{N_C}$ of concept names, a set $\mathsf{N_R}$ of role names, and a set $\mathsf{N_I}$ of individual names. Concept descriptions and role inclusions in $\mathcal{EL}^{++}$ are built with the constructors depicted in Table 1. We write r, s to denote role names and C, D to denote concept descriptions. The semantics of the concept descriptions in $\mathcal{EL}^{++}$ are defined in terms of an interpretation $\mathcal{I} = (\Delta^{\mathcal{I}}, \cdot^{\mathcal{I}})$. The interpretation function $\cdot^{\mathcal{I}}$ is recursively defined as shown in Table 1. A concept C *is subsumed by* a concept D with respect to a CBox $\mathcal{C}$, written $C \sqsubseteq_{\mathcal{C}} D$, if $C^{\mathcal{I}} \subseteq D^{\mathcal{I}}$ in every model $\mathcal{I}$ of $\mathcal{C}$.

A constraint box (CBox) is a finite set of general concept inclusions (GCIs) and role inclusions (RIs). Given a CBox $\mathcal{C}$, we use $\mathsf{BC}_{\mathcal{C}}$ to denote the set of *basic concept descriptions*, that is, the smallest set of concept descriptions consisting of the top concept $\top$, all concept names used in $\mathcal{C}$, and all nominals $\{a\}$ appearing in $\mathcal{C}$. Then, $\mathcal{C}$ is in normal form if all GCIs have one of the following forms, where $C_1, C_2 \in \mathsf{BC}_{\mathcal{C}}$ and $D \in \mathsf{BC}_{\mathcal{C}} \cup \{\bot\}$:

$$C_1 \sqsubseteq D; \qquad C_1 \sqsubseteq \exists r.C_2;$$
$$C_1 \sqcap C_2 \sqsubseteq D; \ \exists r.C_1 \sqsubseteq D$$

and if all role inclusions are of the form $r \sqsubseteq s$ or $r_1 \circ r_2 \sqsubseteq s$. By applying a finite set of rules and introducing new concept and role names, any CBox $\mathcal{C}$ can be turned into a normalized CBox [1]. For any $\mathcal{EL}^{++}$ CBox $\mathcal{C}$ we write $\mathsf{norm}(\mathcal{C})$ to denote the set of normalized axioms that result from the application of the normalization rules to $\mathcal{C}$. A normalized $\mathcal{EL}^{++}$ CBox is *classified* when subsumption relationships between *all* concept names are made explicit. A CBox $\mathcal{C}$ is *coherent* if for all concept names C in $\mathcal{C}$ we have that $C \not\sqsubseteq_{\mathcal{C}} \bot$.

3 Coherent Ontology Alignment

Ontology alignment is the process of inferring correspondences between entities of two ontologies. We begin by formally defining the notions of *correspondence* and *alignment* based on a definition by Euzenat and Shvaiko [6]. In this paper, each *ontology* is equivalent to a $\mathcal{EL}^{++}$ CBox *without* nominals and concrete domains, that is, an OWL 2 EL ontology without nominals and datatype properties. We refer the reader to [9] for a primer of the W3C recommendation for OWL 2 and its profiles.

Definition 1 (Correspondence and Alignment). *Given ontologies $\mathcal{O}_1$ and $\mathcal{O}_2$, let q be a function that defines sets of matchable entities $q(\mathcal{O}_1)$ and $q(\mathcal{O}_2)$. A correspondence between $\mathcal{O}_1$ and $\mathcal{O}_2$ is a triple $\langle e_1, e_2, r\rangle$ such that $e_1 \in q(\mathcal{O}_1)$, $e_2 \in q(\mathcal{O}_2)$, and r is a semantic relation. An alignment between $\mathcal{O}_1$ and $\mathcal{O}_2$ is a set of correspondences between $\mathcal{O}_1$ and $\mathcal{O}_2$.*

The general form of Definition 1 captures a wide range of correspondence types. In the following we focus on equivalence correspondences between concepts and object properties, respectively. The majority of matching systems provide normalized confidence values for each correspondence. Based on these confidence values, an alignment is extracted by applying a threshold $\tau \in [0, 1]$ meaning that only the correspondences with a confidence value greater than or equal to τ are included in the alignment.

In this paper, however, we are interested in solutions to the problem of computing *coherent alignments* between ontologies. An alignment $\mathcal{A}$ is coherent with respect to the coherent ontologies $\mathcal{O}_1$ and $\mathcal{O}_2$ if the ontology $\mathcal{O}_1 \cup \mathcal{O}_2 \cup \mathcal{A}$ is coherent, that is, if the ontology that results from merging $\mathcal{O}_1$ and $\mathcal{O}_2$ under the alignment $\mathcal{A}$ is coherent. Hence, in the remainder of the paper, we assume the existence of confidence values provided by, for instance, state-of-the-art matching systems. We refer to these values as *a-priori* confidence values. The *score* of an alignment is the sum of confidence values of its correspondences. We say that an alignment $\mathcal{A}$ of size k with score s is *optimal* if for every other alignment of size at most k with score s' we have that $s' \leq s$.

3.1 Greedy Coherent Top-*k* Alignment

The first algorithm for generating top-k coherent alignments from a set of correspondences with a-priori confidence values follows a greedy strategy. It appends an initially empty alignment with correspondences according to their a-priori confidence values in descending order. After each addition, it employs a reasoner to check whether the resulting alignment causes incoherences, and if it does, removes the previously added correspondence. The advantage of the approach is its efficiency – classification and, therefore, checking coherence of OWL EL ontologies can be performed in polynomial time. However, the approach does not compute optimal alignments. Once a correspondence has been added it cannot be revoked in later stages of the computation. The following example from the conference domain demonstrates said problem.

Table 2. The first-order theory $\mathcal{F}$. Valid instantiations of the formulas are those compatible with the types of the predicates from Definition 2. The predicates *cmap* and *pmap* model the correspondences between concept and role names, respectively. $\bot$ and $\top$ are constant symbols representing the bottom and top concept.

F_1	$\forall c : sub(c, c)$
F_2	$\forall c : sub(c, \top)$
F_3	$\forall c, c', d : sub(c, c') \wedge sub(c', d) \Rightarrow sub(c, d)$
F_4	$\forall c, c_1, c_2, d : sub(c, c_1) \wedge sub(c, c_2) \wedge int(c_1, c_2, d) \Rightarrow sub(c, d)$
F_5	$\forall c, c', r, d : sub(c, c') \wedge rsup(c', r, d) \Rightarrow rsup(c, r, d)$
F_6	$\forall c, r, d, d', e : rsup(c, r, d) \wedge sub(d, d') \wedge rsub(d', r, e) \Rightarrow sub(c, e)$
F_7	$\forall c, r, d, s : rsup(c, r, d) \wedge psub(r, s) \Rightarrow rsup(c, s, d)$
F_8	$\forall c, r_1, r_2, r_3, d, e : rsup(c, r_1, d) \wedge rsup(d, r_2, e) \wedge pcom(r_1, r_2, r_3) \Rightarrow rsup(c, r_3, e)$
F_9	$\forall c : \neg sub(c, \bot)$
F_{10}	$\forall c_1, c_2 : cmap(c_1, c_2) \Rightarrow sub(c_1, c_2)$
F_{11}	$\forall c_1, c_2 : cmap(c_1, c_2) \Rightarrow sub(c_2, c_1)$
F_{12}	$\forall r_1, r_2 : pmap(r_1, r_2) \Rightarrow psub(r_1, r_2)$
F_{13}	$\forall r_1, r_2 : pmap(r_1, r_2) \Rightarrow psub(r_2, r_1)$

Example 1. Let $\mathcal{O}_1$ contain the axiom $Review \sqcap JournalReviewer \sqsubseteq \bot$ and $\mathcal{O}_2$ the axiom $Reviewer \sqcap PaperReview \sqsubseteq \bot$. Moreover, consider the following correspondences and their associated a-priori confidence values: $\langle Reviewer \equiv Review, 0.9\rangle$, $\langle PaperReview \equiv Review, 0.7\rangle$, $\langle Reviewer \equiv JournalReviewer, 0.6\rangle$. The greedy top-$k$ approach would include the correspondence $\langle Reviewer \equiv Review, 0.9\rangle$ and would not add more correspondences due to the resulting incoherence. While an optimal top-k approach would also add the same correspondence for $k = 1$ it would generate the correct alignment $\{\langle PaperReview \equiv Review, 0.7\rangle, \langle Reviewer \equiv JournalReviewer, 0.6\rangle\}$ for $k = 2$ revoking the previous decision.

In the following we introduce a novel algorithm that computes optimal coherent top-k alignments. It leverages the completion rules for the DL $\mathcal{EL}^{++}$ without nominals and concrete domains [1,12].

3.2 Optimal Coherent Top-*k* Alignment

The *optimal* top-k alignment algorithm which we describe in the remainder of this section computes an *optimal coherent* alignment of size at most k from a given set of a-priori confidence values. The crucial insight is that the optimal alignment problem can be reduced to an optimization problem: Given the a-priori confidence values and the two input ontologies $\mathcal{O}_1$ and $\mathcal{O}_2$, *maximize* the sum of confidence values of correspondences in the alignment *subject to* the coherence of the ontology that results when merging $\mathcal{O}_1$ and $\mathcal{O}_2$ under the alignment. In order to guarantee the coherence of the alignment we map the normalized axioms of the two ontologies to ground predicates and formulate the optimization problem in such a way that all solutions to the problem correspond to coherent ontologies. We achieve this through a set of materialization formulas that capture the underlying DL semantics. We refer the reader to [12] for more details on materialization calculi and to [1,12] for the completeness of a finite set of completion rules for $\mathcal{EL}^{++}$ from which the set of formulas $\mathcal{F}$ (see Table 2) is partially derived. Furthermore, we refer the reader to [16] for the introduction of log-linear description logic which is the foundation of our approach. We begin by defining the mapping φ between ontologies and sets of ground atoms of the theory $\mathcal{F}$.

Definition 2 (Ontology Transformation). *Let $\mathcal{O}_1$ and $\mathcal{O}_2$ be two normalized ontologies, let $\mathsf{N_U} = \mathsf{BC}_{\mathcal{O}_1} \cup \mathsf{BC}_{\mathcal{O}_2}$ be the set of basic concept descriptions of both ontologies, and $\mathcal{H}$ be the set of all valid instantiations of predicates in $\mathcal{F}$ (see Table 2) relative to $\mathsf{N_U}$ (the Herbrand base of $\mathcal{F}$ with respect to $\mathsf{N_U}$ as a set of constant symbols). The function φ maps $\mathcal{O}_1 \cup \mathcal{O}_2$ to a subset of $\mathcal{H}$ as follows.*

$$
\begin{array}{ll}
C_1 \sqsubseteq D & \mapsto sub(c_1, d) \\
C_1 \sqcap C_2 \sqsubseteq D & \mapsto int(c_1, c_2, d) \\
C_1 \sqsubseteq \exists r.C_2 & \mapsto rsup(c_1, r, c_2) \\
\exists r.C_1 \sqsubseteq D & \mapsto rsub(c_1, r, d) \\
r \sqsubseteq s & \mapsto psub(r, s) \\
r_1 \circ r_2 \sqsubseteq r_3 & \mapsto pcom(r_1, r_2, r_3).
\end{array}
$$

All predicates are typed meaning that $r, s, r_i, (1 \leq i \leq 3)$, are role names, C_1, C_2 basic concept descriptions, and D basic concept descriptions or the bottom concept.

Based on the previously defined mapping, we can state the computation of an optimal top-k alignment as an instance of integer linear programming (ILP). Let $\mathcal{O}_1$ and $\mathcal{O}_2$ be two normalized *coherent* ontologies, let $\mathsf{N_U} = \mathsf{BC}_{\mathcal{O}_1} \cup \mathsf{BC}_{\mathcal{O}_2}$, let $\mathcal{F}^{\mathsf{N_U}}$ be the set of all valid instantiations of $\mathcal{F}$ relative to $\mathsf{N_U}$, and let $\mathcal{H}$ be the Herbrand base of $\mathcal{F}$ relative to $\mathsf{N_U}$. Moreover, let $\mathcal{K} = \varphi(\mathcal{O}_1 \cup \mathcal{O}_2)$ and let $\mathcal{L}$ be a set of valid instantiations of the predicates *cmap* and *pmap* modeling correspondences between classes and object properties, respectively, each associated with its a-priori confidence.

For each ground atom g_i occurring at least once in either $\mathcal{L}$ (with a-priori confidence value w_i), $\mathcal{K}$, or in a formula in $\mathcal{F}^{\mathsf{N_U}}$ we associate a variable $x_i \in \{0, 1\}$. Let C^L be the set of indices of ground atoms in $\mathcal{L}$, let C^K be the set of indices of ground atoms in $\mathcal{K}$, and let $C_j^F (\bar{C}_j^F)$ be the set of indices of unnegated (negated) ground atoms in the clause equivalent to $F_j \in \mathcal{F}^{\mathsf{N_U}}$. Then, the *top-k ILP with respect to $\mathcal{L}$* is stated as follows

$$
\textbf{max} \sum_{i \in C^L} w_i x_i \qquad \textbf{subject to} \quad \sum_{i \in C^L} x_i \leq k \ \textbf{ and}
$$

$$\sum_{i \in C^K} x_i \geq |C^K| \text{ and } \sum_{i \in C_j^F} x_i + \sum_{i \in \bar{C}_j^F} (1 - x_i) \geq 1, \ \forall j \quad (2).$$

Theorem 1. *Each solution of the Top-k ILP with respect to $\mathcal{L}$ corresponds to an ontology that results from (a) merging the ontologies $\mathcal{O}_1$ and $\mathcal{O}_2$ under an optimal alignment $\mathcal{A} \subseteq \mathcal{L}$ of size at most k and (b) classifying the merged ontology.*

Thus, the algorithm not only computes an optimal coherent top-k alignment but also classifies the merged ontologies making subsumption relationships between each pair of classes explicit. For a proof concerning the classification and the coherency of Theorem 1 the reader is referred to [16].

The immediate addition of all above constraints, however, would result in a very complex and potentially intractable optimization problem. In order to avoid this problem, we combine variants of the cutting plane inference algorithm [20] and the delayed column generation algorithm [14] both of which were first proposed for computing maximum a-posteriori (MAP) states in Markov logic networks [19]. To compute the solution of a top-k ILP we first construct the top-k ILP *with respect to* the set $\mathcal{L}'$ containing only $m \geq k$ correspondences with highest a-priori confidence values. The ILP is initially solved *without* the constraints of type (2). Given the current solution, the algorithm determines all violated constraints of type (2) in polynomial time, adds those to the ILP, and solves the updated problem. This is repeated until no violated constraints remain. If the solution contains k correspondences we have found an optimal top-k alignment. Otherwise, the set $\mathcal{L}'$ is augmented with m more correspondences with highest a-priori confidence values and the top-k ILP *with respect to* $\mathcal{L}'$ is solved as before. This is repeated until we have found a solution with k correspondences or until $\mathcal{L}'$ contains *all* correspondences.

Due to the extendability of the ILP formulation of the top-k alignment problem it is possible to include additional types of constraints such as constraints enforcing functional and one-to-one alignments and constraints modeling known correct correspondences.

4 Experimental Evaluation

We conducted extensive experiments to evaluate the performance of the greedy and optimal top-k alignment algorithms. In particular, we compared the optimal with the greedy top-k algorithm both in terms of computation time and alignment accuracy. We also assessed the accuracy of the alignments by comparing them to the alignments generated by state-of-the-art matching systems that participated in the latest OAEI of 2010 [7]. Moreover, we analyzed and compared the degree of coherence of each of the alignments computed by the matching systems.

4.1 Experimental Set-Up

For the experimental evaluation we used the ontologies of the conference and anatomy tracks of the OAEI. The availability of reference alignments and recent

Table 3. Number of classes and properties as well as number of normalized EL axioms in the respective ontologies we used for the experiments.

Axiom type	C	P	$C_1 \sqsubseteq D$	$C_1 \sqcap C_2 \sqsubseteq D$	$C_1 \sqsubseteq \exists r.C_2$	$\exists r.C_1 \sqsubseteq D$	$r \sqsubseteq s$	$r_1 \circ r_2 \sqsubseteq s$
Conference ontologies								
cmt	30	49	25	27	0	48	0	0
conference	60	46	56	14	7	47	13	0
confof	39	13	42	43	9	11	0	1
edas	104	30	90	409	3	29	0	0
ekaw	73	33	80	74	6	20	8	3
iasted	141	38	291	3	126	49	0	0
sigkdd	50	17	59	0	15	23	0	0
Anatomy ontologies								
mouse_anatomy	2744	3	4493	0	1637	0	0	0
nci_anatomy	3304	2	5423	17	1662	0	0	1

results from state-of-the-art matching systems make the two tracks particularly suitable. The conference track consists of several expressive ontologies modeling the domain of scientific conferences. The ontologies have been developed by different groups and, therefore, reflect different conceptualizations of the same domain. Reference alignments for seven of these ontologies are made available by the organizers of the OAEI. These 21 alignments contain correspondences between concepts and properties including a reasonable number of non-trivial instances. The two ontologies of the anatomy track are from the medical domain modeling the anatomy of humans and mice, respectively, and consist of over 2500 classes each. Since our matching approach is restricted to EL axioms we used the OWL API to downgrade the more expressive conference ontologies. We applied the set of rules from [1,2] to normalize the ontologies and to also include existing range restrictions. Table 3 lists the resulting conference and the anatomy ontologies along with the number of classes, properties, and normalized EL axioms.

We have argued that the top-k algorithms are not intended to compete with existing matching systems but rather to complement their strengths in generating high-quality a-priori confidence values. Hence, in order to asses the algorithms' ability to compute alignments from given confidence values we used the Levenshtein distance normalized to the range $[-1, 1]$. Using such a naïve algorithm to derive confidence values lets us evaluate the performance of the alignment algorithm without being influenced by highly sophisticated confidence measures. Please note, however, that the strength of the approach is the ability to incorporate confidence values generated by existing matching systems. We employed the reasoner Pellet [22] for the greedy algorithm and the mixed ILP

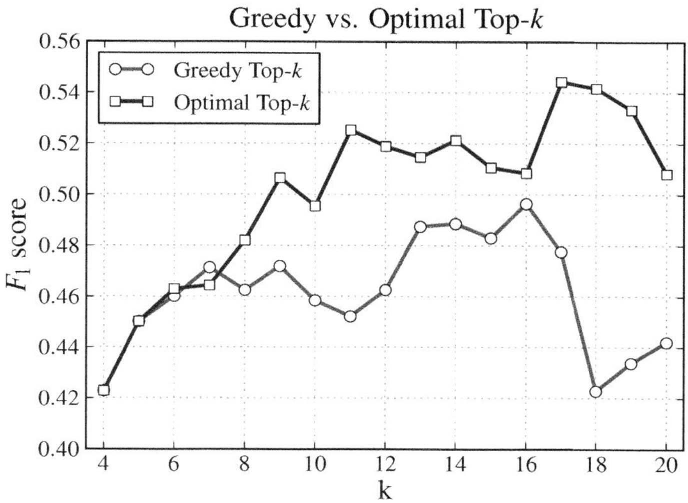

Fig. 1. F_1 scores of the optimal top-k and the greedy top-k algorithms averaged over the 21 alignment problems in the conference ontologies

solver Gurobi[1] for the optimal algorithm. We also augmented the ILP with constraints enforcing functional one-to-one alignments and we set the parameter m to $2k$. The experiments were run on a desktop PC with AMD Athlon Dual Core Processor 5400B with 2.6GHz and 1GB RAM. The source files and supplementary materials are available at `http://code.google.com/p/elmatch/`.

4.2 Results of the Evaluation

We first assessed the relative performance of the two top-k algorithms with respect to their F_1 score. Figure 1 shows the F_1 scores of the optimal and greedy top-k algorithms averaged over the 21 ontology pairs of the conference track. For $k \leq 6$ the F_1 scores are almost identical which is due to the absence of incoherence causing correspondences in the small alignments. With $k = 8$, however, the optimal algorithm starts to outperform the greedy approach as the larger alignments cause incoherences and substitutions of correspondences of the type described in Example 1 are becoming more prevalent.

The runtime of the algorithms is summarized in Table 4. For the conference ontologies and $k \leq 10$ the run time of the optimal algorithm is comparable to the greedy approach. The reason for the increase in runtime of the optimal algorithm for $k = 20$ is caused by the small size of the ontologies – alignments of size 20 exist only between 9 of the 21 pairs of ontologies. Hence, the optimal algorithm has to include *all* correspondences in its ILP formulation thus increasing the complexity of the optimization problem. Interestingly, the effect is reversed for

---

[1] http://www.gurobi.com/

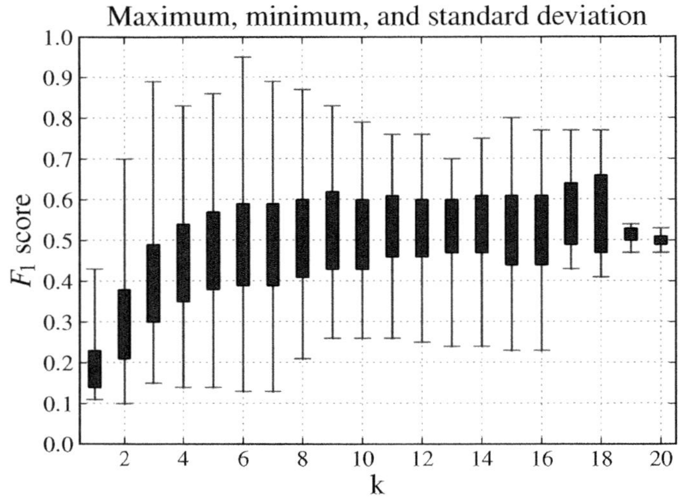

Fig. 2. Minimum, maximum and standard deviation of the F_1 score for the optimal top-k algorithm on the conference ontologies. The decrease in standard deviation for $k \geq 19$ is due to the fact that there are only few pairs of ontologies with functional one-to-one alignments of size k.

the anatomy ontologies. While the optimal algorithm has an overhead of about 40 seconds for classifying the large merged ontology the increase in runtime is smaller compared to the greedy approach. For $k = 1$ the greedy approach is about 10 times faster but only about twice as fast for $k = 20$. Considering that the reasoner Pellet is highly optimized for EL ontologies we find this to be a convincing result.

A suitable choice for the parameter k of the top-k algorithms clearly depends on the number of matchable elements and, therefore, on the size of the involved ontologies. Figure 2 shows the minimum, maximum and standard deviation of the optimal algorithm for the 21 different alignments. The large standard deviation and the discrepancy between the minima and maxima makes it evident that we need to adjust the parameter k *individually* for each alignment instance. We used the following *ad-hoc* heuristic to determine a suitable choice for the parameter k. We first computed the number P of correspondences where both matchable elements have identical labels. We then computed the parameter k with the formula $k = P + \alpha(k_{max} - P)$ where k_{max} is the maximal possible number of correspondences and $\alpha \in [0, 1]$. The parameter α determines the fraction of "non-trivial" correspondences one wants to derive and depends on the heterogeneity of the involved ontologies. In our experiments, we set the parameter α to 0.2. Table 5 depicts precision, recall, and F_1 scores of the optimal coherent top-k algorithm and a selection of matching systems that participated in the OAEI[2]

[2] Please visit `http://oaei.ontologymatching.org/2010/` for a complete list of results and all matching systems that participated at the OAEI 2010.

Table 4. The average time in seconds needed to compute coherent top-k alignments for the benchmark and anatomy ontologies *and* classifying the merged ontologies

k	1	5	10	15	20
Conference ontologies					
Greedy Top-k	0.36	0.41	0.56	0.77	1.21
Optimal Top-k	0.49	0.49	1.21	2.93	14.24
Anatomy ontologies					
Greedy Top-k	4.67	4.96	10.39	17.66	29.68
Optimal Top-k	40.76	42.42	45.10	48.74	53.60

Table 5. Comparison of the optimal top-k algorithm with state-of-the-art matching systems on the conference ontologies. Precision, recall, and F_1 scores are measured relative to the reference alignments. *Coh. Align* is the fraction of coherent alignments and *Coh. Class* is the fraction of coherent classes relative to the number of classes in all ontologies.

Matcher	Top-k	Falcon	AgrMaker	Aroma	ASMOV
Precision	0.78	0.59	0.50	0.36	0.45
Recall	0.44	0.58	0.65	0.49	0.07
F_1 score	0.57	0.58	0.57	0.42	0.12
Coh. Align	1.0	0.29	0.38	0.14	1.0
Coh. Class	1.0	0.95	0.84	0.64	1.0

with standard threshold 0.5. The coherent top-k algorithm has the best precision and competitive F_1 scores.

The main advantage of both top-k approaches compared to other matching systems, however, is the coherence of their alignments for EL ontologies. Table 5 lists the fraction of coherent alignments and classes, respectively, in the merged ontologies. Except for ASMOV, whose incomplete semantic verification algorithm [11] also reduces incoherences, all other matching systems generated incoherent alignments. In summary, only 14% of Aroma's, 29% of Falcon's, and 38% of AgreementMakers alignments were coherent indicating that these systems do not leverage the notion of coherence during the alignment process.

5 Conclusion and Future Work

With this paper, we presented a greedy and a novel optimal algorithm for computing coherent top-k alignments between OWL EL ontologies. The optimal algorithm employs integer linear programming solvers to maximize the sum of confidence values subject to the coherence of the ontology. Our evaluation showed that although we spent no effort on optimizing the confidence values (we used the simple Levenshtein distance), our F_1 scores were competitive compared to the participating systems at the OAEI 2010. The real strength of the top-k algorithms, however, is their ability to existing incorporate a-priori confidence values.

Currently, our approach is limited to the description logic $\mathcal{EL}^{++}$ without nominals and concrete domains but we intend to extend it to more expressive description logic languages such as Horn-$\mathcal{SHIQ}$. Moreover, we will work on supporting class and role assertions, nominals, and concrete domains. Apart from this, we will modify our approach to incorporate confidence values for complex correspondences. To this end, we will express complex matching patterns and their confidence values and integrate them in the optimization problem to compute coherent complex alignments between ontologies.

References

1. Baader, F., Brandt, S., Lutz, C.: Pushing the $\mathcal{EL}$ envelope. In: Proceedings of the 19th International Joint Conference on Artificial Intelligence (2005)
2. Baader, F., Brandt, S., Lutz, C.: Pushing the $\mathcal{EL}$ envelope further. In: Proceedings of the OWLED Workshop (2008)
3. Baader, F., Calvanese, D., McGuinness, D.L., Nardi, D., Patel-Schneider, P.F. (eds.): The Description Logic Handbook. Cambridge University Press, Cambridge (2003)
4. Cruz, I., Stroe, C., Caci, M., Caimi, F., Palmonari, M., Antonelli, F., Keles, U.: Using AgreementMaker to Align Ontologies for OAEI 2010. In: Proceedings of the 5th Workshop on Ontology Matching (2010)
5. David, J., Guillet, F., Briand, H.: Matching directories and OWL ontologies with AROMA. In: Proceedings of the 15th Conference on Information and knowledge management (2006)
6. Euzenat, J., Shvaiko, P.: Ontology matching. Springer, Heidelberg (2007)
7. Euzenat, J., et al.: First Results of the Ontology Alignment Evaluation Initiative 2010. In: Proceedings of the 5th Workshop on Ontology Matching (2010)
8. Gal, A.: Managing uncertainty in schema matching with top-k schema mappings. J. Data Semantics VI (2006)
9. Hitzler, P., Krötzsch, M., Parsia, B., Patel-Schneider, P.F., Rudolph, S. (eds.): OWL 2 Web Ontology Language: Primer. W3C Recommendation (2009)
10. Hu, W., Chen, J., Cheng, G., Qu, Y.: ObjectCoref & Falcon-AO: Results for OAEI 2010. In: Proceedings of the 5th International Ontology Matching Workshop (2010)
11. Jean-Maya, Y.R., Shironoshitaa, E.P., Kabuka, M.R.: Ontology matching with semantic verification. Web Semantics 7(3) (2009)
12. Krötzsch, M.: Efficient inferencing for OWL EL. In: Janhunen, T., Niemelä, I. (eds.) JELIA 2010. LNCS, vol. 6341, pp. 234–246. Springer, Heidelberg (2010)
13. Meilicke, C., Tamilin, A., Stuckenschmidt, H.: Repairing ontology mappings. In: Proceedings of the Conference on Artificial Intelligence (2007)
14. Niepert, M.: A Delayed Column Generation Strategy for Exact k-Bounded MAP Inference in Markov Logic Networks. In: Proceedings of the 25th Conference on Uncertainty in Artificial Intelligence (2010)
15. Niepert, M., Meilicke, C., Stuckenschmidt, H.: A Probabilistic-Logical Framework for Ontology Matching. In: Proceedings of the 24th AAAI Conference on Artificial Intelligence (2010)

16. Niepert, M., Noessner, J., Stuckenschmidt, H.: Log-Linear Description Logics. In: Proceedings of IJCAI (2011)
17. Noessner, J., Niepert, M.: CODI: Combinatorial Optimization for Data Integration–Results for OAEI 2010. In: Proceedings of the 5th Workshop on Ontology Matching (2010)
18. Noy, N., Musen, M.: The PROMPT suite: interactive tools for ontology merging and mapping. International Journal of Human-Computer Studies 59(6), 983–1024 (2003)
19. Richardson, M., Domingos, P.: Markov logic networks. Machine Learning 62(1-2) (2006)
20. Riedel, S.: Improving the accuracy and efficiency of map inference for markov logic. In: Proceedings of the Conference on Uncertainty in Artificial Intelligence (2008)
21. Schlobach, S., Huang, Z., Cornet, R., Harmelen, F.v.: Debugging incoherent terminologies. J. Autom. Reasoning 39(3) (2007)
22. Sirin, E., Parsia, B., Grau, B.C., Kalyanpur, A., Katz, Y.: Pellet: a practical OWL-DL reasoner. Journal of Web Semantics 5(2), 51–53 (2007)

Generalized Possibilistic Logic

Didier Dubois and Henri Prade

IRIT, Université Paul Sabatier, 31062 Toulouse Cedex 09, France
{dubois,prade}@irit.fr

Abstract. Usual propositional possibilistic logic formulas are pairs made of a classical logic formula associated with a weight thought of as a lower bound of its necessity measure. In standard possibilistic logic, only conjunctions of such weighted formulas are allowed (a weighted classical conjunction is equivalent to the conjunction of its weighted conjuncts, due to the min-decomposability of necessity measures). However, the negation and the disjunction of possibilistic logic formulas make sense as well. They were briefly introduced by the authors some years ago, in a multiple agent logic context. The present paper hints at the multi-tiered logic that is thus generated, and discusses its semantics in terms of families of possibility distributions. Its practical interest for expressing higher order epistemic states is emphasized.

1 Introduction

Possibilistic logic formulas [4,5] are pairs made of a classical logic formula p, which as such can be nothing but true or false, associated with a weight $\alpha \in [0,1]$ interpreted in connection with different types of set-functions in possibility theory. They are of interest when modeling uncertainty, preferences, or priorities. In the following, we restrict ourselves to *propositions* estimated in terms of *(strong) necessity* measures N and their dual *(weak) possibility* measures $\Pi(p) = 1 - N(\neg p)$. The possibilistic formula (p, α) is semantically understood as standing for the constraint $N(p) \geq \alpha$. Necessity measures have the characteristic property of being min-decomposable: $N(p \wedge q) = \min(N(p), N(q))$. Thus, it turns out that $N(p \wedge q) \geq \alpha \Leftrightarrow N(p) \geq \alpha$ and $N(q) \geq \alpha$, or if we prefer, $(p \wedge q, \alpha)$ is equivalent to $(p, \alpha) \wedge (q, \alpha)$. So a possibilistic logic base can be put in the form of a conjunction of weighted classical clauses.

Necessity measures are neither decomposable w.r.t. disjunction nor w.r.t. negation. The constraint '$N(p \vee q) \geq \alpha$' represents a piece of information weaker than '$N(p) \geq \alpha$ or $N(q) \geq \alpha$', and '$N(\neg p) \geq \alpha$' is a piece of information much stronger than 'not $N(p) \geq \alpha$' $\Leftrightarrow$ '$N(p) < \alpha$' $\Leftrightarrow$ '$\Pi(\neg p) > 1 - \alpha$' (since $\min(N(p), N(\neg p)) = 0$ and $N(p) = 0 \Leftrightarrow \Pi(\neg p) = 1$). This behavior agrees with the intuition that being sure of $p \wedge q$ requires being sure of both p and of q, while one may be sure of $p \vee q$ without being sure of p nor of q. Similarly, not being fully sure of p is much weaker than being sure of $\neg p$.

In the following, we discuss how to handle the disjunction and the negation of standard possibilistic logic formulas, at the semantic level and at the syntactic

S. Benferhat and J. Grant (Eds.): SUM 2011, LNAI 6929, pp. 428–432, 2011.
© Springer-Verlag Berlin Heidelberg 2011

level in the setting of a multi-tiered logic, in the multiple agent perspective first suggested in [6]. Indeed, while we may not be sure whether another agent is certain of p or certain of q, it is strange to make such a statement about oneself.

2 Semantics of Possibilistic Logic Constraints

A necessity measure N and the dual possibility measure Π are associated with a possibility distribution [8] over interpretations of the language. Namely, $\Pi(p) = \max_{\omega \in [p]} \pi(\omega)$ where $[p]$ denotes the sets of models of p. Thus, $N(p) \geq \alpha$ corresponds semantically to the possibility distribution $\pi_{(p,\alpha)}(\omega) = \max(\mu_{[p]}(\omega), 1-\alpha)$ where $\mu_{[p]}$ is the characteristic function of $[p]$ [4]. In fact, the constraint $N(p) \geq \alpha$ ($\Leftrightarrow \Pi(\neg p) \leq 1 - \alpha$) defines the *set* of possibility distributions $\mathbf{Pi}((p, \alpha)) = \{\pi | \max_{\omega \in [\neg p]} \pi(\omega) \leq 1 - \alpha\} = \{\pi | \forall \omega\ \pi_{(p,\alpha)}(\omega) \geq \pi(\omega)\}$ that are at least as informative as $\pi_{(p,\alpha)}$ (the last equality holds since $\forall \omega \in [p]\ \pi_{(p,\alpha)}(\omega) = 1$).

In this view, the pair (p, α) is both a possibilistic logic formula at the object level, and a classical formula at the meta level. Indeed, since (p, α) is semantically interpreted as $N(p) \geq \alpha$, a possibilistic formula can be manipulated as a formula that is true (if $N(p) \geq \alpha$) or false (if $N(p) < \alpha$). Then possibilistic formulas can be combined with all propositional connectives. For instance, the conjunction '$N(p) \geq \alpha$ and $N(q) \geq \beta$' defines the set of distributions smaller or equal to the min-combination of the largest possibility distributions representing each constraint. Indeed:

$$\mathbf{Pi}((p, \alpha) \wedge (q, \beta)) = \{\pi | \pi \leq \min(\pi_{(p,\alpha)}, \pi_{(q,\beta)})\} = \mathbf{Pi}((p, \alpha)) \cap \mathbf{Pi}((q, \beta)).$$

As for disjunction, the set of possibility distributions representing the disjunctive constraint '$N(p) \geq \alpha$ or $N(q) \geq \beta$' has no longer a unique extremal element in general. Indeed: $\mathbf{Pi}((p, \alpha) \vee (q, \beta)) = \{\pi |\ \pi_{(p,\alpha)} \geq \pi \text{ or } \pi_{(q,\beta)} \geq \pi\} = \mathbf{Pi}((p, \alpha)) \cup \mathbf{Pi}((q, \beta))$, while $\mathbf{Pi}((p \vee q, \alpha)) = \{\pi | \max(\pi_{(p,\alpha)}, \pi_{(q,\alpha)}) \geq \pi\} \supseteq \{\pi | \pi_{(p,\alpha)} \geq \pi \text{ or } \pi_{(q,\alpha)} \geq \pi\}$. For the negation of a possibilistic formula, we get $\mathbf{Pi}(\neg(p, \alpha)) = \{\pi | \max_{\omega \in [\neg p]} \pi(\omega) > 1 - \alpha\} = \{\pi | \exists \omega \in [\neg p]\ \pi(\omega) > 1 - \alpha\} = \{\pi | \pi \not\leq \pi_{(p,\alpha)}\} = \overline{\mathbf{Pi}((p, \alpha))} \supset \mathbf{Pi}((\neg p, \alpha))$, where overline represents complementation.

A possibility-qualified statement in the sense of [9], syntactically denoted by $< p, \alpha >$, stands for a constraint of the form $\Pi(p) \geq \alpha$. It represents the set of possibility distributions $\mathbf{Pi}(< p, \alpha >) = \{\pi | \max_{\omega \in [p]} \pi(\omega) \geq \alpha\}$. Then again:

$$\mathbf{Pi}(< p, \alpha > \vee < q, \beta >) = \mathbf{Pi}(< p, \alpha >) \cup \mathbf{Pi}(< q, \beta >),$$

$$\mathbf{Pi}(< p, \alpha > \wedge < q, \beta >) = \mathbf{Pi}(< p, \alpha >) \cap \mathbf{Pi}(< q, \beta >),$$

and $\mathbf{Pi}(\neg < p,\alpha >) = \overline{\mathbf{Pi}(< p, \alpha >)} \subset \mathbf{Pi}(< \neg p,\alpha >)$. Moreover, $\mathbf{Pi}(< p \wedge q, \alpha >) \subseteq \mathbf{Pi}(< p,\alpha >) \cap \mathbf{Pi}(< q,\alpha >)$. Note that $< p, \alpha > \wedge < \neg p, \beta >$ represents (graded) ignorance about p, since it claims that both p and $\neg p$ are somewhat possible. There is a close connection between $< \neg p, 1 - \alpha >$ and $\neg(p, \alpha)$. If the necessity scale is a finite subset$\{0 < \alpha_1 < \cdots < \alpha_n = 1\}$ of $[0, 1]$, then $\neg(p, \alpha_i)$ is the same as $< \neg p, 1 - \alpha_{i-1} >$. On the whole unit interval, though, one should then distinguish between formulas that express inequalities in the broad sense such as $N(p) \geq \alpha$ or $\Pi(p) \geq \alpha$, and those that express strict inequalities such as $N(p) > \alpha$ or $\Pi(p) > \alpha$, to make this connection at the syntactic level.

A piece of information such as (p, α) is naturally held by an agent or more generally a set of agents $\mathcal{A} \subseteq \mathcal{U}$. This is denoted by $(p, \alpha/\mathcal{A})$. It means that all the agents in $\mathcal{A}$ are certain at level α that p is true. The semantics is given by a collection of possibility distributions $\pi^a_{(p,\alpha)}$, where $a \in \mathcal{U}$. Namely, $\pi^a_{(p,\alpha)} = \pi_{(p,\alpha)}$ if $a \in \mathcal{A}$, and $\pi^a_{(p,\alpha)} = 1$ (total ignorance) if $a \notin \mathcal{A}$.

One can move one step further by considering nested formulas of the form $((p, \alpha), \beta)$, viewing (p, α) as a true or false statement (indeed $\pi \in \mathbf{Pi}((p, \alpha))$ or not) using the *forcing* semantics [3]. One can define the set $\mathbf{Pi}(((p, \alpha), \beta))$ of higher-order possibility distributions π^2 over the π's such that $\pi \leq \pi_{(p,\alpha)}$ (that makes $N(p) \geq \alpha$ true). It possesses a greatest element $\pi^2_{((p,\alpha),\beta)}$ such that $\pi^2_{((p,\alpha),\beta)}(\pi) = 1$ if $\pi \leq \pi_{(p,\alpha)}$ and $\pi^2_{((p,\alpha),\beta)}(\pi) = 1 - \beta$ otherwise. The higher-order formula $((p, \alpha), \beta)$ may be then reduced to $(p, \min(\alpha, \beta))$ via the disjunctive weighted aggregation $\max(\min(\pi_{(p,\alpha)}, 1), \min(1, 1 - \beta))$. Then, $((p, \alpha), \beta)$ is interpreted as: either it is the case that $N(p) \geq \alpha$ with a possibility level equal to 1, or one knows nothing with possibility $1 - \beta$. Similarly, the semantics of $((p, \alpha/\mathcal{A}), \beta/\mathcal{B})$, is obtained by associating, to each agent $b \in \mathcal{B}$, a possibility distribution over a set of possibility distributions π^a (such that $\pi^a \leq \pi^a_{(p,\alpha)}$ or not) for each $a \in \mathcal{U}$. In case $a = b$, the above reduction may be applied.

Now we can consider expressions such as $(\neg(p, \mathcal{A}) \vee (q, \beta/\mathcal{B}), \gamma/\mathcal{C})$ (stating that for agents in $\mathcal{C}$ it is γ-certain that if the agents in $\mathcal{A}$ are certain of p (at level 1), those in $\mathcal{B}$ are certain of q at least at level β), or $((p, \alpha/\mathcal{A}), \beta/\mathcal{B})$ (for expressing that agents in $\mathcal{B}$ are certain at least at level β that agents in $\mathcal{A}$ are certain at least at level α that p is true).

3 Inference in Generalized Possibilistic Logic

To achieve inference in this kind of generalised possibilistic logic, we shall use the forcing approach to entailment [3]. The semantic entailment of standard possibilistic logic defined by an inequality between two maximal possibility distributions can be equivalently expressed by the inclusion between two sets of distributions. But only the latter scales up to generalised possibilistic logic. Namely if Φ and Ψ are generalised possibilistic formulae of the same kind, then $\Phi \models \Psi$ if and only if $\mathbf{Pi}(\Phi) \subseteq \mathbf{Pi}(\Psi)$, which presupposes that both possibility sets refer to the same universe where possibility distributions are defined.

Syntactic inference from nested possibilistic formulas should be handled as a two-layer process (assuming for simplicity that nestedness is not iterated, i.e. the insertion of standard possibilistic logic formulas inside possibilistic logic formulas is not iterated). More precisely, classical resolution may be applied "externally" to the possibilistic logic formulas of the highest level (regarding the possibilistic logic formulas inside (if any) as classical formulas), or "internally" to the possibilistic logic formulas inside, once the "context" has been properly made homogeneous (by weakening) in agreement with the semantics described above.

Namely, the following internal inference rule is clearly valid

$$(\neg p \vee q, \alpha/\mathcal{A}); (p \vee r, \beta/\mathcal{B}) \models (q \vee r, \min(\alpha, \beta)/\mathcal{A} \cap \mathcal{B})$$

When $\mathcal{A} = \mathcal{U} = \mathcal{B}$, we retrieve the standard possibilistic resolution rule. We also have the following weakening and fusion rules

- $\forall\, \beta \leq \alpha\ \forall\, \mathcal{B} \subseteq \mathcal{A}\ (p, \alpha/\mathcal{A}) \vdash (p, \beta/\mathcal{B})$ **(weight weakening)**
- si $p \vdash q$, alors $(p, \alpha/\mathcal{A}) \vdash (q, \alpha/\mathcal{A})$ **(logical weakening)**
- $(p, \alpha/\mathcal{A})\ (p, \beta/\mathcal{A}) \vdash (p, \max(\alpha, \beta)/\mathcal{A})$ **(weight fusion 1)**
- $(p, \alpha/\mathcal{A})\ (p, \alpha/\mathcal{B}) \vdash (p, \alpha/(\mathcal{A} \cup \mathcal{B}))$ **(weight fusion 2)**

This is illustrated on the two following examples.

1. Consider the two following formulas: $\{((p, \alpha/\mathcal{A}'), \rho/\mathcal{C}), \neg(p, \beta/\mathcal{A}) \vee (q, \gamma/\mathcal{B}), \delta/\mathcal{D})\}$. Assume $\alpha > \beta$ and $\mathcal{A}' \supset \mathcal{A}$. Then from the first premise, we get $((p, \beta/\mathcal{A}), \rho/\mathcal{C})$ by weakening; then by "external" resolution with the second expression, we obtain
$$((q, \gamma/\mathcal{B}), \min(\rho, \delta)/\mathcal{C} \cap \mathcal{D}).$$
2. Consider now $\{((p, \alpha/\mathcal{A}'), \rho/\mathcal{C}), ((\neg p \vee q, \beta/\mathcal{B}), \delta/\mathcal{D})\}$. Assume $\rho > \delta$ and $\mathcal{C} \supset \mathcal{D}$. By weakening, we get $((p, \alpha/\mathcal{A}'), \delta/\mathcal{D})$; and by "internal resolution", we finally obtain
$$((q, \min(\alpha, \beta)/\mathcal{A}' \cap \mathcal{B}), \delta/\mathcal{D}).$$

It is clear that the above weakening steps can be always applied by taking the minimum of the certainty levels, and the intersection of the sets of agents, even if the first statement does not involve higher certainty levels and larger sets of agents. Mind that getting an empty set of agents after intersection makes the result trivial.

Lastly, the difference between the formulas $(\neg p \vee q, \alpha)$ and $\neg(p, \alpha) \vee (q, \alpha)$, for $\alpha > 0$, in the presence of (p, α) affects inferences one may draw from them[1]. Indeed, consider the formulas $(\neg p \vee q, \alpha)$ and $\neg(p, \alpha) \vee (q, \alpha)$. The latter means that either (p, α) cannot be ascertained or (q, α) is sure. It is interesting to observe that while the formula $(\neg p \vee q, \alpha)$, enables us to deduce both (q, α) if (p, α) holds, and $(\neg p, \alpha)$ if $(\neg q, \alpha)$ holds, the formula $\neg(p, \alpha) \vee (q, \alpha)$ still enables us to get (q, α) from (p, α) (since (p, α) is taken for granted), but no longer $(\neg p, \alpha)$ in the presence of $(\neg q, \alpha)$. Indeed, $(\neg q, \alpha), \alpha > 0$ expresses that $N(\neg q) \geq \alpha$, which entails $N(q) = 0$. Now the latter along with $\neg(p, \alpha) \vee (q, \alpha)$, that is, $N(p) < \alpha$ or $N(q) \geq \alpha > 0$ entails $N(p) < \alpha$. It differs from the stronger conclusion $(\neg p, \alpha)$, i.e., $N(\neg p) \geq \alpha$, which may be obtained from $(\neg p \vee q, \alpha)$ and $(\neg q, \alpha)$. Moreover $N(p) < \alpha$, i.e., $\neg(p, \alpha)$ is a different kind of possibilistic formula (it is precisely $< \neg p, 1 - \alpha^- >$) where $\alpha^- < \alpha$ is the next smaller element in the necessity scale.

The asymmetric nature of $\neg(p, \alpha) \vee (q, \alpha)$ appears to be in the spirit of logic programming (LP), encoding rules of the form "q is certain provided that p is

[1] We do not indicate the set of agents here since they do not play any role in the point.

certain". However in LP, such rules are only used to deduce q from p, while in possibilistic logic, as shown above, it is allowed to use them in the opposite way. If moreover one wants to introduce negation as failure, as in "q is certain provided that p is certain and that one cannot establish r", this can be expressed as if $N(p) \geq \alpha$ and $\Pi(\neg r) \geq \beta$ then $N(q) \geq \alpha$. It corresponds to the generalized possibilistic formula $\neg(p, \alpha) \vee \neg < \neg r, \beta > \vee (q, \alpha)$. A mixed resolution rule exists (e.g., [5]) for reasoning from such clauses, namely we have $< \neg p \vee q, \alpha >$; $(p \vee r, \beta) \vdash < q \vee r, \beta >$ if $\beta > 1 - \alpha$. One can indeed mimic non-monotonic logic programming as suggested in [7].

4 Concluding Remarks

We have highlighted the possibility to substantially enlarge the framework of possibilistic logic by making possible the manipulation of formulas expressing the epistemic states of agents in its setting, and advocated the interest of such an extension. The development of the whole machinery would require an article much longer than this introductory paper. To this end, one may draw lessons from modal logic MEL [1] where the binary-valued possibility case is handled as a two-tiered classical logic. It should enable us to handle constraints of the form $\alpha \geq N(p) \geq \beta$ (using $\Pi(\neg p) > 1 - \alpha$). It is not to be confused with the situation where the lower bound of $N(p)$ is imprecisely located inside a given interval, as studied in the recent interval-based extension of possibilistic logic [2].

References

1. Banerjee, M., Dubois, D.: A simple modal logic for reasoning about revealed beliefs. In: Sossai, C., Chemello, G. (eds.) ECSQARU 2009. LNCS (LNAI), vol. 5590, pp. 805–816. Springer, Heidelberg (2009)
2. Benferhat, S., Hué, J., Lagrue, S., Rossit, J.: Interval-based possibilistic logic. In: Proc. 22nd Inter. Joint Conf. on Artif. Intellig (IJCAI 2011), Barcelona (July 16-22, 2011)
3. Boldrin, L., Sossai, C.: Local possibilistic logic. Journal of Applied Non-Classical Logics 7(3), 309–333 (1997)
4. Dubois, D., Lang, J., Prade, H.: Possibilistic logic. In: Gabbay, D.M., et al. (eds.) Handbook of Logic in Artificial Intelligence and Logic Programming, vol. 3, pp. 439–513. Oxford University Press, Oxford (1994)
5. Dubois, D., Prade, H.: Possibilistic logic: a retrospective and prospective view. Fuzzy Sets and Systems 144, 3–23 (2004)
6. Dubois, D., Prade, H.: Toward multiple-agent extensions of possibilistic logic. In: Proc. IEEE Inter. Conf. on Fuzzy Systems (FUZZ-IEEE 2007), London (UK), pp. 187–192 (July 23-26, 2007)
7. Dubois, D., Prade, H., Schockaert, S.: Rules and meta-rules in the framework of possibility theory and possibilistic logic. Scientia Iranica,18, Special issue dedicated to the 90th birthday of L. A. Zadeh (to appear, 2011)
8. Zadeh, L.A.: Fuzzy sets as a basis for a theory of possibility. Fuzzy Sets and Systems 1, 3–28 (1978)
9. Zadeh, L.A.: PRUF: A meaning representation language for natural languages. Int. J. of Man-Machine Studies 10, 395–460 (1978)

Transformations around Quantitative Possibilistic Logic

Hadja Faiza Khellaf-Haned

LRIA, Computer Science Department, USTHB,
PB 32 Bab Ezzouar Algiers Algeria
fkhellaf@usthb.dz

Abstract. In the framework of quantitative possibility theory, two representation modes were developed: logical representation in term of quantitative possibilistic base and graphical representation in term of product-based possibilistic network. This article deals with logical and graphical representations of uncertain information around quantitative possibility theory. First, a deep analysis of relationships between these two forms of representational frameworks is provided. Then, in the logical setting, syntactical relations between penalty logic and quantitative possibilistic base are developed. Afterward, the relationship which exists between UCP networks and product-based possibilistic networks is pointed out in the graphical setting. These translations are useful for different applications and are interesting by taking advantage from each format at the inferential level. From these translations, we also exhibit the relation which is deduced, between UCP networks and penalty logic.

Keywords: possibilistic logic, product-based possibilistic network, UCP network, penalty logic.

1 Introduction

Generally, uncertain pieces of information or flexible constraints can be represented in different equivalent formats. In possibility theory, possible formats can be:

- logical-based representations which are simple extensions of classical logic,
- graphical-based representations, viewed as counterparts of probabilistic Bayesian networks [1,2].

In graphical representations [3,4,10], uncertain information is encoded by means of possibilistic networks which are composed of Directed Acyclic Graph (DAG) and conditional possibility distributions. In logical representations [6], uncertain information is encoded by means of possibilistic knowledge bases which are sets of weighted formulas having the form (ϕ_i, α_i) where ϕ_i is a propositional formula and α_i is a positive real number belonging to the unit interval [0,1]. Each possibilistic network (resp. each possibilistic knowledge base) induces a ranking between possible interpretations of a language, called a possibility distribution.

S. Benferhat and J. Grant (Eds.): SUM 2011, LNAI 6929, pp. 433–446, 2011.
© Springer-Verlag Berlin Heidelberg 2011

The possibility degree associated with an interpretation is obtained by combining the satisfaction degrees of this interpretation with respect to each weighted formula of the knowledge base, or with respect to each conditional possibility degree of the possibilistic network. Two combination operators have been generally used [6]: minimum operator and product operator. Therefore, there are two kinds of possibilistic networks: min-based possibilistic networks and product-based possibilistic networks. Similarly, two kinds of possibilistic knowledge bases are defined: min-based possibilistic logic (standard possibilistic logic) and product-based possibilistic logic called also quantitative possibilistic logic. In this paper, we first investigate the syntactic relations which exists between a possibilistic knowledge base and a penalty knowledge base. The penalty logic has interesting proprieties. Indeed, on of the important advantage of penalty logic is its ability to deal with inconsistency. Then, we exhibit relationships between Utility CP-networks and Product-based possibilistic networks.

The rest of this paper is organized as follows. The following section gives logical models for representing uncertain knowledge. Section 3 describes graphical models for representing uncertain knowledge. Section 4 relates the transformation from a product-based possibilistic network to quantitative possibilistic base. Section 5 provides the main translations released around quantitative possibility theory. Section 6 concludes the paper.

2 Logical Frameworks for Uncertain Knowledge

2.1 Quantitative Possibilistic Logic

Let $\mathcal{L}$ be a finite propositional language and Ω be the set of all propositional interpretations. Let $\phi, \psi, \ldots$ be propositional formulas. For interpretation ω and propositional formula $\phi, \omega \models \phi$ means that ω is a model (in the way of propositional logic) of ϕ. A possibility distribution [6] π is a mapping from a set of interpretations Ω into the unit interval [0,1]. A possibility distribution π is said to be normalized if an interpretation ω exists such as $\pi(\omega) = 1$. In this paper, only normalized possibility distributions are considered. Given a possibility distribution π, two dual measures are defined on the set of propositional formulas:

- The possibility measure of a formula ϕ, defined by:

$$\Pi(\phi) = max\{\pi(\omega) : \omega \models \phi \ and \ \omega \in \Omega\}$$

  which evaluates the extent to which ϕ is consistent with the available beliefs expressed by π. For $\phi \equiv \bot$ (a contradiction), we have $\Pi(\bot) = 0$.
- The necessity measure of a formula ϕ, defined by:

$$N(\phi) = 1 - \Pi(\neg\phi)$$

  which evaluates the extent to which ϕ is entailed by the available beliefs. For $\phi \equiv \top$ (a tautology), we have $N(\top) = 1$.

A possibilistic knowledge base Σ is a set of weighted formulas:

$$\Sigma = \{(\phi_i, \alpha_i) : i = 1, ..., n\}$$

where ϕ_i is a propositional formula and $\alpha_i \in]0, 1]$ represents the certainty level of ϕ_i. Each piece of information (ϕ_i, α_i) of a possibilistic knowledge base can be viewed as a constraint that restricts possibility degrees associated with interpretations. The possibility distribution associated with a weighted formula (ϕ_i, α_i) is: $\forall \omega \in \Omega$,

$$\pi_{(\phi_i,\alpha_i)}(\omega) = \begin{cases} 1 - \alpha_i \; if \; \omega \not\models \phi_i \\ 1 \quad \text{otherwise} \end{cases} \tag{1}$$

More generally, the possibility distribution associated with a possibilistic knowledge base Σ is the result of combining possibility distributions associated with each weighted formula (ϕ_i, α_i) of Σ, namely: $\forall \omega \in \Omega$,

$$\pi_\Sigma(\omega) = \oplus\{\pi_{(\phi_i,\alpha_i)}(\omega) : (\phi_i, \alpha_i) \in \Sigma\}. \tag{2}$$

where $\oplus$ is in general either equal to the minimum operator (in standard possibilistic logic), or to the product operator (*). In the rest of the paper, we only focus on the case where $\oplus = *$. The possibilistic base Σ is then called product-based or quantitative possibilistic knowledge base. Equation (2) can then be written as: $\forall \omega \in \Omega$,

$$\pi_\Sigma(\omega) = \begin{cases} 1 & \text{if } \forall(\phi_i, \alpha_i) \in \Sigma, \omega \models \phi_i \\ *\{1 - \alpha_i : (\phi_i, \alpha_i) \in \Sigma, \omega \not\models \phi_i\} & \text{otherwise} \end{cases} \tag{3}$$

2.2 Penalty Logic

Penalty logic introduced by Pinkas [7] and developed in [8], associates to each formula of a knowledge base the price to pay if this formula is violated. More the penalty is higher more the formula is important. Formally, let $\mathbb{R}^{*+}$ be the union of the set of all the strictly positive real numbers and $\{+\infty\}$, reserved to quantify the completely certain formulas which allow no exceptions. A penalty knowledge base PK is a finite multi-set of pairs (ϕ_i,α_i) where $\phi_i \in \mathcal{L}$ and $\alpha_i \in \mathbb{R}^{*+}$, α_i is the penalty associated to ϕ_i. If α_i=$+\infty$ then it is forbidden to remove ϕ_i from PK (ϕ_i is inviolable).

Cost of an Interpretation. Let $PK = \{(\phi_i, \alpha_i) : i = 1, ..., n\}$ be a penalty knowledge base. The cost of an interpretation $\omega \in \Omega$ with respect to PK, denoted by κ_{PK}, is equal to the sum of the penalties of the formulas in PK violated by ω [7]:

$$\kappa_{PK}(\omega) = \begin{cases} 0 & \text{if } \forall(\phi_i, \alpha_i) \in PK, \omega \models \phi_i \\ \sum\{\alpha_i : (\phi_i, \alpha_i) \in PK, \omega \not\models \phi_i\} & \text{otherwise} \end{cases} \tag{4}$$

Then, a minimum cost interpretation corresponds to the most interesting one.

Example 1. Let $PK = \{(\neg a \vee b, 4), (b \vee \neg c, 8), (\neg a \vee c, 2), (c, +\infty)\}$. The corresponding interpretations costs are given by Table 1.

Table 1. Interpretations costs of Example 1

ω	$\kappa_{PK}(\omega)$	ω	$\kappa_{PK}(\omega)$
$a\ b\ c$	0	$\neg a\ b\ c$	0
$a\ b\ \neg c$	$+\infty$	$\neg a\ b\ \neg c$	$+\infty$
$a\ \neg b\ c$	4+8=12	$\neg a\ \neg b\ c$	8
$a\ \neg b\ \neg c$	$+\infty$	$\neg a\ \neg b\ \neg c$	$+\infty$

Cost of Consistency of a Formula. The cost of consistency of a formula ϕ with respect to PK, denoted $K_{PK}(\phi)$, is the minimum cost with respect to PK of an interpretation satisfying ϕ:

$$K_{PK}(\phi) = min_{\omega \in \Omega}\{\kappa_{PK}(\omega) : \omega \models \phi\} \tag{5}$$

3 Graphical Frameworks for Uncertain Knowledge

3.1 Product-Based Possibilistic Networks

A possibilistic network [4,9,10] (which can be viewed as a counterpart of a probabilistic network) is a graphical representation of uncertain pieces of information. Let $V = \{A_1, A_2, .., A_n\}$ be a set of variables (or attributes). We denote by D_i the domain associated with the variable A_i. For Boolean variables, x_i denotes any of the two instances of A_i which can be either $x_i = a_i$ or $x_i = \neg a_i$. In this paper, only binary variables are considered without loss of generality. The Cartesian product of all boolean variable domains in V is simply the set of interpretations. Depending on the context, interpretations are denoted either by tuples: $\omega = (a_1, ..., a_n)$ or by conjunctions: $\omega = a_1 \wedge .... \wedge a_n$. A possibilistic network, denoted by Π_G, is a Directed Acyclic Graph (DAG), where nodes represent variables and edges encode the "causal" (or influence) links between these variables. When a link exists from a node A_i to a node A_j, A_i is called a parent of A_j. The set of parents of a node A_j is denoted by $Par(A_j)$, and an instance of $Par(A_j)$ is denoted by u_j. Uncertainty is represented on each node by means of normalized conditional possibility distributions and expresses the strength of the links between variables. Conditional possibility distributions are associated with the DAG in the following way:

- For root nodes A_i, we specify the prior possibility degrees of $\Pi(a_i)$ and $\Pi(\neg a_i)$ with $max(\Pi(a_i), \Pi(\neg a_i)) = 1$ (the normalization condition).
- For other nodes A_j, we specify, for each u_j an instance of $Par(A_j)$, conditional possibility degrees of $\Pi(a_j \mid u_j)$ and $\Pi(\neg a_j \mid u_j)$ with: $max(\Pi(a_j \mid u_j), \Pi(\neg a_j \mid u_j)) = 1$.

Example 2. Let us consider the product-based possibilistic network Π_G represented by the DAG of Figure 1. The local conditional possibility distributions are given by Tables 2.

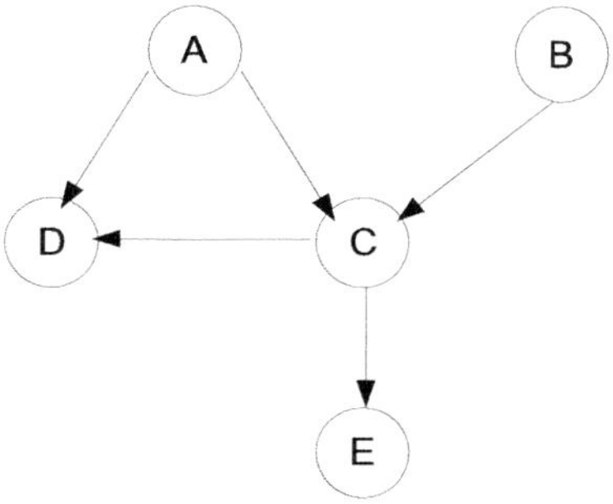

Fig. 1. Example of a DAG

Table 2. Conditional possibility distributions for variables A,B, C, D and E

A	$\Pi(A)$	B	$\Pi(B)$	ABC	$\Pi(C \mid AB)$	ACD	$\Pi(D \mid AC)$	CE	$\Pi(E \mid C)$
a	1	b	0.2	abc	0.3	acd	1	ce	0.3
$\neg a$	0.4	$\neg b$	1	$ab\neg c$	1	$ac\neg d$	0.3	$c\neg e$	1
				$a\neg bc$	0.4	$a\neg cd$	1	$\neg ce$	1
				$a\neg b\neg c$	1	$a\neg c\neg d$	0.4	$\neg c\neg e$	0.5
				$\neg abc$	1	$\neg acd$	0.2		
				$\neg ab\neg c$	0.1	$\neg ac\neg d$	1		
				$\neg a\neg bc$	0.2	$\neg a\neg cd$	0.3		
				$\neg a\neg b\neg c$	1	$\neg a\neg c\neg d$	1		

In possibility theory, different kinds of possibilistic conditioning have been defined (for a detailed discussion on possibilistic conditioning see [11,12,13]). In this section, we only recall the product-based conditionning:

– A product-based conditioning is defined as:

$$\Pi(\psi \mid \phi) = \begin{cases} \frac{\Pi(\psi \wedge \phi)}{\Pi(\phi)} & \text{if } \Pi(\phi) \neq 0 \\ 1 & \text{otherwise} \end{cases} \tag{6}$$

Each product-based possibilistic network Π_G (DAG and local conditional possibility distributions) induces a unique joint possibility distribution using a so-called product-based chain rule similar to the one used in probabilistic Bayesian networks. Let $\omega = (x_1, x_2, ..., x_n)$ be a given interpretation, and x_i is an instance of A_i which can be either $x_i = a_i$ or $x_i = \neg a_i$. The product-based chain rule is defined by:

$$\pi_G(\omega) = *\{\Pi(x_i \mid u_i) : \omega \models x_i \wedge u_i, i = 1, .., n\} \tag{7}$$

where u_i is an instance of the parents of A_i.

3.2 UCP-Networks

A UCP-network, denoted by U_G [14] is a directed graphical representation of utility functions that combines aspects of two preference models: Generalized

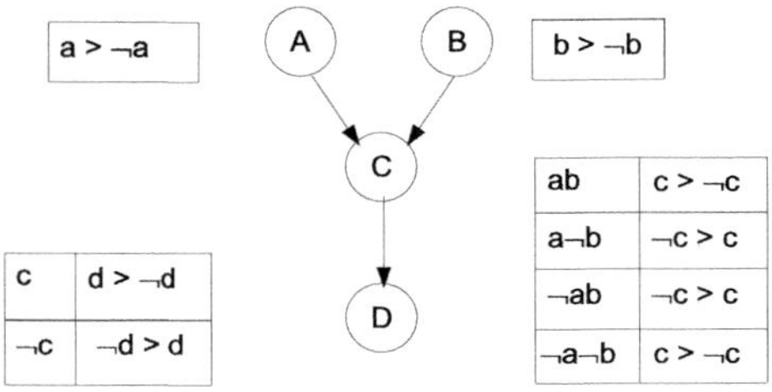

Fig. 2. A CP-Network

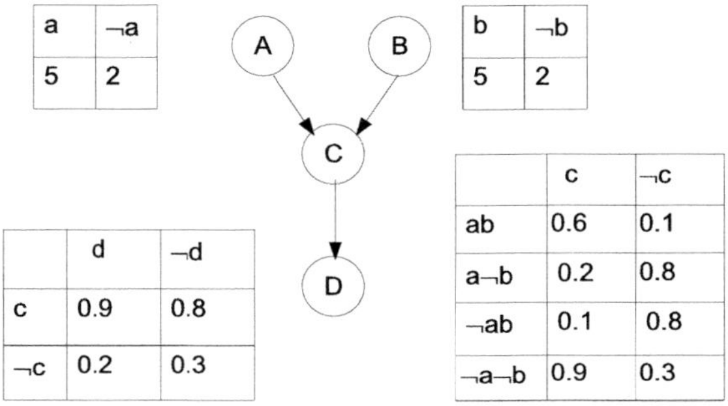

Fig. 3. A UCP-Network

additive models based on the notion of generalized additive independence (GAI) [15] and CP-network [16] which allows to represent qualitative preference functions that captures conditional preference statements under a *cetris paribus* (all else equal) assumption. CP-networks are directed acyclic graphs whose nodes are the variables of V, where a conditional preference table (CPT) is associated to each node X specifying a preference order over $X's$ values given each instantiation, of its parents U.

Figure 2 represents a CP-network, given in [14] defined over four variables, where, for example, the CPT for C specifies that c is preferred to $\neg c$ when a and b hold. Let $X_1, ..., X_k$ be the sets of variables such that $V = \cup_i X_i$. $X_1, ..., X_k$ are generalized additive independent (GAI) for an underlying utility function u if u can be written as [15]:

$$u(V) = \sum_{i=1}^{k} f_i(X_i) \tag{8}$$

A UCP-net extends a CP-net by allowing quantification of nodes with conditional utility information. Semantically, the different factors are treated as generalized additive independent of one another. For example the network in Figure 2 can be extended with utility information by including a factor for each family in the network, specifically, $f_1(A), f_2(B), f_3(A, B, C)$ and $f_4(C, D)$ (see Figure 3). We interpret this network using GAI: $u(A, B, C, D) = f_1(A) + f_2(C) + f_3(A, B, C) +$

$f_4(C, D)$. Each of these factors is quantified by quantitative CPT tables in the network. For exemple, in Figure 3, $f_3(a, b, c) = 0.6$ while $f_3(a, b, \neg c) = 0.1$. Thus, the CPT tables along with the GAI interpretation provide a full specification of the utility function. For example, $u(a, b, \neg c, \neg d) = f_1(a) + f_2(b) + f_3(a, b, \neg c) + f_4(\neg c, \neg d) = 5 + 5 + 0.1 + 0.3 = 10.4$.

4 Logical Encoding of Product-Based Possibilistic Networks

In [18], the logical encoding of product-based possibilistic network was given. Let Π_G be a product-based possibilistic network represented by a set of triples as in [17]:

$$\mathcal{P}_G = \{(x_i, u_i, \alpha_i) : \Pi(x_i \mid u_i) = \alpha_i \neq 1 \in \Pi_G\},$$

where x_i is an instance of the variable A_i and u_i is an instance of $Par(A_i)$. Then, the possibilistic base Σ_G associated with a product-based possibilistic network Π_G is defined as follow:

$$\Sigma_G = \{(\neg x_i \vee \neg u_i, 1 - \alpha_i) : (x_i, u_i, \alpha_i) \in \mathcal{P}_G\}. \quad (9)$$

Example 3. The possibilistic knowledge base associated with $\mathcal{P}_G$ of Example 2 is:
$\Sigma_G = \{(a, 0.6), (\neg b, 0.8), (\neg a \vee \neg b \vee \neg c, 0.7), (\neg a \vee b \vee \neg c, 0.6), (a \vee \neg b \vee c, 0.9), (a \vee b \vee \neg c, 0.8), (\neg a \vee \neg c \vee d, 0.7), (\neg a \vee c \vee d, 0.6), (a \vee \neg c \vee \neg d, 0.8), (a \vee c \vee \neg d, 0.7), (\neg c \vee \neg e, 0.7), (c \vee e, 0.5)\}$.

The equivalence between the possibility distribution $\pi_{\Sigma_G}(\omega)$ associated with a knowledge base Σ_G and the conditional possibility distribution $\pi_G(\omega)$ induced by a product-based possibilistic network $\mathcal{P}_G$ is given by the following equation:

$$\forall \omega \in \Omega, \pi_{\Sigma_G}(\omega) = \pi_G(\omega)$$

where π_{Σ_G} is obtained using Equation (3) and π_G is obtained from Equation (7).

Note that the transformation from product-based possibilistic networks into quantitative knowledge bases is efficient. More precisely, it is linear (with respect to the number of parameters in the product-based possibilistic network ,i.e., number of conditional possibility degrees).

5 Transformations around Quantitative Possibility Theory

This section completes the translations developed around quantitative possibility theory. First, we exhibit the syntactical relations which exist between quantitative possibilistic base and penalty base. Second, we point out the relations between product-based possibilistic networks and UCP-networks. Finally, using the set of the developed procedures, we provide the relation which permits to transform a UCP-network to an equivalent penalty base.

5.1 Encoding Quantitative Possibilistic Base to Penalties

Clearly, under a specific scale, there are strong relationships between penalty logic and quantitative possibilistic base. Indeed, let Σ be a quantitative possibilistic base defined by:

$$\Sigma = \{(\phi_i, \alpha_i) : i = 1, ..., n\}, \alpha_i \in]0, 1].$$

The penalty base PK associated to the quantitative possibilistic base PK is defined by: $PK = \{(\phi_i, k_i) : (\phi_i, \alpha_i) \in \Sigma, k_i = -ln(1 - \alpha_i)\}$.

Proposition 1. *Let Σ be a quantitative possibilistic base defined by:*

$\Sigma = \{(\phi_i, \alpha_i) : i = 1, ..., n, \ -ln(1 - \alpha_i) \in \mathbb{N}\}$.

Let PK be the penalty base associated to the quantitative possibilistic base Σ defined by:

$PK = \{(\phi_i, k_i) : i = 1, ..., n, k_i \in \mathbb{N} \cup \{+\infty\}\}$ *with* $(\phi_i, \alpha_i) \in \Sigma$ *and* $k_i = -ln(1 - \alpha_i)$.

Then, π_Σ is the possibility distributions associated to Σ and κ_{PK} is the distribution associated to the penalty base PK where:

$$\forall \omega \in \Omega, \pi_\Sigma(\omega) = e^{-\kappa_{PK}(\omega)} \quad (10)$$

Proof. Using Equation (3), we consider two cases:

If $\forall (\phi_i, \alpha_i) \in \Sigma, \omega \models \phi_i$, then, $\pi_\Sigma(\omega) = 1$. Using Equation (4), we have $\kappa_{PK}(\omega) = 0$. We obtain then: $\pi_\Sigma(\omega) = e^{-\kappa_{PK}(\omega)} = 1$
Else, $\pi_\Sigma(\omega) = *\{1 - \alpha_i : (\phi_i, \alpha_i) \in \Sigma, \omega \not\models \phi_i\}$
$\pi_\Sigma(\omega) = *\{1 - (1 - e^{-k_i}) : (\phi_i, k_i) \in PK, \omega \not\models \phi_i\}$, as $\alpha_i = 1 - e^{-k_i}$
$\pi_\Sigma(\omega) = *\{e^{-k_i} : (\phi_i, k_i) \in PK, \omega \not\models \phi_i\}$
$\pi_\Sigma(\omega) = e^{\{-\sum k_i : (\phi_i, k_i) \in PK, \omega \not\models \phi_i\}}$
Using Equation (4), we obtain then: $\pi_\Sigma(\omega) = e^{-\kappa_{PK}(\omega)}$

Table 3. The distributions associated to Example 4

ω	$\kappa_{PK}(\omega)$	$\pi_\Sigma(\omega)$
abc	5	0.0068
$ab\neg c$	0	1
$a\neg bc$	4	0.018
$a\neg b\neg c$	1	0.36
$\neg abc$	5	0.0068
$\neg ab\neg c$	4	0.019
$\neg a\neg bc$	3	0.05
$\neg a\neg b\neg c$	0	1

Example 4. Let Σ be the following quantitative possibilistic base:

$\Sigma = \{(\neg a \vee b, 0.6321), (a \vee \neg b \vee c, 0.9816), (\neg b \vee \neg c, 0.8646), (\neg c, 0.9502)\}$

The equivalent penalty base PK is:

$PK = \{(\neg a \vee b, 1), (a \vee \neg b \vee c, 4), (\neg b \vee \neg c, 2), (\neg c, 3)\}$

Using Equations (3) and (4), we obtain the distributions associated to the quantitative possibilistic base and the penalty base, represented by Table 3.
Thus, the relation between the two distributions given by Equation (10) is verified. Indeed, $\forall \omega, \pi_{\Sigma}(\omega) = e^{-\kappa_{PK}(\omega)}$.

5.2 From UCP Networks to Possibilistic Product-Based Graph

There is strong relation between UCP networks and possibilistic product-based graphe. Indeed, each UCP network can be translated into an equivalent product-based possibilistic network:

- The graphical component represented by the DAG is identic,
- The numerical component is obtained by considering for each variable X_i of the DAG, the conditional possibility distributions $\Pi(x_i \mid u_i)$ from $u(x_i, u_i)$ where x_i is an instance of the variable X_i and u_i is an instance of $Par(X_i)$ as follow:

$$\Pi(x_i \mid u_i) = e^{-u(x_i, u_i)} \tag{11}$$

Example 5. Let us consider again the UCP network represented by the Figure 3. The product-based possibilistic graph associated to the UCP network is represented by the same DAG and using the equation (11), we obtain the conditional possibility distributions represented by Table 4.

Table 4. Conditional possibility distributions associated to the Example 5

A	$\Pi(A)$	B	$\Pi(B)$	ABC	$\Pi(C \mid AB)$	CD	$\Pi(D \mid C)$
a	0.0067	b	0.0067	abc	0.5488	cd	0.4065
$\neg a$	0.1353	$\neg b$	0.1353	$ab\neg c$	0.9048	$c\neg d$	0.4493
				$a\neg bc$	0.8187	$\neg cd$	0.8187
				$a\neg b\neg c$	0.4493	$\neg c\neg d$	0.7408
				$\neg abc$	0.9048		
				$\neg ab\neg c$	0.4493		
				$\neg a\neg bc$	0.4065		
				$\neg a\neg b\neg c$	0.7408		

Proposition 2. *Let u be the utility function associated to a UCP network given by Equation (8). Let π_G be the possibility distribution associated to the product-based possibilistic graph equivalent to the UCP-network given by Equation (7). The equivalence between the utility function and the possibility distribution is given by the following equation:*

$$\forall (x_1, ..., x_n) \in \Omega, \pi_G(x_1, ..., x_n) = e^{-u(x_1, ..., x_n)} \tag{12}$$

Proof. By definition, using equation (7), we have:

$\pi_G(x_1, ..., x_n) = *\{\Pi(x_i \mid u_i) : x_1, ..., x_n \models x_i \wedge u_i, i = 1, ..., n\}$

Using Equation (11), we obtain:

$\pi_G(x_1, ..., x_n) = *\{e^{-u(x_i,u_i)} : x_1, ..., x_n \models x_i \wedge u_i, i = 1, ..., n\}$
$\pi_G(x_1, ..., x_n) = e^{\{-\sum_i f_i(x_i,u_i):x_1,...,x_n \models x_i \wedge u_i, i=1,...,n\}}$

Using equation (8), we obtain then: $\pi_G(x_1, ..., x_n) = e^{-u(x_1,...,x_n)}$

Example 6. Let ω_1 be the following interpretation: $\omega_1 = (\neg a \neg b \neg c d)$.
Using Equation (7), we obtain the following possibility distribution associated to ω_1:

$\pi_G(\omega_1) = \Pi(\neg a) * \Pi(\neg b) * \Pi(\neg c \mid \neg a \neg b) * \Pi(d \mid \neg c) = 0.1353 * 0.1353 * 0.7408 * 0.8187 = 0.0111$

Applying equation (8), we obtain the following utility function associated to ω_1:
$u(\neg a \neg b \neg c d) = f_1(\neg a) + f_2(\neg b) + f_3(\neg c \mid \neg a \neg b) + f_4(d \mid \neg c) = 2+2+0.2+0.3 = 4.5$

Then, the equation (11) is verified. Indeed, $\pi_G(\omega_1) = e^{-u(\omega_1)}$.

5.3 From Possibilistic Product-Based Graph to UCP Networks

The converse translation from a product-base possibilistic network to an equivalent UCP network is also possible. The graphical component is the same and the numerical components $u(x_i, u_i)$, for each variable $X_i \in V$ where x_i is an instance of the variable X_i and u_i is an instance of variable X_i, are obtained using the following Equation deduced form Equation (11):

$$u(x_i, u_i) = -ln(\Pi(x_i \mid u_i)) \quad (13)$$

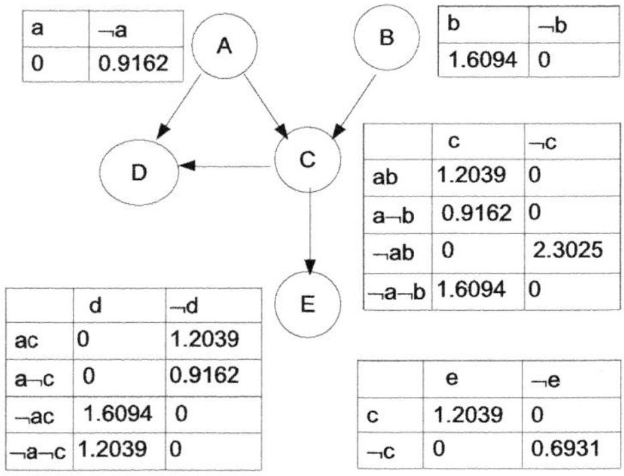

Fig. 4. The UCP-Network equivalent to the product-based possibilistic network of Example 2

Example 7. Let us consider the product-based possibilistic graph of Example 2. Using Equation (13), we obtain the UCP network represented by Figure 4.

5.4 Relating UCP-Networks and Penalty Logic

From these different transformations released around quantitative possibility theory, we establish a corollary which transforms a UCP network to an associated penalty base.

Corollary 1. *Let U_G be a UCP network over $X_1, X_2, ..., X_n$ represented by a DAG and $u(x_i, a_i)$ where x_i is an instance of a variable A_i of the DAG and utility functions a_i is an instance of $Par(A_i)$. The penalty base $PK = \{(\phi_i, k_i), i = 1, ..., n\}$ equivalent to the UCP-network U_G is obtained by:*

1. *Transforming the UCP network U_G to an equivalent product-based network Π_G having the same graphical component represented by the DAG and the numerical component is such that: $\Pi(x_i \mid a_i) = e^{-u(x_i, a_i)}$ where $u(x_i \mid a_i)$ represent the utilities associated to the UCP network with x_i is an instance of a variable A_i of the DAG and a_i is an instance of $Par(A_i)$, as described in Section 5.2.*
2. *Transforming the obtained product-based causal network Π_G to an equivalent quantitative possibilistic base $\Sigma_G = \{(\neg x_i \vee \neg a_i, 1 - \alpha_i)\}$ with $\alpha_i = \Pi(x_i \mid a_i)$ et $\Pi(x_i \mid a_i) \neq 1 \in \Pi_G$, as described in Section 4,*
3. *Transforming the quantitative possibilistic base Σ_G to an equivalent penalty base $PK = \{(\neg x_i \vee \neg a_i, k_i)\}$ with $k_i = -ln(1 - \alpha_i)$ as described in Section 5.1.*

Table 5. Conditional possibility distributions associated to the Example 8

A	$\Pi(A)$	B	$\Pi(B)$	ABC	$\Pi(C \mid AB)$	BCD	$\Pi(D \mid BC)$
a	1	b	0.0498	abc	0.0498	bcd	1
$\neg a$	0.0183	$\neg b$	1	$ab\neg c$	1	$bc\neg d$	0.3679
				$a\neg bc$	1	$b\neg cd$	0.0025
				$a\neg b\neg c$	0.36793	$b\neg c\neg d$	1
				$\neg abc$	1	$\neg bcd$	0.1353
				$\neg ab\neg c$	0.0067	$\neg bc\neg d$	1
				$\neg a\neg bc$	0.0025	$\neg b\neg cd$	1
				$\neg a\neg b\neg c$	1	$\neg b\neg c\neg d$	0.0067

Example 8. Let us consider the UCP network represented by Figure 3.

The corresponding product-based possibilistic network has the same graphical structure and the numerical component is represented by the Table 5. The corresponding quantitative possibilistic base Σ_G is: $\Sigma_G = \{(a, 0.9817), (\neg b, 0.9502),$ $(\neg a \vee \neg b \vee \neg c, 0.9502), (\neg a \vee b \vee c, 0.6321), (a \vee \neg b \vee c, 0.9933), (a \vee b \vee \neg c, 0.9975), (\neg b \vee$

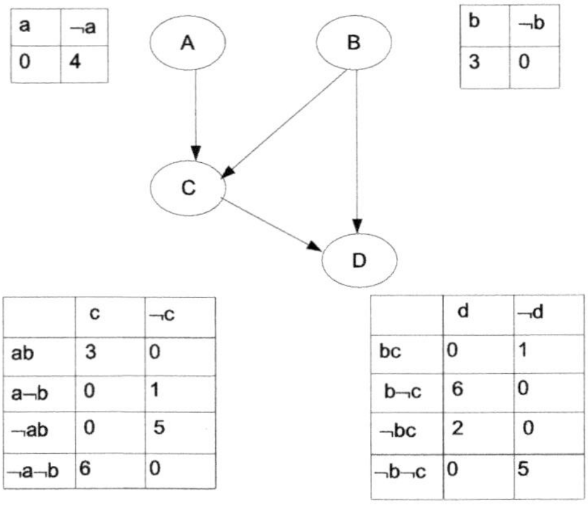

Fig. 5. A UCP-Network of Example 8

$\neg c \vee d, 0.6321), (\neg b \vee c \vee \neg d, 0.9975), (b \vee \neg c \vee \neg d, 0.6847), (b \vee c \vee d, 0.9933)\}$. The penalty base PK equivalent to the quantitative possibilistic base Σ_G is: $PK = \{(a, 4), (\neg b, 3), (\neg a \vee \neg b \vee \neg c, 3), (\neg a \vee b \vee c, 1), (a \vee \neg b \vee c, 5), (a \vee b \vee \neg c, 6), (\neg b \vee \neg c \vee d, 1), (\neg b \vee c \vee \neg d, 6), (b \vee \neg c \vee \neg d, 2), (b \vee c \vee d, 5)\}$.

6 Conclusion

In this paper, some transformations between different formats for representing uncertain knowledge have been proposed. In one hand, the linear transformation from a product-based possibilistic network to a quantitative possibilistic base has been recalled, then, the transformation from a quantitative possibilistic base to a penalty base has been developed. In the other hand, a link which allows bridging the gap between UCP network and product-based possibilistic network has been presented. This established translation may have some impact on inferential issues. Indeed, the propagation algorithm for product based possibilistic network can be exploited on UCP networks. Furthermore, a link between UCP networks and quantitative possibilistic bases was shown. It allows an unified framework for beliefs and preferences.

Nevertheless, it is important to note that in quantitative possibilistic setting, the minimum specificity principle must be verified. This problem can be resolved using fusion techniques developed in [21]. Indeed, in [21] authors showed that the combination of possibility distributions with an operator as the product can be handled in standard possibilistic logic. Thus we can consider each formula of a quantitative possibilistic base as an elementary knowledge base, then we can use the appropriate fusion technic in order to compute the possibility distribution associated to the resulting knowledge base.

A future work would be to study the similarities which exist between the product-based possibilistic networks and Valuations Based Systems (VBS) [19,20].

Indeed, the two graphical representation modes are based on the same definition of the conditioning based on the product.

References

1. Pearl, J.: Probabilistic Reasoning in Intelligent Systems: Networks of Plausible Inference. Morgan Kaufmann Publ. Inc., San Mateo (1988)
2. Jensen, V.F.: An introduction to Bayesian Networks. UCL Press, University College London (1996)
3. Ben Amor, N., Benferhat, S., Mellouli, K.: Anytime propagation algorithm for min-based possibilistic graphs. Soft Computing, A fusion of foundations: methodologies and applications 8(2), 150–161 (2003)
4. Fonk, P.: Réseaux d'inférence pour le raisonnement possibiliste, PHD thesis, Université de liège, Faculté des sciences (1994)
5. Gebhardt, J., Kruse, R.: POSSINFER - A Software Tool for Possibilistic Inference. In: Dubois, D., Prade, H., Yager, R. (eds.) Fuzzy set Methods in Information Engineering, A Guided Tour of Applications (1997)
6. Dubois, D., Lang, J., Prade, H.: Possibilistic Logic. In: Gabbay, D., et al. (eds.) Handbook of Logic in Artificial Intelligence and Logic Programming, vol. 3, pp. 439–513. Oxford University Press, Oxford (1994)
7. Pinkas, G.: Propositional nonmonotonic reasoning and inconsistency in symmetric neural networks. In: 12th IJCAI, Sydney Australia, pp. 525–530. Morgan-Kaufmann, San Francisco (1991)
8. Dupin De Saint-Cyr, F., Lang, J., Schiex, T.: Penalty logic and its link with Dempster-Shafer theory. In: Proceedings of 10th International Conference on Uncertainty in Artificial Intelligence (UAI 1994), pp. 204–211 (1994)
9. Gebhardt, J., Kruse, R.: Background and perspectives of possibilistic graphical models. In: 4th European Conference on Symbolic and Quantitative Approaches to Reasoning and Uncertainty (ECSQARU 1997). LNCS (LNAI), vol. 2143, pp. 108–121. Springer, Heidelberg (1997)
10. Gebhardt, J., Kruse, R.: Background and perspectives of possibilistic graphical models. In: 4th European Conference on Symbolic and Quantitative Approaches to Reasoning and Uncertainty (ECSQARU 1997). LNCS (LNAI), vol. 2143, pp. 108–121. Springer, Heidelberg (1997)
11. Dubois, D., Prade, H.: The logical view of conditioning and its application to possibility and evidence theorie. International Journal of Approximate Reasoning 4(1), 23–46 (1990)
12. De Campos, L.M., Huete, J.F.: Independence concepts in possibility theory. Fuzzy Sets and Systems (1998)
13. Ben Amor, N., Benferhat, S., Dubois, D., Geffner, H., Prade, H.: Independence in Qualitative Uncertainty Frameworks. In: Seventh International Conference on Principles of Knowledge Representation and Reasoning (KR 2000), Breckenridge, Colorado, pp. 235–246. Morgan Kaufmann, San Francisco (April 2000)
14. Boutilier, C., Bacchus, F., Brafman, R.I.: UCP-Networks: A directed Graphical Representation of Conditional Utilities. In: Proceedings of the Seventeenth Annual Conference on Uncertainty Artificial Intelligence (UAI), Seattle, pp. 56–64 (2001)
15. Bacchus, F., Grove, A.: Graphical models for preference and utility. In: Proceedings of the Eleventh Conference on Uncertainty in Artificial Intelligence (UAI), Montreal, pp. 3–10 (1995)

16. Boutilier, C., Brafman, R.I., Hoos, H., Poole, D.: Reasoning with conditional ceteris paribus preference statements. In: Proceedings of 15th Conference on Uncertainty in Artificial Intelligence (UAI), Stockholm, pp. 71–80 (1999)
17. Benferhat, S., Dubois, D., Garcia, L., Prade, H.: On the transformation between possibilistic logic bases and possibilistic causal networks. International Journal of Approximate Reasoning 29, 135–173 (2002)
18. Benferhat, S., Khellaf-Haned, F., Mokhtari, A.: Product-based causal networks and quantitative possibilistic bases. International Journal of Uncertainty, Fuzziness and Knowledge-based Systems 13(5), 469–493 (2005)
19. Shenoy, P.P.: A valuation-based language for expert systems. International Journal of Approximate Reasoning 3, 383–411 (1989)
20. Shenoy, P.P.: Valutions-based systems for propositional logic. Methodologies for Intelligent Systems 5, 305–312 (1990)
21. Benferhat, S., Dubois, D., Prade, H.: From semantic to syntactic approaches to information combinaison in possibilistic logic. In: Bouchon-Meunier, B. (ed.) Aggregation and Fusion of Imperfect Information, pp. 141–161. Physica-Verlag, Heidelberg (1987)

Imprecise Regression Based on Possibilistic Likelihood

Mathieu Serrurier and Henri Prade

IRIT - Université Paul Sabatier
118 route de Narbonne 31062, Toulouse Cedex 9, France

Abstract. Machine learning, and more specifically regression, usually focuses on the search for a precise model, when precise data are available. It is well-known that the model thus found may not exactly describe the target concept, due to the existence of learning biases. So, we are interested in a learning process that accounts also for the uncertainty around the predicted value which should not be illusionary precise. The goal of imprecise regression is to find a model that offers a good trade-off between faithfulness w.r.t. data and (meaningful) precision. The function that is learnt associates, to each input vector, a possibility distribution which represents a family of probability distributions. Based on this interpretation of a possibilistic distribution, we define the notion of possibilistic likelihood. Then, we propose a framework of imprecise regression based on the previous notion and a particle swarm optimization process. This approach takes advantage of the capability of triangular possibility distributions to approximate any unimodal probability distribution from above. We illustrate our approach with a generated dataset.

1 Introduction

Fuzzy regression methods have been proposed for now more than twenty years (e.g. [2,12]). The motivations that have been put forward for such extended forms of regression have been either to generalize regression to fuzzy data, or to describe envelopes for the data by associating each input with an interval covering the output data. This second type of regression (often termed 'possibilistic regression') yields interval representations even when input and output data are non-fuzzy. This suggests that possibilistic regression does not serve exactly the same purpose as classical regression. Still the purpose of possibilistic regression has never been fully laid bare (beyond the informal idea of coverage of the data).

The goal of classical least square regression is to learn a function that associates a precise value to any input vector, from a set of data. Due to the existence of learning biases, especially the limited amount of data available and the necessarily incomplete language used for describing them, the model that is found does not describe exactly the reality. This is particularly true when considering complex concepts. The regression line is supposed to pass through the "middle" of the "cloud" of data points. In the statistical view, the regression curve is interpreted as the mean of a probability distribution, usually a Gaussian one,

S. Benferhat and J. Grant (Eds.): SUM 2011, LNAI 6929, pp. 447–459, 2011.
© Springer-Verlag Berlin Heidelberg 2011

for the output, given an input vector. This interpretation requires that the data variations obey the assumed law. These assumptions allow for an a posteriori description of the uncertainty around the prediction. The analysis of the error can be used for describing the general shape of the uncertainty distribution. A local estimation of the uncertainty can also be done by considering the neighborhood, in the training set, of the input vector considered. However, these approaches suffer from some drawbacks. First, these analyses are done a posteriori and they will be constrained by the model learnt and the assumption made in order to learn it. The global uncertainty estimation requires to know a priori the type of the probability distribution of the error and it supposes that some parameters of the distribution are fixed (variance for Gaussian distributions for instance). The local estimation supposes a high density of data and also some knowledge about the shape of the distribution. Thus, when the data is poorly described (or are too complex), these methods may provide an illusionary precise description of the predictions and the uncertainty associated to it. This may be problematic for risk analysis in particular when strict security constraints should be enforced.

Imprecise regression, whose a preliminary form has been proposed in [10,9], may be considered as being midway between possibilistic regression (due to its coverage concern) and least square regression (due to an uncertainty interpretation). Imprecise regression associates input variables to a possibility distribution over the values of the output. In this paper, we take advantage of the interpretation of a possibility distribution in terms of a family of probability distributions [5]. In this scope, we propose a possibilistic counterpart of the maximum likelihood principle and we consider it as a quality measure for fuzzy functions. We find the optimal function by using a particle swarm algorithm. This allows us to learn both the general tendency of the data and the variation around it. Due to the capability of possibility distributions to upper bound probability distributions, our imprecise regression approach does not need to have knowledge about the type of the probability distribution.

The paper is structured as follows. Section 2 provides some background about possibility distributions and their interpretations in terms of a family of probabilities. In section 3, we propose a definition of a likelihood measure in the context of possibility theory. Section 4 describes the framework of imprecise regression. The section 5 is devoted to comparisons with the related literature. Lastly, we report the results of experiments on a generated dataset, which illustrate the interest of the approach.

2 Background on Possibility Theory

2.1 Possibility Distribution

Possibility theory, introduced by Zadeh [14], was initially created in order to deal with imprecision and uncertainty due to incomplete information as the one provided by linguistic statement. This kind of epistemic uncertainty may not be handled by probability theory, especially when a priori knowledge about the nature of the probability distribution is lacking. A possibility distribution π is a mapping from Ω to $[0,1]$ (Ω may be a discrete universe or a continuous one, i.e. $\Omega = \mathbb{R}$). This value $\pi(x)$ is named possibility degree. For any subset of Ω, the possibility measure is defined as follows :

$$\forall A \subseteq \Omega, \Pi(A) = max\{\pi(x), x \in A\}.$$

If it exists one singleton $x \in \Omega$ for which we have $\pi(x) = 1$, the distribution is normalized. We can distinguish two extreme cases of knowledge situation:

- complete knowledge: $\exists x \in \Omega$ such as $\pi(x) = 1$ and $\forall y \in \Omega, y \neq x, \pi(y) = 0$;
- total ignorance: $\forall x \in \Omega, \pi(x) = 1$.

The necessity is the dual measure of the possibility measure. We have:

$$\forall A \subseteq \Omega, N(A) = 1 - \Pi(\overline{A}).$$

Let us introduce the α-cuts of the distribution π defined y:

$$D_\alpha = \{x \in \Omega, \pi(x) \geq \alpha\}.$$

It can be checked that if the distribution is normalized, continuous, and $\Omega = \mathbb{R}$, we have $\forall \alpha \in [0,1], \Pi(D_\alpha) = 1$ and $N(D_\alpha) = 1 - \alpha$.

2.2 Possibility Distribution as a Family of Probability Distributions

One view of possibility theory is to consider a possibility distribution as a family of probability distributions (see [3] for an overview). Thus, a possibility distribution π will represent the family of the probability distributions for which the measure of each subset of Ω will be bounded by its necessity and its possibility measures. More formally, if $\mathcal{P}$ is the set of all probability distributions defined on Ω, the family of probability distributions $\mathcal{P}_\pi$ associated with π is defined as follows:

$$\mathcal{P}_\pi = \{p \in \mathcal{P}, \forall A \in \Omega, N(A) \leq P(A) \leq \Pi(A)\}. \tag{1}$$

where P is the probability measure associated with p. In this scope, the situation of total ignorance corresponds to the case where all probability distributions are possible. This type of ignorance cannot be described by a single probability distribution. The case of complete knowledge corresponds to the case where only one value is possible and then where there are no randomness nor imprecision. When $\Omega = \mathbb{R}$, this family of probability distribution can also be described in terms of confidence intervals. Given a probability distribution p, a confidence interval I_α is a subset of Ω such as $P(I_\alpha) = \alpha$. We define I^*_α, also referred as quantile, as the smallest confidence interval with probability measure equal to α. Thus, an alternative to Equation 1 is to look for the family of probability function:

$$\mathcal{P}_\pi = \{p \in \mathcal{P}, \forall I^*_\alpha \in \Omega, I^*_\alpha \subseteq D_{1-\alpha}(\pi)\} \tag{2}$$

where $D_{1-\alpha}$ is the $(1-\alpha)$-cut of π. Thus, the possibility distribution π contains the probability distributions for which confidence intervals at level α are upper bounded by its $(1-\alpha)$-cuts.

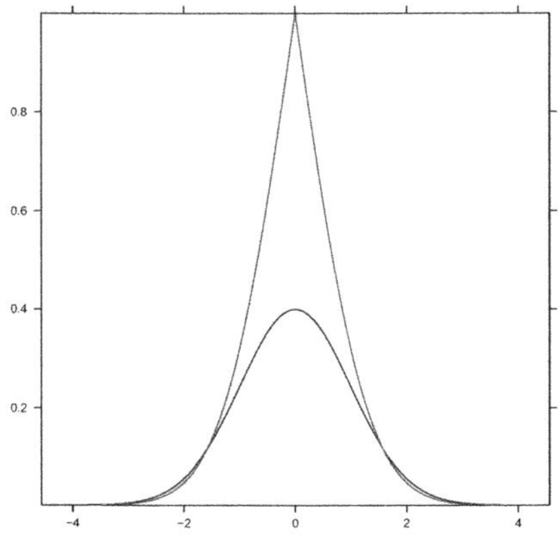

Fig. 1. probability to possibility transformation of a Gaussian distribution

2.3 Probability to Possibility Transformation

According to this probabilistic interpretation, a method from transforming probability distributions to possibility distributions has been proposed in [6]. The idea behind this, is to consider the most informative possibility distribution, i.e. the tightest one, that contains the probability distribution. Let us consider a probability distribution p, the possibility distribution π^* is defined in the following way:

$$\forall x \in \Omega, \pi^*(x) = max_{\alpha, x \in I_\alpha}(1 - \alpha). \tag{3}$$

Then, in the spirit of Equation 2, given p and its transformation π^* we have :

$$D^*_{1-\alpha} = I^*_\alpha$$

where $D^*_{1-\alpha}$ is the $(1 - \alpha)$-cut of π^*. Thus, if p has a finite number of modes, π^* is the possibility distribution for which each $(1 - \alpha)$-cut corresponds to the α-quantile of p. When p is unimodal, the unique value x such that $\pi^*(x) = 1$ is the mode of p.

3 Possibilistic Likelihood

3.1 Definition of a Likelihood Function

Likelihood measures have been introduced in order to evaluate the adequateness of a probability distribution with respect to a set of data. In this section we define a likelihood function for a possibility distribution which supports the interpretation of a possibility distribution in terms of a family of probability distributions (see [11] for details). We first consider the case of a discrete universe, i.e. $\Omega = \{C_1, \dots, C_q\}$. Let us consider a set of data $X = \{x_1, \dots, x_n\}$ belonging to Ω. Let $\alpha_1, \dots, \alpha_q$ be the frequency of the elements of X that belong respectively to $\{C_1, \dots, C_q\}$.

Let us also assume that the frequencies of examples in class C_i are put in increasing order, i.e. $\alpha_1 \geq \ldots \geq \alpha_q$. In this case, the probability distribution p^* that maximizes the likelihood is such that $p^*(x \in C_i) = p_i^* = \alpha_i$. In the following, given a possibility distribution π, we note π_i the value $\pi(x \in C_i)$. It has been shown in [4] that the transformation of p^* into a possibility distribution π^* (see Equation (3)), is:

$$\forall i \in \{1, \ldots, q\}, \pi_i^* = \sum_{j=i}^{q} \alpha_j. \tag{4}$$

This possibility distribution is one of the cumulated functions of p^*. It is worth noticing that it is the tightest one. What we expect from possibility likelihood is that the maximum of this function is reached for π^*. In the following, we assume that $\pi_1 \geq \ldots \geq \pi_q$ (and not necessarily $\alpha_1 \geq \ldots \geq \alpha_q$). We propose the following function:

$$\begin{aligned} \mathcal{L}_{pos}(\pi|x_1, \ldots, x_n) = &\sum_{i=1}^{q}(-\alpha_i * \sum_{j=1}^{i}(1 - \pi_j)) - \sum_{i=1}^{q} \frac{(1 - \pi_i)^2}{2} \\ &+ \sum_{i=1}^{q}(1 - \pi_i). \end{aligned} \tag{5}$$

The following proposition shows that $\mathcal{L}_{pos}$ is an acceptable likelihood function for possibility distributions viewed as families of probabilities.

Proposition 1. *Given a set of data $X = \{x_1, \ldots, x_n\}$ belonging to a discrete universe $\Omega = \{C_1, \ldots, C_q\}$, the possibility distribution π^* that maximizes the function $\mathcal{L}_{pos}$ is the transform of the probability distribution p^* such as $\forall i \in \{1, \ldots, q\}, p_i^* = \alpha_i$.*

Proof: The result is directly obtained by deriving $\mathcal{L}_{pos}$ with respect to the π_i's ■

This likelihood depends on the surface shared between the considered possibility distribution and the optimal one.

It is worth noticing that, when optimal distributions can only be approximated, finding the best approximation with respect to $\mathcal{L}_{pos}$ is not equivalent to finding the best probability approximation with respect to probabilistic likelihood and then turning it into a possibility distribution. This result is fundamental since it illustrates that using a probabilistic likelihood and then the probability-possibility transformation is not an effective approach for constructing a possibility distribution from data. The maximization of $\mathcal{L}_{pos}$ is more adapted in this scope.

We now consider the continuous case where $\Omega = \mathbb{R}$. In the continuous case, the consideration of the values of π in increasing order is naturally replaced by the use of α-cuts. We adapt Equation 5 as follows:

$$\mathcal{L}_{pos}(\pi|x_1,\ldots,x_n) = -(\sum_{i=1}^{n}\int_{D_{\pi(x_i)}}(1-\pi(t))dt) - \int_{\mathbb{R}}\frac{(1-\pi(t))^2}{2}dt + \int_{\mathbb{R}}(1-\pi(t))dt \quad (6)$$

where $D_{\pi(x_i)}$ is the $\pi(x_i)$-cut of π. If we only consider one piece of data, we obtain:

$$\mathcal{L}_{pos}(\pi|x) = -\int_{D_{\pi(x)}}(1-\pi(t))dt + C*(-\int_{\mathbb{R}}\frac{(1-\pi(t))^2}{2}dt + \int_{\mathbb{R}}(1-\pi(t))dt) \quad (7)$$

where C is a constant (usually $\frac{1}{n}$, where n is the number of pieces of data considered). Proposition 1 remains true in the continuous case. The possibilistic counterpart of likelihood being defined, we will now considered the particular case of triangular possibility distributions.

3.2 Triangular Distribution

We define a triangular possibility distribution as the triple $\pi_{tri} = (m, l, r)$ where m is the mode of the triangle and l and r the left and the right spread respectively. Since the 0-cut is infinite for triangular distributions, we assume that X is bounded and have a maximal size equal to the constant MAX_{size}.

We consider a piece of data $x \in X$. We note $\mu = \pi_{tri}(x)$ the possibility degree of x and $[a, b]$ the μ-cut of π_{tri}. There are two cases for the term that depends on $\pi_{tri}(x)$ in (7). We consider the case of $x \in]m-l, m+r[$. We have:

$$MemSurf(\pi_{tri}|x) = -\int_{D_{\pi_{tri}(x)}}(1-\pi_{tri}(t))dt = -(1-\mu)^2 * \frac{l+r}{2}.$$

In the case of $x \notin]m-l, m+r[$, with the bounding assumption, we obtain:

$$MemSurf(\pi_{tri}|x) = -MAX_{size} + \frac{l+r}{2}.$$

Note that a more flexible approach on the bounding can be used, for instance by considering that the weight MAX_{size} depends on the distance between x and the triangle. The other part of Equation 7 is computed such as:

$$-\int_{\mathbb{R}}\frac{(1-\pi_{tri}(t))^2}{2}dt + \int_{\mathbb{R}}(1-\pi(t))dt = -\frac{l+r}{6} - \frac{MAX_{size}}{2}.$$

The terms MAX_{size} neither depends on π_{tri}, nor on x, and can then be omitted. Finally, we obtain :

$$\mathcal{L}_{pos}(\pi_{tri}|x) = MemSurf(\pi_{tri}|x) - C * \frac{l+r}{6}. \quad (8)$$

4 Imprecise Regression Framework

4.1 Definition

Knowing that the representation of the examples corresponds necessarily to an incomplete view of the world, the goal of imprecise regression is to search for imprecise hypotheses that take into account this incompleteness. Thus, given a set of crisp data, we will search for a model that is as precise as possible and which provides a faithful description of the data. When the imprecision tends to 0, we obtain a crisp hypothesis that describes the concept exactly. In a formal way, imprecise regression allows us to represent the imprecision associated with the model by taking into account the incompleteness of the information provided by the data and the chosen representation space for the hypotheses.

A regression database is a set of m pairs $(\overrightarrow{x}_i, y_i)$, $1 \leq i \leq m$, where $\overrightarrow{x}_i \in \mathbb{R}^n$ is a vector of n input variables and $y_i \in \mathbb{R}$ is the real output variable. An imprecise fuzzy function F is a function from $\mathbb{R}^n \rightarrow [0,1]^{\mathbb{R}}$ that associates a distribution on the possible values of the output to the input vector $\overrightarrow{x}$. The goal of imprecise regression is to find the fuzzy function $F(\overrightarrow{x})$ that maximizes the possibilistic likelihood for each piece of data :

$$\mathcal{L}_{pos}(F) = -\sum_{i=1}^{m} \mathcal{L}_{pos}(\pi_i|y_i) \tag{9}$$

where $\pi_i = F(\overrightarrow{x}_i)$. The maximum is reached when the function describes exactly the data without imprecision. Since the learning bias may prevent reaching this maximum, the function will describe both the general tendency of the data and the variations around it. By maximizing the possibility likelihood, the function that we learn will estimate locally the distribution of the data with respect to the input vector. In the next section, we propose an algorithm for imprecise regression with triangular distributions.

4.2 Algorithm

In the following, we consider imprecise regression functions of the form $F_{m,l,r}(\overrightarrow{x}) = T_{f_m(\overrightarrow{x}),f_l(\overrightarrow{x}),f_r(\overrightarrow{x})}$ which associate a triangular fuzzy set to a vector of input variables, although the framework would be applicable to any kind of membership functions. Triangular-shaped possibility distributions are defined as follows:

$$T_{m,l,r}(x) = \begin{cases} 0 & \text{if } x \leq l \text{ or } x \geq r \\ \frac{x-l}{m-l}, & \text{if } x \leq m \text{ and } x > l \\ \frac{r-x}{r-m} & \text{if } x > m \text{ and } x < r \end{cases}$$

The functions f_m, f_l and f_r are independent functions. These can be encoded by affine functions of the form

$$f(\overrightarrow{x} =< x_1, \ldots, x_n >) = a_0 + a_1 * x_1 + \ldots + a_n * x_n,$$

or by kernel functions (e.g. Gaussian kernels in our application)

$$f(\overrightarrow{x}) = a_0 + a_1 * K(s_1, x) + \ldots + a_k * K(s_k, x),$$

where s_1, s_k are support vectors which are computed previously by using a k-means algorithm. Finding optimal f_m, f_l, f_r constitutes a hard problem which is not solvable by classical optimization methods. We propose to solve the problem by using a particle swarm optimization algorithm [7]. The goal of the particle swarm is to determine the function F that maximizes the possibilistic likelihood. One of the advantages of the particle swarm optimization with respect to the other meta-heuristics is that it is particularly suitable for continuous problems. Here, one particle represents the parameters of a fuzzy function (the parameters $a_1, \ldots, a_n$ for each f_m, f_l, f_r in the affine case). At each step the algorithm, each particle is moved along its velocity vector (randomly fixed at the beginning). The velocity vectors are updated at each step by considering the current vectors, the vector from the current particle position to particle best known position and the vector from the current particle position to global swarm's best known position. The second advantage of the algorithm is that it is easy to tune. The three parameters for the updating of the velocity ω, ϕ_p and ϕ_g correspond respectively to the coefficient for the current velocity, the velocity to the particle best known position and the velocity to the global swarm's best known position. The numbers of particle and the number of iteration will depend on the problem, but generic values perform well in most of the case.

4.3 Properties of Triangular Possibility Distributions

Due to the convex nature of the result of the probability-possibility transformation (see Figure 1), triangular possibility distributions offer a convenient way for upper-approximating unimodal probability distributions. This is illustrated by the following proposition, proved in [4]:

Proposition 2. *The triangular symmetric possibility distribution with support* $[x_1, x_2]$ *and with mode* $\frac{x_1+x_2}{2}$ *is the least upper bound of all the possibility transforms of symmetric probability distributions with support* $[x_1, x_2]$ *and with mode* $\frac{x_1+x_2}{2}$.

This proposition shows that triangular distributions may approximate any possibility transformation of a bounded symmetric unimodal distribution. We have shown in Section 3 that approximating a possibility distribution by maximizing $\mathcal{L}_{pos}$ is more efficient that approximating the probability distribution and then turn it into possibility distribution. It validates the use of the possibilistic likelihood for building triangular distributions from data. Even if Proposition 2 is not always true when the distribution is asymmetric, triangular possibilistic distribution performs well in general for any type of unimodal distribution. This allows us to learn fuzzy function that describes the data faithfully without having any a priori knowledge about the shape of the probability distribution.

5 Related Works

There are a large number of methods in statistics for computing confidence intervals. However most of these methods still assume that the type of the distribution is known. Non parametric approaches can then be used for describing the dispersion around the curve. However, it also requires the choice of a particular kind of probability distribution (whose parameters may vary) and a large amount of data. Since it is a local estimation problem with respect to the data in the neighborhood, this approach is not suitable for prediction or for handling sparse data. In quantile regression [8], the goal is to learn a function that associates the quantile interval (of a prefixed probability) for the output variable to the input variables. However, there are some limitations. First, a unique probability value has to be chosen for the quantile. Second it supposes that the error distribution is symmetric. Imprecise regression with possibilistic likelihood allows us to learn the general tendency of the data, together with a local estimation of the error. Contrarily to the probabilistic case where there does not exist a simple probability distribution that can approximate any other one (due to the constraint surface requirement), triangular possibility distributions can upper approximate any unimodal probability distribution. Thus no a priori knowledge about the distribution of the error is required. If we consider that a possibility distribution contains the probability distributions for which each quantile of probability α is a subset of the $(1 - \alpha)$-cut, our method can be viewed as a non linear infinite approximation of the quantile regression. In the same way, imprecise regression is a modal estimator since the top of the triangle corresponds to the mode of the error distribution.

A first type of fuzzy regression approach assumes that we start with fuzzy data, which means that the output values are fuzzy and maybe also the input values. Then, a fuzzy representation is searched for describing such data [1,2]. Diamond's method is based on the extension of least square error minimization using a metrics on fuzzy sets. The major advantage of the least square method is that it appears to be a natural mathematical extension of crisp regression. In this context, when data inputs and output are not fuzzy, fuzzy least square regression reduces to standard least square regression, thus leading to a non fuzzy result. This constitutes a major difference with our approach. In fact, imprecise regression aims at being faithful to the distribution of the data, and associates a fuzzy representation with crisp input and output data.

A second type of approach, named possibilistic regression, has been initially proposed by Tanaka [12], and is reminiscent of quantile regression. The goal of this approach is to associate the data with a pair of upper and lower regression functions, while minimizing the total spread of the output coverage. The main disadvantage of this method is that it is very sensitive to outliers (even if it may be somewhat controlled [13] by using SVM's together with outliers tolerance). Indeed, the optimal upper (resp. lower) bound function is basically a function that is immediately above (resp. below) the whole set of output data. Thus, outliers may affect to a large extent the function that is learnt (even if it may be somewhat controlled [13] by using SVM's together with outliers tolerance). At first glance, imprecise regression may seem to be close to possiblistic regression. First, the two approaches deal with crisp data. Second, they use separate functions in order to represent fuzzy sets or intervals. However, the two approaches differ both at the theoretical level and at the algorithmic level. Possibilistic regression aims at finding the most precise function that is totally accurate with respect to all the examples up to some fixed outliers tolerance. On the contrary, the goal of imprecise regression is to find the function that has the better trade-off between data faithfulness and precision in order to take into account epistemic uncertainty associated with the learning problem. This is why imprecise regression is less sensitive to outliers than possibilistic regression. Moreover, imprecise regression with possibilistic likelihood has an interpretation in terms of imprecise probabilities (family of probability measures).

6 Experimentations

We illustrate our approach with a generated dataset. The data have one variable in input. We assume that for each point in the input space, the output value follows a skewed normal distribution. The skewed normal distribution has three parameters : the location ξ (that affects the position of the distribution), the scale ω (that affects the variance of the distribution) and the shape α (that affects the symmetry of the distribution). The mean of the distribution is $\xi + \omega\delta\sqrt{\frac{2}{pi}}$, the variance is $\omega^2(1 - \frac{2\delta^2}{\pi})$, with $\frac{\alpha}{\sqrt{1+\alpha^2}}$. We generate 3000 pairs of input-output values. The input variable values are uniformly distributed in the range $[-5, 5]$. In a first time, given an input variable x, we randomly take the associated output value in the skewed distribution of mean equals to $cos(x)$, of variance equals to 0.30 and of shape equals to $-4 * cos(\frac{x}{5} * pi)$. It makes that the underlined distribution is symmetric when $x = -2.5$ or $x = 2.5$ or very asymmetric when $x = -5$,

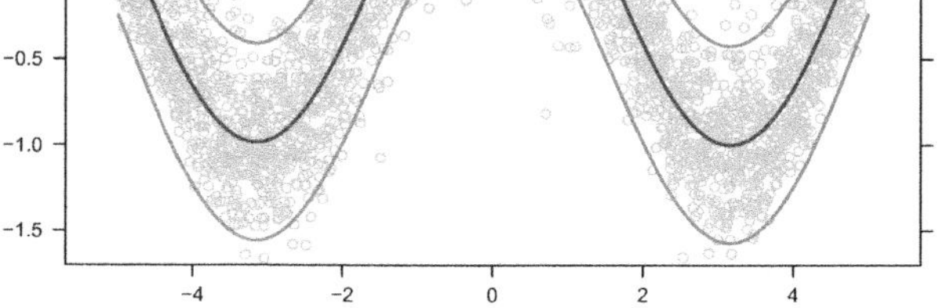

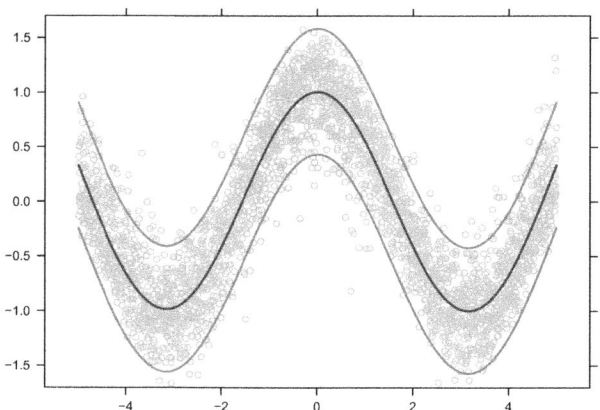

Fig. 2. Non linear least square regression applied to generated data with constant variance

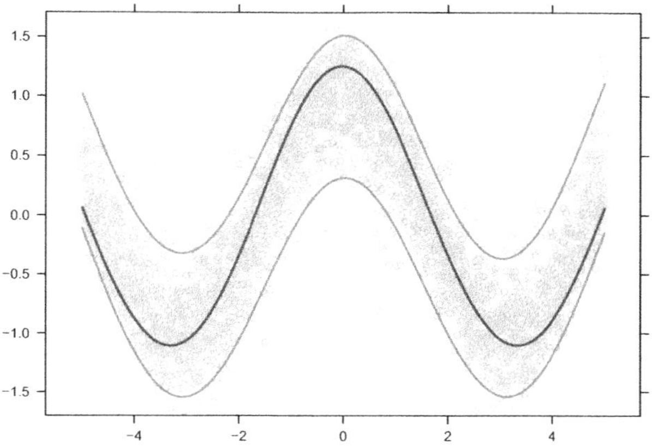

Fig. 3. Imprecise regression applied to generated data with constant variance

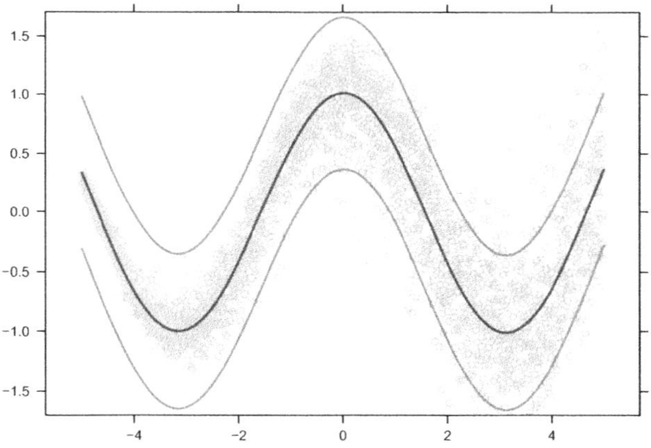

Fig. 4. Non linear least square regression applied to generated data with increasing variance

$x = 0$, or $x = 5$. We apply two methods on these data. The first one is the non linear least square regression with support vector machine. Confidence intervals are 0.95 quantile of the Gaussian distribution centered on the predicted values and with a variance equals to the variance of the error. The second method used is the imprecise regression with possibilistic likelihood. Results are respectively presented in Figures 2 and 3. The estimated mode of the distribution is in red, the estimated 0.95 confidence interval (which corresponds to the 0.05-cuts in the possibility case) is in green.

We can observe that, even that the least square regression predicts well the mean of the distribution, it fails to identify the mode as expected. We can observe that the imprecise regression enables us to overcome this problem. We remark

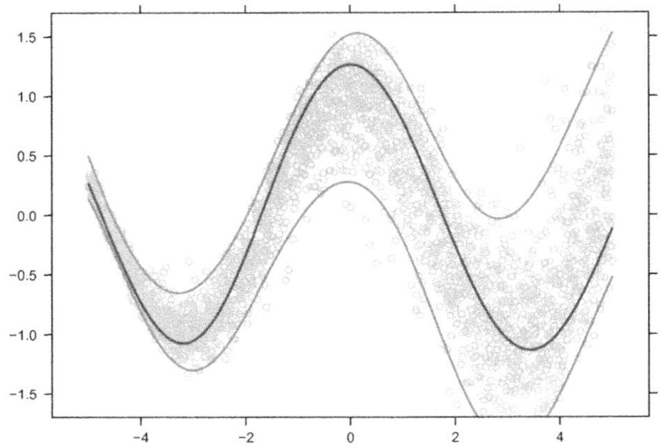

Fig. 5. Imprecise regression applied to generated data with increasing variance

that the triangular possibility distribution provides the best description of the data. Indeed, it clearly identifies the mode and the asymmetrical nature of the data. This is not the case for the distributions based on least square regression since they are centered on the mean.

In the following experiment. We use the same scheme than the previous experiment except that the variance grows linearly following the equation $\frac{x}{5}*0.25+0.30$. Results are respectively presented in Figures 4 and 5. Least square regression still makes a good prediction for the mean, but does not produce faithful confidence intervals and fails again to predict the mode. On the contrary, imprecise regression describes both the mode and the increasing of the variance. These two experiments clearly emphasize the capacity of imprecise regression with possibilistic likelihood to faithfully describe the general tendency of the data and the variation around it without any information on the shape of the distribution of the error.

7 Discussion and Conclusion

In this paper we have proposed an imprecise regression method based on the possibilistic likelihood. This approach allows us to describe both the general tendency and the variation around it. We have shown that by using triangular possibility distributions, we obtain a good approximation of the distribution around the prediction, without any a priori knowledge. More precisely, it describes an upper bound of each quantile of the distribution. When considering the 1-cut we obtain the mode of the distribution. The choice of a particular method depends on the problem at hand. If the shape of the law to be learnt is known then classical regression is enough for the task. On the contrary, as illustrated in the experiments, if the law to be learnt is not completely determined by the input variables available, imprecise regression allows us to capture the epistemic uncertainty associated with the model.

Classical regression when applied to noisy data pertaining to the measurement of some precise law provides an estimate of this law, while imprecise regression looks for a possibility distribution that for each input represents a family of probability distributions which reflects an epistemic uncertainty due to learning biases.

A further line of research is the handling of multi modal distribution of error by the use of trapezoidal possibility distributions. We also plan to take into consideration the quantity of data that are used for determining the possibility distribution at each point in order to re-inject the uncertainty due the estimation based on a sample.

References

1. Celmins, A.: Least squares model fitting to fuzzy vector data. Fuzzy Sets and Systems 22(3), 245–269 (1987)
2. Diamond, P.: Fuzzy least squares. Information Science 46(3), 141–157 (1988)
3. Dubois, D.: Possibility theory and statistical reasoning. Computational Statistics and Data Analysis 51, 47–69 (2006)
4. Dubois, D., Foulloy, L., Mauris, G., Prade, H.: Probability-possibility transformations, triangular fuzzy sets, and probabilistic inequalities. Reliable Computing 10, 273–297 (2004)
5. Dubois, D., Prade, H.: When upper probabilities are possibility measures. Fuzzy Sets and Systems 49, 65–74 (1992)
6. Dubois, D., Prade, H., Sandri, S.: On possibility/probability transformations. In: Proceedings of Fourth IFSA Conference, pp. 103–112. Kluwer Academic Publ., Dordrecht (1993)
7. Kennedy, J., Eberhart, R.: Particle swarm optimization. In: Proceedings of IEEE International Conference on Neural Networks 1995, pp. 1942–1948 (1995)
8. Koenker, R., Hallock, K.F.: Quantile regression. Journal of Economic Perspectives 15, 143–156 (2001)
9. Prade, H., Serrurier, M.: Why Imprecise regression: a discussion. In: International Conference on Soft Methods in Probability and Statistics (SMPS), Oviedo (Espagne), pp. 527–535. Springer, Heidelberg (2010)
10. Serrurier, M., Prade, H.: A general framework for imprecise regression. In: FUZZ-IEEE 2007, London, pp. 1597–1602 (2007)
11. Serrurier, M., Prade, H.: Maximum-likelihood principle for possibility distributions viewed as families of probabilities. In: Proceeding of 2011 International Conference on Fuzzy Systems, FUZZ-IEEE (2011)
12. Tanaka, H.: Fuzzy data analysis by possibilistic linear models. Fuzzy Sets and Systems 24(3), 363–376 (1987)
13. Xu, S.-Q., Luo, Q.-Y., Xu, G.-H., Zhang, L.: Asymmetrical interval regression using extended epsilon-SVM with robust algorithm. Fuzzy Sets and Systems 160(7), 988–1002 (2009)
14. Zadeh, L.A.: Fuzzy sets as a basis for a theory of possibility. Fuzzy sets and systems 1, 3–25 (1978)

Possibilistic Network-Based Classifiers: On the Reject Option and Concept Drift Issues

Karim Tabia

Univ Lille Nord de France, F-59000 Lille, France
UArtois, CRIL UMR CNRS 8188, F-62300 Lens, France
tabia@cril.univ-artois.fr
http://www.cril.fr/~tabia

Abstract. In this paper, we deal with two important issues regarding possibilistic network-based classifiers. The first issue addresses the reject option in possibilistic network-based classifiers. We first focus on simple threshold-based reject rules and provide interpretations for the ambiguity and distance reject then introduce a third reject kind named incompleteness reject occurring when the inputs are missing or incomplete. The second important issue we address is the one of concept drift. More specifically, we propose an efficient solution for revising a possibilistic network classifier with new information.

Keywords: Possibilistic networks, classification, reject option, concept drift, belief revision, Jeffrey's rule of conditioning.

1 Introduction

Classification is an important task in several domains. It consists in predicting the class label of an item described by a set of features. There is in the literature several classifiers based on different approaches and having different algorithmic and performance abilities. For instance, Bayesian network classifiers [12], decision trees [22] and SVM [7] are among the well-known and most efficient classifiers in practice. Unlike Bayesian network-based classifiers, possibilistic ones are studied only in few works [4] [15] [2]. While possibilistic networks allow better handling ignorance and some uncertainty types, some important issues closely related to possibilistic network-based classifiers have not been addressed.

In some applications, the cost of misclassifications (classification errors) is more important than ignoring and rejecting the item to classify. This issue is known as the reject option [6] where a classifier makes predictions only if it is "reliable" or "confident" to a given threshold. The key issue when implementing the reject option in practice is how to measure and interpret a classifier's confidence. More importantly, after a reject decision one often needs to know about the reason why the classifier is not confident enough for making good predictions. In the literature, most reject decision rules are threshold-based and consider two types of reject: ambiguity and distance ones. In [6], the author proposed a rejection mechanism consisting in rejecting any item whose posterior probability

S. Benferhat and J. Grant (Eds.): SUM 2011, LNAI 6929, pp. 460–474, 2011.
© Springer-Verlag Berlin Heidelberg 2011

is less then a predefined threshold t. Several authors proposed more sophisticated rejection rules and strategies. For instance, the author in [14] developed a method allowing to associate the item to classify with several classes. In [19], a class-rejective schema is proposed where some classes are first rejected then the item is associated with the remaining classes. In earlier works [11], the author proposed a unifying approach of rejection schemas for probabilistic, fuzzy and possibilistic classifiers and viewed a classifier as a couple (L, H) of a labeling and hardening functions respectively where the labeling function L provides a measure of typicality of the item to classify with respect to the different classes while the hardening function H assigns a class to the item in hand. In all these works, even those dealing with possibilistic classifiers, the authors adopt a statistical framework[1] and do not address the reject issue from the perspective of the encoded beliefs representing the classifier in case where this latter is elicited from an expert.

In the possibilistic framework, a model's confidence regarding a prediction can naturally be estimated by the necessity measure. However, this measure is unable to identify the reject type. In this paper, we interpreted and analyze some simple threshold-based rejection rules where the used classifiers are possibilistic networks representing an experts beliefs. More specifically, given a possibilistic network classifier and a rejection threshold t, then we are interested in identifying situations where a given item may be rejected with respect to the encoded beliefs. Note that we deal with exclusive classification and we neither consider reject strategies nor ambiguity reject types as in [11]. This paper provides preliminary results on naive possibilistic network-based classifiers using threshold-based reject option. In particular, we analyze the distance and ambiguity reject option then we deal with reject option when the inputs are incomplete.

Over time, the expert's beliefs or the properties of a model or the studied problem change and it becomes important to revise the old model to reflect these changes. For instance, the plausibility order of some events may change over time. This issue is known as the concept drift [24] and the main problems here are (i) how to detect that the current model is no more appropriate and (ii) how to revise it in order to meet the new requirements. In the machine learning field, the detection and update are mostly performed empirically (testing the model with data with known labels, and retraining the model for example [23]). But how can one do this if the system is built on the basis of expert knowledge? We propose in this paper an efficient solution for implementing the concept drift. We propose a method for representing the new beliefs to take into account then revise the old model in order to fully integrate the new knowledge. This revision gives results that are consistent with the results obtained if the revision is performed using Jeffrey's rule of conditioning [17]. Our method can be viewed as a graphical counterpart of Jeffrey's rule as the belief revision operation operates directly on some local distributions associated with the network nodes.

---

[1] For instance, the labeling function in [11] is based on the notion of typicality estimated according to a prototype for each class or class centers, etc.

2 Possibility Theory and Possibilistic Network Classifiers

2.1 Possibility Theory: A Brief Refresher

Possibility theory [9] provides a powerful and simple alternative to probability theory in particular for dealing with some types of uncertainty. It lies on a pair of dual measures (which are the possibility and necessity measures) in order to assess the knowledge/ignorance. One of the fundamental concepts of possibility theory is the one of possibility distribution π which is a mapping from the universe of discourse Ω (possible states of the world) to the unit interval $[0, 1]$. A possibility degree $\pi(w_i)$ expresses to what extent a world $w_i \in \Omega$ can be the actual state of the world. By convention, $\pi(w_i)=1$ means that w_i is totally possible and $\pi(w_i)=0$ denotes an impossible event. The relation $\pi(w_i)>\pi(w_j)$ means that w_i is more possible than w_j. A possibility distribution π is normalized if $max_{w_i \in \Omega}\pi(w_i) = 1$. The normalization condition ensures that the possibility distribution is free from contradictions. The other important concept is the one of possibility measure $\Pi(\phi)$ which evaluates the possibility degree relative to an event $\phi \subseteq \Omega$. It is defined as follows:

$$\Pi(\phi) = \max_{w_i \in \phi}(\pi(w_i)). \tag{1}$$

The necessity measure evaluates the certainty entailed by the current knowledge of the world encoded by the distribution π:

$$N(\phi) = 1 - \Pi(\overline{\phi}) = 1 - \max_{w_i \notin \phi}(\pi(w_i)), \tag{2}$$

where $\overline{\phi}$ denotes the complementary of ϕ in Ω. Note that if $\Pi(\phi)<1$ then $N(\phi)=0$.
In possibility theory, there are several interpretations for the possibilistic scale [0,1]. Accordingly, there are two variants of possibility theory:

1. **Qualitative (or min-based) possibility theory** where the possibility measure is a mapping from the universe of discourse Ω to an "ordinal" scale where only the "ordering" of values is important.
2. **Quantitative (or product-based) possibility theory**: In this case, the possibilistic scale [0,1] is numerical and possibility degrees are like numeric values that can be manipulated by arithmetic operators.

In this work, we only focus on the quantitative possibilistic setting. The other fundamental notion in possibility theory is the one of conditioning which is concerned with updating the current beliefs encoded by a possibility distribution π when a completely sure event (evidence) is observed. Note that there are several definitions of the possibilistic conditioning [16] [9]. The product-based possibilistic conditioning is defined as follows:

$$\pi(w_i|\phi) = \begin{cases} \frac{\pi(w_i)}{\Pi(\phi)} & \text{if } w_i \in \phi; \\ 0 & \text{otherwise.} \end{cases} \tag{3}$$

2.2 Possibilistic Networks

Possibilistic networks [5] [1] are graphical models that allow to compactly and easily encode an agent's beliefs. They are possibilistic counterparts of Bayesian networks [21]. A possibilistic network is composed of two components:

1. **A graphical component:** it consists in a directed acyclic graph (DAG) capturing the direct dependence relationships between domain variables.
2. **A numerical component:** This component is composed of a set of local possibility distributions assessing the plausibility of each domain variable A_i in the context of its parents U_{A_i}.

In order to guarantee that the joint possibility distribution encoded by a possibilistic network is normalized, every local possibility distribution must satisfy the normalization condition expressed as follows:

$$\forall u_{A_i} \in D_{U_{A_i}}, \max_{a_{i_j} \in D_{A_i}} (\pi(a_{i_j}|u_{A_i})) = 1. \quad (4)$$

In Equation 4, $a_{i_j} \in D_{A_i}$ denotes a value a_{i_j} belonging to the domain of the variable A_i, denoted D_{A_i}. The possibility degree associated with an event is computed using the product-based chain rule defined as follows:

$$\forall a_i \in D_{A_i}, \Pi(a_1 a_2..a_n) = \prod_{i=1}^{n} (\pi(a_i|u_{a_i})). \quad (5)$$

2.3 Possibilistic Network Classifiers

Classification consists in predicting the value of a target (non observable) variable given the values of the observed variables. Namely, given $a_1 a_2..a_n$ the values of observed variables $A_1,..,A_n$ describing the items to classify, it is required to predict the right value of the class variable C among a predefined set of class instances $D_C=\{c_1,..,c_m\}$.

Classification based on a possibilistic network classifier is achieved by computing the a posteriori most plausible class instance given the item to classify. Namely,

$$c^* = argmax_{c_k \in D_C}(\Pi(c_k|a_1 a_2..a_n)) = argmax_{c_k \in D_C}(\frac{\Pi(c_k \wedge a_1 a_2..a_n)}{\Pi(a_1 a_2..a_n)}), \quad (6)$$

where the term $\Pi(c_k|a_1 a_2..a_n)$ denotes the a posteriori possibility degree of class instance c_k given the observation $a_1 a_2..a_n$. Note that the denominator in Equation 6 if a normalization term ensuring that the a posteriori possibility distribution relative to the class variable C is normalized. The normalization operation is generally ignored in classification problems since the denominator is the same over all the classes. It is clear that the computational complexity of classifying completely certain observations is linear in the number of attributes and the number of class instances.

A naive possibilistic classifier is a very simple network where the only direct dependence relationships are from the class node C to every attribute node

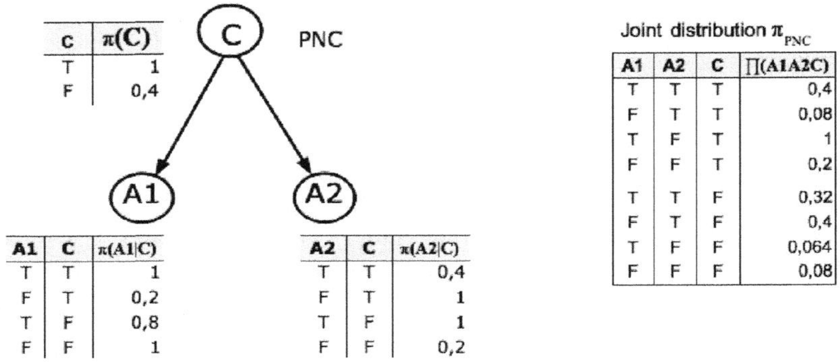

C	π(C)
T	1
F	0,4

A1	C	π(A1|C)
T	T	1
F	T	0,2
T	F	0,8
F	F	1

A2	C	π(A2|C)
T	T	0,4
F	T	1
T	F	1
F	F	0,2

A1	A2	C	∏(A1A2C)
T	T	T	0,4
F	T	T	0,08
T	F	T	1
F	F	T	0,2
T	T	F	0,32
F	T	F	0,4
T	F	F	0,064
F	F	F	0,08

Fig. 1. Example of a naive possibilistic classifier PNC and the corresponding joint distribution π_{PNC} computed using the product-based chain rule of Equation 5

A_i. Figure 1 provides a naive classifier example with a binary class variable C and two binary attributes A_1 and A_2. Naive network classifiers rely on the assumption that attribute variables are independent in the context of the class variable. Hence, naive possibilistic network-based classification in the product-based setting can be achieved as follows:

$$c^* = argmax_{c_k \in D_C}(\prod_{i=1..n} \pi(a_i|c_k) * \pi(c_k)). \tag{7}$$

Naive network classifiers are very easy to build either empirically or by eliciting an expert's beliefs. Moreover, several empirical evaluations show that naive classifiers such as probabilistic ones are very efficient [12][15].

3 Classification with Reject Option

In several domains, the cost of some misclassifications is more important than rejecting the item to classify. For instance, in medical diagnosis, military target identification, etc. it is often better to reject (not classify) an object than misclassifying it. However, when rejecting an item, it is important to know the reason why this item is rejected. This problem is extensively studied for some statistical and machine learning techniques [6][14][19].

A classifier can be seen as the frontiers separating the items composing a same class. In the literature, there are mainly two kinds of classification with reject option. These two reject types refer to situations where the classifier is not "confident". In ambiguity reject the object to classify *belongs* to several classes simultaneously. This may be due to the fact that the modeled classes are not completely disjoint. Distance reject identify situations where the instance to classify *does not belong* to any of the classes represented by the classification model. This may be due to the existence of a class which is not represented or to the fact that the item to classify is an outlier. Distance reject is used to define and delimit the classes modeled by the classifier and can thus reject what is beyond its frontiers.

3.1 Classifiers' Confidence and Reject Option

The classifier's confidence concept is a key issue for the reject option and it denotes how much the model is confident when making a prediction. In our case, we are interested in the confidence of a possibilistic network classifier where classification is of maximum a posteriori (according to Equation 6). Namely, given a possibilistic network classifier PNC encoding the current knowledge or beliefs and given an item $a_1..a_n$ to classify, then we compute $\Pi_{PNC}(c_i|a_1..a_n)$ for every candidate class instance $c_i \in D_C$. Note that the obtained distribution over D_C is normalized. Consequently, the most plausible class instance c^* is the one(s) having $\Pi_{PNC}(c_i|a_1..a_n)=1$. Then the necessity degree $N_{PNC}(c^*|a_1..a_n)$ will provide a kind of confidence degree on the decision. It is inversely proportional to the possibility degree of the second most plausible class instance. In particular, if at least two class instances are totally possible, then the necessity degree will be zero. Hence, computing the necessity degree of a class instance allows to estimate the confidence but one can not identify what type of reject we are facing. This is due to the normalization performed by the conditioning operator. As we will see in the following, it is enough to ignore the normalization operation in order to identify the reject kind. In this work, we deal with exclusive classification and we assume that all the misclassifications have a more important cost than rejecting the item to classify.

3.2 Distance Reject

In practice, distance reject is often used for anomaly and outlier detection and it is implemented by measuring the degree of belonging (or distance) of the object to classify to the different classes. A threshold t_d is set above which the objects to classify are rejected. Intuitively, given an instance to classify $a_1..a_n$, the value of $\Pi(c_i \wedge a_1..a_n)$ can be interpreted as a distance of the item $a_1..a_n$ from the class c_i. Hence, the distance of $a_1..a_n$ from the existing classes can be estimated by

$$distance(a_1..a_n) = 1 - \Pi(c^* \wedge a_1..a_n). \tag{8}$$

It is clear that according to Equation 8, a distance close to 1 reveals an instance to classify which is impossible in any of the existing class instances. Such a case reveals a contradiction between the beliefs encoded by the classifier and the item to classify. This situation may occur when a system has not been revised for a long time or in case there were problems when collecting the inputs (presence of outliers, etc.) or due to problems when eliciting the beliefs. Now, in order to implement the distance option, one can simply set a threshold t_d to implement the distance reject as follows:

$$c^* = \begin{cases} argmax_{c_k \in D_C}(\Pi(c_i \wedge a_1..a_n)) & if\ distance(a_1..a_n) \leq t_d, \\ \emptyset_d & otherwise \end{cases} \tag{9}$$

The value $\emptyset_d$ denotes a distance reject decision. The threshold t_d can be set empirically or according to the encoded beliefs. The value of the distance threshold t_d must be as follows:

$$t_d \geq \min_{c_i a_1 a_2 .. a_n} (\Pi(c_i \wedge a_1..a_n)). \tag{10}$$

It is clear that the distance reject threshold t_d should be greater than the plausibility of the least plausible configuration. Now, given a naive possibilistic network classifier PNC and a distance threshold t_d then we have the following result:

Proposition 1. Let PNC be a naive possibilistic classifier PNC and let t_d be the distance reject threshold then:
if $\exists a_i \in D_{A_i}$ such that $\max_{c_i \in D_C}(\pi_{PNC}(a_i|c_k)) < 1\text{-}t_d$ then every attribute configuration $a_1..a_i..a_n$ where the value of variable A_i is a_i will be rejected.

Proposition 1 provides an immediate result in naive classifiers stating that if an attribute value a_i is less plausible than t_d in all the class instances of the classifier PNC, then all the attribute configurations involving a_i will be rejected.

3.3 Ambiguity Reject

Ambiguity reject is generally implemented by detecting data items that are simultaneously close to several classes. In [6], the author proposed to use the a posteriori probability of the instance to be classified in the different classes. Intuitively, there is ambiguity if

1. $distance(a_1..a_n) \leq t_d$.
2. There exists a class instance $c^{**} \neq c^*$ such that $\Pi(c^* \wedge a_1..a_n)\text{-}\Pi(c^{**} \wedge a_1..a_n) < t_a$.

The first condition checks if there is distance reject while the second condition states that there exists a second class instance c^{**} that is "as plausible as" c^* for the item to classify (the difference is less than the ambiguity threshold t_a). It is easy to show that if the two conditions above are satisfied then the a posteriori necessity degree $N(c^*|a_1..a_n)$ equals 0 and we are in presence of ambiguity reject. We can estimate the ambiguity relative to $a_1..a_n$ denoted as $ambiguity(a_1..a_n)$ by

$$ambiguity(a_1..a_n) = 1 - (\Pi(c^* \wedge a_1..a_n) - \Pi(c^{**} \wedge a_1..a_n)). \tag{11}$$

This measure is used in [20] where the authors study classifiers' confidence evaluation in a probabilistic framework. It estimates the gap between the plausibility of the two most plausible class instances. Hence, a user wanting to reject every item where the classifier's confidence is not very high will use a decision rule taking into account the required confidence level t_a. The a posteriori classification rule of Equation 6 is reformulated as follows:

$$c^* = \begin{cases} argmax_{c_k \in D_C}(\Pi(c_i \wedge a_1..a_n)) & if\ ambiguity(a_1..a_n) \leq t_a \\ \emptyset_a & otherwise \end{cases} \tag{12}$$

The value $\emptyset_a$ denotes the ambiguity reject decision. Let us now see the possible values for t_a with respect to the plausibility of c^* and c^{**}, the two most plausible class instances for the attribute configuration $a_1a_2..a_n$:

$$\max_{a_1a_2..a_n}((\Pi(c^*\wedge a_1..a_n)-\Pi(c^{**}\wedge a_1..a_n)) \geq t_a \geq \min_{a_1a_2..a_n}((\Pi(c^*\wedge a_1..a_n)-\Pi(c^{**}\wedge a_1..a_n)). \tag{13}$$

It is clear that a threshold t_a greater (resp. smaller) than the maximum (resp. minimum) of Equation 13 will reject all (resp. none) of the items to classify.

3.4 Inputs' Incompleteness Reject

One of the main advantages of belief networks in general and possibilistic network classifiers in particular is the ability to reason and perform classification even if some input observations are incomplete or uncertain. By incomplete inputs, we mean that the values of some attributes of the item to classify are not available. The need to know whether a reject decision is due to inputs incompleteness is important since one can ask for more inputs in order to make a more confident decision. The treatment is different if the reject decision is caused by distance or ambiguity reject what ever are the missing inputs. Let $a_i..a_j$ be an instance of the attribute subset $A_i..A_j$ representing the item to classify where some attributes are missing. Hence, there is input incompleteness reject if

- There is ambiguity or distance reject, namely $ambiguity(a_i..a_j)\geq t_a$ or $distance(a_i..a_j)\geq t_d$.
- There exists an instance of the missing attributes such that there is no reject, namely $\exists A'\subseteq A/\{A_i..A_j\}$ and $\exists\ a'\in D_{A'}$ such that $ambiguity(a_i..a_j \wedge a')<t_a$ and $distance(a_i..a_j \wedge a')<t_d$.

The two conditions above identify an input incompleteness reject. The first one allows to detect the incompleteness reject as ambiguity or distance reject while the second allows to identify an instance a' of a subset A' of the missing attributes that can prevent the reject decision. Hence, one can require the missing attributes A'. In the next section, we propose an efficient solution for revising a possibilistic network classifier to efficiently solve the concept drift problem.

4 Concept Drift as a Belief Revision Process

In classification problems, the concept drift [23][24] is the phenomenon denoting the fact that the properties and beliefs of some variables change over time. For example, the plausibility of an event in a given class had decreased. Hence, it becomes important after some time to revise the encoded beliefs in order to meet such changes. Two questions need to be answered regarding this issue. The first question is how to detect that the model suffers from concept drift and needs revision while the second one is how to revise the current model. The first question is out of the scope of this paper[2]. The issue we deal with in this section

[2] One way to do this is to gather new labelled data and classify it using the classifier. If this latter provokes a high reject rate, than this can be due to concept drift.

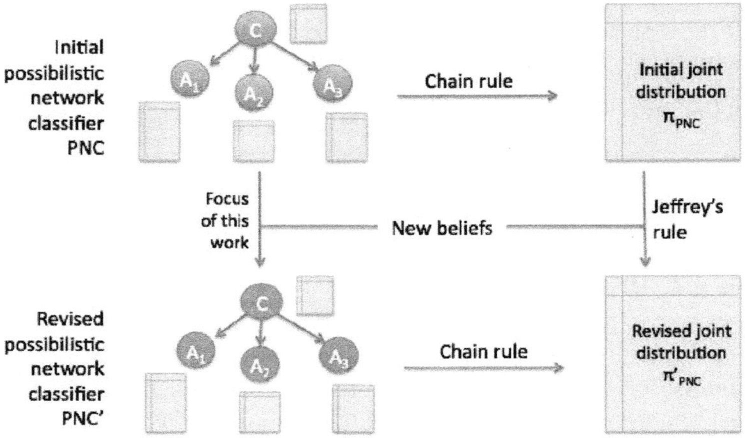

Fig. 2. Our belief revision framework

is how to revise an existing possibilistic network classifier? Namely, given an initial naive possibilistic classifier PNC and given some new beliefs about what should be the classifier, then how to revise PNC in order to accept the new beliefs. There remains to find how to encode the new beliefs and how to perform the revision transformation. The belief revision framework we adopt in this work is the one based on Jeffrey's rule of conditioning [17]. As shown in Figure 2, the old beliefs are compactly encoded by a naive possibilistic classifier PNC. The joint possibility distribution π_{PNC} encoded by PNC can be obtained using the chain rule. In order to revise the beliefs encoded by PNC by the new inputs, one can first compute the joint distribution π_{PNC} then directly use Jeffrey's rule for performing the revision. However, computing a joint distribution and revising it in that way is exponential in the number of variables composing the network. Our contribution consists in a solution allowing to directly revise the initial network PNC with the new beliefs while ensuring that the joint distribution $\pi_{PNC'}$ encoded by the revised network PNC' equals π'_{PNC} obtained by revising the joint distribution π_{PNC} by the new beliefs using Jeffrey's rule. Without loss of generality, we deal in this work only with product-based naive classifiers.

4.1 Jeffrey's Revision Rule

Jeffrey's rule [17] extends the probabilistic conditioning to the case where the evidence is uncertain. In a possibilistic setting, it allows revising the beliefs encoded by a possibility distribution π into a posterior distribution π' given the uncertainty bearing on a set of mutually exclusive and exhaustive events λ_i. In this method, the uncertainty is encoded in the form (λ_i, α_i) with $\alpha_i{=}\Pi'(\lambda_i)$. Jeffrey's rule aims to minimize belief change while fully accepting the new inputs. The revised distribution is the solution satisfying the following conditions:

Condition 1: $\forall \lambda_i \subseteq \Omega$, $\Pi'(\lambda_i)=\alpha_i$.
Condition 2: $\forall \lambda_i \subseteq \Omega$, $\forall \phi \subseteq \Omega$, $\Pi'(\phi|\lambda_i)=\Pi(\phi|\lambda_i)$.

The first condition guarantees that the new inputs are fully accepted while the second condition guarantees that although there is disagreements about the plausibility of the events λ_i in the distributions π and π', the conditional possibility degree of any event $\phi \subseteq \Omega$ given any uncertain event λ_i should remain the same in the original and the revised distributions. As in the probabilistic setting, in the product-based possibilistic one, there always exists a unique solution while in the min-based setting, there exist situations where there is no solution satisfying Condition 1 and Condition 2 at the same time [3]. In the product-based possibilistic setting, the solution satisfying Condition 1 and Condition 2 is computed as follows [10]:

$$\forall \phi \subseteq \Omega, \Pi'(\phi) = \max_{\lambda_i}(\alpha_i * \frac{\Pi(\phi \wedge \lambda_i)}{\Pi(\lambda_i)}). \tag{14}$$

5 Concept Drift and Revising a Naive Possibilistic Classifier

This section presents our solution for revising a naive possibilistic classifier.

Step1: Encoding the inputs (new beliefs) for the revision operation
The aim of our procedure is to revise the beliefs encoded by a possibilistic network *PNC* with some new beliefs. These latter can be of two main kinds:

- **Marginal beliefs:** Here the new beliefs are about single variables (class or attribute variables). For example, a class instance c_i becomes totally possible (namely $\pi'(c_i)=1$).
- **Conditional beliefs:** The new beliefs are on attributes in the context of the class variable. For instance, the possibility degree of a_i in the class c_j becomes equal to 1 (namely $\pi'(a_i|c_j)=1$).

We focus in this work only on the beliefs bearing on single variables as they represent the most common revision situations for naive classifiers in practice (the other situations can be easily reformulated and expressed as new beliefs on single variables). Recall that in naive network classifiers (see Figure 1) the only existing relationships are from the class variable C to each attribute A_i. Recall also that in order to use Jeffrey's rule, the new beliefs should bear on an exhaustive and mutually exclusive set of events. In case where the new beliefs are relative to a non exhaustive and mutually exclusive set of events λ_1 .. λ_i then we complete them with the old beliefs of the missing events λ_{i+1} .. λ_n. Namely, the uncertainty is encoded as follows:

- For j=1 to i, (λ_j, α_j) such that $\Pi'(\lambda_j)=\alpha_j$ (new beliefs)
- For j=$i+1$ to n, (λ_j, α_j) such that $\Pi(\lambda_j)=\alpha_j$. (old beliefs)

The underlying assumption here is that if the new beliefs do not bear a given event, then there is no reason to alter the old belief regarding this event.

Example. Assume that the class variable C is associated with the domain D_C=$\{c_1, c_2, c_3\}$ and with a prior possibility distribution π_C=$\{1, .4, .2\}$. Assume now that we want to revise the old beliefs such that $\pi'_C(c_2)$=.8. Then the complete inputs will be π'_C=$\{1, \mathbf{.8}, .2\}$.

Note that there may exist situations where the new beliefs can be incomplete and lead to sub-normalized distributions. For example, assume that π_C=$\{1, .4, .2\}$ and we want to revise π_C such that $\pi'_C(c_1)$=.6. According to our assumption, the complete inputs would be π_C=$\{\mathbf{.6}, .4, .2\}$. In order to be guarantee that the revised beliefs are normalized then the inputs must be normalized[3]. The inputs (λ_i, α_i) for the revision operation here are of the form $(c_i, \Pi'(c_i))$ or $(a_i, \Pi'(a_i))$ and they are relative to an exhaustive and mutually exclusive set of events. Let us now give an example about conditional beliefs.

Example. Let C be the class variable and A_i be an attribute respectively associated with the domains D_C=$\{c_1, c_2, c_3\}$ and D_{A_i}=$\{a_{i_1}, a_{i_2}\}$. Assume that $\pi_{A_i|c_1}$=$\{1, .4\}$ (a short for the possibility distribution of attribute A_i in case where C takes the value c_1, namely $\pi_{A_i|C}(a_{i_1}|c_1)$=1 and $\pi_{A_i|C}(a_{i_2}|c_1)$=.4). If we have new beliefs stating that $\pi'_{A_i|C}(a_{i_2}|c_1)$=.8 then the complete new inputs are $\pi'_{A_i|c_1}$=$\{1, \mathbf{.8}\}$ and the remaining beliefs of A_i in the class c_2 and c_3, namely the conditional distributions $\pi_{A_i|c_2}$ and $\pi_{A_i|c_3}$.

Note that in this case, the inputs (λ_i, α_i) are of the form $(a_{i_j}|c_k, \Pi'(a_{i_j}|c_k))$ and they are exhaustive and mutually exclusive as they cover each instance of A_i in each class instance.

Step2: Revising the classifier with the new beliefs We present now our solution for revising the beliefs encoded by a naive possibilistic network PNC with new beliefs in the form of (λ_i, α_i) where the events λ_i are exhaustive and mutually exclusive and there is at least one event λ_i associated with α_i=1. Note also that by revising a naive possibilistic network, we mean revising only the necessary local distributions since revising the structure of the network does not make sense here. In order to revise PNC, we distinguish the following cases:

The input is a marginal distribution relative to a single variable
In this case, the new beliefs $\alpha_1..\alpha_n$ are relative either to the class variable C or on an attribute variable A_i. The revision of PNC is done as follows:

Case 1: The new beliefs are relative to the class variable C (namely α_i=$\Pi'(c_i)$ where $c_i \in D_c$): In this case, the revision of PNC is done by replacing the distribution π_C encoding the old beliefs about C by the distribution π'_C where $\forall c_i \in D_C$, $\Pi'(c_i)$=α_i. Hence we have the following proposition:

[3] In case where the new beliefs regarding an attribute A_i are not normalized, then we can easily normalize them by altering the conditional distribution $\pi_{A_i|C}$ and π_C (see for instance [2] for more details). However, if the sub-normalized beliefs bear on the class variable then there is no way to normalize it.

Proposition 2. Let PNC be the naive possibilistic classifier and π_{PNC} be the joint possibility distribution encoded by PNC. Let PNC' be the possibilistic network obtained as follows:

- PNC and PNC' have the same naive structure.
- The new possibility distribution of C is such that $\forall c_i \in D_C$, $\pi'_C(c_i) = \alpha_i$.
- For each value a_{i_j} of attribute A_i in the context of each class c_k of C, $\pi'_{A_i|C}(a_{i_j}|c_k) = \pi_{A_i|C}(a_{i_j}|c_k)$.

Then for every instance $c_i a_1..a_n$ of $CA_1..A_n$ we have $\pi'_{PNC}(c_i a_1..a_n) = \alpha_i * \frac{\pi_{PNC}(c_i a_1..a_n)}{\Pi_{PNC}(c_i)} = \pi_{PNC'}(c_i a_1..a_n)$.

Proposition 2 states that the revised possibility degree of any variable configuration computed from the revised network PNC' or from the revised joint distribution π'_{PNC} are exactly equal. The proof is straightforward from the chain rule and Jeffrey's rule of Equation 14.

Case 2: The new beliefs are relative to an attribute A_i (namely $\alpha_j = \Pi'(a_{i_j})$ where $a_{i_j} \in D_{A_i}$): Here we have marginal beliefs about A_i but the network contains a conditional distribution $\pi_{A_i|C}$ associated with A_i in the context of the class variable C. The following proposition formalizes how to revise PNC:

Proposition 3. Let PNC be the initial naive possibilistic classifier and π_{PNC} be the joint possibility distribution encoded by PNC. Let PNC' be the possibilistic network obtained as follows:

- PNC and PNC' have the same structure.
- The distribution of C remains unchanged, namely $\forall c_k \in D_C$, $\pi'_C(c_k) = \pi_C(c_k)$.
- For each attribute $A_{l \neq i}$ in the context of the class c_k, $\pi'_{A_l|C}(a_{l_j}|c_k) = \pi_{A_l|C}(a_{l_j}|c_k)$.
- For the attribute A_i in the context of each class c_k, $\pi'_{A_i|C}(a_{i_j}|c_k) = \alpha_{i_j} * \frac{\pi_{A_i|C}(a_{i_j}|c_k)}{\max_{c_m \in D_C}(\pi(a_{i_j}|c_m) * \pi(c_m))}$.

Then for every instance $c_k a_1..a_n$ of $CA_1..A_n$ we have $\pi'_{PNC}(c_k a_1..a_n) = \alpha_{i_j} * \frac{\pi_{PNC}(c_k a_1..a_n)}{\Pi_{PNC}(c_k)} = \pi_{PNC'}(c_k a_1..a_n)$.

Proposition 3 revises the initial beliefs encoded by the network PNC given new marginal beliefs about the attribute A_i by revising the conditional distribution $\pi_{A_i|C}$ of A_i in the context of C. With $\Pi_{PNC}(a_i) = \max_{c_k \in D_C}(\pi_{A_i|C}(a_i|c_k) * \pi(c_k))$ one can easily show that the revised possibilistic network PNC' encodes exactly the same joint distribution as the one obtained by revising the initial joint distribution π_{PNC} using Jeffrey's rule with the same inputs.

The input is a conditional distribution relative to a single variable. In this case, the new beliefs are relative to an attribute variable A_i in the context of the class variable (namely $\alpha_{i_j k} = \Pi'(a_{i_j}|c_k)$ where $a_{i_j} \in D_{A_i}$ and $c_k \in D_C$). The revision of the initial network is achieved by simply altering the old conditional distribution $\pi_{A_i|C}$ as formalized in the following proposition:

Proposition 4. Let PNC be the initial naive possibilistic classifier encoding the joint possibility distribution π_{PNC}. Let PNC' be as follows:

- PNC and PNC' have the same structure.
- The possibility distribution of C remains unchanged, namely $\forall c_k \in D_C, \pi'_C(c_k)=\pi_C(c_k)$.
- The new conditional possibility distribution of A_i in the context of C is such that $\pi'_{A_i|C}(a_{i_j}|c_k)=\alpha_{i_j k}$.
- For each attribute $A_{l \neq i}$ in the context of C, $\pi'_{A_l|C}(a_{l_j}|c_k)=\pi_{A_i|C}(a_{l_j}|c_k)$.

Then for every instance $c_k a_1..a_n$ of $CA_1..A_n$ we have $\pi'_{PNC}(c_k a_1..a_n)=\alpha_{i_j k} * \frac{\pi_{PNC}(c_k a_1..a_n)}{\pi_{A_i|C}(a_{i_j}|c_k)}=\pi_{PNC'}(c_k a_1..a_n)$.

In Proposition 4, we state that the revised network PNC' encodes exactly the same possibility distribution as π_{PNC} revised with the new beliefs using Jeffrey's rule. The proof is straightforward by applying the chain rule on PNC'.

6 Summary and Conclusions

This paper addressed two important issues regarding possibilistic network-based classifiers. After analyzing simple threshold-based reject rules for the ambiguity and distance reject and the underlying interpretations, we introduced a new kind of reject due to missing inputs. The ambiguity reject provokes ignorance in the decision phase while distance reject reveals contradictions between the data in hand and the classification model. Finally, the incompleteness reject occurs as ambiguity or distance reject because of lack of evidence. As for complexity issues, it is clear that the computational complexity of the ambiguity and distance reject option rules is exactly the same as the one of classification without the reject option. However, for the incompleteness reject option, finding the instance a' is exponential in the number of missing attributes. The second important issue addressed in this work is the one of revising a possibilistic network for concept drift purposes. Our solution for revising a possibilistic network is simple and intuitive and guarantees the same a posteriori beliefs as Jeffrey's rule.

This paper provided preliminary results on some issues that have not been dealt with before in possibilistic network classifiers. For instance, the solution consisting in dealing with the concept drift problem as a belief revision problem is original. In [8], the authors deal with case-based prediction using possibilistic rules but did not deal with reject option. The authors in [18] only deal with ambiguity and distance reject using the notions of typicality and fuzzy implications. As for revising a belief network, there exist some preliminary works [13] for Markov networks but the problem of revising a possibilistic network to implement the concept drift has never been conducted as far as we know. Some related issues remain open for future works. More specifically, min-based possibilistic counterparts of our contributions as well as generalizing them for general network structures are quite straightforward. There remains also to deal with the commutativity issue of our approach for revising a possibilistic network given that Jeffrey's rule is well-known to be not commutative. The issue of solution existence and uniqueness also represents an interesting problem especially in the min-based possibilistic setting.

References

1. Ben-Amor, N., Benferhat, S., Mellouli, K.: A two-steps algorithm for min-based possibilistic causal networks. In: Benferhat, S., Besnard, P. (eds.) ECSQARU 2001. LNCS (LNAI), vol. 2143, pp. 266–277. Springer, Heidelberg (2001)
2. Benferhat, S., Tabia, K.: An efficient algorithm for naive possibilistic classifiers with uncertain inputs. In: Greco, S., Lukasiewicz, T. (eds.) SUM 2008. LNCS (LNAI), vol. 5291, pp. 63–77. Springer, Heidelberg (2008)
3. Benferhat, S., Tabia, K., Sedki, K.: On analysis of unicity of jeffrey's rule of conditioning in a possibilistic framework. In: 11th International Symposium on Artificial Intelligence and Mathematics, Florida, USA (2010)
4. Borgelt, C., Gebhardt, J.: A naive bayes style possibilistic classifier. In: Proceedings of the 7th European Congress on Intelligent Techniques and Soft Computing, Verlag Mainz, Aachen, Germany (1999)
5. Borgelt, C., Kruse, R.: Graphical Models: Methods for Data Analysis and Mining. John Wiley and Sons, Inc., USA (2002)
6. Chow, C.: On optimum recognition error and reject tradeoff. IEEE Transactions on Information Theory 16(1), 41–46 (1970)
7. Cortes, C., Vapnik, V.: Support-vector networks. Machine Learning 20(3), 273–297 (1995)
8. Dubois, D., Hüllermeier, E., Prade, H.: Flexible control of case-based prediction in the framework of possibility theory. In: 5th European Workshop on Advances in Case-Based Reasoning, London, UK, pp. 61–73 (2000)
9. Dubois, D., Prade, H.: Possibility theory. Plenium Press, New-York (1988)
10. Dubois, D., Prade, H.: A synthetic view of belief revision with uncertain inputs in the framework of possibility theory. Int. J. of Approximate Reasoning 17(2-3), 295–324 (1997)
11. Frélicot, C.: On unifying probabilistic/fuzzy and possibilistic rejection-based classifiers. In: Amin, A., Pudil, P., Dori, D. (eds.) SPR 1998 and SSPR 1998. LNCS, vol. 1451, pp. 736–745. Springer, Heidelberg (1998)
12. Friedman, N., Geiger, D., Goldszmidt, M.: Bayesian network classifiers. Machine Learning 29(2-3), 131–163 (1997)
13. Gebhardt, J.: Knowledge revision in markov networks. Mathware & Soft Computing 11, 93–107 (2004)
14. Ha, T.M.: The optimum class-selective rejection rule. IEEE Trans. Pattern Anal. Mach. Intell. 19, 608–615 (1997)
15. Haouari, B., Ben Amor, N., Elouedi, Z., Mellouli, K.: Naïve possibilistic network classifiers. Fuzzy Sets and Systems 160(22), 3224–3238 (2009)
16. Hisdal, E.: Conditional possibilities independence and non interaction. Fuzzy Sets and Systems, 283–297 (1978)
17. Jeffrey, R.C.: The Logic of Decision. McGraw Hill, NY (1965)
18. Le Capitaine, H., Frelicot, C.: Classification with reject options in a logical framework: a fuzzy residual implication approach. In: Proc. IFSA/EUSFLAT Conference, Lisbonne Portugal, pp. 855–860 (2009)
19. Le Capitaine, H., Frelicot, C.: An Optimum Class-Rejective Decision Rule and Its Evaluation. In: Proceedings of the 2010 20th International Conference on Pattern Recognition, Istanbul, Turquie, pp. 3312–3315. IEEE Computer Society, Los Alamitos (2010)
20. Leray, P., Zaragoza, H., d'Alch-Buc, F.: Pertinence des mesures de confiance en classification. In: 12eme Congres Francophone AFRIF-AFIA Reconnaissance des Formes et Intelligence Articifielle, Paris, France, pp. 267–276 (2000)

21. Pearl, J.: Probabilistic reasoning in intelligent systems: networks of plausible inference. Morgan Kaufmann Publishers Inc., San Francisco (1988)
22. Ross Quinlan, J.: C4.5: Programs for Machine Learning. Morgan Kaufmann, San Francisco (1993)
23. Tsymbal, A.: The problem of concept drift: Definitions and related work. Technical report, TCD-CS-2004-15, Trinity College Dublin, Ireland (2004)
24. Widmer, G., Kubat, M.: Learning in the presence of concept drift and hidden contexts. Mach. Learn. 23, 69–101 (1996)

Weak and Strong Disjunction in Possibilistic ASP

Kim Bauters[1,⋆], Steven Schockaert[1,⋆⋆], Martine De Cock[1], and Dirk Vermeir[2]

[1] Department of Applied Mathematics and Computer Science
Universiteit Gent, Krijgslaan 281, 9000 Gent, Belgium
{kim.bauters,steven.schockaert,martine.decock}@ugent.be
[2] Department of Computer Science
Vrije Universiteit Brussel, Pleinlaan 2, 1050 Brussel, Belgium
dvermeir@vub.ac.be

Abstract. Possibilistic answer set programming (PASP) unites answer set programming (ASP) and possibilistic logic (PL) by associating certainty values with rules. The resulting framework allows to combine both non-monotonic reasoning and reasoning under uncertainty in a single framework. While PASP has been well-studied for possibilistic definite and possibilistic normal programs, we argue that the current semantics of possibilistic disjunctive programs are not entirely satisfactory. The problem is twofold. First, the treatment of negation-as-failure in existing approaches follows an all-or-nothing scheme that is hard to match with the graded notion of proof underlying PASP. Second, we advocate that the notion of disjunction can be interpreted in several ways. In particular, in addition to the view of ordinary ASP where disjunctions are used to induce a non-deterministic choice, the possibilistic setting naturally leads to a more epistemic view of disjunction. In this paper, we propose a semantics for possibilistic disjunctive programs, discussing both views on disjunction. Extending our earlier work, we interpret such programs as sets of constraints on possibility distributions, whose least specific solutions correspond to answer sets.

1 Introduction

Answer Set Programming (ASP) is a form of declarative programming based on the stable model semantics [11] that allows to succinctly formulate and easily solve complex combinatorial problems. Possibilistic logic (PL) [7], which is based on possibility theory [17], allows us to reason about (partial) ignorance or uncertainty in a non-probabilistic way. Possibilistic ASP (PASP) [13,3] unites ASP and PL and provides a single framework for declarative programming under uncertainty. The certainty of a conclusion is then given by the lowest certainty of the rules that were used to establish the conclusion (*i.e.* the strength of the conclusion is determined by the weakest piece of information involved).

⋆ Funded by a joint Research Foundation-Flanders (FWO) project.
⋆⋆ Postdoctoral fellow of the Research Foundation-Flanders (FWO).

S. Benferhat and J. Grant (Eds.): SUM 2011, LNAI 6929, pp. 475–488, 2011.
© Springer-Verlag Berlin Heidelberg 2011

The semantics of both ordinary and possibilistic ASP can be characterized as constraints on possibility distributions [3]. Under this characterization we treat a rule of the form '$c \leftarrow a, not\ b$' intuitively as follows: we can conclude that 'c' is certain when we know that 'a' is certain and when it is consistent to assume that 'b' is false. This characterization of ASP clearly highlights the intuition that underlies ASP and the epistemic flavor of such rules. Indeed, when we know that 'a' is true and we do not know that 'b' is true then we conclude that 'c' should be accepted as true. In the possibilistic case, when we attach certainty degrees to the atoms and rules, we have that the certainty of the conclusion can be no stronger than the certainty of the different pieces of information that were used to deduce the conclusion.

When we consider disjunctions in the head of rules, then reasoning under this epistemic view suggests that when we are certain of the body, we should accept the head of the rule. For example, given the rules

$$a \vee b \leftarrow$$
$$c \leftarrow a$$
$$c \leftarrow b$$

the epistemic reading of the program is that we know that '$a \vee b$' is true and that as soon as we know explicitly whether it is 'a' or 'b' that is true we can conclude c. Hence, without any further information, we cannot conclude 'c' since we only have the underspecified information that '$a \vee b$' is true. We would be able to conclude 'c', however, if we had a rule $c \leftarrow a \vee b$. This particular view of disjunction does not correspond with the intuition in ordinary ASP. Indeed, the semantics of disjunctive rules in ASP say that whenever the body of a rule is satisfied, we should make a non-deterministic choice as to which atom in the head of our rule is chosen to be true (alongside with a minimality requirement on the resulting answer sets). Regardless, as we will see it is often the case that when reasoning under uncertainty we are driven towards this epistemic view of disjunction.

In this paper we examine the differences between these two treatments of disjunction in the head using the framework of possibilistic logic. As we will see, if we treat ASP rules as constraints on possibility distributions we naturally obtain two ways in which we can interpret a disjunctive rule. In one case we retrieve the semantics of ordinary disjunctive ASP (this interpretation of disjunction will be called strong disjunction) and in the other case we retrieve the epistemic view of disjunction (this interpretation of disjunction will be called weak disjunction). The resulting characterizations of disjunctive ASP programs can then naturally be generalized to possibilistic programs, where each rule is labelled with a degree of certainty.

The remainder of this paper is organized as follows. In Section 2 we start by introducing some background on ASP, PL and PASP. In Section 3 we present the strong semantics for possibilistic disjunctive ASP. Then in Section 4 we present the weak semantics for possibilistic disjunctive ASP. We discuss related work in Section 5 and we conclude with Section 6 in which we provide our conclusions.

2 Preliminaries

2.1 Answer Set Programming

To define ASP programs, we start from a finite set of atoms $\mathcal{A}$. A *naf-atom* is either an atom 'a' or an atom 'a' preceded by '*not*' which we call the *negation-as-failure operator*. Intuitively, '$not\ a$' is true when we cannot prove 'a'.

An expression of the form $a_0; ...; a_k \leftarrow a_{k+1}, ..., a_m, not\ a_{m+1}, ..., not\ a_n$ with a_i an atom with $0 \leq i \leq n$ is called a *disjunctive rule*. We call $a_0; ...; a_k$ the *head* of the rule (interpreted as a disjunction) and $a_{k+1}, ..., a_m, not\ a_{m+1}, ..., not\ a_n$ the *body* of the rule (interpreted as a conjunction).

A positive disjunctive rule is a disjunctive rule without negation-as-failure, *i.e.* $n = m$. A rule of the form $a_0; ...; a_k \leftarrow$ is called a *fact* and is used as a shorthand for $a_0; ...; a_k \leftarrow \top$ with $\top$ a special language construct that denotes tautology.

A disjunctive program P is a finite set of disjunctive rules. The *Herbrand base* $\mathcal{B}_P$ of a disjunctive program P is the set of atoms appearing in P. An *interpretation* I of a disjunctive program P is any set of atoms $I \subseteq \mathcal{B}_P$. A *normal rule* is a disjunctive rule with exactly one atom in the head, *i.e.* $k = 0$. A *definite rule* is a normal rule with no negation-as-failure, *i.e.* $k = 0$ and $n = m$. A *normal* (resp. *definite*) *program* P is a finite set of normal (*resp.* definite) rules.

An interpretation I is a *model* of a positive disjunctive rule $r = a_0; ...; a_k \leftarrow a_{k+1}, ..., a_m$, denoted $I \models r$, if $\{a_0, ..., a_k\} \cap I \neq \emptyset$ or $\{a_{k+1}, ..., a_m\} \not\subseteq I$, *i.e.* the body is false or at least one of the atoms in the head is true. An interpretation I of a positive disjunctive program P is a *model* of P iff $\forall r \in P \cdot I \models r$.

The *reduct* [11,10] P^I of a disjunctive program P *w.r.t.* an interpretation I is defined as

$$\begin{aligned} P^I = &\{a_0; ...; a_k \leftarrow a_{k+1}, ..., a_m \mid (\{a_{m+1}, ..., a_n\} \cap I = \emptyset) \\ &\wedge (a_0; ...; a_k \leftarrow a_{k+1}, ..., a_m, not\ a_{m+1}, ..., not\ a_n) \in P\}. \end{aligned}$$

We say that I is an answer set of the disjunctive program P when I is a minimal model *w.r.t.* set inclusion of P^I.

The answer set of a definite program P can also be defined using the *immediate consequence operator* T_P, which is defined *w.r.t.* an interpretation I as:

$$T_P(I) = \{a_0 \mid (a_0 \leftarrow a_1, ..., a_m) \in P \wedge \{a_1, ..., a_m\} \subseteq I\}.$$

We use $P^\star$ to denote the fixpoint which is obtained by repeatedly applying T_P starting from the empty interpretation, *i.e.* the least fixpoint of T_P *w.r.t.* set inclusion, which is guaranteed to exist [16]. An interpretation I is an *answer set of a definite program* P iff $I = P^\star$.

2.2 Possibilistic Logic

At the semantic level, possibilistic logic [7] is defined in terms of a *possibility distribution* π on the universe of interpretations. For $\Omega = 2^{\mathcal{B}_P}$ the set of all interpretations of a program P, we have that the possibility distribution is an $\Omega \rightarrow [0,1]$ mapping which encodes for each interpretation (or world) I to what extent it is plausible that I is the actual world. Rather than using certainty degrees from $[0,1]$, we could use any linearly ordered set, together with an involutive order-reversing mapping. Intuitively, $\pi(I)$ represents the compatibility of the interpretation I with available information. By convention, $\pi(I) = 0$ means that I is impossible and $\pi(I) = 1$ means that no available information prevents I from being the actual world. Note that possibility degrees are mainly interpreted qualitatively: when $\pi(I) > \pi(I')$, I is considered more plausible than I'. For two possibility distributions π_1 and π_2 with the same domain Ω we write $\pi_1 > \pi_2$ when $\forall I \in \Omega \cdot \pi_1(I) \geq \pi_2(I)$ and $\exists I \in \Omega \cdot \pi_1(I) > \pi_2(I)$. The satisfaction relation $\models$ is defined for a set of atoms A as $A \models a$ iff $a = \top$ or $a \in A$, otherwise $A \not\models a$. Furthermore, $A \models \neg a$ iff $A \not\models a$.

A possibility distribution π induces two uncertainty measures that allow us to rank propositions. The *possibility measure* Π is defined by [7]:

$$\Pi(p) = \max \{\pi(I) \mid I \models p\}$$

and evaluates the extent to which a proposition p is consistent with the beliefs expressed by π. The dual *necessity measure* N is defined by:

$$N(p) = 1 - \Pi(\neg p)$$

and evaluates to which extent a proposition p is entailed by available beliefs [7].

An important property that necessity measures have is min-decomposability *w.r.t.* conjunction: $N(p \wedge q) = \min(N(p), N(q))$ for all propositions p and q. However, for disjunction only the inequality $N(p \vee q) \geq \max(N(p), N(q))$ holds. As possibility measures are dual to necessity measures, they have the important property of max-decomposability *w.r.t.* disjunction, whereas for conjunction only the inequality $\Pi(p \wedge q) \leq \min(\Pi(p), \Pi(q))$ holds.

At the syntactic level, a *possibilistic knowledge base* Σ corresponds to a set of constraints $N(p) \geq c$ where p is a propositional formula and $c \in [0,1]$ expresses the certainty that p is the case. Typically, there will be many possibility distributions that satisfy these constraints. In practice, we are usually only interested in the *least specific possibility distribution* of these possibility distributions, which is the possibility distribution that makes minimal commitments, *i.e.* the largest possibility distribution *w.r.t.* the ordering $>$ defined above.

2.3 Possibilistic Normal ASP

Possibilistic ASP combines ASP and possibilistic logic [7] by associating a certainty value with atoms and rules. A possibilistic normal (*resp.* definite) rule is a pair (r, λ) where r is a normal (*resp.* definite) rule and where $\lambda \in [0,1]$ is a certainty attached to r. We also write a pair (r, λ) as 'λ: r'. A possibilistic normal (*resp.* definite) program is a set of possibilistic normal (*resp.* definite) rules.

As we recalled in Section 2.1, in ASP, an answer set of a program P is an interpretation that satisfies some additional requirements. Note that an interpretation I of P can be thought of as a $\mathcal{B}_P \rightarrow \{0,1\}$ mapping. As a generalization of this, in possibilistic ASP, an answer set of a program is a valuation V, which is a $\mathcal{B}_P \rightarrow [0, 1]$ mapping, that satisfies the requirements formally defined in Definition 1. This is the mechanism used to associate certainty values with atoms appearing in a program. The intuition is that for an atom $a \in \mathcal{B}_P$, $V(a) = c$ means that we can derive with certainty c that a is true. For notational convenience, we also use the set notation $V = \{a^c, \ldots\}$. In accordance with this set notation, we write $V = \emptyset$ to denote the valuation in which each atom is mapped to 0. We generally omit atoms and rules with an associated certainty of 0 due to their triviality.

Possibilistic normal programs (and therefore also ordinary normal programs) are interpreted in terms of constraints on possibility distributions. Intuitively, an ordinary rule of the form '$rule = (head \leftarrow body)$' says that we are able to conclude that '*head*' is true when we know that '*body*' is true. When we associate necessities with the information in '*body*' and with '*rule*' itself, then we can only deduce '*head*' with a certainty $\min\{N(body), N(rule)\}$. Indeed, the contribution of a single rule to the necessity with which its head is true cannot be stronger than the necessity of the weakest information used to derive the body of the rule. However, a stronger conclusion could be derived using another set of rules, hence the '$rule = (head \leftarrow body)$' induces the constraint $N(head) \geq \min\{N(body), N(rule)\}$.

In the ordinary case, we tackle negation-as-failure by making certain assumptions about which atoms we will be able to derive (we guess an I) and then checking whether our assumptions are stable (we verify that $I = (P^I)^\star$). When dealing with uncertain information, the assumptions we need to make are not whether an atom is true or not, but rather with what certainty we will be able to derive an atom. We make these assumptions by guessing a valuation V, *i.e.* an association of a necessity with each atom. At the end we verify whether $V(a) = N(a)$, *i.e.* whether our guess is stable. This is the possibilistic counterpart of the ordinary reduct.

Definition 1. *[3] Let P be a possibilistic normal program. Let $V : \mathcal{B}_P \rightarrow [0, 1]$ be a valuation. For every $p \in P$, the constraint $\gamma_V(p)$ induced by $p = (r, \lambda)$ with $r = (a_0 \leftarrow a_1, ..., a_m, not\ a_{m+1}, ..., not\ a_n)$ and V is given by*

$$N(a_0) \geq \min\{N(a_1), ..., N(a_m), 1 - V(a_{m+1}), ..., 1 - V(a_n), \lambda\}.$$

We write $C_{(P,V)} = \{\gamma_V(p) \mid p \in P\}$ to denote the set of constraints imposed by program P. A possibility distribution that satisfies the constraints in $C_{(P,V)}$ is called a possibilistic model of $C_{(P,V)}$. We write $S_{(P,V)}$ for the set of all least specific possibilistic models of $C_{(P,V)}$. V is called a possibilistic answer set of P iff there exists a $\pi \in S_{(P,V)}$ such that $\forall a \in \mathcal{B}_P \cdot V(a) = N(a)$.

The ordinary case can be retrieved if we also require $\forall a \in \mathcal{B}_P \cdot N(a) \in \{0, 1\}$, *i.e.* if for every atom we are entirely sure whether or not the atom is necessary. The above definitions can then be used to characterize the semantics of ordinary normal programs.

Proposition 1. *[3] Let P' be a normal program, let P be a possibilistic normal program such that $P = \{(r', 1) \mid r' \in P'\}$ and let $V : \mathcal{B}_P \rightarrow [0,1]$ be a valuation. If V is a possibilistic answer set of P and $\forall a \in \mathcal{B}_P \cdot V(a) \in \{0,1\}$, then $M = \{a \mid V(a) = 1, a \in \mathcal{B}_P\}$ is an answer set of P'.*

Before we extend the semantics to cover the case of disjunction in Section 3 and 4, we first provide an example of the possibilistic semantics applied to a PASP program in order to further clarify the approach.

Example 1. Two common symptoms associated with fibromyalgia (a medical disorder consisting of pain in muscle and joint tissue) are a feeling of weakness and joint pain, where feeling weak without other causes is a telltale sign of fibromyalgia. Our patient tells us that she is experiencing both symptoms. However, the patient is known as a hypochondriac and is not entirely trustworthy. In the past she sometimes complained about weakness without any grounds, though she hardly ever complains about pain without an actual physical or mental cause. Her sagging eyes hint at an iron deficiency (which might explain the weakness in itself), though it is highly unlikely that sagging eyes by themselves correctly identify an iron deficiency. We have the program P with the rules:

$$\begin{aligned}
&\mathbf{0.2:}\ \mathit{fibro} \leftarrow \mathit{pain}\\
&\mathbf{0.6:}\ \mathit{fibro} \leftarrow \mathit{weak}, \mathit{not}\ \mathit{deficiency}\\
&\mathbf{0.9:}\ \mathit{pain} \leftarrow\\
&\mathbf{0.8:}\ \mathit{weak} \leftarrow\\
&\mathbf{0.1:}\ \mathit{deficiency} \leftarrow
\end{aligned}$$

which induces the set $C_{(P,V)}$ of constraints

$$\begin{aligned}
\{&N(\mathit{fibro}) \geq \min\{N(\mathit{pain}), 0.2\},\\
&N(\mathit{fibro}) \geq \min\{N(\mathit{weak}), 1 - V(\mathit{deficiency}), 0.6\},\\
&N(\mathit{pain}) \geq 0.9 \quad , \quad N(\mathit{weak}) \geq 0.8 \quad , \quad N(\mathit{deficiency}) \geq 0.1\}.
\end{aligned}$$

The set of least specific possibility models $S_{(P,V)}$ is a singleton and $\pi \in S_{(P,V)}$ is defined as $\pi(I) = 0.1$ when $I \models \{\neg p\}$, $\pi(I) = 0.2$ when $I \models \{p, \neg w\}$, $\pi(I) = 0.4$ when $I \models \{\neg f, p, w\}$, $\pi(I) = 0.9$ when $I \models \{f, p, w, \neg d\}$ and $\pi(I) = 1$ when $I \models \{f, p, w, d\}$, where we use the first letter of the atom as abbreviation to save space. The possibilistic answer set of this program is unique and is given by

$$V = \left\{\mathit{pain}^{0.9}, \mathit{weak}^{0.8}, \mathit{deficiency}^{0.1}, \mathit{fibro}^{0.6}\right\}$$

which can readily be verified.

We would also like to point out that, unlike the approach above, the approaches from [13,14] do not take the *extent* of certainty of information into account when determining the reduct of a PASP program. In these other semantics, *any* proof of 'a', no matter how uncertain it is, suffices to eliminate the expression '*not* a'.

Hence, in the example above, rule 2 would be eliminated based on rule 5, and we would only be able to derive *fibro*$^{0.2}$. This clearly is not the intended meaning of the program as the limited certainty of an actual deficiency should not be sufficient to dismiss the certainty we have in diagnosing fibromyalgia.

3 Strong Possibilistic Semantics

In this section we extend the semantics of possibilistic normal ASP [3] to disjunctive programs, in a way which remains faithful both to ordinary disjunctive ASP (see Section 2.1) and to the semantics from [6]. As necessity measures do not have the max-decomposability property, we have a choice of how to interpret the disjunction in the head. This is similar to the choice one has for the semantics of disjunction when characterizing ASP using autoepistemic logic [12] or using meta-rules in possibilistic logic [9]. A *possibilistic disjunctive rule* $p = (r, \lambda)$ with $r = a_0; ...; a_k \leftarrow a_{k+1}, ..., a_m, not\ a_{m+1}, ..., not\ a_n$ can either be interpreted as the constraint

$$\max\{N(a_0), ..., N(a_k)\} \geq \min\{N(a_{k+1}), ..., N(a_m), 1 - V(a_{m+1}), ..., 1 - V(a_n), \lambda\} \quad (1)$$

which we will call *strong disjunction* or as the constraint

$$N(a_0 \vee ... \vee a_k) \geq \min\{N(a_{k+1}), ..., N(a_m), 1 - V(a_{m+1}), ..., 1 - V(a_n), \lambda\} \quad (2)$$

which we will call *weak disjunction*. This is a choice that does not arise for the conjunction in the body since min-decomposability dictates that $N(a \wedge \ldots \wedge z) = \min\{N(a), \ldots, N(z)\}$. However, for the disjunction, we only have $N(a \vee \ldots \vee z) \geq \max\{N(a), \ldots, N(z)\}$.

The choice of how to treat disjunction is an important one that profoundly impacts the nature of the resulting answer sets. The main distinction between strong and weak disjunction has to do with the way that we regard an answer set. If we see an answer set as a solution to a problem, then the non-deterministic nature of strong disjunction provides a useful way to generate different (candidate) solutions. If we take an answer set as a representation of an epistemic state, then weak disjunction models the current state of belief.

In the remainder of this section we consider the characterization of disjunction as (1). In Section 4 we consider the characterization of disjunction as (2).

As it turns out, the characterization of disjunction as (1) makes the disjunction behave as in ordinary ASP (see Section 2.1). Indeed, the interplay between the strong possibilistic semantics of disjunction together with the requirement that we are looking for the least specific possibility distribution ensures that disjunction induces a choice. Similar as in the ordinary case, the constraint (1) will generate a number of possible outcomes. The requirement that we are looking for the least specific possibility distribution behaves similarly as the requirement of trying to find the minimal model. We will first give the general definition and then we will illustrate the semantics using an ordinary disjunctive ASP program.

Definition 2. *Let P be a possibilistic disjunctive program and let $V : \mathcal{B}_P \rightarrow [0,1]$ be a valuation. For every $p \in P$ with $p = (r, \lambda)$, the constraint γ_V^s induced by $r = (a_0; ...; a_k \leftarrow a_{k+1}, ..., a_m, not\ a_{m+1}, ..., not\ a_n)$ and V under the strong possibilistic semantics of disjunction is given by*

$$\max\{N(a_0), ..., N(a_k)\} \geq \min\{N(a_{k+1}), ..., N(a_m), 1 - V(a_{m+1}), ..., 1 - V(a_n), \lambda\}.$$

$C^s_{(P,V)} = \{\gamma_V^s(r) \mid r \in P\}$ is the set of constraints imposed by program P and V. A possibility distribution that satisfies the constraints in $C^s_{(P,V)}$ is called a possibilistic model of $C^s_{(P,V)}$. We write $S^s_{(P,V)}$ to denote the set of all least specific possibilistic models of $C^s_{(P,V)}$. V is called a possibilistic answer set of P iff there exists a $\pi \in S^s_{(P,V)}$ such that $\forall a \in \mathcal{B}_P \cdot V(a) = N(a)$.

Whenever P is a positive disjunctive program we have no need for a specific valuation V to pin down the constraints (since there is no negation-as-failure), and we simplify the notation to γ^s, C^s_P and S^s_P. As before we retrieve the semantics for the ordinary case if we require $\forall a \in \mathcal{B}_P \cdot N(a) \in \{0,1\}$.

Example 2. Consider the program P with the single rule

$$\mathbf{0.7}\text{: } a; b \leftarrow .$$

The set of constraints C^s_P is given by $\{\max\{N(a), N(b)\} \geq 0.7\}$. This constraint induces a choice, *i.e.* we either need to pick $N(a) = 0.7$ or $N(b) = 0.7$ to conform to the principle of least specificity. We can conclude that the two possibilistic answer sets of P are given by $\{a^{0.7}\}$ and $\{b^{0.7}\}$. This corresponds with the non-deterministic intuition of the problem; we choose either 'a' or 'b' and assign a certainty of 0.7 to the atom we choose.

Example 3. We cannot simply simulate disjunction using negation-as-failure, as would be possible in the ordinary case. Indeed, the possibilistic normal program that simulates program P from Example 2 would have the rules

$$\mathbf{0.7}\text{: } a \leftarrow not\ b \qquad\qquad \mathbf{0.7}\text{: } b \leftarrow not\ a.$$

This program has an infinite set of possibilistic answer sets, with certainty degrees ranging from 0.3 to 0.7 for a and $1 - N(a)$ for b. When trying to simulate the rule from Example 2 we clearly do not want a possibilistic answer set such as $\{a^{0.4}, b^{0.6}\}$ as this does not correspond with our intuition. This again highlights the importance of satisfactory semantics for possibilistic disjunctive ASP.

As before, when we impose the additional constraint $\forall a \in \mathcal{B}_P \cdot N(a) \in \{0,1\}$ we retrieve the ordinary semantics for disjunctive ASP.

Proposition 2. *Let P' be a disjunctive program, let P be a possibilistic disjunctive program such that $P = \{(r', 1) \mid r' \in P'\}$. Let $V : \mathcal{B}_P \rightarrow [0,1]$ be a valuation. If V is a possibilistic answer set of P and $\forall a \in \mathcal{B}_P \cdot V(a) \in 0, 1$, then $M = \{a \mid V(a) = 1, a \in \mathcal{B}_P\}$ is an answer set of the disjunctive program P'.*

Example 4. Consider the possibilistic disjunctive program P with the rules

$$\mathbf{1}\colon\ a; b \leftarrow \qquad\qquad \mathbf{1}\colon\ a \leftarrow b.$$

which induces the constraints

$$\max\{N(a), N(b)\} \geq N(\top) = 1 \qquad\qquad N(a) \geq N(b).$$

Intuitively, the first constraint induces a choice. To satisfy the constraint, we need to take either $N(a) = 1$ or $N(b) = 1$. Since $N(a) = 1 = 1 - \Pi(\neg a)$ tells us that $\Pi(\neg a) = 0 = \max\{\pi(I) \mid I \models \neg a\}$, we have for every I with $a \notin I$ that $\pi(I) = 0$ (and obviously when $N(b) = 1$ then $\pi(I) = 0$ whenever $b \notin I$). Depending on our choice of whether we take $N(a) = 1$ or $N(b) = 1$ (and since $N(a) = N(b)$ whenever $N(b) = 1$ due to the last constraint), we obtain two possibility distributions π_1 and π_2 defined by:

$$\pi_1(\{a, b\}) = 1 \qquad \pi_1(\{b\}) = 0 \qquad \pi_1(\{a\}) = 1 \qquad \pi_1(\{\}) = 0$$

and

$$\pi_2(\{a, b\}) = 1 \qquad \pi_2(\{b\}) = 0 \qquad \pi_2(\{a\}) = 0 \qquad \pi_2(\{\}) = 0.$$

It is clear that π_2 cannot be least specific since $\pi_1 > \pi_2$. We then have that S_P^{s} only contains a single element, namely π_1. With N the necessity measure induced by π_1 we obtain $N(a) = 1$ and $N(b) = 0$. The unique answer set of P is thus $\{a^1\}$, which indeed corresponds with the answer set of the ordinary disjunctive program $P' = \{a; b \leftarrow, a \leftarrow b\}$.

4 Weak Possibilistic Semantics

The semantics we discussed thus far have a clear non-deterministic flavor: if the antecedent is known, a rule declares that we should choose one or more consequents to accept (and at the same time choose as few as possible). Under this non-deterministic view the rule '$a; b \leftarrow$' means that a is believed to be true or b is believed to be true. However, there are cases in which we do not want to or cannot make a commitment. In other words, we want a rule '$a \vee b \leftarrow$' to mean that *a or b* is true, without clarifying whether it is a, b or both that are true. In this sense we regard answer sets more as epistemic states than as possible solutions to a problem.

In this section we first define an alternative semantics for disjunctive ASP (different from the one in Section 2.1) and then extend it to possibilistic ASP. Before we give an example of the semantics, we first note that when we use weak disjunction it matters whether we model a sentence like "when it is raining or snowing, you will get wet" as either $wet \leftarrow rain \vee snow$ or as the set of rules $\{wet \leftarrow rain, wet \leftarrow snow\}$. Indeed, the latter implies that we can only derive '*wet*' after we have made the choice between '*rain*' or '*snow*'. To accommodate for this we need to slightly alter the syntax of ASP.

Definition 3. *A* clause e *is a disjunction of one or more atoms. A* clausal rule *is an expression of the form* $e_0 \leftarrow e_1, ..., e_m, not\ e_{m+1}, ..., not\ e_n$ *with* e_i *a clause for every* $0 \leq i \leq n$*. The clause* e_0 *is called the head of the rule and* $e_1, ..., e_m, not\ e_{m+1}, ..., not\ e_n$ *the body of the rule. A* clausal program *is a finite set of clausal rules. A* positive clausal program *is a finite set of* positive clausal rules *which are expressions of the form* $e_0 \leftarrow e_1, ..., e_m$*. The* Herbrand base $\mathcal{B}_P$ *of a clausal program* P *is redefined as being the set of clauses appearing in* P*.*

It is easy to see that disjunctive programs are a special case of clausal programs. We can now take a look at an example.

Example 5. We live in Ohio and we want to book a holiday trip. Either we go to Venice, Rome or Florida (USA). After we have selected our destination, we can book our trip. Also, as soon as we know that we go to either Venice or Rome we need to arrange our visa so that we have it well before our departure date. We have program P with the rules:

$$\begin{aligned} venice \vee rome \vee florida &\leftarrow \\ book\_transportation &\leftarrow venice \\ book\_transportation &\leftarrow rome \\ book\_transportation &\leftarrow florida \\ arrange\_visa &\leftarrow venice \vee rome \end{aligned}$$

Definition 4. *For a positive clausal program* P *we define the immediate consequence operator* T_P^{w} w.r.t. *a set of clauses* E *as:*

$$T_P^{\text{w}}(E) = \{e_0 \mid e_0 \leftarrow e_1, ..., e_m \in P \wedge \forall i \in \{1, ..., m\} \cdot \exists e \in E \cdot e \subseteq e_i\}$$

where we identify a clause with its set of atoms. We use $P_{\text{w}}^{\star}$ *to denote the fixpoint which is obtained by repeatedly applying* T_P^{w} *starting from the empty interpretation,* i.e. *this is the least fixpoint of* T_P^{w} w.r.t. *set inclusion. An interpretation* E *is called an answer set of a positive clausal program* P *iff it is a minimal set of clauses for which* $E \models P_{\text{w}}^{\star}$*.*

Proposition 3. *The operator* T_P^{w} *is monotonic.*

Note that this definition of the immediate consequence operator is a generalization of the immediate consequence operator for a definite program from Section 2.1. Indeed, for a positive clausal program where all clauses contain only a single atom, *i.e.* a definite program, we have that $P^{\star} = P_{\text{w}}^{\star}$. Also note that a positive clausal program always has a unique answer set.

Example 6. We again consider Example 5. It is easy to see that the unique answer set of program P is $\{venice \vee rome \vee florida\}$. Indeed, without any further information we cannot derive anything but the disjunction itself.

Proposition 4. *Let* P *be a positive clausal program. We can compute the unique answer set of* P *in polynomial time.*

Thus we find that weak disjunction has a lower complexity than strong disjunction (which is in NP when we only have rules without negation-as-failure and otherwise in Σ_2^{P} [1]). This is clearly due to the fact that there is no non-determinism and that the unique answer set can be found using an iterative procedure. This does, however, imply that we can reason with certain forms of disjunction in an intuitive way without requiring additional complexity. This is an advantage of weak disjunction since many situations that involve disjunction and have uncertainty tend to lead to an epistemic view, for which weak disjunction offers a lower complexity than strong disjunction.

To define the concept of an answer set of an arbitrary (not necessarily positive) clausal program we also need to redefine the reduct for clausal programs. As in the ordinary case, we want that '*not e*' is removed whenever the clause '*e*' is satisfied, *i.e.* '*not e*' is true when there does not exist an $e' \in E$ such that $e' \subseteq e$. In other words, the guess E reflects what we think that we are capable of deriving from the program and we need to verify that this is indeed the case.

Definition 5. *Given a clausal program P and a set of clauses E, the reduct P^E of P* w.r.t. *E is defined as*

$$P^E = \{e_0 \leftarrow e_1, ..., e_m \mid \forall i \in \{m+1, ..., n\} \cdot \forall e \in E \cdot e \not\subseteq e_i \\ \wedge (e_0 \leftarrow e_1, ..., e_m, not\ e_{m+1}, ..., not\ e_n) \in P\}.$$

We say that E is an answer set of the clausal program P iff $(P^E)^\star_{\mathrm{w}} = E$, *i.e.* if E is the answer set of the reduct P^E.

Example 7. Consider the following clausal program P:

$$a \vee b \leftarrow \qquad c \leftarrow \qquad d \leftarrow not\ (a \vee b \vee d) \qquad e \leftarrow not\ c.$$

The reduct P^E with $E = \{a \vee b, c, e\}$ is then:

$$a \vee b \leftarrow \qquad c \leftarrow$$

since $\{a, b\} \subseteq \{a, b, d\}$ and $\{c\} \subseteq \{c\}$. The answer set of the reduct P^E is given by $(P^E)^\star_{\mathrm{w}} = \{a \vee b, c\}$, hence E is not an answer set of P since $(P^E)^\star_{\mathrm{w}} \neq E$.

We now extend the semantics to the case of *possibilistic clausal programs* which are finite sets of *possibilistic clausal rules* $p = (r, \lambda)$ with r a clausal rule.

Definition 6. *Let P be a possibilistic clausal program. Let $V : \mathcal{B}_P \to [0, 1]$ be a valuation. For every $p \in P$ where we have that $p = (r, \lambda)$ and $r = (e_0 \leftarrow e_1, ..., e_m, not\ e_{m+1}, ..., not\ e_n)$ the constraint γ_V^{w} induced by r under the weak possibilistic semantics is given by*

$$N(e_0) \geq \min\{N(e_1), ..., N(e_m), 1 - V(e_{m+1}), ..., 1 - V(e_n), \lambda\}. \tag{3}$$

$C^{\mathrm{w}}_{(P,V)} = \{\gamma_V^{\mathrm{w}}(r) \mid r \in P\}$ is the set of constraints imposed by program P and V. A possibility distribution that satisfies the constraints in $C^{\mathrm{w}}_{(P,V)}$ is called a possibilistic model of $C^{\mathrm{w}}_{(P,V)}$. We write $S^{\mathrm{w}}_{(P,g)}$ to denote the set of all least specific possibilistic models of $C^{\mathrm{w}}_{(P,V)}$. V is called a possibilistic answer set of P iff there exists a $\pi \in S^{\mathrm{s}}_{(P,V)}$ such that $\forall e \in \mathcal{B}_P \cdot V(e) = N(e)$.

Example 8. Let us consider the program P from Example 2, yet this time using the weak semantics for disjunction. We have the program Q with the single rule

$$\mathbf{0.7:}\ a \vee b \leftarrow .$$

The set of constraints C_Q^{w} is this time around given by $\{N(a \vee b) \geq 0.7\}$. The unique possibilistic answer set of Q is given by $\left\{(a \vee b)^{0.7}\right\}$, which is also an intuitively satisfactory result.

We now direct our attention to the discussion of how one should treat a constraint, which are rules of the form '$\leftarrow a_{k+1}, ..., a_m, not\ a_{m+1}, ..., not\ a_n$', in a (disjunctive) possibilistic ASP program.[1] If we look at the intuition that underlies constraints in ordinary ASP, then it seems natural to treat a constraint such as '$\leftarrow a, b$' as $N(a \wedge b) = 0$, *i.e.* it is not possible that both atoms are true at the same time. The next definition extends Definition 6 by formalizing possibilistic constraints.

Definition 7. *Let P be a possibilistic clausal program. Let $V : \mathcal{B}_P \rightarrow [0,1]$ be a valuation. For every* possibilistic constraint *p with $p \in P$, $p = (r, \lambda)$ and $r = (\leftarrow e_1, ..., e_m, not\ e_{m+1}, ..., not\ e_n)$ we define the associated constraint γ_g^{w} induced by r by*

$$N(e_1 \wedge ... \wedge e_m) \leq 1 - \min\{\lambda, 1 - V(e_{m+1}), ..., 1 - V(e_n)\}. \tag{4}$$

Example 9. If we once again consider Example 5 we can extend the program P with the rule '$\leftarrow$ *florida*'. The unique answer set of the program is then $\{rome \vee venice, arrange\_visa\}$. Indeed, with the additional knowledge that we will not be traveling to Florida we can readily conclude that we should arrange our visa. However, as desired, we still cannot conclude that we should book our transportation, which we can only do as soon as we pick either Venice or Rome as the actual destination.

It is important to note, however, that adding constraints to a positive clausal program affects its complexity. Indeed, finding whether an atom belongs to an answer set of a positive clausal program which, in addition, has possibilistic constraints, is NP-complete.

Proposition 5. *Let P be a possibilistic positive clausal program with possibilistic constraints. Finding whether P has a possibilistic answer set is* NP*-complete.*

Proof. (sketch) We can readily simulate 3SAT using positive clausal programs and possibilistic constraints. Let $\phi = (l_{11} \vee ... \vee l_{13}) \wedge \ldots \wedge (l_{n1} \vee ... \vee l_{n3})$ be an expression in conjunctive normal form where all clauses have 3 literals (*i.e.* either

[1] Note that one way to treat such constraints in ordinary ASP is to simulate these constraints as $\theta \leftarrow not\ \theta, a_{k+1}, ..., a_m, not\ a_{m+1}, ..., not\ a_n$ with θ a fresh atom. However, similar as to Example 3 we would end up with many possibilistic answer sets, *i.e.* the simulation does not succeed in eliminating undesired solutions.

an atom or the negation of an atom). Let P be a positive clausal program with a rule $p = (r, 1)$ and $r = (l'_{i1} \vee ... \vee l'_{i3} \leftarrow)$ for every clause in ϕ and where l' is l when $l = a$ is an atom and l' is a fresh atom $\bar{l}$ when $l = \neg a$. For every literal l in ϕ we also add the possibilistic constraint $p = (r, 1)$ with $r = (\leftarrow l, \bar{l})$. It is then easy to see that ϕ has a model if and only if P has an answer set. □

The same complexity result hold for clausal programs without certainty weights.

5 Related Work

A large body of research has been devoted to combining uncertainty with ASP. Different approaches are proposed in the literature depending on whether the uncertainty is treated in a qualitative or quantitative way. When uncertainty is treated in a quantitative way, probability theory seems to be the most often used. For example, in [2] uncertain information is encoded as a probabilistic atom which, intuitively, describes the probability that the atom will take on a certain value in some random selection given some other known evidence.

The most popular approach for dealing with uncertainty in a qualitative way is possibility theory. Combining possibility theory with logic programming was an idea first proposed in [7]. The work in [13] was one of the first papers to explore the idea of combining possibility theory with ASP. This work was later extended to also cover the case of disjunctive ASP in [14]. It was, however, noted in [3] that the semantics from [13,14] offer unintuitive results in certain cases since neither approach takes the certainty into account when dealing with negation-as-failure. This problem was discussed in [3] and a new characterization of normal ASP programs was established based on constraints on possibility distributions, which can naturally be generalized to cover possibilistic normal ASP programs.

Alternative semantics for PASP exist in the form of pstable models [15,4]. Yet these models are closer to the intuition of classical models than they are to the intuition of stable models as used in ASP. Hence the intuition that is captured by pstable models is different, where the focus is more on finding reasonable results in programs faced with uncertainty and which are inconsistent. There is a formal connection between the approach from [3] and the work on residuated logic programs [5] under the Gödel semantics. Both approaches are different in spirit, however, in the same way that possibilistic logic (which deals with uncertainty or priority) is different from Gödel logic (which deals with graded truth). The formal connection is due to the fact that necessity measures are min-decomposable. The work in this paper clearly differs from the work on residuated logic programs since necessity measure are not max-decomposable, which highlights that possibilistic logic is not truth-functional in general [8].

6 Conclusion

In this paper we have introduced the semantics of possibilistic disjunctive ASP in terms of constraints on possibility distributions. This provides us with natural semantics for dealing with possibilistic disjunctive ASP pervaded by uncertainty. We explored two different views of disjunction, the non-deterministic view as found in ordinary disjunctive ASP as well as a more epistemic view of disjunction. These two views are unearthed by the two distinct ways in which we can interpret a disjunctive rule as a constraint on possibility distributions. Due to the epistemic nature of possibilistic logic we find that the epistemic view of disjunction is oftentimes the one that offers the most intuitive understanding of the problem. Finally, we also examined the complexity of weak disjunction.

References

1. Baral, C.: Knowledge, Representation, Reasoning and Declarative Problem Solving. Cambridge University Press, Cambridge (2003)
2. Baral, C., Gelfond, M., Rushton, N.: Probabilistic reasoning with answer sets. Theory and Practice of Logic Programming 9(1), 57–144 (2009)
3. Bauters, K., Schockaert, S., De Cock, M., Vermeir, D.: Possibilistic answer set programming revisited. In: Proc. of UAI 2010 (2010)
4. Confalonieri, R., Nieves, J.C., Vázquez-Salceda, J.: Pstable semantics for logic programs with possibilistic ordered disjunction. In: Proc. of AI*IA 2009, pp. 52–61 (2009)
5. Damásio, C.V., Pereira, L.M.: Monotonic and residuated logic programs. In: Benferhat, S., Besnard, P. (eds.) ECSQARU 2001. LNCS (LNAI), vol. 2143, pp. 748–759. Springer, Heidelberg (2001)
6. Dubois, D., Lang, J., Prade, H.: Towards possibilistic logic programming. In: Proc. of ICLP 1991, pp. 581–595 (1991)
7. Dubois, D., Lang, J., Prade, H.: Possibilistic logic. Handbook of Logic for Artificial Intelligence and Logic Programming 3(1), 439–513 (1994)
8. Dubois, D., Prade, H.: Can we enforce full compositionality in uncertainty calculi? In: Proc. of AAAI 1994, pp. 149–154 (1994)
9. Dubois, D., Prade, H., Schockaert, S.: Rules and meta-rules in the framework of possibility theory and possibilistic logic. Scientia Iranica (to appear, 2011)
10. Gelfond, M., Lifschitz, V.: Classical negation in logic programs and disjunctive databases. New Generation Computing 9, 365–385 (1991)
11. Gelfond, M., Lifzchitz, V.: The stable model semantics for logic programming. In: Proc. of ICLP 1988, pp. 1081–1086 (1988)
12. Lifschitz, V., Schwarz, G.: Extended logic programs as autoepistemic theories. In: Proc. of LPNMR 1993, pp. 101–114 (1993)
13. Nicolas, P., Garcia, L., Stéphan, I., Lefèvre, C.: Possibilistic uncertainty handling for answer set programming. Annals of Mathematics and Artificial Intelligence 47(1–2), 139–181 (2006)
14. Nieves, J.C., Osorio, M., Cortés, U.: Semantics for possibilistic disjunctive programs. In: Baral, C., Brewka, G., Schlipf, J. (eds.) LPNMR 2007. LNCS (LNAI), vol. 4483, pp. 315–320. Springer, Heidelberg (2007)
15. Osorio, M., Pérez, J.A.N., Ramírez, J.R.A., Macías, V.B.: Logics with common weak completions. Journal of Logic and Computation 16(6), 867–890 (2006)
16. Tarski, A.: A lattice-theoretical fixpoint theorem and its applications. Pacific Journal of Mathematics 5(2), 285–309 (1955)
17. Zadeh, L.A.: Fuzzy sets as a basis for a theory of possibility. Fuzzy Sets and Systems, 3–28 (1978)

t-DeLP: A Temporal Extension of the Defeasible Logic Programming Argumentative Framework

Pere Pardo and Lluís Godo

Institut d'Investigació en Intel·ligència Artificial (IIIA - CSIC)
Campus UAB, E-08193 Bellaterra, Catalonia, Spain

Abstract. The aim of this paper is to offer an argumentation-based defeasible logic that enables temporal forward reasoning. We extend the DeLP logical framework by associating temporal parameters to literals. A temporal logic program is a set of temporal literals and durative rules. These temporal facts and rules combine into durative arguments representing temporal processes, that permit us to reason defeasibly about future states. The corresponding notion of logical consequence, or warrant, is defined slightly different from that of DeLP, due to the temporal aspects. As usual, this notion takes care of inconsistencies, and in particular we prove the consistency of any logical program whose strict part is consistent. Finally, we define and study a sub-class of arguments that seem appropriate to reason with natural processes, and suggest a modification to the framework that is equivalent to restricting the logic to this class of arguments.

1 Introduction

In this contribution, we present an argumentation-based temporal defeasible logic, t-DeLP, with temporal literals (for facts) and durative strict or defeasible rules defining temporal logical programs. The main motivation is to reason about interacting processes modeling them as arguments (combinations of facts and rules). An argument expresses some delay between each premise (cause) and the conclusion (effect), thus suggesting how a process might evolve. Since different arguments (process descriptions) might conflict, a dialectical procedure is proposed that decides which arguments are undefeated, thereby defining the set of logical consequences of a logical program.

An important feature of *defeasible* logics is the logical parsimony one obtains both at the level of representing knowledge bases (like for the family of non-monotonic logics), as well as regarding the associated logical machinery. This parsimony is in accordance with everyday causal reasoning, where it is standard practice to list only those causes that are uncommon or specific to the process (e.g. a *spark*, rather than *oxygen-in-air* is listed among the causes of fire). Other causes are only mentioned if their actual absence explains the non-occurrence of the effect (in absence of oxygen, a spark does not cause a fire).

Among non-monotonic logics, those based on an argumentation process present several advantages. First, logical consequence relations built upon arguments are based on a preference relation (between conflicting arguments) which is more expressive than the priority relation of purely rule-based approaches. Another advantage of argumentation-based logics is that these mirror the inference mechanisms of a deliberating human

S. Benferhat and J. Grant (Eds.): SUM 2011, LNAI 6929, pp. 489–503, 2011.
© Springer-Verlag Berlin Heidelberg 2011

agent (or set of agents), thus producing logical formalisms that appear more natural and conceptually transparent.

An important contribution along these lines is García and Simari's [7]. The authors present an argumentation-based defeasible logic, called DeLP, and discuss several issues related to application domains. For instance, the question of which criteria the preference relation should be based upon is discussed at length. These criteria play a central role in argumentation-based logics, since they determine which relation of defeasible logical consequence, or *warrant*, one obtains. Among other alternatives, a formal criteria based on a preference for more (or more specific) information is suggested in [7]: the *more premises* criterion and the *more direct rules* criterion. But since the latter criterion conflicts with durative rules, it will not be considered here.[1] Other genuinely temporal or causal features cannot be modeled as in DeLP either, and must be adapted from [7]. These are mainly due to the temporal asymmetry (past vs. future) of causation. Thus, persistence, the attack relation and the formal criteria for preference are among the elements deserving special attention for the temporal case. As a consequence, the notion of warrant for (temporal) literals is slightly different than that studied in [7].

Another reason to use defeasible logic, more to the point of the present contribution, is its ability to reason with interactions between different processes or between different aspects of a given process. In t-DeLP, the value of a parameter in a process evolves according to the conclusions of the sequence of undefeated arguments representing this process at increasing time-points. If an internal cause or influence is just part of the support of an argument, an external influence can be seen, from a logical point of view, as a contradiction obtaining between two arguments. These interactions between processes reflect the fact that causal laws occurring in their description are conceived in idealized or isolated conditions, so in practice instances of laws may contradict each other. The temporal argumentation framework t-DeLP permits to naturally address the problem posed by these contradictions in a compact way. This system does detect and remove all the contradictions that exist in a given logical program. Moreover this logic can compute the positive facts resulting from these interactions, if the logical program is supplied with sufficient knowledge.

The paper is structured as follows: after some preliminaries, we present t-DeLP logic and study some of its logical properties. In particular we prove that any temporal logical program outputs a consistent set of warranted literals. As a consequence, when logical programs restrict its strict information to facts (i.e. all rules are defeasible), the resulting notion of warrant is a consequence operator. Finally, we focus on the study of a particularly interesting sub-class of arguments that presumably capture natural processes. We produce a counter-example showing that if we do not restrict temporal arguments to this class some unintuitive consequences may occur. Then we revise some of the definitions to prune these counterexamples, and show that this revised notion of warrant coincides with warrant restricted to arguments in the class. We also discuss if these problematic cases can be prevented by faithful representations.

---

[1] This criterion captures the preference for {*penguins do not fly*} over {*penguins are birds*, *birds fly*}. It might still apply between non-durative arguments in t-DeLP, though evidence-based reasoning lies out of the scope of this paper.

Notation. We will use the following conventions: strong negation is denoted $\sim p$, for a propositional variable $p \in \mathsf{Var}$. Given two sets X, Y we denote the size of X as $|X|$, set-theoretic difference as $X \smallsetminus Y$, the power set of X as $\mathcal{P}(X)$, and the Cartesian product of X and Y as $X \times Y$, or X^2 for $X \times X$; $X^{<\omega}$ is the set of finite sequences of elements of X. If f is a function $f : X \rightarrow Y$ and $X' \subseteq X$, we define $f[X'] = \{f(a) \in Y \mid a \in X'\}$. Given a family of sets $\mathbf{M}$, its union is denoted $\bigcup \mathbf{M}$.

2 Knowledge Representation

Our language builds upon a set of temporal literals, consisting of a pair $\langle$*literal, time*$\rangle$. Literals are expressions of the form $p, \sim p$ from a given set of variables $p \in \mathsf{Var}$. Strong negation $\sim$ cannot be nested, so we will use the following notation over literals: if $\ell = p$ then $\sim\ell$ will denote $\sim p$, and if $\ell = \sim p$ then $\sim\ell$ will denote p. These literals, though, might rather be seen as ground predicates, of the form *literal* = (*object, property*) or also *literal* = (*object, parameter, value*).

Time is relevant to determine whether a pair of temporal literals contradict each other: for this contradiction to exist, the literals expressed must be the negation of each other *and* they must be claimed to hold at the same time. A temporal or causal statement (possibly an instance of some general law) is represented as a rule: *a set of tuples* $\langle$(*object, parameter, value*), *time*$\rangle$ imply *a tuple* $\langle$(*object, parameter, value*), *time*$\rangle$; rules with no duration (delay) stand for static or structural constraints.

In any case, literals of the form $\langle$(*object, parameter, value*), *time*$\rangle$ tacitly require some constraints to be satisfied: an object cannot have different values of a given parameter at a given time. These absolute constraints, represented by strict rules, can be seen as induced by a family of sets of pairwise incompatible literals $X = \{(o, p, v), (o, p, v'), \ldots\}$, for fixed p and o (these literals are also called *mutex* in the literature, for mutual exclusion). Similarly, defeasible rules may represent contingent constraints.

Example 1. Let $\mathcal{O}$ and $\mathcal{L}$ be the sets of objects o and locations l; and let $@(o, l) \in \mathsf{Var}$ denote: *o is at l*;

- the *at most one location per object* policy is defined by a set $\{o\} \times \mathcal{L}$ for each $o \in \mathcal{O}$; this set induces rules of the form $\langle \sim@(o, l), t\rangle \leftarrow \langle @(o, l'), t\rangle$, if $l \neq l'$.
- the *at most one object per location* policy is defined by a set $\mathcal{O} \times \{l\}$ for each $l \in \mathcal{L}$; this set induces rules $\langle \sim@(o, l), t\rangle \leftarrow \langle @(o', l), t\rangle$, if $o \neq o'$.

The *orientation* $\nearrow (o, y)$ of an object o can be used to constrain the location of bound parts o_0, o_1 w.r.t. each other and reason with the motion of rigid objects[2]:

$$\langle @(o_0, x_0), t\rangle \prec \langle @(o_1, x_1), t\rangle, \langle \nearrow (o_1, y_1), t\rangle, \langle \mathsf{bound}(o_0, o_1), t\rangle.$$

3 t-DeLP: Defeasible Logic with (discrete) Time

We take the set of natural numbers $\mathbb{N}$ as our working set of discrete time points. The logic t-DeLP is based on temporal literals $\langle \ell, t\rangle$, where ℓ is a literal and $t \in \mathbb{N}$, denoting

[2] For reasons of space, though, the examples presented below to illustrate t-DeLP rather involve qualitative reasoning.

ℓ holds at t. In order to solve conflicts between arguments the preference (or defeat) relation between arguments is based on: a preference for arguments with *more premises* and for *longer* arguments over its parts (so an argument can defeat the persistence of its subarguments' conclusions, if inconsistent with them). Arguments that make only use of strict information are also preferred to arguments conflicting with them. A final criterion, *less durative rules*, is not considered here.[3]

Definition 1. *Given a finite set of propositional variables* Var, *we define* $\mathsf{Lit} = \mathsf{Var} \cup \{\sim p \mid p \in \mathsf{Var}\}$. *The define set of temporal literals* $\mathsf{TLit} = \{\langle \ell, t\rangle \mid \ell \in \mathsf{Lit}, t \in \mathbb{N}\}$. *If* $\Gamma \subseteq \mathsf{Lit}$, *we say that* Γ *is* consistent *if there is no* $p \in \mathsf{Var}$ *such that* $p, \sim p \in \Gamma$. *If* $\Gamma^* \subseteq \mathsf{TLit}$, *we say that* Γ^* *is* consistent *if each* $\Gamma^*_t := \{\ell \mid \langle \ell, t\rangle \in \Gamma^*\}$ *is consistent.*

Definition 2. *A* temporal defeasible (resp. strict rule) *is an expression* δ *of the form* $\langle \ell, t\rangle \prec \langle \ell_0, t_0\rangle, \ldots, \langle \ell_n, t_n\rangle$ *(resp.* $\langle \ell, t\rangle \leftarrow \langle \ell_0, t_0\rangle, \ldots, \langle \ell_n, t_n\rangle$*), where* $t \geq \max\{t_0, \ldots t_n\}$. *We write* $\mathsf{head}(\delta) = \langle \ell, t\rangle$, $\mathsf{body}(\delta) = \{\langle \ell_0, t_0\rangle, \ldots, \langle \ell_n, t_n\rangle\}$ *and* $\mathsf{literals}(\delta) = \{\mathsf{head}(\delta)\} \cup \mathsf{body}(\delta)$.

A strict rule with an empty body, e.g. $\langle \ell, t\rangle \leftarrow$, also denoted $\langle \ell, t\rangle$, represents a basic fact that holds at time t. (As usual in DeLP -see [7], [4]- basic defeasible facts, or presumptions, $\langle \ell, t\rangle \prec$ are not considered). A strict rule $\delta \in \Pi$ preserves the truth from $\mathsf{body}(\delta)$ to $\mathsf{head}(\delta)$ (plus it preserves its being strictly derived). A defeasible rule $\delta \in \Delta$ states a weaker claim: if the premises are true, this is in principle a reason for believing that the conclusion is also true (though this conclusion may be withdrawn for other reasons). A special subset of rules is that of *persistence* rules, of the form $\langle \ell, t+1\rangle \leftarrow \langle \ell, t\rangle$ or $\langle \ell, t+1\rangle \prec \langle \ell, t\rangle$, stating that ℓ is preserved from t to $t+1$ (if true at t) and, resp., that *ceteris paribus* it will persist at $t+1$. The set of defeasible (strict) persistence rules will be denoted Δ_{p} (resp. Π_{p}).

Example 2. $\langle \sim\text{tuesday}, t+24\rangle \leftarrow \langle \text{tuesday}, t\rangle$ and $\langle \text{wednesday}, t+24\rangle \leftarrow \langle \text{tuesday}, t\rangle$, where time units are hours, are examples of strict rules with a temporal delay.[4]

Definition 3. *A* temporal DeLP program, *or t-de.l.p., is a pair* (Π, Δ), *where* Π *is a set of temporal strict rules,* Δ *a set of temporal defeasible rules and the set of derivable literals from* Π *is consistent.*

Temporal rules as above can be seen as instances of general rules of the form $\delta^* = \ell \leftarrow (\ell_0, d_0), \ldots, (\ell_n, d_n)$ -and similarly for defeasible rules with $\prec$ -, where each d_i expresses how much time in advance must ℓ_i hold for the rule to apply and produce a derivation of ℓ. Such a general rule is to be understood as a shorthand for the set of rules $\{\langle \ell, t\rangle \leftarrow \langle \ell_0, t-d_0\rangle, \ldots, \langle \ell_n, t-d_n\rangle \mid t \in \mathbb{N}, t \geq \max\{d_0, \ldots, d_n\}\}$. For example, the rule $\langle p, 4\rangle \prec \langle q, 3\rangle$ would be an instance of the general rule $p \prec (q, 1)$. Persistence rules can therefore be expressed as general rules of the form $\ell \leftarrow (\ell, 1)$ or $\ell \prec (\ell, 1)$; the latter defeasible general persistence rule for ℓ will be denoted δ_ℓ. The formal definitions do make use only of instances of general rules, i.e. temporal rules only.

[3] This is important, since rules with long duration might fail to detect conflicts, (so, e.g. the program might fail to predict that two balls running into each will collide). Instead, we will assume rules are precise enough.

[4] Note the condition that $\mathsf{body}(\delta)$ occurs no later than $\mathsf{head}(\delta)$ in Def. 2 prevents strict rules with positive delay (like in this example) to be closed under transposition, cf. [3].

Example 3. Consider the next example (with arguments). Lars, a tourist visiting the Snake Forest, has been bitten by a venomous snake. The poison of this type of snake does kill a person in 3 hours ($\mathcal{A}$). But since our subject, Lars, is experienced (it has been bitten and cured a few times before), he may resist up to 5 hours ($\mathcal{B}_0, \mathcal{B}_1$). We decide to take him to the nearest hospital, which in normal conditions this would take 2 hours ($\mathcal{C}$), but since today is sunday, the traffic jam makes it impossible to reach the hospital in less than 4 hours ($\mathcal{D}_0, \mathcal{D}_1$). The antidote takes less than an hour to become effective ($\delta^\star$). This scenario is modeled by the following temporal facts and general rules:

$$\begin{aligned}\Pi = \{ & \langle \text{@forest(Lars)}, 0\rangle, \langle \text{bitten(Lars)}, 0\rangle, \langle \text{exp(Lars)}, 0\rangle, \\ & \langle \sim\text{dead(Lars)}, 0\rangle, \langle \text{sunday}, 0\rangle \}\end{aligned}$$

$$\begin{array}{rll}
\Delta = \{ \quad \text{dead(Lars)} & \prec (\text{bitten(Lars)}, 3) & \mathcal{A} \\
\sim\text{dead(Lars)} & \prec (\text{bitten(Lars)}, 3), (\text{exp(Lars)}, 3), (\sim\text{dead(Lars)}, 3) & \mathcal{B}_0 \\
\text{dead(Lars)} & \prec (\text{bitten(Lars)}, 5), (\text{exp(Lars)}, 5), (\sim\text{dead(Lars)}, 5) & \mathcal{B}_1 \\
\text{@hospital(Lars)} & \prec (\text{bitten(Lars)}, 2), (\text{@forest(Lars)}, 2), (\sim\text{dead(Lars)}, 2) & \mathcal{C} \\
\sim\text{@hospital(Lars)} & \prec \left\{ \begin{array}{l} (\text{traffic.jam}, 2), \ (\text{bitten(Lars)}, 2), \ (\text{exp(Lars)}, 2), \\ (\sim\text{dead(Lars)}, 2), \ (\sim\text{@hospital(Lars)}, 2) \end{array} \right\} & \mathcal{D}_0 \\
\text{@hospital(Lars)} & \prec \left\{ \begin{array}{l} (\text{traffic.jam}, 4), \ (\text{bitten(Lars)}, 4), \ (\text{exp(Lars)}, 4), \\ (\sim\text{dead(Lars)}, 4), \ (\sim\text{@hospital(Lars)}, 4) \end{array} \right\} & \mathcal{D}_1 \\
\text{traffic.jam} & \prec (\text{sunday}, 0) & \mathcal{D}_x \\
\sim\text{dead(Lars)} & \prec (\text{@hospital(Lars)}, 1), (\text{bitten(Lars)}, 1), (\sim\text{dead(Lars)}, 1) \quad \} & \delta^\star
\end{array}$$

For this example, Δ_p contains persistence rules for all literals except those of the form $\sim$@location(Lars). We prove below in t-DeLP that Lars survives the snake attack.

Derivability in t-DeLP, denoted by $\vdash$, is defined (as in DeLP) by closure under the *modus ponens* rule: $(\Pi, \Delta) \vdash \langle \ell, t\rangle$ if $\langle \ell, t\rangle \in \Pi$; and $(\Pi, \Delta) \vdash \mathsf{head}(\delta)$ if $(\Pi, \Delta) \vdash \langle \ell', t'\rangle$ for each $\langle \ell', t'\rangle \in \mathsf{body}(\delta)$ and some $\delta \in \Pi \cup \Delta$. As it happens in DeLP, the set of derivable literals in (Π, Δ) will not in general be consistent.

Definition 4. *Given a t-de.l.p.* (Π, Δ), *an* argument *for* $\langle \ell, t\rangle$ *is a set* $\mathcal{A} \subseteq \Pi \cup \Delta$, *such that*

(1) $\mathcal{A} \vdash \langle \ell, t\rangle$,
(2) the set of derivable literals from $\Pi \cup \mathcal{A}$ *is consistent,*
(3) $\mathcal{A}$ *is* $\subseteq$*-minimal satisfying (1) and (2).*

Thus, arguments are consistent minimal derivations (i.e. without redundant information). In Example 3, each argument (e.g. $\mathcal{D}_0$) is made of the rules in Δ labeled by this argument ($\mathcal{D}_0, \mathcal{D}_x$ in this case). Observe that, although Π and Δ may be infinite (due to the coding of general rules as an infinite set of temporal rules), an argument for a t-de.l.p. (Π, Δ) will be always a finite subset of $\Pi \cup \Delta$. We also define for an argument $\mathcal{A}$ for $\langle \ell, t\rangle$:

$$\mathsf{concl}(\mathcal{A}) = \langle \ell, t\rangle \qquad \mathsf{base}(\mathcal{A}) = \{\delta \in \mathcal{A} \mid \mathsf{body}(\delta) = \emptyset\}$$
$$\mathsf{literals}(\mathcal{A}) = (\bigcup \mathsf{body}[\mathcal{A}]) \cup \mathsf{head}[\mathcal{A}] \qquad \|\mathcal{A}\| = t - t(\mathcal{A})$$

where $t(\mathcal{A}) = \min\{t' \in \mathbb{N} \mid \langle \ell', t'\rangle \in \mathsf{base}(\mathcal{A})\}$.

In contrast to DeLP, we make explicit in argument $\mathcal{A}$ which is the strict information used, to facilitate the detection of inconsistencies with an intermediate step of $\mathcal{A}$ in its strict part. The reason is that there exist many possible ways to complete defeasible rules in $\mathcal{A}$ into a derivation for $\mathsf{concl}(\mathcal{A})$. And these different ways may be attacked by different arguments. For example, the sets $\{\langle p, 4\rangle \leftarrow \langle q, 2\rangle, \langle q, 2\rangle \leftarrow \langle r, 1\rangle, \langle r', 0\rangle\}$ and $\{\langle p, 4\rangle \leftarrow \langle s, 3\rangle, \langle s, 3\rangle \leftarrow \langle r, 1\rangle, \langle r', 0\rangle\}$ may both complete the set of defeasible rules $\{\langle p', 5\rangle \prec \langle p, 4\rangle, \langle r, 1\rangle \prec \langle r', 0\rangle\} \subseteq \Delta$ into an argument (derivation) for $\langle p', 5\rangle$, but only the latter is attacked by an argument concluding $\langle \sim s, 3\rangle$.

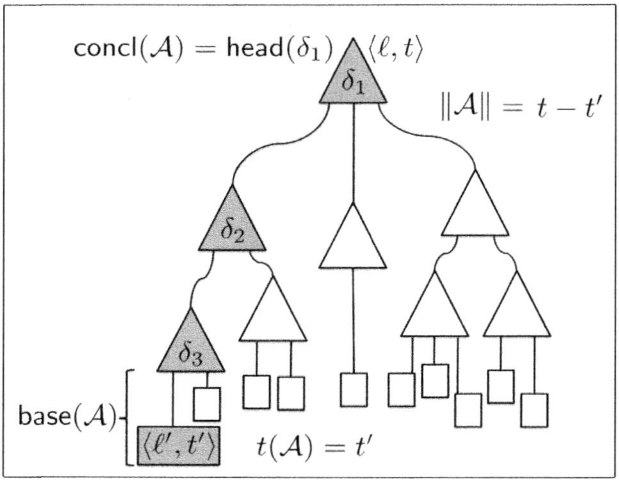

Fig. 1. Facts (rectangles) and rules (triangles) in grey define the delay $\|\mathcal{A}\|$ of argument $\mathcal{A}$

Now we define a sub-argument of $\mathcal{A}$. A sub-argument will be the actual target of an attack by another argument.

Definition 5. *Let (Π, Δ) be a t-de.l.p. and let $\mathcal{A}$ be an argument for $\langle \ell, t\rangle$ in (Π, Δ). Given some $\langle \ell_0, t_0\rangle \in \mathsf{literals}(\mathcal{A})$, a* sub-argument *for $\langle \ell_0, t_0\rangle$ is a subset $\mathcal{B} \subseteq \mathcal{A}$ such that $\mathcal{B}$ is an argument for $\langle \ell_0, t_0\rangle$.*

Notice that each literal $\langle \ell_0, t_0\rangle$ in an argument $\mathcal{A}$ uniquely determines its corresponding subargument, that we will denote by $\mathcal{A}(\langle \ell_0, t_0\rangle)$. For example, in Figure 1, $\mathcal{A}(\mathsf{head}(\delta_2)) = \{\delta_2, \delta_3, \delta_4, \langle \ell', t'\rangle, \ldots\}$.

Definition 6. *Given a t-de.l.p. (Π, Δ), let $\mathcal{A}_0$ an argument for $\langle \ell_0, t_0\rangle$ and let $\mathcal{A}_1$ an argument for $\langle \ell_1, t_1\rangle$. We say $\mathcal{A}_1$* attacks *$\mathcal{A}_0$ if there exists a subargument $\mathcal{B}$ of $\mathcal{A}_0$ for $\langle \sim\ell, t_1\rangle$ and $\Delta \cap \mathcal{B} \neq \emptyset$. In this case we say that $\mathcal{A}_1$ attacks $\mathcal{A}_0$ at $\mathcal{B}$.*

Notice that if $\mathcal{A}_1$ attacks $\mathcal{A}_0$ at $\mathcal{B}$, $\mathcal{B}$ cannot only consist of strict information, in particular of a strict fact: $\mathcal{B} \neq \{\langle \ell, t'\rangle\}$.

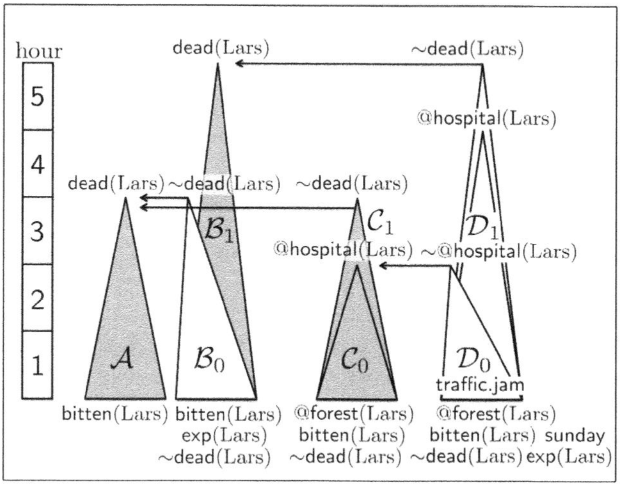

Fig. 2. The Snake Bites Lars scenario

Example 4. (Cont'd) See Figure 2 for an illustration of Example 3. The arguments related by an arrow attack each other: $\mathcal{C}_1$, $\mathcal{B}_0$ attack $\mathcal{A}$ and viceversa. But there are asymmetries in the quantity of information supporting each argument. Intuitively, we have (1) argument $\mathcal{B}_0$ should prevail over $\mathcal{A}$ since it is based on more information (premises); (2) the argument $\mathcal{D}_1$ for $\langle$@hospital(Lars), 4$\rangle$ should be preferred over the argument $\mathcal{E}$ for $\langle$@forest(Lars), 4$\rangle$ that merely uses persistence, from of $\langle$@forest(Lars), 0$\rangle$; the idea is that $\mathcal{D}_1$ contains more recent information: $\mathcal{E}$ is just a*sub-argument* of $\mathcal{D}_1$ *extended with* defeasible persistence rules. Finally, rephrase Example 3 with these rules: black-spotted snakes are generally poisonous, while green snakes are generally harmless. (3) If Lars was bitten by a green black-spotted snake, we would not be able to decide whether he has been poisoned or not, since reasons for and against would not dominate each other.

As in DeLP, one refines the relation of attack relation into a defeat relation to decide which argument prevails in case of an attack. This relation can be in principle specified by the user, but in this paper we adopt the following definition in order to meet the above intuitive preferences in Example 4 above.

Definition 7. *Let* $\mathcal{A}_1$ *attack* $\mathcal{A}_0$ *at* $\mathcal{B}$, *where* $\mathsf{concl}(\mathcal{B}) = \langle \ell, t \rangle$. *We say:*

- $\mathcal{A}_1$ *is a* proper defeater for $\mathcal{A}_0$ *iff* $\mathcal{A}_1 \subseteq \Pi$, *or* $\mathsf{base}(\mathcal{A}_1) \supsetneq \mathsf{base}(\mathcal{B})$, *or*
  for some $t' < t$, $\mathcal{A}_0 = \mathcal{A}_1(\langle \ell, t' \rangle) \cup \{\langle \ell, t''+1 \rangle \prec \langle \ell, t'' \rangle \mid t' \leq t'' < t\}$.

Otherwise,

- $\mathcal{A}_1$ *is a* blocking defeater for $\mathcal{A}_0$ *iff* $\mathsf{base}(\mathcal{A}_1) \not\subseteq \mathsf{base}(\mathcal{B})$ *or* $\mathsf{base}(\mathcal{A}_1) = \mathsf{base}(\mathcal{B})$

Blocking and proper defeat relations are denoted, resp., $\mathcal{A}_1 \prec\succ \mathcal{A}_0$, *and* $\mathcal{A}_1 \succ \mathcal{A}_0$.

Thus, a strict set of rules is a proper defeater for any argument attacked by it. In the other cases, a properly defeated argument $\mathcal{A}_0$ either has *less premises than* (case (1) of Example 4) or is a *sub-argument of* its defeater $\mathcal{A}_1$, *extended with* a sequence of

$\|\mathcal{A}_1\| - \|\mathcal{B}\|$ instances of persistence rule $\delta_{\mathsf{concl}(\mathcal{B})}$ (case (2) of Example 4). In the latter case we say that $\mathcal{A}_1$ is a *longer* argument than $\mathcal{A}_0$. (Observe $\mathcal{A}_1$ does not defeat its sub-argument, only the δ_ℓ-extension of it. See Figure 3 (top left) for an illustration.) Finally, note that since Π is assumed to produce a consistent set of derivable literals and the other two conditions for being a proper defeater are asymmetric relations, no pair of arguments can be a proper defeater for each other. Blocking defeaters are defined to satisfy case (3) of Example 4. In Figure 2, arguments defeated by some argument are depicted as grey.

An argument $\mathcal{B}$ defeating $\mathcal{A}$ can at its turn have its own defeaters $\mathcal{C}, \ldots$ and so on. These give rise to *argumentation lines* where each argument defeats its predecessor. Intuitively, the notion of defeat in an argumentation line $[\ldots, \mathcal{A}, \mathcal{B}, \mathcal{C}, \ldots]$ should exclude a blocking defeater $\mathcal{C}$ for $\mathcal{B}$ as a defeater, *provided that* $\mathcal{B}$ is already blocking defeater for $\mathcal{A}$. (The reason is that otherwise, we could have cycles $[\ldots, \mathcal{A}, \mathcal{B}, \mathcal{A}, \mathcal{B}, \ldots]$.) Other forms of cyclic defeats are also excluded in the definition.

Definition 8. *(Adapted from [7]) Let $\mathcal{A}_1$ be an argument in (Π, Δ). An* argumentation line *for $\mathcal{A}_1$ is a sequence $\Lambda = [\mathcal{A}_1, \mathcal{A}_2, \ldots]$ where*

(i) supporting arguments, i.e. in odd positions $\mathcal{A}_{2i+1} \in \Lambda$ are jointly **M***-consistent, and similarly for interfering arguments $\mathcal{A}_{2i} \in \Lambda$*
(ii) a sub-argument of $\mathcal{A}_i$ can occur later in Λ, i.e. as $\mathcal{A}_{i+2j}$ only if $\|\mathcal{A}_{i+2j}\| < \|\mathcal{A}_i\|$ (i.e. its duration is stricly less than that of $\mathcal{A}_i$) [5]
(iii) $\mathcal{A}_{i+1}$ is a proper defeater for $\mathcal{A}_i$ if $\mathcal{A}_i$ is a blocking defeater for $\mathcal{A}_{i-1}$

The union of maximal argumentation lines Λ for $\mathcal{A}_1$ is the dialectical tree *for $\mathcal{A}_1$:*
$$\mathcal{T}_{(\Pi,\Delta)}(\mathcal{A}_1) = \bigcup\{\Lambda \in (\Pi \cup \Delta)^{<\omega} \mid \Lambda \text{ is a maximal argumentation line for } \mathcal{A}_1\}$$

Example 5. (Cont'd) Define $\mathcal{E}^+ = \mathcal{E} \cup \{\langle \sim@\mathsf{hospital(Lars)}, 4\rangle \leftarrow \langle @\mathsf{forest(Lars)}, 4\rangle\}$; and $\mathcal{D}_1^+ = \mathcal{D}_1 \cup \{\langle \sim@\mathsf{forest(Lars)}, 4\rangle \leftarrow \langle @\mathsf{hospital(Lars)}, 4\rangle\}$. Note that $\mathcal{D}_1^+$ properly defeats $\mathcal{E}^+$ at $\mathcal{E}$. Both $[\mathcal{E}, \mathcal{D}_1, \mathcal{E}^+, \mathcal{D}_1^+]$ and $[\mathcal{D}_1, \mathcal{E}^+, \mathcal{D}_1^+]$ are maximal arg. lines.

The next bottom-up marking procedure on the tree $\mathcal{T}_{(\Pi,\Delta)}(\mathcal{A}_1)$ decides whether $\mathcal{A}_1$ is undefeated in (Π, Δ).

Definition 9. *(From [7]) Let $\mathcal{T} = \mathcal{T}_{(\Pi,\Delta)}(\mathcal{A}_1)$ be the dialectical tree for $\mathcal{A}_1$. Then,*

(1) mark all terminal nodes of $\mathcal{T}$ with a U (for undefeated);
(2) mark a node $\mathcal{B}$ with a D (for defeated) if it has a children node marked U;
(3) mark $\mathcal{B}$ with U if all its children nodes are marked D.

See Figure 3 (right) for an example of a dialectical tree with root $\mathcal{A}_1$. Arguments marked U are represented white, and those marked D are represented black.

Definition 10. *Given a t-de.l.p. (Π, Δ), we say $\langle \ell, t\rangle$ is* warranted *in (Π, Δ) iff there exists an argument $\mathcal{A}_1$ for $\langle \ell, t\rangle$ in (Π, Δ) such that $\mathcal{A}_1$ is undefeated in $\mathcal{T}_{(\Pi,\Delta)}(\mathcal{A}_1)$. We will denote by* $\mathsf{warr}(\Pi, \Delta)$ *the set of warranted literals in (Π, Δ).*

---

[5] This is a weaker condition that in DeLP, where no sub-argument *at all* can occur later than an argument in Λ. In our temporal case, a sub-argument (of $\mathcal{A}$) talking about a previous time may offer legitimate reasons to the defense of $\mathcal{A}$.

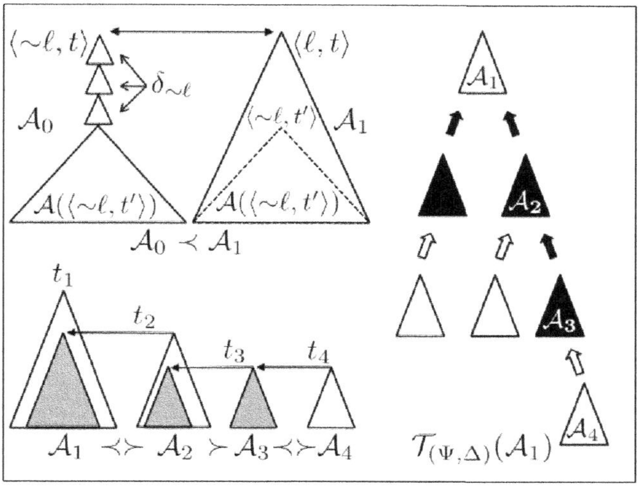

Fig. 3. (Top Left) $\mathcal{A}_1$ is a proper defeater of $\mathcal{A}_0$, a sub-argument of $\mathcal{A}_1$ extended with $\delta_{\sim\ell}$. (Bottom Left) An argumentation line. (Right) The dialectical tree $\mathcal{T}_{(\Pi,\Delta)}(\mathcal{A}_1)$.

Example 6. (Cont'd) In Example 4, the argument $\mathcal{E}$ is defeated, while $\mathcal{D}_1$ is undefeated. Hence, $\langle$@hospital(Lars), 4$\rangle$ is warranted, and so is $\langle\sim$dead(Lars), 5$\rangle$; i.e. Lars survives.

Example 7. (Yale Shooting Scenario) Even if able to prove that a turkey dies if shot just after loading a gun, early logical formalisms failed to prove the same if some waiting occurred between load and shoot. To prove this in t-DeLP, represent the action shoot (at t) by rules: $\delta_0(t) = \langle\sim\mathsf{loaded}, t+1\rangle \prec \langle\mathsf{loaded}, t\rangle$ and $\delta_1(t) = \langle\sim\mathsf{alive}, t+1\rangle \prec \langle\mathsf{loaded}, t\rangle$; and action load (at t) by $\delta(t) = \langle\mathsf{loaded}, t+1\rangle \prec \langle\sim\mathsf{loaded}, t\rangle$. Instead of a wait action we use persistence rules. In the t-de.l.p. $(\{\langle\sim\mathsf{loaded}, 0\rangle\}, \{\delta(3), \delta_0(8), \delta_1(8)\})$, the literal $\langle\sim\mathsf{alive}, 9\rangle$ is warranted.

The next results show t-DeLP ensures the consistency of a t-de.l.p. (Π, Δ), provided that its strict part Π outputs a consistent set of derivable literals.

Lemma 1. *Given some (Π, Δ), let $\mathcal{A}$ be an argument in (Π, Δ) for $\langle\ell, t\rangle$. Also, let $\mathcal{B}$ be an argument for $\langle\sim\ell, t\rangle$, with $\mathcal{A}$ a defeater for $\mathcal{B}$ at $\mathcal{B}$. If $\mathcal{A}$ is defeated in $\mathcal{T}_{(\Pi,\Delta)}(\mathcal{B})$, then $\mathcal{A}$ is defeated in $\mathcal{T}_{(\Pi,\Delta)}(\mathcal{A})$.*

Proof. First, let $\Lambda = [\mathcal{D}_1, \ldots, \mathcal{D}_n]$ be a maximal argumentation line, with $\mathcal{D}_i \in \Lambda$ defeated. Define a witness for the defeat of $\mathcal{D}_i$ as some argumentation line $\Lambda' = [\mathcal{D}_i, \ldots, \mathcal{D}_m]$ of even length $m - i$. Such a witness exists: set e.g. $\Lambda' = \Lambda$ for $i = 1$; or Λ' a terminal segment $[\mathcal{D}_i, \ldots, \mathcal{D}_n]$ of Λ).

Let then $\mathcal{A}$ a defeater for $\mathcal{B}$ at $\mathcal{B}$. This implies the existence of some $\Lambda = [\mathcal{B}, \mathcal{A}, \ldots] \subseteq \mathcal{T}_{(\Pi,\Delta)}(\mathcal{B})$. Assuming $\mathcal{A}$ is defeated in $\mathcal{T}_{(\Pi,\Delta)}(\mathcal{B})$, we have in particular some $\Lambda^* = [\mathcal{B}, \mathcal{A}, \mathcal{B}_3, \ldots, \mathcal{B}_{2m+1}]$ witnessing the defeat of $\mathcal{A}$ (i.e. with $\mathcal{B}_{2i+1}$ undefeated in $\mathcal{T}_{(\Pi,\Delta)}(\mathcal{B})$, for each $1 \leq i \leq m$).

We show first that any arg. line $[\mathcal{B}, \mathcal{A}, \mathcal{B}_3, \ldots]$ in $\mathcal{T}_{(\Pi,\Delta)}(\mathcal{B})$ contains a sequence $[\mathcal{A}, \mathcal{B}_3, \ldots]$ that is an arg. line in $\mathcal{T}_{(\Pi,\Delta)}(\mathcal{A})$. Let then $[\mathcal{B}, \mathcal{A}, \mathcal{B}_3, \ldots] \subseteq \mathcal{T}_{(\Pi,\Delta)}(\mathcal{B})$. Clearly $\Lambda = [\mathcal{A}, \mathcal{B}_3, \ldots]$ satisfies the 3 conditions for an arg. line for $\mathcal{A}$: (1) its first element is $\mathcal{A}$; (2) the set of even (odd) members is jointly consistent (since, otherwise, the set of odd (resp. even) members of $[\mathcal{B}, \mathcal{A}, \mathcal{B}_3, \ldots]$ would also be inconsistent). (3) a sub-argument of some argument $\mathcal{B}_i$ (with the same duration than $\mathcal{B}_i$) does not occur after argument $\mathcal{B}_i$ in the line (otherwise, the same would be true of $[\mathcal{B}, \mathcal{A}, \mathcal{B}_3, \ldots]$ in $\mathcal{T}_{(\Pi,\Delta)}(\mathcal{B})$). Finally, (4) for no three consecutive elements $[\ldots, \mathcal{D}_i, \mathcal{D}_{i+1}, \mathcal{D}_{i+2}, \ldots]$ we have $\mathcal{D}_{i+2}$ is a blocking defeater for $\mathcal{D}_{i+1}$ and $\mathcal{D}_{i+1}$ a blocking defeater for $\mathcal{D}_i$ (since otherwise the same would occur in arg. line $[\mathcal{B}, \mathcal{A}, \mathcal{B}_3, \ldots]$).

Now, assume, towards a contradiction, that $\mathcal{A}$ is undefeated in $\mathcal{T}_{(\Pi,\Delta)}(\mathcal{A})$. We show the previous inclusion, namely that any $[\mathcal{B}, \mathcal{A}, \ldots] \subseteq \mathcal{T}_{(\Pi,\Delta)}(\mathcal{B})$ is such that $[\mathcal{A}, \ldots] \subseteq \mathcal{T}_{(\Pi,\Delta)}(\mathcal{A})$, plus both assumptions ($\mathcal{A}$ is defeated in $\mathcal{T}_{(\Pi,\Delta)}(\mathcal{B})$ and $\mathcal{A}$ is undefeated in $\mathcal{T}_{(\Pi,\Delta)}(\mathcal{A})$) imply the existence of an increasing sequence of argumentation lines of arbitrarily finite length, which is impossible.

The previous inclusion shows in particular that witness Λ^*-minus-$\mathcal{B}$ is an arg. line in $\mathcal{T}_{(\Pi,\Delta)}(\mathcal{A})$. Since $\mathcal{A}$ is undefeated in $\mathcal{T}_{(\Pi,\Delta)}(\mathcal{A})$, some $\mathcal{B}_3$ must be defeated in $\mathcal{T}_{(\Pi,\Delta)}(\mathcal{A})$. Let $\Lambda_0 = [\mathcal{A}, \mathcal{B}_3, \mathcal{C}_1^0, \ldots, \mathcal{C}_{2n_0+1}^0]$ witness the defeat of $\mathcal{B}_3$ in $\mathcal{T}_{(\Pi,\Delta)}(\mathcal{A})$; i.e. $\mathcal{C}_{2n_0'+1}^0$ is undefeated in this tree, for any $n_0' \leq n_0$. By assumption on the original witness Λ^* in $\mathcal{T}_{(\Pi,\Delta)}(\mathcal{B})$, if $\mathcal{C}_1^0$ occurs in $\mathcal{T}_{(\Pi,\Delta)}(\mathcal{B})$, then $\mathcal{C}_1^0$ must be defeated in $\mathcal{T}_{(\Pi,\Delta)}(\mathcal{B})$. To see this $\mathcal{C}_1^0$ will effectively occur in $\mathcal{T}_{(\Pi,\Delta)}(\mathcal{B})$ it suffices to prove $\mathcal{C}_1^0$ is not a sub-argument of $\mathcal{B}$ with $\|\mathcal{C}_1^0\| = \|\mathcal{B}\|$. For this, assume the contrary. Then, by Def. of arg. line, we have $\|\mathcal{B}\| = \|\mathcal{A}\| = \|\mathcal{B}_3\| = \|\mathcal{C}_1^0\|$. But then, $\mathcal{C}_1^0 = \mathcal{B}(\mathsf{concl}(\mathcal{C}_1^0))$, $\|\mathcal{C}_1^0\| = \|\mathcal{B}\|$ and $\mathcal{C}_1^0$ a defeater for $\mathcal{B}_3$ (hence inconsistent with it) jointly imply that $\mathcal{B}$ and $\mathcal{B}_3$ are not consistent (contradiction). Moreover, this $\mathcal{C}_1^0$ satisfies in the tree for $\mathcal{B}$ the restriction against two consecutive blocking defeaters, since it satisfies this restriction in the tree for $\mathcal{A}$ (this preservation is automatic since $\mathcal{C}_1^0$ is not the 2nd element in Λ_0).

Let then $\Lambda_1 = [\mathcal{B}, \mathcal{A}, \mathcal{B}_3, \mathcal{C}_1^0, \mathcal{C}_1^1, \ldots, \mathcal{C}_{2n_1+1}^1]$ be a witness to the defeat of $\mathcal{C}_1^0$. By the former inclusion, this latter witness Λ_1-minus-$\mathcal{B}$ is in the tree for $\mathcal{A}$. By the assumption that $\mathcal{A}$ is undefeated in $\mathcal{T}_{(\Pi,\Delta)}(\mathcal{A})$, the element $\mathcal{C}_1^1$ of this witness must be defeated in $\mathcal{T}_{(\Pi,\Delta)}(\mathcal{A})$, since $\mathcal{C}_1^0$ is undefeated in it. Let $\Lambda_2 = [\mathcal{B}, \mathcal{A}, \mathcal{B}_3, \mathcal{C}_1^0, \mathcal{C}_1^1, \mathcal{C}_1^2 \ldots, \mathcal{C}_{2n_2+1}^2]$ be a witness to the defeat of $\mathcal{C}_1^1$.

This procedure can be continued *ad infinitum* with analogous reasonings from $\mathcal{T}_{(\Pi,\Delta)}(\mathcal{B})$ to $\mathcal{T}_{(\Pi,\Delta)}(\mathcal{A})$ and viceversa. Thus, there exists an infinite sequence of arg. lines (witnesses) Λ_n of the form $[\mathcal{B}, \mathcal{A}, \mathcal{B}_3, \mathcal{C}_1^0, \mathcal{C}_1^1, \ldots, \mathcal{C}_1^n, \ldots]$ (for n odd) or of the form $[\mathcal{A}, \mathcal{B}_3, \mathcal{C}_1^0, \mathcal{C}_1^1, \ldots, \mathcal{C}_1^n, \ldots]$ (for n even). Thus, arg. lines of arbitrarily finite length must exist, and elements of the form $\mathcal{C}_1^n$ form an infinite sequence $\Lambda_\omega = [\mathcal{A}, \mathcal{B}_3, \mathcal{C}_1^0, \mathcal{C}_1^1, \ldots, \mathcal{C}_1^n, \mathcal{C}_1^{n+1}, \ldots]$ satisfying: any initial segment of Λ_ω is an arg. line.

We show such an infinite sequence Λ_ω cannot exist. Since $t(\mathcal{A}) + \|\mathcal{A}\|$ is finite, and arguments $\mathcal{C}$ in Λ_ω must satisfy $t(\mathcal{C}) + \|\mathcal{C}\| \leq t(\mathcal{A}) + \|\mathcal{A}\|$, we have that rules in these $\mathcal{C}$ are finite sequences of literals in the finite set $\mathsf{Lit} \times \{0, \ldots, t(\mathcal{A}) + \|\mathcal{A}\|\}$. Hence, the number of these rules is finite. Hence, there are only finitely many different arguments which can occur in Λ_ω. But since Λ_ω is infinite, we will have some repetition $\mathcal{C}_1^j = \mathcal{C}_1^{j+i}$. Then, the sequence $[\mathcal{A}, \mathcal{B}_3, \ldots, \mathcal{C}_1^j, \ldots, \mathcal{C}_1^{j+i}]$ will violate the corresponding condition of Definition 8. Thus, such an infinite sequence Λ_ω cannot exist (contradiction). Then, $\mathcal{A}$ must be defeated in $\mathcal{T}_{(\Pi,\Delta)}(\mathcal{A})$, if it is defeated in $\mathcal{T}_{(\Pi,\Delta)}(\mathcal{B})$. □

Theorem 1. *Given a t-de.l.p.* (Π, Δ)*, the set of literals* $\mathsf{warr}(\Pi, \Delta)$ *is consistent.* [6]

Proof. Let $\langle \ell, t\rangle \in \mathsf{warr}(\Pi, \Delta)$. Thus, some argument $\mathcal{A}$ for $\langle \ell, t\rangle$ in (Π, Δ) exists that is undefeated in $\mathcal{T}_{(\Pi,\Delta)}(\mathcal{A})$. It suffices to show that $\langle \sim\ell, t\rangle \notin \mathsf{warr}(\Pi, \Delta)$. The reason is that if, instead, an attack occurred at a previous time, i.e. $\mathcal{A}$ was attacked at some $\mathcal{A}(\langle \ell_0, t_0\rangle)$, and defeated by some $\mathcal{B}$, the same reasoning given next would apply for $\mathcal{A}(\langle \ell_0, t_0\rangle)$ and the corresponding defeater $\mathcal{B}$ (i.e. that $\langle \sim\ell_0, t_0\rangle \notin \mathsf{warr}(\Pi, \Delta)$.

Thus, assume -towards a contradiction- that $\langle \sim\ell, t\rangle \in \mathsf{warr}(\Pi, \Delta)$. Then some argument $\mathcal{B}$ for $\langle \sim\ell, t\rangle$ exists in (Π, Δ), undefeated in $\mathcal{T}_{(\Pi,\Delta)}(\mathcal{B})$. Observe first that if $\mathcal{A} \subseteq \Pi$, then either each such argument $\mathcal{B}$ contains some rule in Δ, in which case $\mathcal{A}$ will attack and defeat any such $\mathcal{B}$ (contradicting that $\langle \sim\ell_0, t_0\rangle \in \mathsf{warr}(\Pi, \Delta)$); or, also some such $\mathcal{B}$ for $\langle \sim\ell_0, t_0\rangle$ is a subset of Π, contradicting the assumption that the set of derivable literals from Π alone is consistent. Thus, we may assume that $\mathcal{A} \cap \Delta \neq \emptyset$. Now, consider again the possibility that some such argument $\mathcal{B}$ for $\langle \sim\ell_0, t_0\rangle$ is a subset of Π. Then, $\mathcal{A}$ is defeated by an undefeated argument, contradicting the initial assumption: $\langle \ell, t\rangle \in \mathsf{warr}(\Pi, \Delta)$. Thus, we may also assume that $\mathcal{B} \cap \Delta \neq \emptyset$.

This implies that $\mathcal{A}$ attacks $\mathcal{B}$ at $\mathcal{B}$, and $\mathcal{B}$ attacks $\mathcal{A}$ at $\mathcal{A}$. Consider next the following cases. (Case) $\mathcal{A}$ is not a defeater for $\mathcal{B}$. Then, since the only possibilities are $\mathsf{base}(\mathcal{A}) \not\subseteq \mathsf{base}(\mathcal{B})$ and $\mathsf{base}(\mathcal{A}) \neq \mathsf{base}(\mathcal{B})$ we conclude that $\mathsf{base}(\mathcal{B}) \supsetneq \mathsf{base}(\mathcal{A})$, so $\mathcal{B}$ is a (proper) defeater for $\mathcal{A}$. Thus, $[\mathcal{A}, \mathcal{B}, \ldots]$ is in $\mathcal{T}_{(\Pi,\Delta)}(\mathcal{A})$. From this and the assumption that $\mathcal{B}$ is undefeated in $\mathcal{T}_{(\Pi,\Delta)}(\mathcal{B})$, we can apply Lemma 1 to show that $\mathcal{B}$ is undefeated in $\mathcal{T}_{(\Pi,\Delta)}(\mathcal{A})$. Hence, $\mathcal{A}$ is defeated in $\mathcal{T}_{(\Pi,\Delta)}(\mathcal{A})$ (contradiction). (Case) If $\mathcal{A}$ is a defeater for $\mathcal{B}$, then $[\mathcal{B}, \mathcal{A}, \ldots]$ is in $\mathcal{T}_{(\Pi,\Delta)}(\mathcal{B})$, so by assumption on $\mathcal{B}$, $\mathcal{A}$ is defeated in $\mathcal{T}_{(\Pi,\Delta)}(\mathcal{B})$. Then, by Lemma 1, we obtain that $\mathcal{A}$ is defeated in $\mathcal{T}_{(\Pi,\Delta)}(\mathcal{A})$ (contradiction). Hence $\langle \sim\ell, t\rangle \notin \mathsf{warr}(\Pi, \Delta)$. Thus, $\mathsf{warr}(\Pi, \Delta)$ is consistent. □

Next we consider the restriction of strict information in t-de.l.p.s to strict facts. This is called ODeLP in [4] in the case of DeLP programs. Under the proposed restriction, the t-DeLP programming framework can be proved to be a *logic* in the sense of Tarski, and to satisfy the Rationality Postulates for argumentation frameworks stated by Caminada and Amgoud in [3], that may fail if we have strict rules.

Corollary 1. *For any t-de.l.p.* (Π, Δ)*, with* $\Pi \subseteq \mathsf{TLit}$*, we have* $\mathbf{C}(\Pi, \Delta) = (\mathsf{warr}(\Pi, \Delta) \cup \Pi, \Delta)$ *is a logical consequence operator: i.e. it satisfies* inclusion $\Pi \subseteq \mathsf{warr}(\Pi, \Delta) \cup \Pi, \Delta \subseteq \Delta$*;* idempotence $\mathbf{C}(\Pi, \Delta) = \mathbf{C}(\mathbf{C}(\Pi, \Delta))$*; and* coherence $\mathbf{C}(\emptyset, \emptyset) = (\emptyset, \emptyset)$ *is consistent.*

4 Nature Does Not Wait: Eager Arguments

We study in this section a sub-class of t-DeLP arguments, called *eager*, for reasoning with natural processes. In law-governed processes *as soon as* all conditions hold,

[6] Recall that, according to Defintion 1, $\mathsf{warr}(\Pi, \Delta)$ is consistent iff there is no p such that both p and $\sim p$ belong to $\mathsf{warr}(\Pi, \Delta)$. This differs from stronger notions of consistency [3] requiring that $\mathsf{warr}(\Pi, \Delta) \cup \Pi$ does not derive any pair of contradictory literals (see Corollary 1 below).

the process can do nothing else than start. This would exclude from the class of arguments that model some natural process those constructible arguments that unnecessarily postpone the (start of an) application of a rule after its body holds (i.e. arguments that introduce some unnecessary delay after the rule becomes *applicable*). No natural process corresponds to these t-DeLP arguments, so they should be excluded from reasoning about natural processes.

Interestingly, any argument $\mathcal{A}$ can be transformed into an eager argument $\mathcal{A}^*$ by following an iterative procedure. The idea is that $\mathcal{A}^*$ orders non-persistence (resp. persistence) rules in $\mathcal{A}$ to occur as early (resp. late) as possible while keeping the same base. To obtain $\mathcal{A}^*$, let initially $n = 0$ and $\mathcal{A}_0 = \mathcal{A}$, and apply iteratively the following transformation $\mathcal{A}_n \mapsto \mathcal{A}_{n+1}$ until it cannot be applied any longer:

1. Select a rule $\delta^* \in \mathcal{A}_n$ satisfying the condition: for each $\langle \ell_i, t_i \rangle \in \mathsf{body}(\delta^*)$, there exists (at least) an instance of the persistence rule δ_{ℓ_i} supporting this $\langle \ell_i, t_i \rangle$. If there is no such a rule let $\mathcal{A}^* = \mathcal{A}_n$ and STOP. Otherwise follow to the next step.
2. Let $\{\langle \ell_i, t_i \rangle \prec \langle \ell_i, t_i - 1 \rangle, \ldots, \langle \ell_i, t_i - k_i + 1 \rangle \prec \langle \ell_i, t_i - k_i \rangle\} \subseteq \mathcal{A}_n$ be the set of persistence rules for each $\langle \ell_i, t_i \rangle \in \mathsf{body}(\delta^*)$, i.e. such that $\langle \ell_i, t_i - k_i \rangle \in \mathsf{literals}(\mathcal{A}_n)$ is supported by some non-persistence rule. Let k_j be minimal among those $\langle \ell_i, t_i - k_i \rangle$ supported by non-persistence rules.
3. Define a new rule, denoted $\delta^* - k_j$, as the rule where the temporal parameter of each literal $\langle \ell, t \rangle \in \mathsf{literals}(\delta^*)$ is subtracted k_j. Then, we
4. In $\mathcal{A}_n$ replace δ^* by $\delta^* - k_j$
5. For each $\langle \ell_i, \cdot \rangle \in \mathsf{body}(\delta^*)$, delete from $\mathcal{A}_n$ the k_j instances of persistence rules δ_{ℓ_i} of the form: $\langle \ell_i, t_i \rangle \prec \langle \ell_i, t_i - 1 \rangle, \ldots, \langle \ell_i, t_i - k_j + 1 \rangle \prec \langle \ell_i, t_i - k_j \rangle$
6. If $\mathsf{head}(\delta^*) = \langle \ell, t \rangle$, add to $\mathcal{A}_n$ the k_j instances of persistence rule δ_ℓ of the form: $\langle \ell, t \rangle \prec \langle \ell, t - 1 \rangle, \ldots, \langle \ell, t - k_j + 1 \rangle \prec \langle \ell, t - k_j \rangle$.
7. Let $\mathcal{A}_{n+1}$ be the resulting argument. Set $n \leftarrow n + 1$ and START again at step 1.

The ouput of the above procedure, $\mathcal{A}^*$, is an argument sharing many properties with $\mathcal{A}$: $\mathsf{base}(\mathcal{A}^*) = \mathsf{base}(\mathcal{A})$, $\mathsf{concl}(\mathcal{A}^*) = \mathsf{concl}(\mathcal{A})$, hence $\|\mathcal{A}^*\| = \|\mathcal{A}\|$ and $t(\mathcal{A}^*) = t(\mathcal{A})$. It can be observed that these transformations define an equivalence relation $\equiv_{\mathsf{p}}$ on the set of arguments in (Π, Δ): $\mathcal{B}, \mathcal{C}$ are $\equiv_{\mathsf{p}}$-equivalent iff $\mathcal{B}^* = \mathcal{C}^*$.

Example 8. For a counterexample to the defeat relation in Definition 7, see Figure 4. Read ru = *the agent runs* and ti = *the agent is tired.* Consider the rules: if the agent starts running when untired, after 2 hours she is tired (this is sub-argument $\mathcal{A}(\langle \mathsf{ti}, 2 \rangle)$); and in this case she also stops running, i.e. $\langle {\sim}\mathsf{ru}, 2 \rangle$. Then, after 3 hours the agent is fresh again $\langle {\sim}\mathsf{ti}, 5 \rangle$. This is argument $\mathcal{A}$ at the center of Fig. 4. A problem occurs with argument $\mathcal{B}$, obtained by applying persistence to $\mathsf{base}(\mathcal{A})$ so as to postpone $\mathcal{A}$'s sub-argument for an hour, denoted $\mathcal{A}(\langle \mathsf{ti}, 2 \rangle) + 1$. But then $\mathcal{A}$ is only a blocking defeater for $\mathcal{B}$ (and viceversa), according to Def. 7. Thus, we cannot conclude $\langle {\sim}\mathsf{ru}, 2 \rangle$ or $\langle {\sim}\mathsf{ti}, 5 \rangle$.

Counterexamples like these of Example 8 can be pruned either by

(a) restricting to the set of eager arguments of a given t-de.l.p., given by the previous procedure $Args(\Pi, \Delta) \mapsto Args^*(\Pi, \Delta)$. This move redefines $\mathcal{T}^*_{(\Pi,\Delta)}(\mathcal{A})$ and $\mathsf{warr}^*(\Pi, \Delta)$ when only eager arguments are considered. Or,

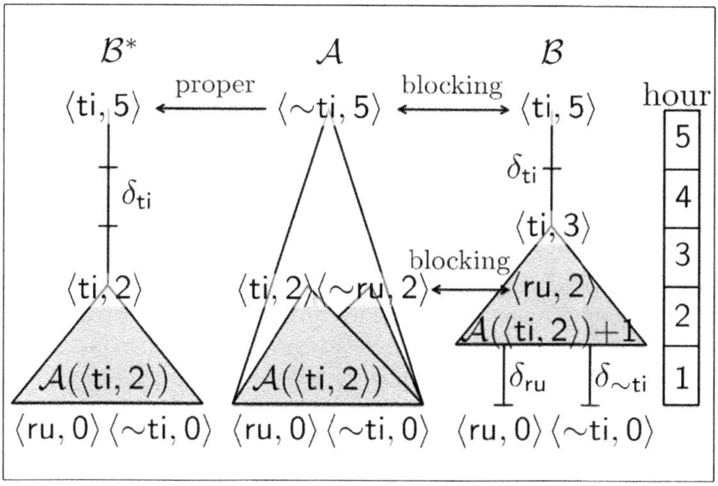

Fig. 4. The argument $\mathcal{A}$ should be a proper defeater for $\mathcal{B}$, as it is for $\mathcal{B}^*$

(b) by redefining the notion of *longer argument* so that $\mathcal{A}$ is a longer argument than $\mathcal{B}$ iff $\mathcal{A}$ is longer than $\mathcal{B}^*$ (in the old sense of Definition 7), where $\mathcal{B}^* \in [\mathcal{B}]_{\equiv_\mathsf{p}}$ is eager. This modification to Def. 7 propagates to new notions $\mathcal{T}^\bullet_{(\Pi,\Delta)}(\mathcal{A})$, $\mathsf{warr}^\bullet(\Pi, \Delta)$.

In the following we prove the two options are equivalent. With the new definition of *defeater* given by (b), we have the next result.

Proposition 1. *Let* (Π, Δ) *be a t-de.l.p. and let* $\mathcal{A}^*$ *be an eager argument in* (Π, Δ). *Then, for any other arguments* $\mathcal{A} \in [\mathcal{A}^*]_{\equiv_\mathsf{p}}$ *and* $\mathcal{B}$ *in* (Π, Δ) *we have:*

(1) if $\mathcal{B}$ *is a proper (blocking) defeater for* $\mathcal{A}^*$ *then so is* $\mathcal{B}$ *for* $\mathcal{A}$.
(2) if $\mathcal{A}$ *is a proper (blocking) defeater for* $\mathcal{B}$, *so is* $\mathcal{A}^*$.

Claim (1) shows eager arguments are the safest among their $\equiv_\mathsf{p}$ equivalence class in (Π, Δ). Define $\ell \in \mathsf{warr}^*(\Pi, \Delta)$ iff there exists an eager argument $\mathcal{A}^*$ in (Π, Δ) undefeated in $\mathcal{T}^*_{(\Pi,\Delta)}(\mathcal{A}^*)$; and similarly define $\mathcal{T}^\bullet_{(\Pi,\Delta)}(\mathcal{A}^*)$, $\mathsf{warr}^\bullet(\Pi, \Delta)$ as in (b).

Corollary 2. *Fix a t-de.l.p.* (Π, Δ). *Let* $\mathcal{A}_0$ *be an eager argument. Then* $\mathcal{A}_0$ *is undefeated in* $\mathcal{T}^*_{(\Pi,\Delta)}(\mathcal{A}_0)$ *iff it is undefeated in* $\mathcal{T}^\bullet_{(\Pi,\Delta)}(\mathcal{A}_0)$. *As a consequence,* $\mathsf{warr}^*(\Pi, \Delta) = \mathsf{warr}^\bullet(\Pi, \Delta)$.

A final remark: in Example 3-6, we forbid persistence rules $\delta_\ell \in \Delta_\mathsf{p}$ when ℓ is of the form $\sim$@location(Lars). This prevents to derive, say, the intermediate step $\langle$@highway(Lars), 1$\rangle$, then $\langle\sim$@hospital(Lars), 1$\rangle$, and then apply persistence on the latter up to $t = 4$. Because of the strict rule used $\langle\sim$@hospital(Lars), 1$\rangle \leftarrow \langle$highway(Lars), 1$\rangle$, argument $\mathcal{D}_1$ is not longer than this argument (according to the modified Def. 7), and moreover it is eager. Thus, the restriction to eager arguments does not suffice. Restricting Δ_p seems to work as well for Example 8: if we forbid persistence of activities ru, argument $\mathcal{B}$ does not occur. We wonder whether the restrictions upon Δ_p or with eager arguments suffice to prevent unintuitive results in t-DeLP.

5 Conclusions and Future Work

We have presented t-DeLP a temporal extension of DeLP with temporal literals and rules with duration. Indeed one can think of the DeLP framework to correspond to t-DeLP (with strict rules made explicit in arguments and with only one time-point, e.g. when all temporal literals are of form $\langle \ell, 0 \rangle$). We proved the (weak) consistency of any program and that the restriction to strict facts makes the logic programming framework a consequence operator in the sense of Tarski. Then, we studied particular questions arising in temporal reasoning with persistence.

In the literature, other rule-based defeasible logics [2], [10] exist as well, but they lack the reasoning power and conceptual transparency of argumentation-based logics. The same goes for defeasible logics extended with temporal parameters associated to literals [8]. On the other hand, argumentation-based logical approaches (inspired by the work of [6]) do not in general take the particularities involved in temporal reasoning into account. Among works that do consider argumentation and time, we find some proposals associating time intervals to arguments [1], [5] and [9]. Our approach differs from these works in that the *interval where an event or argument holds*, rather than being a primitive notion, derives from the argumentation process on temporal literals . Thus, our time-point based approach accommodates features from [8], [1] (expiring literals, persistence), though in the present paper these features are subject to argumentation processes, rather than having them fixed from start.

For future work, we would like to expand temporal reasoning with evidence-based (backward) reasoning, among other improvements on the generality of our language or the results we obtained.

Acknowledgements. The authors acknowledge partial support of the Spanish MICINN project CONSOLIDER-INGENIO 2010 Agreement Technologies CSD2007-00022; LoMoReVI FFI2008-03126-E/FILO (FP006); ARINF TIN2009-14704-C03-03 and the Generalitat de Catalunya grant 2009-SGR-1434.

References

1. Augusto, J., Simari, G.: Temporal Defeasible Reasoning Knowledge and Information Systems, vol. 3, pp. 287–318 (2001)
2. Billington, D.: Defeasible logic is stable. Journal of Logic and Computation 3, 379–400 (1993)
3. Caminada, M., Amgoud, L.: On the evaluation of argumentation formalisms. Artificial Intelligence 171, 286–310 (2007)
4. Capobianco, M., Chesñevar, C., Simari, G.: Argumentation and the Dynamics of Warranted Beliefs in Changing Environments. JAAMAS 11, 127–151 (2005)
5. Cobo, L., Martínez, D., Simari, G.: On Admissibility in Timed Abstract Argumentation Frameworks. In: Proc. of European Conference on AI, ECAI 2010, pp. 1007–1008 (2010)
6. Dung, P.: On the acceptability of arguments and its fundamental role in nonmonotonic reasoning, logic programming and n-person games* 1. Artificial Intelligence 77(2), 321–357 (1995)

7. García, A., Simari, G.: Defeasible logic programming: An argumentative approach. Theory and Practice of Logic Programming 4(1+2), 95–138 (2004)
8. Governatori, G., Terenziani, P.: Temporal Extensions to Defeasible Logic. In: Proc. of Australian Joint Conf. on AI, AI 2007, pp. 1–10 (2007)
9. Mann, N., Hunter, A.: Argumentation Using Temporal Knowledge. In: Proc. of Computer Models of Argumentation (COMMA 2008), pp. 204–215. IOS Press, Amsterdam (2008)
10. Nute, D.: Defeasible Logic. In: Handbook of Logic in Artificial Intelligence and Logic Programming, vol. 3, pp. 353–395. Oxford Univ. Press, Oxford (1994)

Learning to Act Optimally in Partially Observable Markov Decision Processes Using Hybrid Probabilistic Logic Programs

Emad Saad

Department of Computer Science
Gulf University for Science and Technology
Mishref, Kuwait
saad.e@gust.edu.kw

Abstract. We present a probabilistic logic programming framework to reinforcement learning, by integrating reinforcement learning, in POMDP environments, with normal hybrid probabilistic logic programs with probabilistic answer set semantics, that is capable of representing domain-specific knowledge. We formally prove the correctness of our approach. We show that the complexity of finding a policy for a reinforcement learning problem in our approach is NP-complete. In addition, we show that any reinforcement learning problem can be encoded as a classical logic program with answer set semantics. We also show that a reinforcement learning problem can be encoded as a SAT problem. We present a new high level action description language that allows the factored representation of POMDP. Moreover, we modify the original model of POMDP so that it be able to distinguish between knowledge producing actions and actions that change the environment.

1 Introduction

Reinforcement learning is the problem of learning to act by trial and error interaction in dynamic environments. Reinforcement learning problems can be represented as Markov Decision Processes (MDP), under the assumption that accurate and complete model of the environment is known. This assumption requires the agent to have perfect sensing and observation abilities.

However, complete and perfect observability is unrealistic for many real-world reinforcement learning applications, although necessary for learning optimal policies in MDP environments. Therefore, different model is needed to represent and solve reinforcement learning problems with partial observability. This model is Partially Observable Markov Decision Processes (POMDP). Similar to MDP, POMDP requires the model of the environment to be known, however states of the world are not completely known. Consequently, the agent perform actions to make observations about the states of the worlds. These observations can be noisy due to imperfect agent's sensors. Similar to MDP, dynamic programming methods, by value iteration, have been used to learn the optimal policy for a reinforcement learning problem in POMDP environment.

S. Benferhat and J. Grant (Eds.): SUM 2011, LNAI 6929, pp. 504–519, 2011.
© Springer-Verlag Berlin Heidelberg 2011

A logical framework to reinforcement learning in MDP environment has been developed in [23], which relies on techniques from probabilistic reasoning and knowledge representation by normal hybrid probabilistic logic programs [26]. The normal hybrid probabilistic logic programs framework of [23] has been proposed upon observing that dynamic programming methods to reinforcement learning in general and value iteration in particular are incapable of exploiting domain-specific knowledge of the reinforcement learning problem domains to improve the efficiency of finding the optimal policy. In addition, these dynamic programming methods use primitive representation of states and actions as this representation does not capture the relationship between states [18] and makes it difficult to represent domain-specific knowledge. However, using richer knowledge representation frameworks for MDP and POMDP allow efficiently finding optimal policies in more complex stochastic domains and lead to develop methods to find optimal policies with larger domains sizes [18].

The choice of normal hybrid probabilistic logic programs (NHPLP) to solve reinforcement learning problems in MDP environment is based on that; NHPLP is nonmonotonic, therefore more suitable for knowledge representation and reasoning under uncertainty; NHPLP subsumes classical normal logic programs with classical answer set semantics [7], a rich knowledge representation and reasoning framework, and inherits its knowledge representation and reasoning capabilities including the ability to represent and reason about domain-specific knowledge; NHPLP has been shown applicable to a variety of fundamental probabilistic reasoning problems including probabilistic planning [25], contingent probabilistic planning [22], the most probable explanation in belief networks, the most likely trajectory in probabilistic planning, and Bayesian reasoning [24].

In this view, we integrate reinforcement learning in POMDP environment with NHPLP, providing a logical framework that overcomes the representational limitations of dynamic programming method to reinforcement learning in POMDP and is capable of representing its domain-specific knowledge. In addition, the proposed framework extends the logical framework of reinforcement learning in MDP of [23] with partial observability. We show that any reinforcement learning problem in POMDP environment can be encoded as a SAT problem. The importance of that is reinforcement learning problems in POMDP environment can be now solved as SAT problems.

2 Syntax and Semantics of NHPLP

We introduce a class of NHPLP [26], namely NHPLP$_{\mathcal{PO}}$, that is sufficient to represent and reason about POMDP.

2.1 The Language of NHPLP$_{\mathcal{PO}}$

Let $\mathcal{L}$ be a first-order language with finitely many predicate symbols, constants, and infinitely many variables. The Herbrand base of $\mathcal{L}$ is denoted by $\mathcal{B}_{\mathcal{L}}$. Probabilities are assigned to atoms in $\mathcal{B}_{\mathcal{L}}$ as values from $[0, 1]$. An *annotation*, μ, is either a constant in $[0,1]$, a variable (*annotation variable*) ranging over $[0,1]$, or $f(\mu_1,\ldots,\mu_n)$ (called *annotation function*) where f is a representation of a computable total function $f : ([0,1])^n \rightarrow [0,1]$ and $\mu_1,\ldots,\mu_n$ are annotations. Let $a_1, a_2 \in [0,1]$. Then we say that $a_1 \leq_t a_2$ iff $a_1 \leq a_2$. A probabilistic logic program (*p-program*) in NHPLP$_{\mathcal{PO}}$ is a pair $P = \langle R, \tau \rangle$, where R is a finite set of normal probabilistic rules (p-rules) and τ is a mapping $\tau : \mathcal{B}_{\mathcal{L}} \rightarrow S_{disj}$, where S_{disj} is a set of disjunctive probabilistic strategies (p-strategies) whose composition functions, c, are mappings $c : [0,1] \times [0,1] \rightarrow [0,1]$. A composition function of a disjunctive p-strategy returns the probability of a disjunction of two events given the probability values of its components. A p-rule is an expression of the form

$$A : \mu \leftarrow A_1 : \mu_1, \ldots, A_n : \mu_n, not\ (B_1 : \mu_{n+1}), \ldots, not\ (B_m : \mu_{n+m})$$

where A, $A_1,\ldots,A_n, B_1,\ldots,B_m$ are atoms and μ, μ_i $(1 \leq i \leq m+n)$ are annotations. Intuitively, the meaning of a p-rule is that if for each $A_i : \mu_i$, the probability value of A_i is at least μ_i (w.r.t. $\leq_t$) and for each $not\ (B_j : \mu_j)$, it is not *believable* that the probability values of B_j is at least μ_j, then the probability of A is μ. The mapping τ associates to each atom A a disjunctive p-strategy that will be employed to combine the probability values obtained from different p-rules having A in their heads. A p-program is ground if no variables appear in any of its p-rules.

2.2 Probabilistic Answer Set Semantics of NHPLP$_{\mathcal{PO}}$

A probabilistic interpretation (p-interpretation) is a mapping $h : \mathcal{B}_{\mathcal{L}} \rightarrow [0,1]$. Let $P = \langle R, \tau \rangle$ be a ground p-program, h be a p-interpretation, and r be

$$A : \mu \leftarrow A_1 : \mu_1, \ldots, A_n : \mu_n, not\ (B_1 : \beta_1), \ldots, not\ (B_m : \beta_m) \in R.$$

Then, we say

- h satisfies $A_i : \mu_i$ (denoted by $h \models A_i : \mu_i$) iff $\mu_i \leq_t h(A_i)$.
- h satisfies $not\ (B_j : \beta_j)$ (denoted by $h \models not\ (B_j : \beta_j)$) iff $\beta_j \nleq_t h(B_j)$.
- h satisfies $Body \equiv A_1 : \mu_1, \ldots, A_n : \mu_n, not\ (B_1 : \beta_1), \ldots, not\ (B_m : \beta_m)$ (denoted by $h \models Body$) iff $\forall (1 \leq i \leq n), h \models A_i : \mu_i$ and $\forall (1 \leq j \leq m), h \models not\ (B_j : \beta_j)$.
- h satisfies $A : \mu \leftarrow Body$ iff $h \models A : \mu$ or h does not satisfy $Body$.
- h satisfies P iff h satisfies every p-rule in R and for every atom $A \in \mathcal{B}_{\mathcal{L}}$, we have

$$c_{\tau(A)} \{\!\{ \mu | A : \mu \leftarrow Body \in R\ such\ that\ h \models Body \}\!\} \leq_t h(A).$$

The probabilistic reduct P^h of P w.r.t. h is a p-program $P^h = \langle R^h, \tau \rangle$ where:

$$R^h = \left\{ A : \mu \leftarrow A_1 : \mu_1, \ldots, A_n : \mu_n \;\middle|\; \begin{array}{l} A : \mu \leftarrow A_1 : \mu_1, \ldots, A_n : \mu_n, \\ \quad not\ (B_1 : \beta_1), \ldots, not\ (B_m : \beta_m) \in R\ and \\ \forall (1 \leq j \leq m),\ \beta_j \nleq_t h(B_j) \end{array} \right\}$$

A probabilistic model (*p-model*) of a p-program P is a p-interpretation of P that satisfies P. A p-interpretation h of a p-program P is said to be a probabilistic answer set of P if h is the minimal p-model of the probabilistic reduct of P w.r.t. h.

3 Partially Observable Markov Decision Processes

In this section we review finite-horizon POMDP [11] with stationary transition functions, stationary bounded reward functions, and stationary policies. POMDP is a tuple of the form $\mathbf{M} = \langle S, S_0, A, T, \lambda, \mathcal{R}, \Omega, O\rangle$ where: S is a finite set of states; S_0 is the initial state distribution; A is a finite set of stochastic actions; T is stationary transition function $T : S \times A \times S \rightarrow [0,1]$, where for any $s \in S$ and $a \in A$, $\sum_{s' \in S} T(s,a,s') = 1$; $\lambda \in [0,1)$ is the discount factor; $\mathcal{R} : S \times A \times S \rightarrow \mathbb{R}$ is a stationary bounded reward function; Ω is a finite set of observations that the agent observes in the environment; and O is observation function $O : S \times A \times \Omega \rightarrow [0,1]$, where for any $s \in S$ and $a \in A$ where $\sum_{o\in\Omega} O(s,a,o) = 1$. A stationary policy is a mapping from states to actions of the form $\pi : S \rightarrow A$. The value function of a policy π with respect to an initial state $s_0 \in S_o$, with finite horizon of n steps remaining, $V_n^\pi(s_0)$, is calculated by

$$V_n^\pi(s_0) = \sum_{s_1\in S} T(s_0,\pi(s_0),s_1) \sum_{o_i\in\Omega} O(s_1,\pi(s_0),o_i) \left[\mathcal{R}(s_0,\pi(s_0),s_1) + \lambda\, V^\pi_{i,n-1(s_1)}\right]$$

which determines the expected sum of discounted rewards resulting from executing the policy π starting from s_0.

Discussion
The original model of POMDP does not distinguish between knowledge producing (sensing) actions and actions that affects and change the environment (non-sensing actions). This means that it treats sensing and non-sensing actions equally in the sense that, like non-sensing actions, a sensing action affects and change the environment as well as producing knowledge resulting from observing the environment. However, [27] proved that sensing actions produce knowledge (make observations) and does not change the state of the world. Therefore, actions that change the state of the world are different from the knowledge producing actions. In addition, the value function described above makes the agent observing the environment at every step of its life with each action it takes. However, this is not necessary to be always the case, since it is possible for the agent to start with observing the environment then performing a sequence of actions, or the agent could start with performing a sequence of actions then observing the environment. To overcome these limitations, we define the value function of n-step finite horizon POMDP with respect to an initial state $s_0 \in S_o$ as:

- if $\pi(s_0)$ is a non-sensing action then

$$V_n^\pi(s_0) = \sum_{s_1\in S} T(s_0,\pi(s_0),s_1)[\mathcal{R}(s_0,\pi(s_0),s_1) + \lambda\, V^\pi_{n-1}(s_1)]$$

- if $\pi(s_0)$ is sensing action then

$$V_n^\pi(s_0) = \sum_{s_1 \in S} O(s_0, \pi(s_0), s_1)[\mathcal{R}(s_0, \pi(s_0), s_1) + \lambda\, V_{n-1}^\pi(s_1)]$$

where $O(s_0, \pi(s_0), s_1)$ is the probability of observing the state s_1, where for some $o \in \Omega$, o is observed in s_1. Notice that O is treated as a mapping $O : S \times A \times S \rightarrow [0,1]$, where A is the set of sensing actions. For any $s \in S$ and $a \in A$, $O(s, a, .)$ is the probability distribution over states resulting from executing a in s, such that $\sum_{s' \in S} O(s, a, s') = 1$. As in the original model of POMDP, T is a mapping $T : S \times A \times S \rightarrow [0,1]$, where A is the set of non-sensing actions. Extension to infinite horizon POMDP can be achieved in a similar manner. This definition of POMDP distinguishes between knowledge producing actions and actions that change the environment. In this view, the optimal policy V_n^* is given by: $V_n^*(s_0) = \max_\pi V_n^\pi(s_0)$.

4 $\mathcal{A}_{PO}$ an Action Language for POMDP

We introduce an action language for POMDP, namely $\mathcal{A}_{PO}$, that extends both the action languages, $\mathcal{A}_{MD}$ [23] and $\mathcal{P}$ [22] for representing and reasoning about MDPs and imperfect sensing actions respectively.

4.1 Syntax of $\mathcal{A}_{PO}$

A fluent is a predicate, which may contain variables. Given that $\mathcal{F}$ is a set of fluents and $\mathcal{A}$ is a set of actions that can contain variables, a fluent literal is either a fluent $f \in \mathcal{F}$ or $\neg f$. A conjunction of fluent literals of the form $l_1 \wedge \ldots \wedge l_n$ is conjunctive fluent formula, where $l_1, \ldots, l_n$ are fluent literals. Sometimes we abuse the notation and refer to a conjunctive fluent formula as a set of fluent literals ($\emptyset$ denotes *true*). An action theory, **PT**, in $\mathcal{A}_{PO}$ is a tuple $\mathbf{PT} = \langle S_0, \mathcal{D}, \lambda \rangle$, where S_0 is a proposition of the form (1), $\mathcal{D}$ is a set of propositions from (2-4), and $0 \leq \lambda < 1$ is a discount factor as follows:

$$\textbf{initially} \, \{ \, \psi_i \; : \; p_i, \quad 1 \leq i \leq n \tag{1}$$

$$\textbf{executable} \; a \; \textbf{if} \; \psi \tag{2}$$

$$a \; \textbf{causes} \, \{ \, \phi_i \; : \; p_i \; : \; r_i \; \textbf{if} \; \psi_i, \quad 1 \leq i \leq n \tag{3}$$

$$a \; \textbf{observes} \, \{ \, o_i \; : \; p_i \; : \; r_i \; \textbf{sensing} \; \psi_i, \quad 1 \leq i \leq n \tag{4}$$

where $\psi, \psi_i, \phi_i, o_i, (1 \leq i \leq n)$ are conjunctive fluent formulas, $a \in \mathcal{A}$, and $p_i \in [0,1]$. The set of all ground ψ_i and o_i must be exhaustive and mutually exclusive.

The *initial agent's belief state*—a probability distribution over the possible initial states, is represented by (1), that says that each possible initial state ψ_i holds with probability p_i. *Executability condition* is represented by (2). A non-sensing action, a, is represented by (3), which says that for each $1 \leq i \leq n$, a

causes ϕ_i to hold with probability p_i and reward r_i is received in a successor state to a state in which a is executed and ψ_i holds. A sensing action, a, is represented by (4), which says that for each $1 \leq i \leq n$, whenever a correlated ψ_i is known to be true, a causes any of o_i to be known true with probability p_i and reward r_i is received in a successor state to a state in which a is executed, where the literals in ψ_i determine what the agent is observing and literals in o_i determine what the sensor reports on. Similar to [5], when a property of the world cannot be directly sensed by the sensor, another correlated property of the world, that can be sensed by the sensor, can be used instead. An action theory is ground if it does not contain any variables.

In the sequel, we represent an action a in (3) as a set of the form $a = \{a_1, \ldots, a_n\}$, where each a_i corresponds to ϕ_i, p_i, r_i, and ψ_i. For each $1 \leq i \leq n$, (3) can be represented as a_i **causes** $\phi_i \ : \ p_i \ : \ r_i$ **if** ψ_i. Similarly, (4) can be represented as a_i **observes** $o_i \ : \ p_i : \ r_i$ **sensing** ψ_i.

Example 1. Consider the tiger domain from [17]. A tiger is behind left (tl) or right ($\neg tl$) door with equal probability 0.5. If left door is opened and tl, punishment of -100 is received, but a reward of 10 is received if $\neg tl$ and the other way around. The sensing action *listen* used for hearing the tiger behind left door (htl), a correlated property to tl. But, the agent's hearing is not perfect and costs -1. If the agent listens for htl, then it reports tl with 0.85 and erroneously reports $\neg tl$ with 0.15. Similarly for listening to the right door. This is represented by the action theory $\mathbf{PT} = \langle S_0, \mathcal{D}, \lambda \rangle$, where **executable** AC **if** $\emptyset$, where $AC \in \{openL, openR, listen\}$ and

$$S_0 = \textbf{initially} \left\{ \begin{array}{ll} \{tl, htl\} & : 0.5 \\ \{\neg tl, \neg htl\} & : 0.5 \end{array} \right\} \quad listen \ \textbf{observes} \left\{ \begin{array}{l} \{tl\} \ \ : 0.85 : -1 \ \textbf{sensing} \ \{htl\} \\ \{\neg tl\} : 0.15 : -1 \ \textbf{sensing} \ \{htl\} \\ \{\neg tl\} : 0.85 : -1 \ \textbf{sensing} \ \{\neg htl\} \\ \{tl\} \ \ : 0.15 : -1 \ \textbf{sensing} \ \{\neg htl\} \end{array} \right\}$$

$$openL \ \textbf{causes} \left\{ \begin{array}{lll} \{\neg tl\} : 1 : -100 & \textbf{if} \ \{\neg tl\} \\ \{tl\} \ \ : 1 : 10 & \textbf{if} \ \{tl\} \end{array} \right\} \quad openR \ \textbf{causes} \left\{ \begin{array}{lll} \{\neg tl\} : 1 : -100 & \textbf{if} \ \{\neg tl\} \\ \{tl\} \ \ : 1 : 10 & \textbf{if} \ \{tl\} \end{array} \right\}$$

4.2 Semantics of $\mathcal{A}_{PO}$

A set of ground literals ϕ is consistent if it does not contain a pair of complementary literals. If a literal l belongs to ϕ, then we say l is true in ϕ, and l is false in ϕ if $\neg\, l$ is in ϕ. A set of literals σ is true in ϕ if σ is contained in ϕ. A state s is a complete and consistent set of literals that describes the world at a certain time point.

Definition 1. *Let* $\mathbf{PT} = \langle S_0, \mathcal{D}, \lambda \rangle$ *be a ground action theory in* $\mathcal{A}_{PO}$*,* s *be a state,* a_i **causes** $\phi_i \ : \ p_i : \ r_i$ **if** ψ_i $(1 \leq i \leq n)$ *be in* $\mathcal{D}$*, and* $a = \{a_1, \ldots, a_n\}$ *be an action, where each* a_i *corresponds to* ϕ_i, p_i, r_i*, and* ψ_i *for* $1 \leq i \leq n$ *(similarly for* a_i **observes** $\phi_i \ : \ p_i \ : \ r_i$ **sensing** ψ_i*). Then, the state resulting from executing* a *in* s*, denoted by* $\Phi(a_i, s)$*, is given by:*

- $l \in \Phi(a_i, s)$ *and* $\neg\, l \notin \Phi(a_i, s)$ *iff* $l \in \phi_i$ *and* $\psi_i \subseteq s$.
- $\neg\, l \in \Phi(a_i, s)$ *and* $l \notin \Phi(a_i, s)$ *iff* $\neg\, l \in \phi_i$ *and* $\psi_i \subseteq s$.
- *Otherwise,* $l \in \Phi(a_i, s)$ *iff* $l \in s$ *and* $\neg\, l \in \Phi(a_i, s)$ *iff* $\neg\, l \in s$.

Definition 2. *Let s be a state, and a_i* **causes** $\phi_i \; : \; p_i \; : \; r_i$ **if** ψ_i *(similarly a'_i* **observes** $o_i \; : \; p'_i \; : \; r'_i$ **sensing** ψ_i*) $(1 \leq i \leq n)$. Then, the transition probability distribution after executing a (a') in s is given by:*

$$T(s, a, s') = \begin{cases} p_i & if s' = \Phi(a_i, s) \\ 0 & otherwise \end{cases}$$
$$O(s, a', s') = \begin{cases} p'_i & if s' = \Phi(a'_i, s) \\ 0 & otherwise \end{cases}$$

The reward received in a state s' after executing a (a') in s is $\mathcal{R}(s, a, s') = r_i$ if $s' = \Phi(a_i, s)$, $\mathcal{R}(s, a', s') = r'_i$ if $s' = \Phi(a'_i, s)$, otherwise $\mathcal{R}(s, a, s') = \mathcal{R}(s, a', s') = 0$.

Definition 3. *Let s_0 be an initial state, s, s' be states, and π be a policy in* **PT**. *Then, the value function of n-step remaining, V^π_n, of π is given by:*

- *if $\pi(s_0)$ is a non-sensing action and then*
  $V^\pi_n(s_0) = \sum_{s_1 \in S} T(s_0, \pi(s_0), s_1) \left[\mathcal{R}(s_0, \pi(s_0), s_1) + \lambda\, V^\pi_{n-1}(s_1)\right]$
- *if $\pi(s_0)$ is sensing action then*
  $V^\pi_n(s_0) = \sum_{s_1 \in S} O(s_0, \pi(s_0), s_1) \left[\mathcal{R}(s_0, \pi(s_0), s_1) + \lambda\, V^\pi_{n-1}(s_1)\right]$

where after n steps, $V^\pi_0(s_n) = \mathcal{R}(s_{n-1}, \pi(s_{n-1}), s_n)$.

Executing sensing or non-sensing action, $\pi(s)$, in s causes a transition to a set of states, $\sigma = \{s'_1, s'_2, \ldots, s'_m\}$. Let $\pi(\sigma)$ denotes the set of actions $\pi(s'_1), \pi(s'_2), \ldots, \pi(s'_m)$ executed in the states $s'_1, s'_2, \ldots, s'_m$ respectively. Notice that if $\pi(\sigma)$ is a singleton, i.e., the same action is executed in every state in σ, then this corresponds to executing an action in a belief state $\sigma = \{s'_1, s'_2, \ldots, s'_m\}$. Since executing $\pi(\sigma)$ in σ produces another set of states σ', then executing $\pi(\sigma)$ causes a transition from a belief state to another belief state.

For finite horizon POMDP, a policy $\pi : S \rightarrow \mathcal{A}$ can be represented as a set of ordered pairs, starting from the initial belief state σ_0 (the set of initial states in S_0), as $\pi = \{(\sigma_0, \pi(\sigma_0)), (\sigma_1, \pi(\sigma_1)), \ldots, (\sigma_{n-1}, \pi(\sigma_{n-1}))\}$, where for $1 \leq i \leq n$, σ_i represents a belief state (a set of states) resulting from executing $\pi(\sigma_{i-1})$ in σ_{i-1}. This set representation of finite horizon policies in POMDP leads to view a policy as a set of trajectories, where each trajectory takes the form $j(n) \equiv s_0, \pi(s_0), s_1, \pi(s_1), \ldots, s_{n-1}, \pi(s_{n-1}), s_n$ where s_0 is an initial state in S_0 and for all $1 \leq i \leq n$, $s_i \in \sigma_i$ and $\pi(s_i) \in \pi(\sigma_i)$, such that for any $1 \leq i \leq n$, $s_i = \Phi(s_{i-1}, \pi(s_{i-1}))$. Let π be a policy for a finite horizon POMDP and T_π be the set of trajectories representation of π, given the trajectory view of π, the value function of π can be now described as:

$$V^\pi_n(s_0) = \sum_{j(n) \in T_\pi} \left[\sum_{t=0}^{n-1} \lambda^t \left[\prod_{i=0}^{t} X(s_i, \pi(s_i), s_{i+1})\right] \mathcal{R}(s_t, \pi(s_t), s_{t+1})\right] \quad (5)$$

where

$$X(s_i, \pi(s_i), s_{i+1}) = \begin{cases} T(s_i, \pi(s_i), s_{i+1}), \pi(s_i) \text{ is nonsensing} \\ O(s_i, \pi(s_i), s_{i+1}), \pi(s_i) \text{ is sensing} \end{cases}$$

Thus, the optimal policy V_n^*, the maximum value function among all policies, is given by $V_n^*(s_0) = \max_\pi V_n^\pi(s_0)$

5 Reinforcement Learning in NHPLP$_{\mathcal{PO}}$

The p-program encoding of an action theory in $\mathcal{A}_{\mathcal{PO}}$ follows related encoding described in [23,22,28]. We assume that the length of the optimal policy that we are looking for is known and finite. We use the following predicates: $holds(L,T)$ for literal L holds at time moment T, $occ(A,T)$ for action A executes at time T, $state(T)$ for a state of the world at time T, $reward(T,r)$ for the reward received at time T is r, $value(T,V)$ for the value function of a state at time T is V, and $factor(\lambda)$ for the discount factor λ. If an atom appears in a p-rule in R with no annotation it is assumed to be associated with the annotation 1. We use $p(\psi)$ to denote $p(l_1), \ldots, p(l_n)$ for p is a predicate and $\psi = \{l_1, \ldots, l_n\}$.

Let $\Pi_{\mathbf{PT}} = \langle R, \tau \rangle$ be the p-program encoding $\mathbf{PT} = \langle S_0, \mathcal{D}, \lambda \rangle$, where R is the set of the following p-rules.

- Each action $a = \{a_1, \ldots, a_n\} \in \mathcal{A}$, is encoded as

$$action(a_i) \leftarrow \tag{6}$$

for all $1 \leq i \leq n$. Each fluent $f \in \mathcal{F}$ is encoded as a fact of the form $fluent(f)$. Fluent literals are encoded as

$$literal(F) \leftarrow fluent(F) \tag{7}$$

$$literal(\neg F) \leftarrow fluent(F) \tag{8}$$

To specify that fluents F and $\neg F$ are contrary literals, we use the following p-rules.

$$contrary(F, \neg F) \leftarrow fluent(F) \tag{9}$$

$$contrary(\neg F, F) \leftarrow fluent(F) \tag{10}$$

- The initial belief state **initially** $\{\psi_i \ : \ p_i, 1 \leq i \leq n$ is represented in R as follows. Let $s_1, s_2, \ldots, s_n$ be the set of possible initial states, where for each $1 \leq i \leq n$, $s_i = \{l_1^i, \ldots, l_m^i\}$, and the initial probability distribution be $Pr(s_i) = p_i$. Moreover, let $s = s_1 \cup s_2 \cup \ldots \cup s_n$, $s' = s_1 \cap s_2 \cap \ldots \cap s_n$, $\widehat{s} = s - s'$. Let $s^{report} = \{l \,|\, l \in s_i$ and l is a sensor report literal $\}$ be the set of all sensor report literals in all s_i. We denote $s'' = \{l | l \in (\widehat{s} - s^{report}) \vee \neg l \in (\widehat{s} - s^{report})\}$. Let s^{sense} be the set of all pairs (δ_i, γ_i), where δ_i and γ_i are sets of literals contained in s_i, such that δ_i is the set of sensor reading literals and γ_i is the set of sensor report literals appearing in s_i. The set of all possible initial states are generated as follows: for each $l \in s'$, we include in R

$$holds(l, 0) \leftarrow \tag{11}$$

which says that l holds at time 0. For each $l \in s''$,

$$holds(l, 0) \leftarrow not\ holds(\neg l, 0) \tag{12}$$

$$holds(\neg l, 0) \leftarrow not\ holds(l, 0) \tag{13}$$

These p-rules say l (similarly $\neg l$) holds at time moment 0, if $\neg l$ (similarly l) does not hold at the time moment 0. For each $(\delta, \gamma) \in \psi^{sense}$, let $\gamma = \{l_1, \ldots, l_m\}$, then for each $1 \leq i \leq m$, R includes

$$holds(l_i, 0) \leftarrow holds(\delta, 0) \tag{14}$$

The initial probability distribution over the initial states is encoded as follows, which says that the probability of a state at time 0 is p_i, if $l^i_1, \ldots, l^i_m$ hold at the time 0.

$$state(0) : p_i \leftarrow holds(l^i_1, 0), \ldots, holds(l^i_m, 0) \tag{15}$$

- Each executability condition of an action of the form (2) is encoded for each $1 \leq i \leq n$ as

$$exec(a_i, T) \leftarrow holds(\psi, T) \tag{16}$$

- For each non-sensing action proposition a_i **causes** $\phi_i\ :\ p_i\ :\ r_i$ **if** $\psi_i, 1 \leq i \leq n$, in $\mathcal{D}$, let $\phi_i = \{l^1_i, \ldots, l^m_i\}$. Then, $\forall (1 \leq j \leq m)$, R includes

$$holds(l^j_i, T+1) \leftarrow occ(a_i, T), exec(a_i, T), holds(\psi_i, T) \tag{17}$$

If a occurs at time T and ψ_i holds at the same time moment, then the l^j_i holds at the time $T+1$. Then, we have

$$state(T+1) : p_i \times U \leftarrow state(T) : U, occ(a_i, T), exec(a_i, T),$$
$$holds(\psi_i, T), holds(\phi_i, T+1) \tag{18}$$

where U is an annotation variable ranging over $[0, 1]$ acts as a place holder. This p-rule states that if ψ_i holds in a state at time T, whose probability is U, and in which a is executable, then the probability of a successor state at time $T+1$ is $p_i \times U$, in which ϕ_i holds.

- For each sensing action proposition a_i **observes** $o_i\ :\ p_i\ :\ r_i$ **sensing** ψ_i, $1 \leq i \leq n$, in $\mathcal{D}$, let $o_i = \{l^1_i, \ldots, l^m_i\}$ and $\psi_i = \{l'^1_i, \ldots, l'^m_i\}$. Then, $\forall (1 \leq j \leq m)$, R includes

$$observed(l'^j_i, T) \leftarrow occ(a_i, T), exec(a_i, T), holds(\psi_i, T) \tag{19}$$

$$holds(l^j_i, T+1) \leftarrow occ(a_i, T), exec(a_i, T), observed(\psi_i, T) \tag{20}$$

where (19) says that executing the sensing action a at time T in which ψ_i holds causes ψ_i to be observed to be known true at the same moment T, and (20) states that if a occurs at time T and the literals in ψ_i are observed to be known true at the same moment, then the literals $l_i^j \in o_i$ are known to hold at the time moment $T+1$.

$$state(T+1) : p_i \times U \leftarrow state(T) : U, occ(a_i, T), exec(a_i, T), observed(\psi_i, T), holds(o_i, T+1) \quad (21)$$

The above p-rule says that the probability of a state at time $T+1$ is $p_i \times U$ if o_i become known true at the same moment, after executing a in a state at time T, whose probability is U, in which the literals in ψ_i are observed true.

- The reward r_i received at time $T+1$ after executing a in a state at T is encoded as

$$reward(r_i, T+1) \leftarrow occ(a_i, T), exec(a_i, T) \quad (22)$$

- The value function $T+1$ steps away from the initial state given the value function T steps away from the initial state is encoded as
  - if a is a non-sensing action

$$value(V + \lambda^T * U * r_i, T+1) \leftarrow value(V, T), factor(\lambda), state(T+1) : U, reward(r_i, T+1), occ(a_i, T), exec(a_i, T), holds(\psi_i, T), holds(\phi_i, T+1) \quad (23)$$

  - if a is a sensing action

$$value(V + \lambda^T * U * r_i, T+1) \leftarrow value(V, T), factor(\lambda), state(T+1) : U, reward(r_i, T+1), occ(a_i, T), exec(a_i, T), observed(\psi_i, T), holds(o_i, T+1) \quad (24)$$

  where the variables $V \in \mathbb{R}$, $\lambda \in [0, 1)$, $U \in [0, 1]$, and $factor(\lambda)$ is a fact in R. These p-rules state that the value function at time $T+1$ is equal to the value function at time T added to the product of the reward r_i received in a state at $T+1$ and the probability of a state at time $T+1$ discounted by λ^T.

- The following p-rule asserts that a literal L holds at $T+1$ if it holds at T and its contrary does not hold at $T+1$.

$$holds(L, T+1) \leftarrow holds(L, T), not\ holds(L', T+1), contrary(L, L') \quad (25)$$

- The literal, A, and its negation, $\neg A$, cannot hold at the same time, where $inconsistent$ is a literal that does not appear in **PT**.

$$inconsistent \leftarrow not\ inconsistent, holds(A, T), holds(\neg A, T) \quad (26)$$

- Actions are generated once at a time by the p-rules:

$$occ(AC^i, T) \leftarrow action(AC^i), not\ abocc(AC^i, T) \quad (27)$$
$$abocc(AC^i, T) \leftarrow action(AC^i), action(AC^j), occ(AC^j, T), AC^i \neq AC^j \quad (28)$$

- The goal expression $\mathcal{G} = g_1 \wedge \ldots \wedge g_m$ is encoded as

$$goal \leftarrow holds(g_1, T), \ldots, holds(g_m, T) \quad (29)$$

Example 2. The p-program encoding of the tiger domain presented in Example 1 is given by $\Pi = \langle R, \tau \rangle$, where τ is arbitrary and R consists of the following p-rules, in addition to the p-rules (7), (8), (9), (10), (25), (26), (27), (28):

$$action(openL_i) \leftarrow \quad action(openR_i) \leftarrow \quad action(listen_j) \leftarrow$$

for $1 \leq i \leq 2$ and $1 \leq j \leq 4$. Properties of the world are described by the fluents tl and htl which are encoded in R by the p-rules

$$fluent(tl) \leftarrow \qquad fluent(htl) \leftarrow$$

The set of possible initial states are encoded by the p-rules:

$$\begin{array}{ll} holds(tl, 0) & \leftarrow not\ holds(\neg tl, 0) \\ holds(\neg tl, 0) & \leftarrow not\ holds(tl, 0) \\ holds(tl, 0) & \leftarrow holds(htl, 0) \\ holds(\neg tl, 0) & \leftarrow holds(\neg htl, 0) \end{array}$$

The initial probability distribution over the possible initial states is encoded by the p-rules

$$\begin{array}{l} state(0) : 0.5 \leftarrow holds(tl, 0), holds(htl, 0) \\ state(0) : 0.5 \leftarrow holds(\neg tl, 0), holds(\neg htl, 0) \end{array}$$

The executability conditions of actions are encoded by the following p-rules

$$exec(openL_i) \leftarrow \quad exec(openR_i) \leftarrow \quad exec(listen_j) \leftarrow$$

for $1 \leq i \leq 2$ and $1 \leq j \leq 4$. Effects of the $openL$ action are encoded by the p-rules

$$\begin{array}{ll} holds(tl, T+1) & \leftarrow occ(openL_1, T), exec(openL_1, T), holds(tl, T) \\ holds(\neg tl, T+1) & \leftarrow occ(openL_2, T), exec(openL_2, T), holds(\neg tl, T) \end{array}$$

Effects of the $openR$ action are encoded by the p-rules

$$\begin{array}{ll} holds(\neg tl, T+1) & \leftarrow occ(openR_1, T), exec(openR_1, T), holds(\neg tl, T) \\ holds(tl, T+1) & \leftarrow occ(openR_2, T), exec(openR_2, T), holds(tl, T) \end{array}$$

Effects of the $listen$ action are encoded by the p-rules

$$\begin{array}{ll} observed(htl, T) & \leftarrow occ(listen_1, T), exec(listen_1, T), holds(htl, T) \\ observed(htl, T) & \leftarrow occ(listen_2, T), exec(listen_2, T), holds(htl, T) \\ observed(\neg htl, T) & \leftarrow occ(listen_3, T), exec(listen_3, T), holds(\neg htl, T) \\ observed(\neg htl, T) & \leftarrow occ(listen_4, T), exec(listen_4, T), holds(\neg htl, T) \\ holds(tl, T+1) & \leftarrow occ(listen_1, T), exec(listen_1, T), observed(htl, T) \\ holds(\neg tl, T+1) & \leftarrow occ(listen_2, T), exec(listen_2, T), observed(htl, T) \\ holds(\neg tl, T+1) & \leftarrow occ(listen_3, T), exec(listen_3, T), observed(\neg htl, T) \\ holds(tl, T+1) & \leftarrow occ(listen_4, T), exec(listen_4, T), observed(\neg htl, T) \end{array}$$

The probability distribution resulting from executing the *listen* action is given by

$$
\begin{aligned}
state(T+1) : 0.85 \times V \leftarrow\ & occ(listen_1, T), exec(listen_1, T), state(T) : V, \\
& observed(htl, T), holds(tl, T+1) \\
state(T+1) : 0.15 \times V \leftarrow\ & occ(listen_2, T), exec(listen_2, T), state(T) : V, \\
& observed(htl, T), holds(\neg tl, T+1) \\
state(T+1) : 0.85 \times V \leftarrow\ & occ(listen_3, T), exec(listen_3, T), state(T) : V, \\
& observed(\neg htl, T), holds(\neg tl, T+1) \\
state(T+1) : 0.15 \times V \leftarrow\ & occ(listen_4, T), exec(listen_4, T), state(T) : V, \\
& observed(\neg htl, T), holds(tl, T+1)
\end{aligned}
$$

The rewards received from executing the actions are encoded by

$$
\begin{aligned}
reward(-100, T+1) &\leftarrow occ(openL_1), exec(openL_1) \\
reward(10, T+1) &\leftarrow occ(openL_2), exec(openL_2) \\
reward(-100, T+1) &\leftarrow occ(openR_1), exec(openR_1) \\
reward(10, T+1) &\leftarrow occ(openR_2), exec(openR_2) \\
reward(-1, T+1) &\leftarrow occ(listen_1), exec(listen_1) \\
reward(-1, T+1) &\leftarrow occ(listen_2), exec(listen_2) \\
reward(-1, T+1) &\leftarrow occ(listen_3), exec(listen_3) \\
reward(-1, T+1) &\leftarrow occ(listen_4), exec(listen_4)
\end{aligned}
$$

The value function is encoded in R by the p-rules:

$$
\begin{aligned}
value(V + \lambda^T * U * -100, T+1) \leftarrow\ & value(V, T), factor(\lambda), state(T+1) : U, \\
& reward(-100, T+1), occ(openL_1, T), exec(openL_1, T), holds(tl, T), \\
& holds(tl, T+1) \\
value(V + \lambda^T * U * 10, T+1) \leftarrow\ & value(V, T), factor(\lambda), state(T+1) : U, \\
& reward(10, , T+1), occ(openL_2, T), exec(openL_2, T), holds(\neg tl, T), \\
& holds(\neg tl, T+1) \\
value(V + \lambda^T * U * -100, T+1) \leftarrow\ & value(V, T), factor(\lambda), state(T+1) : U, \\
& reward(-100, T+1), occ(openR_1, T), exec(openR_1, T), holds(\neg tl, T), \\
& holds(\neg tl, T+1) \\
value(V + \lambda^T * U * 10, T+1) \leftarrow\ & value(V, T), factor(\lambda), state(T+1) : U, \\
& reward(10, T+1), occ(openR_2, T), exec(openR_2, T), holds(tl, T), \\
& holds(tl, T+1) \\
value(V + \lambda^T * U * -1, T+1) \leftarrow\ & value(V, T), factor(\lambda), state(T+1) : U, \\
& reward(-1, T+1), occ(listen_1, T), exec(listen_1, T), \\
& observed(htl, T), holds(tl, T+1) \\
value(V + \lambda^T * U * -1, T+1) \leftarrow\ & value(V, T), factor(\lambda), state(T+1) : U, \\
& reward(-1, T+1), occ(listen_2, T), exec(listen_2, T), observed(htl, T), \\
& holds(\neg tl, T+1) \\
value(V + \lambda^T * U * -1, T+1) \leftarrow\ & value(V, T), factor(\lambda), state(T+1) : U, \\
& reward(-1, T+1), occ(listen_3, T), exec(listen_3, T), observed(\neg htl, T), \\
& holds(\neg tl, T+1) \\
value(V + \lambda^T * U * -1, T+1) \leftarrow\ & value(V, T), factor(\lambda), state(T+1) : U, \\
& reward(-1, T+1), occ(listen_4, T), exec(listen_4, T), observed(\neg htl, T), \\
& holds(tl, T+1)
\end{aligned}
$$

6 Correctness

Let the domain of T be $\{0, \ldots, n\}$. Let Φ be a transition function associated with **PT**, s_0 be a possible initial state, and $a_0, \ldots, a_{n-1}$ be a set of actions in $\mathcal{A}_{PO}$. Recall, any action a_i can be represented as $a_i = \{a_{1_i}, \ldots, a_{m_i}\}$. Therefore, a trajectory $s_0, \pi(s_0), s_1, \pi(s_1), \ldots, s_{n-1}, \pi(s_{n-1}), s_n$ in **PT** can be also represented as $s_0\ a_{j_0}\ s_1 \ldots a_{j_{n-1}}\ s_n$ for $(1 \leq j \leq m)$ and $(0 \leq i \leq n)$, such that $\forall(0 \leq i \leq n)$, s_i is a state, a_i is an action, $a_{j_i} \in a_i = \{a_{1_i}, \ldots, a_{m_i}\}$, $a_{j_i} = \pi(s_i)$, and $s_i = \Phi(a_{j_{i-1}}, s_{i-1})$.

Theorem 1. *Let* **PT** *be an action theory in* $\mathcal{A}_{PO}$, π *be a policy in* **PT**, *and* T_π *be the set of trajectories in* π. *Then,* $s_0, \pi(s_0), s_1, \pi(s_1), \ldots, s_{n-1}, \pi(s_{n-1}), s_n$ *is a trajectory in* T_π *iff* $occ(\pi(s_0), 0), \ldots, occ(\pi(s_{n-1}), n-1)$ *is true in a probabilistic answer set of* $\Pi_{\mathbf{PT}}$.

Intuitively, an action theory, **PT** in $\mathcal{A}_{PO}$, can be encoded to a p-program, $\Pi_{\mathbf{PT}}$, whose probabilistic answer sets correspond to trajectories in **PT**.

Theorem 2. *Let* h *be a probabilistic answer set of* $\Pi_{\mathbf{PT}}$, π *be a policy in* **PT**, *and* T_π *be the set of trajectories in* π. *Let* $\mathcal{OCC}$ *be a set that contains* $h \models \tau = occ(\pi(s_0), 0)$,
$\ldots, occ(\pi(s_{n-1}), n-1)$ *iff* $s_0, \pi(s_0), s_1, \pi(s_1), \ldots, s_{n-1}, \pi(s_{n-1}), s_n \in T_\pi$. *Then,*
$\sum_{h \models value(n,v) \text{ and } h \models \tau \in \mathcal{OCC}} v = V_n^\pi(s_0)$

The p-program encoding of the reinforcement learning problems, in finite-horizon POMDP, finds optimal policies using the flat representation of the problem domains [19]. Hence, Theorem 4 follows directly from Theorem 3.

Theorem 3 ([19]). *The stationary policy existence problem for finite-horizon POMDP in the flat representation is NP-complete.*

Theorem 4. *The policy existence problem for a reinforcement learning problem in POMDP environment using* $NHPLP_{\mathcal{PO}}$ *with probabilistic answer set semantics is NP-complete.*

7 Reinforcement Learning Using Answer Set Programming

Excluding the p-rules (15), (18), (21) – (24) from the p-program encoding, $\Pi_{\mathbf{PT}}$, of **PT**, results p-program, denoted by $\Pi_{\mathbf{PT}}^{normal}$, with only annotations of the form 1. As shown in [26], the syntax and semantics of this class of p-programs is equivalent to classical normal logic programs with classical answer set semantics.

Theorem 5. *Let* $\Pi_{\mathbf{PT}}^{normal}$ *be the* normal logic program *resulting after deleting the p-rules (15), (18), (21) – (24) from* $\Pi_{\mathbf{PT}}$ *and* π *be a policy in* **PT**. *Then,* $s_0, \pi(s_0), s_1, \pi(s_1)$
$, \ldots, s_{n-1}, \pi(s_{n-1}), s_n$ *is a trajectory in* π *iff* $occ(\pi(s_0), 0), \ldots, occ(\pi(s_{n-1}), n-1)$ *is true in an answer set of* $\Pi_{\mathbf{PT}}^{normal}$.

Any reinforcement learning problem in POMDP can be encoded as a SAT problem. Any normal logic program, Π, can be translated into a SAT formula, $\mathcal{S}$, where the models of $\mathcal{S}$ are equivalent to the answer sets of Π [16].

Theorem 6. *Let* **PT** *be an action theory and* $\Pi_{\mathbf{PT}}^{normal}$ *be the normal logic program encoding of* **PT**. *Then, the models of the SAT encoding of* $\Pi_{\mathbf{PT}}^{normal}$ *are equivalent to valid trajectories in* **PT**.

Corollary 1. *Let* **PT** *be an action theory. Then,* **PT** *can be directly encoded as a SAT formula* $\mathcal{S}$ *where the models of* $\mathcal{S}$ *are equivalent to valid trajectories in* **PT**.

However, encoding reinforcement learning problems in NHPLP$_{\mathcal{PO}}$ has advantages over normal logic program and SAT encoding. These include, the explicit representation of probabilities, the explicit assignment of probabilities to states, and the direct propagation of probabilities through states, which are naturally present in NHPLP$_{\mathcal{PO}}$ with probabilistic answer set semantics.

8 Conclusions and Related Work

The difference between $\mathcal{A}_{PO}$ and the action languages [1], [3], [6], [10], and [15] is that $\mathcal{A}_{PO}$ is a high level language and allows the factored specification of POMDP.

The approaches for solving POMDP to find the optimal policies can be categorized into two main approaches; dynamic programming approaches and the search-based approaches (a detailed survey on these approaches can be found in [3]). However, dynamic programming approaches use primitive domain knowledge representation. Moreover, the search-based approaches mainly rely on search heuristics which have limited knowledge representation capabilities to represent and use domain-specific knowledge. In [18], a logical approach for solving POMDP, for probabilistic contingent planning, has been presented which converts a POMDP specification of a probabilistic contingent planing problem into a stochastic satisfiability problem and solving the stochastic satisfiability problem instead. Our approach is similar in spirit to [18] in the sense that both approaches are logic based approaches. However, it has been shown in [24] that NHPLP is more expressive than stochastic satisfiability from the knowledge representation point of view. In [14], based on first-order logic programs without nonmonotonic negation, a first-order logic representation of MDP has been described. Similar to the first-order representation of MDP in [14], $\mathcal{A}_{MD}$ allows objects and relations. However, unlike $\mathcal{A}_{PO}$, [14] finds policies in the abstract level. But, NHPLP allows objects and relations. [4] presented a more expressive first-order representation of MDP than [14] that is a probabilistic extension to Reiter's situation calculus. However, it is more complex than [14].

References

1. Baral, C., Tran, N., Tuan, L.C.: Reasoning about actions in a probabilistic setting. In: AAAI 2002 (2002)
2. Bagnell, J., Kakade, S., Ng, A., Schneider, J.: Policy search by dynamic programming. In: Neural Information Processing Systems, vol. 16. MIT Press, Cambridge (2003)
3. Boutilier, C., Dean, T., Hanks, S.: Decision-theoretic planning: structural assumptions and computational leverage. Journal of AI Research 11, 1–94 (1999)
4. Boutilier, C., Reiter, R., Price, B.: Symbolic dynamic programming for first-order MDPs. In: 17th IJCAI (2001)
5. Draper, D., Hanks, S., Weld, D.: Probabilistic planning with information gathering and contingent execution. In: 2nd ICAIPS (1994)
6. Eiter, T., Lukasiewicz, T.: Probabilistic reasoning about actions in nonmonotonic causal theories. In: 19th Conference on Uncertainty in Artificial Intelligence (2003)
7. Gelfond, M., Lifschitz, V.: The stable model semantics for logic programming. In: ICSLP. MIT Press, Cambridge (1988)
8. Gelfond, M., Lifschitz, V.: Classical negation in logic programs and disjunctive databases. New Generation Computing 9(3-4), 363–385 (1991)
9. Gelfond, M., Lifschitz, V.: Representing action and change by logic programs. Journal of Logic Programming 17, 301–321 (1993)
10. Iocchi, L., Lukasiewicz, T., Nardi, D., Rosati, R.: Reasoning about actions with sensing under qualitative and probabilistic uncertainty. In: 16th ECAI (2004)
11. Kaelbling, L., Littman, M., Cassandra, A.: Planning and acting in partially observable stochastic domains. Artificial Intelligence 101, 99–134 (1998)
12. Kaelbling, L., Littman, M., Moore, A.: Reinforcement Learning: A Survey. Journal of Artificial Intelligence Research 4, 237–285 (1996)
13. Kautz, H., Selman, B.: Pushing the envelope: planning, propositional logic, and stochastic search. In: 13th National Conference on Artificial Intelligence (1996)
14. Kersting, K., De Raedt, L.: Logical Markov decision programs and the convergence of logical TD(λ). In: 14th International Conference on Inductive Logic Programming (2004)
15. Kushmerick, N., Hanks, S., Weld, D.: An algorithm for probabilistic planning. Artificial Intelligence 76(1-2), 239–286 (1995)
16. Lin, F., Zhao, Y.: ASSAT: Computing answer sets of a logic program by SAT solvers. Artificial Intelligence 157(1-2), 115–137 (2004)
17. Littman, M., Cassandra, A., Kaelbling, L.: Learning policies for partially observable environments: scaling up. In: 12th ICML (1995)
18. Majercik, S., Littman, M.: Contingent planning under uncertainty via stochastic satisfiability. Artificial Intelligence 147(1–2), 119–162 (2003)
19. Mundhenk, M., Goldsmith, J., Lusena, C., Allender, E.: Complexity of finite-horizon Markov decision process problems. Journal of the ACM (2000)
20. Niemela, I., Simons, P.: Efficient implementation of the well-founded and stable model semantics. In: Joint ICSLP, pp. 289–303 (1996)
21. Saad, E.: Incomplete knowlege in hybrid probabilistic logic programs. In: 10th European Conference on Logics in Artificial Intelligence (2006)
22. Saad, E.: Probabilistic planning with imperfect sensing actions using hybrid probabilistic logic programs. In: Godo, L., Pugliese, A. (eds.) SUM 2009. LNCS, vol. 5785, pp. 206–222. Springer, Heidelberg (2009)

23. Saad, E.: A logical framework to reinforcement learning using hybrid probabilistic logic programs. In: 2nd International Conference on Scalable Uncertainty Management (2008)
24. Saad, E.: On the relationship between hybrid probabilistic logic programs and stochastic satisfiability. In: 2nd International Conference on Scalable Uncertainty Management (2008)
25. Saad, E.: Probabilistic planning in hybrid probabilistic logic programs. In: 1st International Conference on Scalable Uncertainty Management (2007)
26. Saad, E., Pontelli, E.: A new approach to hybrid probabilistic logic programs. Annals of Mathematics and Artificial Intelligence Journal 48(3-4), 187–243 (2006)
27. Scherl, R., Levesque, H.: The frame problem and knowledge producing actions. In: AAAI 1993 (1993)
28. Son, T., Baral, C., Nam, T., McIlraith, S.: Domain-dependent knowledge in answer set planning. ACM Transactions on Computational Logic 7(4), 613–657 (2006)
29. Sutton, R., Barto, A.: Reinforcement Learning: An Introduction. MIT Press, Cambridge (1998)

ChaseT: A Tool for Checking Chase Termination

Andrea De Francesco, Sergio Greco, Francesca Spezzano, and Irina Trubitsyna

DEIS, Università della Calabria, 87036 Rende, Italy
{adefrancesco,greco,fspezzano,irina}@deis.unical.it

Abstract. Consistency problems arise in many fundamental database applications as data exchange, data integration, data warehouse and many others. The chase algorithm is a fundamental and useful tool fixing inconsistencies of database instances with respect to a set of data dependencies. It is well known that the chase algorithm may be non-terminating and several techniques and criteria for checking chase termination have been recently proposed. This paper presents *ChaseT*, a tool that allows users to design data dependencies and combine different criteria and rewriting algorithms for checking chase termination.

1 Introduction

In the design of databases an important role is played by the definition of data dependencies. Data dependencies (also called integrity constraints) are used to define data properties which must be satisfied by data. Databases not satisfying constraints are said to be in an inconsistent state and they need to be repaired through the application of update operations [1,2]. An alternative solution is to leave databases in an inconsistent state and to answer queries distinguishing tuples (in the answers) which are true from those which are undefined (missing tuples are false). This alternative solution considers all possible repaired databases: true (resp. undefined) answers are those derived from all (resp. a proper subset of) repaired databases. The presence of inconsistent data arise in several practical contexts including the database integration, data exchange, data warehouses, querying federated databases, and many others [3,8].

To repair databases several contexts, based on different sets of update operations, have been considered. In the case of correct, but incomplete databases (i.e. only tuple insertion is taking into account to restore the consistency), the algorithm which is often used to make databases consistent is the the well-known *chase* fixpoint, consisting in the insertion of tuples in order to fulfill an unsatisfied tuple generating dependency (TGD), and the modification of nulls, to satisfy an equality generating dependency (EGD) which is violated by the current database instance. The chase algorithm could i) terminate successfully, if a consistent state is reached in a finite number of steps, ii) terminate unsuccessfully, if an EGD cannot be satisfied by the current database instance iii) be non-terminating, as shown by the following example.

S. Benferhat and J. Grant (Eds.): SUM 2011, LNAI 6929, pp. 520–524, 2011.
© Springer-Verlag Berlin Heidelberg 2011

Example 1. Consider the set of constraints Σ_1:

$$\begin{array}{l}\forall x\ Employee(x) \rightarrow \exists y\ WorksFor(x,y)\\ \forall x \forall y\ WorksFor(x,y) \rightarrow \exists z\ Managed(y,z)\\ \forall x \forall y\ Department(x), Managed(x,y) \rightarrow Employee(y)\end{array}$$

and the database instance $D = \{Employee(john), WorksFor(john, cs), Department(cs)\}$. Since the database does not satisfy the second constraint, a tuple $Managed(cs, n_1)$ should be inserted, where n_1 is a new labeled null value. At this point the third constraint is not satisfied and the tuple $Employee(n_1)$ should be added to the database. The chase ends by inserting the tuples $WorksFor(n_1, n_2)$ and $Managed(n_2, n_3)$ to satisfy the first two constraints, because of the presence of $Employee(n_1)$ in the database. Moreover, observe that, by removing the atom $Department(x)$ in the second constraint, the chase algorithm does not terminate and an infinite number of tuples are inserted. □

Thus, it is important to design data dependencies where the chase terminates, independently from the database instance.

This paper presents *ChaseT*, a tool for testing whether, given a set of data dependencies Σ, the application of the chase terminates, for all database instances. *ChaseT* checks chase termination for two different types of chase algorithms known as *standard* and *oblivious* chase [10,9]. Since the problem of checking whether the chase terminates is undecidable [4], several criteria, defining sufficient conditions have been recently proposed. The first and simplest criterium proposed is *Weak Acyclicity* (WA) [5] checking whether the set of constraints does not present cyclic conditions for which a new null value forces (directly or indirectly) the introduction of another null in the same position. For instance, in Example 1 we have that $Employee_1$ (denoting position 1 in relation $Employee$) forces the introduction of a new null value in position $WorkFor_2$ and $WorkFor_2$ forces the introduction of a new null value in position $Managed_2$ (denoted as $Employee_1 \xrightarrow{*} WorkFor_2$ and $WorkFor_2 \xrightarrow{*} Managed_2$ respectively); this value is then propagated in position $Employee_1$ (denoted as $Managed_2 \rightarrow Employee_1$) owing to the third constraint. The cycle going through the special edge, i.e. edge marked with "*", means that an infinite number of nulls could be introduced. The WA criterium has been generalized in several ways: *Safety* (SC) [10] and *Super-weak Acyclicity* (SwA) [9] analyze the propagation of nulls trough positions and their check requires a polynomial time complexity, whereas *C-Stratification* ($CStr$) and its extensions analyze how constraints may activate each other and their check is in co-$\mathcal{NP}$ [4,10,7]. SwA extends SC, but is not comparable with $CStr$.

In order to enlarge the class of terminating constraints, rewriting techniques, denoted by Adn and Adn^+, have been proposed in [6]. The idea consists in rewriting the original set of constraints Σ into an 'equivalent' set Σ^α, where predicate symbols are adorned, and verifying the structural properties for chase termination on Σ^α. In particular, if Σ satisfies a chase termination criterium C, then the rewritten set Σ^α satisfies C as well, but the vice versa is not true, that is there are significant classes of constraints for which Σ^α satisfies C and Σ does not. More specifically, in the Adn technique, an adorned predicate is of the form

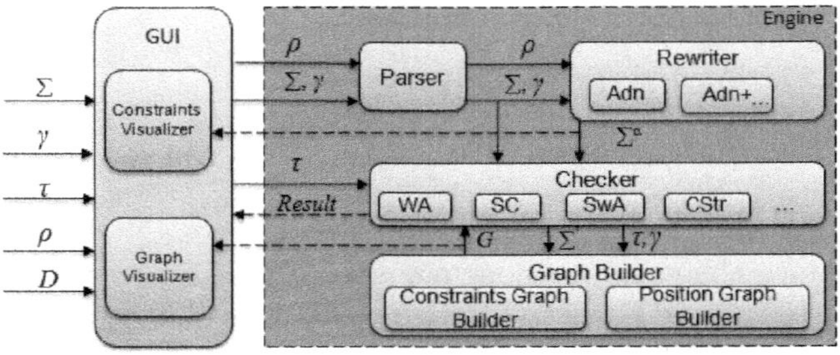

Fig. 1. *ChaseT* architecture

$p^{\alpha_1 \dots \alpha_m}(x_1, ..., x_m)$ where $\alpha_i = b$ means that the variable x_i is bounded, i.e. can take values from a finite domain, otherwise ($\alpha_i = f$) we say that x_i is free. The Adn^+ technique performs more precise analysis: instead of simply using f to denote that a position may contain null values, it uses adornments of the form f_i, where i is a fresh subscript. This allows to recognize that different nulls are generated. The rewriting of constraints allows us to recognize larger classes of constraints for which chase termination is guaranteed, although in some case the number of adorned constraints could be exponential.

Example 2. Consider the set of constraints Σ_1. (For the sake of space limitation, we will use abbreviations instead of the full predicate names.) The *Adn* technique starts by rewriting constraints, associating to body predicates strings of b symbols and to the head predicates the same body adornments for universally quantified variables and f symbols for existentially quantified variables (dependencies on the left column):

$$\forall x\ E^b(x) \rightarrow WF^{bf}(x,y) \qquad \forall x \forall y\ WF^{bf}(x,y) \rightarrow M^{ff}(y,z)$$
$$\forall x \forall y\ WF^{bb}(x,y) \rightarrow M^{bf}(y,z) \qquad \forall x \forall y\ D^b(x), M^{bf}(x,y) \rightarrow E^f(y)$$
$$\forall x \forall y\ D^b(x), M^{bb}(x,y) \rightarrow E^b(y) \qquad \forall x\ E^f(x) \rightarrow WF^{ff}(x,y)$$
$$\forall x \forall y\ WF^{ff}(x,y) \rightarrow M^{ff}(y,z)$$

Subsequently, due to the new predicates $WorksFor^{bf}(x,y)$ and $Managed^{bf}(x,y)$ the rewriting continues by producing the constraints on the right column.

At this point, the rewriting terminates, since the predicate $Department^b(x)$ cannot be joined with $Managed^{ff}(x,y)$, to produce a new adorned constraints, because the adornment of the variable x is not coherent. The rewritten set of constraints is weakly acyclic, whereas the original set Σ_1 is not recognized by *SwA* and *CStr* criteria. □

2 System Description

ChaseT implements the above criteria and techniques for checking chase termination. Its architecture is depicted in Fig. 1 and consists of five main modules which allow users to define data dependencies, check chase termination properties and visualize explanations. The Graphical User Interface (GUI) allows the user to provide the set Σ of data dependencies and three parameters: i) γ, denoting the type of chase she/he is interested in (standard, skolem oblivious, naive oblivious), ii) τ, denoting the selected termination criterium (*WA*, *SC*, *SwA* or *CStr*), and iii) ρ, denoting the possible rewriting technique she/he would apply (Adn, Adn+ or none). If the user wants to check the termination of the selected chase algorithm by applying a rewriting technique, the input set of dependencies Σ is rewritten (by the module *Rewriter*) into a set of adorned dependencies Σ^α. Since the rewriting output also depends on the particular chase procedure the user wants to check, the Rewriter module receives in input Σ and the parameters ρ and γ and gives in output the rewritten set of constraints Σ^α. The system also allows users to check termination conditions without indicating any specific criterium. In such a case all techniques are applied and the system returns the properties of the input dependencies (see the bottom right window in Fig. 2). Fig. 2 shows how the user interacts with the system through the GUI. The left window shows the dependencies input set while the rewritten set of dependencies is showed in the right window; parameters are introduced through check boxes.

It is worth noting that the data dependencies defined by the user are first parsed (by the module *Parser*) to check syntactic errors and inconsistencies (e.g. the use of predicates having the same name and different arity).

For the analysis of the structural properties of a set of dependencies Σ, the *Checker* builds two specific graphs for the selected criterium: the *constraints graph* shows how constraints may activate each other, while the *position graph* shows how nulls may propagate through positions. The construction of the graphs is performed by the module *Graph Builder* which receives in input the set of dependencies Σ, the criterium τ and the type of chase γ. The graphs can be visualized using a graph visualization tool for a better understanding of data dependency properties (see Fig. 3). It is important to observe that the system allows user to select a specific termination criterium for a better understanding of the source of possible cycles generating tuples with nulls.

The system has been developed in Java using *Eclipse IDE* and is downloadable from *wwwinfo.deis.unical.it/chaset*. The GUI has been written using the *Swing Java* libraries and the open source library *JGraphX* for the visualization of graphs. The interactions among the different modules are carried out through interfaces so that each module can be easily modified without any inference on the other modules.

3 Application Scenario

In the following, a typical use-case scenario of *ChaseT* is shown. Suppose, for instance, that the user wants to check the termination of the standard chase procedure for the set of constraints of Example 1. As shown in Fig. 2, the user introduces the set of constraints Σ in the "*Input data dependency*" window, selects "*standard*" as chase type and tries to test the known termination conditions (*Run test* button) by taking into account that more general techniques require greater computational effort and the explanation is more complex. For the example shown in Fig. 2 the application of all termination conditions to the

Fig. 2. *ChaseT* User Interface

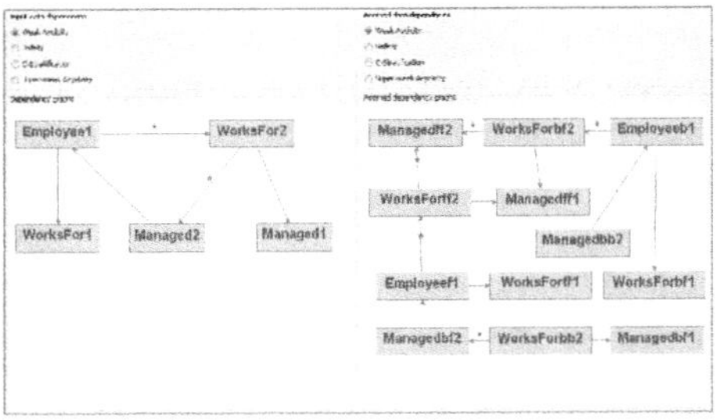

Fig. 3. Graph Visualizer

original set of dependencies produces a negative result. However, the application of the simplest rewriting generates the equivalent set $Adn(\Sigma)$, which is weakly acyclic. The rewritten set of constraints can be visualized in the "*Adorned data dependency*" window (using the *Adorning* button). The user can also visualize on the screen the graphs generated by the *Graph Builder* in order to analyze graphs and understand the behaviors of the different termination criteria. As shown in Fig 3, the dependency graph of Σ contains a cycle going through a special edge ("*Dependency graph*" window) instead the dependency graph of $Adn(\Sigma)$ is acyclic ("*Adorned dependency graph*" window).

4 Conclusion

This paper has presented a tool for testing whether, given a set of data dependencies, the chase terminates, for all database instances. We described the architecture of the system and the use-case scenarios. As future work, we plan to extend the set of chase termination criteria with new ones proposed in [7].

References

1. Bertossi, L.E.: Consistent query answering in databases. SIGMOD Record 35(2), 68–76 (2006)
2. Chomicki, J.: Consistent query answering: Five easy pieces. In: Schwentick, T., Suciu, D. (eds.) ICDT 2007. LNCS, vol. 4353, pp. 1–17. Springer, Heidelberg (2006)
3. De Giacomo, G., Lembo, D., Lenzerini, M., Rosati, R.: On reconciling data exchange, data integration, and peer data management. In: PODS (2007)
4. Deutsch, A., Nash, A., Remmel, J.B.: The chase revisited. In: PODS (2008)
5. Fagin, R., Kolaitis, P.G., Miller, R.J., Popa, L.: Data exchange: semantics and query answering. Theor. Comput. Sci. 336(1), 89–124 (2005)
6. Greco, S., Spezzano, F.: Chase termination: A constraints rewriting approach. PVLDB 3(1), 93–104 (2010)
7. Greco, S., Spezzano, F., Trubitsyna, I.: Stratification criteria and rewriting techniques for checking chase termination. PVLDB 4(11) (2011)
8. Lenzerini, M.: Data integration: A theoretical perspective. In: PODS (2002)
9. Marnette, B.: Generalized schema-mappings: from termination to tractability. In: PODS (2009)
10. Meier, M., Schmidt, M., Lausen, G.: On chase termination beyond stratification. CoRR, abs/0906.4228 (2009)

Swapping-Based Partitioned Sampling for Better Complex Density Estimation: Application to Articulated Object Tracking

Séverine Dubuisson, Christophe Gonzales, and Xuan Son Nguyen

Laboratoire d'Informatique de Paris 6 (LIP6/UPMC)
4 place Jussieu, 75005 Paris, France
firstname.name@lip6.fr

Abstract. In this paper, we propose to better estimate high-dimensional distributions by exploiting conditional independences within the Particle Filter (PF) framework. We first exploit Dynamic Bayesian Networks to determine conditionally independent subspaces of the state space, which allows us to independently perform propagations and corrections over smaller spaces. Second, we propose a *swapping* process to transform the weighted particle set provided by the update step of PF into a "new particle set" better focusing on high peaks of the posterior distribution. This new methodology, called *Swapping-Based Partitioned Sampling*, is successfully tested and validated for articulated object tracking.

1 Introduction

Dealing with high-dimensional state and observation spaces is a major concern for many research communities. There exist essentially two ways to tackle high-dimensional problems: either reduce the dimension of the state space/search space by approximation or exploit conditional independences naturally arising in the state space to partition the latter into low-dimensional spaces. In this paper, we chose the latter and focus on articulated object tracking. Actually, it is an important computer vision task for a wide variety of applications including gesture recognition, human tracking and event detection. However, tracking articulated structures with accuracy and within a reasonable time is challenging due to the high dimensionality of the state and observation spaces. In the optimal filtering context, the goal of tracking is to estimate a state sequence $\{\mathbf{x}_t\}_{t=1,\ldots,T}$ whose evolution is specified by a dynamic equation $\mathbf{x}_t = \mathbf{f}_t(\mathbf{x}_{t-1}, \mathbf{n}_t^{\mathbf{x}})$ given a set of observations. These observations $\{\mathbf{y}_t\}_{t=1,\ldots,T}$, are related to the states by $\mathbf{y}_t = \mathbf{h}_t(\mathbf{x}_t, \mathbf{n}_t^{\mathbf{y}})$. Usually, $\mathbf{f}_t$ and $\mathbf{h}_t$ are vector-valued and time-varying transition functions, and $\mathbf{n}_t^{\mathbf{x}}$ and $\mathbf{n}_t^{\mathbf{y}}$ are noise sequences, independent and identically distributed. All these equations are usually considered in a probabilistic way and their computation is decomposed in two main steps. First the prediction of the density function $p(\mathbf{x}_t|\mathbf{y}_{1:t-1}) = \int_{\mathbf{x}_{t-1}} p(\mathbf{x}_t|\mathbf{x}_{t-1})p(\mathbf{x}_{t-1}|\mathbf{y}_{1:t-1})d\mathbf{x}_{t-1}$ with $p(\mathbf{x}_t|\mathbf{x}_{t-1})$ the prior density related to transition function $\mathbf{f}_t$, and then a correction step $p(\mathbf{x}_t|\mathbf{y}_{1:t}) \propto p(\mathbf{y}_t|\mathbf{x}_t)p(\mathbf{x}_t|\mathbf{y}_{1:t-1})$ with $p(\mathbf{y}_t|\mathbf{x}_t)$ the likelihood density

S. Benferhat and J. Grant (Eds.): SUM 2011, LNAI 6929, pp. 525–538, 2011.
© Springer-Verlag Berlin Heidelberg 2011

related to the measurement function $\mathbf{h}_t$. When functions $\mathbf{f}_t$ and $\mathbf{h}_t$ are linear, or linearizable, and when distributions are Gaussian or mixtures of Gaussians, sequence $\{\mathbf{x}_t\}_{t=1,\ldots,T}$ can be computed analytically by Kalman, Extended Kalman or Unscented Kalman Filters [4]. Unfortunately, most vision tracking problems involve nonlinear functions and non-Gaussian distributions. In such cases, tracking methods based on Particle Filters (PF) [4,6], also called Sequential Monte Carlo Methods (SMC), can be applied under very weak hypotheses: their principle is not to compute the parameters of the distributions, but to approximate these distributions by a set of N weighted samples $\{\mathbf{x}_t^{(i)}, w_t^{(i)}\}$, also called *particles*, corresponding to hypothetical state realizations. As optimal filtering approaches do, PF consists of two main steps: (i) a prediction of the object state in the scene (using previous observations), that consists of propagating the set of particles $\{\mathbf{x}_t^{(i)}, w_t^{(i)}\}$ according to a *proposal function* $q(\mathbf{x}_t|\mathbf{x}_{0:t-1}^{(i)}, \mathbf{y}_t)$, followed by (ii) a correction of this prediction (using a new available observation), that consists of *weighting* the particles w.r.t. a *likelihood function*, so that $w_t^{(i)} \propto w_{t-1}^{(i)} p(\mathbf{y}_t|\mathbf{x}_t^{(i)}) \frac{p(\mathbf{x}_t^{(i)}|\mathbf{x}_{t-1}^{(i)})}{q(\mathbf{x}_t|\mathbf{x}_{0:t-1}^{(i)}, \mathbf{y}_t)}$, with $\sum_{i=1}^N w_t^{(i)} = 1$. Particles can then be resampled, so that those with highest weights are duplicated, and those with lowest weights are removed. The estimation of the posterior distribution is then given by $\sum_{i=1}^N w_t^{(i)} \delta_{\mathbf{x}_t^{(i)}}(\mathbf{x}_t)$, where $\delta_{\mathbf{x}_t^{(i)}}$ are Dirac masses centered on particles $\mathbf{x}_t^{(i)}$. There exist many models of PF, each having its own advantages. Unfortunately, the computational cost of PF highly depends on the number of dimensions of the state space and, for large state and observation spaces, it may be unrealistically high due to the large number of particles needed to approximate the distributions and to the costs of computing weights $w_t^{(i)}$.

In this paper, we propose to exploit conditional independences in the state space to transform by *swapping* processes the weighted particle set provided by the correction step of PF into a "new particle set" better focusing on high peaks of the posterior distribution. This enables to deal with high-dimensional state spaces by reducing the needed number of particles while increasing the accuracy of the estimation of the probability distribution of the tracked object's state. This paper is organized as follows. Section 2 gives a short overview of the existing approaches that try to solve the high-dimensionality problem by exploiting conditional independences to decompose probabilistic computations. Section 3 recalls the Partitioned Sampling approach and, then, details our approach. Section 4 gives experimental results on challenging synthetic video sequences. Finally, concluding remarks and perspectives are given in Section 5.

2 Exploiting Conditional Independences for Tracking

It has been shown in [12] that the number of particles needed to track an object grows exponentially with the dimension of the state space of this object. For problems of articulated object or multiple object tracking, the state space may have very high dimensions, which makes PF unusable for real-time tracking.

Several methods aim at reducing the number of necessary particles by exploiting conditional independences in the state space to divide it into small parts.

Partitioned Sampling (PS) [11] is one of the most popular among these methods. It exploits the fact that, in many problems, both the system dynamics and the likelihood function are decomposable over small subspaces. The key idea, that will be described in Section 3, is then to substitute the application of one PF over the whole state space by a sequence of applications of PF over these small subspaces, thus significantly speeding-up the process. However, for the articulated object tracking purpose, PS suffers from numerous resampling steps that increase noise and decrease the tracking accuracy over time.

The same kind of decomposition is exploited in [9] in the context of a general PF for Dynamic Bayesian Networks (DBN). Here, the proposal distributions of the prediction step is decomposed as the product of the conditional distributions of all nodes of the current time slice in the network. The prediction step then follows the topological order of the nodes of the current time slice of the DBN and uses for each node its conditional probability as the proposal distribution. This allows to integrate the current observations into the proposal distribution.

In [15], the sampling idea of [9] is combined with that of resampling proposed in [11] to create a PF algorithm fitted for DBNs. This algorithm can be seen as a generalization of PS. By following a DBN topological order for sampling and by resampling the particles each time an observed node is processed, particles with low likelihood for one subspace are discarded just after the instantiation of this subspace due to the resampling step, whereas particles with high likelihood are multiplied. This has the same effect as *weighted resampling* in PS.

One of the most recent and promising approach that uses a decomposition technique is the nonparametric Belief Propagation algorithm [17,8]. It combines the PF framework with the well-known Loopy Belief Propagation algorithm [14] for speeding-up computations (but at the expense of approximations). It has been successfully applied on many problems of high dimensions [16,2,7]

Another popular approach is the Rao-Blackwellized Particle Filter for DBN (RBPF) [5]. By using a natural decomposition of the conditional probability, RBPF decomposes the state space into two parts that fulfill the following condition: the conditional distribution of the second part given the first part can be estimated using classical techniques such as Kalman filter. The distribution of the first part is then estimated using PF and the conditional distribution of the second part given the first one is estimated using Kalman filter. As the dimension of the first part is smaller than that of the whole state space, the sampling step of particle filter for the first part needs fewer particles and the variance of the estimation can be reduced. Though RBPF is very efficient for reducing the high dimension of the problem, it can not be applied on all DBNs because the state space cannot always be decomposed into two parts fulfilling the condition.

The framework introduced in [3] is somewhat related to ours. This is a parallel PF for DBNs that uses the same decomposition of the joint probability as a Bayesian Network (BN) to reduce the number of particles required for tracking. The state space is divided into several subspaces that are in some respect relatively independent. The particles for these subspaces can then be generated independently using different proposal densities. This approach offers a very flexible way of choosing the proposal density for sampling each subspace. However the definition of different subspaces requires the DBN to have a particular independence structure, limiting the generalization of this algorithm. In our paper, we address more general problems where no such independences hold. We focus on PS [11,12] for its simplicity and generalization potential. In [1], PS was proved to be one of the best algorithm for tracking problems of high dimension. We believe that PS can be improved by better exploiting the independences in DBNs. This idea will be presented in the next section.

3 Proposed Approach

3.1 Partitioned Sampling (PS)

PS is an effective Particle Filter (PF) designed for tracking complex objects with large state space dimensions using only a reduced number of particles. Its key idea is to divide the state space into an appropriate set of partitions and to apply sequentially a PF on each partition, followed by a specific "weighted resampling" ensuring that the sets of particles represent the joint distribution of the whole state space and are focused on its peaks.

Let $g : \mathcal{X} \mapsto \mathbb{R}$ be any strictly positive continuous function, with $\mathcal{X}$ the state space. Given a set of particles $\mathcal{P}_t = \{\mathbf{x}_t^{(i)}, w_t^{(i)}\}_{i=1}^N$ with weights $w_t^{(i)}$, weighted resampling proceeds as follows: let ρ_t be defined as $\rho_t(i) = g(\mathbf{x}_t^{(i)}) / \sum_{j=1}^N g(\mathbf{x}_t^{(j)})$ for $i = 1, \ldots, N$. Select independently indices $k_1, \ldots, k_N$ according to probability ρ_t. Finally, construct a new set of particles $\mathcal{P}'_t = \{\mathbf{x'}_t^{(i)}, w'^{(i)}_t\}_{i=1}^N$ defined by $\mathbf{x'}_t^{(i)} = \mathbf{x}_t^{(k_i)}$ and $w'^{(i)}_t = w_t^{(k_i)} / \rho_t(k_i)$. MacCormick [10] shows that $\mathcal{P}'_t$ represents the same probability distribution as $\mathcal{P}_t$ while focusing on the peaks of g.

PS's key idea is to exploit some decomposition of the system dynamics w.r.t. subspaces of the state space in order to apply PF only on those subspaces. This leads to a significant reduction in the number of particles needed for tracking. So, assume that state space $\mathcal{X}$ can be partitioned as $\mathcal{X} = \mathcal{X}^1 \times \cdots \times \mathcal{X}^P$ as well as observation space $\mathcal{Y} = \mathcal{Y}^1 \times \cdots \times \mathcal{Y}^P$. For instance, a system representing a hand could be defined as $\mathcal{X}^{\text{hand}} = \mathcal{X}^{\text{palm}} \times \mathcal{X}^{\text{thumb}} \times \mathcal{X}^{\text{index}} \times \mathcal{X}^{\text{middle}} \times \mathcal{X}^{\text{ring}} \times \mathcal{X}^{\text{little}}$. Assume in addition that the dynamics of the system follows this decomposition, i.e., that:

$$f_t(\mathbf{x}_{t-1}, n_t^{\mathbf{x}}) = f_t^P \circ f_t^{P-1} \circ \cdots \circ f_t^2 \circ f_t^1(\mathbf{x}_{t-1}), \tag{1}$$

where $\circ$ is the usual function composition operator and where each function $f_t^i : \mathcal{X} \mapsto \mathcal{X}$ modifies the particles' states only on subspace $\mathcal{X}^i$ [1].

The PF scheme consists of resampling particles, of propagating them using proposal function f_t and, finally, of updating their weights using the observations at hand. Here, the same result is achieved by substituting the f_t propagation

[1] Note that, in [10], functions f_t^i are more general since they can modify states on $\mathcal{X}^i \times \cdots \times \mathcal{X}^p$. However, in practice, particles are often propagated one $\mathcal{X}^j$ at a time.

step by the sequence of applications of the f_t^i as given in Eq. (1), each one followed by a weighted resampling that produces new particles sets focused on the peaks of a function g. To be effective, PS thus needs g to be peaked with the same region as the posterior distribution restricted to $\mathcal{X}^i$. When the likelihood function decomposes as well on subsets $\mathcal{X}^i$, i.e., when:

$$p(\mathbf{y}_t|\mathbf{x}_t) = \prod_{i=1}^{P} p^i(\mathbf{y}_t^i|\mathbf{x}_t^i), \tag{2}$$

where $\mathbf{y}_t^i$ and $\mathbf{x}_t^i$ are the projections of $\mathbf{y}_t$ and $\mathbf{x}_t$ on $\mathcal{Y}^i$ and $\mathcal{X}^i$, weighted resampling focusing on the peaks of the posterior distribution on $\mathcal{X}^i$ can be achieved by first multiplying the particles' weights by $p^i(\mathbf{y}_t^i|\mathbf{x}_t^i)$ and, then, by performing a usual resampling. Note that Eq. (2) naturally arises when tracking articulated objects. This leads to the condensation diagram given in Fig. 1, where operations "$*f_t^i$" refer to propagations of particles using proposal function f_t^i as defined above, "$\times p_t^i$" refer to the correction steps where particle weights are multiplied by $p^i(\mathbf{y}_t^i|\mathbf{x}_t^i)$ (see Eq. (2)), and "$\sim$" refer to usual resamplings. MacCormick and Isard show that this diagram produces mathematically correct results [12].

3.2 Swapping-Based Partition Sampling (SBPS)

The hypotheses used by PS can best be explained on a dynamic Bayesian network (DBN) representing the conditional independences between random variables of states and observations [13]. Assume for instance that an object to be tracked is composed of 3 parts: a torso, a left arm and a right arm. Let $\mathbf{x}_t^1, \mathbf{x}_t^2, \mathbf{x}_t^3$ represent these parts respectively. Then, the probabilistic dependences between these variables and their observations $\mathbf{y}_t^1, \mathbf{y}_t^2, \mathbf{y}_t^3$, can be represented by the DBN of Fig. 2. In this figure, Eq. (2) implicitly holds because, conditionally to states $\mathbf{x}_t^i$, observations $\mathbf{y}_t^i$ are independent of the other random variables. In addition, the probabilistic dependences between substates $\mathbf{x}_t^1, \mathbf{x}_t^2, \mathbf{x}_t^3$ suggest that the dynamics of the system is decomposable on $\mathcal{X}^1 \times \mathcal{X}^2 \times \mathcal{X}^3$. As a consequence, the condensation diagram of Fig. 1 can be exploited to track the object.

Through the d-separation criterion [14], DBNs offer a strong framework for analyzing probabilistic dependences among sets of random variables. By this criterion, it can be remarked that, on Fig. 2, $\mathbf{x}_t^2$ and $\mathbf{x}_t^3$ are independent conditionally to $(\mathbf{x}_t^1, \mathbf{x}_{t-1}^2)$ and $(\mathbf{x}_t^1, \mathbf{x}_{t-1}^3)$ respectively. As a consequence, for each particle, PS's propagations/corrections over subspaces $\mathcal{X}^2$ and $\mathcal{X}^3$ can be performed independently since, in this case, $\mathbf{x}_t^1$ and $\mathbf{x}_{t-1}$ are known and fixed. This suggests the new condensation diagram of Fig. 3.

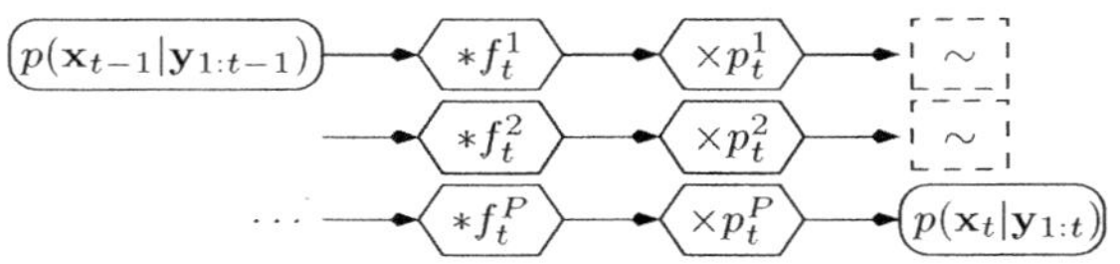

Fig. 1. Partitioned Sampling condensation diagram

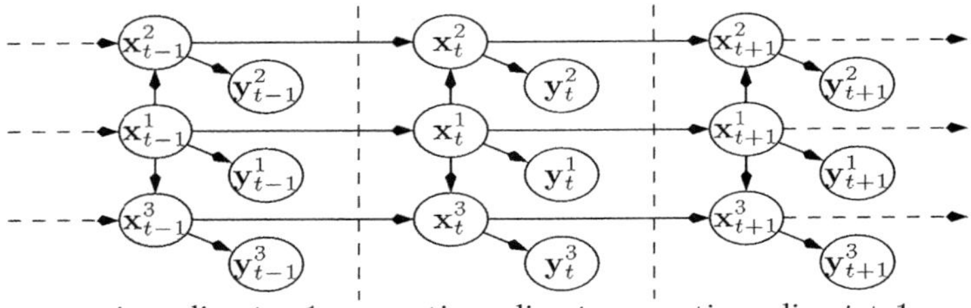

Fig. 2. A Dynamic Bayesian network for body tracking

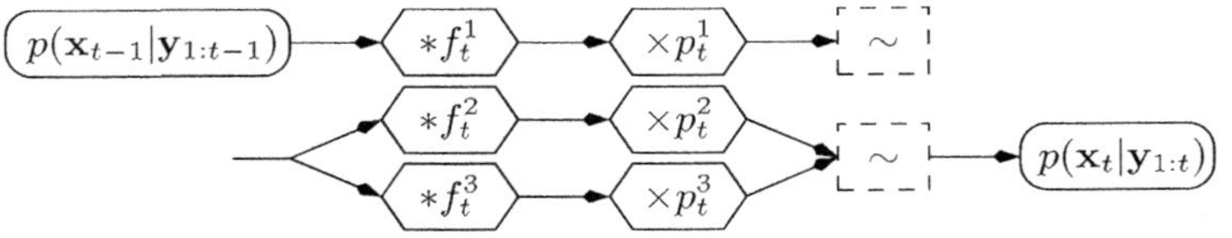

Fig. 3. Condensation diagram exploiting conditional independences

Proposition 1. *The set of particles resulting from the Particle Filter of Fig. 3 represents probability distribution* $p(\mathbf{x}_t|\mathbf{y}_{1:t})$.

Proof. Propagations performed in parallel concern subspaces that are probabilistically independent. So, they produce the same result as if they were performed sequentially. Hence, the only difference between PS and the PF of Fig. 3 is that fewer resamplings are performed. But resamplings do not change the probability distributions represented by the particle sets. Hence the result. □

There are two major differences between PS and the PF of Fig. 3: the latter performs fewer resamplings, thus introducing less noise in the particle set and, more importantly, it enables to produce better fitted particles by swapping their subparts. Actually, consider again our body tracking example and assume that we generated the 3 particles $\mathbf{x}_t^{(i)}$ of Fig. 4.a where $\mathcal{X}^1$ is the middle part of the object and $\mathcal{X}^2$ and $\mathcal{X}^3$ are its left and right parts respectively, and where the shaded areas represent the object's true state. According to the DBN of Fig. 2, for fixed values of $\mathbf{x}_{1:t}^1$, the sets of left and right parts of the particles represent $p(\mathbf{x}_t^2, y_{1:t}^2|\mathbf{x}_{1:t}^1)$ and $p(\mathbf{x}_t^3, y_{1:t}^3|\mathbf{x}_{1:t}^1)$ respectively (summing out variables $\mathbf{x}_j^2, \mathbf{x}_j^3$ from the DBN). Hence, after permuting the values of the particles on $\mathcal{X}^2$ (resp. $\mathcal{X}^3$) for a fixed value of $\mathbf{x}_{1:t}^1$, distribution $p(\mathbf{x}_t^2, y_{1:t}^2|\mathbf{x}_{1:t}^1)$ (resp. $p(\mathbf{x}_t^3, y_{1:t}^3|\mathbf{x}_{1:t}^1)$) remains unchanged. A fortiori, this does not affect the representation of the joint posterior distribution $\int p(\mathbf{x}_{1:t}^1, y_{1:t}^1)p(\mathbf{x}_t^2, y_{1:t}^2|\mathbf{x}_{1:t}^1)p(\mathbf{x}_t^3, y_{1:t}^3|\mathbf{x}_{1:t}^1)d\mathbf{x}_{1:t-1}^1 = p(\mathbf{x}_t, \mathbf{y}_{1:t})$. On Fig. 4.a, particles $\mathbf{x}_t^{(1)}$ and $\mathbf{x}_t^{(3)}$ have the same state on $\mathcal{X}^1$. Thus their right parts can be permuted, resulting in the new particle set of Fig. 4.b. Remark that we substituted 2 particles, $\mathbf{x}_t^{(1)}$ and $\mathbf{x}_t^{(3)}$, which had low weights due to their bad estimation of the object's right or left part states, by one particle

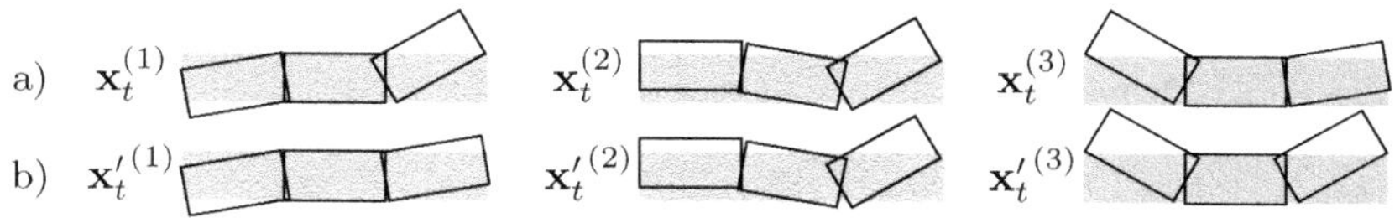

Fig. 4. The particle swapping scheme: a) before swapping; b) after swapping

$\mathbf{x}_t'^{(1)}$ with a high weight and another one $\mathbf{x}_t'^{(3)}$ with a very low weight. After resampling, the later will most probably be discarded and, therefore, swapping will have focused particles on the peaks of the posterior distribution. Note however that not all permutations are allowed: for instance, none can involve particle $\mathbf{x}_t^{(2)}$ because its center part differs from that of the other particles.

Let us now formulate SBPS. Assume again that state space $\mathcal{X}$ is decomposed as $\mathcal{X} = \mathcal{X}^1 \times \cdots \times \mathcal{X}^P$ and that the probabilistic dependences between all random variables $\mathbf{x}_t^i$ and $\mathbf{y}_t^i$, $i = 1, \ldots, P$, are represented by a DBN $\mathcal{G}$. Let $\{P_1, \ldots, P_K\}$ be a partition of $\{1, \ldots, P\}$ such that, for all i, $\{\mathbf{x}_t^j\}_{j \in P_i}$ are mutually independent conditionally to $(\cup_{h=1}^{i-1} \cup_{k \in P_h} \mathbf{x}_t^k) \cup \mathbf{x}_{t-1}$. Such sets are easily identified by d-separation on DBN $\mathcal{G}$ [14]. By definition, after processing $P_1, \ldots, P_{i-1}$, all the variables of each set P_i can be processed independently. Denote the elements of P_i by $\{i_{P_i}^1, \ldots, i_{P_i}^{k_i}\}$. Then, the SBPS algorithm can be described by the condensation diagram of Fig. 5, where operations "$\rightleftharpoons^{P_i}$" refer to the particle subpart swappings briefly described previously. Remark that, after the resampling operation of part P_i, the high weighted particles will be duplicated, which will enable swapping when processing next part P_{i+1}. Swappings need however to be further formalized. Let $\mathbf{pa}(X_t^i)$ denote the parents of node X_t^i in $\mathcal{G}$ in time slice t and let $\text{Link}(X_t^i)$ be the set of nodes in all time slices such that there exists an undirected path in $\mathcal{G}$ linking them to X_t^i while not passing through any node in $(\mathbf{pa}(X_t^i))_{1:t}$. Assume now that SBPS was executed up to (but not including) operation $\rightleftharpoons^{P_k}$. Let $r \in P_k$ be some part of the object. Then, for each value of $(\mathbf{pa}(\mathbf{x}_t^r))_{1:t}$, substates $\mathbf{x}_t^r$ of the particles represent $p(\mathbf{x}_t^r, \mathbf{y}_{1:t}^r | \mathbf{pa}(\mathbf{x}_t^r)_{1:t})$. Thus, permuting substates $\mathbf{x}_t^r$ among particles with the same value of $(\mathbf{pa}(\mathbf{x}_t^r))_{1:t}$ does not change this distribution. However, if $\mathbf{x}_t^s$ is a child of $\mathbf{x}_t^r$, then not permuting similarly substates $\mathbf{x}_t^s$ of these particles changes $p(\mathbf{x}_t^s, \mathbf{y}_{1:t}^s | \mathbf{pa}(\mathbf{x}_t^s)_{1:t})$, hence resulting in incorrect computations. More generally, it is easily shown that all the values of the substates in $\text{Link}(\mathbf{x}_t^r)$ (and only those values) need be permuted to ensure that no conditional probability is affected by the swapping. By conditional independences w.r.t. $\mathbf{pa}(\mathbf{x}_t^r)_{1:t}$, the product of all these distributions is the joint probability. So, operation $\rightleftharpoons^{P_k}$ refers to permuting some values of $\text{Link}(\mathbf{x}_t^r)$ for some $r \in P_k$ and among particles having the same substate $\mathbf{pa}(\mathbf{x}_t^r)_{1:t}$. In practice, whenever $\mathbf{pa}(\mathbf{x}_t^r)_t$ is identical for two particles, the continuous nature of the state space make it highly probable that one particle is a copy of the other due to resampling. Hence, their $\mathbf{pa}(\mathbf{x}_t^r)_{1:t}$ values should also be identical. So, our implementation approximates the posterior distributions by performing swapping for fixed values of $\mathbf{pa}(\mathbf{x}_t^r)_t$ instead of $\mathbf{pa}(\mathbf{x}_t^r)_{1:t}$.

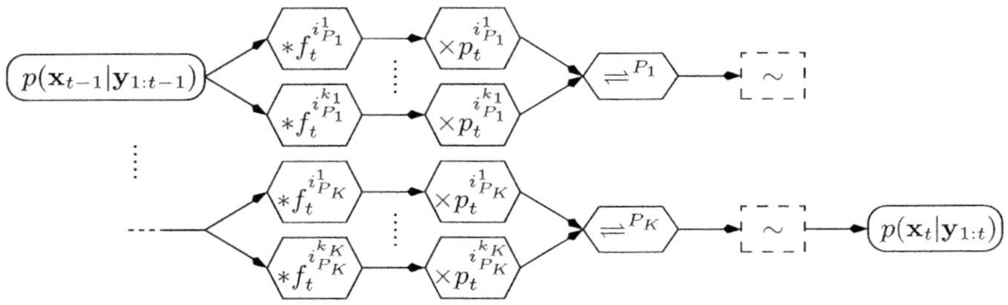

Fig. 5. Swapping-based Partitioned Sampling condensation diagram

Proposition 2. *The set of particles resulting from SBPS represents* $p(\mathbf{x}_t|\mathbf{y}_{1:t})$.

Proof. If we do not swap particles, the proof follows from Proposition 1. If some swapping occurs, say on substate $\mathbf{x}_t^r$, then, as mentioned previously, distribution $p(\mathbf{x}_t^r, \mathbf{y}_{1:t}^r|\mathbf{pa}(\mathbf{x}_t^r)_{1:t})$ remains unchanged. By definition, swapping also occurs on the neighbors $\mathbf{x}_t^s$ of $\mathbf{x}_t^r$ and their neighbors, so that the conditional distribution of $\mathbf{x}_t^s$ remains unchanged. By induction on neighborhoods in $\mathcal{G}$, no conditional distribution is ever affected by swapping and the result follows. □

Finally, let us show how $\rightleftharpoons^{P_k}$ can determine attractive swappings, i.e., how high-peaked regions can be discovered. For simplicity, we will only describe swapping on the example of Fig. 6, but the principle can easily be generalized. Assume that $\mathcal{X} = \mathcal{X}^1 \times \mathcal{X}^2 \times \mathcal{X}^3 \times \mathcal{X}^4$, where parts are defined from left to right. In addition, assume that $P_1 = \{2\}$, $P_2 = \{3\}$ and $P_3 = \{1, 4\}$, i.e., P_3 corresponds to the extremal sides of the object. Let us describe operation $\rightleftharpoons^{P_3}$. Just before propagating part P_3, SBPS has constructed particles with equal weights (due to its resamplings). In the rectangles of parts $\mathcal{X}^2$ and $\mathcal{X}^3$, identical letters indicate identical substate values (e.g., particles 1 and 2 have the same value on $\mathcal{X}^2$). Just before executing $\rightleftharpoons^{P_3}$, particles have been propagated on the extremal sides of the object (resulting in Fig. 6.a) and operations $\times p_t^1$ and $\times p_t^4$ have updated their weights. These weights (unormalized for clarity reasons) are displayed inside the rectangles corresponding to $\mathcal{X}^1$ and $\mathcal{X}^4$ and the total weights of the particles are shown on their right side (According to the DBN, these weights are equal to $p^1(\mathbf{y}_t|\mathbf{x}_t^{1,(i)}) \times p^4(\mathbf{y}_t|\mathbf{x}_t^{4,(i)})$). To find the best swappings, we exploit the data structure given in Fig. 6.b: the circle nodes correspond to the values of the particles on $\mathbf{pa}(\mathbf{x}_t^1)_t$ and $\mathbf{pa}(\mathbf{x}_t^4)_t$. The values within rectangles correspond to the set of values and weights of the particles on $\mathbf{x}_t^1$ and $\mathbf{x}_t^4$ for each value of their parents $\mathbf{pa}(\mathbf{x}_t^1)_t$ and $\mathbf{pa}(\mathbf{x}_t^4)_t$. Finally, there exists an edge between two circles if and only if there exists at least one particle with the values of both circles. For instance, edge (A, D) is induced by particle 2. By definition, all the rectangle values that are attached to a given circle (e.g., 5 and 2) can be swapped since they have the same parent values in the DBN. As there exists an edge between two circles if and only if there exists a particle with both circle values, we can conclude that any value attached to one such circle can be combined by swapping with any value attached to the other circle. For instance, 5, which is attached

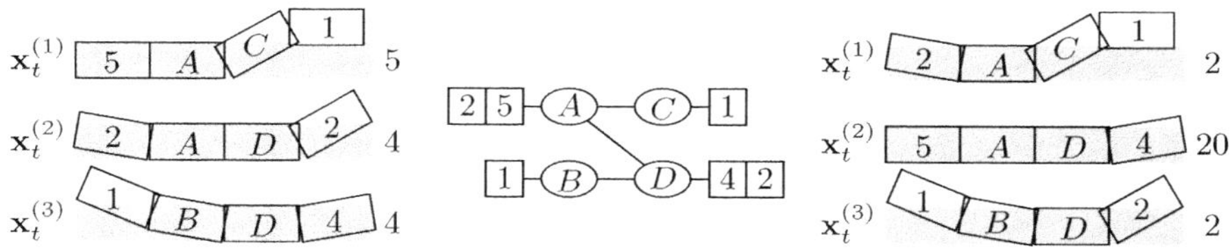

a) before swapping b) swapping data structure c) after swapping

Fig. 6. Details on the swapping process

to A, can be combined with 1 (attached to C) or 4 or 2 (attached to D). To get the best particle, we shall only consider combinations with the highest rectangle values, hence, for the pair (A, D), we shall only consider combining 5 with 4. The same shall be done for nodes attached to B and the first new particle constructed is the one with the highest product (here 5×4 from pair (A, D)). Once this combination has been used, remove values 5 and 4 from the graph and iterate. When no more rectangle is attached to a circle, this one is removed from the graph. This process is fast and very effective to produce high-weight particles.

4 Experimental Results

We have tested our method and compared it to PS on synthetic video sequences because we wanted to highlight its interest in terms of dimensionality reduction and tracking accuracy without having to take into account specific properties of images (noise, *etc.*). Moreover, it is possible to simulate specific motions and then to test and compare with accuracy our method with PS. For that, we have generated our own synthetic video sequences, each one containing 300 frames, showing two different kinds of articulated objects: chains or squids. A chain is the concatenation of P colored rectangles (P is also called the length of the object), and a squid is made of two chains crossing in their middle part: in a sense, a chain can be defined by a central rectangle, and two tentacles starting from this central part, while a squid has four tentacles starting from its central part. Chains and squids are translating and distorting over time, see examples of squids in Fig. 10 and 11, and of a chain in Fig. 8. The goal, here, is to observe the capacity of PS and SBPS to deal with articulated objects composed of a varying number of parts and subject to weak or strong motions.

The tracked articulated object is modeled by a set of P rectangles whose corners are labeled $C_1, \ldots, C_4$. The state space contains parameters describing each rectangle, and is defined by $\mathbf{x}_t = \{\mathbf{x}_t^1, \mathbf{x}_t^2, \ldots, \mathbf{x}_t^P\}$, with $\mathbf{x}_t^i = \{x_t^i, y_t^i, \theta_t^i\}$, where (x_t^i, y_t^i) denotes the coordinates of the center of the ith rectangle, and θ_t^i is its orientation, $i = 1, \ldots, P$. A particle $\mathbf{x}_t^{(j)} = \{\mathbf{x}_t^{1,(j)}, \mathbf{x}_t^{2,(j)}, \ldots, \mathbf{x}_t^{P,(j)}\}$, $j = 1, \ldots, N$, is thus a possible configuration of an articulated object. In the first frame, particles are uniformly generated around the object. During the prediction step, particles are propagated following a random walk whose variance

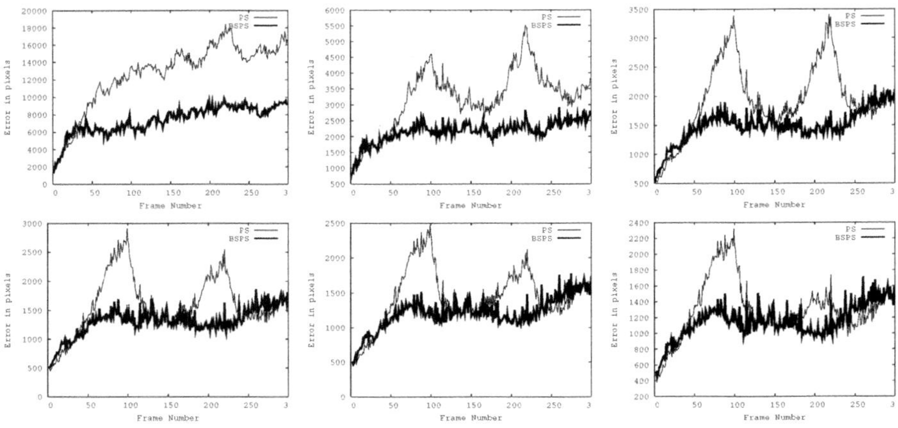

Fig. 7. Tracking errors for PS and SBPS approaches for a chain object of length $P = 11$ with, from left to right and from top to bottom, $N = 5, 10, 20, 30, 40, 50$ particles. Motions in frames [50, 100] and [150, 200] are stronger than in the other frames.

has been manually chosen. The weights of the particles are then computed using the current observation (i.e. the current frame). Finally, the particle's weights are given by $w_t^{(j)} = w_{t-1}^{(j)} p(\mathbf{y}_t | \mathbf{x}_t^{(j)}) \propto w_{t-1}^{(j)} e^{-\lambda d^2}$, with d the Bhattacharyya distance between the histograms of the target (prior) and the reference (previously estimated) regions. Parameter λ was set to 50 in all our tests.

In both approaches, the articulated object's joint distribution is estimated by starting from its center part. PS then propagates and corrects particles part after part to derive a global estimation of the object. SBPS considers tentacles of objects as totally independent, and thus propagates, swaps and corrects simultaneously in all tentacles. PS and SBPS are compared by measuring the tracking error as the distance between the ground truth and the estimated articulated object at each instant. This distance is given by the sum of the Euclidean distances between each corner C_i of each estimated rectangle and its corresponding corner C_i of the same rectangle in the ground truth. All the results presented in this paper correspond to a mean over 100 runs.

Our first test concerns the tracking of a chain object of length $P = 11$. To test the stability of our approach, we have generated video sequences in which motions during two specific temporal intervals (frames [50, 100] and [150, 200]) are strong. Comparative results of tracking errors of PS and SBPS are reported in Fig. 7, for different numbers N of particles. We can see on these graphs that SBPS always outperforms PS, which shows its stability especially when the motion becomes strong: the tracking error drastically increases for PS whereas that of SBPS is relatively stable. Visual results of tracking are shown on Fig. 8 with $N = 30$ particles for this object: the estimation of the articulated object is represented by the concatenated white rectangles. Over 100 runs, the tracking error resulting from our approach was reduced by 19% as compared to PS.

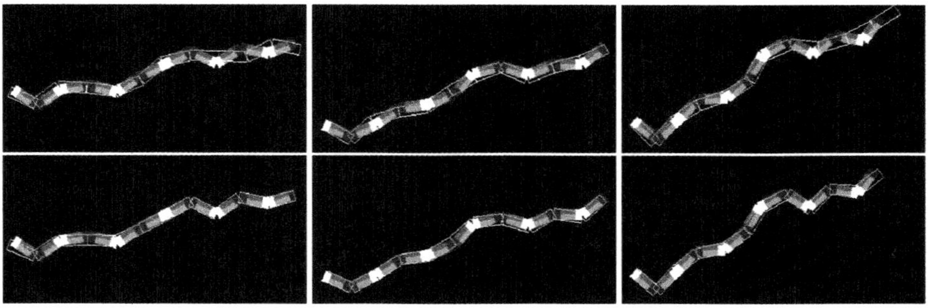

Fig. 8. Zooms on tracking results obtained for PS (top line) and SBPS (bottom line) on frames 100, 150 and 200, for a chain object of length $P = 11$ with $N = 30$ particles. White articulated objects represent the mean estimations of the articulated object. Mean tracking error: 1670 pixels for PS, and 1286 pixels for SBPS.

Table 1. Comparison of tracking mean errors (in pixels) over all the sequences obtained by PS and SBPS depending on the object (chain or squid), its length P, and the number N of particles. $\searrow$ %$= \left(1 - \frac{SBPS}{PS}\right) \times 100$ is the error reduction percentage using SBPS.

		Chain					Squid				
		$P=3$	$P=5$	$P=7$	$P=9$	$P=11$	$P=5$	$P=9$	$P=13$	$P=17$	$P=21$
$N=5$	PS	514	1565	3440	8546	12666	1999	18473	37710	66659	77864
	SBPS	469	1480	1706	6638	7374	812	6408	9125	12397	13075
	$\searrow$ %	**-9%**	**-6%**	**-50%**	**-33%**	**-42%**	**-60%**	**-76%**	**-76%**	**-82%**	**-84%**
$N=10$	PS	187	315	1302	1528	3199	289	894	1862	2339	7786
	SBPS	167	293	1044	1215	2161	193	627	746	1407	2225
	$\searrow$ %	**-11%**	**-7%**	**-20%**	**-21%**	**-33%**	**-34%**	**-30%**	**-60%**	**-40%**	**-72%**
$N=20$	PS	136	193	949	1529	1944	153	596	519	1046	2610
	SBPS	125	185	813	1192	1495	114	428	405	696	1374
	$\searrow$ %	**-9%**	**-5%**	**-15%**	**-23%**	**-24%**	**-26%**	**-29%**	**-22%**	**-34%**	**-48%**
$N=30$	PS	123	164	819	1313	1606	120	510	404	772	1919
	SBPS	112	160	706	1069	1309	97	377	327	527	1127
	$\searrow$ %	**-9%**	**-3%**	**-14%**	**-19%**	**-19%**	**-20%**	**-27%**	**-20%**	**-32%**	**-42%**
$N=40$	PS	115	159	768	1199	1440	108	467	349	666	1615
	SBPS	108	147	671	997	1211	91	351	287	460	1016
	$\searrow$ %	**-7%**	**-8%**	**-13%**	**-17%**	**-16%**	**-16%**	**-25%**	**-18%**	**-31%**	**-38%**
$N=50$	PS	112	141	735	1109	1306	102	426	317	592	1534
	SBPS	105	138	648	956	1151	88	337	265	428	943
	$\searrow$ %	**-7%**	**-3%**	**-12%**	**-14%**	**-12%**	**-14%**	**-21%**	**-17%**	**-28%**	**-39%**

Table 1 summarizes all the tests performed on different video sequences showing chains of length $P = 3, 5, 7, 9, 11$ or squid of length $P = 5, 9, 13, 17, 21$, for values $N = 5, 10, 20, 30, 40, 50$. Tracking errors (in pixels) over all the sequences are reported for both approaches, and the percentage of reduction of tracking error obtained with our approach, denoted by $\searrow$ %, is computed as $\left(1 - \frac{SBPS}{PS}\right) \times 100$. As another example, we also reported the tracking errors for a squid object of size $P = 17$ in Fig. 9. We can see that our approach always decreases the tracking

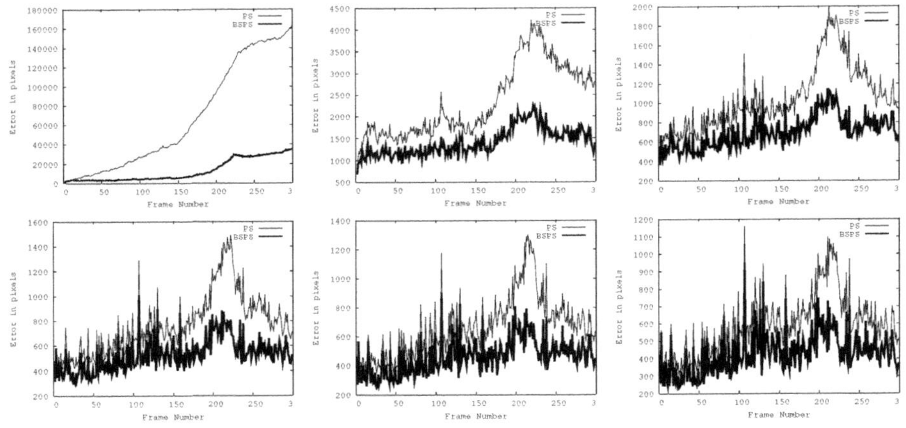

Fig. 9. Tracking errors for PS and SBPS approaches for a squid object of length $P = 17$ with, from left to right and from top to bottom, $N = 5, 10, 20, 30, 40, 50$ particles

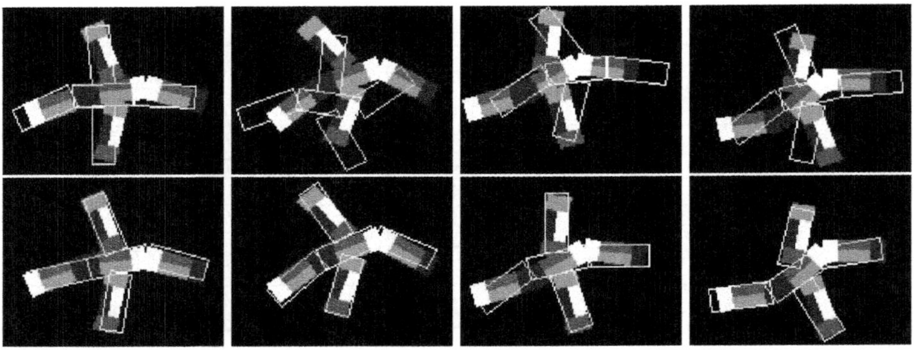

Fig. 10. Zooms on tracking results obtained for PS (top line) and SBPS (bottom line) on frames 50, 100, 200 and 250, for a squid of length $P = 5$ with $N = 5$ particles. White articulated objects represent the mean estimations of the articulated object. Mean tracking error: 1454 pixels for PS, and 403 pixels for SBPS.

error. This is especially noticeable for high values of P. Even for small objects ($P = 3$) and a large number of particle ($N = 50$), which should be highly sufficient to provide good tracking results, SBPS outperforms PS, decreasing the error by 7%. The visual tracking results for a squid object of size $P = 5$ using $N = 5$ particles are given in Fig. 10: mean tracking error is 1454 pixels for PS , and 503 pixels for SBPS. All these results show why exploiting both independence between the different object's parts and subpart swapping is highly efficient: "much better" particles are constructed which, in turn, allows to better estimate the joint probability of the articulated object. This advantage is illustrated by Fig. 11 for a squid object of length $P = 13$ (frames 50, 100, 150 and 200): we have voluntarily

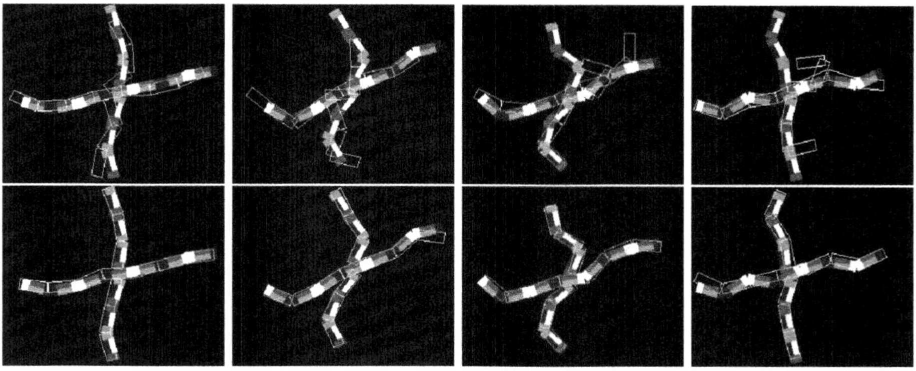

Fig. 11. Zooms on the best particles (i.e. with the highest weight) of PS (top line) and SBPS (bottom line) approaches, drawn in white for a squid object of length $P = 13$, with $N = 5$. From left to right, frames 50, 100, 150 and 200

Table 2. Computation times (in sec.) of PS and SBPS for the tracking of a chain and of a squid of lengths $P = 9$ and $P = 17$ respectively, for different values of N

		$N = 5$	$N = 10$	$N = 20$	$N = 30$	$N = 40$	$N = 50$
Chain: $P = 9$	PS	0.70	1.21	2.18	3.22	4.2	5.15
	SBPS	0.72	1.30	2.31	3.45	4.43	5.41
Squid: $P = 17$	PS	1.18	2.12	4.01	6.02	8.16	10.28
	SBPS	1.20	2.15	4.11	6.21	8.37	10.69

used a small number of particle for tracking ($N = 5$) and have drawn into the frames the "best" particle, i.e. the one with the highest weight. If we compare PS (top line) and SBPS (bottom line), we can see the benefit of swapping: unlike the best particle of SBPS, that of PS totally misses the articulated object.

Finally, Table 2 reports the mean computation times (in seconds), over 100 runs, for tracking two different objects: a chain of length $P = 9$ and a squid of length $P = 17$, depending on the number N of particles. Of course, we can see that PS is faster than SBPS, but previous tests show that SBPS requires fewer particles to provide a tracking as good as PS. For instance, for a chain of length $P = 9$, using SBPS and $N = 20$, in 2.31 seconds, we get similar tracking results than those with PS using $N = 30$ (in 3.22 seconds). Similarly, for a squid of length $P = 17$, using SBPS and $N = 20$, in 4.11 seconds, we get similar tracking results than those with PS using $N = 40$ (in 8.16 seconds).

5 Conclusion

We have introduced a new framework, *Swapping-Based Partitioned Sampling*, exploiting conditional independences to simultaneously propagate, correct and swap particles in independent subspaces. As a result, the particle sets produced are more concentrated on high peaks of the posterior distribution than in the classical Partition Sampling. Thus, our estimations of the probability densities of the tracked object are more accurate. Empirical tests have shown that SBPS always outperforms PS, especially in cases where the object motion is strong and when the dimension of the state space is high (i.e., the number of parts is large). There still remains to validate our approach on real video sequences.

References

1. Bandouch, J., Engstler, F., Beetz, M.: Evaluation of Hierarchical Sampling Strategies in 3D Human Pose Estimation. In: BMVC, pp. 925–934 (2008)
2. Bernier, O., Cheung-Mon-Chan, P., Bouguet, A.: Fast nonparametric belief propagation for real-time stereo articulated body tracking. Computer Vision and Image Understanding 113, 29–47 (2009)
3. Besada-Portas, E., Plis, S.M., Cruz, J.M., Lane, T.: Parallel subspace sampling for particle filtering in dynamic Bayesian networks. In: ECML PKDD, pp. 131–146 (2009)
4. Chen, Z.: Bayesian filtering: from Kalman filters to particle filters, and beyond (2003)
5. Doucet, A., de Freitas, N., Murphy, K.P., Russell, S.J.: Rao-Blackwellised particle filtering for dynamic Bayesian networks. In: UAI, pp. 176–183 (2000)
6. Gordon, N., Salmond, D.J., Smith, A.: Novel approach to nonlinear/non-Gaussian Bayesian state estimation. IEE Proceedings of Radar and Signal Processing 140(2), 107–113 (1993)
7. Ihler, A., Fisher, J., Iii, J.W.F., Willsky, A., Moses, R.: Nonparametric belief propagation for self-calibration in sensor networks. In: ISIPSN, pp. 225–233 (2004)
8. Isard, M.: PAMPAS: real-valued graphical models for computer vision. In: CVPR, pp. 613–620 (2003)
9. Kanazawa, K., Koller, D., Russell, S.: Stochastic simulation algorithms for dynamic probabilistic networks. In: UAI, pp. 346–351 (1995)
10. MacCormick, J.: Probabilistic modelling and stochastic algorithms for visual localisation and tracking. Ph.D. thesis. Oxford University (2000)
11. MacCormick, J., Blake, A.: A probabilistic exclusion principle for tracking multiple objects. In: ICCV, pp. 572–587 (1999)
12. MacCormick, J., Isard, M.: Partitioned sampling, articulated objects, and interface-quality hand tracking. In: Vernon, D. (ed.) ECCV 2000. LNCS, vol. 1843, pp. 3–19. Springer, Heidelberg (2000)
13. Murphy, K.: Dynamic Bayesian Networks: Representation, Inference and Learning. Ph.D. thesis, UC Berkeley, Computer Science Division (2002)
14. Pearl, J.: Probabilistic Reasoning in Intelligent Systems: Networks of Plausible Inference. Morgan Kaufman Publishers, San Francisco (1988)
15. Rose, C., Saboune, J., Charpillet, F.: Reducing particle filtering complexity for 3D motion capture using dynamic Bayesian networks. In: AAAI, pp.1396–1401 (2008)
16. Sigal, L., Isard, M., Sigelman, B.H., Black, M.J.: Attractive people: Assembling loose-limbed models using non-parametric belief propagation. In: NIPS, pp. 1539–1546 (2003)
17. Sudderth, E.B., Ihler, A.T., Isard, M., Freeman, W.T., Willsky, A.S.: Nonparametric belief propagation. Commununications of ACM 53, 95–103 (2010)

A Fuzzy-Based Approach to the Value of Information in Complex Military Environments

Timothy Hanratty[1], Robert J. Hammell II[2], and Eric Heilman[3]

[1] Computational Information, Science Directorate, US Army Research Laboratory
timothy.hanratty@us.army.mil
[2] Department of Computer & Information Sciences, Towson University
rhammell@towson.edu
[3] Computational Information Science Directorate US Army Research Laboratory
Eric.g.heilman@us.army.mil

Abstract. The last several decades have seen an unprecedented increase in the types and amount of information pertaining to the military environment. For the military commander and his staff, separating the important information from the routine has become a primary challenge in calculating the Value of Information (VOI). Wrought with uncertainty and contradiction, new methodologies are required to confront this significant issue. This paper presents an approach for calculating the VOI in complex military environments using fuzzy logic as a method for managing uncertain and imprecise information.

1 Introduction

In a broad sense, the last several decades have seen an unprecedented increase in the types and amount of information pertinent to the military environment. From sophisticated unmanned ground acoustic sensors to open-source RSS news feeds, the military commander and his staff are challenged not only by the established information overload dilemma, but more importantly with separating the important information from the routine. Calculating information importance, termed the value of information (VOI) metric, is a daunting task that is highly dependent upon its application to dynamic situations [1]. Solution flexibility is paramount since VOI understanding must be readily applicable across a disparate range of information types and situational states of affairs.

Towards this end, this paper presents an approach for tackling the calculation of VOI in complex military environments. Specifically, outlined is the use of fuzzy logic as method for managing uncertainty and contradictory information with respect to calculating VOI in complex military environments. The paper will address an understanding of the complexity of the military information domain with emphasis on VOI (section 2); examples of addressing VOI challenges (section 3); an outline of the fuzzy approach along with an associated use case example (section 4); and concluding remarks and future directions (section 5).

S. Benferhat and J. Grant (Eds.): SUM 2011, LNAI 6929, pp. 539–546, 2011.
© Springer-Verlag Berlin Heidelberg 2011

2 Understanding the Domain Challenge

On today's battlefield, information drives action. Personnel must know details about important persons, places and events within their area of operations to address issues ranging from kinetic fights to adjudicating legal disputes to revitalizing a depleted economy. Soldiers at the edge of conflict gather data to support their mission. As Major General Michael Flynn points out:

> "At the battalion level and below, intelligence officers know a great deal about their local Afghan districts but are generally too understaffed to gather, store, disseminate, and digest the substantial body of crucial information that exists outside traditional intelligence channels." [2]

Several factors contribute to the challenge of analyzing VOI within military environments. As depicted in Table 1, bottom-most echelons containing fewer assets and executing more immediate tactical missions require an increasing timeliness of the information.

Table 1. Military Echelons with typical Operational Times and Areas [3]

Echelon	Planning	Execution	Reports / Hr	Area of Operation
Division	Weeks	Week/ days		Province
Brigade	Days	Days	170K	Province/ district
Battalion	Day/hours	Day	56K	District
Company	Hours	Hours	18K	Village
Platoon	Hour/Min	Hour/ Min	6K	Village/ hamlet

At the company level, for example, the decision making cycle is measured in hours. Information quickly becomes irrelevant at this level. On the other hand, as the echelon increases, the scope of military operations and number of information reports grows tremendously. The ability to manage information effectively at higher echelon levels becomes exceedingly difficult.

Today, each of these echelons can supplement human information collection by using a host of automated data gathering devices. For example, the addition of a single unmanned aerial vehicle into the reconnaissance effort with sensors such as full motion video cameras or light detection and ranging (LIDAR) equipment will generate voluminous data feeds measured in terabytes per hour of operation. While this does not impose much of an increase in number of collection entities, the volume of generated data increases rapidly and the ability to merge automated and manually collected data is not yet fully developed. Since "the value of information is largely subjective" [4] [5], estimation of data usefulness remains with the analyst who may not be able to sufficiently fathom the large volume of varied data. Providing commanders and analysts with VOI tools to filter an increasing number of data feeds holds the potential of quickly generating useful intelligence at all echelons.

US Army Field Manual (FM) 3-0, Operations, defines the "instruments of national power" in terms of diplomatic, informational, military, and economic activities, normally referred to as the DIME [6]. Consequently, missions within a counter insurgency operation concentrate on actions supportive of DIME concepts. These activities are a part of reestablishing civil authority within a country by positively increasing its civil government's Political, Military, Economic, Social, Infrastructure, and Information (PMESII) capabilities. Tactical requirements to support these varied operations further increases the volume of collected information and expands the nature of intelligence analysis.

Finding a method of accurately determining the value of information generated from gathered data will take some of the burden from human analysts. Yet the nature of intelligence requires flexible methods and tools that incorporate measurements covering many topics. Attempting to qualify and quantify VOI measures within these varying military situations is an open topic for exploration.

3 Related VOI Challenges

As a concept, VOI is the subject of investigations within many different fields and for many differing purposes. For instance, placing a monetary VOI on intellectual property is part of legal inquiries while assigning VOI to medical data can save lives. Within almost every instance of science, determining the VOI for gathered data is a significant part of new technology development. Within the military context, much like the medical, VOI can save lives and optimize the conduct of field operations. As an illustration, two examples will follow of challenges in the field of military VOI estimation.

Example 1. In a project to determine the flow of valued, but classified, information within the spectra of military operations, researchers are "developing a scientific basis for valuing information for use at the tactical level and ... [demonstrating] technologies for flowing information among network nodes to increase the value of information available for command decisions [3]." A value assignment for a piece of classified information will relate directly to its congruence with commander's intent statements for mission execution. Tactical situation dynamics, a sufficient trust measurement for the receiving unit, and, collectively, a "chain of trust" in interconnecting network communication nodes, will combine to enable release of classified information, thus increasing the VOI available to a unit's commander.

Example 2. The prevalence of information gathering and reporting technologies is causing an overload situation for personnel responsible for monitoring, filtering, and analyzing incoming data. In fact, Wilkins, et al., found that "algorithms that alert on constraint violations and threats in a straight forward manner inundate users in dynamic domains [7]." In their development of a value of alerts (VOA) measurement, these researchers found several contributing VOI criteria. Included are: the plan, polices, user's awareness of the situation, system's view of current situation, user's cognitive load, resources (particularly time) available for analysis and response, information about adversarial agents, characterization of uncertainty, age of information and age of user's awareness, and source of information [7]. While these criteria form a good set of VOI modifiers, further research may reveal additional criteria. However, this set is useful as a point of departure for the development of fuzzy descriptive measures in the determination of VOI.

From these and other efforts, three points become clear: namely, 1) VOI of military data is an important determiner of tactical action; 2) one or more analysts subjectively determine VOI for most military applications; and 3) short timeframes for usefulness, high information density, and the amount of analytical experience available all influence the quality of VOI determination. The authors propose that the application of fuzzy parameters, based on the assignment of significant VOI factors, to capture efficient analytical practice will result in improved military information utilization within all echelon levels.

4 Approach

A fuzzy-based methodology will be used to develop a system for assisting with the problem of determining VOI. This section first provides a very brief description of fuzzy logic, then describes the rationale for choosing to use a fuzzy system, and finally discusses the general approach that will be followed.

Fuzzy Logic Background. In 1965, Lotfi Zadeh wrote his famous paper formally defining multivalued, or "fuzzy" set theory [8]. He extended the two-valued *indicator function* of traditional set theory to a multivalued *membership function.* The membership function is used to assign a grade of membership, ranging from 0 to 1, to each object in the fuzzy set. Zadeh formally defined fuzzy sets, their properties, and various mathematical operations on fuzzy sets. In a later paper [9] he introduced the concept of linguistic variables which have values that are linguistic in nature (i.e. speed = {slow, medium, fast}).

Fuzzy logic extends conventional Boolean (two-valued) logic so that it can handle truth values other than 0 (completely false) and 1 (completely true). That is, fuzzy logic can work with values that indicate partial truth. Fuzzy logic is built upon fuzzy sets and the basic concept is easy to grasp. In reality, we input, process, and output vague and imprecise information every day. Suppose you are teaching your child to drive and are discussing rules for how to handle the approach to an intersection. Would you tell the child "If the light has been green for *30 seconds*, release the accelerator *75 feet* from the intersection"? Or would it be better to say "If the light has been green for a *long time*, let off the gas pedal as you *get near* the intersection"? The precision in the first rule makes it impossible to follow; the more vague, or fuzzy, second rule can be easily understood and applied.

One use of fuzzy logic is to develop fuzzy inference systems; these systems provide the ability to perform approximate, or fuzzy, reasoning. Zadeh [10] defines approximate reasoning as "the process or processes by which a possibly imprecise conclusion is deduced from a collection of imprecise statements." His idea of approximate reasoning uses fuzzy logic which contains linguistic truth values (true, somewhat true, false, etc.) and approximate rules of inference. *Linguistic variables* are an important concept in fuzzy inference. A linguistic variable is used to approximately characterize relationships and values. For example, numbers can be used to characterize a person's height, but using words instead might provide categories such as tall, quite tall, more or less tall, not very tall, more or less small, and so on. The imprecision introduced by using words may or may not be by choice. That is, the imprecision may be intentional based on not needing to be more precise. More often, however, the imprecision is dictated by the lack of a means to quantitatively specify the attributes of an object. [11]

Fuzzy rules of inference encapsulate the approximate relationships between the input and output, or in the terminology of rules, the antecedent and consequent, domains. A fuzzy rule with two antecedents has the form "If *X* is *A* and *Y* is *B* then *Z* is C" where *A* and *B* are fuzzy sets over the input domains *U* and *V*, respectively and *C* is a fuzzy set over the output domain *W*. When using fuzzy sets in fuzzy inference, a domain is typically decomposed into overlapping fuzzy sets; each fuzzy set represents a classification. An element in the domain has some grade of membership, from 0 to 1 inclusive, in each fuzzy set in the domain. The membership function determines the grade of membership; the shape of the fuzzy sets determines the membership function.

Rationale for Choosing Fuzzy Logic. The literature reveals that significant work has been done with respect to "imperfect" data, especially in the context of data and knowledge bases [12, 13, 14, 15, 16]. Several types of imperfection have been noted; while the exact classification categories, and their definitions, vary somewhat from author to author it seems that a reasonable categorization might include:

- Uncertainty: cannot determine for sure if some statement is true or not ("The enemy will attack tomorrow")
- Incompleteness: lack of relevant information ("The attacking force is comprised of several key combat elements", without saying what the elements are)
- Imprecision: granularity problem ("The enemy will attack tomorrow" is imprecise if we need to know the exact hour the attack will occur)
- Vagueness: fuzzy imprecision ("The size of the attacking force is small")
- Inconsistency: conflicting values ("The enemy will attack at 0200 hours" and "The enemy will attack after dawn") [16]

Most of this work has been done from the viewpoint of the "quality" of information; our focus is on the "value" of information which is perhaps a subtle, but distinct, difference. A generally accepted definition for *information quality* seems to be as a "fitness for use" measure of the information [17]. In contrast, the *value* of information hinges more on how important a piece of information should be in a given decision-making context.

A fuzzy logic-based approach to solving the VOI problem was chosen for several reasons. First, fuzzy systems are known to be good at approximate reasoning where information is uncertain, incomplete, imprecise, and/or vague (four of the five data imperfections presented above) [18, 19]. Additionally, a fuzzy knowledge-based system was developed in [19] to model situation and threat assessment in a littoral environment; preliminary testing demonstrated good performance. The author states that before deciding to use a fuzzy approach several other methods for handling uncertainty were studied "in depth", including Bayesian methods, Dempster-Shafer

Theory, probability, artificial neural networks, and others. A fuzzy logic approach was also proven effective in detecting temporal aspects of data in data mining activities [20], which seems to relate quite well to the "timeliness" characteristic of information. Further, we presume that the successful development of the proposed system will rely heavily on integrating knowledge from subject matter experts (SMEs). Numerous works have discussed the efficacy and potential of using fuzzy logic in knowledge acquisition efforts [21, 22, 23].

General Approach. As a proof-of-concept effort, a fuzzy system will be developed to consider a simplified scenario in which VOI will be computed based on two factors: the timeliness of the information and the pedigree of the source. As such, the system will have two inputs (timeliness and pedigree), and one output (VOI). Timeliness is related to the availability of the information with respect to the organizational decision cycle. If the information arrives during the planning and execution phase, then it will have the highest timeliness rating; information arriving before or after this period will have a decreasing timeliness aspect. The pedigree of the data is a function of the believability associated with the data, and is directly related to the source's history of being trustworthy [24].

Figure 1 depicts a possible decomposition of the input and output domains for the system. The timeliness domain is broken into three fuzzy sets, while the pedigree and VOI domains are each decomposed into five fuzzy sets. The x-axis for each domain would comprise the *universe* of the domain; that is, the set of all values within the domain. The y-axis is the membership function value in the range [0,1]. Note that the shape of the fuzzy sets can take on many forms (triangles, trapezoids, or even beta, Gaussian, or other curves) and they all do not have to be the same (the uniformity of the pedigree and VOI domain decompositions was simply done for convenience). Also, the fuzzy sets

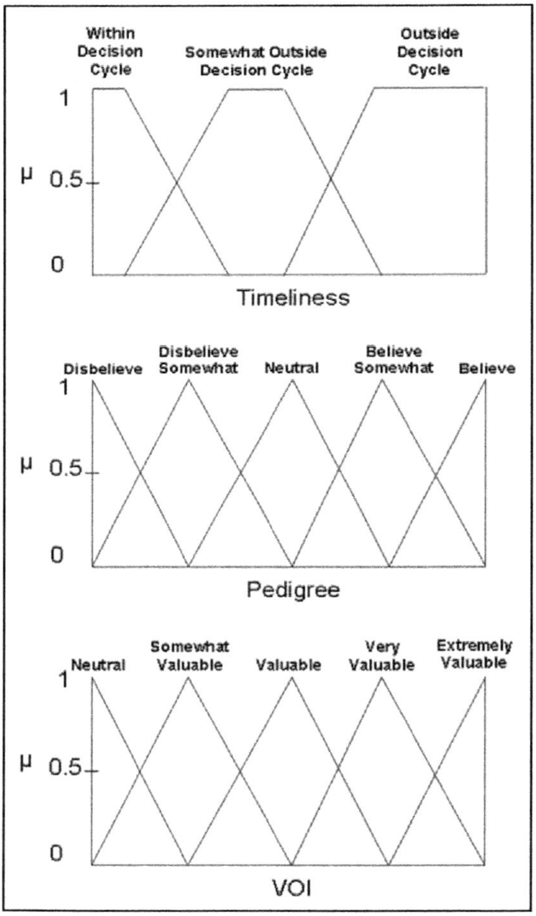

Fig. 1. Example Membership Functions

decomposing a domain do not have to overlap in a regular pattern as shown in Figure 1, nor does the sum have to be 1 for the membership values for all sets in which an element belongs (which would be the case in all three of the decompositions in Figure 1).

The membership functions depicted in Figure 1 are only shown to demonstrate an example potential decomposition structure. We anticipate that the "rules" for determining VOI will have to come from subject matter experts (SMEs). As such, the number of fuzzy sets in each domain will be driven by the granularity of the rules derived from the SMEs. An example rule that could be captured by the membership functions in Figure 1 might be "If the timeliness is within the decision cycle and the pedigree is neutral, then the VOI is somewhat valuable." However, the number of fuzzy sets needed in a domain will be determined by how finely or coarsely (the granularity) an SME chooses to divide it. After the system is constructed, it will be tested by devising multiple scenarios which will be submitted to a SME and the system. The answer from the system will be compared with that from the SME to judge the system's performance.

5 Conclusion

Information within complex military environments is often of uncertain pedigree, imprecise, and time sensitive. Within this context, fuzzy logic methodology may offer a superior metric for forming a VOI solution. Contributions of this research will include a proof-of-concept system as well as initial research in obtaining SMEs quantification of their judgments with respect to context-dependent VOI determination. Logical extension of this effort includes working with SMEs to decompose the information space into finer characterization and exploring methods for tuning the membership functions depending on the operational tempo of a given scenario. As the program matures, the capability to accommodate inconsistent or contradictory information will be investigated. For the military, the ability to efficiently and effectively calculate VOI and separate the wheat from the chaff is paramount. This program is an important step towards that goal.

References

[1] Alberts, D.S., Garstka, J.J., Hayes, R.E., Signori, D.T.: Understanding Information Age Warfare. CCRP, Washington (2001)

[2] Flynn, M.T., et al.: Fixing Intel: A Blueprint for Making Intelligence relevant in Afghanistan, US Army (January 5, 2010)

[3] James, J.: "Military Data", presentation, Network Science Center, West Point (October 2010)

[4] Ahituv, N.: Assessing the value of information: Problems and approaches. Paper presented at the Proceedings of the Tenth International Conference on Information Systems, Boston, MA (1989)

[5] Rafaeli, S., Raban, D.R.: Experimental investigation of the subjective value of information in trading. Journal of the Association for Information Systems 4(5), 119–139 (2003)

[6] Anonymous, US Army Field Manual (FM) 3-0, Operations, US Army (June 2001)
[7] Wilkins, D.E., et al.: Interactive Execution Monitoring of Agent Teams. Journal of Artificial Intelligence Research 18 (March 2003)
[8] Zadeh, L.A.: Fuzzy sets. Information and Control 8, 338–353 (1965)
[9] Zadeh, L.A.: Outline of a new approach to the analysis of complex systems and decision processes. IEEE Transactions on Systems, Man, and Cybernetics 3, 28–44 (1973)
[10] Zadeh, L.A.: A theory of approximate reasoning. In: Yager, R., Orchinnikov, S., Tong, R., Nguyen, H. (eds.) Fuzzy Sets and Applications, pp. 367–412. John Wiley & Sons, New York (1987)
[11] Zadeh, L.A.: The Concept of a Linguistic Variable - I. Information Sciences 8, 199–249 (1975)
[12] Agrawal, P., Sarma, A., Ullman, J., Widom, J.: Foundations of Uncertain-Data Integration. In: Proceedings of the VLDB Endowment, vol. 3(1-2), pp. 1080–1090 (September 2010)
[13] Magnani, M., Montesi, D.: A Survey on Uncertainty Management in Data Integration. Journal of Data and Information Quality 2(1), 5:1–5:33 (2010)
[14] Helfert, M., Foley, O.: A Context Aware Information Quality Framework. In: Proceedings of the Fourth International Conference on Cooperation and Promotion of Information Resources in Science and Technology, pp. 187–193 (November 2009)
[15] Yu, B., Kallurkar, S., Vaidyanathan, G., Steiner, D.: Managing the Pedigree and Quality of Information in Dynamic Information Sharing Environments. In: Proceedings of the 6th International Joint Conference on Autonomous Agents and Multiagent Systems, pp. 1248–1250 (May 2007)
[16] Parsons, S.: Current Approaches to Handling Imperfect Information in Data and Knowledge Bases. IEEE Transactions on Knowledge and Data Engineering 8(3), 353–372 (1996)
[17] Wang, R.Y., Strong, D.: Beyond Accuracy. What Data Quality Means to Data Consumers. Journal of Management Information Systems 12(4), 5–34 (1996)
[18] Yen, J., Langari, R.: Fuzzy Logic: Intelligence, Control, and Information. Prentice Hall, Upper Saddle River (1999)
[19] Liang, Y.: An Approximate Reasoning Model for Situation and Threat Assessment. In: Proceedings of the Fourth International Conference on Fuzzy Systems and Knowledge Discovery, pp. 246–250 (November 2007)
[20] Vincenti, G., Hammell II, R.J., Trajkovski, G.: Scouting for Imprecise Temporal Associations to Support Effectiveness of Drugs During Clinical Trials. In: Proceedings of the Annual Conference of the North American Fuzzy Information Processing Society (NAFIPS 2005), Ann Arbor, MI (June 22-25, 2005)
[21] Barnes, A., Hammell II, R.J.: Employing Intelligent Decision Systems to Aid in Information Technology Project Status Decisions. In: Nag, B. (ed.) Intelligent Systems in Operations: Models, Methods, and Applications, pp. 1–26. IGI Global, Hershey (2010)
[22] McQuighan, J., Hammell II, R.J.: Computational Intelligence for Project Scope. In: Proceedings of the 22nd Midwest Artificial Intelligence and Cognitive Science Conference, Cincinnati, OH, April 16-17, pp. 47–53 (2011)
[23] Tolosa, J., Guadarrama, S.: Collecting Fuzzy Perceptions from Non-expert Users. In: Proceedings of the 19th IEEE International Conference on Fuzzy Systems (FUZZ-IEEE 2010), Barcelona, (Spain), pp. 1–8 (July 2010)
[24] Cerruti, M., Das, S., Ashenfelter, A., Raven, G., Brooks, R., Sudit, M., Chen, G., Wright, E.: Pedigree Information for Enhanced Situation and Threat Assessment. In: Proceedings of the Ninth International Conference on Information Fusion, pp. 1–8 (July 2006)

Handling Sequential Observations in Intelligent Surveillance

Jianbing Ma, Weiru Liu, and Paul Miller

School of Electronics, Electrical Engineering and Computer Science,
Queen's University Belfast, Belfast BT7 1NN, UK
{jma03,w.liu}@qub.ac.uk, p.miller@ecit.qub.ac.uk

Abstract. Demand for intelligent surveillance in public transport systems is growing due to the increased threats of terrorist attack, vandalism and litigation. The aim of intelligent surveillance is in-time reaction to information received from various monitoring devices, especially CCTV systems. However, video analytic algorithms can only provide static assertions, whilst in reality, many related events happen in sequence and hence should be modeled sequentially. Moreover, analytic algorithms are error-prone, hence how to correct the sequential analytic results based on new evidence (external information or later sensing discovery) becomes an interesting issue. In this paper, we introduce a high-level sequential observation modeling framework which can support revision and update on new evidence. This framework adapts the situation calculus to deal with uncertainty from analytic results. The output of the framework can serve as a foundation for event composition. We demonstrate the significance and usefulness of our framework with a case study of a bus surveillance project.

Keywords: Intelligent Surveillance, Active System, Situation Calculus, Belief Change, Sequential Observation.

1 Introduction

Recently, more and more attention has been paid by governments and transport operators to protect vehicles and passengers with surveillance cameras, e.g., the Florida School Bus Surveillance project [2], the First Glasgow Bus Surveillance [22], Federal Intelligent Transportation System Program in the US [18], Washington rail corridor surveillance [17], Airport Corridor Surveillance in the UK [16], etc. These applications require deployment of large-scale CCTV systems giving rise to unique problems. For example, in a reasonable sized provincial city there may be several hundred buses, each of which has 12-14 cameras, giving a total of several thousand cameras. The large amount of cameras for monitoring passengers/vehicles makes it almost impossible to detect possible incidents manually without delay. For this reason, video analytics and event reasoning are being introduced to CCTV systems in order to ensure in-time reaction.

The aim of video analytics is single-event recognition. Recently, however, developers have realized that it is necessary to manage the events generated by video analysis software. For instance, to prevent anti-social behaviors on public transport systems, one

S. Benferhat and J. Grant (Eds.): SUM 2011, LNAI 6929, pp. 547–560, 2011.
© Springer-Verlag Berlin Heidelberg 2011

has to make a decision based on a sequence of detected events. Ideally, this would be straightforward if the recognized events are correct and certain. Unfortunately, in reality, imperfection and mistakes frequently occur in practical applications. For example, in the case of a person entering the bus doorway, the person may be classified as male with a certainty of 85% by the classification analytics, rather than with 100% certainty. Even worse, the analytic algorithm may classify a person as female at one time instant and male at a later time. In addition to inaccurate analytics, a large amount of mistakes are caused by the unreliability of the data sources. For example, in the classification example above, the camera may have been tampered with, illumination could be poor, or the classifier training set may be unrepresentative. Any, or all of these, can result in imperfection and errors.

In this paper, we introduce a sequential observation modeling framework which is able to deal with uncertainty, and can support revision and update when new evidence is received, thereby removing the influence of past errors. This framework, which operates at a higher level than analytic algorithms, deploys a situation calculus foundation with the ability to deal with uncertainty in analytic results. It is also able to handle belief revision and update properly. We demonstrate the significance and usefulness of our framework with a case study of a bus surveillance project [12,7,10,15]. Our approach provides a sound framework for surveillance applications, such as CCTV for buses, airports, etc. The output of the framework, i.e., primitive events, can be used as a starting point of event composition.

Usually in situation calculus, there are sensing and non-sensing actions, or epistemic and ontic actions [4]. A sensing/epistemic action senses a property of the domain and does not change the environment. A non-sensing/ontic action is an action done by the agent which changes the environment. A major difference between real-world situations, such as those encountered in surveillance applications, and situation calculus approaches is that in the former, even the result of an ontic action, e.g., a passenger changes their position from standing to seated, is observed by cameras and analyzed by video analytic algorithms (and hence sometimes we call the agent of interest an *observable*). Therefore, situation calculus should be significantly adapted to make it suitable for intelligent surveillance purposes.

In intelligent surveillance applications the results of both epistemic and ontic actions are provided by video analytics, therefore, we must differentiate between both kinds of actions, since they respond differently upon new evidence being obtained. The properties sensed by epistemic actions are generally invariable, e.g., the gender of a person, etc., whilst properties related to ontic actions are those that can be changed at will, e.g., the position of a person, etc. We also allow external information to be handled in this framework. External information, when received and used, can be seen as a kind of epistemic or ontic action, according to the information properties. For instance, if a piece of information tells us a person is a male, then it can be seen as an epistemic action; if it tells us a person is standing, it can be seen as an ontic action. In addition, a property related to an epistemic action will be called an **epistemic** or an **invariable** property, and similarly, a property related to an ontic action is called an **ontic** or a **variable** property. In summary, if a property indicates an intrinsic character of an observable of interest, and hence is invariable, then it is an epistemic property. But we also need to point out that this property could be mis-classified or even intentionally disguised, which seemingly makes it variable. For example, although a person is a male, he could be wrongly classified as a female. He could even disguise himself as a female if he wants to. However, this superficial variability should not cause any confusion. Instead, ontic properties are usually *external* properties between an observable of interest and the environment, and hence are variable, e.g., a passenger can move from the drivers cabin area to the saloon area on a bus. This differentiation between epistemic and ontic properties also applies to properties obtained by video analytics. For instance, if the gender of a person is in fact estimated from the captured video, it is still called an epistemic property since gender is an intrinsic character of a person.

The rest of the paper is organized as follows. Section 2 provides the preliminaries on the situation calculus. In Section 3, formal approaches to deal with uncertain observations are presented, including the ways in which to handle epistemic and ontic actions. Section 4 shows how belief revision is adequately handled in our framework. We then provide a case study, which is a simplified bus surveillance scenario, in Section 5. Finally, we conclude the paper in Section 6.

2 Preliminaries

Situation calculus, introduced by John McCarthy [13,14], has been applied widely to model and reason about actions and changes in dynamic systems. It was reinterpreted in [19] as *basic action theories* which are comprised of a set of foundational axioms defining the space of situations, unique-name axioms for actions, action preconditions and effects axioms, and the initial situation axioms [5]. The well known frame problem is solved by a set of special action effects axioms called *successor state axioms*.

Since actions carried out by agents cause constant changes of the agents' beliefs, developing strategies of managing belief changes triggered by actions is an important issue. The problem of *iterated belief change* within the framework of situation calculus has been investigated widely, e.g., [20,21,24,11]. In [24], a new framework extending previous approaches was proposed, in which a plausibility value is attached to every situation. This way, the framework is able to deal with nested beliefs, belief introspection, mistaken beliefs, and it can also handle belief revision and update together in a seamless way. The framework in [24] is based on an extension of action theory [19] stemming from situation calculus [13,14]. Here we introduce the notion of situation calculus from [24] which includes a belief operator [20,21].

According to [24], the situation calculus is a predicate calculus language for representing dynamically changing domains. A situation represents a snapshot of the domain. There is a set of initial situations corresponding to what the agents believe the domain might be initially. The actual initial state of the domain is represented by a distinguished initial situation constant, S_0, which may or may not be among the set of initial situations believed by an agent. The term $do(a, s)$ denotes the unique situation that results from the agent performing action a in situation s.

Predicates and functions whose values may change from situation to situation (and whose last argument is a situation) are called *fluents*. For instance, we use the fluent $\mathsf{InR1}(s)$ to represent that the agent is in room $R1$ in situation s. The effects of actions on fluents are defined using successor state axioms [19], which provide a succinct representation for both effect axioms and frame axioms [13,14]. For instance, if there are two rooms ($R1, R2$) and an action Leave takes the agent from the current room to the other room. Then, the successor state axiom for InR1 is [24]:

$$\mathsf{InR1}(do(a,s)) \equiv \big((\neg\mathsf{InR1}(s) \wedge a = \mathsf{Leave}\big) \vee \big(\mathsf{InR1}(s) \wedge a \neq \mathsf{Leave})\big).$$

This axiom says that the agent will be in Room 1 after doing action a in s iff either it is in Room 2 and leaves for Room 1 or is currently in Room 1 and does not leave.

Levesque [6] introduced a predicate, $\mathsf{SF}(a,s)$, to describe the result of performing the binary-valued epistemic action a. $\mathsf{SF}(a,s)$ holds (returns *true*) iff the sensor associated with a returns the sensing value 1 in situation s. Each epistemic action senses some property of the domain. The property sensed by an action is associated with the action using a *guarded sensed fluent axiom* [3]. For example, the following two axioms

$$\mathsf{InR1}(s) \rightarrow \big(\mathsf{SF}(\mathsf{SenseLight}, s) \equiv \mathsf{Light1}(s)\big)$$

$$\neg\mathsf{InR1}(s) \rightarrow \big(\mathsf{SF}(\mathsf{SenseLight}, s) \equiv \mathsf{Light2}(s)\big)$$

can be used to specify that SenseLight senses whether the light is on in the room the agent is currently located.

In this paper, we adopt the following conventions about *guarded action theories* Σ consisting of: (A) successor state axioms for each fluent, and guarded sensed fluent axioms for each action; (B) unique names axioms for actions, and domain-independent foundational axioms; and (C) initial state axioms which describe the initial state of the domain and the initial beliefs of agents. A *domain-dependent fluent* means a fluent other than the probability fluent p, and a *domain-dependent formula* is one that only mentions domain-dependent fluents. However, since this is a paper focusing on applications, we will not introduce the axioms here. Interested readers can refer to [24,11]. We further assume that there is only one agent acting in a chosen domain, although the framework is capable of accommodating multiple agents.

3 The Revised Situation Calculus Framework

In this section, we extend the situation calculus to include a probability operator to account for iterated belief changes and deal with uncertainty.

Usually in situation calculus, the result of all actions are accurate. In recent work, e.g., [1,23,24,11], noisy epistemic actions have been proposed and studied. However, in intelligent surveillance applications, not only can epistemic actions be noisy, but ontic actions can also be subject to noise (recall an epistemic action senses some property of the domain, but leaves the environment unchanged, while an ontic action changes the actual environment). That is, in these applications, the results of ontic actions are also reported by video analytic algorithms, which may (and in fact usually, if not always) present uncertain results. For example, if a passenger takes a seat, then from a normal situation calculus point of view, its status certainly changes from "standing" to "seated". However, if the scenario is analyzed by an algorithm, due to the imperfection of the algorithm, it can only give 90% degree of certainty that the passenger is seated, leaving the remaining 10% still standing. That is, ontic actions can also bring uncertainty, or noise. In fact, in a few scenarios (e.g., light changes suddenly outside the window), the analytic results could be very inaccurate. It may conclude that with 60% degree of certainty the passenger is seated and 40% degree of certainty the passenger is standing. These cases cannot be handled in classical situation calculus. Hence, we need to adapt the situation calculus to deal with such cases.

For convenience, we denote SA the set of all epistemic actions and hence for each action a, $a \in SA$ means that a is an epistemic action while $a \notin SA$ means that a is a ontic action. Since an ontic action a can bring up more than one possible result (e.g., "seated" or "standing"), the corresponding $do(a, s)$ may also give rise to more than one successive situation. An epistemic action a can also bring up more than one possible result if it is not accurate.

Example 1. *Let a situation $s = M \wedge S$ (the passenger is male and standing), then a noisy (inaccurate) epistemic action presents a result as the passenger is male with probability 0.4 and female with probability 0.6, then two successive situations $s^1 = M \wedge S$ with probability value 0.4 and $s^2 = \neg M \wedge S$ with probability 0.6 should be expected.*

Hence, subsequently in this paper, we assume $do(a, s)$ is a set of situations instead of a single situation. Moreover, in this paper, we assume $\mathsf{SF}(a, s)$ (different from the definition in [6] where $\mathsf{SF}(a, s)$ returns a boolean value) gives the tuple-valued sensing result $(x_1, \cdots, x_k)$ where each x_i stands for the probability that the epistemic action returns result X_i. For instance, in the above example, $\mathsf{SF}(a, s) = (0.4, 0.6)$ means that the passenger is male with probability 0.4 and female with probability 0.6 when a is an epistemic action returning the gender of the passenger. For convenience, we also write $\mathsf{SF}(X, a, s)$ to denote the probability that the epistemic action a returns X, e.g., $\mathsf{SF}(M, a, s) = 0.4, \mathsf{SF}(\neg M, a, s) = 0.6$. Similarly, we write $\mathsf{NSF}(a, s)$ (NS is short for Non-Sensing) to present the tuple-valued ontic action result $(x_1, \cdots, x_t)$ where each x_i stands for the probability that the ontic action returns result X_i. For instance, if a is an ontic action changing the behavior of a passenger, then $\mathsf{NSF}(a, s) = (0.2, 0.8)$ means that there is a probability 0.2 such that a passenger is standing and a probability 0.8 such that it is seated ($\neg$S). Similarly, we also denote $\mathsf{NSF}(X, a, s)$ the probability that the ontic action a returns X.

In this paper, for simplicity, we assume that all actions, regardless of whether they are epistemic or ontic, can only provide two possible results. In fact, if they could return more than two possible results, no essential changes are needed for the framework, but a more cumbersome description of the scenarios, e.g., $\mathsf{SF}(a, s)$ will be a n-tuple value where $n > 2$ and the number of successive situations will become greater, etc..

In this paper, we use ordinals as time points to indicate the sequence of situations. More precisely, all the initial situations will have a subscript 0, denoted as s_0^i (where i indicates the i-th possible situation), and the successive situation of a situation s_n will be s_{n+1}. Let $\mathcal{S}_n$ denote the set of all situations with subscript n, i.e., the set of situations in the n-th run. Note that if s and s' are both in $\mathcal{S}_n$, then we should have $\mathsf{SF}(a, s) = \mathsf{SF}(a, s')$ (resp. $\mathsf{NSF}(a, s) = \mathsf{NSF}(a, s')$) since the action is taken in the real world, it should return only one result, no matter what we *think* the real-world situation might be (e.g., s or s'). From this sense, we can write $\mathsf{SF}_n(a)$ or $\mathsf{NSF}_n(a)$ to denote the action result for situations in the n-th run. Furthermore, we use s/X to denote a situation that the property corresponding to X in s is changed to X, e.g., for $s = M \wedge S$, we have $s/M = s$ and $s/\neg M = \neg M \wedge S$.

The belief set of $\mathcal{S}_n$ is defined as follows. Let $\phi[s]$ denote that ϕ is assessed in s. For example M (the passenger is a male) is assessed in $s = M \wedge S$, hence $M[s]$ holds. Let $p_n(\phi) = \sum_{s:s\in\mathcal{S}_n\wedge\phi[s]} p(s)$ indicate the total probability of ϕ in $\mathcal{S}_n$.

Definition 1. $Bel_n(\phi) \stackrel{def}{=} p_n(\phi) > p_n(\neg\phi)$.

That is, ϕ is believed in the n-th run if it is more probable than its negation. Since this definition is not closed under deduction, i.e., $Bel_n(\phi) \wedge Bel_n(\psi) \nrightarrow Bel_n(\phi \wedge \psi)$, we usually only consider probabilities (and hence beliefs) on atoms (e.g., $Male$, $Stand$, etc.), while probabilities (and hence beliefs) of other formulae are computed from probabilities of atoms (with independence assumptions) [9].

Based on the above notations and definitions, we can define a probability function p for each situation s to measure how possible an agent considers s is. The p functions for initial situations are provided with a normalization condition that the sum of probabilities of all initial situations is 1. This is expressed as follows:

Axiom 1. *(Initial State Axiom)* $\sum_{s:\mathsf{Init}(s)} p(s) = 1$.

Probabilities of successor situations are defined as follows.

If a is an epistemic action and $\mathsf{SF}(a, s_n) = (t, 1-t)$, or $\mathsf{SF}(X, a, s_n) = t$ and $\mathsf{SF}(\neg X, a, s_n) = 1 - t$, then in general it induces two successive situations for s_n, i.e., $s_{n+1} = s_n/X$ and $s'_{n+1} = s_n/\neg X$ with probability $p(s_n)t$ and $p(s_n)(1-t)$ respectively. Note that if $t = 0$ or $t = 1$, it in fact only induces one successive situation (situations with probability 0 will be ignored).

However, it is not always reasonable to simply change the current situation to the successive situations as stated above. In some scenarios, we must keep the probabilities of the beliefs induced by the current situations. For instance, assume that at $\mathcal{S}_k$, a passenger is classified as a male with probability 0.8, and at $\mathcal{S}_{k+1}$, this passenger is classified as a male with probability 0.6, then do we need to change the probability to 0.6? The answer is no. In real-world applications such as intelligent surveillance, we observe that if a video analytic algorithm is used to continuously check the gender of a person based on a video, then the probability of that person being a male will fluctuate. Hence in practice, if at some time point, it is classified as a male with probability 0.9, and later with probability 0.85, we can just keep the probability 0.9. A more persuasive scenario is that at some time point we have external information (e.g., an analyst views the person on a monitor) which indicates that the passenger is 100% a male, but later the algorithm still classifies it as a male with probability 0.85, then it is obvious that we do not need to change the probability from 1 to 0.85.

An exception to the above statement, is that the change of probability leads to a change of beliefs. For example, if at $\mathcal{S}_k$, a passenger is classified as male with probability 0.8 (hence $Bel_k(M)$ holds), but at $\mathcal{S}_{k+1}$, it is classified as female with probability 0.75 (hence $Bel_{k+1}(\neg M)$ holds), then this major change should not be ignored. It may indicate that an interesting event has happened. In real systems, such belief changes with respect to an invariable property, may, by themselves, justify alerting an analyst.

In short, the probabilities of the current belief are only changed when the sensing result overwhelms the current belief. Here by overwhelm we mean either the belief induced by the sensing result is the same as the current belief but with a greater probability or the belief becomes different.

More precisely, let a be an epistemic action and $\mathsf{SF}(a, s_n) = (t, 1-t)$. Without loss of generality, we assume that $t > 0.5$[1] and hence $Bel_{n+1}(X)$ holds. For each situation $s_n \in \mathcal{S}_n$, it induces two successive situations, i.e., $s_{n+1} = s_n/X$ and $s'_{n+1} = s_n/\neg X$. The probabilities of these two situations are defined as follows:

- if $Bel_n(\neg X)$ holds, then $s_{n+1} = s_n/X$ and $s'_{n+1} = s_n/\neg X$ are assigned with probability $p(s_n)t$ and $p(s_n)(1-t)$ respectively,
- if $Bel_n(X)$ holds, then $s_{n+1} = s_n/X$ and $s'_{n+1} = s_n/\neg X$ are assigned with probability $p(s_n)max(p_n(X), t)$ and $p(s_n)(1 - max(p_n(X), t))$ respectively.

For this we have the following result which shows that the assignment of probabilities satisfy the statements we argued before.

Proposition 1. *For any epistemic property X and $n > 0$, if both $Bel_n(X)$ and $Bel_{n+1}(X)$ hold, then $p_{n+1}(X) = max(p_n(X), \mathsf{SF}(X, a, s_n))$. If both $Bel_n(\neg X)$ and $Bel_{n+1}(X)$ hold, then $p_{n+1}(X) = \mathsf{SF}(X, a, s_n)$.*

There might be some equivalent situations in $\mathcal{S}_{n+1}$ (in terms of all fluents). For convenience, they can be merged together.

Example 2. *Assume that a video analyzer detects a passenger on board but it does not know whether the passenger is male or female. So it considers two possible situations S_0^1 and S_0^2 at the beginning where*

$$S_0^1 = \mathsf{Male} \wedge \mathsf{Stand},\ S_0^2 = \mathsf{Female} \wedge \mathsf{Stand}$$

The video analytic algorithm gives S_0^1 with probability 0.8 and S_0^2 with probability 0.2. The bottom-half of Fig. 1 illustrates these two situations.

After some seconds, the camera does a second detection from which the video analytic algorithm asserts that the passenger is male with probability 0.9 and female with probability 0.1. In Fig. 1, Sensing Gender *is abbreviated as* SG. *Hence each situation induces two successive situations (in Fig. 1, 0.18=0.2*0.9, etc.) and then equivalent situations are merged together, which finally forms two situations S_1^1 and S_1^2 in Fig. 1.*

It seems that we can let each initial situation in Fig. 1 only induce one successive situation and simply change the probabilities of the successive situations to 0.9 and 0.1, respectively. The reason why we do not follow this way is that the latter approach is not applicable on some occasions. For instance, if there is only one initial situation S_0^1, then from the second detection, probability 0.1 should be assigned to a successive situation that the passenger is a female but no initial situations can induce such a successive situation.

[1] In practice, we can always change $(0.5, 0.5)$ to $(0.5+\epsilon, 0.5-\epsilon)$ for a small positive real ϵ. It does not make much difference.

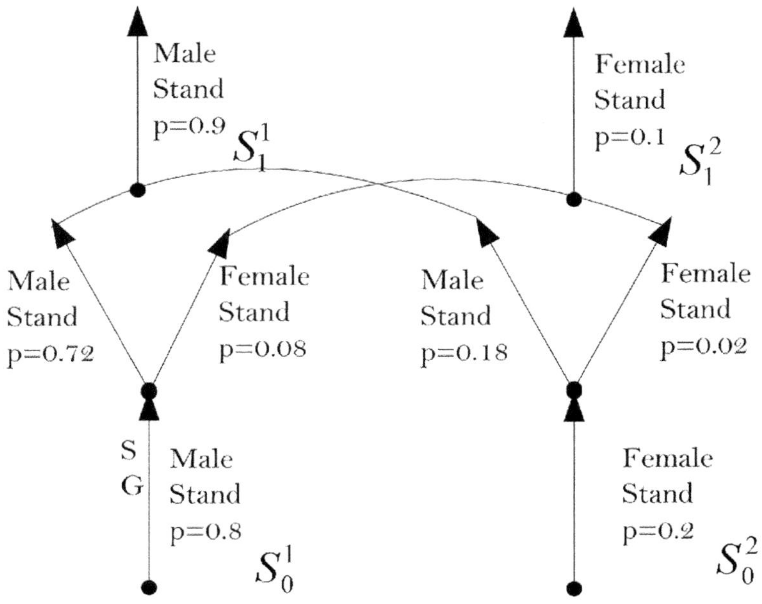

Fig. 1. Situations after Sensing the Gender of a Passenger

If a is an ontic action and $\mathsf{NSF}(a, s) = (t, 1-t)$, then for each situation $s_n \in \mathcal{S}_n$, it induces two successive situations $s_{n+1} = s_n/X$ and $s'_{n+1} = s_n/Y$ with probability $p(s_n)t$ and $p(s_n)(1-t)$ respectively. Similarly, if there are equivalent induced situations, then we can merge them.

We have the following result.

Proposition 2. *For any ontic property X and $n > 0$, $p_{n+1}(X) = \mathsf{NSF}(X, a, s_n)$.*

Example 3. *Assume we have two possible initial situations S_0^1 and S_0^2 at the beginning where*

$$S_0^1 = \mathsf{Male} \wedge \mathsf{Stand}, S_0^2 = \mathsf{Male} \wedge \neg\mathsf{Stand}$$

The video analytic algorithm gives S_0^1 with probability 0.8 and S_0^2 with probability 0.2. The bottom-half of Fig. 2 illustrates these two situations (Note that this figure is similar to Fig. 1 except that the action is an ontic action).

After some seconds, the camera does a second detection from which the video analytic algorithm asserts that the passenger takes a seat with probability 0.9 and it is standing with probability 0.1. In Fig. 2, Sensing Position *is abbreviated as* SP. *Using the above method, finally we get two situations S_1^1 and S_1^2 in Fig. 2.*

For any epistemic or ontic actions, the revised probabilities always sum up to 1.

Proposition 3. *For any $n > 0$, $\sum_{s \in \mathcal{S}_n} p(s) = 1$.*

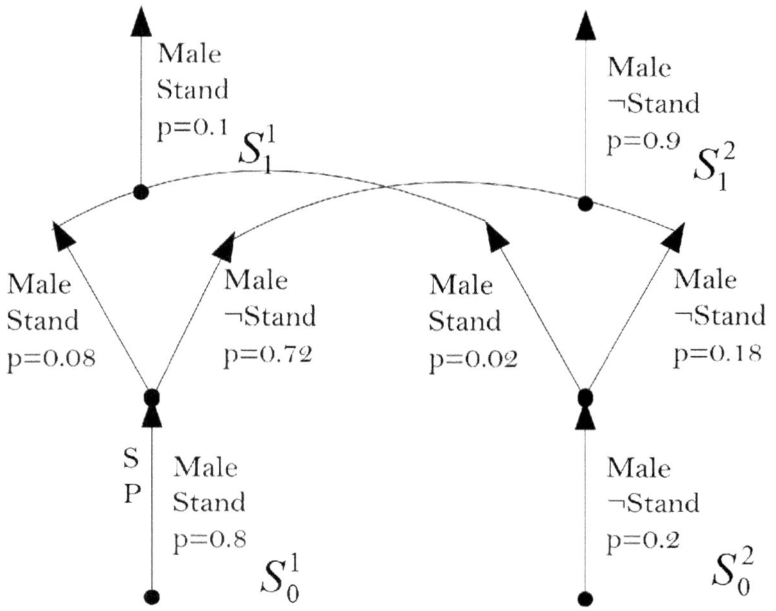

Fig. 2. Situations after Sensing the Position of a Passenger

4 Belief Revision

Belief revision studies how an agent's beliefs can be changed based on some new information if the new information must be believed. Any property of interest (no matter epistemic properties or ontic properties) could be revised when obtaining certain new information on that property of the observable. Studying belief revision in situation calculus is a natural course for managing an agent's beliefs. In the following, we assume that for each formula ϕ to be revised, there is a corresponding action that obtains information on that property.

Definition 2. *(Uniform formula, adapted from [24,11]) A formula is* uniform *if it contains no unbound variables.*

Definition 3. *A uniform formula ϕ is called obtainable from an action A with regard to a situation s, denoted: $(A, s) \to \phi$, if*

$$\begin{cases} SF(\phi, A, s) > SF(\neg\phi, A, s), & A \text{ is an epistemic action,} \\ NSF(\phi, A, s) > NSF(\neg\phi, A, s), & \text{otherwise.} \end{cases}$$

A is called a revision *action for ϕ w.r.t. Σ, if for any s, $(A, s) \to \phi$.*

Note that here the meaning of *revision* is in fact extended to updating as it also handles changes of ontic properties as ontic actions changes the environment[2].

Now by abuse of notation, we use $Bel(\phi, s)$ to denote $Bel_n(\phi)$ where s is a situation in the n-th run.

Theorem 1. *Let ϕ be a domain-dependent, uniform formula, and A be a revision action for ϕ w.r.t. Σ, then we have:*
$\Sigma \models \big[\forall s, \phi[s] \rightarrow Bel(\phi, do(A, s))\big] \wedge \big[\forall s, \neg\phi[s] \rightarrow Bel(\neg\phi, do(A, s))\big].$

This theorem proves that revision (as well as updating) in our framework is handled adequately. That is, if new information indicates that ϕ holds, then the agent will believe that ϕ holds after performing A. Conversely, if new information shows that ϕ does not hold, then the agent will believe $\neg\phi$ after performing A. This theorem is also consistent with the framework in [20,21,24,11].

Theorem 2. *Let A be a revision action for domain-dependent, uniform formula ϕ w.r.t. Σ, then the following sentence is satisfiable:*
$\Sigma \cup \{Bel(\neg\phi, S_0), Bel(\phi, do(A, S_0)), \neg Bel(FALSE, do(A, S_0))\}.$

This theorem shows that even if the agent believes $\neg\phi$ in S_0, it will believe ϕ after performing A when action A provides that ϕ is true, and still maintains consistent beliefs ($\neg Bel(FALSE, do(A, S_0))$).

5 Example

The advantage of the methods proposed in this paper is that it can tolerate the existence of errors and correct errors, hence keeps a well established track of video analytics. Error correction can be done by either internal inspections or external inferences. In this section, we use a surveillance example to illustrate this advantage.

Example 4. *Now we are going to model a simplified scenario that a passenger boards a bus. We use Fig. 3 to illustrate the situation pedigree. Multiple passengers can be modeled by multiple situation pedigrees.*

Similar to the previous examples, we use SP *to denote Sensing Position and* SG *to denote Sensing Gender. In addition to the internal actions* SP *and* SG*, we also allow external instructions as external actions into the system, e.g., $P(M) = 1$ (resp. $P(F) = 1$) in Fig. 3 which means that the external instructions suggesting the passenger is definitely a male (resp. a female).*

Now we give the explanations of the process depicted by Figure 3.

The two initial situations are:

$$S_0^1 = \mathsf{Male} \wedge \mathsf{Stand}, S_0^2 = \mathsf{Female} \wedge \mathsf{Stand}$$

That is, initially the video analytics tell us that the passenger is standing but it does not know accurately whether it is male or female, only providing probability values 0.8 for male and 0.2 for female.

[2] Revision receives (and accepts) information about a static world while updating means that the world itself has been changed [8].

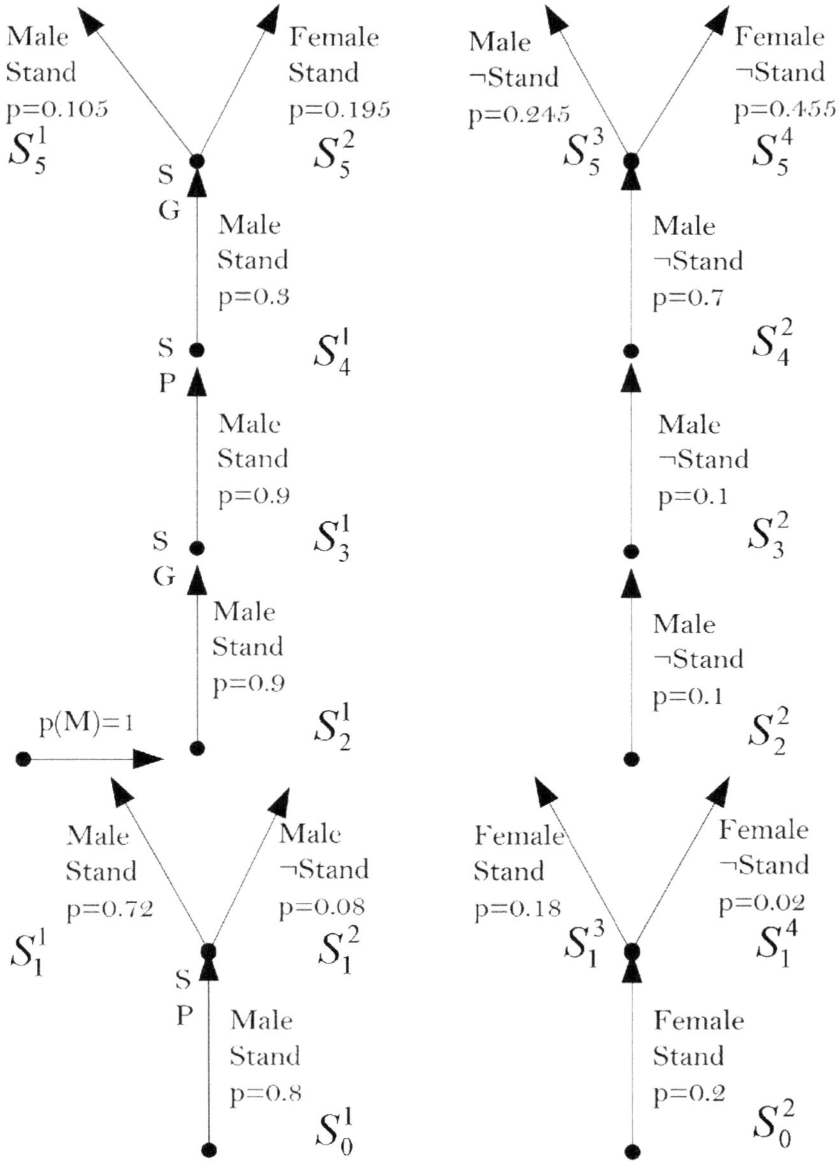

Fig. 3. Surveillance Example

*After a while, the sensor re-examines the position of the passenger (*SP *in the bottom of Fig. 3) and tells us it is now standing with probability 0.9 and seated with probability 0.1. Hence our method gives four possible successive situations* S_1^1, S_1^2, S_1^3, *and* S_1^4 *as shown in Fig. 3.*

Now the monitor in the control room provides a piece of information that this passenger is definitely a male ($p(M) = 1$ *in Fig. 3). Hence we get two successive situations* S_2^1 *and* S_2^2*. We can see that the possibility of the passenger being a female is eliminated.*

*Then the sensor re-examines the gender of the passenger (*SG *in the middle of Fig. 3), and tells us it is a male with probability 0.85 and female with probability 0.15. However, since it does not change the belief that this person is a male and the probability of this person being a male (0.85) is less than the one in the current situation (where the probability of the person being a male is 1), according to our procedure, we do not need to change the probability, hence the two successive situations* S_3^1 *and* S_3^2 *are just the same to their predecessors. Note that here new information from* SG *shows the passenger is male, and after performing* SG*, the proposition* the passenger is male *is believed. It verifies Theorem 1.*

*After that the sensor checks the position of the passenger (*SP *in the top half of Fig. 3). This time the passenger might have taken a seat, hence the video analytics tell us that the passenger is standing with probability 0.3 and seated (*$\neg Stand$*) with probability 0.7. Then two successive situations are induced as* S_4^1 *and* S_4^2 *in Fig. 3.*

Finally, the passenger accidentally removes some of her disguise and the video analytics tell us that it is a female with probability 0.65 *and a male with probability* 0.35*. Then we obtain four successive situations* S_5^1, S_5^2, S_5^3 *and* S_5^4*. Since there is a big change in epistemic properties (*$Male \rightarrow Female$*), an alert is triggered and reported to the control room. Also note that here Theorem 1 is verified.*

This example clearly shows how the beliefs are smoothly maintained or changed with uncertain internal (video analytics) and external information, whilst video analytics just tell what is what *at each time point, without continuity.*

6 Conclusion

In this paper, we have proposed a framework to deal with uncertain observations. This framework is based on a revised version of situation calculus. It allows external instructions as well as internal actions. It is able to tackle with uncertain epistemic actions and uncertain ontic actions. Early errors can be corrected in this framework by revision and updating.

In the literature, probabilistic methods (e.g. dynamic bayesian networks, [26], etc.) and other AI techniques (e.g. Bilattice reasoning, [25], etc.) have been applied in computer vision/video surveillance. Comparing to these approaches, our framework considers and easily handles external information. In addition, the ability of belief revision and updating makes it easy to correct past mistakes.

For future work, we are implementing this framework as a part of an on-going intelligent surveillance project (CSIT) to enhance the power of event reasoning, especially for error correction. The full implementation includes a thorough set of ontic actions, epistemic actions for a set of properties of interest. It also can be naturally extended to allow for multiple agents (passengers). In addition, the situations can serve as a foundation for event inference proposed in [12]. Another interesting issue is to study the properties that the framework satisfy.

Acknowledgement. This research work is sponsored by the EPSRC projects EP/G034303/1 and EP/H049606/1 (the CSIT project).

References

1. Bacchus, F., Halpern, J., Levesque, H.: Reasoning about noisy sensors and effectors in the situation calculus. Artificial Intelligence 111(1-2), 171–208 (1998)
2. Bsia: Florida school bus surveillance, `http://www.bsia.co.uk/LY8VIM18989_action;displaystudy_sectorid;LYCQYL79312_caseid;NFLEN064798`
3. De Giacomo, G., Levesque, H.: Progression using regression and sensors. In: Procs. of IJCAI, pp. 160–165 (1999)
4. Herzig, A., Lang, J., Marquis, P.: Action representation and partially observable planning using epistemic logic. In: Procs. of IJCAI, pp. 1067–1072 (2003)
5. Lakemeyer, G., Levesque, H.: A semantic characterization of a useful fragment of the situation calculus with knowledge. Artificial Intelligence 175(1), 142–164 (2011)
6. Levesque, H.: What is planning in the presence of sensing. In: Procs. of AAAI, pp. 1139–1146 (1996)
7. Liu, W., Miller, P., Ma, J., Yan, W.: Challenges of distributed intelligent surveillance system with heterogenous information. In: Procs. of QRASA, Pasadena, California, pp. 69–74 (2009)
8. Ma, J., Liu, W., Benferhat, S.: A belief revision framework for revising epistemic states with partial epistemic states. In: Procs. of AAAI, pp. 333–338 (2010)
9. Ma, J., Liu, W., Hunter, A.: Inducing probability distributions from knowledge bases with (in)dependence relations. In: Procs. of AAAI, pp. 339–344 (2010)
10. Ma, J., Liu, W., Miller, P.: Event modelling and reasoning with uncertain information for distributed sensor networks. In: Deshpande, A., Hunter, A. (eds.) SUM 2010. LNCS, vol. 6379, pp. 236–249. Springer, Heidelberg (2010)
11. Ma, J., Liu, W., Miller, P.: Belief change with noisy sensing in the situation calculus. In: Procs. of UAI (2011)
12. Ma, J., Liu, W., Miller, P., Yan, W.: Event composition with imperfect information for bus surveillance. In: Procs. of AVSS, pp. 382–387. IEEE Press, Los Alamitos (2009)
13. McCarthy, J.: Situations, Actions and Causal Laws. Stanford University, Stanford (1963)
14. McCarthy, J., Hayes, P.: Some philosophical problems from the standpoint of artificial intelligence. In: Machine Intelligence, vol. 4, pp. 463–502. Edinburgh University Press, Edinburgh (1969)
15. Miller, P., Liu, W., Fowler, F., Zhou, H., Shen, J., Ma, J., Zhang, J., Yan, W., McLaughlin, K., Sezer, S.: Intelligent sensor information system for public transport: To safely go.. In: Procs. of AVSS (2010)
16. ECIT Queen's University of Belfast. Airport corridor surveillance (2010), `http://www.csit.qub.ac.uk/Research/ResearchGroups/IntelligentSurveillanceSystems`
17. US Defense of the Homeland. Washington rail corridor surveillance (2006), `http://preview.govtsecurity.com/news/Washington-rail-corridor-surveillance/`

18. US Department of Transportation. Rita - its research program (2010), http://www.its.dot.gov/ITS_ROOT2010/its_program/ITSfederal_program.htm
19. Reiter, R.: The frame problem in the situation calculus: A simple solution (sometimes) and a completeness result for goal regression. In: Aritificial Intelligence and Mathematical Theory of Computation: Papers in Honor of John McCarthy, pp. 359–380. Academic Press, London (1991)
20. Scherl, R., Levesque, H.: The frame problem and knowledge-producing actions. In: Procs. of AAAI, pp. 689–695 (1993)
21. Scherl, R., Levesque, H.: Knowledge, action, and the frame problem. Artificial Intelligence 144(1-2), 1–39 (2003)
22. Gardiner Security. Glasgow transforms bus security with ip video surveillance, http://www.ipusergroup.com/doc-upload/Gardiner-Glasgowbuses.pdf
23. Shapiro, S.: Belief change with noisy sensing and introspection. In: Procs. of NRAC, pp. 84–89 (2005)
24. Shapiro, S., Pagnucco, M., Lespérance, Y., Levesque, H.: Iterated belief change in the situation calculus. Artificial Intelligence 175(1), 165–192 (2011)
25. Shet, V.D., Neumann, J., Ramesh, V., Davis, L.S.: Bilattice-based logical reasoning for human detection. In: IEEE Conf. on Computer Vision and Pattern Recognition, CVPR 2007, pp. 1–8 (2007)
26. Wang, T., Diao, Q., Zhang, Y., Song, G., Lai, C., Bradski, G.: A dynamic bayesian network approach to multi-cue based visual tracking. In: Procs. of Pattern Recognition, ICPR, pp. 167–170 (2004)

Author Index

GPSR Compliance

The European Union's (EU) General Product Safety Regulation (GPSR) is a set of rules that requires consumer products to be safe and our obligations to ensure this.

If you have any concerns about our products, you can contact us on ProductSafety@springernature.com

In case Publisher is established outside the EU, the EU authorized representative is:

Springer Nature Customer Service Center GmbH
Europaplatz 3
69115 Heidelberg, Germany

Batch number: 09490872

Printed by Printforce, the Netherlands